T0181673

Trace Environmental
Quantitative Analysis

Trace Environmental Quantitative Analysis

Including Student-Tested Experiments

Third Edition

Paul R. Loconto

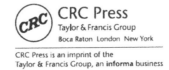

CRC Press
Taylor & Francis Group
Boca Raton London New York

CRC Press is an imprint of the
Taylor & Francis Group, an **informa** business

About the Cover: *Plot of the logarithm of the octanol-water partition coefficient vs. the logarithm of the aqueous solubility for selected organic compounds of enviro-chem/enviro-health interest. (Adapted from Chiou, G. et al.,* Environmental Science and Technology, *11(5): 475–478, 1977)*

Third Edition published 2020
by CRC Press
6000 Broken Sound Parkway NW, Suite 300, Boca Raton, FL 33487-2742

and by CRC Press
2 Park Square, Milton Park, Abingdon, Oxon OX14 4RN

© 2021 Taylor & Francis Group, LLC
First edition published by CRC Press 2001
Second edition published by CRC Press 2005
CRC Press is an imprint of Taylor & Francis Group, LLC

Library of Congress Cataloging-in-Publication Data
Names: Loconto, Paul R., 1947– author.
Title: Trace environmental quantitative analysis: including
student-tested experiments / Paul R. Loconto.
Description: Third edition. | Abingdon, Oxon; Boca Raton, FL : CRC Press, 2020. |
Includes bibliographical references and index.
Identifiers: LCCN 2020035599 (print) | LCCN 2020035600 (ebook) |
ISBN 9780367445331 (hardback) | ISBN 9780367631062 (paperback) |
ISBN 9781003010609 (ebook)
Subjects: LCSH: Environmental chemistry. | Trace analysis. |
Chemistry, Analytic–Quantitative.
Classification: LCC TD193 .L63 2020 (print) |
LCC TD193 (ebook) | DDC 628.5028/7–dc23
LC record available at https://lccn.loc.gov/2020035599
LC ebook record available at https://lccn.loc.gov/2020035600

ISBN: 978-0-367-44533-1 (hbk)
ISBN: 978-0-367-63106-2 (pbk)
ISBN: 978-1-003-01060-9 (ebk)

DOI: 10.1201/9781003010609

Typeset in Times
Typeset by Newgen Publishing UK

To
Patricia Ann (Loconto) Stevens
1940–2014

She read the Chemistry Set instructions while helping her kid brother safely per-form experiments in the basement of our home in south central Massachusetts during the late 1950s.

She very much wanted to be (and became) a superb 6th grade school teacher. I'd like to think that I became, down in that basement, her first sixth grade pupil!

Contents

Preface to the Third Edition

A recent article in a weekly periodical for those who maintain membership in the *American Chemical Society* titled "The Lurking Contaminant"* is as good a topic as any to begin a book with a title such as that shown on the front cover. The article begins by stating that a research associate was using the company's gas chromatography-mass spectrometry instrument to test one of the first drugs on the firm's list: an acid reflux baby syrup prescribed to the company cofounder's daughter. The syrup contained *rantidine,* a drug commonly prescribed for heartburn. The peak in the mass spectrum had a mass to charge ratio (*m/z*) of 74. A mass spectral library search was undoubtedly done and the result of the search identified the chromatographically resolved peak as N-nitrosodimethylamine (NDMA). The molecular structure of NDMA is shown below:

If you add up the atomic weights of the chemical elements that comprise NDMA, you get a molecular weight of 74! It has been believed for a long time that the chief sources of environmental exposure to NDMA by humans are due to tobacco; cured meats, such as bacon, where the salt, sodium nitrite, is routinely added; fermented foods, such as beer and cheese; and shampoos, cleansers, detergents, and pesticides. It is believed that one possible mechanism for the formation of NDMA involves the generation of a nitrosonium ion (NO^+) from a reaction of the nitrite ion in the presence of acid (H^+) followed by a subsequent reaction with dimethylamine to form NDMA!

This is a book that introduces the topic of trace quantitative organic and inorganic chemical analysis from samples of the environment (air, water, and soil) or from human body fluids written from an analytical chemist's point of view. Those who take the time to study the book's content might someday be the ones who discover yet another chemical contaminant like NDMA in a pharmaceutical preparation, or a toxic chemical from a hazardous waste site, or a chemical warfare agent from a specimen of human serum, etc. This book provides the necessary learning tools to accomplish these noteworthy objectives.

Since the 2nd Edition appeared in 2006, let's briefly catalog just a few of the major changes in the science of trace environmental quantitative analysis:

- Organochlorine pesticides that were once separated, confirmed, and quantitated by dual capillary column/dual detector have been replaced by single capillary column GC-MS.
- GC-MS has emerged as the dominant determinative technique for separating, identifying, and quantitating volatile and semivolatile analytes of environmental concern.
- Autosampler technology has evolved from stand-alone dedicated units to dual-rail robotic multipurpose samplers.
- Gas and liquid chromatographs and atomic absorption and atomic emission spectrometers are instruments that, today, are entirely computer controlled and the data generated acquired by computer software. Often an analytical chemist must first learn how to use and understand the commercial software before he/she can obtain useful trace analytical data from such instruments.
- Ninety-six-well plate solid-phase extraction has become a dominant sample preparation technique in the trace analytical laboratory.
- Analytical results must satisfy strict QA/QC criteria before being released to the client.

* The Lurking Contaminant. *Chemical & Engineering News,* April 20, 2020 p. 27.

- Reports from the contemporary analytical laboratory are routinely entered into a laboratory information management system (LIMS) and sent to clients on-line.
- LC-MS/MS has become *the* dominant determinative technique with respect to separating, identifying, and quantitating trace nonvolatile polar analytes of environmental concern.
- Advances in LC-MS and LC-MS/MS have almost completely eliminated the need for chemical derivatization with subsequent quantitative analysis by GC or GC-MS.
- GC-MS/MS has brought an unprecedented level of specificity to the GC-MS determinative technique.
- How to measure the instrument detection limit still remains controversial.

It has been an honor for this author to have had the opportunity after well over a decade to envision how the 2nd Edition could be revised and even improved upon while at the same time sustaining what was of value in the first place! Time and maturation on the part of the author does have its rewards! I'd like to think of this book as a *3 in 1 resource*: text, reference, and handbook!

The overall style for the 3rd Edition has remained the same. Section headings continue to be cast in the form of a question. New terms have been italicized when they appear for the first time. Key concepts within a given paragraph have also been italicized for emphasis. Each chapter starts with a *Chapter at a Glance* so that the reader can more easily visualize what will be introduced in a given chapter. In addition, for reference purposes, the reader can more quickly find topics of immediate interest.

Unlike the 2nd Edition, all figures have been numbered and provided with a figure legend to assist in clarification. I have eliminated in-text drawings and abandoned the use of the word scheme to describe flow charts, etc. The reader will find many more molecular structures drawn when compared to the 2nd Edition. I have made extensive use of *Google* in finding correctly drawn molecular structures. I have eliminated an appendix that listed useful internet links found in the 2nd Edition since any keyword entered into a Google search seems to take you to where you want to go! Graphs continue to be sketches that I drew or carried over from earlier editions. I have tried very hard to make this book readable, comprehensible, interesting and relevant, and, at the same time, introduce sound scientific principles and practices. I have been careful to draw a distinction between samples taken from air, water, and soil (so-called *enviro-chemical*) and human body fluids (so-called *enviro-health*) samples.

I sincerely appreciate the early efforts by Hilary Lafoe at Taylor & Francis Group in turning the author's desire to write a 3rd Edition into a reality. I sincerely appreciate the efforts of Jessica Poile from Taylor & Francis Group. She guided the specifics providing the author with an electronic version of the 2nd Edition from which he could begin the revision. She was the recipient of the revised chapters, one by one. Traditionally, retirement does not usually mean *working* tirelessly on either a desktop or laptop computer *daily to* revise an electronic version of the 2nd Edition for a book of this size and content. I could only accomplish such a task with the loving support and understanding of my wife, Priscilla. She has been my base of support for each and every intellectual effort "beyond the call of duty" that her husband has engaged himself in for nearly closing in on five decades. This one has been no exception!

It's not what you gather, but what you scatter
That tells what kind of life you've lived.

—**Helen Walton**

About the Author

Paul R. Loconto holds a Ph.D. in analytical chemistry and a M.S. in physical chemistry. He has published 35 peer-reviewed papers in analytical chemistry and in chemical education. He has given over 40 talks and poster presentations at various workshops, meetings, and conferences.

After brief stints at the American Cyanamid Co. (Stamford, CT) and the Dow Chemical Co. (Midland, MI) and beginning in 1974, Dr. Loconto taught introductory, environmental, general, and organic chemistry at Dutchess Community College (Poughkeepsie, NY) for 12 years. He then joined NANCO Environmental Services (Wappingers Falls, NY) as R&D manager in 1986. He joined the Michigan Biotechnology Institute (Lansing, MI) in 1990.

In 1992 he became the laboratory manager for the graduate program in Environmental Engineering at Michigan State University (East Lansing, MI) where he conducted analytical method development for both the NIEHS analytical core and the EPA Hazardous Substance Research Center while coordinating the development of an instructional analytical laboratory for the graduate school.

In 2001, he joined the Michigan Department of Community Health, Bureau of Laboratories (Lansing, MI) as a Laboratory Scientist Specialist. Here, in addition to training new employees on how to use GC-FID, GC-TSD, GC-MS, GC-MS/MS, GC-AED, HPLC-UV and FTIR instruments as well as SPE sample prep techniques, he taught co-workers how to satisfy QA/QC requirements. He focused on developing analytical methods for biomonitoring while conducting trace quantitative analysis in support of the Laboratory Response Network for the U.S. Centers for Disease Control and Prevention (CDC).

He retired in late 2013 yet continues as a consultant, educator, and writer.

Contributors

Michael C. Stagliano, from outside Philadelphia, PA, resides in East Lansing, MI, where he is a Laboratory Scientist Specialist for the Michigan Department of Health and Human Services, Chemistry and Toxicology Division, Analytical Chemistry Section. He oversees the validation of LC-MS methods for the CDC's Chemical Threat Laboratory Response Network, and develops and validates new LC-MS methods for both emerging environmental contaminants as well as clinically relevant small molecules. Prior to joining the State of Michigan, he spent five years as a Research Assistant at Michigan State University, where he focused on developing rapid, sensitive LC-MS methods for lipid detection and quantitation. Michael holds a doctorate degree in chemistry from The Pennsylvania State University and baccalaureate degrees in both chemistry and biochemistry from The University of Michigan. In his free time, he enjoys traveling and gardening, and is an avid hockey and college football fan.

Scott Forsyth is a laboratory scientist at the Michigan Department of Health and Human Services, Bureau of Laboratories, Chemistry and Toxicology Division, Analytical Chemistry Section, Trace Metals Unit. He has been active with routine testing methods and method development in trace metals since 2011. He has run methods for lead determination in human whole blood, lead determination in environmental samples, biomonitoring for various metals in human hair, and biomonitoring for various metals in human whole blood and urine, as well as running various CDC methods for determination of various metals associated with potential chemical terrorism events. He is familiar with several atomic spectrometry techniques, such as atomic absorption, ICP-AES, ICP-MS, and ICP-MS/MS, HPLC-ICP-MS, and HPLC-ICP-MS/MS. Prior to working in trace metals, he worked on the organic side of analytical chemistry from 2004 to 2010, where he primarily worked on GC-ECDs for detection of PCBs and pesticides in fish tissue, biomonitoring for the same class of compounds in human serum, helped develop a method for determining various toxaphene parlars in fish tissue utilizing GC-SIM-MSD instrumentation, and ran methods associated with the CDC national chemical terrorism/chemical warfare agent determination using various techniques, such as SPME-GC-MSD, GC-MS/MS, HPLC-MS/MS, and Headspace-GC-MSD. Prior to this, Scott worked in two hospital laboratories—Tallahassee Memorial Hospital (2000), where he worked as a Generalist clinical laboratory scientist, and Sparrow Health System from 2001 to 2004, where he worked as a clinical laboratory scientist as Generalist in the Serology and Molecular Diagnostics departments. Mr. Forsyth graduated from Michigan State University in 1993 with a BS degree in Zoology. He went to graduate school for a year and realized that plant parasitic nematology was not a wise career choice. Mr. Forsyth ended up taking some medical technology courses and was accepted at St Vincent's Medical Center, School of Medical Technology, where he received his certificate in 1999. In 2003, Scott earned a certificate in Molecular Diagnostics. His interests include, but are not limited to and in no particular order, scuba diving, sailing, computer gaming, hiking, fishing, amateur astronomy, and eating copious amounts of Thai food. He also spends as much quality time as he can with his two daughters.

1 Introduction to Trace Environmental Quantitative Analysis (TEQA)

CHAPTER AT A GLANCE

DDT, oil slicks, the PBB mix-up, and Pb in drinking water
Trace environmental (chemical vs. health) quantitative analysis
To what extent do environmental chemicals enter humans?
Analytical chemistry approaches to biomonitoring
Environmental chemistry: a contemporary definition
EPA regulations
Analytical methods that satisfy EPA regulations
Physical/chemical basis for EPA's methods protocols
References

If you teach a person what to learn, you are preparing him for the past. If you teach him how to learn, you are preparing him for the future.

—Anonymous

1.1 WHAT CAUSED THE BIRTH OF THE ENVIRONMENTAL MOVEMENT IN THE U.S.?

The environmental movement in the United States began with the publication in 1962 of the book *Silent Spring* written by biologist Rachel Carson. The need to protect and sustain the environment became very focused when the Cuyahoga River near Cleveland, Ohio caught fire on June 22, 1969.[1,2] Carson was the first "whistleblower" who sounded the alarm concerning the negative environmental consequences of indiscriminately dispersing the chemical DDT. No understanding existed at that time of the persistent nature of DDT and neither was it known that its degradation product DDE would bio-accumulate and bio-magnify into living things including humans. Molecular structures and nomenclature for DDT and DDE are shown below:

dichlorodiphenyltrichloroethane (pp'DDT)   dichlorodiphenyldichloroethylene (pp'DDE)

The molecular structure differences reveal a loss of HCl (hydrogen chloride) via environmental degradation. This awareness eventually led to the first Earth Day on April 22, 1970 and subsequent establishment of the U.S. Environmental Protection Agency (EPA) in 1972. For the scientific world this awareness led to advances in analytical chemistry that enabled organochlorine pesticides to be identified and quantified in ways that today seem unimaginable back then!

DDT was not the only culprit! Listed below are numerous organochlorine pesticides that the analyst had to identify and quantitate. These chemical substances appear in a 1988 version of EPA Method 508: *Determination of chlorinated pesticides in waters by Gas Chromatography with an Electron Capture Detector:*[2]

Aldrin	**Dieldrin**	HCH-∂	**Aroclor 1016**
Chlordane-α	Endosulfan I	HCH-γ (Lindane)	Aroclor 1221
Chlordane-γ	Endosulfan II	Heptachlor	Aroclor 1232
Chlorneb	Endosulfan sulfate	Heptachlor epoxide	Aroclor 1242
Chlorobenzilate (a)	Etridiazole	Hexachlorobenzene	Aroclor 1248
Chlorothalonil	Endrin	Methoxychlor	Aroclor 1254
DCPA	Endrin aldehyde	cis-Permethrin	Arolcor 1260
4,4'-DDD	Etridiazole	trans-Permethrin	Toxaphene
4,4'-DDE	HCH-α	Propachlor	Chlordane
4,4-'DDT	HCH-β	Trifluralin	

Organochlorine pesticides (insecticides) consist of semivolatile organic compounds and therefore are amenable to trace analysis using gas chromatography (GC). The electron capture detector (ECD) is very selective and very sensitive to organic compounds in the vapor state that contain molecules with chlorine covalently bonded to carbon. To prepare drinking water samples for analysis by GC-ECD in EPA Method 508, the analyst must use liquid–liquid extraction (LLE). This author was fortunate early on to explore the use of solid-phase extraction (SPE), as an alternative to LLE, to isolate and recover three representative organochlorine pesticides listed above.[3]

Lindane   Endrin   Methoxyclor

This book provides the means within which to better understand how to identify and quantitate organochlorine pesticides that have been extracted from the environment, i.e. air, water and soil. It's not only organic compounds with chlorine bonded to carbon that have polluted the environment; bromine bonded to carbon has a unique history and more recently fluorine bonded to carbon.

1.2   WHAT HAPPENED WHEN FLAME RETARDANT CHEMICALS WERE INADVERTENTLY SUBSTITUTED IN CATTLE FEED IN MICHIGAN?

Fries has identified four major factors that may have led to the contamination of the food chain and related public health effects[4]. First, fire safety legislation in the 1960s stimulated research and development of flame retardant chemicals called polybrominated biphenyls (PBBs). Second, increasing size coupled to growing agribusiness interests in the 1960s–1970s led to the use of magnesium oxide, MgO, as an additive to cattle feed to compensate for the increased acidity of corn silage. Third, technological advances in analytical instrumentation in the early 1970s that enabled detection of trace concentration levels of chemical contaminants in complex matrices were realized. Finally, increased public awareness and concern by the early 1970s (as discussed earlier) fueled the investigation.

The Michigan Chemical Corporation (MCC) was making FireMaster BP-6® in St. Louis, MI with a yearly production of 5 million pounds.[4] The major component in BP-6 was the PBB congener whose molecular structure is shown below:

2,2'4,4',5,5'-hexabromobiphenyl

A new product that might have been the most significant contributor to the mix-up that eventually led to widespread environmental distribution of PBBs was the introduction of Firemaster FF-1® in 1972 by the MCC. FF-1 was made by grinding BP-6 and adding enough calcium silicate, $CaSiO_3$ to comprise ~2% of the total weight of FF-1.

Another chemical produced by the MCC was magnesium oxide, MgO, marketed as Nutrimaster® which as stated earlier had emerged as an additive to cattle feed. FF-1 and MgO were packaged in paper bags with color-coded labeling. However during a paper shortage, both products were packaged in plain bags with black stenciled labeling and were stored in the same warehouse. During May, 1973, some FF-1 was mistakenly included in a shipment of MgO to the Farm Bureau Services feed mill at Climax, MI. The amount of FF-1 is estimated not to exceed 650 lb (295 kg). Some "MgO" (suspected now to be FF-1) was used for mixing feeds at the Climax mill. This feed was sent directly to farms or to retail outlets. This feed was also shipped to affiliate feed mills in the state and used in mixing operations at those locations. Feeds that were not formulated to contain MgO also became contaminated because of carryover of PBBs from batch to batch in the mixing equipment. "MgO" was used for formulating feed #402 as early as July 1973.

Although the Halbert Dairy Farm was not the first farm to receive feed with high concentration levels of PBBs, severe health problems in their herd led directly to the identification of the toxicant and the actions that followed. Approximately 350 lactating cows at the Halbert farm received Farm Bureau feed #402 in 9-ton lots in 1973. Among other clinical signs, a 50% drop in feed consumption and milk production in late September, 1973 was observed. Feed #402 began to be suspect as the cause of the problem. However, the PBB culprit proved elusive to find at first. The feed did not contain excess urea or nitrates while heavy metal concentrations were within acceptable limits. Mycotoxins were not detected while organo-chlorine pesticides (OCs) and polychlorinated biphenyl (PCB) residues were present at the usual background levels. As luck would have it, PBBs were discovered only when a gas chromatograph was inadvertently allowed to operate several hours longer than usual.[5] Since atoms of the chemical element bromine are much heavier than atoms of the chemical element chlorine, it appeared that either a longer elution time and/or a higher GC column temperature was required to elute PBBs when compared to OCs or PCBs.

1.3 WHAT WAS THE GOVERNMENT RESPONSE TO THE MIX-UP?

The U.S. Food and Drug Administration (FDA) established a temporary guideline of 1 ppm PBBs in milk and tissue fat for contaminated herds. Contaminated animals were killed including 9,400 head of cattle, 2,000 swine, 400 sheep, and over two million chickens! It was then suspected that farm families who used milk or slaughtered animals for home consumption might have high exposures to PBBs. In June, 1974, the Michigan Department of Public Health (MDPH) surveyed farm residents who were potentially exposed and found no consistent symptoms or adverse health effects that could be associated with exposure to PBBs. A cohort of 4,000 Michigan farmers, their families, and others were enrolled for further study and observation for chronic effects of PBB exposure[6]. A control group of 1,000 Iowa famers and their families was also enrolled. Following the initial governmental response, controversies involving dairy farmers who became concerned about human health issues ensued. A PBB present in human breast milk during the summer of 1976 ignited media interest and led to bills introduced in the state legislature to reduce tolerances. Michigan Public Act 77 lowered the tolerance to 0.02 ppm in body fat of all culled dairy cows offered for slaughter. Also, the finding of a single violated animal did not lead to quarantine and disposal of the entire herd. Passage of Act 77 reduced public concern but did not dampen media coverage of PBB contamination. In 1978, only when the first lawsuits brought against Farm Bureau Services and the MCC were completed and ruled in favor of the defendants did the controversy begin to subside. Act 77 was allowed to expire in 1982.

1.4   WAS ANALYTICAL TESTING FOR PBBS ESTABLISHED WITHIN THE STATE?

Yes indeed! The Division of Environmental Epidemiology (DEE) was established by the MDPH in response to the PBB crisis. The Division contained field staff to interview and draw specimens from the cohort members. The Division also included laboratory staff to analyze specimens and quantify exposure, as well as administrative staff of epidemiologists, physicians, toxicologists, and clerical support to design and refine study parameters. The DEE worked closely with the Centers for Disease and Prevention (CDC) to improve analytical and information-gathering methods. Also, they worked with the MDPH Vital Statistics group to store, monitor, and assess the large amount of data generated. Harold Humphrey, PhD, PBB project director at the time stated that the "Division was created as a direct outgrowth of the PBB problem. It was the last nail in the coffin." Irving Selikoff, PhD, a prominent environmental scientist at Mount Sinai Hospital in New York City, stated several years following the disaster that "we have made a Faustian bargain with our chemical world. Now we have to learn enough so that we can cut our losses."[7] Figure 1.1 shows a typical gas chromatogram using a packed column with an electron capture detector (PC-GC-ECD) and nicely illustrates the need to increase the temperature program range so that PBBs will also elute and get detected while Figure 1.2 displays a packed column GC-ECD chromatogram that was typical of the 1970s to mid-1980s era. Note the various PBB congeners associated with different chromatographically separated peaks. Contrast this chromatogram with a more contemporary one, while injecting an aliquot of the original FF-1. This chromatogram, shown in Figure 1.3, used a capillary or open-tubular GC column interfaced to a mass spectrometer that employed electron-capture negative ion detection (C-GC-ECNI-MS).

1.5   WHAT WERE THE MOST SIGNIFICANT ANALYTICAL RESULTS FROM THE PBB CRISIS?

A comprehensive study of the distribution of PBBs into the human population was not conducted until 1978[4]. The mean concentration of PBB in human adipose tissue, detected in 97% of human specimens was 400 parts per billion (ppb). Mean concentrations of PBB in human serum, detected in only 69% of adults and 73% of children, were 1.3 ppb (adult) and 1.8 ppb (children).[8] The detection of PBB in human serum was deemed less sensitive when compared to human adipose.[1] Results from applying a pharmacokinetic model to the mean serum concentrations for adults

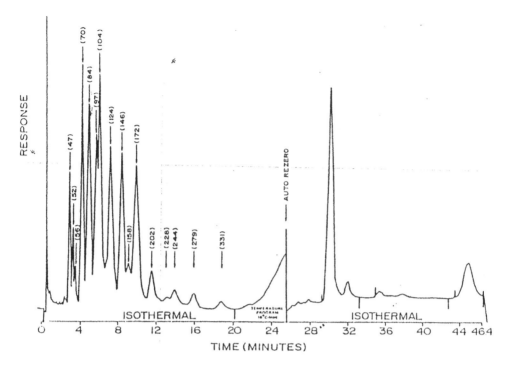

FIGURE 1.1

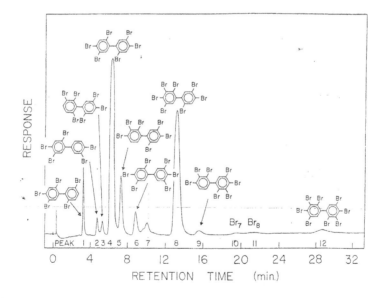

FIGURE 1.2

indicate that total exposure was 9–10 mg for an average male with an average adipose content. In some individuals, the highest amount accumulated was 800–900 mg.[9] Human PBB accumulation in Michigan was associated with regions near where the contaminated feed had been used and marketed. Higher than average concentrations occurred in an area centered in Muskegon County in western Michigan. This area also had the largest number of quarantined farms.[1] Groups with

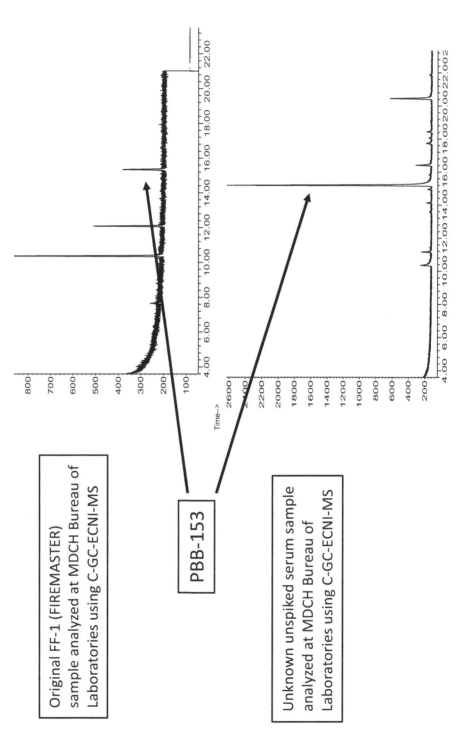

Original FF-1 (FIREMASTER) sample analyzed at MDCH Bureau of Laboratories using C-GC-ECNI-MS

PBB-153

Unknown unspiked serum sample analyzed at MDCH Bureau of Laboratories using C-GC-ECNI-MS

FIGURE 1.3

the highest exposure to PBBs were chemical plant workers involved in the manufacturer of PBB, farmers on contaminated farms who consumed their own dairy and meat products and a group who purchased products directly from contaminated farms. Of the 55 chemical plant workers out of a pool of 270 who agreed to be tested, all had detectable levels of PPBs in their serum. Four out of ten production workers had serum concentrations greater than 500 ppb with a mean of ~600 ppb. The mean concentration level of PBBs in non-production workers was only 16 ppb.[1] The Michigan Department of Health Cohort Study initiated as early as June, 1974 contained at the time the most comprehensive data.[1]

This book provides the means within which to better understand how to identify and quantitate analytes like PBBs that have been extracted from the environment, i.e. from air, water and soil and humans.

1.6 WHAT HAPPENED IN FLINT, MICHIGAN? IS IT IMPORTANT TO LOWER DETECTION LIMITS FOR QUANTITATING BLOOD PB?

In 2014, the City of Flint, MI then under direction of a governor-appointed emergency manager made a fateful decision to change the source of the city's drinking water from Lake Huron to the Flint River. To add insult to injury, the routine addition of so-called corrosion inhibitor chemicals at the drinking water treatment plant was eliminated! Two factors contributed to significantly increase the concentration of lead in the city's drinking water: 1) a change in the source water and 2) a failure to treat the source water! Students of general chemistry learn that if you mix an aqueous solution containing a water-soluble lead (Pb^{2+}) salt, a precipitate is likely according to:

$$3Pb^{2+} + 2\ PO_4^{3-} \longrightarrow Pb_3\ (PO_4)_{2(s)}$$

Students also know that the solubility product constant (K_{SP}) for lead phosphate where

$$Ksp = [Pb^{2+}]^3 \times [(PO_4)^{3-}]^2 = 8.0 \times 10^{-43}$$

is extremely low. This fact suggests that if excess phosphate ion is available, the dissolved concentration of Pb^{2+} in water remains very low.[10] In the absence of phosphate ion continuously precipitating Pb^{2+}, it is feasible that significant increases in the concentration of Pb^{2+} are realized.

The current blood lead reference value (BLRV) of 5 µg/dL (50 ppb) for children is an important parameter in public health. It means that a measured blood Pb concentration level that is above the BLRV in children is considered elevated.[11] The fact that we can measure as low as 50 ppb in a human blood sample is worth considering here.

Advances in atomic spectroscopy since 1960 have continually served to lower the BLRV because with each advance in atomic spectroscopy instrumentation, the method detection limit (MDL) has got lower and lower. Figure 1.4 compares what the CDC considers safe concentration levels of Pb and how these standards have decreased as MDLs have declined with advances in analytical instrumentation over the years. Using a flame atomic absorption spectrophotometer (FlAA) with an MDL that might be ~120 µg/dL for Pb in blood in the year 1960, a BLRV back then was established at 60 µg/dL. Compare that MDL to that afforded today using an inductively coupled plasma mass spectrometer (ICP-MS) which can easily reach 0.01 or even 0.001 µg/dL Pb! A recent study used laser ablation ICP-MS and reported traces of Pb and As in baby teeth from children living near a Pb battery recycling center in Los Angeles, CA.[12]

Every decade in the late 20th and early 21st centuries brought a profound realization that industrial chemicals cannot continue to pollute the environment. This understanding led to huge scientific initiatives to identify and quantitate these pollutants and employed hundreds of analytical chemists and chemical technicians to do the work! During the 1950s and 1960s,

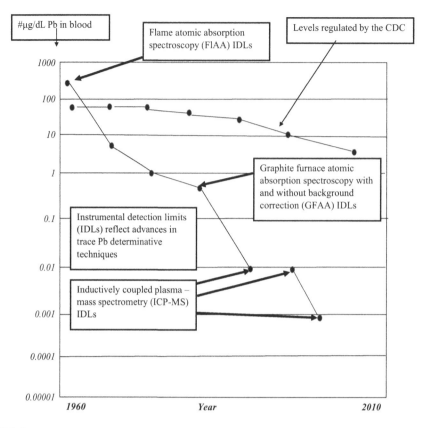

FIGURE 1.4

organochlorine pesticides became front and center, during the 1970s through to the 1990s polychlorinated and then polybrominated biphenyls emerged, while the 2000s through the 2010s found polybrominated diphenyl ethers emerging. Today we hear about perfluoroalkyl substances, PFAS, having potentially contaminated the water we drink.

1.7   WHAT OTHER ENDEAVORS REQUIRE TRACE ANALYSIS?

State public health laboratories with considerable support from the CDC are responding to the potential threat from chemical warfare agents that result from so-called bio-terrorist activities. These laboratories now have an analytical chemistry capability in trace environmental health quantitative analysis to identify and to quantitate metabolites from human exposure to a number of chemical warfare agents such as organic phosphonate nerve agents. In addition, numerous state public health laboratories have a capability to conduct biomonitoring. These methods are designed to measure trace concentration levels of chemical substances that either persist (persistent organic pollutants (POPs)) or are eliminated rather quickly by the body, i.e., nonpersistent organic pollutants (NPOPs).

Bioterrorism and biomonitoring are key initiatives that are currently driving the changing nature of trace quantitative organics and inorganics analysis. The third edition of this book attempts to reflect these changes. This new emphasis, when combined with the more established methods of trace environmental quantitative analysis, has led this author to adopt a new term: *trace enviro-chemical / enviro-health quantitative analysis*, whose acronym is also TEQA. I have tried to add

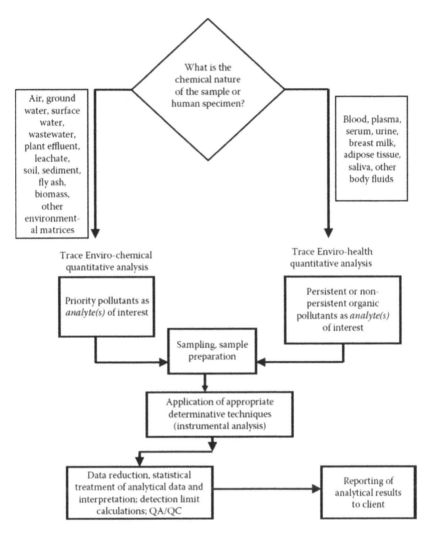

FIGURE 1.5

those analytical concepts that are most relevant to conducting trace enviro-health quantitative analysis. Sampling, sample preparation, determinative technique, and data reduction/interpretation are very similar to both trace enviro-chemical and trace enviro-health quantitative analysis. Figure 1.5 depicts both the enviro-chemical and the enviro-health aspects of trace quantitative organics and inorganics analysis while discerning the similarities and differences in both. One starts with an understanding of the chemical nature of the sample or human or animal specimen received. A client needs to understand just what chemical substances (analytes) are to be measured and how these two pathways lead to the four steps in the process as shown in Figure 1.5. There is no substitute for effective communication between the client and the analytical laboratory. Sampling (introduced in Chapter 2), sample preparation (introduced in Chapter 3), determinative techniques, often referred to as *instrumental analysis* (introduced in Chapter 4), and data reduction, statistical treatment, and interpretation of analytical data (introduced in Chapter 2) comprise the important aspects of successfully implementing TEQA.

This third edition updates the principles and practices of trace enviro-health and trace enviro-chemical quantitative analysis as introduced earlier by the author.[13]

1.8   CAN EXAMPLES PROVIDE INSIGHT INTO TRACE ENVIRONMENTAL HEALTH AND ENVIRONMENTAL CHEMICAL QUANTITATIVE ANALYSIS?

Yes indeed! Recent work at the CDC demonstrates how exposure to the plant-based toxic proteins ricin (derived from the castor bean) and abrin (derived from the rosary pea) can be evaluated in human plasma based on quantifying their metabolites ricinine and abrine.[14] Molecular structures for these analytes are shown below:

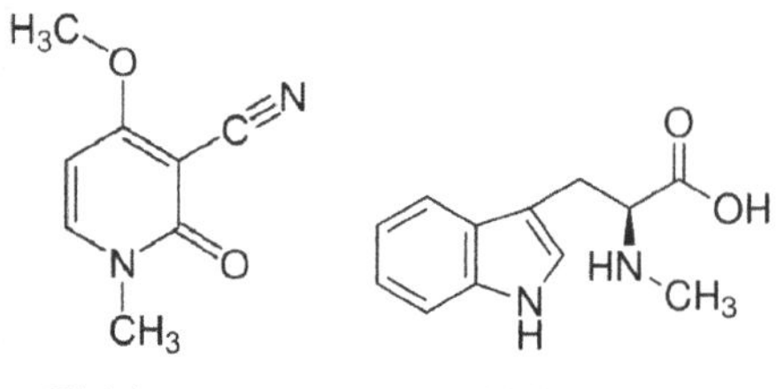

Ricinine   Abrine

Plasma samples are initially diluted and the analytes (shown above) are isolated and recovered using reversed-phase solid-phase extraction in a 96-well plate configuration followed by protein precipitation. Eluents were injected into a liquid chromatograph with tandem mass spectrometric detection. Ricinine and abrine (N-methyl-L-tryptophan) can be identified and detected in one sample by this isotope dilution method.[14]

Polychlorinated biphenyls (PCBs) once manufactured in huge quantities (Aroclors) between 1929 and 1975 in the U.S. and used in electrical transformers and in various ink formulations among other uses have been continuously monitored particularly in drinking water, wastewater effluents, and contaminated soil. PCBs bio-accumulate in fish. There are 209 possible PCB congeners whose general molecular structure is shown below:

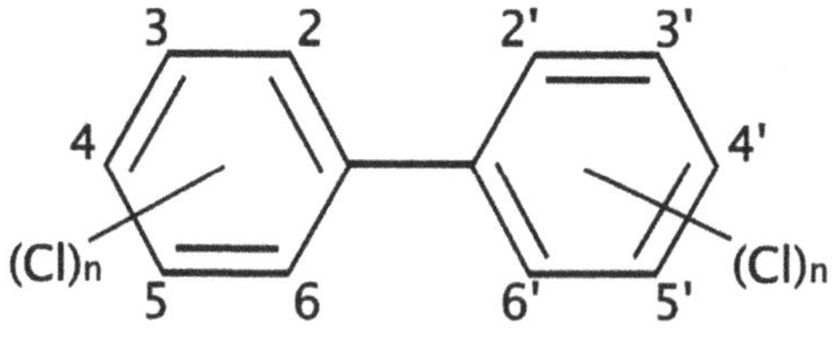

Gas chromatography using capillary or open tubular columns interfaced to either electron capture (ECD) or mass spectrometric (MS) detectors is the preferred determinative technique. The more complex the sample matrix, the more sample preparation steps are required to isolate and recover PCBs. For example, during the packed column days, EPA Method 505 enabled various organochlorine pesticides as well as various Aroclors to be quantitated in drinking water[15]. Today, congener-specific PCBs can be isolated and recovered as capillary columns replaced packed columns in gas chromatography. GC-ECD and GC-MS determinative techniques will be introduced in Chapter 4.

PCBs are called legacy priority pollutants while perfluoroalkyl substances (PFAS) have recently arrived on the environmental toxicant radar screen. Molecular structures for two major categories of PFAS as represented specifically by perfluorooctanesulfonic and perfluorooctanoic acids are shown below:

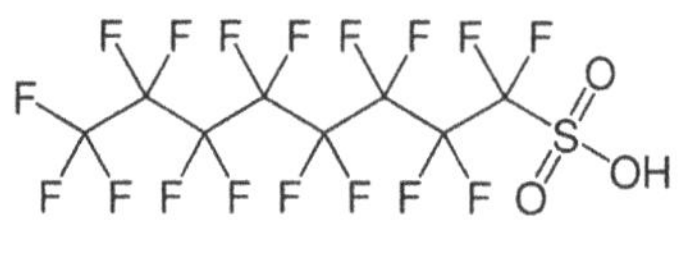

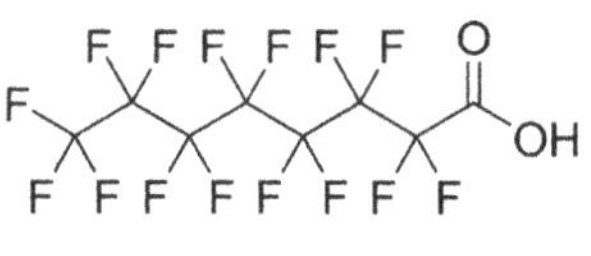

EPA Method 537.1 isolates and recovers some 14 PFAS from drinking water using solid-phase extraction (SPE) and liquid chromatography tandem mass spectrometry. These sample preparation and determinative techniques will be introduced in Chapters 3 and 4. A recent note illustrates how EPA Method 537 can be modified to incorporate automated SPE coupled to ultra-high pressure liquid chromatography.[16]

1.9 TO WHAT EXTENT DO ENVIRONMENTAL CONTAMINANTS ENTER HUMANS?

Tables 1.1 through 1.5 highlight selected trace quantitative analytical results for the 2001–2002 survey years of the *Third National Report on Human Exposure to Environmental Chemicals* conducted by the U.S. Department of Health and Human services, Centers for Disease Control and Prevention (CDC). The analytical work was performed by the National Center for Environmental Health, Division of Laboratory Sciences in Atlanta, GA. This report provides exposure information about people who participated in an ongoing national survey of the U.S. population. Of the 15 categories of inorganic and organic chemical compounds quantitated, metals and four

TABLE 1.1
Geometric Mean of Blood and Urine Levels of Environmental Metals

Metal	Age Range (Years)	Human Specimen	No. of People Sampled	Units	Geometric Mean (95% Confidence Interval)
Cd	All ages 1 and older (1+)	Blood	8945	µg/L	< LOD
Pb	1+	Blood	8945	µg/L	1.45(1.39 – 1.51)
Pb	1–5	Blood	898	µg/L	1.70(1.55 – 1.87)
Pb	6–11	Blood	1044	µg/L	1.25(1.14 – 1.36)
Hg	1–5	Blood	872	µg/L	0.318(0.268 – 0.377)
Hg	16–49 (females)	Blood	1928	µg/L	0.833(0.738 – 0.94)
Sb	All ages 6 and older (6+)	Urine	2690	µg/L	0.134(0.126 – 0.142)
Ba	6+	Urine	2690	µg/L	1.52(1.41 – 1.65)
Cd	6+	Urine	2690	µg/L	0.210(0.189 – 0.235)
Cs	6+	Urine	2690	µg/L	4.81(4.40 – 5.26)
Co	6+	Urine	2690	µg/L	0.379(0.355 – 0.404)
Pb	6+	Urine	2690	µg/L	0.677(0.637 – 0.718)
Mo	6+	Urine	2690	µg/L	45.0(42.1 – 48.0)
Pt	6+	Urine	2690	µg/L	<LOD
Th	6+	Urine	2653	µg/L	0.165(0.154 – 0.177)
W	6+	Urine	2652	µg/L	0.082 (0.073 – 0.092)
U	6+	Urine	2690	µg/L	0.009(0.007 – 0.10)

Source: Adapted from the Third National Report on Human Exposure to Environmental Chemicals, 2005, using only the 2001–2002 survey. Chemicals or their metabolites were measured in blood and urine samples from a random sample of participants from the National Health and Nutrition Examination Survey (NHANES), Department of Health and Human Services, Centers for Disease Control and Prevention, National Center for Environmental Health, Division of Laboratory Sciences, Atlanta, GA 30341–3724

TABLE 1.2
Geometric Mean of Urine Levels of Environmental Phthalates, Phosphates, and Thiophosphates

Organic Metabolite	Age Range (Years and Older +)	Human Specimen	No. of People Sampled	Units	Geometric Mean (95% Confidence Interval)
monomethylphthalate	6+	Urine	2782	μg/L	1.15(0.985 – 1.34)
monoethylphthalate	6+	Urine	2782	μg/L	178(159 – 199)
mono-n-butylphthalate	6+	Urine	2782	μg/L	18.9(17.4 – 20.6)
monoisobutylphthalate	6+	Urine	2782	μg/L	2.71(2.49 – 2.94)
Mono-2-ethylhexylphthalate	6+	Urine	2782	μg/L	4.27(3.80 – 4.79)
Dimethylphosphate	6+	Urine	2519	μg/L	<LOD
Dimethylthiophosphate	6+	Urine	2518	μg/L	<LOD
Diethylthiophosphate	6+	Urine	2519	μg/L	0.457(0.353 – 0.592)
3,5,6-trichloro-2-pyridinol	6+	Urine	2509	μg/L	1.76(1.52 – 2.03)

Source: Adapted from the Third National Report on Human Exposure to Environmental Chemicals, 2005, using only the 2001–2002 survey. Chemicals or their metabolites were measured in blood and urine samples from a random sample of participants from the National Health and Nutrition Examination Survey (NHANES). Department of Health and Human Services, Centers for Disease Control and Prevention, National Center for Environmental Health, Division of Laboratory Sciences, Atlanta, GA 30341–3724

organics categories are the focus here. These five categories of environmental contaminants found in humans were all above the analytical method's so-called limit of detection (LOD).

Table 1.1 lists the specific metal contaminant, whether the human specimen was blood or urine, the number of people whose body fluids were sampled, and the concentration of the contaminant in the units specified. With respect to lead (Pb) the concentration of Pb in #μg/dL (a commonly used unit to report the concentration of Pb in human blood), differences are seen among the three age categories cited. The higher concentration of Mo is noted; however, it was mentioned in the report that Mo is an essential nutrient and enters the body primarily from dietary sources.

Table 1.2 lists human exposure to various organic compounds known as phthalates or esters of phthalic acid. Phthalates are industrial chemicals that can act as plasticizers, which when added to polymers, impart flexibility and resilience. We frequently touch plastic materials. Our food is packaged in plastic materials. Finding a measurable amount of one or more phthalate metabolite in urine does not mean that they cause an adverse health effect.

Table 1.3 aptly called in the report, the tobacco smoke study, lists the concentration of cotinine, a metabolite of nicotine that tracks exposure to environmental tobacco smoke (ETS) in human serum. Measuring cotinine is preferred over measuring nicotine because cotinine persists longer in the body (half-life of about 16 hours).[17] Molecular structures for both cotinine and its precursor, nicotine, are shown below:

Nicotine   Cotinine

TABLE 1.3
Geometric Mean of Serum Levels of Cotinine. Cotinine as Metabolite from Exposure to Nicotine from Environmental Tobacco Smoke

Age Range (Years and Older +)	Human Specimen	No. of People Sampled	Units	Geometric Mean (95% Confidence Interval)
3+	Serum	6813	ng/mL	0.062(0.050 – 0.077)
3–11	Serum	1414	ng/mL	0.110(0.076 – 0.160)
12–19	Serum	1902	ng/mL	0.086(0.059 – 0.126)
20+	Serum	3497	ng/mL	0.052(<LOD – 0.063)
Males	Serum	3149	ng/mL	0.075(0.059 – 0.095)
Females	Serum	3664	ng/mL	0.053(<LOD – 0.066)
Mexican Americans	Serum	1877	ng/mL	0.060(<LOD – 0.084)
Non-Hispanic Blacks	Serum	1599	ng/mL	0.164(0.136 – 0.197)
Non-Hispanic Whites	Serum	2845	ng/mL	0.052(<LOD – 0.068)

Source: Adapted from the Third National Report on Human Exposure to Environmental Chemicals, 2005, using only the 2001–2002 survey. Chemicals or their metabolites were measured in blood and urine samples from a random sample of participants from the National Health and Nutrition Examination Survey (NHANES). Department of Health and Human Services, Centers for Disease Control and Prevention, National Center for Environmental Health, Division of Laboratory Sciences, Atlanta, GA 30341–3724

From 1988 through 1991, as part of NHANES III, CDC determined that the median level (50th percentile) of cotinine among nonsmokers in the United States was 0.2ng/mL (ppb).[18] Since that 1988–1991 survey period, median levels of cotinine (as measured in the NHANES 1999–2002) have decreased 68% in children, 69% in adolescents, and about 75% in adults. This reduction in cotinine levels suggests a major reduction in exposure of the general U.S. population to ETS since the period 1988–1991.[19]

Table 1.4 focuses on one metabolite among many from exposure to polycyclic aromatic hydrocarbons (PAHs), 2-Hydroxynaphthalene or 2-naphthol. PAHs are a class of chemicals that result from the incomplete burning of coal, oil, gas, wood, garbage or other organic substances such as tobacco and charbroiled meat. Exposure can occur through air, water, soil or food. PAHs enter the air from motor vehicle exhaust, residential and industrial furnaces, tobacco smoke, volcanoes, agricultural burning, residential wood burning, and wildfires.[20] When interpreting the magnitude of the concentration of 2-hydroxynaphthalene in urine in Table 1.4 note that the units refer to parts per trillion (ppt). For example, for the 20 and older age group, 2,620 ng/L = 2.60 μg/L or 2.60 ppt.

Table 1.5 focuses on quantitating p,p'-DDE or 1.1'-(2,2-dichloroethenylidene)-bis(4-chlorobenzene) in human serum. Analytical results are reported in #ng/g of lipid since these organic compounds are predominantly found in fatty tissue. This lipophilic organic compound arises from the environmental degradation over time of DDT. Molecular structures and common nomenclature for DDT and DDE were shown earlier. DDT is an insecticide that was used in the 1940s by the military against mosquitoes that carried vector-borne diseases such as malaria. The U.S. EPA banned the use of DDT in the United States in 1973, and it is no longer being applied in this country. However, DDT is still used in other countries. Commercially available DDT (technical grade) contains three forms of DDT: p,p'DDT, o,o'-DDT, and o,p-DDT. In the general

TABLE 1.4
Geometric Mean of Urine Levels of 2-Hydroxynaphthalene. 2-Hydroxynaphthalene is a Metabolite from Exposure to Environmental Polycyclic Aromatic Hydrocarbons

Age Range (Years and Older +)	Human Specimen	No. of People Sampled	Units	Geometric Mean (95% Confidence Interval)
6+	Urine	2748	ng/L	2470(2110 – 2890)
6–11	Urine	387	ng/L	1690(1560 –1840)
12–19	Urine	735	ng/L	2220(1700 – 2900)
20+	Urine	1626	ng/L	2620(2220 – 3100)
Males	Urine	1349	ng/L	2750(2360 – 3210)
Females	Urine	1399	ng/L	2220(1860 – 2660)
Mexican Americans	Urine	665	ng/L	2700(2360 – 3080)
Non-Hispanic Blacks	Urine	692	ng/L	3970(3470 – 4540)
Non-Hispanic Whites	Urine	1207	ng/L	2190(1760 – 2720)

Source: Adapted from the Third National Report on Human Exposure to Environmental Chemicals, 2005 using results for only the 2001–2002 survey. Chemicals or their metabolites were measured in blood and urine samples from a random sample of participants from the National Health and Nutrition Examination Survey (NHANES). Department of Health and Human Services, Centers for Disease Control and Prevention, National Center for Environmental Health, Division of Laboratory Sciences, Atlanta, GA 30341–3724

TABLE 1.5
Geometric Mean of Serum Levels of p,p'-DDE. p,p'-DDE or 1,1'-(2,2-Dichloroethenylidene)-bis[4-Chlorobenzene] is the Stable Environmental Degradation Product of DDT (Dichlorodiphenyltrichloroethane)

Age Range (Years and Older +)	Human Specimen	No. of People Sampled	Units	Geometric Mean (95% Confidence Interval)
Total 12+	Serum	2298	ng/g lipid	295(267–327)
12–19	Serum	758	ng/g lipid	124(106–146)
20+	Serum	1540	ng/g lipid	338(303–376)
Males	Serum	1069	ng/g lipid	285(252–323)
Females	Serum	1229	ng/g lipid	305(273–341)
Mexican Americans	Serum	566	ng/g lipid	652(569–747)
Non-Hispanic Blacks	Serum	515	ng/g lipid	324(262–400)
Non-Hispanic Whites	Serum	1053	ng/g lipid	253(226–284)

Source: Adapted from the Third National Report on Human Exposure to Environmental Chemicals, 2005 using results for only the 2001–2002 survey. Chemicals or their metabolites were measured in blood and urine samples from a random sample of participants from the National Health and Nutrition Examination Survey (NHANES). Department of Health and Human Services, Centers for Disease Control and Prevention, National Center for Environmental Health, Division of Laboratory Sciences, Atlanta, GA 30341–3724

population, food is the primary source of exposure to DDT. Many foods and commodities contain detectable residues of DDT or its degradation products. The estimated food intake of DDT in the U.S. had decreased since the 1950s. However, food imported into the U.S. from other countries that still use DDT may contain DDT or DDE residues. In addition, local spraying with DDT can add greatly to body burdens. For example, after a single application of DDT for malaria control, DDT levels were sevenfold higher in people tested 1 year after the application than in a comparison population.[20] After DDT enters the body, it is metabolized to DDE. Table 1.5 shows that those in the population sampled who are 20 years and older have over twice the concentration of DDE versus those in the 12–19 years age group. Finding a measurable amount of p,p'DDE does not mean that this finding will result in an adverse health effect. This data also provides physicians with a reference range so they can determine whether or not other people have been exposed to higher levels of DDT or DDE than levels found in the general population.

Let us summarize some regulatory issues, first from this emerging trace enviro-health quantitative analysis arena. We then complete this introductory chapter with an emphasis on the well-established trace enviro-chemical quantitative analysis arena, largely reviewing the significant environmental regulations. We then show just how the EPA methods fit in. A significant question is before us with respect to enviro-health.

Trace enviro-health quantitative analysis, also abbreviated TEQA, is, in this author's opinion, an evolving subdiscipline of trace environmental quantitative analysis. The Clinical Laboratory Improvement Act of 1988 (CLIA 88) regulates the chemical laboratory and addresses those aspects of traditional clinical chemistry, such as determining the concentration of creatinine in blood. Toxicological chemistry also includes blood alcohol, digoxin, lithium, primidone, and theophylline assays. The concentrations in the blood and urine of these analytes are significantly higher than those that would be considered at a trace level. Our focus in this book is to discuss how environmental pollutants can be quantitatively determined in human specimens. However, environmental-priority pollutants found in human specimens may have entered the human domain via the various routes of exposure. Refer to Figure 1.6 which depicts a biomonitoring scenario centered with respect to a person's blood supply and shows routes of exposure to environmental priority pollutants along with the possible kinds of body fluids, shown as ovals, that could be defined as suitable human specimens for biomonitoring.[20] Three routes of exposure include inhalation to the respiratory tract, ingestion to the gastrointestinal tract, and absorption through the skin, often termed dermal exposure. The development of so-called biological markers (*biomarkers*) represents a very active research area involving toxicologists and epidemiologists. TEQA has a vital role to play in this research today. A biomarker can be either cellular, biochemical, or molecular in nature and can be measured analytically in biological media such as tissues, cells, or fluids. A suitable biomarker could be an exogenous substance or its metabolite. It could also be a product of an interaction between the xenobiotic agent and some target molecule. *Exposure* and *dose* are two terms that are further elaborated below:

- Exposure is contact of a biological, chemical, or physical agent with the surface of the human body.
- Dose is the time integral of the concentration of the toxicologically active form of the agent at the biological target tissue.
- Dose links exposure to risk of disease.
- Exposure ≈ dose ≈ effect.

The relationship between exposure, dose, and potential health effects is summarized in Figure 1.7. This model does not include adipose or other human tissue. Blood and urine have emerged as the most convenient human specimens to collect and conduct biomonitoring.

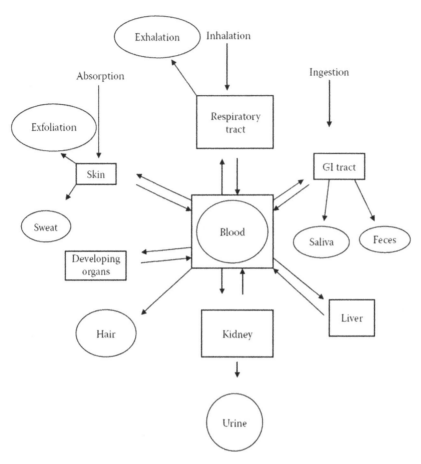

FIGURE 1.6

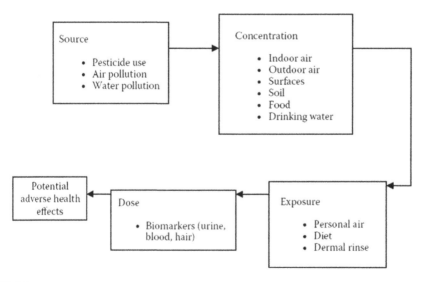

FIGURE 1.7

Let us return to the concept of a biomarker as a key ingredient in biomonitoring. Biomarkers provide evidence of both exposure and uptake. The concentration level of a given biomarker is directly related to tissue dose. Biomarkers account for all possible routes, as shown in Figure 1.6. Biomarkers account for differences in people. Factors that limit the usefulness of biomarkers include:[21]

- The fact that many biomarkers are still being developed
- The need for standardized protocols in both collection and analysis
- Variability in relationship with exposure
- Timing—each biomarker has a characteristic half-life
- Expense
- Difficulty in interpreting.

1.10   WHAT MIGHT AN ANALYTICAL CHEMISTRY APPROACH TO BIOMONITORING LOOK LIKE?

One answer to this question can be found in Figure 1.8. The scenario "from human specimen to analytical result" is listed in terms of five essential and sequential steps, each linked by a *chain-of-custody* protocol. The arrows show that the relationship between steps must include a chain-of-custody protocol. This protocol might take the form of a written document. If, however, a Laboratory Information Management System (LIMS) is in place, the protocol would include data entry into a computer that utilizes an LIMS. Referring to Figure 1.8, the sample prep lab may give to the analyst a complete sample extract along with a signed chain-of-custody form to provide evidence as to where the sample extract is headed next. This five-step approach to biomonitoring is also applicable to trace enviro-chemical quantitative analysis.

Consider the need for biomonitoring as a result of a terrorist-related event; let's work through the scenario presented in Figure 1.8 with this in mind. Consider this hypothetical event. A group of about 20 employees gather in a typical conference room. They sit at a large conference table as they begin to conduct their monthly meeting. One disgruntled employee brings two containers to the meeting. Each container holds a liter of liquid of unknown chemical composition. He waits for the intended moment and then pours the contents from the two containers toward the same location on the large table. Those humans sitting at the table begin to smell a very unpleasant odor and eventually get quite ill. Fatalities are possible. The local authorities are called in to assess the situation. It has become a local crime scene as well as a public health bio-terrorist event. With Figure 1.8 in mind, sampling might begin very quickly. Blood, urine and possibly other body fluids might be sampled, collected and appropriately labeled. Sample/specimen preservation becomes an important consideration. We all know what it is like to get blood and urine withdrawn for medical considerations. It becomes crucial in the case of a potential bio-terrorist event to properly sample, label and store blood and urine specimens while adhering to strict protocols. In the United States, the Centers for Disease Control and Prevention (CDC) has primary jurisdiction in the case of a bio-terrorist event. After chemical identification is made, the CDC will direct any and all further testing from the site of the bio-terrorist event to appropriate state public health laboratory facilities. Exactly what chemicals were found in the biological samples obtained from humans dictates the type of sample preparation approach and determinative analytical technique to be used to conduct trace environmental health quantitative analysis. For example, if nerve agent metabolites were found in affected urine samples, the appropriate determinative technique might be liquid chromatography while gas chromatography would be preferred if cyanide was found in affected blood samples. Results would be reported out from the specific state laboratory assigned the project to the appropriate federal agencies. This author was involved early on in developing several analytical methods at a state laboratory for the CDC as part of its Laboratory Response Network during the early years of the new millennium.

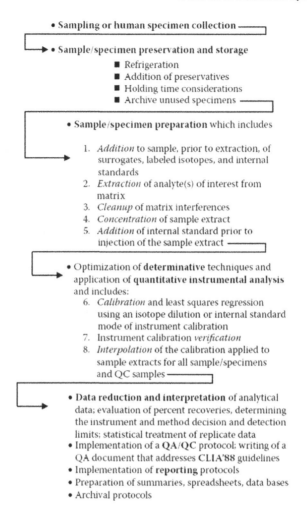

• Sampling or human specimen collection

• Sample/specimen preservation and storage
  ■ Refrigeration
  ■ Addition of preservatives
  ■ Holding time considerations
  ■ Archive unused specimens

• Sample/specimen preparation which includes

  1. *Addition* to sample, prior to extraction, of surrogates, labeled isotopes, and internal standards
  2. *Extraction* of analyte(s) of interest from matrix
  3. *Cleanup* of matrix interferences
  4. *Concentration* of sample extract
  5. *Addition* of internal standard prior to injection of the sample extract

• Optimization of **determinative** techniques and application of **quantitative instrumental analysis** and includes:

  6. *Calibration* and least squares regression using an isotope dilution or internal standard mode of instrument calibration
  7. Instrument calibration *verification*
  8. *Interpolation* of the calibration applied to sample extracts for all sample/specimens and QC samples

• **Data reduction and interpretation** of analytical data; evaluation of percent recoveries, determining the instrument and method decision and detection limits; statistical treatment of replicate data
• Implementation of a **QA/QC** protocol; writing of a QA document that addresses **CLIA'88** guidelines
• Implementation of **reporting** protocols
• Preparation of summaries, spreadsheets, data bases
• Archival protocols

FIGURE 1.8

As an example of biomonitoring not related to bioterrorism, consider human exposure to the controversial chemical Bisphenol A (recall the concern over polycarbonate baby bottles). Bisphenol A (BPA) or 2,2 -bis(4-hydroxyphenyl)propane, is an industrial chemical used in a variety of plastic materials, some of which are used in packaging and hence is believed to leach out into consumable items such as food and dental fillings. Earlier, Brock and coworkers and the CDC developed a quantitative analytical method to determine just how much BPA might be present in a urine sample.[21] BPA is apparently excreted from a person believed to be exposed to BPA. Urinary BPA glucuronide seems to be a longer-lived biomarker (12 to 48 h). After de-glucuronidation using β-glucuronidase, BPA was isolated and recovered by reversed-phase solid-phase extraction. The isolate was converted to its pentafluorobenzyl diether. This is a classic example of the use of chemical derivatization. Long before liquid chromatography mass spectrometry was well developed, polar analytes like BPA had to be chemically derivatized so as to make the analyte amenable to quantitative analysis using gas chromatography. The pentafluorobenzyl diether of BPA was quantitated using isotope dilution gas chromatography mass spectrometry (GC-MS). A method detection limit (MDL) was reported to be 120 parts per trillion (ppt). Figure 1.9 shows two different mass spectra for the

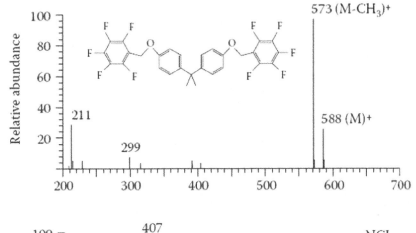

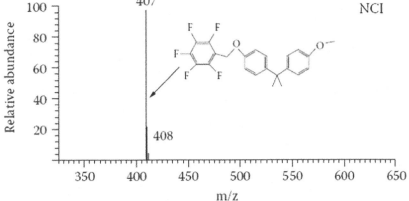

FIGURE 1.9

pentafluorobenzyl ether of BPA that eluted from the gas chromatographic column at ~26.4 min. The top mass spectrum in Figure 1.9 was obtained via electron-impact mass spectrometry and reflects positive fragment ions, while the bottom mass spectrum was obtained via negative chemical ionization mass spectrometry and reflects negative fragment ions. Pooled human urine samples showed no detectable BPA before the urine was treated, while BPA concentration levels varied from 0.11 to 0.51 parts per billion (ppb) for the treated urine. Molecular structures for Bisphenol A and for pentafluorobenzyl bromide (α-bromo-2,3,4,5,6 -pentafluorotoluene) are shown below:

Bisphenol A   Pentafluorobenzyl bromide

We leave for the moment this introductory topic of trace *enviro-health* quantitative analysis and pick up the introduction to trace *enviro-chemical* quantitative analysis. Let us first define what we mean by environmental chemistry.

1.11   WHAT KIND OF CHEMISTRY IS THIS?

Environmental chemistry can be defined as a systematic study of the nature of matter that exists in the air, water, soil, and biomass. This definition could be extended to the plant and animal domains where chemicals from the environment are likely to be found. This discipline, whose origins go back as far as the late 1960s, requires a knowledge of the traditional branches of organic, inorganic, physical, and analytical chemistry. Environmental chemistry is linked to biotechnology as well as to chemical, environmental, and agricultural engineering practices.

Environmental analytical chemistry can be further defined as a systematic study that seeks to answer two fundamental questions: *What* and *how* much matter exists in the air, water, soil, and biomass? This definition could also be extended to the plant and animal domains just discussed. This discipline, which developed in the 1970s, spearheaded by the first Earth Day in 1970 and the establishment of the U.S. EPA, requires a knowledge of traditional quantitative analysis, contemporary instrumental analysis, and selected topics, such as statistics, electronics, computer software, and experimental skill. Environmental analytical chemistry represents *the* fundamental measurement science to biotechnology and to chemical, environmental, and agricultural engineering practices. That portion of environmental analytical chemistry devoted to rigorously quantifying the extent to which chemical substances have contaminated the air, water, soil, and biomass is the subject of this book.

In its broadest sense, environmental chemistry might be considered to include the chemistry of everything outside of the synthetic chemist's flask. The moment that a chemical substance is released to the environment, its physical-chemical properties may have an enormous impact on ecological systems, including humans. Researchers have identified 51 synthetic chemicals that disrupt the endocrine system. Hormone disrupters include some of the 209 polychlorinated biphenyls (PCBs) and some of the 75 dioxins and 135 furans that have a myriad of documented effects (p. 81).[22] The latter half of the 20th century has witnessed more synthetic chemical production than any other period in world history. Between 1940 and 1982, the production of synthetic chemicals increased about 350 times. Billions of pounds of synthetic materials were released into the environment during this period. U.S. production of carbon-based synthetic chemicals topped 435 billion pounds in 1992, or 1,600 pounds per capita (p. 137).[22]

The concept of environmental contaminants as estrogenic "mimics" serves to bring attention to the relationship between chemicals and ecological disruption. The structural similarity between DDT and diethyl stilbestrol is striking. The former chemical substance was released into the environment decades ago, whereas the latter was synthesized and marketed to pregnant women during the 1950s and then used as a growth promoter in livestock until it was banned by the Food and Drug Administration (FDA) in 1979.[23]

At levels typically found in the environment, hormone-disrupting chemicals do not kill cells or attack DNA. Their target is hormones, the chemical messengers that move about constantly within the body's communication. They mug the messengers or impersonate them. They jam signals. They scramble messages. They sow disinformation. They wreak all manner of havoc. Because messages orchestrate many critical aspects of development, from sexual differentiation to brain organization, hormone-disrupting chemicals pose a particular hazard before birth and early in life (pp. 203–204).[22]

A more recent controversy has arisen around the apparent leaching of Bisphenol A from various sources of plastics that are in widespread use among consumers. Earlier, the isolation and recovery of Bisphenol A from human urine was discussed. How could that method be changed to enable Bisphenol A to be isolated and recovered from an environmental matrix such as plastic wrap? Molecular structures for p,p'-DDT and diethyl stilbestrol are shown below. Compare these structures to that shown earlier in this chapter for Bisphenol A. The similarities in molecular structure are striking.

p, p'-DDT   Diethylstilbestrol

Much earlier the EPA released its plan for testing 15,000 chemicals for their potential to disrupt hormone systems in humans and wildlife. These chemicals were chosen because they are produced in volumes greater than 10,000 pounds per year.[24]

One usually hears about environmental catastrophes through the mass media and increasingly through social media today. A name is assigned to the environmental disaster that also includes a geographic connotation. In a chapter titled "Controversies and Issues," the author cites the following as major controversies and issues addressed by the EPA over the years: Love Canal and the Hooker Chemical Company, the Exxon Valdez: Oil Spilled, Oil Drilling in the Arctic National Wildlife Refuge: Oil Drilled?, Three Mile Island: Radiation Leaked, Disposal of Nuclear Waste at Yucca Mountain: Radiation Disposed?[25] What is not so newsworthy, yet may have as profound an impact on the environment, is the ever-so-subtle pollution of the environment day in and day out. Both catastrophic pollution as well as the more subtle aspects of day-to-day pollution of our air, water and soil require the techniques of TEQA to obtain data that enables society to continuously monitor the environment to ensure minimal ecological and toxicological disruption. It is the combination of sophisticated analytical instruments (Chapter 4), conventional and novel sample preparation strategies (Chapter 3), mathematical treatment of analytical data (Chapter 2), and practical procedures as illustrated by student experiments (Chapter 5) that enables a student or practicing analyst to effectively conduct TEQA.

This book provides insights and tools that enable an individual who either works in an environmental testing lab or public health lab or anticipates having a career in the environmental science or environmental health field to make a contribution. Individuals are thus empowered and can begin to deal with the problems of monitoring and sometimes finding the extent to which chemicals have contaminated the environment or entered the human body.

1.12 WHO NEEDS ENVIRONMENTAL TESTING?

It is too easy to answer this question with "everyone." The industrial sector of the U.S. economy is responsible for the majority of chemical contamination released to the environment. Since the early 1970s, industry has been under state and federal regulatory pressures not to exceed certain maximum contaminant levels (MCLs) for a variety of so-called priority pollutant organic and inorganic chemical substances. However, one of the more poignant examples of small-time pollution is that of dry cleaning establishments located in various shopping plazas throughout the U.S. These small businesses would follow the practice of dumping their dry cleaning fluid into their septic systems. It was not unusual, particularly during the 1980s, for labs to analyze drinking water samples drawn from an aquifer that served the shopping plaza and find parts per billion (ppb) or greater concentration levels of chlorinated volatile organics such as trichloroethylene (perchloroethylene or PCE) whose molecular structure is shown below:

$$\begin{array}{ccc} Cl & & Cl \\ \diagdown & & \diagup \\ & C{=}C & \\ \diagup & & \diagdown \\ Cl & & H \end{array}$$

The necessary sample preparation needed to modify a sample taken from an aquifer that is expected to contain PCE, so as to enable the sample to become compatible with the appropriate analytical instrument, will be described in Chapter 3. The identification and quantitative determination of priority pollutants like PCE in drinking water require sophisticated analytical instrumentation. These so-called determinative techniques will be described in Chapter 4. A laboratory exercise that might introduce a student to the technique involved in sample preparation and instrumental analysis to quantitatively determine the presence or absence of a chlorinated volatile organic like PCE will be described in Chapter 5.

1.13 WHO REQUIRES INDUSTRY TO PERFORM TEQA?

Demand for trace enviro-chemical quantitative analysis is largely regulatory-driven, with the exception of the research done in methods development by both the private sector and federal, state, and academic labs. The major motivation for a company to conduct TEQA is to demonstrate that its plant's effluent falls within the MCLs for the kinds of chemical contaminants that are released. There exists a myriad of laws that govern discharges, and these laws also specify MCLs for targeted chemical contaminants. The following outline is a brief overview of the regulations, and it incorporates the abbreviations used by practitioners in this broad category of environmental compliance and monitoring (pp. 1–32).[26]

1.14 HOW DOES ONE MAKE SENSE OF ALL THE "REGS"?

The following outline summarizes the federal regulations responsible for environmental compliance:

A Title 40 Code of Federal Regulations (40 CFR): This is the ultimate authority for environmental compliance. New editions of 40 CFR are published annually and are available on the World Wide Web. This resource includes chapters on air, water, pesticides, radiation protection, noise abatement, ocean dumping, and solid wastes. Superfund, Emergency Planning and Right-to-Know, effluent guidelines and standards, energy policy, and toxic substances are among other topics.

B Government regulations administered by the Environmental Protection Agency (EPA)
   1 Resource Conservation and Recovery Act (RCRA): Passage of the RCRA in 1976 gave the EPA authority to oversee waste disposal and hazardous waste management. Identification of a waste as *hazardous* relied on specific analytical tests. Analytical methods that deal with RCRA are found in a collection of four volumes titled *Test Methods for Evaluating Solid Waste Physical/Chemical Methods,* commonly referred to as SW-846. Three major subtitles deal with hazardous waste management, solid waste management, and underground storage tanks.
   2 Comprehensive Environmental Response, Compensation, and Liability Act (Superfund) (CERCLA): This authority granted to the EPA enables the agency to take short-term or

emergency action to address hazardous situations that affect health. The release of the toxic chemical isocyanate in the Bhopal, India, community that left over 3,000 dead might have fallen under CERCLA if it had occurred in the U.S. In addition, the CERCLA contains the authority to force the cleanup of hazardous waste sites that have been identified based on environmental analytical results and placed on the National Priority List. The EPA also has authority to investigate the origins of waste found at these sites and to force the generators and other responsible parties to pay under CERCLA. Analytical methods that deal with CERCLA are provided through the Contract Laboratory Program (CLP). The actual methods are found in various Statements of Work (SOWs) that are distributed to qualified laboratories.

3 Drinking Water and Wastewater: The Safe Drinking Water Act, amended in 1986, and then amended again in 1996, gives EPA the authority to regulate drinking water quality. The law requires many actions to protect drinking water and its sources: rivers, lakes, reservoirs, springs and groundwater wells. (The act does not regulate private wells that serve fewer than 25 individuals).[25] Two broad categories are considered within which to place numerous priority pollutants, called *targeted analytes*, in this act. The first category is the National Primary Drinking Water Standards. These chemical substances affect human health, and all drinking water systems are required to reduce their presence to below a certain analyte concentration in drinking water established for each compound by the federal government. The second group of targeted analytes is the National Secondary Drinking Water Standards. These analytes include chemical substances that affect the taste, odor, color, and other non-health-related qualities of water. A given chemical compound may appear on both lists at different levels of action. A selection of the most recent primary drinking water monitoring contaminants are listed in Tables 1.6 through 1.8 by category. The most recent secondary drinking water monitoring contaminants are listed in Table 1.9. The maximum contaminant level goal (MCLG) as well as the current maximum contaminant level (MCL) in #mg of analyte per liter of drinking water are found in these tables. Toxicological considerations govern the decision to estimate what makes for an environmentally acceptable MCL. The more toxic the chemical, the lower is the value of both the MCLG as well as the MCL. Both parameters depend upon analytical chemistry methods that have method detection limits that must be lower than the reported values for both the MCLG and the MCL. For some analytes, the MCLG has been met, while for other analytes, the MCLG has not been met.

4 The Clean Water Act, which was last amended in 1987, provides for grants to municipalities to build and upgrade treatment facilities. The act also establishes a permit system known as the National Pollutant Discharge and Elimination System (NPDES) for discharge of natural water bodies by industry and municipalities. Over two thirds of the states have accepted responsibility for administration of the act. The act and its amendments are based on the fact that no one has the right to pollute the navigable waters of the U.S. Permits limit the composition and concentration of pollutants in the discharge. Wastewater effluents are monitored through the NPDES, and this analytical testing has been a boon for commercial analytical laboratories. Every industry and wastewater treatment facility that directs into a receiving stream or river has either a federal or state NPDES permit. Methods of analysis are given in 40 CFR 136 for the following:

(1) Conventional pollutants such as biological oxygen demand (BOD), chemical oxygen demand (COD), pH, total suspended solids, oil and grease, and fecal coliforms.

(2) Nonconventional pollutants such as nitrogen, phosphorous, and ammonia.

(3) The 129 so-called priority pollutants.

A select number of these pollutants are listed in Tables 1.6 through 1.8.

TABLE 1.6
Primary Drinking Water Monitoring Requirements for Inorganics

Contaminant	MCLG (mg/L)	MCL (mg/L)
Antimony (Sb)	0.006	0.006
Arsenic (As)	Zero	0.01
Barium (Ba)	2	2
Beryllium	0.004	0.004
Bromate (BrO_3^-)	Zero	0.01
Cadmium (Cd)	0.005	0.005
Chloramine	4	4
Chlorine	4	4
Chlorine dioxide	0.8	0.8
Chromium total (Cr)	0.1	0.1
Copper (Cu at tap)	1.2	1.3
Cyanide (CN^-)	0.2	0.2
Fluoride (F^-)	4	4
Lead (Pb at tap)	Zero	0.015
Manganese (Mn)	–	–
Mercury (Hg)	0.002	0.002
Molybdenum (Mo)	–	–
Nickel (Ni)	---	---
Nitrate (NO_3^-) as N	10	10
Nitrite (NO_2^-) as N	1	1
Nitrate + Nitrite ($NO_3^- + NO_2^-$) as N	10	10
Perchlorate (ClO_4^-)	---	---
Selenium (Se)	0.05	0.05
Silver (Ag)	---	---
Strontium (Sr)	---	---
Thallium (Tl)	0.0005	0.002
White Phosphorous (P)	---	---
Zinc (Zn)	---	---

MCLG – maximum contamination level goal in milligrams per liter
MCL – maximum contaminant level in milligrams per liter

Source: 2018 Edition of the Drinking Water Standards and Health Advisory Tables,epa.gov/sites/production/files/ 2018-03/documents/dw tables.2018.pdf, EPA 822-f-18-01. Office of Water, U.S. Environmental Protection Agency, Washington, D.C., March, 2018.

TABLE 1.7
Primary Drinking Water Monitoring Requirements for Semivolatile Organics

Contaminant	MCLG (mg/L)	MCL (mg/L)
Aldicarb	0.001	0.003
Aldrin	---	---
Anthracene (PAH)	---	---
Atrazine	0.003	---
Benzo[a]anthracene (PAH)	---	---
Benzo[a]pyrene (PAH)	---	---
Carbofuran	0.04	0.04
Chlordane	Zero	0.002
Chloropyrifos	---	---
Chrysene (PAH)	---	---
2,4-Dichlorophenoxyacetic acid (2,4-D)	0.07	0.07
Di(2-ethylhexyl) adipate	0.4	
Di(2-ethylhexyl) phthalate	Zero	0.006
Diazinon	---	---
Dicamba	---	---
Dinoseb	0.007	0.007
Diquat	0.02	0.02
Endothall	0.1	0.1
Ethylenedibromide (EDB)	Zero	0.0005
Glyphosate	0.7	0.7
Heptachlor	Zero	0.0004
Hexachlorobenzene	Zero	
Hexachlorocyclopentadiene	0.05	0.05
Lindane	0.0002	0.0002
Methoxychlor	0.04	0.04
Oxamyl	0.2	0.2
Polychlorinatedbiphenyls (PCBs)	Zero	0.001
Pentachlorophenol	Zero	0.001
Picloram	0.5	0.5
Simazine	0.004	0.004
Toxaphene	Zero	
2,3,7,8-TCDD (Dioxin)	Zero	
2,4,5-trichlorophenoxyacetic (Silvex)	0.05	0.05

MCLG – maximum contamination level goal in milligrams per liter
MCL – maximum contaminant level in milligrams per liter
PAH – polycyclic aromatic hydrocarbon

Source: 2018 Edition of the Drinking Water Standards and Health Advisory Tables,epa.gov/sites/production/files/ 2018-03/documents/dw tables.2018.pdf, EPA 822-f-18-01. Office of Water, U.S. Environmental Protection Agency, Washington, D.C., March, 2018.

TABLE 1.8
Primary Drinking Water Monitoring Requirements for Volatile Organics

Contaminant	MCLG (mg/L)	MCL (mg/L)
Carbon Tetrachloride	Zero	0.005
Benzene	Zero	0.005
o-dichlorobenzene	0.6	0.6
p-dichlorobenzene	0.075	0.075
1,2-dichloroethane	Zero	0.005
1,1-dichloroethane	0.007	0.007
Cis 1,2-dichloroethylene	0.07	0.07
Trans-1,2-dichloroethylene	0.1	0.1
Dichloromethane	Zero	0.005
1,2-dichloropropane	Zero	0.005
Ethylbenzene	0.7	0.7
Styrene	0.1	0.1
Tetrachloroethylene (PCE)	Zero	0.005
Tolulene	1	1
1,2,4-trichlorobenzene	0.07	0.07
1,1,2-trichloroethane	0.2	0.2
Trichloroethylene (TCE)	Zero	0.005
Vinyl chloride	zero	0.002
Xylenes	10	10

MCLG – maximum contamination level goal in milligrams per liter
MCL – maximum contaminant level in milligrams per liter

Source: 2018 Edition of the Drinking Water Standards and Health Advisory Tables,epa.gov/sites/production/files/2018-03/documents/dw tables.2018.pdf, EPA 822-f-18-01. Office of Water, U.S. Environmental Protection Agency, Washington, D.C., March, 2018.

5 Federal Insecticide, Fungicide, and Rodenticide Act (FIFRA): This act requires the EPA to oversee the manufacture and use of chemical substances that directly terminate insects, fungi, and rodents. Analytical methods are required for the analysis of residues in air, soil, and water. Specifications for labeling and warnings are also required under this act.

6 Food, Drug, and Cosmetic Act: Administered by the Food and Drug Administration (FDA), this act governs all chemicals added to foods, drugs, and cosmetics. These substances are not directly under EPA oversight; however, the wastes and by-products generated from the manufacture of these substances are controlled by EPA.

7 Toxic Substance Control Act (TSCA): Since 1976, the EPA has been given the authority to gather information on the toxicity and hazardous nature of individual chemicals. Chemical producers are required to supply information dealing with risk assessment of proposed products 90 days before proposed manufacture or import. Included in this information are chemical fate testing, environmental effects testing, and health effects testing.

8 Superfund Amendments and Reauthorization Act (SARA): Passed in 1986, this act extends the lifetime of the legislation begun in CERCLA and gives the EPA the authority

TABLE 1.9
Secondary Drinking Water Monitoring Requirements

Contaminant	SDWR
Aluminum (Al)	0.05 – 0.2 mg/L
Chloride (Cl⁻)	250 mg/L
Color	15 color units
Copper (Cu)	1.0 mg/L
Corrosivity	Non-corrosive
Fluoride (F⁻)	2.0 mg/L
Foaming agents	0.5 mg/L
Iron	0.3 mg/L
Manganese	0.05 mg/L
Odor	3 Threshold odor numbers
pH	6.5 – 8.5
Silver	0.1 mg/L
Sulfate	250 mg/L
Total dissolved solids (TDS)	500 mg/L
Zinc	5 mg/L

SDWR – Secondary Drinking Water Regulations, non-enforceable Federal guidelines regarding cosmetic or aesthetic effects

Source: 2018 Edition of the Drinking Water Standards and Health Advisory Tables,epa.gov/sites/production/files/2018-03/documents/dw tables.2018.pdf, EPA 822-f-18-01. Office of Water, U.S. Environmental Protection Agency, Washington, D.C., March, 2018

to remediate a site if there are no responsible parties that can be found to pay the cost for remediation.

9 Clean Air Act (CAA): The CAA amendments passed in 1990 gave the EPA authority to regulate many hazardous air pollutants (HAPs). Over 100 of these HAPs must be regulated; they include well-known organic compounds such as acetaldehyde, benzene, carbon tetrachloride, 1,4-dioxane, hexane, methyl methacrylate, and so forth. The CAA also requires the EPA to establish permits for the regulation of the maximum amounts of emissions by various industries in a manner similar to that of the NPDES program for wastewater effluents.

As if the number of regulations is not enough, many environmental testing labs also have to satisfy state requirements. Most states run their own private lab certification programs. Also, some of the method details differ between state and federal programs. One example from the author's own experience working in an EPA contract lab during the late 1980s involves the implementation of EPA Method 8270 for semivolatile organics in solid waste. EPA Method 3640A, Gel Permeation Chromatography (GPC), is added to any one of the 3500 series of sample preparation methods, particularly if soil samples are saturated with oil. To satisfy the New Jersey Department of Environmental Protection requirements when this method is implemented requires that the GPC cleanup step *not be used* even if some soil samples that arrive to the laboratory are indeed saturated with oil. On the other hand, if similar oil-laden soil samples were analyzed for the New York Department of Environmental Conservation, GPC cleanup is acceptable. What is an analyst to do?

1.15   WHAT ANALYTICAL METHODS SATISFY THESE REGULATIONS?

Today, myriads of analytical methods exist that the above-cited regulations use to demonstrate compliance. The two parameters cited in Tables 1.6 through 1.8, namely, the MCLG and the MCL, are only obtained with good accuracy and precision by applying all of the skills and techniques of TEQA. Analysts must also have a good idea of how low a concentration of a specific chemical analyte can be accurately and precisely measured in a given analytical method. Each analytical method should have a known instrument (IDL) and known method detection limit (MDL). The author will discuss his approach to calculating IDLs and then MDLs in Chapter 2. The analyst who is about to work in the environmental testing laboratory must have access to these written methods in order to become familiar with their content before he or she can effectively execute the methods themselves.

The EPA during the past 30 years has operated within two broad mental frameworks or paradigms. The first paradigm, which began in the early 1970s and lasted until the early 1990s, suggested that the methods written by the agency were the only ones that a laboratory could use in order to satisfy the "regs." This mind-set led to the proverbial "EPA approved" stamp of approval. Instrument manufacturers during the late 1980s would incorporate into their advertisements that their product was EPA approved. During these years, instrument vendors sought collaborations with various EPA offices in an attempt to accelerate the more than 10-year delay between proposal of a new method and its acceptance by the EPA.

For example, the development of an instrumental method that would determine more than one inorganic anion in a drinking water sample was announced in 1975.[26] It took until around 1985 for this technique to become incorporated into EPA Method 300.0 as applied to drinking water. Prior to the development of ion chromatography, individual anionic species such as chloride, nitrate, phosphate, and sulfate were determined individually by applying specific colorimetric methods. Each colorimetric method would require separate sample preparation and determinative steps culminating with the use of a spectrophotometer. The idea that a sample could be injected into some kind of chromatograph and all of the ions separated, detected, and quantitated in a reasonable period represented a bit of a revolutionary concept. It was also thought that there would be a serious matrix effect that would render this new method useless when it came time to analyze "dirty" samples. Today, after some 20 or more years of instrumentation refinement, the advantages and limitations of the method that requires ion chromatography are well understood. This technique has even advanced to the point where one can conduct analyses without having to prepare either an eluent (mobile phase passed through the ion exchange column) or a regenerate (mobile phase passed through the suppressor), thus further simplifying the implementation of this methodology.[28] Should it take a decade or two for a newly developed method to become EPA approved? The future appears not to be emulating the past, as the second broad mental framework is taking hold; the agency calls it the performance-based method (PBM).

The new paradigm places more of the responsibility for quality assurance on the individual analyst and the laboratory rather than on the method itself. It used to be that as long as an analyst followed the recipes (sounds a lot like cooking) in the various methods, then quality data was assured. The analyst's input in the process was not required, nor was it requested. This point of view took the analyst out of the picture and served to define the analyst more as a robot-technician than as an involved and motivated professional. This point of view left no room for analyst intervention. If an analyst wished to skip a step here or there in the method, it was assumed that the method was not followed and, therefore, the data could not be assured or, in the parlance of the time, EPA approved. PBMs hopefully have and will change this somewhat myopic view of TEQA and, furthermore, begin to place the responsibility on the analyst to yield quality data that is assured based on method performance, as opposed to blindly adopting the recipe.

For example, instead of an EPA-approved method specifying that a particular gas chromatographic (GC) column be used to measure analyte A in matrix B, the laboratory is free to choose what column to use, as long as an equivalent degree of quality assurance is obtained. In other words, analytical results for the concentration of analyte A in matrix B from the approved method vs. the PBM approach should be nearly identical. In addition, the laboratory would not have to conduct the side-by-side comparison of a more conventional sample preparation against a proposed alternative in this new paradigm. For example, it was once considered heresy for an analyst to analyze wastewater samples to determine the slew of organochlorine pesticides using anything except a glass column that contained Supelcoport coated with 1.5% SP-2250/1.95% SP-2401 packed into a 1.8m-long × 4mm-inner-diameter tube (the essence of Method 608). However, during the late 1980s, labs, including the one this author worked in, began to investigate megabore capillary columns as alternatives to packed columns. Megabore capillary columns made from fused silica could be easily connected to the common ¼-in. injection port used for packed columns via megabore adapter kits. When combined with element-specific detectors, many new organics could be separated and identified using megabore capillary GC columns.[29]

Performance-based methods are an empowering concept. PBMs enable analysts and the laboratories that they work in to become more creative and encourage attempts to develop alternative methods to perform TEQA. Many of these alternative approaches are much more cost effective while yielding the same outcome. In other words, the performance of the method is the same, deliverable outcome, and the inputs are much reduced in cost and labor. Alternative methods of sample preparation will be discussed in Chapter 3.

1.16   HOW ARE THE REGULATORY METHODS FOR TEQA IDENTIFIED?

All EPA methods are classified based on a number, whereas other methods are categorized based on the chemical composition of the analytes of interest. It is useful at this point to briefly outline these EPA method numbers while providing a brief description of their origin (pp. 33–36).[26] All of the major offices within the EPA have had a role in developing analytical methods that bear a number. Series 1 to 28 are air monitoring methods found in 40 CFR Appendix A, whereas series 101 to 115 are air monitoring methods in 40 CFR Appendix B. Methods that have three digits refer to methods for the chemical analysis of water and wastes and represent some of the earliest EPA methods. The 100 series of methods characterizes the physical properties of water and wastes. The 200 series is devoted to metals, and the 300 series deals with inorganics and nonmetals. This is why the ion chromatographic technique discussed earlier is found in this series. The 400 series deals with organics where the emphasis is on nonspecific organics that are primarily found in wastewater. Method 413 for oil and grease, Method 415 for total organic carbon, Method 420 for total recoverable phenols, and Method 425 for methylene blue active surfactants are examples of nonspecific analytical methods that give good indications of whether a waste sample contains organics without the need to be specific. These are workhorse methods that are usually found in most environmental testing labs.

Method 413 has recently become controversial in that the manufacture of the common extraction solvent used in the method, 1,1,2-trichlorotrifluoroethane (TCTFE), itself a fluorocarbon (Freon 113), is banned. The attributes of TCTFE are such that it enabled this method to move from a purely gravimetric approach (EPA Method 413.1, Gravimetric, Separatory Funnel Extraction) to an instrumental approach, whereby the carbon-to-hydrogen stretching frequency in the infrared spectrum enabled a significant reduction in the MDL to occur. The infrared approach was assigned its own method number (EPA Method 413.2, Spectrophotometric, Infrared). This author's personal opinion is that the more cost-effective infrared approach coupled with the lower MDLs should be maintained. Labs should be strongly encouraged to recycle the waste TCTFE solution so that it can be reused. A simple distillation should suffice to return TCTFE in sufficient purity to use to re-extract samples. However, the EPA has chosen to establish Method 1664 for oil and grease analysis. TCTFE has been replaced by hexane, and the determination is by weighing the residue after solvent evaporation.

One good thing about this new method is that it is performance based. The analyst can use an alternative approach to oil and grease isolation and recovery. One approach is to use reversed-phase solid-phase extraction (RP-SPE) to isolate the oil and grease, whereas the other approach is to find a way to eliminate use of the solvent. The recently introduced use of infrared cards by 3M Corporation, whereby the oil and grease remain on a thin film of Teflon, is an example of the latter alternative to the conventional liquid–liquid extraction (LLE). The thin film that now offers a fixed path length is then inserted into the sample holder of a conventional infrared spectrophotometer, and the absorbance of the C–H stretching vibration at 2900 cm^{-1} is directly related to concentration via the Beer–Lambert law of spectrophotometry, discussed in Chapter 4.

The 500 series methods refer to measurement of trace organics in drinking water. Method 508 was introduced early in this chapter. This is a collection of methods first published in the late 1980s and incorporating a number of the more innovative techniques introduced at that time. The 600 series methods describe how to isolate, recover, and measure trace organics in wastewaters. These methods were first promulgated in the 1970s. Both series of methods measure specific organic compounds and require a means to separate, identify, and quantify these organics. Thus, samples that may contain two or more compounds require a chromatographic separation. Hence, almost all methods in these two series are GC methods. The scope of both methods is also limited to organics that are amenable to GC. Sample preparation for trace volatile organics (commonly abbreviated VOCs) will be introduced in Chapter 3, and the principles of GC will be introduced in Chapter 4.

The 900 series of methods refer to the measurement of radioactivity in drinking water. These methods were first promulgated in 1980.

The four-digit series, methods 0000 to 9999, involve the analysis of solid wastes. This series was promulgated in the early 1980s and has undergone several major revisions up to the present. This series, when compared to all of the others, seems to be the most dynamic and is considered by the EPA as a general guide. These viewpoints imply that these methods were written within the framework of the second paradigm. In other words, the methods themselves are performance based. The series is periodically updated and revised, and new methods are continuously added. A novice to the field of TEQA would do well to initially focus on this series as a learning tool. A table of contents for this series of analytical methods, which is collectively known as SW-846, 3rd edition, is given in Table 1.10.[30] To illustrate the dynamic nature of the SW-846 series of analytical methods for TEQA, consider the very recent development of Method 7473 for determining mercury in various solids of environmental interest.[31] A solid sample such as apple leaves, oyster tissue, coal fly ash, or river sediment is dried and thermally decomposed at 750°C in an oxygenated furnace. Mercury vapor is released, swept through, and trapped on a gold amalgamator. The amalgam is thermally desorbed, and the Hg vapor is swept into a cold vapor atomic absorption spectrophotometer, where the absorbance is measured at 254 nm. Again, the Beer–Lambert law is used to relate Hg absorbance to concentration. An instrument detection limit (IDL) is reported to be 0.01 ng of total Hg.

One of the most challenging methods to implement in the environmental testing laboratory is EPA Method 8270. This is a method that separates, identifies, and quantifies over 100 priority pollutant semivolatile and nonvolatile organics in various solid wastes. The method is a determinative one and is used in combination with one of the sample preparation methods found in the 3000 series of SW-846. A brief discussion of this method follows in order to introduce the reader to the challenges of implementing such a method. The most current SW-846 series lists a third revision of this method, identified as 8270C. When Method 8270 was first promulgated, a

TABLE 1.10
Method Series 0000–9999: Test Methods for Evaluating Solid Wastes Physical/Chemical Methods

Method	Description
0000s	Air emission sampling methods from stationary sources
1000s	Methods to determine hazardous waste characteristics: ignitibility, corrosivity, and reactivity; includes the toxicity characteristic leaching procedure (TCLP)
2000s	Unused
3000s	Sample preparation methods for inorganics and organics
4000s	Soil screening methods by immunoassay
5000s	Sample preparation methods for volatile organics and miscellaneous sample preparation methods
6000s	Methods to determine more than one metal at a time
7000s	Methods to determine one metal at a time
8000s	Methods to determine organics
9000s	Miscellaneous test methods; includes cyanides, sulfides, phenols, oil and grease, chlorine, total coliform, potentiometric halides and cyanide, colorimetric chloride and radium-228 isotope

Source: Test Methods for Evaluating Solid Wastes, Physical/Chemical Methods, SW-846, 3rd ed., Supplement III Update, July 1997.

number of analytes from the targeted list found in the method either did not yield a high recovery or did not chromatograph well. This resulted in the analyst being unable to satisfy the criteria and led to a rejection of the analytical results in the validation process. Method 8270C clearly states at the onset that certain analytes are non-detected by this method or have an unfavorable distribution coefficient (to be discussed in Chapter 3) or adsorb to the inner walls of glassware or hydrolyze during sample preparation. The three revisions of this method serve to illustrate nicely how the EPA has been able to readjust its promulgated methods under the new PBM paradigm to address early flaws. Under the old paradigm, this realization on the part of the EPA was not evident. The method itself is succinctly summarized as follows:

> Method 8270 can be used to quantitate most neutral, acidic or basic organic compounds that are soluble in methylene chloride and capable of being eluted, without derivatization, as sharp peaks from a gas chromatographic fused-silica capillary column coated with a slightly polar silicone. Such compounds include poly-nuclear aromatic hydrocarbons, chlorinated hydrocarbons and pesticides, phthalate esters, organophosphate esters, nitrosamines, halo-ethers, aldehydes, ethers, ketones, anilines, pyridines, quinolines, aromatic nitro compounds and phenols including nitrophenols.[30]

1.17 WHY IS IT CONSIDERED A CHALLENGE TO IMPLEMENT AN EPA METHOD?

The most tedious and time-consuming aspects of Method 8270 are involved in the preparation and organization of the various referenced chemical standards. This is in addition to the use of reference standards that are required in the various sample preparation methods (i.e., the 3000 series). Method 8270C requires that the following categories of reference standards be prepared, maintained, and replenished, when necessary, by the analyst:

TABLE 1.11
List of Priority Pollutant Semivolatile Organics Quantitated with Respect to the Internal Standard Phenanthrene-d$_{10}$

4-Aminobiphenyl	Hexachlorobenzene
Anthracene	N-Nitrosodiphenylamine
4-Bromophenyl phenyl ether	Pentachlorophenol
Di-n-butyl phthalate	Pentachloronitrobenzene
4,6-Dinitro-2-methyl phenol	Phenancetin
Diphenylamine	Phenanthrene
Fluoranthene	Pronamide

1 Stock reference standards prepared in the laboratory from the dissolution of neat forms of the individual targeted chemical compounds or purchased from various suppliers. The EPA used to provide these from its repository located in Research Triangle Park, NC, but no longer does.

2 Internal standard reference solutions that contain, with the exception of 1,4-dichlorobenzene-d$_8$, five deuterated polycyclic aromatic hydrocarbons which must be prepared. Table 1.11, adapted from the method, shows which of the semivolatile organics are to be quantitated against phenanthrene-d$_{10}$ as the internal standard. The analytes listed in Table 1.11 all elute from the capillary column over a range of retention times that are close to that of phenanthrene-d$_{10}$.

3 GC-MS tuning standard for semivolatiles that contains the tuning compound decafluorotriphenyl phosphine (DFTPP) and also 4,4′-DDT, pentachlorophenol, and benzidine. DFTPP is used to verify that the hardware tune was carried out previously, and the other three analytes are used to verify injection port inertness and GC column performance. Polar compounds like phenols tend to adsorb to contaminated surfaces and result in the complete loss of response or poor peak shape.

4 Dilution of stock solutions to prepare the series of working calibration standards that are injected into the gas chromatograph mass spectrometer (GC-MS) to calibrate the instrument for each of the 100 or more semivolatile organics. This series of standards should also contain at least one initial calibration verification (ICV) standard to assess the accuracy and precision of the calibration curve.

5 Surrogate standard solutions whose analytes are chemically different from any of the targeted analytes and are added to each sample to assess percent recovery.

6 Matrix spike standard solutions and laboratory control standards that are added to one sample per batch to assess matrix effects.

7 System performance check standards and calibration check standards are used to ensure that minimum response factors are met before the calibration curve is used.

8 Blanks that represent standards that contain no targeted analytes, surrogates, or matrix spikes and are used to assess the degree to which a laboratory is contaminated.

All of these standards must be available to the analyst in the laboratory. A most important aspect of an analyst's job is to keep track of all of these standards.

The requirement that the analytes be quantitated using an internal standard mode of instrument calibration is due to the fact that mass spectrometers are intrinsically unstable; that is, their response factor varies with time when compared to other GC detectors, such as the flame ionization detector (FID). The internal standard technique of instrument calibration is discussed in Chapter 2, and the principles of mass spectrometry are introduced in Chapter 4.

1.18 WHAT MADE EPA METHOD 625 SO UNIQUE?

EPA Method 625 discusses exactly how to isolate, recover, and *quantitate* the various priority pollutant semivolatile organic compounds (SVOCs) from a wastewater matrix. The 600 series of EPA methods include both sample prep and determinative techniques in contrast to the SW-846 methods. It was realized back in the early 1970s that priority pollutant analyte identification (answers the trace qualitative analysis question) based only on analyte retention time (the time it takes for a given analyte or chemical compound to pass through the chromatographic column after being injected; retention time is usually given the symbol t_R) was inadequate for regulatory purposes relying only on gas chromatographs with element-specific or so-called standard detectors. According to Budde (p. 171):[32]

The proposal of Method 625 for regulatory use was very significant and had a major impact on the subsequent practice of environmental analyses and the development of commercial GC-MS instrumentation. The prospect of using a single analytical method with no required extract fractionation for 83 analytes was attractive even though it required a more costly analytical instrument. The alternative was the implementation of eight different GC columns and detectors. The economic implications were clear to the regulated industries and the emerging environmental testing industry.

Figure 1.10 introduces the sample preparation rationale that was used to isolate and recover the original 83 priority pollutants from wastewater. Prior to the development of capillary GC columns, the use of packed GC columns necessitated that two fractions be prepared. A *base-neutral* fraction was prepared and injected into a GC-MS, followed by an *acid* fraction. The organic phase was further evaluated as shown in Figure 1.11. Based on how "colored" the extract is, a preliminary screen uses a GC incorporating a flame ionization detector (GC-FID). The FID will be described in detail in Chapter 4; suffice to say here that this detector is universally selective for organic compounds whose molecules contain a carbon–hydrogen covalent bond. The aqueous phase from Figure 1.10 that contains the conjugate base for the organic acids, such as phenols, is acidified, extracted, and also evaluated based on the color of the extract, as shown in Figure 1.12. *Flowcharts* will be used throughout this book to briefly describe sample preparation techniques. Their purpose is to help facilitate a quicker understanding of the many approaches to sample preparation. Budde summarizes and further comments on Method 625 from a contemporary perspective, as follows (p. 173):[32]

Method 625 was promulgated in wastewater regulations on 26 October 1984. The method requires the pH adjustments before extraction and specifies two packed GC columns. However, the method does allow the use of capillary GC columns, but does not provide information to assist the analyst in the application of these columns. As of 1999, Method 625 has not been revised to modify the pH adjustments before extraction, provide information about fused-silica capillary columns, or incorporate other improvements in analytical techniques.

1.19 WHAT ABOUT METHODS FOR THE ANALYSIS OF AIR AND OTHER METHODS?

This book clearly focuses on sample matrices drawn from condensed states of matter. However, there is a set of established methods chiefly to monitor targeted VOCs in ambient air. These are the EPA's toxic organics (TO) methods. Table 1.12 lists eleven TO methods along with the required sampling and determinative techniques. These methods are conventional air sampling techniques. A recent and exciting addition to ambient air sampling is to use a solid-phase microextraction fiber as a sampling device (this topic is found in Chapter 3). The fiber needs to be carefully used, since it can pick up all kinds of atmospheric pollutants. Methods IP1A to IP10B are a compendium of methods for determining indoor air pollutants. There is a National Institute

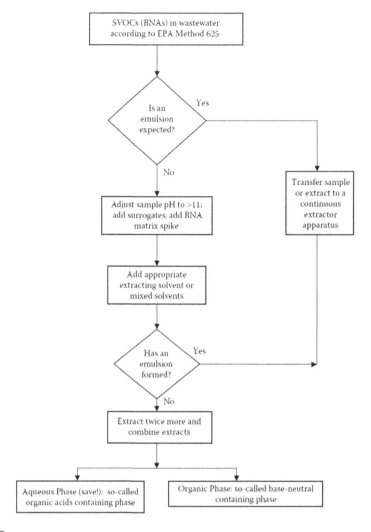

FIGURE 1.10

for Occupational Safety and Health (NIOSH) *Manual of Analytical Methods* that covers air analysis. Two methods are found in the Code of Federal Regulations: 40 CFR 50 for lead, ozone, particulates, and NO_X and 40 CFR 80 for phosphorous and lead in gasoline. A draft method for air analysis exists under the EPA Contract Laboratory Program (CLP).

Standard Methods for the Examination of Water and Wastewater, published once every 5 years by the American Public Health Association, American Water Works Association, and the Water Environment Federation is a well-regarded "bible" of methods for water and wastewater. Table 1.13 is a much abbreviated version of the Table of Contents for the 23rd edition (current as of this writing) and provides a concise overview of its content. There is a strong emphasis on microbiology that is not necessarily found in EPA's SW-846 series.

Official Methods of AOAC International is the other large compendium of analytical methods that includes discussions pertinent to TEQA. The Association of Official Analytical Chemists (AOAC) has been a pioneer in conducting so-called round-robin studies among participating laboratories as a minimum requirement for a validated analytical method to appear in this collection.

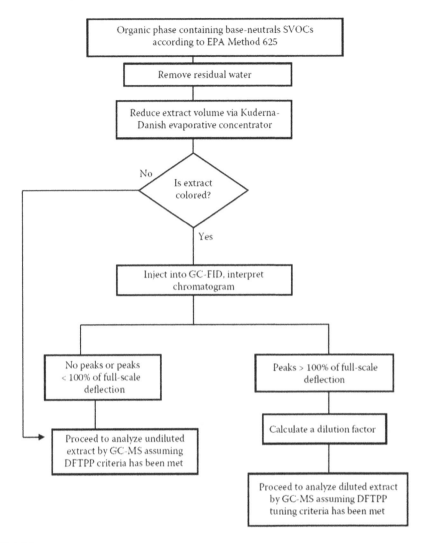

FIGURE 1.11

The primary and secondary literature of analytical chemistry are major sources of new methods, techniques, variations on older methods, and innovation. The American Chemical Society (ACS) publishes two premier journals, *Analytical Chemistry* and *Environmental Science and Technology*. The biennial reviews in *Analytical Chemistry* cover both principles and applications in alternate years. There exists a host of other very relevant scientific journals and reviews, such as *Chromatographia, International Journal of Environmental Analytical Chemistry, Analytical Chimica Acta, The Analyst, Journal of Chromatography A and B, Journal of Chromatographic Science, Journal of Analytical Toxicology, Trends in Analytical Chemistry,* and *CRC Critical Reviews in Analytical Chemistry,* among others journals that are pertinent to TEQA.

Trade journals that arrive for free to qualified professionals include the following:

- *LC-GC: Solutions for Separation Scientists* is a must read for the practicing chromatographer, irrespective of whether the reader's interest is environmental, pharmaceutical, industrial, academic, or some combination (as is the case of most of us).

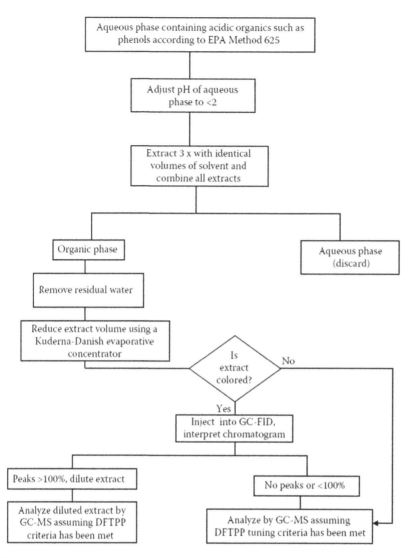

FIGURE 1.12

- *American Laboratory* and *American Laboratory News Edition* introduce readers to diverse areas of applied analytical science. This trade journal terminated printing issues to subscribers in 2019. It is readily available online.
- *Spectroscopy* is focused on the practical aspects of atomic and molecular spectroscopy and is, at times, also pertinent to TEQA.

Each of the above trade journals maintains valuable online archives of past issues. Instrumentation manufacturers such as Agilent Technologies, Thermo Fisher, Shimadzu, Waters, among others, as well as consumables suppliers such as Restek, Phenomenex, among others, provide timely newsletters, websites, and comprehensive catalogs.

As of this writing, online webinars are usually free to the viewer and have become very popular! This is a great way to listen and learn from experienced professionals in the field of analytical chemistry.

TABLE 1.12
EPA Analytical Methods for Sampling and Quantitating Air Pollutants

Method	Target Analytes	Sampling/Chemical Derivatization Technique	Determinative Technique
TO-1	VOCs	Tenax® adsorption	GC-MS
TO-2	VOCs	Carbon molecular sieve adsorption	GC-MS
TO-3	VOCs	Cryogenic preconcentration	GC-FID/ECD
TO-4	Pesticides, PCBs	High volume PUF adsorption	GC-multi-detector
TO-5	Aldehydes, ketones	Impinger/ DNPH derivatization	HPLC-UV
TO-6	Phosgene	Impinger/ reacts with aniline to form carbanile (1,3-diphenylyurea)	HPLC-UV(@254nm)
TO-7	N-nitroso-dimethylamine	Thermosorb® adsorption	GC-MS
TO-8	Phenols, cresols	Impinger/NaOH	HPLC-UV(@254nm)
TO-9A	Polybrominated/ polychlorinated Dibenzo-p-dioxins /benzofurans/PCBs	PUF adsorption	GC- HRMS
TO-10A	Pesticides and Polychlorinated. Biphenyls (PCBs)	Low Volume PUF adsorption	Gas Chromatographic-Multi-Detector Detection (GC/MD)
TO-11A	Formaldehyde and other aldehydes and ketones	DNPH adsorbent cartridge	HPLC-UV (@ 350nm)
TO-12	Non-methane Organic Compounds (NMOC))	Cryogenic preconcentration/ thermal desorption directly to a flame ionization detector	Direct Flame Ionization Detection (FID)
TO-13A	Polycyclic Aromatic Hydrocarbons (PAHs)	PUF or XAD-2 adsorbent cartridge	GC-MS
TO-14A	Volatile Organic Compounds (VOCs)	Collected in specially prepared canisters	GC using nonspecific and specific detectors
TO-15	Volatile Organic Compounds (VOCs)	Collected in specially prepared canisters	GC-MS
TO-16:	Carbon monoxide	Long-path/open path tube	FTIR absorption
TO-17:	VOCs	Ambient air by pulling air through various sorbent tubes	Thermal desorption/ GC-MS

Note: PUF-polyurethane foam, HPLC-high performance liquid chromatography, UV-ultraviolet absorption spectrophotometry, GC-gas chromatography, FID-flame ionization detection, ECD-electron capture detection, MS-mass spectrometry, DNPH-dinitrophenyl-hydrazine, HRMS-high resolution mass spectrometry, FTIR-Fourier Transform infrared spectroscopy.

Source: Compendium of Methods for the Determination of Toxic Organic Compounds in Ambient Air. 2nd Edition, January, 1999 U.S. Environmental Protection Agency.

TABLE 1.13
Standard Methods for the Examination of Water and Wastewater, 23rd Edition: Table of Contents

Part #	Title	Abbreviated Content
1000	Introduction	Quality assurance, data quality, method development, evaluation and expression of results, collection and preservation of samples, reagent water, laboratory occupational health and safety, waste minimization and disposal, and more
2000	Physical and aggregate properties	QA/QC, appearance, color, turbidity, odor, taste, flavor profile analysis, acidity, alkalinity, and more
3000	Metals	QA/QC, preliminary treatment of samples, atomic spectroscopy (AA): flame, cold vapor, electrothermal; plasma emission; methods for specific metals, and more
4000	Inorganic nonmetallic constituents	QA/QC, determination of anions by ion chromatography, inorganic nonmetals by flow injection analysis, inorganic anions by capillary electrophoresis, specific inorganic methods, cyanide amenable to chlorination after distillation, and more
5000	Aggregate organic constituents	QA/QC. biological oxygen demand, chemical oxygen demand, total organic carbon, oil and grease, aquatic humic substances, phenols, surfactants, and more
6000	Individual organic compounds	QA/QC, volatile organic compounds, trihalomethanes, extractable base-neutrals and acids, polychlorinated biphenyls, polycyclic aromatic hydrocarbons, glyphosate herbicide, and more
7000	Radioactivity	QA/QC, counting instruments, gross alpha and gross beta radioactivity, radioactive cesium, radon, radioactive strontium and strontium-90, and more
8000	Toxicity	QA/QC, mutagenesis, bacterial bioluminescence, and more
9000	Microbiological examination	QA/QC, preparation of culture media, membrane filter techniques for coliform, and more
10000	Biological examination	Plankton, periphyton, macrophytes, and more

Books on any and all aspects of analytical chemistry should be frequently consulted. A number of contemporary books either directly or indirectly related to TEQA are widely available.[33–42]

1.20   WHAT IS THE PHYSICAL/CHEMICAL BASIS OF THE EPA'S ORGANICS PROTOCOL?

A fundamental question can be asked to the newcomer to all of this: On what basis do all of the 200 or so priority pollutant organics get organized so that a systematic approach to TEQA can be accomplished? In other words, is there a simple and unifying theme that can be introduced to make some sense of all of this? The answer is yes. A flowchart is developed for the protocol as a whole and is depicted in Figure 1.13. Any sample matrix obtained from the environment

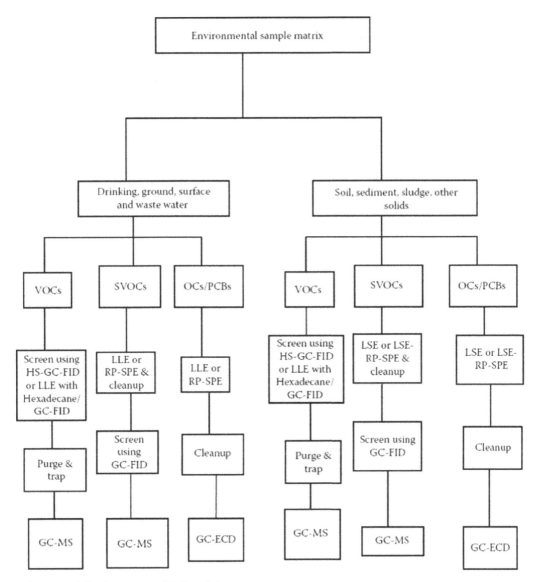

Acronyms defined that are used in the scheme

VOCs—volatile organics
SVOCs—semi-volatile organics
OCs/PCBs—organochlorine pesticides/polychlorinated biphenyls
HS-GC-FID—static headspace coupled to gas chromatography incorporating flame ionization detection
LLE—liquid-liquid extraction (separatory funnel, countercurrent apparatus)
LSE—liquid-solid extraction (Soxhlet apparatus) or solvent leaching from solid matrices
RP-SPE—reversed-phase solid-phase extraction using chemically bonded silica sorbents
LSE-RP-SPE—liquid-solid leaching followed by isolation on reversed-phase sorbents
GC-ECD—gas chromatography incorporating electron-capture detection
GC-MS—gas chromatography interfaced to low-resolution quadrupole mass spectrometry

FIGURE 1.13

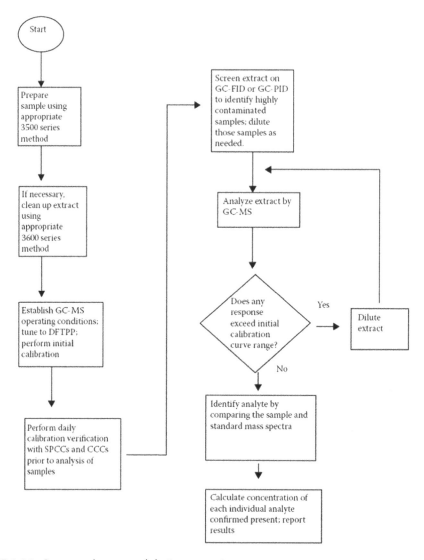

FIGURE 1.14 Erroneously removed during corrections.

is first categorized as being either water or soil/sediment, and in turn, appropriate methods are implemented based in large part on the degree of volatility of the respective targeted analytes. Ideally, one would wish to screen samples to obtain a preliminary indication as to whether one must dilute the extract before injection into the determinative instrument. For example, EPA Method 8270C fits very nicely into this scheme. A sample of soil taken from a Superfund hazardous waste site would be analyzed according to the path for semivolatile organics. This path assumes that an initial screen is considered, and based on this screen, an appropriate dilution of an extract is prepared from the sample made. The sample preparation techniques of liquid–liquid extraction (LLE) and sample cleanup are used and the quantitative determination accomplished using GC-MS. Figure 1.14 shows a flowchart for this method and directs the analyst in implementing the method. Note that, along the way, there is ample opportunity for analyst intervention and decision making. Chapters 2, 3 and 4 introduce those principles, techniques and applications responsible for implementing the flowcharts shown in Figures 1.13 and 1.14, not only for this particular method, but, in general, for all analytical methods.

Before we plunge into the details of analytical methods to measure environmental contaminants, it is important first to consider just how to interpret analytical data. This topic is usually either placed near the end of most books or added in the form of an appendix.[43] Increasing emphasis on a more rigorous treatment of analytical data has emerged.[44]

Often you have to decide when the data is not as good as you would like.

—**Donald Kennedy**

REFERENCES

1. Carson, R., *Silent Spring*, Houghton Mifflin, 1962.
2. *Chemical & Engineering News*, June 17, 2019, p. 30.
3. Loconto, P.R., "Interactions on hydrophobic surfaces as a means of isolating environmentally significant analytes at trace concentrations in water." Ph.D. Thesis, University of Massachusetts at Lowell (formerly The University of Lowell), 1985.
4. Fries, G., "The PBB episode in Michigan: an overall appraisal." *CRC Crit Rev Toxicol.* 16: 105–156, 1985.
5. Halbert, F. and S., *Bitter Harvest, Investigation of the PBB Contamination: A Personal Story,* 1978, Grand Rapids, MI: Erdmans Pub.Co.
6. Landrigan, P. et al. "Cohort study of Michigan residents exposed to PBBs: epidemiologic and immunologic findings." Ann NY Acad Sci. 320: 284, 1979.
7. *Enquirer and News*, Battle Creek, Michigan, March 26th, 1978.
8. Wolff, M., H. Anderson, I. Selikoff, "Human tissue burdens of halogenated aromatic chemicals in Michigan." *J Am Med Assoc.* 247: 2112, 1982.
9. Tuey, D. and H. Mathews. "Distribution of excretion of 2,2'3,3'4,4'-hexachlorobiphenyl in rats and man: pharmacokinetic model predictions." *Toxicol Appl Pharmacol.* 53: 420, 1980.
10. J. Dean (ed.) *Lange's Handbook of Chemistry*, 13th ed. 1985, McGraw-Hill, New York.
11. Thomas, R. "The critical role of atomic spectroscopy in understanding the links between lead toxicity and human disease" *Spectroscopy*, 33(10): 12–21. 2018.
12. Johnston, J. et al. "Lead and Arsenic in shed deciduous teeth of children living near a lead-acid battery smelter" *Environmental Science and Technology*, 53(10): 6000–6006, 2019.

13. Loconto, P.R. *Trace Environmental Quantitative Analysis*, 2nd edition, 2006, CRC Press, Taylor and Francis.
14. Isenberg, S. et al. *J Anal Toxicol*. 42: 630–636, 2018.
15. Methods for the Determination of Organic Compounds in Drinking Water, EPA 600 4.88 039, U.S. Environmental Protection Agency, December 1988.
16. Addink, R. and T. Hall, Automated low background solid-phase extraction of perfluorinated compounds in water, *Am Lab*. 51(2): 31–33, 2019.
17. Benowitz, N.L. and P. Jacob, *Clin Pharmacol Ther*. 56: 483–394, 1994.
18. Pirkle, J.L., et al. *J Am Med Assoc*. 275: 1233–1240, 1996.
19. *Third National Report on Human Exposure to Environmental Chemicals*, DHHS, CDC, NCEH Pub. No. 05-057, p. 74, 324.
20. Egeghy, P., Biological Monitoring in Exposure Assessment. Paper presented at the workshop at the 11th Annual Meeting of the International Society of Exposure Analysis, Charleston, SC 2001.
21. Brock, J. et al. *J Expo Anal Environ Epidemiol*. 11: 323–328, 2001.
22. Coburn, T., D. Dumanoski, J. Myers. *Our Stolen Future: Are we threatening our Fertility, Intelligence, and Survival? – A Scientific Detective Story*. New York: Penguin Books, 1996.
23. Merck and Company. *The Merck Index*. 12th ed. Rathway, NJ: Merck and Company. 1996, p. 529.
24. Chemical Testing Plan Draws Praise. *Chemical and Engineering News*. October 12, 1998.
25. Collin, R.W. *The Environmental Protection Agency: Cleaning Up America's Act*, Wesport, CT, Greenwood Press, 2006.
26. Smith, R.K. *Handbook of Environmental Analysis*, 2nd ed., Genium Publishing, Amsterdam, NY, 1994.
27. Weiss, J. *Handbook of Ion Chromatography*, 2nd ed. New York: VCH, 1994.
28. Small, H. et al. *Anal Chem*. 70: 3629–3635, 1998.
29. Loconto, P.R., A.K. Gaind. *J Chromatogr Sci*. 27: 569–573, 1989.
30. *Test Methods for Evaluating Solid Wastes, Physical/Chemical Methods*, SW-846, 3rd ed. Final Update III, Method 8270C, July 1997.
31. Fordham O. *Environ Test Anal*. 7:10–13, 1998.
32. Budde W. *Analytical Mass Spectrometry*. Washington, DC: Oxford University Press, 2001.
33. Csuros, M. *Environmental Sampling and Analysis*, 2000, Lewis Publishers, CRC Press, 373pp.
34. Kebbekus, B., S. Mitra, *Environmental Chemical Analysis*, 1998, Chapman & Hall/CRC Press, 330pp.
35. Patnaik, P. *Handbook of Environmental Analysis*, 3rd ed., 2018, CRC Press, 608pp.
36. Overway, K. *Environmental Chemistry: an analytical approach*, 2017, Wiley, 352pp.
37. Fifield, F., P. Haines, (editors) *Environmental Analytical Chemistry*, 2nd ed., 2000, Wiley, 512pp.
38. Mita, S., P. Patnaik, B. Kebbekos, *Environmental Chemical Analysis*, 2nd ed., 2019, CRC Press, 424pp.
39. Bpkdyk, G., *Environmental Analysis: The Ultimate Step by Step*, 2018, 5TARCooks, 142pp, $79.00 (paperback).
40. Popek, E. *Sampling and Analysis of Environmental Chemical Pollution*, 2018, Elsevier, 436pp.
41. Dean, J. *Environmental Trace Analysis Techniques and Applications*, 2013, Wiley, 278pp.
42. Nollet, L., L. DeGelder, *Handbook of Water Analysis*, 3rd ed., 2013, CRC Press, 995pp.
43. Skoog D., J. Leary. *Principles of Instrumental Analysis,* 4th ed. Philadelphia: Saunders, 1992.
44. Loconto, P.R., "Use of weighted least squares and confidence band calibration statistics to find reliable instrument detection limits for trace organic chemical analysis" *Am Lab*. 47(7): 34–39, 2015.

2 Calibration, Verification, Quantification, Statistical Treatment of Analytical Data, Detection Limits, and Quality Assurance/Quality Control

If you can measure that of which you speak, and can express it by a number, you know something of your subject, but if you cannot measure it, your knowledge is meager and unsatisfactory.

—Lord Kelvin

Chromatographic and spectroscopic analytical instrumentation are the key determinative tools to quantitate the presence of chemical contaminants in biological fluids and in environmental matrices such as air, soil, and water. These instruments generate electrical signals that are related to the amount or concentration of a chemical contaminant (analytical chemists refer to these chemical contaminants as analytes) of *enviro-health* and *enviro-chemical* significance. Body fluids include blood, urine, saliva, etc. while environmental matrices include groundwater, surface water, air, soil, wastewater, sediment, sludge, and so forth. Computer technology has aided immensely in the conversion of an analog signal from the analytical instrument's transducer to the digital domain. This achievement enabled computer software to play a central role in the contemporary practice of TEQA. It is the relationship between the analog or digital output from the instrument and the amount or concentration of an analyte of *enviro-health* or *enviro-chemical* origin that is discussed in this chapter. The process by which an electrical signal is transformed to an amount or concentration is called instrument calibration. Chemical analysis based on measuring the mass or volume obtained from chemical reactions is stoichiometric. Gravimetric (where the analyte of interest is weighed) and volumetric (where the analyte of

43

interest is titrated) techniques are methods that are stoichiometric. Such methods do not require calibration. Most instrumental determinative methods are nonstoichiometric and thus require instrument calibration.

This chapter introduces *the* most important aspect of TEQA: how to interpret data generated by an analytical instrument and how to report accurate and precise quantitative results. After the basics of what constitutes good laboratory practice are discussed, the concept of instrumental calibration is introduced and the mathematics used to establish such calibrations is developed. The uncertainty present in the interpolation of the calibration is then discussed. A comparison is made between the more conventional approach to determining instrument detection limits and the more contemporary approaches that have recently been discussed in the literature.[1-6] These more contemporary approaches use least squares regression and incorporate relevant concepts from statistics applicable to analytical chemistry.[7] The use of weighted least squares regression is introduced and applied to trace organics quantitative analysis to calculate instrument detection limits (IDLs). The relationship between IDLs and method detection limits (MDLs) is discussed. The importance of sound quality assurance / quality control (QA/QC) practices in the analytical laboratory is discussed. Readers can compare QA/QC practices from two fictitious environmental testing laboratories. The principles that enable a detector's analog signal to be digitized via analog-to-digital converters is briefly introduced. A brief introduction to the basic principles of environmental sampling is presented.

Every employer wants to hire an analytical chemist who knows of and practices good laboratory behavior!

2.1   WHAT IS GOOD LABORATORY PRACTICE?

Good laboratory practice (GLP) requires that a quality control (QC) protocol for trace environmental analysis be put in place. A good laboratory QC protocol for any laboratory attempting to achieve precise and accurate TEQA requires the following considerations:

- Deciding whether an external standard, internal standard, or standard addition mode of instrument calibration is most appropriate for the intended quantitative analysis application.
- Establishing a calibration plot (still referred to as a calibration curve) that relates instrument response to analyte amount or concentration by preparing reference standards and measuring their respective instrument responses.
- Performing a least squares regression analysis on the experimental calibration data to evaluate instrument linearity over a range of concentrations of interest and to establish the best relationship between instrument response and amount or concentration of the analyte of interest.
- Computing the statistical parameters that assist in specifying the uncertainty of the least squares fit to the experimental data points.
- Quantifying one or more reference standards referred to as initial calibration verification (ICV) standards.
- Computing the statistical parameters for the ICV that assist in specifying the precision and accuracy of the least squares fit to the experimental data points.
- Determining the instrument detection limits (IDLs) of the specific method.
- Determining the method detection limits (MDLs), which requires establishing the percent recovery for a given analyte in both a clean matrix and the sample matrix. With some techniques, such as static headspace gas chromatography (HS-GC), the MDL cannot be determined independently from the instrument's IDL.
- Preparing, running, and quantifying QC reference standards (often called continuing calibration verification (CCV) standards) at a frequency of once every 5 or 10 samples. This QC standard serves to monitor instrument precision and accuracy during a batch run. This assumes that both calibration and ICV criteria have been met. A mean value for the QC

reference standard should be obtained over all QC standards run in the batch. The standard deviation, s, and the relative standard deviation (RSD) should be calculated.

- Preparing, running and quantifying QC surrogates, matrix spikes, and, in some cases, matrix spike duplicates per batch of samples. A batch is defined in EPA methods to be approximately 20 samples. These reference standard spikes serve to assess extraction efficiency where applicable. Matrix spikes and duplicates are often required in EPA methods.
- Preparing and running laboratory blanks, laboratory control samples, and field and trip blanks. These blanks serve to assess whether samples may have become contaminated during sampling and sample transport.

It has been stated many times by experienced analysts that in order to achieve GLP, close to one QC sample must be prepared and analyzed for nearly each and every real-world environmental sample.

2.2 CAN DATA REDUCTION, INTERPRETATION, AND STATISTICAL TREATMENT BE SUMMARIZED BEFORE WE PLUNGE INTO CALIBRATION?

Yes, indeed. Figure 2.1, adapted and modified, while drawing on recently published International Union of Pure and Applied Chemistry (IUPAC) recommendations, as discussed by Currie,[1] is this author's attempt to do just that! The true amount that is present in the unknown sample can be expressed as an *amount* such as a #ng analyte, or as a *concentration* [#µg analyte/kg of sample (weight/weight) or #pg analyte/L of sample (weight/volume)]. The amount or concentration of true unknown present in either an environmental sample (*enviro-chemical TEQA*) or human/animal specimen (*enviro-health TEQA*) and represented by τ that is shown in Figure 2.1 being transformed to an electrical signal y. Chapters 3 and 4 describe how the *six steps* from sampling

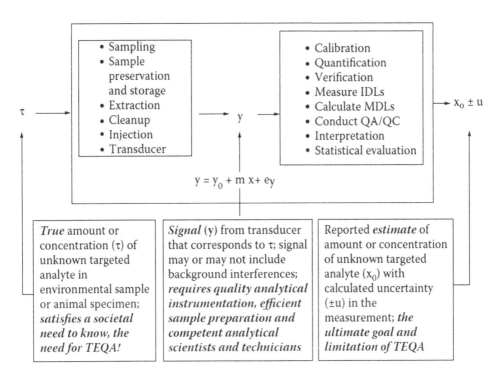

FIGURE 2.1

to transducer are accomplished. The signal y, once obtained, is then converted to the reported estimate x_0, as shown in Figure 2.1. This chapter describes how the *eight steps* from calibration to statistical evaluation are accomplished. The ultimate goal of TEQA is then realized, i.e., a reported estimate x_0 with a calculated *uncertainty* using statistics in the measurement expressed as $\pm u$. We can assume that the transduced signal varies linearly with x, where x is the known analyte amount or concentration of a standard reference. This analyte in the standard reference must be chemically identical to the analyte in the unknown sample represented by its true value τ. x is assumed to be known with certainty since it can be traced to accurately known certified reference standards, such as that obtained from the National Institute of Standards and Technology (NIST). We can then realize that

$$y = y_0 + mx + e_y$$

where

- y_0 = the y intercept, the magnitude of the signal in the absence of analyte.
- m = slope of the best-fit *regression line* (what we mean by regression will be taken up shortly) through the experimental data points. The slope also defines the sensitivity of the specific determinative technique.
- e_y = the error associated with the variation in the transduced signal for a given value of x. We assume that x itself (the amount or concentration of the analyte of interest) is free of error. This assumption is used throughout the mathematical treatment in this chapter and serves to simplify the mathematics introduced.

Referring to Figure 2.1, we can, at best, only estimate τ and report a result for the amount or concentration at a trace level, represented by x_0, with an uncertainty u such that numerical values for x_0 could have a range from a low of $(x_0 - u)$ to a high of $(x_0 + u)$. Let us focus a bit more on the concept of error in measurement.

2.2.1 How Is Measurement Error Defined?

Let us digress a bit and discuss measurement error. Each and every measurement includes error. The length and width of a page from this book cannot be measured without error. There is a true length of this page, yet at best we can only estimate its length. We can measure length only to within the accuracy and precision of our measuring device, in this case, a ruler or straightedge. We could increase our precision and accuracy for measuring the length of this page if we used a digital caliper. Currie has defined x_0 as the statistical estimate derived from a set of observations.

The error in x_0 represented by e is shown in Figure 2.2 to consist of two parts, systematic or bias error represented by Δ and random error represented by δ.[1, 8, 9] Δ is defined as the *absolute difference* between a population mean represented by μ (assuming a Gaussian or normal distribution) and the true value τ. δ is defined as the *absolute difference* between the estimated analytical result for the unknown sample x_0 and the population mean μ. δ can also be viewed in terms of a multiple z of the population standard deviation σ, σ being calculated from a Gaussian or normal distribution of x values from a population.

2.2.2 Are There Laboratory-Based Examples of How Δ and δ are Used?

Yes, indeed. Bias, Δ, reflects systematic error in a measurement. Systematic error *can be instrumental, operational, or personal.*

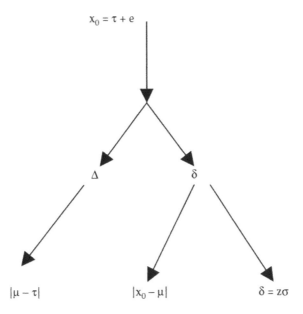

$x_0 = \tau + e$

Δ

δ

$|\mu - \tau|$

$|x_0 - \mu|$

$\delta = z\sigma$

FIGURE 2.2

Instrumental errors arise from a variety of sources such as:[10]

- Poor design or manufacture of instruments
- Faulty calibration of scales
- Wear of mechanical parts or linkages
- Maladjustment
- Deterioration of electrical, electronic, or mechanical parts due to age or location in a harsh environment
- Lack of lubrication or other maintenance.

Errors in this category are often the easiest to detect. They may present a challenge in attempting to locate them. Use of a certified reference standard might help to reveal just how large the degree of inaccuracy as expressed by a *percent relative error* really is. The percent relative error (%error), i.e., the *absolute difference* between the mean or average of a small set of replicate analyses, $\bar{x}$, and the true or accepted value, τ divided by τ and multiplied by 100, is mathematically stated (and used throughout this book) as follows:

$$\% \, error = \frac{|\bar{x} - \tau|}{\tau} \times 100$$

It is common to see the expression "the manufacturer states that its instrument's accuracy is better than 2% relative error." The analyst should work in the laboratory with a good idea as to what the percent relative error might be in each and every measurement that he or she must make. It is often difficult if not impossible to know the true value. This is where certified reference standards such as those provided by the NIST are valuable. It has been said that *good precision* in measurement may or may not mean *acceptable accuracy*!

Operational errors are due to departures from correct procedures or methods. These errors often are time dependent. One example is that of drift in readings from an instrument before the instrument has had time to stabilize. A dependence of instrument response on temperature can be eliminated by waiting until thermal equilibrium has been reached. Another example is the failure to set scales to zero or some other reference point prior to making measurements. Interferences can cause either positive or negative deviations. One example is the deviation from Beer's law at higher concentrations of the analyte being measured. However, in trace analysis, we are generally confronted with analyte concentration levels that tend toward the opposite direction.

Personal errors result from bad habits and erroneous reading and recording of data. Parallax error in reading the height of a liquid in a buret from titrimetic analysis is a classic case in point. One way to uncover personal bias is to have someone else repeat the operation. Occasional random errors by both persons are to be expected, but a discrepancy between observations by two persons indicates bias on the part of one or both.[9]

Consider the preparation of reference standards using an analytical balance that reads a larger weight than it should. This could be due to a lack of adjusting the zero within a set of standard masses. What if an analyst, who desires to prepare a solution of a reference standard to the highest degree of accuracy possible, dissolves what he thinks is 100 mg of standard reference (the solute), but really is only 89 mg, in a suitable solvent using a graduated cylinder and then adjusts the height of the solution to the 10 mL mark? Laboratory practice would suggest that this analyst use a 10 mL volumetric flask. Use of a volumetric flask would yield a more accurate measurement of solution volume. Perhaps 10 mL turns out to be really 9.6 mL when a graduated cylinder is used. We now have inaccuracy, i.e., bias, in both mass and in volume. Bias has direction, i.e., the true mass is always lower *or* higher. Bias is usually never lower for one measurement and then higher for the next measurement. The mass of solute dissolved in a given volume of solvent yields a solution whose concentration is found from dividing the mass by the total volume of solution. The percent relative error in the measurement of mass and the percent relative error in the measurement of volume propagate to yield a combined error in the reported concentration that can be much more significant than each alone. Here is where the cliché "the whole is greater than the sum of its parts" has some meaning.

Random error, δ, occurs among replicate measurement without direction. If we were to weigh 100 mg of some chemical substance, such as a reference standard, on the most precise analytical balance available and repeat the weighing of the same mass additional times while remembering to re-zero the balance after each weighing, we might get data such as that shown below:

Replicate No.	Weight (mg)
1	99.98
2	100.10
3	100.04
4	99.99
5	100.02

Notice that the third replicate weighing yields a value that is less than the second. Had the values kept increasing through all five measurements, systematic error or bias might be evident.

Another example that clearly shows the difference between systematic vs. random error, this time using analytical instrumentation, is to make repetitive 1 μL injections of a reference standard solution into a gas chromatograph (GC). A GC with an atomic emission detector (GC-AED) was used by this author to evaluate whether systematic error was evident for triplicate injection of a 20 ppm reference standard containing tetra-chloro-m-xylene (TCMX) and decachlorobiphenyl (DCBP) dissolved in the solvent iso-octane. Both analytes are used as surrogates in EPA organochlorine pesticide / polychlorinated biphenyl (PCB)-related methods such as EPA Methods 608 and 8080. The atomic emission from microwave-induced plasma excitation of chlorine atoms, monitored at a wavelength of 837.6 nm, formed the basis for the transduced electrical signal. Both analytes are separated chromatographically (refer to Chapter 4 for an introduction to the principles underlying chromatographic separations) and appear in a *chromatogram* as distinct peaks, each with an instrument response. The emitted intensity is displayed graphically in terms of a peak whose *area beneath the curve* is given in units of counts-seconds. This data is shown below:

	TCMX (counts-seconds)	DCBP (counts-seconds)
1st injection	48.52	53.65
2nd injection	47.48	52.27
3rd injection	48.84	54.46

The drop between the first and second injections in the peak area along with the rise between the second and third injections suggests that systematic error has been largely eliminated. A few days before this data was generated a similar set of triplicate injections was made using a somewhat more diluted solution containing TCMX and DCBP into the same GC-AED. The following data was obtained:

	TCMX (counts-seconds)	DCBP (counts-seconds)
1st injection	37.83	41.62
2nd injection	38.46	42.09
3rd injection	37.67	40.70

The rise between the first and second injections in peak area followed by the drop between the second and third injections again suggests that systematic error has been largely eliminated. One of the classic examples of systematic error, and one that is most relevant to TEQA, is to compare the bias and percent relative standard deviations in the peak area for five identical injections using a liquid-handling autosampler against a manual injection into a graphite furnace atomic absorption spectrophotometer using a common 10 µL glass liquid-handling syringe. It is almost impossible for even the most skilled analyst around to achieve the degree of reproducibility afforded by most automated sample delivery devices.

Good laboratory practice suggests that it should behoove the analyst to eliminate any bias, Δ, so that the population mean equals the true value. Mathematically stated:

$$\Delta = 0 = \mu - \tau$$

$$\therefore \mu = \tau$$

Eliminating Δ in the practice of TEQA enables one to consider only random errors. Mathematically stated:

$$\delta = |x_0 - \mu|$$

Random error alone becomes responsible for the absolute difference between the reported estimate x_0 and the statistically obtained population mean μ. Random error can never be completely eliminated. Referring again to Figure 2.1, let us proceed in this chapter to take a more detailed look at those factors that transform y to x_0. We focus on those factors that transform τ to y in Chapters 3 and 4.

2.3 HOW IMPORTANT IS INSTRUMENT CALIBRATION AND VERIFICATION?

It is very important and *the* most important task for the analyst who is responsible for operation and maintenance of analytical instrumentation. Calibration is followed by a verification process in which specifications can be established and the analyst can evaluate whether the calibration is verified or refuted. A calibration that has been verified can be used in acquiring data from samples for quantitative analysis. A calibration that has been refuted must be repeated until verification is achieved, e.g., if, after establishing a multipoint calibration for benzene via a gas chromatographic determinative method, an analyst then measures the concentration of benzene in a certified reference standard. The analyst expects no greater than a 5% relative error and discovers to his surprise a 200% relative error! In this case, the analyst must reconstruct the calibration and measure the certified reference standard again. Close attention must be paid to those sources of systematic error in the laboratory that would cause the relative error to greatly exceed the minimally acceptable relative error criteria previously developed for this method.

An analyst who expects to implement TEQA and begins to use any one of the various chromatography data acquisition and processing software packages available in the marketplace today is immediately confronted with several calibration modes available. Most software packages will contain most of the *modes of instrumental calibration* that appear in Table 2.1. For each calibration mode, the general advantages as well as the overall limitations are given. Area percent and normalization percent (norm%) are not suitable for quantitative analysis at the trace concentration level. This is due to the fact that a concentration of 10,000 ppm is only 1% (parts per hundred), so that a 10 ppb concentration level of, for example, benzene, in drinking water is only 0.000001% benzene in water. Weight% and mole% are subsets of norm% and require response factors for each analyte in units of peak area or peak height per gram or per mole, respectively. Table 2.2 relates each calibration mode with its corresponding quantification equation. Quantification follows calibration and thus achieves the ultimate goal of TEQA, i.e., to perform a quantitative analysis of a sample of *enviro-chemical* or *enviro-health* interest in order to determine the concentration of each targeted chemical analyte of interest at a trace concentration level. Table 2.1 and Table 2.2 are useful as reference guides.

We now proceed to focus on the most suitable calibration modes for TEQA. Referring again to Table 2.1, these calibration modes include *external standard* (ES), *internal standard* (IS), to include its more specialized *isotope dilution mass spectrometry* (IDMS) calibration mode, and *standard addition* (SA). Each mode will be discussed in sufficient detail to enable the reader to acquire a fundamental understanding of the similarities and differences among all three. Correct execution of calibration on the part of the analytical chemist on a given instrument is a major factor in achieving GLP.

2.3.1 How Does the External Mode of Instrument Calibration Work?

The ES mode uses an external reference source for the analyte whose concentration in an unknown sample is sought. A series of working calibration standards are prepared that encompass the entire range of concentrations anticipated for the unknown samples and may include one or more orders of magnitude. For example, let us assume that a concentration of 75 ppb of a trihalomethane (THM) is anticipated in chlorinated drinking water samples. A series of working calibration standards should be prepared whose concentration levels start from a minimum of 5 ppb to a maximum of 500 ppb each THM. The range for this calibration covers two orders of magnitude. Six standards that are prepared at 5, 25, 50, 100, 250, and 500 ppb for each THM, respectively, would be appropriate in this case. Since these standards will not be added to any samples, they are considered external to the samples, hence defining this mode as ES. The calibration curve is established by plotting the instrument response against the concentration of analyte for each THM.

TABLE 2.1
Advantages and Limitations of the Various Modes of Instrument Calibration Used in TEQA

Calibration Mode	Advantages	Limitations
Area%	No standards needed; provides for a preliminary evaluation of sample composition; injection volume precision not critical	Need a nearly equal instrument response for all analytes so peak heights/areas all uniform; all peaks must be included in calculation; not suitable for TEQA
Norm%	Injection volume precision not critical; accounts for all instrument responses for all peaks	All peaks must be included; calibration standards required; all peaks must be calibrated; not suitable for TEQA
ES	Addresses wide variation in GC detector response; more accurate than area%, norm%; not all peaks in a chromatogram of a given sample need to be quantitated; compensates for recovery losses if standards are taken through sample prep in addition to samples; does not have to add any standard to the sample extract for calibration purposes; ideally suited to TEQA	Injection volume precision is critical; instrument reproducibility over time is critical; no means to compensate for a change in detector sensitivity during a batch run; needs a uniform matrix whereby standards and samples should have similar matrices
IS	Injection volume precision not critical; instrument reproducibility over time not critical; compensates any variation in detector sensitivity during a batch run; ideally suited to TEQA	Need to identify a suitable analyte to serve as an IS; bias is introduced if the IS is not added to the sample very carefully; does not compensate for percent recovery losses during sample preparation since IS is usually added after both extraction and cleanup are performed
IDMS	Same as for IS; injection volume precision not critical; instrument reproducibility over time not critical; compensates for analyte percent recovery losses during sample preparation since isotopes are added prior to extraction and cleanup; eliminates variations in analyte vs. internal standard recoveries; ideally suited to TEQA	Need to obtain a suitable isotopically labeled analog of each target analyte; isotopically labeled analogs are very expensive; bias is introduced if the labeled isotope is not added to the sample very carefully; needs a mass spectrometer to implement; mass spectrometers are expensive in comparison to element-selective GC detectors or non-MS LC detectors
SA	Useful when matrix interference cannot be eliminated; applicable where analyte-free matrix cannot be obtained; commonly used to measure trace metals in "dirty" environmental samples	Need two aliquots of same sample to make one measurement; too tedious and time consuming for multiorganics quantitative analysis

Source: Modified and adapted from Agilent Technologies GC-AED Theory and Practice, Training Course from Diablo Analytical, Inc., 2001.

The external standard is appropriate when there is little to no matrix effect between standards and samples. To illustrate this elimination of a matrix effect, consider the situation whereby an aqueous sample is extracted using a nonpolar solvent. The reference chemical standard used to construct the ES calibration is usually dissolved in an organic solvent such as methanol, hexane, or iso-octane. The analytes of interest are now also in a similar organic solvent. ES is

TABLE 2.2
**Summary of Important Quantification Equations for Each Mode of Instrument
Calibration Used in TEQA**

Calibration Mode	Quantification Equation for
Area%	$$C_{unk}^i = \dfrac{A_{unk}^i}{\displaystyle\sum_i A_i} \times 100$$

C_{unk}^i—concentration of analyte i in the unknown sample (the ultimate goal of TEQA)
A_{unk}^i—area of ith peak in unknown sample
N—total number of peaks in chromatogram

Norm%	$$C_{unk}^i = \dfrac{A_{unk}^i RF_i}{\displaystyle\sum_i A_{unk}^i RF_i} \times 100$$

Weight%/Mole%

RFi—response factors for ith analyte; peak area or peak height per unit amount (grams or moles)

ES	$C_{unk}^i = A_{unk}^i / RF_i$

RFi—response factor for the ith analyte; peak area or peak height per unit concentration

IS	$$C_{unk}^i = \left[\dfrac{A_{unk}^i}{A_{IS}^i}\right]\left[\dfrac{C_{IS}^i}{RRF_i}\right]$$

RRF_i—**relative response factor for the ith analyte; peak area or peak height per unit concentration**

IDMS	$$C_{unk}^i = \left[\dfrac{C_{spike}^i W_{spike}}{W_{spike}}\right]\left[\dfrac{f_{spike}^{i,1} - R_m f_{spike}^{i,2}}{R_m f_{unk}^{i,2} - f_{unk}^{i,1}}\right]$$

Refer to text for definition of each term used in the above equation

SA	$$C_{unk}^i = \left[\dfrac{R_{unk}^i - R_{bl-unk}^i}{\left(R_{SA}^i - R_{unk}^i\right) - \left(R_{bl-spike}^i - R_{bl-unk}^i\right)}\right] C_{spike}^i$$

C_{spike}^i—**concentration of analyte i after analyte i (standard) is added to the unknown sample**
R_x^i, R_{XB}^i—**response of unknown analyte and blank, both associated with unknown sample**
R_S^i, R_{SB}^i—**response of unknown analyte and blank, in spiked or standard added known sample**

also appropriate when the instrument is stable and the volume of injection of a liquid sample such as an extract can be reproduced with good precision. A single or multipoint calibration curve is usually established when using this mode.

For a single-point calibration, the concept of a response factor, R_F, becomes important. The use of response factors is valid provided that it can be demonstrated that the calibration curve is, in fact, a straight line. If so, the use of R_F values serves to greatly simplify the process of calibration. R_F is fixed and is independent of the concentration for its analyte for a truly linear calibration. A response factor for the ith analyte would be designated as R_F^i. For example, if twelve analytes are to be calibrated and we are discussing the seventh analyte in this series, i would then equal 7. The magnitude of R_F^i. does indeed depend on the chemical nature of the analyte and on the sensitivity of the particular instrument. The definition of R_F for ES is given as follows (using notation from differential calculus):

$$\lim_{\Delta C_S^i \to 0} \frac{\Delta A_S^i}{\Delta C_S^i} \equiv R_F^i \tag{2.1}$$

A response factor for each analyte (i.e., the ith analyte) is obtained during the calibration and is found by finding the limit of the ratio of the incremental change in peak area for the ith analyte, ΔA_S^i, to the incremental change in concentration of the ith analyte in the reference standard, ΔC_S^i as ΔC_S^i approaches zero. Quantitative analysis is then carried out by relating the instrument response to the analyte concentration in an unknown sample according to

$$A^i = R_F^i C_{unknown}^i \tag{2.2}$$

Equation (2.2) is then solved for the concentration of the ith analyte, $C_{unknown}^i$ in the unknown environmental sample. Refer to the quantification equation for ES in Table 2.2.

Figure 2.3 graphically illustrates the ES approach to multipoint instrument calibration. Six reference standards, each containing Aroclor 1242 (AR 1242), were injected into a gas chromatograph that incorporates a capillary or open tubular column appropriate to the separation and an electron-capture detector. The instrument used to generate the calibration is a capillary gas chromatograph incorporating an electron-capture detector (C-GC-ECD). Aroclor 1242 is a commercially produced mixture of 30 or more polychlorinated biphenyl (PCB) congeners. The

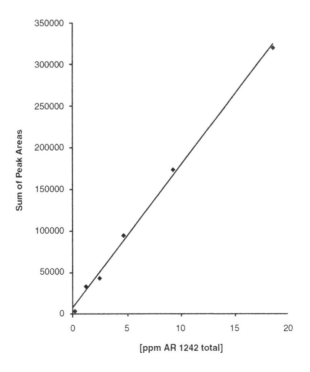

FIGURE 2.3

peak areas under the chromatographically resolved peaks were integrated and reported as an area count in units of microvolts-seconds (μV-sec). The more than 30 peak areas are then summed over all of the peaks to yield a total peak area, A_T^{1242} according to

$$A_T^{1242} = \sum_i A^i$$

The total peak area is then plotted against the concentration of Aroclor 1242 expressed in units of parts per million (ppm). The experimental data points closely approximate a straight line. This closeness of fit demonstrates that the summation of AR 1242 peak areas varies linearly with AR 1242 concentration expressed in terms of a total Aroclor. This data was obtained in the author's laboratory and nicely illustrates the ES mode. However, today with the high chromatographic resolution afforded by contemporary open tubular GC columns, PCBs are quantitated congener by congener and reported by PCB congener number. There are 209 PCB congeners. Chromatography processing software is essential to accomplish such a seemingly complex calibration in a reasonable time frame. This author would not want to undertake such a task armed with only a slide rule!

2.3.2   How Does the is Mode of Instrument Calibration Work and Why is it Increasingly Important to TEQA?

The IS mode is most useful when it has been determined that the injection volume cannot be reproduced with good precision. This mode is also preferred when the instrument response for a given analyte at the same concentration will vary over time. Both the analyte response and the IS analyte response will vary to the same extent over time; hence, the ratio of analyte response to IS response will remain constant. The use of an IS thus leads to good precision and accuracy in construction of the calibration curve. The calibration plot is usually established by plotting the increasing ratio of the analyte response to the fixed IS response versus the increasing concentration of analyte. In our THM example, 1,2-dibromopropane (1,2-DBP) is often used as a suitable IS. The molecular formula for 1,2-DBP is similar to each of the THMs, and this results in an instrument response factor that is near to that of the THMs. The concentrations of IS in all standards and samples must be identical so that the calibration curve can be correctly interpolated for the quantitative analysis of unknown samples. Refer to the THM example above and consider the concentrations cited above for the six-point working calibration standards. 1,2-DBP is added to each standard so as to be present at, for example, 200 ppb. This mode is defined as such since 1,2-DBP must be present in the sample or is considered *internal* to the sample. A single-point or multipoint calibration curve is usually established when using this mode.

The IS mode to instrument calibration has become increasingly important over the past decade as the mass spectrometer (MS) has replaced the element-selective detector as the principal detector coupled to gas chromatographs in the contemporary practice of TEQA. The mass spectrometer is somewhat unstable over time. The IS mode of GC-MS calibration quite adequately compensates for this instability.

Consider the determination of clofibric acid (CF) in wastewater. Clofibric acid or [2-(4-chlorophenoxy)-2-methyl-propanoic] acid is the bioactive metabolite of various lipid-regulating prodrugs. After chemically derivatizing CF to its methyl ester, a plot of the ratio of the CF methyl ester peak area to that of the internal standard 2, 2′, 4, 6, 6′-pentachlorobiphenyl (22′466′PCBP) against the concentration of CF methyl ester in ppm is shown in Figure 2.4. An ordinary (or unweighted) least squares regression line was established and drawn as shown (we will take up

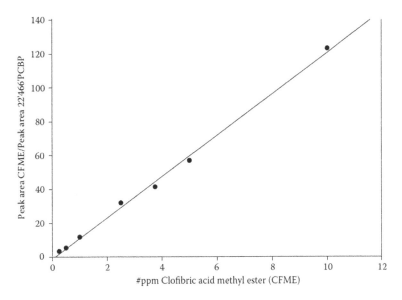

FIGURE 2.4

least squares regression shortly). The line shows a goodness of fit to the experimental data points. This plot demonstrates adequate linearity over the range of CF methyl ester concentrations shown. Any instability of the GC-MS instrument during the injection of these calibration standards is not reflected in the calibration. Therein lies the value and importance of the IS mode of instrument calibration.

For a single-point calibration approach, a relative response factor is used:

$$\lim_{\Delta\left(\frac{C_S^i}{C_{IS}^i}\right)\to 0} \frac{\Delta\left(\dfrac{A_S^i}{A_{IS}^i}\right)}{\Delta\left(\dfrac{C_S^i}{C_{IS}^i}\right)} = RR_F^i \tag{2.3}$$

Quantitative analysis is then carried out by relating the ratio of analyte instrument response for an unknown sample to that of IS instrument response to the ratio of unknown analyte concentration to IS concentration according to

$$\frac{A_{unknown}^i}{A_{IS}^i} = RR_F^i \frac{C_{unknown}^i}{C_{IS}^i} \tag{2.4}$$

Equation (2.4) is then solved for the concentration of analyte i in the unknown sample, $C_{unknown}^i$. Refer to the quantification equation for IS in Table 2.2 and A_{IS}^i are allowed to vary with time. This is what one expects when using high-energy detectors such as mass spectrometers. The ratio $A_{unknown}^i / A_{IS}^i$ remains fixed over time. This fact establishes a constant RR_F^i and hence preserves the linearity of the internal standard mode of instrument calibration. Equation (2.4) suggests that if RR_F^i is constant, and if we keep the concentration of IS to be used with the ith analyte, C_{IS}^i, constant, the ratio $A_{unknown}^i / A_{IS}^i$. varies linearly with the concentration of the ith analyte in the unknown, $C_{unknown}^i$.

The manner in which one uses internal standards in preparing calibration standards, ICVs, matrix spiked samples, and other QC samples will have a significant impact on the analytical result. Three strategies, shown in Figure 2.5, have emerged when considering the use of the IS mode of calibration.[11] In the first strategy, internal standards are added to the final extract after sample prep steps are complete. The quantification equation for IS shown in Table 2.2 would yield an analytical result for $C^i_{unknown}$ that is lower than the true concentration for the ith analyte in the original sample since percent recovery losses are not accounted for. This strategy is widely used in analytical method development. The second strategy first calibrates the instrument by adding standards and ISs to appropriate solvents, and then proceeds with the calibration. ISs are then added in known amounts to samples prior to extraction and cleanup. According to Budde:[11]

The measured concentrations will be the true concentrations in the sample if the extraction efficiencies of the analytes and ISs are the same or very similar. This will be true even if the actual extraction efficiencies are low, for example, 50%.

The third strategy depicted in Figure 2.5 corrects for percent recovery losses. Again, according to Budde:[11]

The system is calibrated using analytes and ISs in a sample matrix or simulated sample matrix, for example, distilled water, and the calibration standards are processed through the entire analytical method … [this strategy] is sometimes referred to as calibration with procedural standards.

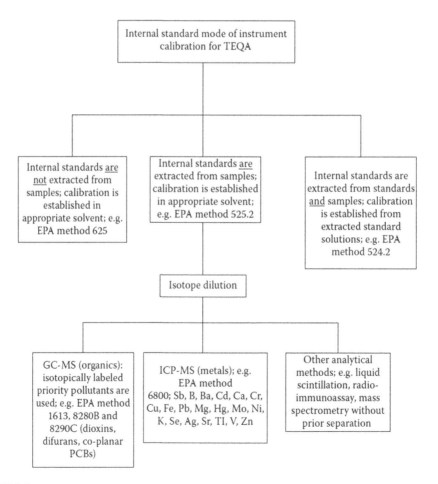

FIGURE 2.5

2.3.2.1   What Is Isotope Dilution?

Figure 2.5 places isotope dilution under the second option for using the IS mode of instrument calibration. The principal EPA methods that require *isotope dilution mass spectrometry* (IDMS) as the means to calibrate determinative techniques such as: GC-MS, GC-MS-MS, LC-MS-MS, and ICP-MS are shown in Figure 2.5. Other analytical methods that rely on isotope dilution as the chief means to calibrate and to quantitate are liquid scintillation counting and various radio-immunoassay techniques that are outside the scope of this book.

TEQA can be implemented using isotope dilution. The unknown concentration of an element or compound in a sample can be found by knowing only the natural isotopic abundance (atom fraction of each isotope of a given element) *and*, after an enriched isotope of this element has been added and equilibrated, by measuring this altered isotopic ratio in the spiked or diluted mixture. This is the simple yet elegant conceptual framework for isotope dilution as a quantitative tool.

2.3.2.2   Can a Fundamental Quantification Equation Be Derived from Simple Principles?

Yes, indeed, and we proceed to do so now. The derivation begins by first defining this altered and measured ratio of isotopic abundance after the enriched isotope (spike or addition of labeled analog) has been added to a sample and equilibrated. Only two isotopes of a given element are needed to provide quantification. Fassett and Paulsen showed how isotope dilution is used to determine the concentration at trace levels for vanadium in crude oil.[12] We use their illustration to develop the principles that appear below.

Let us start by defining R_m as the measured ratio of each of the two isotopes of a given element in the spiked unknown. The contribution made by ^{50}V appears in the numerator, and that made by ^{51}V appears in the denominator. Fassett and Paulsen obtained this measured ratio from mass spectrometry. Mathematically stated,

$$R_m = \frac{amt^{50}V(unknown) + amt^{50}V(spike)}{amt^{51}V(unknown) + amt^{51}V(spike)} \tag{2.5}$$

The amount of ^{50}V in the unknown sample can be found as a product of the concentration of vanadium in the sample as the ^{50}V and the weight of sample. This is expressed as follows:

$$amt^{50}V(unknown) = [atomfraction^{50}V][concV(unknownsample)][weight(unknownsample)] \tag{2.6}$$

The natural isotopic abundances for the element vanadium are 0.250% as ^{50}V and 99.750% as ^{51}V, so that $f^{51} = 0.9975$[12] for the equations that follow.

Equation (2.6) can be abbreviated and is shown rewritten as follows:

$$amt^{50}V = \left[f_{native}^{50}\right]\left[C_{unk}^{V}\right]\left[W_{unk}\right] \tag{2.7}$$

In a similar manner, we can define the amount of the higher isotope of vanadium in the unknown as follows:

$$amt^{51}V = \left[f_{native}^{51}\right]\left[C_{unk}^{V}\right]\left[W_{unk}\right] \tag{2.8}$$

Equation (2.7) and Equation (2.8) can also be written in terms of the respective amounts of the 50 and 51 isotopes in the enriched spike. This is shown as follows:

$$amt^{50}V = \left[f_{enriched}^{50} \right]\left[C_{spike}^{V} \right]\left[W_{spike} \right]$$

(2.9)

$$amt^{51}V = \left[f_{enriched}^{51} \right]\left[C_{spike}^{V} \right]\left[W_{spike} \right]$$

(2.10)

Equation (2.5) can now be rewritten using the symbolism defined by Equation (2.7) to Equation (2.10) and generalized for the first isotope of the ith analyte (i, 1) and for the second isotope of the ith analyte (i, 2) according to

$$R_m = \frac{\left[f_{unk}^{i,1} \right]\left[C_{unk}^{i} \right]\left[W_{unk} \right] + \left[f_{spike}^{i,1} \right]\left[C_{spike}^{i} \right]\left[W_{spike} \right]}{\left[f_{unk}^{i,2} \right]\left[C_{unk}^{i} \right]\left[W_{unk} \right] + \left[f_{spike}^{i,2} \right]\left[C_{spike}^{i} \right]\left[W_{spike} \right]}$$

(2.11)

where

R_m = isotope ratio (dimensionless number) obtained after an aliquot of the unknown sample has been spiked and equilibrated by the enriched isotope mix. This is *measurable* in the laboratory using a determinative technique such as mass spectrometry. The ratio could be found by taking the ratio of peak areas at different quantitation ions (quant ions or Q ions) if GC-MS was the determinative technique used.

$\left[f_{unk}^{i,1} \right]$ = natural abundance (atom fraction) of the ith element of the first isotope in the unknown sample. This is known from tables of isotopic abundance.

$\left[f_{unk}^{i,2} \right]$ = natural abundance (atom fraction) of the ith element of the second isotope in the unknown sample. This is known from tables of isotopic abundance.

$\left[f_{spike}^{i,1} \right]$ = natural abundance (atom fraction) of the ith element of the first isotope in the spiked sample. This is known from tables of isotopic abundance.

$\left[f_{spike}^{i,2} \right]$ = natural abundance (atom fraction) of the ith element of the second isotope in the spiked sample. This is known from tables of isotopic abundance.

C_{unk}^{i} = concentration [μmol/g, μg/g] of the ith element or compound in the unknown sample. This is unknown; *the goal of isotope dilution is to find this value.*

C_{spike}^{i} = concentration [μmol/g, μg/g] of the ith element or compound in the spike. This is known.

W_{unk} = weight of unknown sample in g. This is measurable in the laboratory.

W_{spike} = weight of spike in g. This is measurable in the laboratory.

Equation (2.11), the more general form, can be solved algebraically for C_{unk}^{i} to yield the quantification equation:

$$C_{unk}^{i} = \left[\frac{C_{spike}^{i} W_{spike}}{W_{unk}} \right]\left[\frac{f_{spike}^{i,1} - R_m f_{spike}^{i,2}}{R_m f_{unk}^{i,2} - f_{unk}^{i,1}} \right]$$

(2.12)

Equation (2.12) also appears as the quantification equation for IDMS in Table 2.2. We proceed now to consider the use of isotopically labeled organic compounds in IDMS. Returning again to Figure 2.5, we find the use of IDMS as a means to achieve TEQA when a GC-MS is the determinative technique employed. Methods that determine polychloro-dibenzo-dioxins (PCDDs), polychloro-dibenzo-difurans (PCDFs), and coplanar polychlorinated biphenyls (cp-PCBs) require IDMS. IDMS coupled with the use of high-resolution GC-MS represents the most rigorous and highly precise trace organics analytical techniques designed to conduct TEQA known today. The author recently adapted Equation (2.12) as it relates to more accurately quantifying trace concentration levels of cyanide ion (CN⁻) in whole blood.[13] The determinative technique (see Chapter 4) used in this application was automated static headspace GC-MS. This paper describes an application of isotope dilution applied to *enviro-health* TEQA.

2.3.2.3 What Is Organics IDMS?

Organics IDMS exploits the excellent specificity afforded by MS and particularly MS-MS determinative techniques by utilizing 2H-, 13C-, or 37Cl- isotopically labeled organic compounds as internal standards. The identical unlabeled organic compound is quantitated against its isotopic analog. These labeled analogs are also added to environmental samples or to human specimens to conduct *enviro-chemical* or *enviro-health* TEQA.

Isotopically labeled analogs are structurally identical except for the substitution of ^{2}H for 1H, ^{13}C for 12C, or ^{37}Cl for 35Cl. A plethora of isotopically labeled analogs are now available for most priority pollutants or persistent organic pollutants (POPs) that are targeted analytes in various EPA methods. To illustrate, molecular structures for the priority pollutant phenanthracene and its deuterated isotope, symbolized by 2H, or D, are shown below:

Phenanthracene   Phenanthracene-d10

Polycyclic aromatic hydrocarbons (PAHs), of which phenanthracene is a member, have abundant molecular ions in electron-impact MS. The molecular weight for phenanthracene is 178, while that for the deuterated isotopic analog is 188 (phen-d10). If phenanthracene is diluted with itself, and if an aliquot of this mixture is injected into a GC-MS, the native and deuterated forms can be distinguished at the same gas chromatographic retention time by monitoring the mass-to-charge ratio, abbreviated m/z at 178 and then at 188. Refer back to Table 1.11 whereby phen-d10 is used as an IS to quantitate all of the analytes listed when implementing EPA Method 8270C. Contrast this with the ultimate goal of IDMS, just discussed, in which an isotopically labeled organic compound is needed for each and every targeted organic compound! Isotopically labeled organic reference standards are very expensive!

2.3.3 How Does the SA Mode of Instrument Calibration Work?

The SA mode is used primarily when there exists a significant matrix interference and where the concentration of the analyte in the unknown sample is appreciable. SA becomes a calibration mode of choice when the analyte-free matrix cannot be obtained for the preparation of standards for ES. However, for each sample that is to be analyzed, a second so-called standard added or spiked sample must also be analyzed. This mode is preferred when trace metals are to be determined in complex sample matrices such as wastewater, sediments, and soils. If the analyte response is linear within the range of concentration levels anticipated for samples, it is not necessary to construct a multipoint calibration. Only two samples need to be measured—the unspiked and spiked samples.

2.3.3.1 Can We Derive a Quantification Equation for SA?

Yes, indeed. We proceed to do so now. Assume that C_{unk}^i represents the ultimate goal of TEQA, i.e., the concentration of the ith analyte, such as a metal in the unknown environmental sample or human specimen. Also assume that C_{spike}^i represents the concentration of the ith analyte in a spike solution. After an aliquot of the spike solution has been added to the unknown sample, an instrument response of the ith analyte for the standard added sample, denoted as R_{SA}^i whose concentration must be C_{SA}^i is measured. Knowing only the instrument response for the unknown, R_{unk}^i and the instrument response for the standard added, R_{SA}^i, C_{unk}^i can be found. Mathematically, let us prove this. The proportionality constant k must be the same between the concentration of the ith analyte and the instrument response, such as a peak area in atomic absorption spectroscopy. The following four relationships must be true:

$$C_{unk}^i = kR_{unk}^i \qquad (2.13)$$

$$C_{spike}^i = kR_{spike}^i \qquad (2.14)$$

$$R_{SA}^i = R_{unk}^i + R_{spike}^i \qquad (2.15)$$

$$C_{SA}^i = k\left[R_{unk}^i + R_{spike}^i \right] \qquad (2.16)$$

Solving Equation (2.15) for R_{spike}^i and substituting this into Equation (2.14) leads to the following ratio:

$$\frac{C_{unk}^i}{C_{spike}^i} = \frac{R_{unk}^i}{R_{SA}^i - R_{unk}^i} \qquad (2.17)$$

Solving Equation (2.17) for C_{unk}^i yields the quantification equation

$$C_{unk}^i = \left[\frac{R_{unk}^i}{R_{SA}^i - R_{unk}^i} \right] C_{spike}^i \qquad (2.18)$$

For real samples that may have nonzero blanks, the concentration of the ith analyte in an unknown sample, C_{unk}^i can be found knowing only the measurable parameters R_{SA}^i and R_{unk}^i and instrument responses in blanks along with the known concentration of single standard added or spike concentration R_{spike}^i according to

$$C_{unk}^i = \left[\frac{R_{unk}^i - R_{bl-unk}^i}{\left(R_{SA}^i - R_{unk}^i \right) - \left(R_{bl-unk}^i - R_{bl-unk}^i \right)} \right] C_{spike}^i \qquad (2.19)$$

where R_{bl-unk}^i represents the instrument response for a blank that is associated with the unknown sample. $R_{bl-spike}^i$ is the instrument response for a blank associated with the spike solution and accounts for any contribution that the spike makes to the blank. Equation (2.19) is listed in Table 2.2 as the quantification equation for SA.

If a multipoint calibration is established using SA, the line must be extrapolated across the ordinate (*y*-axis) and terminate on the abscissa (*x*-axis). The value on the abscissa that corresponds to the amount or concentration of unknown analyte yields the desired result. Students are asked to create a multipoint SA calibration to quantitate both Pb and anionic surfactants in Chapter 5. Contemporary software for graphite furnace atomic absorption spectroscopy (GFAA) routinely incorporates SA as well as ES modes of instrument calibration. Autosamplers for GFAA can be programmed to add a precise aliquot of a standard solution containing a metal to an aqueous portion of an unknown sample that contains the same metal at an unknown concentration.

Most comprehensive treatments of various analytical approaches utilizing SA as the principal mode of calibration can be found in an earlier published paper.[14]

2.4 WHAT DOES LEAST SQUARES REGRESSION REALLY MEAN?

Ideally, a calibration curve that is within the linearity range of the instrument's detector exhibits a straight line whose slope is constant throughout the range of concentration of interest. By minimizing the sum of the squares of the residuals, a straight line with a slope *m* and a *y* intercept *b* is obtained. This mathematical approach is called a least squares (LS) fit of a regression line to the experimental data. The degree of fit expressed as a goodness of fit is obtained by the calculation of a correlation coefficient. The degree to which the least squares fit reliably relates detector response to analyte amount or concentration can also be determined using statistics. Upon interpolation of the least squares regression line, the amount or concentration of analyte is obtained. The extent of uncertainty in the interpolated amount or concentration of analyte in the unknown sample is also found. In the next section, equations for the least squares regression will be derived and treated statistically to obtain equations that state to what degree of confidence this can be achieved in an interpolated value. These concepts are at the heart of what constitutes GLP as well as being essential to achieving good precision and accuracy in the experimental aspects of TEQA.

2.4.1 How Do You Derive the Least Squares Regression Equations?

The concept starts with a definition of a residual for the *i*th calibration point. The residual Q_i is defined to be the square of the difference between the experimental data point y_i^e. and the calculated data point from the best-fit line y_i^c. Figure 2.6 illustrates a residual from the author's laboratory where a least squares regression line is fitted from the experimental calibration points for *N,N*-dimethyl-2-amino-ethanol (N,N-DM-2AE) using a gas chromatograph with an ion trap mass spectrometer.[15] Expressed mathematically,

$$Q_i = \left| y_i^e - y_i^c \right|^2$$

where y^c is found according to

$$y_i^c = mx_i + b$$

with *m* being the slope for the best-fit straight line through the data points and *b* being the *y* intercept for the best-fit straight line. x_i is the amount of analyte *i* or the concentration of analyte *i*. x_i is obtained from a knowledge of the analytical reference standard used to prepare the calibration

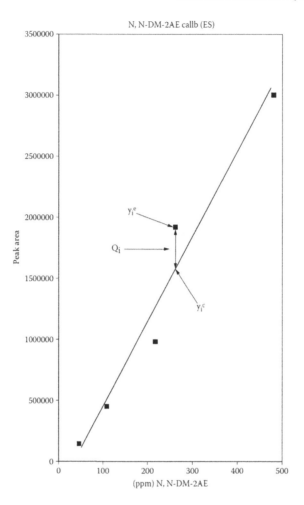

FIGURE 2.6

standards and is assumed to be free of error. There are alternative relationships for least squares regression that assume x_i is not free of error. This approach is outside the scope of this book. To obtain the least squares regression slope and intercept, the sum of the residuals over all N calibration points, defined as Q, is first considered:

$$Q = \sum_{i}^{N} \left| y_i^e - y_i^c \right|^2$$

$$Q = \sum_{i}^{N} \left[y_i^e - \left(mx_i + b \right) \right]^2$$

The total residual is now minimized with respect to both the slope m and the intercept b:

$$\frac{\partial Q}{\partial b} = 0 = -2 \sum_{i}^{N} \left[y_i^e - \left(mx_i + b \right) \right] \tag{2.20}$$

$$\frac{\partial Q}{\partial m} = 0 = -2\sum_{i}^{N} x\left[y_i^e - \left(mx_i + b \right) \right] \tag{2.21}$$

Rearranging Equation (2.20) for b,

$$b = \frac{1}{N}\left[\sum_i y_i - m\sum_i x_i \right] \tag{2.22}$$

Rearranging Equation (2.21) for m,

$$m = \frac{\sum_i x_i y_i - b\sum_i x_i}{\sum_i x_i^2} \tag{2.23}$$

Next, substitute for b from Equation (2.22) into Equation (2.23):

$$m = \frac{\sum_i x_i y_i - (1/N)\left(\sum_i y_i - m\sum_i x_i \right)\sum_i x_i}{\sum_i x_i^2}$$

Upon simplifying, we obtain

$$m = \frac{N\sum_i x_i y_i - \sum_i x_i \sum_i y_i}{N\sum_i x_i^2 - \left(\sum_i x_i \right)^2} \tag{2.24}$$

Recall the definition of a mean:

$$\overline{x} = \frac{\sum_i x_i}{N}, \overline{y} = \frac{\sum_i y_i}{N}$$

Rearranging in terms of summations gives

$$\sum_i x_i = N\overline{x} \tag{2.25}$$

$$\sum_i y_i = N\overline{y} \tag{2.26}$$

Upon substituting Equations (2.25) and (2.26) into Equation (2.24), we arrive at an expression for the least squares slope m in terms of only measurable data points:

$$m = \frac{\sum_i x_i y_i - N\overline{xy}}{\sum_i x_i^2 - N\overline{x}^2} \tag{2.27}$$

Defining the sum of the squares of the deviations in x and y calculated from all N pairs of calibration points gives

$$SS_{xx} = \sum_{i}^{N} (x_i - \bar{x})^2$$

$$SS_{yy} = \sum_{i}^{N} (y_i - \bar{y})^2$$

The sum of the products of the deviation s is given by

$$SS_{xy} = \sum_{i}^{N} (x_i - \bar{x})(y_i - \bar{y})$$

The slope for the least squares regression can then be expressed as

$$m = \frac{SS_{xy}}{SS_{xx}} \tag{2.28}$$

and the y intercept can be obtained by knowing only the slope m and the mean value of all of the x data and the mean value of all of the y data according to

$$b = \bar{y} - m\bar{x} \tag{2.29}$$

Equations (2.28) and (2.29) enable the best-fit calibration line to be drawn through the experimental x, y points. Once the slope m and the y intercept b for the least squares regression line are obtained, the calibration line can be drawn by any one of several graphical techniques. The instrument manufacturer today provides the software tools to perform a least squares (LS) fit of a regression line to the experimental data. For example, if a GC-MS manufactured by Agilent Technologies Inc. is used, the analytical chemist can use the calibration software incorporated in ChemStation® or MassHunter®. Alternatively, a commonly used spreadsheet such as Excel can achieve the same objectives.[16] These techniques certainly beat use of a ruler and graph paper!

A quantitative measure of the degree to which the dependent variable (i.e., the analytical signal) depends on the independent variable, the concentration of analyte i, is denoted by the *correlation coefficient r* according to

$$r = \frac{\sum_{i}\left[(x_i - \bar{x})(y_i - \bar{y})\right]}{\sqrt{\left[\sum_{i}(x_i - \bar{x})^2\right]\left[\sum_{i}(y_i - \bar{y})^2\right]}}$$

Using the previously defined terms, the correlation coefficient can be described as

$$r = \frac{SS_{xy}}{SS_{xx}SS_{yy}} \tag{2.30}$$

As r approaches 1, the correlation is said to approach perfection. A good linear calibration plot can achieve values for r that easily reach and exceed 0.9900, and suggests that good laboratory technique as well as optimum instrument performance have been achieved. A good correlation coefficient should not be confused with the notion of calibration linearity. A curve that upon visual inspection appears to be a nonlinear curve can exhibit a correlation coefficient that is calculated to be quite close to 1. The square of r is called the *coefficient of determination*. Several commercially available chromatography software packages such as Total Chrom® (PerkinElmer) or ChemStation® (Agilent), are programmed to calculate only the coefficient of determination following the establishment of the least squares regression calibration curve. Equation (2.28), which expresses the least squares slope m in terms of a ratio of sums of the square, can be compared to Equation (2.30). If this is done, it becomes obvious that the true nature of r is merely a scaled version of the slope (i.e., the slope estimate multiplied by a factor to keep r always between -1 and $+1$),[7]

$$r = m \sqrt{\frac{S_{xx}}{S_{yy}}}$$

2.4.2 To What Extent Are We Confident in the Analytical Results?

How reliable are these least squares parameters? To answer this question, we first need to find the standard deviation about the least squares best-fit line by summation over N calibration points of the square of the difference between experimental and calculated detector responses according to

$$S_{y/x} = \sqrt{\frac{\Sigma \left(y_i^e - y_i^e \right)^2}{N-2}} \tag{2.31}$$

where $S_{y/x}$ represents the *standard deviation of the vertical residuals* and $N-2$ represents the number of degrees of freedom. N is the number of x, y data points used to construct the best-fit calibration line less a degree of freedom used in determining the slope and a second degree of freedom used to determine the y intercept.

The uncertainty in both the slope and intercept of the least squares regression line can be found using the following equations to calculate the standard deviation in the slope, S_m, and the standard deviation in the y intercept, S_b:

$$S_m = \frac{S_{y/x}}{\sqrt{\sum_i^N |x_i - x|^2}}$$

$$S_b = \frac{S_{y/x} \Sigma_i x_i^2}{\sqrt{N \sum_i^N |x_i - \bar{x}|^2}}$$

2.4.3 How Confident Are We of an Interpolated Result?

For a given instrumental response such as for the unknown, y_0, the corresponding value x_0 from interpolation of the best-fit calibration is obtained and the standard deviation in x_0 can be found according to

$$S_{x_0} = \frac{S_{y/x}}{m} \sqrt{\frac{1}{L} + \frac{1}{N} + \frac{(y_0 - \bar{y})^2}{m^2 \sum_i^N |x_i - \bar{x}|^2}}$$

where s_{x_0} represents the standard deviation in the interpolated value x_0 and L represents the number of replicate measurements of y_0 for a calibration having N, x, y data points.

Upon replacing the summate with S_{xx}, the standard deviation in the interpolated value x_0 yields a most useful expression:

$$S_{x_0} = \frac{S_{y/x}}{m} \sqrt{\frac{1}{L} + \frac{1}{N} + \frac{(y_0 - \bar{y})^2}{m^2 S_{xx}}} \tag{2.32}$$

Equation (2.32) shows that the uncertainty, S_{x0}, in the interpolated value, x_0, is largely determined by minimizing the ratio of $S_{y/x}$ to m. A small value for $S_{y/x}$ implies good to excellent precision in establishment of the least squares regression line. A large value for m implies good to excellent detector sensitivity. The standard deviation in the interpolated value (S_{x_0}) is also reduced by making L replicate measurements of the sample. The standard deviation can also be reduced by increasing the number, N, of calibration standards used to construct the calibration curve.

The determination of x_0 for a given detector response, y_0, is, of course, *the* most important outcome of trace quantitative analysis. x_0, together with an estimate of its degree of uncertainty, represents *the ultimate goal of trace quantitative analysis*; that is, it answers the questions, how much of analyte i is present, and how reliable is this number in a given environmental sample? For TEQA, it is usually unlikely that the population mean for this interpolated value $\mu(x_0)$ as well as the standard deviation in this population mean, $\sigma(x_0)$, can ever be known. TEQA requires the following:

- A determinative technique whereby, for example, the concentration of a priority pollutant VOC, such as vinyl chloride, can be measured by an instrument. It can be assumed that for a given analyte such as vinyl chloride, repeatedly injected into a gas chromatograph, $\mu(x_0)$ and $\sigma(x_0)$ are known.
- However, the concentration of vinyl chloride in the environmental sample may involve some kind of sample preparation to get the analyte into the appropriate chemical matrix for the application of the appropriate determinative technique, and we can assume that both $\mu(x_0)$ and $\sigma(x_0)$ are unknown.

These two constraints require that the confidence limits be found when the standard deviation in the population mean is known. This is where the Student's t statistics have a role to play. Who was "Student"? Anderson has introduced a little history (pp. 70–72):[7]

Unfortunately, we do not know the true standard deviation of our set; we have only an estimate s based on L observations. Thus, the distribution curve for x is an estimate of the true distribution curve. A research chemist, W.S. Gosset, worked out the relationship between these estimates of the distribution curve and the true one so that we can use s in place of σ. Gosset was employed by a well-known Irish brewery which did not want its competition to know it was using statistical methods. However, he was permitted to publish his work under the pen name "Student," and the test for differences in averages based on his work is known as Student's t-test. Student's t-test assumes that the average is based on data from a normal population.

Because we know s_{x_0} (the standard deviation in the interpolated value x_0), we can use *Student's t statistics* to estimate to what extent x_0 estimates the population mean μ_{x_0} according to:

$$\text{Confidence Interval for } \mu_{x_0} = x_0 \pm t_{1-\alpha/2,df} \, S_{x_0}$$

where $t_{1-\alpha/2,df}$ represents the value for a two-tailed Student's t at the significance level of α for $N - 2$ degrees of freedom (df), where N is the number of x, y data points used to construct the calibration curve. The term $\left[\pm t_{1-\alpha2,df} s_{x_0} \right]$ is called the confidence interval and represents the range of x values within which one can be $(1 - \alpha/2)$ 100% confident that the mean value for x_0 will approximate the population mean μ_{x_0}.

Figure 2.7. is this author's attempt to graphically represent the uncertainty present in an interpolated instrument response y_0. A segment of what might be a typical ES or IS calibration plot shrouded with its corresponding confidence limits both above and below the regression line is shown. The confidence interval is shown to be equidistant from the regression line. This assumption represents the *homoscedastic* case, i.e., a regression line having a constant variance, over the range of analyte concentration used to construct the linear regression line. Those sections of the calibration that are highlighted in bold reveal that horizontal movement to either confidence limit at y_0 results in equal confidence intervals. The confidence interval can be related to the product of Student's t and the standard deviation in the interpolated value, s_{x_0}. Mathematically,

$$\left| y_0 - y_{lower} \right| \leftrightarrow t_{1-\alpha/2,df} \, s_{x_0}$$

$$\left| y_0 - y_{upper} \right| \leftrightarrow t_{1-\alpha/2,df} \, s_{x_0}$$

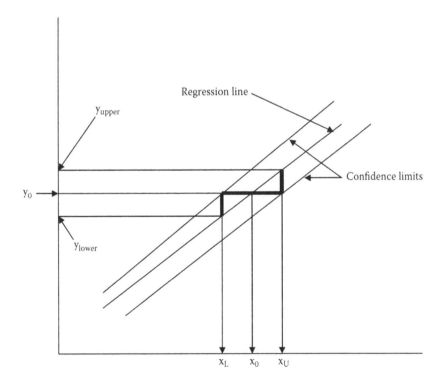

FIGURE 2.7

Figure 2.7 is a good starting point from which to consider a derivation proposed earlier by MacTaggert and Farwell.[17] The prediction interval about x_L (or x_{lower}, as shown in Figure 2.7) can be mathematically defined as follows:

$$y_L \pm tS_{yL} = b + mx_L \pm t_{(1-\alpha/2, df=n-2)} S_{y/x} \sqrt{1 + \frac{1}{N} + \frac{(x_L - \bar{x})^2}{S_{xx}}}$$

where

$b = y$ intercept of the linear regressed line
$m = $ slope of the linear regressed line
$t = $ two-tailed value for Student's t for $N - 2$ degrees of freedom, where N is the number of
    x, y data points used to construct the calibration

Refer again to Figure 2.7. We see that if the value for y_L is taken and the height of this interval added to it, y_0 is obtained. This enables two relationships to emerge that can both be set equal to y_0. These two relationships are given below for y_0:

$$y_0 = b + mx_0$$

$$y_0 = b + mx_L + ts_{y/x} \sqrt{1 + \frac{1}{N} + \frac{(x_L - \bar{x})^2}{S_{xx}}}$$

Both equations being equal to y_0 can be set equal to each other such that

$$b + mx_0 = b + mx_L + ts_{yx} \sqrt{1 + \frac{1}{N} + \frac{(x_L - \bar{x})^2}{S_{xx}}}$$

Eliminating b, setting this equation equal to zero, squaring both sides, and then solving for x_L yields the following quadratic equation:

$$x_L^2 (1 - g) + 2x_L (g\bar{x} - x_0) + x_0^2 - g\bar{x}^2 - gS_{xx} \left(1 + \frac{1}{N}\right) = 0$$

Here g is used to collect terms and greatly simplify the expression

$$g = \frac{t^2 s_{yx}^2}{m^2 S_{xx}}$$

Solving the above quadratic equation via the quadratic formula and incorporating the result for x_U yields a *discrimination interval* about the interpolated value for x_0:

$$\{x_U, x_L\} = \frac{x_0 - g\bar{x}}{1 - g} \pm \frac{ts}{m(1 - g)} \sqrt{\left(1 + \frac{1}{N}\right)(1 - g) + \frac{(x_0 - \bar{x})^2}{S_{xx}}}$$

If L replicates are made of the y_0 value, the following equation may be used to define discrimination intervals about the interpolated value for x_0 according to

$$\{x_U, x_L\} = \frac{x_0 - gx}{1 - g} \pm \frac{ts_{yx}}{m(1-g)} \sqrt{\left(\frac{1}{L} + \frac{1}{N}\right)(1-g) + \frac{(x_0 - \bar{x})^2}{S_{xx}}}$$

The expression for g above can be further simplified by substituting for S_{xx} using Equation (2.28) so that

$$g = \frac{t^2 s_m^2}{m^2}$$

This relationship shows that g is a measure of the statistical significance of the slope value. Further use of g can be made to derive a simpler equation for the discrimination interval. Let us assume that $g \ll 1$ such that

$$\{x_U, x_L\} \approx x_0 \pm \frac{ts_{yx}}{m} \sqrt{\left(\frac{1}{L} + \frac{1}{N}\right) + \frac{(x_0 - \bar{x})^2}{S_{xx}}}$$

This equation shows a symmetrical interval about x_0 and is comparable to a prediction interval, as introduced earlier. We have thus mathematically shown that the standard deviation in the vertical residual, $s_{y/x}$, can approximate the discrimination interval. The following quote reinforces this notion:[19]

Analytical chemists must always emphasize to the public that *the single most characteristic of any result obtained from one or more analytical measurements is an adequate statement of its uncertainty interval.* Lawyers usually attempt to dispense with uncertainty and try to obtain unequivocal statements; therefore, an uncertainty interval must be clearly defined in cases involving litigation and/or enforcement proceedings. Otherwise, a value of 1.001 without a specified uncertainty, for example, may be viewed as legally exceeding a permissible level of 1.

The concept of a confidence interval that shrouds the least squares calibration line will again become important in the calculation of an instrument's detection limit (IDL) for a given analyte. In other words, how reliable is the IDL for a given analyte?

2.5 HOW DO YOU DERIVE EQUATIONS TO FIND INSTRUMENT DETECTION LIMITS?

The IDL for a specific chemical analyte is defined to be the lowest possible concentration that can be reliably measured. Experimentally, if lower and lower concentrations of a given analyte are measured, the smallest concentration that is barely detectable constitutes a rough estimate of the IDL for that analyte using that particular instrument. For years, EPA has required laboratory analysts to first measure a blank replicate a number of times. The most recent guidelines appeared in 40 Code of Federal Regulations (CFR) Part 136, Appendix B.[19] Steps that are recommended to calculate the IDL are listed below:

1. Prepare a calibration curve for the test with standards.
2. Analyze seven laboratory water blanks.
3. Record the response of the test for the blanks.
4. Prepare the mean and standard deviation of the results from the blanks as above.
5. The IDL is three times the standard deviation on the calibration curve.

Fortunately, a bit more thought has been given to the calculation of the IDL than the meager guidelines given above. The average signal that results from these replicate measurements yields a mean signal, S_{blank}. The analyst then calculates the standard deviation of these replicate blanks, S_{blank}, and finds the sum of the mean signal and a multiple k (often $k = 3$) of the standard deviation to obtain a minimum detectable signal S_{IDL} according to[20]

$$S_{IDL} = S_{blank} + k s_{blank}$$

If a least squares calibration curve has been previously established, the equation for this line takes on the common form, with S being the instrument response, m being the least squares slope, and x being the concentration according to

$$S = mx + S_{blank} \tag{2.33}$$

Solving Equation (2.33) for x, denoting the IDL as X_{IDL}, and replacing S with this minimum signal, S_{IDL}, yields

$$x_{IDL} = \frac{S_{IDL} - S_{blank}}{m} \tag{2.34}$$

Equation (2.34) represents a step up from the meager guidelines introduced earlier; it incorporates the slope of the calibration as a means to address detector sensitivity. It becomes obvious that minimizing x_{IDL} requires that the signal at the detection limit, S_{IDL}, be maximized while the noise level remains minimized, and that the steepness of the calibration curve be maximized.

The use of Equation (2.34) to quantitatively estimate the IDL for chromatographs and for spectrometers has been roundly criticized over the years. There have been reported numerous attempts to find alternative ways to calculate IDLs. This author will comment on this most controversial topic in the following manner. The approach encompassed in Equation (2.34) clearly lacks a statistical basis for evaluation, and hence is mathematically found to be inadequate. As if this indictment is not enough, IDLs calculated based on Equation (2.34) also ignore the uncertainty inherent in the least squares regression analysis of the experimental calibration, as presented earlier. In other words, what if the analyte is reported to be absent when, in fact, it is present (a false negative)? In the subsections that follow, a more contemporary approach to the determination of IDLs is presented and starts first with the concept of confidence intervals about the regression line.

2.5.1 Can We Derive Equations for Confidence Intervals about the Regression? Will These Equations Lead to Calculating IDLs?

Yes and yes—however, after much mathematical thought! The least squares regression line is seen to be shrouded within confidence limits as shown in Figure 2.7. The derivation of the relationship introduced below was first developed by Hubaux and Vos[2] some 50 years ago! The upper and lower confidence limits for y that define the confidence interval for the calibration are obtained for any x (analyte amount or concentration) and are given as follows:

$$y = \left[\bar{y} + m(x - \bar{x}) \right] \pm t \sqrt{V_y} \tag{2.35}$$

where $\bar{x}$ and $\bar{y}$ are the respective mean values over all x and y data points, and m represents the slope of the linear least squares regression. t (Student's t) corresponds to a probability of $1 - \alpha$ for the upper limit and $1 - \beta$ for the lower limit. V_y represents the variance in the instrument response y for a normal distribution of instrument responses at a given value of x. This equation represents another way to view the least squares regression line and its accompanying confidence interval. The term in brackets is the calculated response based on least squares regression. The variance in y, V_y, is composed of two contributions. The first is the variance in y calculated from the least squares regression, and the second is the residual variance σ^2. Expressed mathematically,

$$V_y = V_{\bar{y}-m[x-\bar{x}]} + \sigma^2 \tag{2.36}$$

and the variance in the least squares calculated y can be viewed as being composed of a variance in the mean value for y and a variance in the x residual according to

$$V_{\bar{y}-m[x-\bar{x}]} = V_{\bar{y}} + V_m |x-\bar{x}|^2$$

The variance in the least squares regression line can further be broken down into the variance in the mean y expressed as $V\bar{y}$ and the variance in the slope m expressed in terms of V_m according to

$$V_{\bar{y}} = \frac{\sigma}{N} \tag{2.37}$$

$$V_m = \frac{\sigma^2}{\sum_i^N |x-\bar{x}|^2} \tag{2.38}$$

Hence, substituting Equations (2.37) and (2.38) into Equation (2.36) gives

$$V_y = \sigma^2 + \frac{\sigma^2}{N} + \frac{\sigma^2 |x-\bar{x}|^2}{\sum_i^N |x_i-\bar{x}|^2} \tag{2.39}$$

Factoring out σ^2 gives

$$V_y = \sigma^2 \left[1 + \frac{1}{N} + \frac{|x-x|^2}{\sum_i^n |x_i-\bar{x}|^2} \right] \tag{2.40}$$

The residual variance σ^2 may be replaced by its estimate s^2, and upon substituting Equation (2.40) into Equation (2.35), it gives

$$y = \left[\bar{y} + m(x-\bar{x}) \right] \pm ts \sqrt{1 + \frac{1}{N} + \frac{|x-\bar{x}|^2}{\sum_i^N |x_i-\bar{x}|^2}} \tag{2.41}$$

Equation (2.41) is an important relationship if one desires to plot the confidence intervals that surround the least squares regression line. The upper confidence limit for the particular case where $x = 0$ is obtained from *Equation (2.41)*, in which a $1 - \alpha$ probability exists that the normal distribution of instrument responses falls to within the mean y at $x = 0$. Mathematically, the instrument response y_C, often termed a critical response, is defined as

$$y_c = [\bar{y} + m\bar{x}] + t_{1-\pm}s\sqrt{1 + \frac{1}{N} + \frac{x^2}{\sum_i^N |x_i - \bar{x}|^2}} \tag{2.42}$$

Equation (2.42) can be viewed as being composed of two terms. This is expressed as follows:

$$y_c = y_0 + [P][s]$$

The y intercept of the least squares regression line is given by

$$y_0 = \bar{y} - m\bar{x}$$

The value for y_0 cannot be reduced because it is derived from the least squares regression; however, the second term, $[P][s]$, may be reduced and hence lead to lower IDLs. P is defined as

$$P = t_{1-\alpha}\sqrt{1 + \frac{1}{N} + \frac{\bar{x}^2}{SS_{xx}}}$$

Figure 2.8 provides a graphical view of the terms used to define the decision limit, x_C, and detection limit, x_D. The decision limit, x_C, is a specific concentration level for a targeted analyte above which one may decide whether the result of analytical measurement indicated detection. The detection limit, x_D, is a specific concentration level for a targeted analyte above which one may rely upon it to lead to detection. A third limit, known as a determination limit, or using more

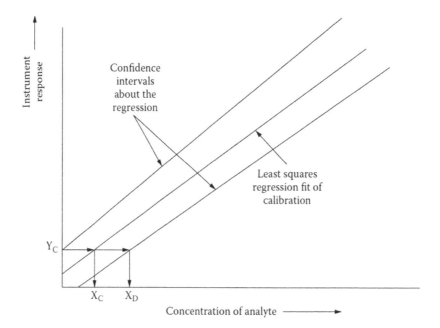

FIGURE 2.8

recent jargon from the EPA, the practical quantitation limit, is a specific concentration at which a given procedure is sufficiently precise to yield satisfactory quantitative analysis.[5] The determination limit will be introduced in more detail later.

Referring to Figure 2.8, y_C can merely be drawn horizontally until it meets with the lower confidence limit in which a $1 - \beta$ probability exists that the normal distribution of instrument responses falls to within the mean at this value of x. The ordinate that corresponds to this value of x, y_D, can be found according to

$$y_D = y_c + t_{1-\beta}s\sqrt{1 + \frac{1}{N} + \frac{|x_D - \bar{x}|^2}{\sum_i^N |x_i - \bar{x}|^2}} \tag{2.43}$$

Equation (2.43) can be viewed as comprising three terms, expressed as follows:

$$y_D = y_0 + [P][s] + [Q][s]$$

where

$$Q = t_{1-\beta}\sqrt{1 + \frac{1}{N} + \frac{|x_D - \bar{x}|^2}{\sum_i^N |x_i - \bar{x}|^2}}$$

The IDL is thus seen to be composed of a fixed value, y_0, which is derived from the nature of the least squares calibration line and the residual standard deviation, s. Factors P and Q can be varied so as to minimize y_D.

Graphically, it is straightforward to interpolate from y_C to the IDL x_D. Numerically, Equations (2.42) and (2.43) can be solved for the critical limit x_C and for the IDL, x_D. The equation for the calibration line can be solved for x as follows:

$$x = \frac{y - y_0}{m}$$

When the instrument response y is set equal to y_C, the concentration that corresponds to a critical concentration limit is defined. This critical level x_C, defined almost 50 years ago,[5,19] is a decision limit at which one may decide whether the result of an analysis indicates detection. The critical level has a specifically defined false positive (type I) error rate, α, and an undefined false negative (type II) error rate, β; x_C is found according to

$$x_c = \frac{y_c - y_0}{m} = \frac{[P][s]}{m} \tag{2.44}$$

In a similar manner, the IDL that satisfies the statistical criteria discussed is obtained according to

$$x_D = \frac{y_D - y_0}{m} = \frac{[P][s] + [Q][s]}{m} = \frac{s(P + Q)}{m} \tag{2.45}$$

The IDL denoted by x_D and given in Equation (2.45) is the true concentration at which a given analytical procedure may be relied upon to lead to detection.[5,19] A false negative rate can now be

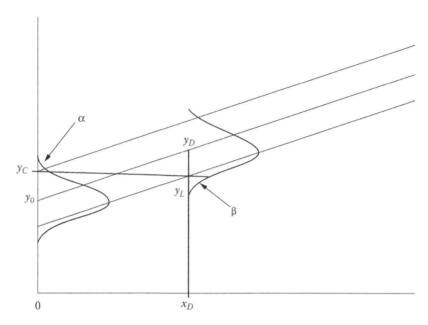

FIGURE 2.9

defined. A normal distribution of instrument responses at a concentration of zero and at a concentration x_D overlaps at x_C and gives rise to α and β. Figure 2.9 graphically depicts this Gaussian distribution and stems from application of Equation (2.41), whereas confidence intervals can be defined that shroud the linear least squares regression line. The intersection of the upper confidence band with the y-axis at y_C, where $x = 0$, corresponds to the highest signal that could be attributed to a blank $100\,(1 - \alpha)\%$ of the time. This is represented by a t distribution with a one-sided tailed α for y at $x = 0$. The intersection of y_C with the lower confidence band at y_L in Figure 2.9 corresponds to the lowest signal that could be attributed to an analyte concentration x_D $100\,(1 - \beta)\%$ of the time. This is represented by the t distribution with a one-tailed β for y at $x = x_D$. The mean for the distribution of signals at x_D is y_D. Equation (2.45) leads to a quadratic equation whose root is given below:[20]

$$x_D = \left[\frac{2ts_{xy}}{m(1-g)}\right]\left[\sqrt{\frac{1}{L} + \frac{1}{N} + \frac{\bar{x}^2}{S_{xx}}} - \frac{2g\bar{x}}{1-g}\right] \qquad (2.46)$$

We have just connected the concept of a confidence interval to an instrument detection limit (IDL)! Let's explore this relationship further using graphical concepts to develop a deeper understanding of the overlay of a probability distribution function atop the calibration plot as illustrated in Figure 2.9. Figure 2.10 depicts the blank statistics model and explains why this model fall short in establishing the correct IDL! Figure 2.10 plots the frequency distribution of instrument responses versus the amount or concentration (x-axis) of analyte. This model defines the LOD (or IDL) as 3x standard deviation in the mean blank signal. A limit of quantitation is defined as 10x standard deviation in the mean blank signal. Let's examine Figure 2.10 more closely as we tie in the mathematics introduced above. Figure 2.11 shows the blank statistics model that satisfies the Type 1 error criteria ($\alpha = 0.05$) and not the Type 2 error criteria ($\beta = 0.5$) while Figure 2.12 shows the confidence band statistics model that satisfies both Type 1 and Type 2 errors ($\alpha = 0.05$ and $\beta = 0.05$). Explanations are also provided in the upper right box for both

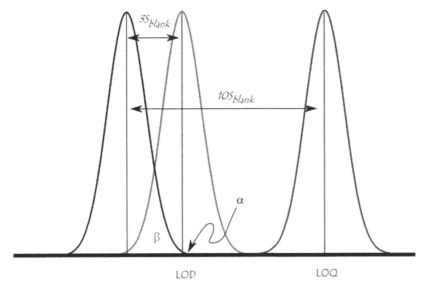

FIGURE 2.10

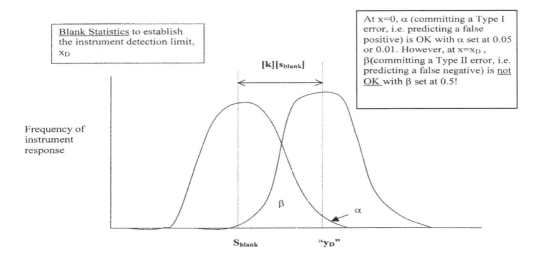

FIGURE 2.11

graphs. Figure 2.13 summarizes in graphical form a calibration plot of instrument response vs. analyte amount or concentration for the ordinary or unweighted least squares regression (OLS) while incorporating the mathematical concepts just discussed. We are still not done! What if we need to weight the least squares regression in the first place?

2.5.2 WHAT IS WEIGHTED LEAST SQUARES AND HOW DOES THIS INFLUENCE IDLS?

The previous derivation of the least squares regression equations assumed that the variance about the regressed line was constant throughout the range of analyte concentrations used to construct the calibration and is referred to as *ordinary least squares* (OLS). Many calibration curves, particularly those at concentration levels close to the IDL, have non-constant variances about the

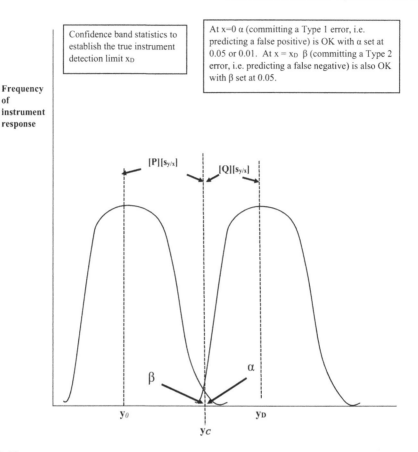

FIGURE 2.12

linear regressed line. Chromatographic processing software queries the user as to whether or not a *weighted least squares regression* (WLS) or OLS regression is to be used. It behooves the analyst to have at least a cursory understanding of WLS.

Burdge et al.[20] put it this way:

> Application of OLS to data with non-constant variance (heteroscedasticity) yields confidence bands and thus detection limits that will not accurately reflect measurement capability. WLS facilitates the determination of realistic detection limits for heteroscedastic data by yielding confidence bands that directly reflect the changing variance … Application of OLS to significantly heteroscedastic data results in the construction of unnecessarily broad confidence bands at the low end of the calibration curve. The detection limit derived from OLS treatment of such data will not be a fair representation of a measurement method's detection capability. *y*

Zorn et al. showed that polychlorinated biphenyls (PCBs) exhibit a non-constant variance (heteroscedasticity).[6] They compared unweighted or ordinary least squares (OLS) regression parameters with weighted least square (WLS) regression parameters. Upon comparing the two approaches, they found that the *y* intercept was moderately affected, the slope was unaffected and that the residual standard deviation decreased significantly. Plots of instrument response

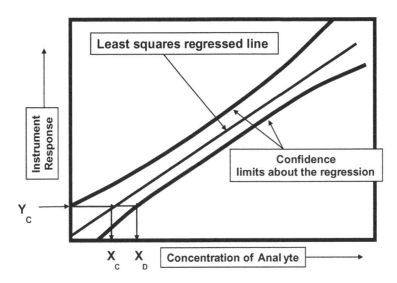

FIGURE 2.13

versus analyte concentration (pg/µL or ppb) showed that the widths of prediction intervals were much smaller at or near the low end of the calibration and much larger near the high end of the calibration. This resulted in lower and more accurate x_C (critical concentration) and x_D (concentration at the lowest level of detection or IDL)[6] Hence, weighting factors or simply *weights* are used to give more emphasis or weight to calibration data points, usually at the lower end of the calibration. This can be done without changing the raw data. However, implementing WLS requires that we modify the previously developed mathematical relationships used to derive OLS.

Let us return to Equation (2.31) for the standard deviation of the vertical residuals for $N - 2$ degrees of freedom, as introduced earlier. A weighted residual would look like the following:

$$S_{wy/x} = \sqrt{\frac{\sum_i wi\left(y_i^e - y_{wi}^c\right)^2}{N - 2}} \qquad (2.47)$$

where y_{wi}^c is the predicted weighted instrument response for the ith data point and is obtained from the WLS regressed line according to:

$$y_{wi}^c = m_w x_{wi} + b_w$$

The WLS slope can similarly be viewed and compared to Equation (2.28):

$$m_w = \frac{S_{wxy}}{S_{wxx}}$$

where

$$S_{wxy} = \frac{\sum w_i x_i y_i - \sum w_i x_i \sum \sum w_i y_i}{\sum w_i}$$

and

$$S_{wxx} = \sum w_i x_i^2 - \frac{\left(\sum w_i x_i\right)^2}{\sum w_i}$$

These three relationships according to Zorn et al.[6] can be used to yield *weighted prediction intervals* around a predicted response, y_{wj}^c, at concentration x_j according to

$$y_j = y_{wj}^c \pm t_{(1-\alpha/2, N-p-2)} s_{wy/x} \left[\frac{1}{w_j} + \frac{1}{\sum w_i} + \frac{\left(x_j - x_w\right)^2}{S_{wxx}}\right]^{1/2} \qquad (2.48)$$

The weighted parameters have replaced the unweighted parameters (s, x_{ave}, S_{xx}), the sum of the weights has replaced N, and $t_{(t-\alpha/2, N-p-2)}$ is the $(1 - \alpha/2)$ 100 percentage point of the Student's t distribution on $N - p - 2$ degrees of freedom (where p is the number of parameters used to model the weights). Also, the inverse weight ($1/w_j$) at x_j has replaced 1 in the unweighted equation [Equation (2.41)]. The reader, at this point, should compare Equation (2.48) with Equation (2.41) while noting similarities and differences between OLS and WLS approaches to instrument calibration.

Three general approaches are introduced to find appropriate weights:

- Define a weight as being inversely proportional to the standard deviation for the ith calibration data point:

$$w_i = \frac{1}{s_i}$$

- Define a weight as being inversely proportional to the variance for the ith calibration data point:

$$w_i = \frac{1}{s_i^2}$$

- Plot the standard deviation as a function of analyte concentration and fit the data to either a quadric or exponential of a two-component model.

Let's address the second question posed at the beginning of this section before we plunge into the mathematics of weighted least squares (WLS) regression of an experimental calibration plot. Consider, for example, the analyte (4-hydroxy-3-nitro) phenyl acetic acid (HNPAA). HNPAA is a known metabolite from human exposure to chemical explosives such as TNT (trinitrotoluene) whose molecular structure is shown below:

TABLE 2.3
Calibration Data and Comparison of OLS/CBCS vs. WLS/CBCS for HNPAA in Urine Using LC-MS-MS

X_i ppb	A_i/A_{ISTD}	OLS/CBCS	WLS/CBCS
5	0.017792	b = 0.003698	b_{1w} = 0.003681
20	0.071242	a = -0.006182	b_{0w} = -0.001559
50	0.17871	s_r = 0.01252	s_r^w = 0.00082
100	0.355414	r^2 = 0.9999	r^2 = 0.9999
200	0.745098	x_C = 7.3 ppb	x_C = 0.98 ppb
500	1.821656	x_D = 15.1 ppb	x_D = 2.3 ppb
1000	3.701299		

TABLE 2.4
Calibration and Comparison of OLS/CBCS vs. WLS/CBCS for HNPAA in Urine Using LC-MS-MS

X_i ppb	A_i/A_{ISTD}	OLS/CBCS	WLS/CBCS
5	0.019995	b = 0.004266	b_{1w} = 0.004234
20	0.080687	a = -0.01066	b_{0w} = -0.002181
50	0.210782	s_r = 0.01418	s_r^w = 0.0009964
100	0.418296	r^2 = 0.9999	r^2 = 0.9998
200	0.818882	x_C = 7.2 ppb	x_C = 1.0 ppb
500	2.110067	x_D = 14.8 ppb	x_D = 2.4 ppb
1000	4.265323		

This author compared calibration data results from applying ordinary least squares regression using confidence band calibration statistics (OLS/CBCS) versus weighted least squares regression using confidence band calibration statistics (WLS/CBCS).[21] OLS/CBCS and WLS/CBCS for two different calibration sets of data for HNPAA are compared in Tables 2.3 and 2.4. Comparable least squares slopes and r^2 values are evident while y-intercepts and s_r^w (standard deviation in the y residuals) are much lower than equivalents using OLS/CBCS. This difference translates into significantly lower x_C and x_D values when using a WLS/CBCS versus a OLS/CBCS approach to instrument calibration. Results from Tables 2.3 and 2.4 reveal nearly identical values for x_C and x_D drawn from calibration data run at two different timeframes and reflects the utmost in accurate and precise TEQA particularly at the very low concentration levels reported. Figure 2.14 shows a WLS regression calibration plot for HNPAA. The calibration plot is shrouded within confidence, often referred to as prediction limits. Referring to Figure 2.14, note the relative "tightness" of the prediction interval that covers the low end of the WLS calibration. Also note the gradual widening of the prediction interval at the high end of the calibration. This is consistent with the *heteroscedastic* behavior exhibited by many organic compounds. HNPAA was diluted with water and filtered using a 96-well plate reversed-phase solid-phase extraction disk (see Chapter 3) to remove interferences. The analyte now in the filtrate was adjusted to a precise final volume and quantitated using LC-MS-MS. Calibration standards including internal standards for HNPAA were prepared in the blank sample matrix and also underwent sample preparation. Has the

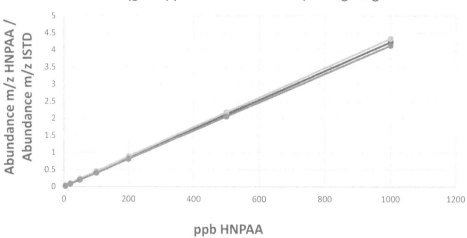

FIGURE 2.14

mathematics been developed for the application of weighted least squares regression to develop algebraic equations that enable x_C and x_D to be realized?

2.5.3 IS THERE A "ROADMAP" WE CAN USE TO FIND X_C AND X_D BASED ON WLS-CBCS?

Yes indeed! Of all the papers published in the analytical chemistry literature on IDLs over the years, and notwithstanding the ensuing controversy at conferences and in the literature, this author has followed and applied the mathematics first articulated by Burdge, MacTaggert, and Farwell[20]. Their paper builds upon earlier work of MacTaggert and Farwell.[17,22] Let's begin by defining what we mean by a *weighted standard deviation in the y residuals*, given earlier in Equation (2.47). A graphical representation with the appropriate mathematical relationships is shown in Figure 2.15. In this "expanded" view of a typical calibration plot, we can easily note the extent to which the experimental (e) points are not aligned with the calculated (c) WLS plot! Included in Figure 2.15 are equations for the WLS calibration, WLS slope and *weighted standard deviation in the y residuals*.

IDLs based on either OLS/CBCS or WLS/CBCS are based on first establishing confidence or prediction intervals that shroud the OLS or WLS regressed calibration curve and then identifying the y intercept y_0, the critical instrument response y_C as determined with a t distribution with $\alpha = 0.05$. The horizontal line is drawn from y_C to the calibration plot (whose x=0 value is at y_0) to establish the analyte's critical decision level x_C. Continuing along the horizontal line to the lower confidence limit y_L where β can be set to 0.05 serves to establish the analyte's x_D or the analyte's IDL. The sketch in Figure 2.16 graphically depicts the important parameters. Figure 2.16 is identical to Figure 2.9 except for superimposing the vertical line defined as the critical concentration (x_C).

2.5.3.1 Finding the Critical Instrument Response y_C and Critical Concentration x_C While Minimizing Type 1 (False Positive) Errors for WLS/CBCS

The intersection of the upper confidence limit with the y-axis at y_C, where x = 0, corresponds to the highest signal that could be attributed to a blank 100 $(1-\alpha)\%$ of the time. In other words,

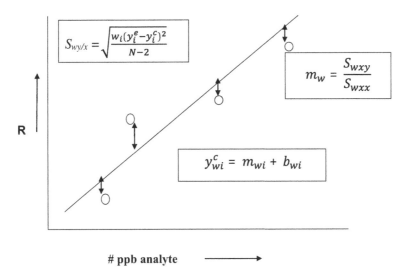

FIGURE 2.15

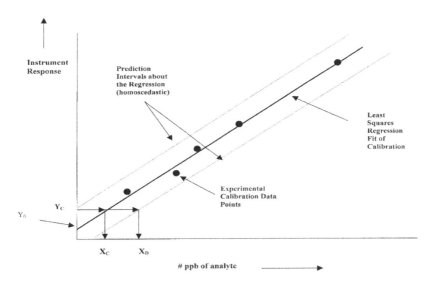

FIGURE 2.16

the critical instrument response y_C is set at $\alpha = 0.05$ such that 95% of all measurements are to be found below this critical value. This is represented by a one-sided tailed α for y at $x = 0$. It is evident that the critical instrument response y_C is established such that the probability of making a Type 1 (false positive) error is minimized at $\alpha = 0.05$.[20]

Equations for the critical instrument response for WLS/CBCS are given below:

$$Y_{crit} = b_{0w} + t_{(1-\alpha, df)} s_w \left[\frac{1}{(m)w_0} + \frac{1}{\Sigma w_i} + \frac{\bar{X}_w^2}{S_{xxw}} \right]^{1/2}$$

(2.49)

Where

y_{crit}—critical instrument response

b_{0w}—y intercept from the WLS regression

$t_{(1-\alpha, df)}$—Student's t for a one-sided hypothesis test for a specific number of degrees of freedom, df, where df = N-2 with N = the number of x,y data points used in the calibration

s_w—standard deviation in the weighted y residuals for a WLS regression and is found as follows:

$$s_w = \sqrt{\frac{\Sigma w_i (y_i - \bar{y}_{wi})^2}{N-2}}$$

m—number of post calibration sample replicates

w_0—weight factor at the y intercept of the calibration y_0

w_i—weight factor for the ith data point in the calibration

y_i—instrument response for the ith data point

N—number of x,y data points in the calibration

$\hat{y}_w$—calculated weighted instrument response that corresponds to the ith data point obtained from application of the weighted least squares equation:

$$\hat{Y}_w = b_{1w}x + b_{0w}$$

$\hat{y}_w$—instrument response for the WLS regression

b_{1w}—slope of the WLS/CBCS regression

x—known amount (e.g. ng) or concentration (e.g. ng/mL) for a given analyte (assumed free of error)

The corresponding critical decision level x_C is calculated according to:

$$X_{crit} = \frac{t_{(1-\alpha, df)} s_w \left[\frac{1}{(m)w_0} + \frac{1}{\Sigma w_i} + \frac{\bar{X}_w^2}{S_{xxw}} \right]^{1/2}}{b_{1w}}$$

(2.50)

Where

x_{crit}—critical amount (e.g. ng) or concentration (e.g. ng/mL); also noted as x_C

b_{1w}—WLS/CBCS regression slope

m—number of post calibration sample replicates

w_0—weight at x=0

Σw_i—sum of the weights over i x,y data points used to construct the WLS regression

$X_{w (ave)}$—mean of N x values used to construct the WLS regression

S_{xxw}—weighted sum of x squared mean deviations

The correlation coefficient r for the WLS/CBCS regression is found according to:

$$r = b_{1w}\sqrt{\frac{Sxxw}{Syyw}}$$

Where

S_{yyw}—weighted sum of y squared mean deviations

Calculation of x_C establishes a critical amount or concentration level. This critical level x_C first defined over 50 years ago is a decision limit at which one may decide whether the result of an analysis indicates detection.[2,20] X_C accounts only for Type 1 (false positives) error and implies that the analyte of interest can be detected but not reliably quantitated. Type 2 error (false negative) has not been minimized at this point. Let's continue on developing these ideas!

2.5.3.2 Estimating the Instrument Response Limit y[D], the Instrument Detection Limit x[D] and Calculating the Instrument Detection Limit x[D] While Minimizing Type 2 (False Negative) Errors

Refer again to Figure 2.16. The intersection of y[C] with the lower confidence limit at y_L corresponds to the lowest signal that could be attributed to an analyte amount or concentration x[D] 100(1-β)% of the time. This is represented by the t distribution with a one-tailed β for y at $x = x[D]$. Equations for the instrument detection limit (IDL) denoted as x_D and obtained from the literature for WLS/CBCS are shown below:[20]

$$x_D = \frac{A \pm \sqrt{B + g_w S_{wxx}\left(C + \frac{1-g_w}{(m)(w_D)}\right)}}{1-g_w} \tag{2.51}$$

Where

$$g_w = \frac{(t_{1-\alpha,df} s_w)^2}{b_w^2 S_{xxw}} \quad B = \bar{x}_w g_w\left(\bar{x}_w g_w - \frac{2ts_w Q}{b_w}\right) \quad Q = \left[\frac{1}{(m)w_0} + \frac{1}{\sum w_i} + \frac{\bar{x}^2}{S_{wxx}}\right]^{1/2}$$

$$A = \frac{(t)(s_w)(Q)}{b_{1w}} - (\bar{x}_w g_w) \quad C = \frac{1}{\sum w_i} + \frac{\bar{x}_w^2}{S_{wxx}} + \frac{g}{(m)(w_D)}$$

These foreboding mathematical equations shown above have been utilized by the author to develop an Excel spreadsheet whose objective it is to calculate x_C and $x_{D!}$[21] An example of applying the spreadsheet to a specific calibration is shown in the Appendix.

To conclude this discussion on IDLs, it is useful to go back and to compare Equations (2.34) and (2.45). Both equations relate x_D to a ratio of a standard deviation to the slope of a least squares regression line multiplied by a factor. In the *blank statistics* case, Equation (2.34), the standard deviation refers to detector noise found in the baseline of a blank reference standard, whereas in the *confidence band calibration statistics* case, Equation (2.45), the standard deviation refers to the uncertainty in the least squares regression itself. This is an important conceptual difference in estimating IDLs! It behooves future analytical chemists to pay attention to this difference!

2.5.4 How Do MDLs Differ from IDLs?

Obtaining an instrument's IDL represents only one aspect of the overall estimate of a method's detection limit. The only situation whereby an IDL is the same as an MDL is when there is no transfer of analyte from the environmental phase to a second phase. Direct aqueous injection of a groundwater sample into a GC in which there is no extraction step involved is one example of where MDLs equal IDLs. The removal of trace organic volatiles from an aqueous sample either by the purge-and-trap technique or by static headspace sampling at elevated temperatures results in nearly 100% efficiency and can also be considered as having MDL equal IDL. Such is not the case for semivolatile to nonvolatile organics in environmental samples when some type of sample extraction is employed. In these situations, the MDL is often much less than the IDL because sample preparation methods are designed to concentrate the analyte from the environmental matrix to the solvent matrix. The efficiency of this concentration must also be taken into account. One cannot always assume that a given sample preparation is 100% efficient. In fact, a key ingredient in GLP is to conduct an efficiency study by adding an accurately known amount of analyte to the sample matrix, and then measure to what extent the method recovers the analyte. The extent of recovery is often expressed as a *percent recovery*.

2.5.5 How Do I Obtain MDLs for My Analytical Method?

The MDL can be found only after the IDL, the concentration factor, i.e., the ratio of *final extract* or *eluent* volume (*extract*, if LLE sample prep techniques were used or *eluent*, if SPE sample prep techniques were used) V_e to the sample volume, V_S, and the percent recovery [$\%R_i = (100)\,(R_i)$], where R_i is the fraction of the ith analyte recovered, have been quantitatively determined. For a groundwater sample, $V_e = 1$ mL and $V_S = 1000$ mL. For a soil sample, $V_e = 10$ mL and the sample weight might be 30 g. For serum, the respective volumes might be: $V_e = 1$ mL and $V_S = 1$ mL.

The important mathematical relationship that answers the question above will now be derived. Mass balance requires that the amount of recovered ith analyte present in the final extract as measured equal that fraction of ith analyte recovered originally present in the groundwater, soil, or serum sample using your method. R must be experimentally measured in the sample matrix in a separate set of experiments called *Percent Recovery Studies*. Expressed mathematically,

$$amt^i_e = amt^i_s \left(R^i \right)$$

so that

$$V_e C^i_e = V_S C^i_S \left(\frac{\%R^i}{100} \right)$$

Solving for C^i_S gives:

$$C^i_S = C^i_e \left(\frac{V_e}{V_S} \right) \left(\frac{100}{\%R^i} \right) \tag{2.52}$$

where

C_S^i = concentration of ith analyte in the original sample

C_e^i = concentration of ith analyte in the final volume of extract (if LLE is used) or final volume of eluent (if SPE is used)

V_e = volume of extract or eluent in mL

V_S = volume of sample in mL

$\%R_i$ = percent recovery for ith analyte measured independently.

If the concentration of ith analyte in the final extract or eluent, C_e^i is low enough so as to represent the IDL, then the concentration of ith analyte in the original sample, C_S^i would yield the MDL for that method. For example, consider the author's work in isolating and recovering various PBDE congeners from human serum using stir bar sorptive extraction (SBSE)[23] Let's say we've established that the IDL for BDE-47 (a specific PBDE congener) is 0.68 ppb. Since volume of hydrophobic coating on the stir bar is ~0.1 mL and assuming we drop the stir bar into ~4.0 mL serum, the phase ratio $V_e / V_S = 0.025$. If we assume a 75% recovery, we can estimate a MDL for BDE-47 using Equation (2.52) (above) of 0.0227 ppb! This MDL of 22.7 ppt illustrates how contemporary analytical instruments coupled to specialized sample preparation techniques can offer ultratrace quantitative analysis of chemicals that contaminate the environment!

The percent recovery ($\%R^i$) can be calculated from measuring the instrument response for the analyte in the spiked matrix denoted by A^i, with uncertainty characterized by a standard deviation s^i and a reference control that is known to be 100% recovered, with uncertainty s_C^i. This is mathematically defined according to

$$\%R^i\left(s_R^i\right) = \frac{A^i\left(s^i\right)}{A_C^i\left(s_C^i\right)} \times 100 \tag{2.53}$$

The calculation of R^i involves dividing A^i by A_C^i; this division requires a propagation of error. A relative standard deviation can be calculated by taking into account the propagation of error for random errors using the equation

$$\frac{s_R^i}{R^i} = \sqrt{\left(\frac{s^i}{A^i}\right)^2 + \left(\frac{s_C^i}{A_C^i}\right)^2} \tag{2.54}$$

Equation (2.50) shows that it is relative errors expressed in terms of variances that must be added to enable a determination of the relative error in the percent recovery calculation using Equation (2.49).

The uncertainty in a percent recovery determination is often expressed in terms of a percent relative standard deviation that is also called the *coefficient of variation*. The coefficient of variation is calculated according to

$$\text{Coeff Var} = \left(\frac{s_R^i}{R^i}\right) \times 100 \tag{2.55}$$

If sufficient replicate analytical data is available, the variance and standard deviation for a percent recovery study can be found by pooling the data. For example, consider a thorough percent recovery study using reversed-phase solid-phase extraction (RP-SPE) for the isolation and recovery of the pesticide methoxychlor from groundwater. If j replicate injections are made for each of g SPE extractions conducted, an overall standard deviation for this pooled data, S_{pooled}, can be calculated according to

$$S_{\text{pooled}} = \sqrt{\frac{1}{N-g} \sum_g \sum_j \left(x_j - \bar{x}\right)^2} \tag{2.56}$$

The author has developed an interesting application of Equation (2.56). Please refer to Appendix G.

The analyst is now ready to run samples using the selected method because instrument calibration, instrument verification, % recoveries, as well as IDLs and MDLs have all been completed.

2.6   WHY SO MANY REPLICATE MEASUREMENTS?

There is no other measurement science that insists on so many replicate measurements as EPA method-driven TEQA. Earlier, the notion of analyzing seven lab blanks was cited as a guideline. In fact, the number 7 appears quite frequently throughout EPA methods. This author has been known to quip to students that since the Beatles used number 9 in their music, the EPA likes to use number 7 in its methods. The answer goes to the heart of the meaning of the central limit theorem of statistics and attempts to apply its meaning to the above-posed question. A number of texts discuss the basics of probability and statistics; this author frequently refers to Mode's text on probability and statistics.[24] How many replicate measurements, L, of a given environmental contaminant such as vinyl chloride are needed to be confident 99% of the time that the estimate of the sample variance is less than three times the true variance (i.e., population variance)? Anderson (pp. 44 and 298) has considered this, and sure enough, in his Table 4A in his Appendix D, titled *Sample Size Required for a probability that s2 will be less than k σ2* there is a table that gives the answer—and lo and behold, for k=3 the answer is seven![7]

2.6.1   How Do Analytical Chemists Deal with Replicate Analytical Data?

They resort to the principles and practice in mathematics known as probability and statistics. A few words about the Gaussian or normal distribution are in order at this point, along with some commonly used statistical definitions. First, consider a hypothetical sellout crowd for a concert held in a large auditorium that seats as many as 1,000 people. If everyone upon purchasing a ticket were asked to list their height, and if everyone did so, we would have a data base of 1,000 heights measured in feet and inches (U.S.) or meters (rest of the world). If we combine identical values for a person's height and proceed to plot the frequency of a given height versus the range of human height, we can expect the variable, a person's height, to be normally distributed across the range of heights available. The largest value at the apex of the peak represents the highest frequency of human heights within this data set. We can then begin to ask questions like those that follow. How likely is it for a National Basketball Association (NBA) player who plays the position of center to be present at the concert? How close is the average height of these concert goers to the average height of people (over 300 million) who live in the United States? The mathematics of probability and statistics can address these questions.

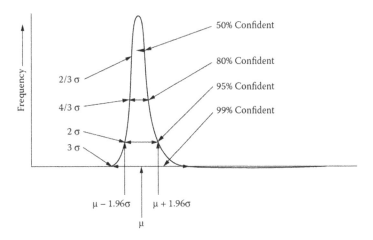

FIGURE 2.17

Replicate measurement (data) that is assumed free of systematic error follows a Gaussian (normal) distribution. Figure 2.17 shows a hypothetical normal distribution with the horizontal (x-axis) defined as the measurement axis—the measurements plotted left to right, low to high.[25] The unlabeled vertical axis in Figure 2.17 indicates the probability distribution. This is related to the probability of finding a given value along the horizontal axis. Clearly, values in the middle of the Gaussian distribution are more probable than values far to the left or far to the right.[25] It is important to distinguish between a *confidence level* or level of significance and a *confidence interval*. This is difference is shown in Figure 2.17. The statement "to be 95% confident that a single analytical result such as x_0, in the absence of any systematic error (i.e., 95 of every subsequent 100 measurements), will fall to within $\pm 1.96\sigma$, where σ is the standard deviation in the population mean" addresses both the confidence level and confidence interval. Confidence intervals are labeled to the left of the Gaussian distribution in Figure 2.17 and indicate width or spread, whereas confidence levels, labeled to the right of the Gaussian distribution, refer to heights. The fraction of a population of observations dL/L, whose values lie in the region x to $x + dx$, is given by

$$\frac{dL}{L} = \frac{1}{2\pi\sigma} e^{-(x-\mu)^2/2\sigma^2} dx \tag{2.57}$$

Deviations of individual values of x from the mean are conveniently expressed in terms of a dimensionless unit of standard deviation z according to

$$z = \frac{x-\mu}{\sigma}$$

If we differentiate z with respect to x and substitute this into Equation (2.57), the Gaussian equation looks like

$$\frac{dL}{L} = \frac{1}{2\pi} e^{-(z)^2/2\sigma^2} dz$$

The distinction between systematic vs. random error in laboratory measurement was introduced at the beginning of this chapter. The normal distribution shown in Figure 2.17 enables a set of confidence levels to be established in the absence of systematic error. Mathematical relationships that enable a confidence interval to be calculated that shrouds the linearly regressed calibration line were also given earlier in the chapter. These relationships enabled a more correct definition of what constitutes an IDL to emerge, as well as to predict an uncertainty in the interpolated value of a concentration x_0 taken from the regression.

Here, we consider the basic statistics of replicate measurements. For example, consider making repeated injections of an initial calibration verification (ICV) reference standard using an autosampler atop a gas chromatograph. The degree of automation essentially eliminates any systematic error in the measurement of each analyte that may comprise the ICV reference standard. Confidence levels (refer again to Figure 2.17) are usually selected at 95 or 99%. The true or population mean μ is fixed for replicate measurement and remains unknown. The normal distribution enables one to set limits such that the measurable sample mean, represented by the x bar below, can be expected to lie within a given degree of probability such that 95% of all measurements lie to within 1.96σ of the true population mean.

Two approaches emerge. A confidence limit can be found when the sample standard deviation, s, is a good approximation of the population standard deviation, σ, such that for a single measurement, the confidence interval for μ is

$$\text{Confidence Interval for } \mu = \bar{x} \pm z\sigma$$

The confidence interval for μ when a sample mean has been found for L replicate measurements is

$$\text{Confidence Interval for } \mu = \bar{x} \pm \frac{z\sigma}{\sqrt{L}}$$

When σ is unknown, the standard deviation, s, of the small number of samples that may have considerable uncertainty requires broader confidence intervals. This is the most likely scenario for conducting TEQA, in that the analyst usually does not know the true value for σ and has in his possession a very limited number of samples. A two-tailed Student's t is then substituted for z. The value of t selected depends on two factors: the desired degree of confidence and the number of degrees of freedom (df), where $df = L - 1$. The confidence interval for the mean of L replicate measurements is found according to

$$\text{Confidence Interval for } \mu = \bar{x} \pm \frac{t_{(1-\alpha/2, df)} s}{\sqrt{L}} \tag{2.58}$$

Short computer programs years ago written in computer programming languages such as BASIC to implement Equation (2.54) have been supplanted today by computerized spreadsheets such as Excel as well as many other software packages as routine tools used by analytical chemists. A brief refresher course on the fundamentals of probability and statistics as this relates to measurement can be found in the Appendix.

2.7   HOW DO I FIND THE LIMIT OF QUANTITATION?

Again, there are two ways, the simpler and the more contemporary. Consider how an established textbook on chemometrics has addressed this difference as follows (p. 66):[26]

In principle, all performance measures of an analytical procedure … can be derived from a certain critical signal value, yC. *These performance measures are of special interest in trace analysis.* The approaches to estimation of these measures may be subdivided into "methods of blank statistics," which use only blank measurement statistics, and "methods of calibration statistics," which in addition take into account calibration confidence band statistics.

The simpler approach, in a manner similar to finding the IDL and discussed earlier in this chapter (and the one suggested by EPA), is to first find the standard deviation in the blank signal, add 10 times this value to the blank signal according to

$$S_{LOQ} = S_{blank} + 10s_{blank} \tag{2.59}$$

and, in a manner similar to that shown in Equation (2.34), interpolate the concentration from S_{LOQ} that can be defined at the limit of quantitation or practical limit, x_{LOQ}, according to

$$x_{LOQ} = \frac{S_{LOQ} - S_{blank}}{m} \tag{2.60}$$

The subscript LOQ is the limit of quantitation. This author has attempted to use the standard deviation in the y intercept of the least squares regression of a given calibration for a given analyte of interest [refer to Equation (2.31) and the equation for s_b]) to establish the standard deviation for the blank, s_{blank}, in Equation (2.55). The y intercept itself, b from Equation (2.29), is used to represent S_{blank}. These two parameters are then substituted into Equation (2.55) to yield S_{LOQ}. This elevated signal, S_{LOQ}, can be substituted into Equation (2.56) to find the limit of quantitation.

Oppenheimer et al.[3] have proposed, for the case of the unweighted least squares regression, a relationship similar to that shown in Equation (2.42) to calculate y_Q according to

$$y_Q = \frac{S_{y/x}}{C} \sqrt{1 + \frac{1}{N} + \frac{\bar{x}^2}{\sum_i^N \left| x_i - \bar{x}^2 \right|}}$$

where C is the ratio of the standard deviation in the signal corresponding to x_{LOQ} to the signal itself and is usually set equal to 0.10. If y_Q is substituted for S_{LOQ} in Equation (2.56), x_{LOQ} is found. This is a "calibration statistical" approach to finding the limit of quantitation, x_{LOQ}. Zorn et al. have defined y_Q as equal to 10 times the standard deviation at the critical level, and this term is added to a weighted y intercept to the weighted least squares regression. In addition, a confidence interval is added to yield an alternative minimum level.[6]

2.7.1 Is There a Way to Couple the Blank and Calibration Approaches to Find X_{LOQ}?

To illustrate the author's approach, consider the determination of the element Cr at low ppb concentration levels using a graphite furnace atomic absorption spectroscopic determinative technique (determinative techniques are introduced in Chapter 4). The calibration data is as follows:

Concentration (ppb)	Absorbance
1	0.0134
5	0.1282
10	0.2260
15	0.3276
20	0.4567

Entering this calibration data into, for example, an Excel spreadsheet using regression statistics to calculate prediction limits about the regression[27] or reverting back to the author's BASIC program LSQUARES,[28] the following statistical parameters are calculated:

Outcome from LSQUARES	Result
Standard deviation in y intercept, $s(b)$	0.0101
Correlation coefficient, r	0.9980
Student's t for one-tail null hypothesis, at a significance level of 95% for three degrees of freedom	2.353
Critical response level, y_C	0.0377
Critical concentration level, x_C	1.67 ppb
Detection concentration level, x_D	3.27 ppb

The measured signal from a blank is 0.0620 absorbance units, and using the author's program LSQUARES to obtain a value for the standard deviation in the least squares regression line, $s_b = 0.0101$. Ten times s_b added to 0.0620 gives $S_{LOQ} = 0.163$. Entering 0.163 into LSQUARES gives 7.22 ± 1.96 ppb Cr as the limit of quantitation. The critical concentration was previously found to be 1.67 ppb Cr, whereas the limit of detection, the IDL for the instrument, was previously found to be 3.27 ppb Cr.

Figure 2.18 is a seemingly complex-looking graph that seeks to show how blank statistics and confidence band statistics can be viewed in an attempt to make sense out of all of this. (p. 65)[26] The confidence interval that surrounds the least squares regression fit through the experimental calibration points is indicated along with extensions to the various Gaussian distributions that appear to the left of the calibration curve. These Gaussian distributions can be interpreted as arising out of the plane of the paper, thus providing for a third dimension. The possibility of committing a significant type II error, represented by β when accepting a detection limit of three times the standard deviation in the blank, S_{blank}, is clearly evident in Figure 2.18. The small type I error at the decision or critical level is evident as well. Note that $y(Q)$ is 10 times the standard deviation in the blank (s_{blank}), and this gives a high degree of certainty that neither type I nor type II errors will be committed at the limit of quantitation. Figure 2.18 obtained from a different source with the exception of the type I error concept serves to confirms many of the ideas presented in this chapter. This concludes our discussion on the statistical treatment of analytical data for TEQA. We now proceed to a discussion of quality in measurement. Before we do this, let us introduce some definitions and clarify some overused words in the lexicon of TEQA.

2.8   CAN IMPORTANT TERMS BE CLARIFIED?

Yes, they can. Let us start by stating an old adage: Samples are analyzed, constituents in samples are identified (qualitative analysis) or determined (quantitative analysis). According to Erickson (p. 107), the following definitions need clarification:[29]

- Technique: A scientific principle or specific operation (e.g., gas chromatography–electron-capture detector (GC-ECD), Florisil column cleanup, or so-called Webb and McCall quantitation).
- Method: A distinct adaptation of a technique for a selected measurement purpose (e.g., a specific GC-ECD operating mode for analysis of PCBs, including column specifications and sample preparation).

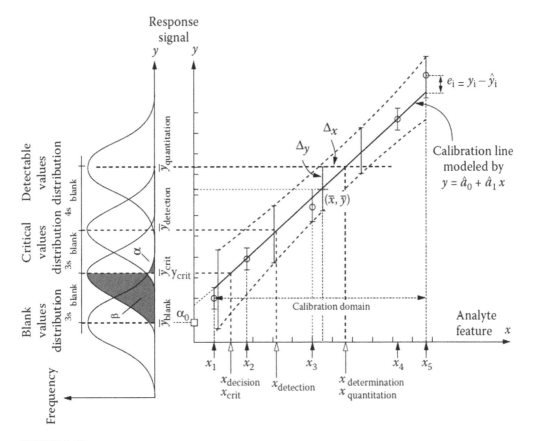

FIGURE 2.18

- Procedure: The written directions to use a method or series of methods and techniques; many authors and agencies blur the distinction between *method* and *procedure*.
- Protocol: A set of definitive directions that must be followed without deviation to yield acceptable analytical results. Analytical protocols may be found as the cookbooks provided within a laboratory derived from common procedures.

The next topic in this chapter introduces the two tenets of good laboratory practice (GLP): quality assurance and quality control. This is a topic of utmost importance, particularly to laboratory managers and corporate executives. A good lab with a lousy GLP is a sad use of human and facility resources. Each batch of samples analyzed must include, in addition to the samples themselves, a matrix spike, a matrix spike duplicate, and one or more laboratory blanks that verify the cleanliness of the laboratory. For the nonstoichiometric chromatographic and spectroscopic determinative techniques (to be discussed in Chapter 4), chemical reference standards must be available and be in high purity as well. These standards are part of GLP and indicate that a laboratory is implementing a quality assurance program.

2.9 WHAT DOES IT MEAN FOR A LABORATORY TO HAVE A QUALITY ASSURANCE PROGRAM?

An important aspect of GLP is for each laboratory to have a document titled "Quality Assurance" in its possession.

One of the most confusing issues in TEQA involves the need to clearly define the difference between quality assurance (QA) and quality control (QC). QA is a management strategy. It is an integrated system of management activities involving the following five areas:

1. Planning
2. Implementation
3. Assessment
4. Reporting
5. Quality improvement.

This plan ensures that a process, item, or service is of the type and quality needed and expected by the client or customer.

Quality control is defined as an overall system of technical activities that measures the attributes and performance of a process, item, or service against defined standards to verify that they meet the stated requirements established by the client or customer. *QA is a management function, whereas QC can be viewed as a laboratory function.* A laboratory must have a written QA document on file, whereas a record of certified reference standards that are properly maintained implies that QC is in place. Much of what has been discussed in this chapter falls under the definition of QA.

Finally, it is necessary to implement an effective QC protocol when all other aspects of the quantitative analysis have been completed. Two topics emerge from a consideration of EPA-related methods. The first comprises a clear set of definitions of QC reference standards to be used in conducting the quantitative analysis, and the second is a set of QC performance criteria and specifications.

2.10   ARE THERE DEFINITIONS OF QC STANDARDS USED IN THE LAB?

Yes indeed! A number of QC standards must be available in the laboratory. These QC standards all have specific roles to play in the QC protocol. The following is a list of the various QC standards, including the role of each QC standard as well.

Surrogate spiking standard: A solution that contains one or more reference compounds used to assess analyte recovery in the method chosen. Surrogates cannot be targeted analytes. For example, tetrachloro-*m*-xylene (TCMX) and decachlorobiphenyl (DCBP) dissolved in MeOH are surrogates used in some EPA methods to determine trace organochlorine (OC) pesticide residue analysis. These two analytes are highly chlorinated and are structurally very similar to OC- and PCB-targeted analytes. TCMX elutes prior to and DCBP elutes after the OCs and PCBs, thus eliminating any co-elution interferences. A fixed volume (aliquot) of this reference solution is added to every sample, matrix spike, blank, blank spike, and control in the protocol.

Matrix spike standard: Consists of a representative set of the targeted analytes whose percent recoveries are evaluated in the sample matrix. For example, if THMs were the targeted analytes, this reference standard would consist of one or more THMs, such as chloroform, bromodichloromethane, and chlorodibromomethane. These compounds would be dissolved in a matrix-compatible solvent such as MeOH. A precise aliquot of this solution is added to a sample so that the effect of the sample matrix on the percent recovery can be evaluated. A second standard, called a *matrix spike duplicate*, is often required in EPA methods and is used to assess matrix recovery precision.

Control standard: Consists of the same representative set of targeted analytes used in the matrix spike. The control standard does not undergo sample extraction. The amount of targeted analyte or surrogate is set equal to the amount injected for the surrogate spike and matrix spikes. This amount is dissolved in the exact final volume of solvent used in sample preparation. In this way, a control standard that ensures a 100% recovery is obtained. The ratio of the amount of surrogate analyte to the amount of control analyte is used to determine the analyte percent recovery.

Stock reference standard: The highest concentration of target analyte either obtained by dissolving a chemically pure form of the analyte in an appropriate solvent or obtained commercially as a certified reference solution.

Primary dilution standard: Results from dilution of the stock reference standard.

Secondary dilution standard: Results from dilution of the primary dilution standard.

Tertiary dilution standard: Results from dilution of the secondary dilution standard.

Calibration or working standards: A series of diluted standards prepared from dilutions made from the tertiary dilution standard. These standards are injected directly into the instrument and used to calibrate the instrument as well as to evaluate the range of detector linearity. The range of concentration should cover the anticipated concentration of targeted analytes in the unknown samples.

Initial calibration verification (ICV) standard: Prepared in a similar manner as the working calibration standards; however, at a different concentration to any of the working standards. Used to evaluate the precision and accuracy for an interpolation of the least squares regression of the calibration curve. This standard is run immediately after the calibration of the working standards.

Continuous calibration (CC) standard: Prepared in a manner similar to the ICV. The CC is run in place of the calibration standards when subsequent batches of samples are anticipated. The initial batch of samples usually includes a series of calibration standards and ICVs. CCs are used to verify that the initial calibration remains analytically viable to quantitate subsequent batches of samples.

Laboratory method blank: A sample that contains every component used to prepare samples except for the analyte(s) of interest. The method blank is taken through the sample preparation process. It is used to evaluate the extent of laboratory contamination for the targeted analytes.

Trip or field blank: A sample that is obtained in the field or prepared during the trip from the field to the laboratory. This blank should contain no chemical contamination, yet should be obtained in a manner similar to the environmentally significant samples themselves.

2.11   WHAT QC SPECS MUST A LAB MEET?

The following criteria represent five separate QC *specifications* that a lab should be provided to each and every client who requests analytical services.

2.11.1   MINIMUM DEMONSTRATION OF CAPABILITY

Before samples are prepared and analytes quantitatively determined by instrumental analysis, the laboratory must demonstrate that the instrumentation is in sound working order and that targeted analytes can be separated and quantitated. For example, GC operating conditions must be established and reproducibility in the GC retention times, t_R, for the targeted analytes must be achieved. Competent analytical scientists coupled with laboratory resources of high quality will greatly help a lab meet this minimum specification.

2.11.2   LABORATORY BACKGROUND CONTAMINATION

Before samples are prepared and analytes quantitatively determined by instrumental analysis, the laboratory must demonstrate that the sample preparation bench-top area instrument itself and all reagents and solvents used are essentially free of traces of the targeted analyte of interest. This is accomplished by preparing method blanks using either distilled or deionized water or solvents of ultrahigh purity. This is particularly important for ultratrace analysis (i.e., concentration levels that are down to the low ppb or high ppt range). The use of pesticide residue analysis-grade solvents for conducting trace organics analysis and the use of ultratrace nitric and hydrochloric acids for trace metals analysis are strongly recommended. This author believes that there is a considerable number of organics in organic solvents at concentrations in the low ppt level! Nonzero blanks prevent true IDLs from ever being obtained. Nonzero blanks even affect analyte percent recoveries!

2.11.3 Assessing Targeted and Surrogate Analyte Recovery

Before samples are prepared and quantitatively determined by instrumental analysis, the laboratory must demonstrate that the surrogate and targeted analytes can be isolated and recovered to a degree by the method selected. A rugged analytical method will have an established range of percent recoveries. In TEQA, the value of the percent recovery is said to be analyte and matrix dependent. For example, if EPA Method 625 is used to isolate and recover phenanthracene from wastewater, the percent recovery is acceptable anywhere between 54 and 120%, whereas that for phenol is between 5 and 112%. These percent recoveries are seen to depend on the chemical nature of the analyte (phenanthracene vs. phenol) and on the matrix, wastewater. Provost and Elder have put forth a mathematics-based argument stating that the range of percent recoveries that can be expected also depends on the ratio of the amount of added spike to the background concentration present in the environmental sample.[30] Table 2.5 applies several mathematical equations discussed by Provost and Elder and clearly shows the impact of spike-to-background ratios on the variability in the percent recovery. Using a spike-to-background ratio of 100 does not change the range of percent recoveries that can be expected. This changes as the ratio is reduced and begins to significantly increase this range when the ratio is reduced to 1 and lower.

This author has conducted more percent recovery studies than he cares to remember. A valid percent recovery study incorporates Equation (2.53) and enlarges upon it for $i = 1, 2, \ldots,$ up to the maximum number of peaks in a multicomponent separation such as that which can be accomplished by applying capillary GC techniques. Using summation notation, Equation (2.61) is enlarged to encompass j replicate injections of an extract that contains the ith analyte from a spiked blank or spiked sample. A control reference standard that contains the ith analyte for k replicate injections of this standard is also prepared such that the percent recovery is found according to

---

TABLE 2.5
Influence of the Spike-to-Background Ratio on Percent Recoveries

		Expected Range in Percent Recoveries		
Spike/Background	Variance in Mean Percent Recovery	p = 1.0, RSD = 0.1	p = 1.0, RSD = 0.2	p = 0.5, RSD = 0.2
Zero background	(100p) 00*(RSD)	80, 120	60, 140	30, 70
100	1.02 (100p)*(RSD)	80, 120	60, 140	30, 70
50	1.04 (100p)*(RSD)	80, 120	59, 141	30, 70
10	1.22 (100p)*(RSD)	78, 122	56, 144	28, 72
5	1.48 (100p)*(RSD)	76, 124	51, 149	26, 74
1	5.00 (100p)*(RSD)	55, 145	10, 190	5, 95
0.5	13.0 (100p)*(RSD)	28, 170	−44, 240	−22, 122
0.1	221 (100p)*(RSD)	−200, 400	−500, 700	−247, 347

[a] 95% tolerance interval for percent recoveries with assumed values for p and RSD; tolerance limits

$$\%R_i = \frac{\sum\limits_{j}^{L} A_j^i / L}{\sum\limits_{k}^{M} A_k^{i,c} / M} \times 100 \qquad (2.61)$$

where L is the total number of replicate injections for the ith analyte in the spiked blank or spiked sample whose peak area is A_j^i and M is the total number of replicate injections for the ith analyte in the control reference standard (this defines a 100% recovery) whose peak area is $A_k^{i,c}$.

The nature of the analytical method largely determines whether a percent recovery study can be conducted in the first place. If sample prep is directly interfaced to the determinative technique, i.e., the analytical instrument, without any opportunity for analyst intervention, a control reference standard from which a 100% recovered analyte can be measured cannot be prepared. This is the case for VOCs since sample preparation is most commonly done by purge-and-trap or static headspace sampling. These sample preparation devices are directly interfaced to the injection port of gas chromatographs. Semivolatile organic compounds (SVOCs) must be extracted via application of phase distribution equilibria (to be introduced in considerable detail in Chapter 3). Some analyte is invariably lost despite the application of excellent lab technique with minimization of systematic error. Percent recoveries for SVOCs are found to be less than 100%. The analyst can prepare a control reference standard in the case of SVOCs by taking a precise aliquot of a more concentrated reference standard and dissolving this aliquot in extracting solvent and adjusting to a precise final volume. This final volume should be identical to the final extract volume used to isolate and recover all spiked blanks, spiked samples, and unspiked samples, and hence, can be compared in Equation (2.61) (i.e., comparing apples to apples) since the same amount of the ith analyte is added to both sample and control. Both sample and control are brought to the same final extract volume. To illustrate, consider spiking a serum sample with 335 ng of the organochlorine pesticide dieldrin and taking it through an appropriate sample preparation method. This sample preparation approach yields a 1 mL hexane extract that would contain dieldrin. Consider also adding 335 ng of dieldrin to 1.0 mL of a hexane extract. This important technique, used repeatedly by this author over the years to conduct precise and accurate percent recovery studies, is illustrated below:

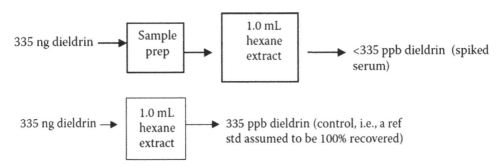

A calibration of the analytical instrument is not necessary to conduct a percent recovery study.

A second example, this time drawing on the author's own research, involves the calculation of the percent recovery of Aroclor (AR) from rat plasma.[31] AR 1248 consists of 20 to 30 PCB congeners and was manufactured for many years in the U.S. by the Monsanto Chemical Company. Aroclors had numerous uses, ranging from an insulating medium for large electrical capacitors to newspaper ink. This extensive use was due to the stable and nonflammable properties of PCBs. Knowledge was minimal years ago about the environmental persistence of PCBs coupled to their lipophilic physico-chemical properties. Three spiking scenarios are presented below with the corresponding instrument outputs in terms of a number of counts. The number of counts is proportional to the area under a chromatographically resolved peak. In this study, the integrated area beneath each peak was summed over the entire 20 to 30 fully or partially resolved peaks for AR 1248.

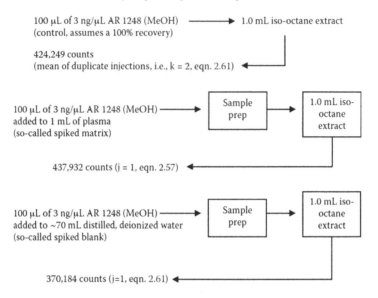

Equation (2.61) addresses only one extraction of one sample. In this case of replicate extractions, the reader should refer back to Equation (2.56). This equation addresses j replicate GC injections for each of g extractions performed.

Once the $\%R_i$ is found, assuming that the IDL, x_D, is known, a method detection limit x_{MDL} can be calculated for the ith analyte. This equation requires a percent recovery expressed as a fraction, R_i, and a knowledge of the phase ratio, β. β is the ratio of extract volume to sample volume. All of this is summarized below:

$$x^i_{MDL} = \frac{x^i_D}{R_i}\left[\frac{V_e}{V_S}\right]$$

(2.62)

$$\%R_i/100 \qquad\qquad \beta$$

Equation (2.62) suggests that in order to reach the lowest MDL, analysts should first achieve as low an IDL as possible. We already discussed how to minimize x_{IDL} [Equation (2.34)] or x_D [Equation (2.45)] from a mathematical point of view. A most efficient extraction also serves to maximize R_i. However, minimizing the phase ratio, β, is only possible within the physical constraints of the experiment. Consider the difference in β between *enviro-chemical* vs. *enviro-health* TEQA. One liter of groundwater that winds up as 1 mL of an organic extract yields $\beta = 0.001$, while a 2-mL serum sample that winds up as 1 mL of an organic extract yields $\beta = 0.5$!

2.11.4   Assessing Goodness of Fit of the Experimental Calibration Data and the Range of Linearity

A correlation coefficient, r, of 0.9900 is the goal that achieves "goodness of fit" between instrument response and analyte amount or concentration for both chromatographic and spectroscopic determinative techniques. An interesting question arises. Over how many orders of magnitude in amount or concentration does it take to yield an r? The answer to this question serves to define what is meant by the range of linearity TEQA. This range will differ among determinative techniques. The majority of column chromatographic used in TEQA occurs at analyte concentration levels from low ppb to low ppm. Both linear (or first-order least squares regression) and polynomial (second-order and even third-order least squares regression) calibration covering one or more orders of magnitude above the instrument's IDL are all useful ways to address the degree of goodness of fit. The least squares slope, called a response factor in some EPA methods, and the uncertainty in the slope are also useful criteria in assessing linearity. The coefficient of determination, r^2, is a measure of the amount of variation in the dependent variable (instrument response) that is accounted for by the independent variable (analyte amount or concentration). (pp. 118–120)[7]

Let us delve into the concept of calibration linearity or lack thereof a bit more. The *linear dynamic range* of an instrument is defined as that range of concentration from the analyte's *limit of quantitation* x_{LOQ}, on up to where a departure from linearity is observed. This concentration is the called the analyte's *limit of linearity*, x_{LOL}. This concept is best understood graphically as shown in Figure 2.19.

The linear dynamic range for most analytical instruments designed to perform TEQA covers at least two orders of magnitude. For example, a range of analyte concentration from 10 to 1,000 ppb covers two orders of magnitude and is quite suitable for TEQA. Some methods have a linear dynamic range over five or six orders of magnitude. Nonlinear least squares regression analysis does exist, can be found in most contemporary statistical software packages, and can be applicable to TEQA.

Consider the construction of a seven-point calibration for the determination of lead, Pb, in a human blood specimen as shown in Figure 2.20. This is an example of *environ-health* TEQA. Declines in blood Pb concentration levels in children over the past 20 years represent one of the real triumphs in public health in the U.S. Refer to the discussion in Chapter 1. Figure 2.20 shows

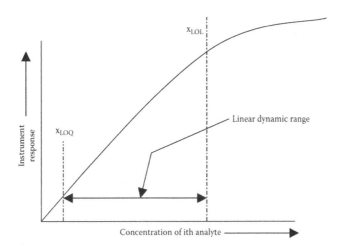

FIGURE 2.19

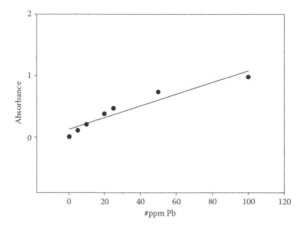

FIGURE 2.20

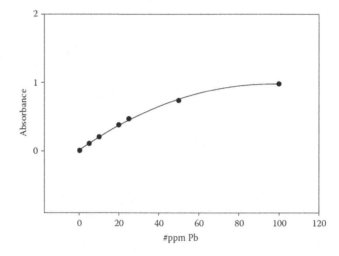

FIGURE 2.21

a calibration plot for quantitating Pb in human blood using flame atomic absorption spectroscopy (FlAA). FLAA was once the dominant determinative technique to quantitate metals. The desire for much lower IDLs brought the graphite furnace atomic absorption spectroscopy (GFAA) to the public health laboratory. Today, inductively coupled plasma atomic emission spectroscopy (ICP-AES) is more commonly found. For laboratories that quantitate trace metals or ultratrace metals concentration levels of detection, ICP-mass spectrometry (ICP-MS) is preferred. GFAA, ICP-AES and ICP-MS determinative techniques are introduced in Chapter 4. Referring to Figure 2.20, note that the concentration levels are in the ppm range. This range is accommodated nicely by FlAA as the determinative technique of choice (the FlAA determinative technique is introduced in Chapter 4). It becomes almost immediately evident that applying a linear least squares fit to the experimental calibration data is not optimum. Figure 2.21 shows that a quadratic (second-order) least squares regression line is a much better fit. Figure 2.22 shows that a cubic (third-order) least squares regression line makes a slight improvement over the quadratic fit. All three least squares regressed lines or curves are polynomials that can be described in terms of a set of coefficients, as shown below:

Polynomial	Type of Fit
$y = a0 + a1x$	Linear
$y = a0 + a1x + a2x2$	Quadratic
$y = a0 + a1x + a2x2 + a3x3$	Cubic

Recall from our previous discussion in this chapter that for the linear fit, a_0 corresponds to the y intercept b while a_1 corresponds to the slope of the least squares regression line m.

It is instructive to examine the sign of the highest-order coefficients in the linear and quadratic equations introduced above. Consider assigning real numbers to each coefficient and plotting these equations as a function of x, $f(x)$. This can be accomplished easily using a graphics calculator. Figure 2.23 shows three superimposed $f(x)$ (a quick review of classic analytic geometry). This result is shown in Figure 2.23. Superimposing curves for the first quadrant for all three polynomials clearly shows just what effect a positive or negative coefficient for x^2 has on the direction of curves 2 and 3. Most nonlinear regressed quadratic curves plotted in the first quadrant

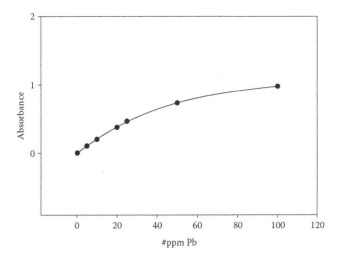

FIGURE 2.22

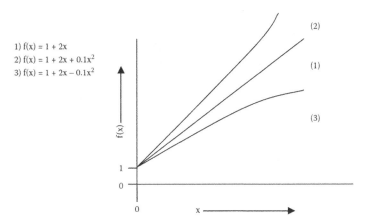

1) f(x) = 1 + 2x
2) f(x) = 1 + 2x + 0.1x²
3) f(x) = 1 + 2x − 0.1x²

FIGURE 2.23

for all positive values for x tend to level off, as shown by curve 3. This curve is quite common for instruments used in TEQA when the linear dynamic range of the detector has been exceeded. A third-order polynomial (not shown in Figure 2.23) tends toward more of a sigmoid or slanted curve shape.

For example, to mathematically conduct a polynomial evaluation of calibration linearity for quantitating Pb in human blood by FlAA, refer to guidelines such as that published by the National Committee for Clinical Laboratory Standards (NCCLS) *EP6-A, vol. 21, no. 20. Evaluation of the Linearity of Quantitative Measurement Procedures: A Statistical Approach: Approved Guideline.* The author had describes the mathematics for this earlier.[28]

2.11.5 Assessing ICV Precision and Accuracy

Precision can be evaluated for replicate measurement of the ICV following establishment of the multipoint calibration by calculating the confidence interval for the interpolated value for the ICV concentration. Triplicate injection of an ICV ($L = 3$) with a concentration well within the linear region of the calibration range enables a mean concentration for the ICV to be calculated. A standard deviation s can be calculated. One ICV reference standard can be prepared closer to the low end of the calibration range, and a second ICV reference standard can be prepared closer to the high end of this range. A confidence interval for the mean of both ICVs can then be found as follows:

$$\text{Confidence Interval for } \mu_{\text{ICV}} = \overline{x}_{ICV} \pm \frac{t\, s_i^{ICV}}{\sqrt{L}}$$

A relative standard deviation, for the ith component, expressed as a percent (%RSD), also called the *coefficient of variation*, can be calculated based on the ICV instrument responses according to

$$\%\text{RSD}_i^{\text{ICV}} = \frac{s_i^{ICV}}{\overline{x}_i^{ICV}} \times 100$$

where s_i is the standard deviation in the interpolated concentration for the ith-targeted component based on a multipoint calibration and $\overline{x}$ is the mean concentration from replicate measurements. For example, one can expect a precision for replicate injection into a GC to have a coefficient of variation of about 2%. As another example, one can expect a precision for replicate solid-phase extractions to have a coefficient of variation of between 10 and 20%. The coefficient of variation itself is independent of the magnitude of the amount or concentration of analyte i measured, whereas reporting a mean ICV (along with a confidence interval at a level of significance) can only be viewed in terms of the amount or concentration measured. The coefficient of variation is thus a more appropriate parameter when comparing the precision between methods.

Accuracy can be evaluated based on a determination of the percent relative error provided that a known value is available. A certified reference standard for at least one analyte or for a mix of analytes constitutes such a standard. Accuracy is calculated and reported in terms of a percent relative error according to

$$\%\text{Error} = \frac{\left| x_{i(unknown)} - x_{i(known)} \right|}{x_{i(known)}} \times 100$$

Statements of precision and accuracy should be established for the ICV, and as long as subsequent measurements of the ICV remain within the established confidence limits, no new calibration curve needs to be generated.

2.11.6 ASSESSING SAMPLE RESULTS

Following the establishment of a calibration and the evaluation of the precision and accuracy for one or more ICVs, we can apply additional statistical assessments on a batch of replicate analytical results from real samples. The Q test can be applied to identify any *outliers* within a series of sample results. The rejection quotient Q is defined as a ratio of the *gap* (difference between the questionable value and its nearest neighbor) to the *range* (difference between questionable value and lowest value in series). (p. 57)[32] For L replicate samples where $L = 8$ to 10, we have:

$$Q_{calc} = \frac{x_n - x_{n-1}}{x_n - x_2}$$

If we have an even larger number of replicate samples such as $L = 11$ to 13, a different equation for Q is used, as shown below:

$$Q_{calc} = \frac{x_n - x_{n-2}}{x_n - x_2}$$

where x_n is a questionable result in the set $x_1, x_2, ..., x_n$ that are arranged in order of increasing value such that $x_1 < x_2 < x_n$. x_{n-1} designates a result that is nearest x_n, and x_{n-2} is a result second nearest x_n. x_1 and x_2 are results furthest and second furthest from x_n. Q_{calc} is compared to $Q_{critical}$ at the 90% confidence level. If Q_{calc} exceeds $Q_{critical}$, then the result is considered an outlier and can be eliminated from the results. A table of Q values is found in Appendix I.

An initial statistical assessment of replicate samples can be accomplished by calculating a *mean* analyte concentration from L replicate samples and a *standard deviation* using previously stated equations found in this chapter.

For two separate batches of replicate sample results, assumed free of systematic error and drawn from the sample population,

$$\sigma_A = \sigma_B$$

The F-test can be used to compare variance from both sets. The F value is calculated according to

$$F_{calc} = \frac{s_A^2}{s_B^2} \quad \begin{array}{l} \nearrow \text{Larger variance put here} \\ \searrow \text{Smaller variance put here} \end{array}$$

F_{calc} is compared to F_{crit} from a tabulated list. Refer to the F table in Appendix I. The number of degrees of freedom for each variance is used to locate the specific F_{crit}. If $F_{calc} < F_{crit}$, then there is no significant difference in two variances.

If a theoretical or established reference whose mean μ is known, we can proceed to find a mean x (average) drawn from L replicate observations and $df = L - 1$ degrees of freedom such that

$$t_{calc} = \frac{|\bar{x} - \mu|}{s / \sqrt{L}}$$

Depending on the outcome of the F-test, two independent means can be compared. There are two approaches to this. If the F-test finds both variances not to be significantly different, then we have

$$t_{calc} = \frac{|\bar{x}_1 - \bar{x}_2|}{s_p\sqrt{\dfrac{1}{L_1} + \dfrac{1}{L_2}}} = \frac{|\bar{x}_1 - \bar{x}_2|}{s_p}\sqrt{\frac{L_1 L_2}{L_1 + L_2}}$$

where s_p is a pooled standard deviation from both sets and is calculated according to

$$s_p = \sqrt{\frac{(L_1 - 1)s_1^2 + (L_2 - 1)s_2^2}{(L_1 - 1) + (L_2 - 1)}}$$

Also,

$$df = L_1 + L_2 - 2$$

If the F-test finds both variances to be significantly different, then we have

$$t_{calc} = \frac{|\bar{x}_1 - \bar{x}_2|}{\sqrt{\dfrac{s_1^2}{L_1} + \dfrac{s_2^2}{L_2}}}$$

where df is computed as shown below:

$$df = \frac{\left(\dfrac{s_2^2}{L_1} + \dfrac{s_2^2}{L_2}\right)^2}{\dfrac{\left(s_1^2 / L_1\right)^2}{L_1 + 1} + \dfrac{\left(s_2^2 / L_2\right)^2}{L_2 + 1}} - 2$$

where $x_{1(ave)}$ and $x_{2(ave)}$ are means from respective batches of replicate sample results for L_1 replicates having a variance of s_1^2 and a variance of s_2^2 is compared to t_{crit} values from a tabulated array of Student's t values. Selective values for Student's t are found in the Appendix. Both df and the confidence level are necessary when obtaining a value for t_{crit}. In both cases, if $t_{calc} < t_{crit}$, then both means that are being compared are *not significantly different*.

Our journey across the "jungle" of data reduction and interpretation has been completed. Topics whose underlying principles also relate to data reduction and interpretation are now introduced. These include:

- The nature of a trace analysis laboratory
- From analog transducer to spreadsheet
- Chromatographic software peak integration
- Sampling for enviro-health and enviro-chemical trace quantitative analysis.

Having discussed what constitutes GLP in terms of QA and QC, let us introduce the very important and not to be neglected *people factor* in all of this.

2.12   HOW IS AN ENVIRONMENTAL TESTING LAB ORGANIZED?

Laboratories are comprised of people in a workplace using sophisticated technology to generate and report analytical data to clients who have a need for the data. Clearly, people function most effectively under an organization in which each individual knows just what is expected of him or her. Figure 2.24 is a flowchart that depicts the organization of a typical environmental testing laboratory. An individual analyst might work within a lab section that is devoted to the analysis of solid waste samples for the determination of trace levels of semivolatile organics and report to a section manager or supervisor who has had 5 or more years of experience in this specialty. A secondary line of responsibility represented by the dashes in Figure 2.24 refers to the QA coordinator's role in guaranteeing that the lab has adopted a QA plan. For example, our individual analyst may have to report some QC data to the QA coordinator from time to time. The chart in Figure 2.24 facilitates QA and offers the hope that a high level of quality in obtaining analytical data can be maintained. A client can then easily recognize which of two labs, hypothetical Lab X or hypothetical Lab Y, should be offered the contract for environmental monitoring of, perhaps, a hazardous waste site.

2.13   WHICH LAB WOULD YOU CHOOSE?

Let us suppose that you are a representative working for an environmental consulting firm and you are asked to choose which company, Lab X or Lab Y, should be hired to analyze your

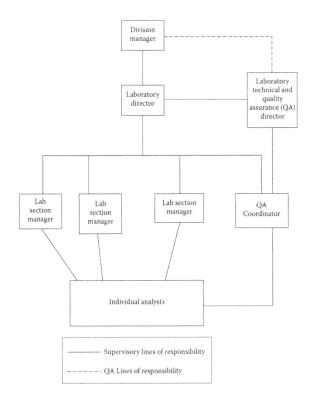

FIGURE 2.24

TABLE 2.6

Analytical Results and QC Reports from Labs X and Y

Analyte	ppb	Confidence Interval at 95%	Correlation Coefficient	% Recovery	%RSD in the % Recovery
Company X Analytical Report for the Hazardous Waste Site					
4-Chloro-3-methyl phenol	100	NA	NA	NA	NA
Bis-2-ethylhexyl phthalate	50	NA	NA	NA	NA
Dibenzo(a,h)anthracene	100	NA	NA	NA	NA
QC results: none					
Company Y Analytical Report for the Hazardous Waste Site					
4-Chloro-3-methyl phenol	56	10	0.9954	67	15
Bis-2-ethylhexyl phthalate	47	9	0.9928	84	13
Dibenzo(a,h)anthracene	73	12	0.9959	73	17
QC results:					
Instrument detection limit (IDL) = 10 ppm					
Method detection limit (MDL) = 10 ppb					
Laboratory blank < MDL					
Field blank < MDL					

Note: NA = not available.

contaminated soil samples from a hypothetical hazardous waste site. Let us also assume that the samples come from a Superfund site and that you are interested in the lab providing you with analytical results and interpretation for semivolatile organics using EPA Method 8270C. You receive analytical results from both labs, as shown in Table 2.6. Based on your interpretation of the reports from both companies, which lab would you choose to contract with for the analysis? It should be obvious.

Lab X lacks any evidence of QC. Although the lab reported concentration levels of three semivolatile organics in ppb, no statement about the precision of the analysis is provided. No correlation coefficient is provided. The lab's efficiency in extracting the analytes of interest from the environmental sample matrix as shown by a percent recovery is not evident in the report. No other QC is reported for this laboratory. This author would immediately ask upon receiving such a report if the QA document is available for review.

Lab Y, on the other hand, not only reported detectable levels of the three organics, but also gave a confidence interval for each reported concentration. Lab Y demonstrates that reference standards were run to establish a least squares regression line, and the goodness of fit is expressed by the value of the correlation coefficient. The lab also conducted the required percent recovery studies and demonstrated acceptable precision in the reported percent recoveries for each of the three analytes. The lab also reports that it conducted a study of the IDLs and MDLs for the requested method. Figure 2.25 summarizes the ideal environmental testing lab and needs no further explanation!

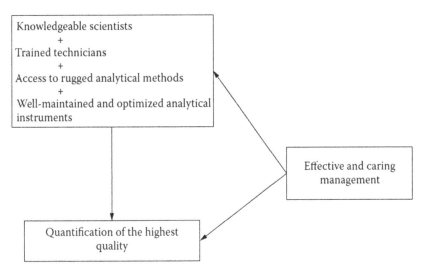

FIGURE 2.25

2.14   HOW DOES THE CDC EVALUATE A METHOD VALIDATION?

This author spent over a decade working as an analytical chemist for a CLIA (Clinical Laboratory Improvement Act) laboratory.[13] QA/QC guidelines for clinical laboratories engaged in TEQA differ from those laboratories engaged in TEQA that must adhere to EPA guidelines. This section will introduce the approach taken by the Centers for Disease Control and Prevention (CDC). Readers should see similarities and differences in QA/QC requirements as we transition from *enviro-chemical* TEQA to *enviro-health* TEQA. Let's suppose that you are the analyst given the task of conducting a validation for a new analytical method. This method if executed properly will provide trace quantitative analysis of various chemicals in blood, serum or urine. This author was assigned to a trace analytical method that sought to quantitate cyanide (CN^-) in whole blood. Here is what was expected. Each laboratory must complete *20 independent analytical runs* and correctly format and submit to the CDC's Quality Assurance Program validation data using the agency's reporting format.

The rules are listed below:

- Each analytical run must include a full calibration curve, one of each QC level including a QC blank.
- Quality control (QC) samples must be quantitated using a unique calibration curve prepared and analyzed.
  - No two sets of QC samples should be quantitated using the same calibration curve.
  - No set of QC samples should be quantitated using more than one calibration curve.
- A minimum of two analysts in the laboratory must perform all aspects of the analysis method.
  - Both a primary and a secondary analyst must develop competence in performing all parts of transferred CT (CDC's Laboratory Response Network, Chemical Terrorism) methods during the validation exercise.
- A maximum of two analytical runs per day may be prepped or analyzed.
  - A minimum timeframe of 10 days is needed to ensure that a reasonable approximation of laboratory and method variation (between run) is incorporated into the QC means and standard deviations.
- Runs must be prepped and run on the same day.
  - To ensure that the age of the samples does not contribute to the variation between analytical runs.

- Observe criteria for dropping standards.
  - The standard dropped is clearly not correct.
  - The same standard cannot be dropped in >25% of the validation runs.
  - The same standard cannot be dropped in three consecutive runs.
- A correlation coefficient of 0.9900 is considered the minimum for acceptability.
  - The origin should be ignored as a possible data point and either a linear or a 1/X weighting can be applied to optimize low-level results.
- Create Levey–Jennings Plots (Excel is recommended to accomplish this).
- Apply Multi-Rule (Westgard) QC as introduced in Appendix D.

2.14.1 EVALUATION CRITERIA

The following statistical criteria should be applied to all laboratory validation data before submission of to the LRN-C Quality Assurance Program where Si refers to the standard deviation in the individual result:

2.14.2 MULTI-RULE QC (WESTGARD)

- 1 3S Rule—Run result is outside a 3 Si limit, where Si refers to the standard deviation (Random Error).
- 2 2S Rule—Current and previous run results are outside the same 2Si limit (Systematic Error).
- 10X-bar Rule—Current and previous nine run results are on the same side of the characterization mean (Systematic Error).
- R 4S Rule—The current and the previous run results differ by more than 4Si (Systematic Error).

The systematic logic described by applying the Westgard Rules is described in Figure 2.26.

2.14.3 ADDITIONAL CDC VALIDATION CRITERIA

1. Individual QC Values. Using a minimum bias, target value, and standard deviation determined during method validation / QC characterization at CDC, a range for acceptable performance, accuracy and precision is applied to individual QC values. This rule does not supersede the 1 3S Rule. The following equation is applied:

$$(\mu - \text{bias}) - 3*s \le \text{Individual QC values} \le (\mu + \text{bias}) + 3*s$$

Where μ- population mean or target value (according to CDC)
Bias—considered as being equal to 4% and substituting into the abov equation

Example

Assume Bias = 4%, μ = 25, s = 5.0 and substituting into the above equation yields

2. QC Mean. A coefficient of 2.093 (determined by applying a t-test) is used to calculate the acceptable range for the QC mean for each QC level. The following equation is applied:

$$(\mu - \text{bias}) - t_{(1-\alpha/2,df)}\frac{s}{\sqrt{L}} \le \text{QC mean value} \le (\mu + bias)t_{(1-\frac{\alpha}{2},df)}\frac{s}{\sqrt{L}}$$

Where t(1-α/2,df) = 2.903 for α = 0.05, two-tailed hypothesis testing and 19 degrees of freedom for 20 replicate QCs for a given level (i.e. L = 20)

3. Standard Deviation. A coefficient of 1.5912 (determined by applying an F-Test) is used to calculate the upper limit for an acceptable standard deviation for each QC level. The following equation is applied:

$$S_{\text{Lab}} \le (s)*1.5912$$

Example

Assume $\mu = 25$, $s = 5.0$ and substituting into the above equation yields:

$$S_{lab} \leq (5)*(5.912) = 7.96$$

2.15   FROM TRANSDUCERS TO SPREADSHEETS: WHAT DOES THIS MEAN?

Envision a trace analytical laboratory of ~1975 vintage. A wire would run from the lab's gas chromatograph to a *strip-chart recorder*. The analog signal from the GC detector (we will address the various detectors used in gas chromatography in Chapter 4) would be recorded on rolled graph paper. The strip-chart recorder's pen would monitor a GC baseline over time. The distance traveled by the graph paper would be related to chromatographic run time based on the speed of the motor in turning the roll. Since 1975, a revolution in data acquisition as well as in instrument control has occurred thanks to advances in solid-state electronics and computer technology.

The contemporary trace analytical laboratory will still have a GC; however, more sophisticated-looking cables emanate out of the instrument to a stand-alone personal computer (PC). An RS232 interface cable is necessary for data transmission. PCs were called *microcomputers* when they first arrived on the scene in the 1980s. Microcomputers were much smaller in size than existing so-called *mainframe* computers. The GC detector's analog signal becomes digitized, read, and stored. The GC chromatogram can then be called up from storage, manipulated, interpreted, restored, and printed out in the form of reports. The analog signal emanating from the GC detector, the transducer, must be converted into the digital domain.

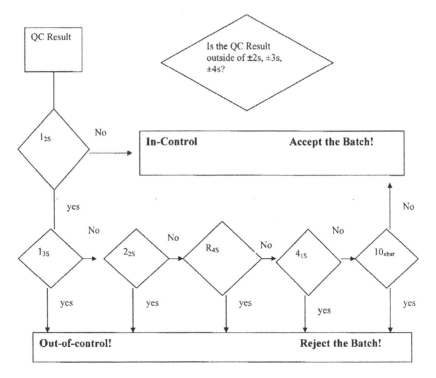

FIGURE 2.26

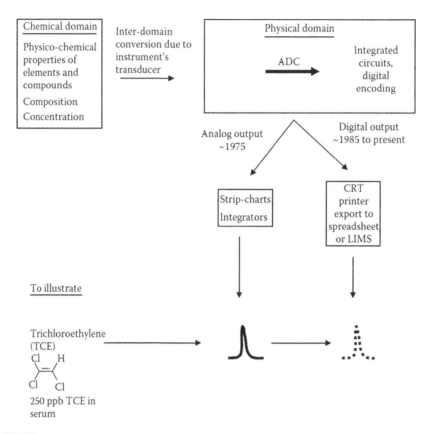

FIGURE 2.27

This is accomplished by adding an analog-to-digital converter (ADC) as an interface between the instrument and PC. This interface can be found external to the PC as a stand-alone box. This interface can also be a board that inserts into the input/output (I/O) peripheral of the PC. Figure 2.27 attempts to conceptually outline the inter-domain conversion from the chemical to the physical, and then within the physical domain, the evolution from analog to digital electronics. To illustrate, trichloroethylene (TCE), a volatile, priority pollutant organic compound, is suspected to be present in a human serum specimen, possibly due to occupational exposure to organic solvents, and is displayed in the GC chromatogram as a detector or transducer response that takes the shape of a Gaussian peak whose area under the curve is proportional to the concentration of TCE injected into the GC. In 1975, the peak would be displayed as analog output on a strip-chart recorder. Analysts of that era would use a ruler and pencil to measure peak height or triangulation to measure peak area. Or earlier they would cut and weigh to measure peak area. Then along came analog integrators. A popular one was made by the company *Spectra Physics*. What a contrast from then until now! Most online webinars that this author has participated in seem to feature presenters who much prefer to introduce their listeners to the many software features that digitized data has afforded. The world that analytical chemists now have to perform their data reduction and interpretation in reflects these changes. However, this author still likes to write with pencil and paper!

Analytical chemists and technicians are not accustomed to delving into electronics (and this includes the author). However, a bit (no pun intended) of dabbling does not hurt as we proceed to introduce those essential principles that take the reader from transducer to spreadsheet. Of the plethora of texts on the subject of electronics, the essential principle for most analysts involves just enough knowledge of the ADC to be able to understand interface specifications and the

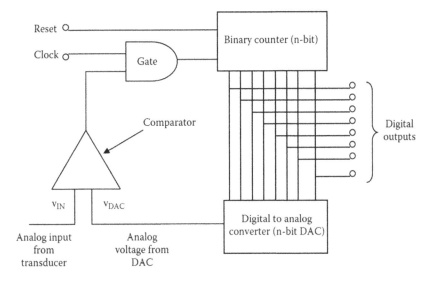

FIGURE 2.28

principle of ADC resolution. This author has chosen to focus on those principles underlying ADCs as deemed most usefully interfaced to gas chromatographs. ADCs are found in almost every kind of analytical instrument today. The major challenge for analysts at the bench is to master whatever data processing software is associated with a particular analytical instrument.

2.15.1 How Does Analog Become Digital?

A simple form of an ADC is shown in the schematic drawing of Figure 2.28.[33] Malmstadt et al., who have written the seminal text on electronics relative to the interest of scientists, have described a staircase ADC in the following manner:[34]

> The conversion cycle begins when a start pulse clears the binary counter. Since v_{DAC} is then less than v_{IN}, the comparator goes to 1 and opens the counting gate. Counts are accumulated until v_{DAC} just exceeds v_{IN}. At this time, the comparator goes to 0 and closes the counting gate. The parallel digital output of the counter is thus the digital equivalent of the analog input voltage.

> Referring again to Figure 2.28 the scenario shown below summarizes how a simple ADC works:

Voltage to Comparator	What Happens to ADC
$VDAC < VIN$	Logical 1; AND gate opens; clock pulses allowed to pass to counter; v_{DAC} continues to increase
$vDAC = vIN$	Logical 0; AND gate closes; v_{DAC} ceases to increase; end-of-conversion flag (EOC) is produced at auxiliary output; counter must be reset to zero before starting next measurement

The most common ADCs that are interfaced to gas chromatographs are voltage-to-frequency converters and dual-slope integrators. Both ADCs measure the signal throughout the entire chromatographic run time. Ouchi said it best:[35]

> Digital values are then used in data analysis and display and in many cases are stored for further analysis. By graphing the digital values, a computer can generate a reconstructed chromatogram, will be as accurate as the original (analog) chromatogram … Converting

one type of data into another creates opportunities for errors to occur. When an ADC digitizes the analog signal from a detector, the signal's continuous voltage and time values are transformed into discrete digital values [as shown in Figure 2.28]. The accuracy of the voltage values is determined by the resolution of the ADC, and the accuracy of the time values is determined by the system's data acquisition rate.

ADC *resolution* is defined by either (1) the number of bits available, (2) the dynamic range, or (3) the amount of uncertainty or error for a given full-scale analog input signal. Typical resolutions for ADCs are 24, 20, 16, and 12 bits. For an n-bit ADC, there are 2^n different combinations (configurations or different ways to represent information) of 1s and 0s. To illustrate, three bits can be configured in 2^3 or 8 different ways, four bits can be configured in 2^4 or 16 different ways, and five bits can be configured in 2^5 or 32 different ways. For example, for $n = 3$, we can arrange three bits in the following manner:

000	001	010
011	100	101
110	111	

Up to eight different values for the analog voltage from the transducer could be digitally measured and stored.

The ADC *dynamic range* is found from the number of different configurations less one (where zero is excluded), or $2n - 1$ different ways. The ADC's resolution is thus calculated for a 1-V full-scale analog output (a 1.0 V signal is typical of a transducer's full range of output) according to

$$R_S^{n-bitADC} = \frac{\# volts\left(full-scale\right)}{2^n - 1}$$

To illustrate, a *20-bit ADC* would have a resolution of 1.0 V divided by 1,048,575 different ways or steps, or 0.954 µV per step. A change of only 0.954 µV is required to register a new configuration for a 20-bit ADC. Contrast this with a 12-bit ADC whose resolution for the same 1.0 V output signal is 244 µV per step. A small peak in a chromatogram whose peak height is 150 µV would be picked up by the 20-bit ADC and missed by the 12-bit ADC.

The *data acquisition rate* (often called a sampling rate) as programmable on most PCs that utilize a chromatographic processing software package defines just how long the analog signal is monitored for each data point used or stored by the computer. As a rule of thumb, between 10 and 20 data points should be taken across a chromatographically resolved Gaussian-shaped peak to achieve the more accurate digitized peak area. Peak widths in capillary gas chromatography (C-GC) are of the order of 1 to 2 sec (1 sec = 0.0166 min) while peak widths in packed column high-performance liquid chromatography (HPLC) are of the order of 5 to 10 sec. These requirements indicate that the data acquisition rate for C-GC should be set somewhere between 5 and 10 points/sec, while the rate for HPLC should be set between 1 and 2 points/sec.

Bunching is yet another parameter found in most chromatographic processing software. Instead of just one data point at a time being sampled, digitized, and stored, a number of data points can be bunched, averaged, digitized, and stored. Bunching enables a much higher data acquisition rate to be set, and this increase has certain advantages. For example, increasing a 10 points/sec sampling rate to 60 points/sec while setting the bunching to 6 sums the value for every 6 data points and, in effect, is the equivalent to a data acquisition rate of 10 points/sec. Data bunching enables a filtering out of unwanted noise, particularly noise due to that caused by 60 Hz alternating current.

Let us look at a recent product release for an ADC as advertised over the internet from *Laboratory Network* (www.laboratorynetwork.com):

The hardware unit interfaces your computer to your detectors, injector, autosampler, and/or fraction collection. It contains a high speed analog-to-digital converter and amplifier which accepts two channels of data simultaneously. Each detector input has independent controls and settings ... is suitable for use with HPLC, GC, ion chromatography (IC) ... capillary electrophoresis and microdialysis systems ... record and analyze output from GC, HPLC, IC, size exclusion, and preparative chromatography.

Specifications for this specific ADC directly from the product release are given in the table below:

Specification Sheet for a Typical ADC

Input Amplifiers	
Analog inputs	Two single-ended or differential
Input impedance	1 MΩ
Input range	±2 mV to ±10 V full scale in 12 steps
Frequency response (−3 dB)	20 kHz at ±10 V full scale, all ranges
Low-pass filter	20 kHz fixed second-order filter; further digital filtering in software
DC offset	Software-corrected
Output	
Analog output	Single-ended
Output ranges	±200 mV to ±10V (software selectable)
Output current	±15 mA maximum
Output resolution	16 bits
System	
ADC resolution	16 bits
Signal resolution	20–24 bits with oversampling
Linearity error	±2 LSB (least significant) (from 0 to 70°C)
System accuracy	0.1% of full scale range ± 1 bit
Sampling rates	200 Hz down to 0.2 Hz (12/min)
Digital outputs	Four contact closure or TTL
Injection signal point	Contact closure (available through instrument connection port)
Processor	PPC403GCX (60 MHz)
Memory	4 MB DRAM
Data communication	USB up to 800 kB/sec
Instrument connection port	20-pin port; accepts strip terminal

Note: TTL = transistor–transistor logic.

One can see in this spec sheet how important the concept of ADC resolution really is.

2.15.2 How Does the 900 Series® Interface Convert Analog Signals to Digital Values?

This author is most familiar with TotalChrom® (upgraded from Turbochrom® PE-Nelson) from PerkinElmer Instruments. TotalChrom is a complete software package for non-mass spectrometric chromatographic data acquisition, processing, and reporting, and for instrument control. Most leading analytical instrument manufacturers have their own complete software packages. Agilent has ChemStation®, etc. Some software packages such as EZ Chrom Elite® (ESA, Inc.) can be downloaded onto a PC, and if this PC can be properly interfaced, the software can be used with a variety of manufacturers' chromatographs.

According to PerkinElmer Instruments:[36]

> Each 900 Series Interface can be connected to one or two chromatographic detectors usually on a single instrument … The 900 Series Interface converts an analog voltage signal to a frequency that varies in proportion to the signal voltage. The interface then counts the pulses and records a value every 0.01 second. The count accumulated during the interval is called a *time slice*. The value of each time slice, or the sum of two or more time slices, becomes a data point on the chromatogram … Because the interface always records a count every 0.01 seconds, its fundamental sampling rate is 100 points per second. However, you can define a lower sampling rate in the method … If you use a rate that is slower than the fundamental rate, the interface sums the appropriate number of slice values taken at the fundamental rate. This integrated value becomes a data point. The number of time slices that are summed to derive a data point depends on the desired sampling rate. For example, if the method calls for a rate of 10 points/sec, the interface sums 10 time slices taken at the fundamental rate.

Nine distinct tasks are identified below if a GC has been downloaded with TotalChrom. Each task is executed starting at the top and proceeding to the bottom.

- Baseline subtraction
- Peak detection
- Peak integration
- Component identification (optional)
- Calibration
- Quantitation
- Report generation (optional)
- Replot generation (optional)
- Post analysis programs (optional).

Let us focus on peak integration while leaving the other eight topics to the user who must operate a given instrument. This assumes that the software has identified a chromatographically resolved peak in a plot of instrument response against run time, i.e., the time after injection of the sample. A sketch of a chromatographically resolved peak is shown in Figure 2.29.

Referring to Figure 2.29, the peak area is found by first dividing the area beginning at a raw data point that corresponds to the *peak stop* and extending horizontally backward to the data point that corresponds to the *peak start*. In the above sketch, eight area slices of equal width are shown. The start point of the peak does not contribute to the peak's area. To integrate the area under this curve, the software first sums the area slices from the peak start to the peak end. Initially, these slices extend vertically from the level of a data point to the zero-microvolt level. Next, the software corrects this sum of the height of the baseline by subtracting the baseline area. This baseline area is the area of a trapezoid between the baseline and the zero-microvolt level. A correct peak area results. This peak is proportional to the amount of analyte injected into the chromatograph,

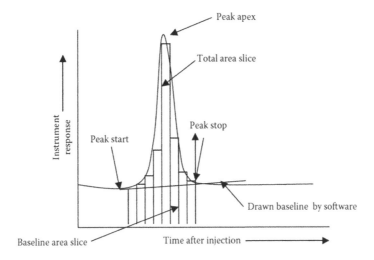

FIGURE 2.29

and this forms the most fundamental basis of TEQA. In the above sketch, note that the peak is somewhat *skewed*; i.e., the peak is slightly unsymmetrical or is said to be not perfectly Gaussian.

2.15.3 WHAT ARE INTEGRATION PARAMETERS?

One other aspect of using chromatographic software that the author wishes to address in this chapter is that of integration parameter settings. Every software package has some sort of integration parameter settings protocol. The settings should be established after a preliminary chromatogram has been obtained and reviewed. Much of a chromatogram will have unnecessary and even unwanted noise and peaks. The integration parameter settings assist in eliminating these undesirable portions of the baseline. These parameter settings also assist the analyst who wants to properly integrate a pair of partially chromatographically resolved peaks. Tabulated below are those parameters as found in Agilent's ChemStation® software.

Integration Parameter Settings	
Initial area reject	Initial peak width
Shoulder detection	Initial threshold
Area reject	Area sum off and/or on
Baseline all valleys off and/or on	Baseline back
Baseline hold off and/or on	Baseline next valley
Baseline now	Integrator off and/or on
Negative peak off and/or on	Peak width
Solvent peak off and/or on	Tangent skim
Threshold	

Integration parameters whose entry includes the word *initial* are default settings and are initiated without user intervention whatsoever. Integrator enables integration to start and stop at preselected time events across the chromatogram. The area sum parameter enables the analyst to sum multiple peaks across a specific time interval in the chromatogram. The user of such software must understand how most of these parameters can be used to yield an analytical method

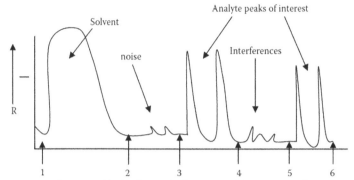

FIGURE 2.30

in the software that is efficient as well as to minimize computer memory and disk storage. The manner by which some of these parameters are employed is seen in Figure 2.30 for a simulated chromatogram that reveals undesirable regions along the time axis. The arrows pointing upward represent those timed events that should be programmed in to eliminate unwanted peak integration due to the presence of baseline noise, interfering peaks, and certainly the initial solvent peak (sometimes called a *solvent front*) that is present when more universal GC detectors are used.

2.15.4   How Are Analytical Results Reported?

The organization of sample results that follows data reduction and interpretation into a tabular format for reporting purposes has evolved significantly over the past 30 years. Laboratory PC software that controls and acquires data all has some type of reporting protocols. The analyst today has numerous options with regard to reporting and presenting analytical results. Reports can be printed out on the accompanying printer right in the laboratory. Various summary versions of sample batches can be printed out in tabular formats. Raw data and results can be transferred to a thumbnail or flash drive and transported to a PC (earlier 3½-in. disks were used and even earlier PCs used *floppy disks* that were 5¼ in. in length and did indeed flop). Older flash drives can store up to 15 GB while newer ones can reach 30 GB! Software such as MassHunter® (Agilent Technologies) provide numerous unique opportunities for analysts to report results! Most chromatography data acquisition and control software commercially available has algorithms that enable a specific raw data or result file to be converted to an *ASCII* (American Standard Code for Information Interchange) file. This ASCII file can be imported into an Excel spreadsheet or an Access database. In recent years, the computer architecture has changed so as to make archiving on hard or portable disk obsolete. The *client–server* architecture in which the laboratory PC becomes the client will succeed in making even flash drives obsolete in the future. Interested readers may find an earlier resource by Dyson of interest.[37] McDowell and colleagues have addressed this topic and continue to write a continuous column on this subject.[38, 39]

2.16   WHAT DO I NEED TO KNOW ABOUT SAMPLING?

Sampling for *enviro-chemical* TEQA differs from sampling for *enviro-health* TEQA due largely to the very different sample matrices. Samples for *enviro-chemical* TEQA are drawn from air, surface, ground- and wastewater, soil of various types, sediment from river bottoms, sludge from water treatment plants, industrial effluent, leachate from landfills and hazardous waste sites, etc. However, samples for *enviro-health* TEQA are limited to human and animal specimens such as whole blood, serum and plasma, adipose and various organ tissue, breast milk, urine, saliva, semen, etc. Sample volumes for *enviro-health* TEQA are limited to anywhere from 0.5 to 10 mL per specimen. In contrast, a typical sample volume to conduct *enviro-chemical* TEQA can range from 100 to 2,000 mL! Sampling 10 L of groundwater is not unusual. These stark differences in sample volume between *enviro-chemical* TEQA and *enviro-health* TEQA result in significant differences in MDLs.

Good sampling technique must seek to preserve sample integrity. A groundwater sample that is known to contain trichloroethylene (TCE) at a concentration level of 125 ppb when sampled and then analyzed might yield an analytical result of 2.5 ppb! Analyte loss is significant, especially for those analytes that are volatile nonpolar organics dissolved in an aqueous matrix such as TCE in groundwater. There exists a spontaneous fugacity or tendency for volatile organic compounds to escape from being dissolved in water into the atmosphere. This is why proper sampling of 1L of groundwater requires a large glass container with a large lid filled "to the brim" with zero headspace. Analyte loss involving environmental water samples also can occur for nonvolatile, semivolatile, as well as for volatile organic compounds during sample storage. Loss in storage due to biological activity can be prevented by:

* Acidifying the sample of pH ~1.5
* Cooling or freezing the sample after collection
* Adding complexing agents
* Adding oxidizing agents such as hydrogen peroxide to destroy organic matter
* Irradiating with ultraviolet light

2.16.1 What Happened When Frozen Blood Specimens Were Thawed Repeatedly?

Blood specimens must be kept cold! Otherwise blood will chemically break down and release gases (so-called off-gassing). The author unexpectedly discovered this fact in the laboratory! Blood samples that were initially stored at -20°C (for an anticipated quantitation of cyanide ion) were repeatedly thawed at room temperature due to faulty refrigeration control! The author was not aware of the problem until he began to realize that the QC samples had violated the Westgard 1_{3s} Rule. The quantitated results for all QC samples were much lower than expected! This was most unusual since all prior QC samples had not violated any of the Westgard Rules! The reason for the QCs violating the Westgard 1_{3s} Rule remained unknown until the author attempted to increase GC resolution by abandoning the temperature program run and operating under isothermal conditions. To his surprise, two partially resolved chromatographic peaks emerged! This chromatogram is shown in Figure 2.31.[13] Propane gas was then purged into a glass flask from a portable source. A sample of the gas was injected into the same HS-C-GC-MS. This chromatogram is shown in Figure 2.32.[13] The electron-impact (EI) mass spectrum for the Total Ion Chromatogram (TIC) shown in Figure 2.33 revealed a dominant abundance of the m/z 29 due to an ethyl moiety![13] It was only then that it was realized that the m/z 29 due to the co-eluted propane was identical and therefore interfered with the isotopic internal standard $H^{13}C^{15}N$! The increased abundance (A) of the m/z 29 altered (decreased) the ratio of A^{27}/A^{29} and thus led to the biased QC result!

2.16.2 Can We Predict How Many Samples Must Be Taken When Sampling the Environment?

To conduct *enviro-chemical* TEQA samples must be collected from either air, water, soil, etc. Since samples obtained from the environment are intrinsically heterogeneous, a large number of samples must be collected and analyzed. A large number of samples can be collected and

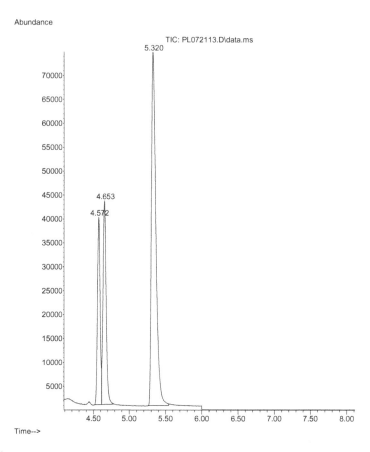

FIGURE 2.31

combined to yield a blend that is reasonably homogeneous. From this composite blend, a sufficient number of subsamples can be analyzed. Provided that the error tolerance and confidence interval are defined, we can view the earlier equation for the absolute difference between a population mean and a sample mean in which systematic error has been eliminated:

$$|\mu - \overline{x}| = \frac{z\sigma_p}{\sqrt{L}}$$

Solving this equation for L and defining a number of samples as L_S gives

$$L_S = \left| \frac{z\sigma_p}{|\mu - \overline{x}|} \right|^2$$

To illustrate, consider the determination of N-methyl carbamates, which is usually accomplished with a direct aqueous injection into a reversed-phase high-performance liquid chromatograph with post-column derivatization and fluorescence detection (RP-HPLC-Der-FL). Carbamates are used chiefly in agriculture as insecticides, fungicides, herbicides, nematocides, and sprout inhibitors. Let us assume that surface water runoff from a particular agricultural field in the Midwest U.S. yields a mean concentration level for carbaryl (Sevin®) of 0.1 ppm, with a known standard deviation of 0.025 ppm. If we accept an error tolerance in the mean concentration of

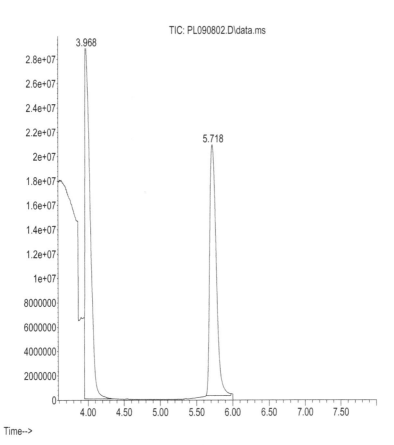

FIGURE 2.32

20% (20% of 0.1 ppm carbaryl is 0.02 ppm) at a 95% confidence interval ($z = 1.96$), using the equation above, we find that it is necessary to run six samples. However, most environmental analyses are done on samples whose mean concentration levels are not known and whose standard deviation for a population is not known as well.[40]

If the standard deviation among replicate measurements can be obtained, the above equation can be used to estimate the number of replicate samples to take to yield a given confidence interval. Measurement for TEQA consists of applying replicate sample preparations such as SPE and replicate determinative techniques such as repeated injection into a GC. There is random error associated with each of these activities. For example, spiking 100 mL of groundwater with chrysene, isolating and recovering the chrysene from the sample matrix, preparing the extract of chrysene in an organic solvent, and injecting the organic solvent into a GC to measure chrysene can be repeated five times to yield the necessary replicate data in order to estimate a standard deviation in the measurement, σ_m. Again, if we specify a given error tolerance, e, we can use Student's t to estimate the number of replicate environmental samples, L_M, to take through the sample preparation and determination steps according to

$$L_M = \left[\frac{t_{(1-\alpha/2, df)} \sigma_M}{e} \right]^2$$

Abundance

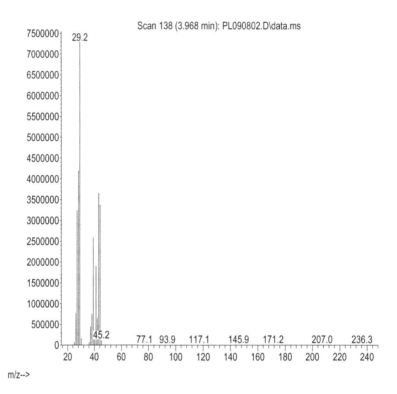

m/z-->

FIGURE 2.33

2.16.3 How Might We Obtain A Representative Sample from a Large Environmental Location?

Imagine you are the environmental engineer who is sent to sample the soil within the currently fenced-off area located on a buff of the Pine River in St. Louis, MI! This EPA hazardous waste superfund site was the location of the former MCC (see Chapter 1). Figure 2.34 nicely depicts all the steps necessary to obtain an environmental sample while introducing appropriate nomenclature. In addition, Figure 2.34 shows how an environmental compartment or bulk material can eventually become a test sample which then leads to an analytical result. This is in accordance with the International Union of Pure and Applied Chemistry (IUPAC) requirements.

2.16.4 What Factors Need to Be Considered in Obtaining Representative Samples?

Eight factors are listed below that answer this question (pp. 95–97):[26]

- Samples must be replicated within each combination of time, location, or other variables of interest.
- An equal number of randomly allocated replicate samples should be taken.
- Samples should be collected in the presence and absence of conditions of interest in order to test the possible effect of the condition.
- Preliminary sampling provides the basis for the evaluation of sampling design and options for statistical analysis.

Sampling operations

Analytical operations

(without sampling errors)

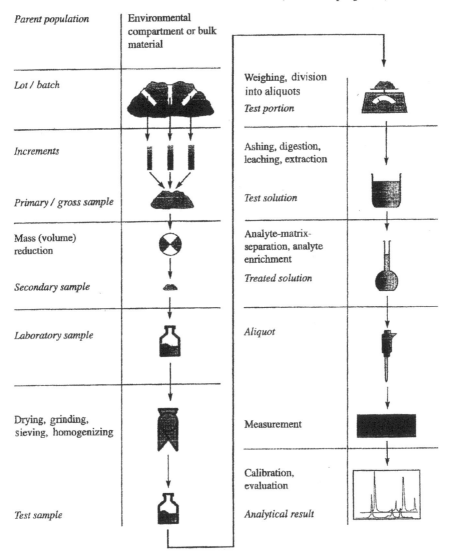

Parent population	Environmental compartment or bulk material
Lot / batch	Weighing, division into aliquots *Test portion*
Increments	Ashing, digestion, leaching, extraction
Primary / gross sample	*Test solution*
Mass (volume) reduction	Analyte-matrix-separation, analyte enrichment
Secondary sample	*Treated solution*
Laboratory sample	*Aliquot*
Drying, grinding, sieving, homogenizing	Measurement
	Calibration, evaluation
Test sample	*Analytical result*

FIGURE 2.34

- Efficiency and adequacy of the sampling device or method over the range of conditions must be verified.
- Proportional focusing of homogeneous subareas or subspaces is necessary if the entire sampling area has high variability as a result of the environment within the features of interest.
- Sample unit size (representative of the size of the population), densities, and spatial distribution of the objects being sampled must be verified.
- Data must be tested in order to establish the nature of error variation to enable a decision on whether the transformation of the data, the utilization of distribution-free statistical analysis procedures, or the test against simulated zero-hypothesis data is necessary.

An optimized sampling program must take the following aspects into account:

- Place, location, and position of sampling
- Size, quantity, and volume of the sample
- Number of samples
- Date, duration, and frequency of sampling
- Homogeneity of the sample
- Contamination of the sample
- Decontamination of the sample
- Sample conservation and storage.

A readily accessible article on sampling for *enviro-chemical* TEQA was recently published.[42] In this paper, the author points out that solid samples are far more heterogeneous when compared to gases and liquids and thus require the use of *random numbers* in planning how to sample a large tract of land. Also cited were references to IUPAC published general sampling nomenclature for analytical chemistry and a terminology recommendation for soil sampling.[41,43]

Well, what is left? Analytical chemists must be able not only to process, manipulate, interpret, and report on analytical data and results, but actually acquire the data. Effectively performed TEQA requires that once sampling is completed, sample preparation techniques must be implemented. This important facet of TEQA comprises the next chapter.

When you develop a measurement method, the resulting method needs to be fit for use. If you discover that your developed method isn't fit for use, you'll need to do further work. This gets you into expensive and time-consuming product development cycles. Thus, it is important to talk with the client *before* development, agree on all of the performance requirements, and focus on meeting those requirements *during* development. Method validation then becomes simple method confirmation.

—**Stan Deming**

REFERENCES

1. Currie L. *Pure Appl Chem* 67: 1699–1723, 1995.
2. Hubaux A., G. Vos. *Anal Chem* 42: 840–855, 1970.
3. Oppenheimer L., T. Capizzi, R. Weppelman, H. Mehta. *Anal Chem* 55: 638–643, 1983.
4. Clayton C., J. Hines, P. Elkins. *Anal Chem* 59: 2506–2514, 1987.
5. Currie L. *Anal Chem* 40: 586–593, 1968.
6. Zorn M., R. Gibbons, W. Sonzogni. *Anal Chem* 69: 3069–3075, 1997.
7. Anderson R. *Practical Statistics for Analytical Chemists.* New York: Van Nostrand Reinhold, 1987.
8. Currie L., *Anal Chim Acta* 391: 105–126, 1999.
9. Currie L., *Anal Chim Acta* 391: 127–134, 1999.
10. Oelke W., Members of the Midwestern Association of Chemisty Teachers in Liberal Arts College. *Laboratory Physical Chemistry.* Cincinnati: Van Nostrand Reinhold, 1969, pp. 29–31.
11. Budde W. *Analytical Mass Spectrometry.* New York: Oxford University Press, 2001, pp. 82–85.
12. Fassett J., P. Paulsen. *Anal Chem* 61: 643A–648A, 1989.
13. Loconto P.R., A decade of quantitating cyanide in aqueous and blood matrices using automated cryotrapping isotopic dilution static headspace GC-MS. *LC-GC North America*, 33(7): 490–505, 2015.
14. Bader M. *J Chem Educ* 579: 703–707, 1980.
15. Loconto P.R., Y. Pan and P. Kamdem, Isolation and recovery of 2-Aminoethanol (2-AE), N-Methyl-2-AE and N,N-Dimethyl-2-AE from aqueous matrices and from treated sawdust using liquid-liquid extraction and liquid-solid extraction combined with capillary gas chromatography / ion-trap mass spectrometry. *J ChromSci* 36: 299–305, 1998.
16. Liengme B. *A Guide to Microsoft Excel for Scientists and Engineers*, London: Arnold, 1997.

17. MacTaggert D., S. Farwell. *J AOAC Int* 75: 594–608, 1992.
18. Keith L., et al. *Anal Chem* 55: 2210–2218, 1983.
19. Smith R. *Handbook of Environmental Analysis,* 2nd ed. Amsterdam, NY: Genium Publishing, 1994, pp. 89–90.
20. Burdge J., D. MacTaggart, S. Farwell. *J Chem Educ* 76: 434–439, 1999.
21. Loconto P.R., Use of weighted least squares and confidence band calibration statistics to find reliable instrument detection limits for trace organic chemical analysis. *Am Lab*, 47 (7): 34–39, 2015. See also: Loconto P.R., Tutorial and full length article supplements to "Use of weighted least squares and confidence band calibration statistics to find reliable instrument detection limits for trace organic chemical analysis" are found in the archives to the September, 2015 issue of *American Laboratory*.
22. MacTaggart, D. and S. Farwell, *JAOAC*, 75: 608–614, 1992.
23. Loconto, P.R., Evaluation of automated stir bar sorptive extraction-thermal desorption-gas chromatography electron capture negative ion mass spectrometry for the analysis of PBDEs and PBBs in sheep and human serum. *J ChromSci*, 47: 656–669, 2009.
24. Mode E. *Elements of Probability and Statistics*. Englewood Cliffs, NJ: Prentice Hall, 1966, pp. 194–200.
25. Deming S. *American Laboratory* 48(8): 35–37, 2016.
26. Einax J.W., W.H. Zwanziger, S. Geiss. *Chemometrics in Environmental Analysis.* New York: VCH, 1997.
27. Billo E.J., *Excel® for Chemists, A Comprehensive Guide*, 2nd Ed., New York: Wiley-VCH, 2001.
28. Loconto P.R. *Trace Environmental Quantitative Analysis, 2nd edition*, 2006, Boca Raton, London, New York: CRC Press, Taylor and Francis, Appendix C.
29. Erickson M. *Analytical Chemistry of PCBs,* 2nd ed. Boca Raton, FL: CRC/Lewis, 1997, p. 107.
30. Provost L., R. Elder. *American Laboratory* 15: 57–63, December 1983.
31. Loconto P.R. *LC-GC* 20: 1062–1068, 2002.
32. Harris D. *Quantitative Chemical Analysis,* 3rd ed. New York: Freeman, 1991.
33. Vassos B., G. Ewing. *Analog and Digital Electronics for Scientists,* 3rd ed. New York: Wiley Interscience, 1985, pp. 340–350.
34. Malmstadt H., C. Enke, S. Crouch. *Electronics and Instrumentation for Scientists*. Menlo Park, CA: Benjamin/Cummings, 1981, p. 382.
35. Ouchi G. *LC-GC* 9: 747–777, 1991.
36. *Total Chrom Workstation User's Guide, Vol. II*. Shelton, CT: PerkinElmer Instruments, 2001, p. A-2.
37. Dyson N. *Chromatographic Integration Methods,* 2nd ed. Cambridge, U.K.: Royal Society of Chemistry, 1998.
38. McDowell R. *LC-GC* 18(1): 56–67, 2000.
39. McDowell R. *LC-GC* 18(2): 180–198, 2000.
40. ACS Committee on Environmental Improvement. *Anal Chem* 55: 2210–2218, 1983.
41. Horwitz W. *Pure Appl.Chem* 62: 1193, 1990.
42. Meyer V.R. Sampling the Ghost in front of the Laboratory Door. *LC-GC North America* 37: 768–774, 2019.
43. deZorzi P. et al. *Pure Appl.Chem* 77: 827, 2005.

3 Sample Preparation Techniques to Isolate and Recover Organics and Inorganics

Separation methods form the basis of chemistry, and the definition of a pure chemical substance ultimately depends on separative operations.

—Arne Tiselius

The importance of sample preparation to TEQA is clearly indicated in the following story. This author was once approached by a student during the era when it became apparent that in the 1970s polychlorinated biphenyls (PCBs) had contaminated the striped bass that migrate up the Hudson River in New York to spawn every spring. Once the student learned that a gas chromatograph (GC) is used to measure the extent that fish are contaminated with PCBs and noticed the instrument on the bench in the corner of the laboratory, the student was curious as to exactly how a fish the size of a striped bass could be put into the injection port of the GC. The diameter of the injection port of the GC was less than 1 mm, which, of course, is minuscule in comparison to the size of the fish. The student thought that all that was necessary was to find a way to get the fish into the injection port and the data, which at that time was displayed on a strip-chart recorder, would indicate the extent of this PCB contamination. The student speculated that it might be easier to cut the fish up and attempt to stuff it into the injection port on the GC. Ah, we see for the first time, in this student, a glimpse into the need for sample preparation.

Indeed, the fish must be transformed in some manner prior to measurement by a determinative technique—in this case, by gas chromatography. Determinative techniques utilize instrumental analysis approaches and are discussed in Chapter 4. The removal of the PCB from fish tissue (known as the sample matrix) to a form that is compatible with the determinative technique or specific analytical instrument—in this case, the GC—is the basis for sample preparation. The GC requires the introduction of a solvent that contains the dissolved solute—in this case, PCBs. A gas can also be injected into the GC. However, it is much more convenient to get the PCBs from the sample matrix to the liquid state. The liquid is quickly vaporized under the elevated temperature of the GC injection port and undergoes GC separation. The number of molecules of each chemically different substance now present in the vapor causes a perturbation in the GC detector. This perturbation results in an electrical signal whose magnitude becomes proportional to the number of molecules present in the liquid.

This chapter introduces the various techniques that are commonly used to prepare environmental samples and animal and human specimens and comprises an important component of TEQA. The laboratory approach used to "get the striped bass into the machine" to achieve the utmost goal of TEQA (i.e., to isolate, identify, and quantitate the PCBs in the sample matrix) defines sample preparation. This chapter starts out with the most common and most conceptually simplistic form of sample preparation, whereby a liquid such as water or a solid such as a soil is placed in a beaker or equivalent container. To this container is added an organic solvent that is immiscible with water. The mixture is shaken and allowed to remain stationary for a period, such as 15 min. The analytes originally dissolved in the water or adsorbed onto soil particulates are partitioned into the organic solvent. The organic solvent that now contains the dissolved analyte as a solute is referred to as the extractant. After the principles of liquid–liquid extraction (LLE) are introduced and developed, the practice of LLE in its various forms will be discussed.

In addition to LLE, there are, today, other types of analyte isolation and recovery methods that are well established. These *sample prep* methods include: reversed-phase, normal-phase, and ion-exchange solid-phase extraction (SPE), supercritical fluid extraction (SFE), stir-bar sorbent extraction (SBSE), solid-phase microextraction (SPME), matrix solid-phase dispersion (MSPD), and QUECHERS (Quick, Easy, Cheap, Effective, Rugged, Safe). SPE refers to those techniques that isolate the analyte from a sample matrix and partition the analytes of interest onto a chemically bonded silica, or polymeric surface. SFE refers to those techniques that isolate the analyte from a sample matrix and partition it into a supercritical fluid (CO_2) that has been heated and pressurized beyond its critical temperature and pressure. Since the early 1980s, *sample prep* method development in TEQA has involved significant experimental research to find alternatives to LLE. It is indeed quite over-simplistic to think that a striped bass can be stuffed into a GC as a means to conduct *enviro-chemical* TEQA!

3.1   WHAT ARE THE PRINCIPLES UNDERLYING LLE?

A good grounding in the basic principles of LLE is a useful way to begin a chapter that focuses on sample preparation for TEQA. LLE was historically the first sample preparation technique used in analytical chemistry. Organic chemists have used LLE techniques for over 150 years for isolating organic substances from aqueous solutions. A good definition of LLE has been given earlier in the literature and is stated here:

A substance distributes between contacting immiscible liquids—water and a suitable organic solvent, for example—roughly in the ratio of its solubility in each if it does not react with either and if it exists in the same form in both. If, at equilibrium, its concentration is much greater in the organic solvent phase than in the aqueous phase, the distribution behavior may be put to analytical use in concentrating the substance into a small volume of the organic liquid and, more importantly, in separating it from substances that do not distribute similarly.[1]

This definition of LLE is concise yet profound in that it covers all ramifications. The first sentence establishes two conditions: compounds that react with the extractant do not obey the rules, and the chemical nature of the compound needs to remain the same throughout the extraction. Mathematical relationships have also been developed to account for the fact that the chemical form may change. This has been called secondary equilibrium effects, and this topic will also be introduced in this chapter. The second sentence implies that a concentration factor can be realized. The concentrating nature of LLE is most important to TEQA. The fact that different chemical substances will distribute differently between immiscible liquids also forms the theoretical basis for separation among two or more organic substances that might be initially dissolved in the aqueous solution. These differences are exploited in the design of sample preparation schemes as well as providing for the fundamental basis to explain analyte separation by chromatography. Aqueous solutions are of prime importance to *enviro-chemical* TEQA because our sample matrix, if a liquid, consists of drinking water, surface (i.e., rivers) water, groundwater, or wastewater obtained from the environment. The sample matrix closest to that of aqueous solutions used in *enviro-health* TEQA is human or animal urine. Blood, serum, and plasma are, although partially aqueous, much more complex due to their biochemical origins. Proteins must be precipitated out of blood before implementing LLE. Refer to the *Glossary* for definitions of serum and plasma and the distinction between them. The fact that the chemical form can change

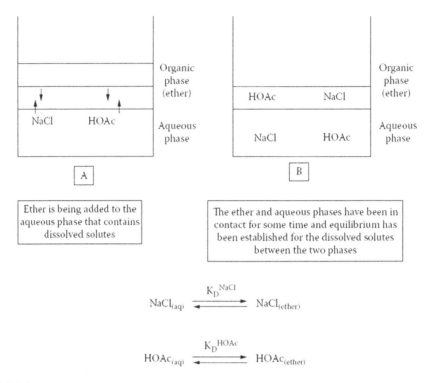

FIGURE 3.1

during the extraction process can be exploited in analytical chemistry toward the development of new methods to separate and isolate the analyte of interest.

To understand the most fundamental concept of liquid–liquid extraction, consider placing 100 mL of an aqueous solution that contains 0.1 M NaCl and 0.1 M acetic acid (HOAc) into a piece of laboratory glassware known as a separatory, or commonly abbreviated as a *sep funnel*.[2] Figure 3.1 shows a conceptually simplified LLE process. Figure 3.1A shows this process just prior to mixing the two immiscible phases. Next, 100 mL of diethyl ether, a moderately polar organic solvent that is largely immiscible with water, is added to the funnel. Indeed, some ether will dissolve in water to the extent of 6.89% at 20°C, while some water dissolves in the ether to the extent of 1.26% at 20°C.[3] Upon shaking the contents of the funnel and allowing some time for the two phases to separate and become stationary, the solute composition of each phase is depicted in Figure 3.1B. The lower layer is removed from the *sep funnel*, thus physically separating the two phases. Taking an aliquot (portion thereof) of the ether phase and separately taking an aliquot of the water phase while subjecting the aliquot to chemical analysis reveals a concentration of NaCl, denoted as [NaCl], at $1.0 \times 10^{-11}\ M$, and that in water, $[NaCl]_{aq} = 0.10\ M$. Analysis of each phase for acetic acid reveals $[HOAc]_{ether} = [HOAc]_{aq} = 5 \times 10^{-2}\ M$. Upon combining both phases again, a second chemical analysis of the composition of each phase reveals exactly the same concentration of HOAc and NaCl in each phase. As long as the temperature of the two phases in contact with each other in the sep funnel remain fixed, the concentration of each chemical species in both phases will not change with time. A dynamic chemical equilibrium has been reached. The significant difference in the extent of partitioning of NaCl and HOAc between diethyl ether and water-immiscible phases can be explained by introducing a thermodynamic viewpoint.

3.2   DOES THERMODYNAMICS EXPLAIN DIFFERENCES IN NaCl VS. HOAc PARTITIONING?

For spontaneous change to occur, the entropy of the universe must increase. The entropy of the universe continues to increase with each and every spontaneous process. LLE represents an ideally closed thermodynamic system in which solutes originally dissolved in an aqueous sample taken from the environment can diffuse across a solvent–water interface and spontaneously partition into the solvent phase. These concepts are succinctly defined in terms of the change in Gibbs free energy, G, for system processes that experience a change in their enthalpy H and a change in the entropy of the system S. The criterion for spontaneity requires that the Gibbs free energy, G, decrease. In turn, this free-energy change is mathematically related to a system's enthalpy H and entropy S. All three depend on the state of the system and not on the particular pathway, so a change in free energy at constant temperature can be expressed as a difference in the exothermic or endothermic nature of the change and the tendency of the matter in the system to spread according to

$$\Delta G = \Delta H - T \Delta S$$

This equation suggests that for spontaneous physical or chemical change to occur, the process proceeds with a decrease in free energy. As applied to phase distribution, equilibrium is reached when the infinitesimal increase in G per infinitesimal increase in the number of moles of solute i added to each phase becomes equal. Hence, the chemical potential of solute i is defined as

$$\mu = \left(\frac{\partial G}{\partial n_i} \right)_{T,P}$$

The chemical potential can also be expressed in terms of a chemical potential under standard-rate conditions μ^0 and the activity a for a solute in a given phase. Recognizing that a phase has an activity equal to unity (i.e., $a = 1$ defines the standard state at a given temperature and pressure), the equation for the chemical potential μ for an activity other than $a = 1$ is found according to

$$\mu = \mu^0 + RT \ln a \qquad (3.1)$$

Once equilibrium is reached, the net change in μ for the transfer of solute i between phases must be zero, so that for our example of NaCl or HOAc in the ether/water-immiscible phase illustration, the chemical potentials are equal:

$$\mu_{ether}^{NaCl} = \mu_{aq}^{NaCl} \qquad (3.2)$$

Hence, upon substituting Equation (3.1) into Equation (3.2) for solute i,

$$\mu_{ether}^0 + RT \ln a_{ether} = \mu_{aq}^0 + RT \ln a_{aq}$$

the above equality which rearranges to

$$RT \ln \left(\frac{a_{ether}}{a_{aq}} \right) = \mu^{0aq} - \mu_{ether}^0 \qquad (3.3)$$

The change in standard-state chemical potential, $\Delta\mu^0$, is usually expressed as the difference between the organic phase and the aqueous phase according to

$$\Delta\mu^0 = \mu^0_{ether} - \mu^0_{aq}$$

Solving Equation (3.3) for the ratio of solute activities gives

$$\frac{a_{ether}}{a_{aq}} = e^{-\Delta\mu^0}$$

Because $\Delta\mu^0$ is the difference of two constant standard-state chemical potentials, it must be a constant. The ratio of activities of NaCl or HOAc is fixed provided that the temperature and pressure are held constant.

A thermodynamic approach has just been used to show what is important analytically; that is, LLE enables an analyte to be transferred from the sample to the extracting solvent and remain at a fixed concentration over time in the extractant. This ratio of activities is defined as the thermodynamic distribution constant, K^0, so that

$$K^0 \equiv \frac{a_{ether}}{a_{aq}} \tag{3.4}$$

3.3   WHAT ARE SOLUTE ACTIVITIES ANYWAY?

A solute dissolved in a solvent such as water is only partly characterized by its concentration. Solute concentration can be expressed in one of any number of units. The most commonly used units include the following: moles solute per liter solution or molarity (M), moles solute/1000 g water or molality (m), and millimoles solute per liter solution or millimolarity (mM). Those units that have greater relevance to TEQA include the following: milligrams of solute per liter solution or parts per million (ppm), micrograms of solute per liter solution or parts per billion (ppb), and picograms of solute per liter solution or parts per trillion (ppt). Note that TEQA relies exclusively on expressing solute concentration in terms of a weight per unit volume basis. The fact that equilibrium constants in chemistry depend not only on solute concentration but also on solute activities serves to explain why any discussion of distribution equilibria must incorporate solute activities. Solute activities are introduced in any number of texts. (chapter 3)[1,4*] Activities become important when the concentration of an electrolyte in an aqueous solution becomes appreciable (i.e., at solute concentrations of 0.01 M and higher).

The extent to which a solution whose concentration of solute i contributes to some physical/chemical property of this solution (i.e., its activity, a_i) is governed by the solute's activity coefficient γ_i according to

$$a_i = \gamma_i c_i$$

For *enviro-chemical* TEQA, it is important to note that neutral molecules dissolved in water do not affect ionic strength and very dilute aqueous solutions are most likely found.

However, one aspect of TEQA that is strongly influenced by ionic strength, and hence provides an opportunity for activity coefficients to play a role, is the concept of *salting out*. The

---

* The concept of activity and activity coefficients is found in most physical and analytical chemistry texts that consider ionic equilibria. The texts listed in reference 1–7 are part of the author's personal library.

solubility of one chemical substance in another, like K^0 [Equation (3.4)] in LLE, is also governed by the need for the substance to lower its Gibbs free energy by dissolving in a solvent. Isopropyl alcohol (IPA) or 2-propanol is infinitely soluble in water, as is true for most lower-molecular-weight alcohols. However, for a solution that might consist of 50% IPA and 50% water, the alcohol can be separated out as a separate phase if enough NaCl is added to almost saturate the system. This is a direct influence of ionic strength in an extreme case. The fact that polar solvents can be separated as an immiscible phase opened up new sample preparation opportunities, e.g. refer to the QUECHERS technique later in this chapter. For example, Loconto and coworkers earlier demonstrated that the homologous series of polar 2-aminoethanols could be efficiently partitioned into IPA from an aqueous sample of interest to wood chemists.[8] The sample was saturated with NaCl, then extracted using IPA.

Two important relationships must be discussed that relate activity coefficients to ionic strength. Ionic equilibria are influenced by the presence of all ions in an aqueous solution. The most useful indicator of the total concentration of ions in a solution is the ionic strength, I. The ionic strength can be calculated if the concentration c_i of an ion whose charge is z_i is known according to

$$I = \frac{1}{2}\sum_i c_i z_i^2 \tag{3.5}$$

The summation is extended over all ions in solution. For example, consider two aqueous solutions, one containing 0.01 M NaCl and the other one containing 0.01 M K$_2$SO$_4$. Using Equation (3.5), the ionic strength for the former solution is calculated to be 0.01 M and that for the latter is 0.03 M. Assume that a solution is created that consists of 0.01 M in each salt. The ionic strength of such a mixture is calculated according to Equation (3.5) to be 0.04 M.

Knowledge of a solution's ionic strength enables a determination of the activity coefficient to be made. This can occur through the application of the Debye–Hückel equation according to

$$\log \gamma = \frac{-0.51 z^2 \sqrt{I}}{1 + \alpha \sqrt{I}/305}$$

where α refers to the size of the hydrated radius of the ion, and z is the charge of the ion. This equation gives good approximations for ionic strengths below or equal to 0.1 M. For ionic strengths less than 0.01 M, the following relationship suffices:

$$\log \gamma = 0.51 z^2 \sqrt{I}$$

3.4 CAN THE DIFFERENCE BETWEEN K^0 VALUES FOR NaCl AND HOAc BE SHOWN GRAPHICALLY?

The thermodynamic relationship between standard-state chemical potential differences and the position of chemical equilibrium can be shown graphically. Figure 3.2 illustrates what happens to the Gibbs free energy G when the solute is partitioned between an aqueous phase in contact with an immiscible organic phase, diethyl ether in this example. The hypothetical plots of G vs. the mole fraction, denoted by X_i, of solute i dissolved in the ether phase, are superimposed for comparison. When there is no solute in the ether phase, a standard-state chemical potential, μ_{aq}^0, can be realized. In the other extreme, when 100% of all of the mass solute is in the ether phase (i.e., having a mole fraction $X_i^{ether} = 1$), a standard-state chemical potential μ_{ether}^0, can also be defined. The situation at $X_i^{ether} = 1$ is a hypothetical one in that for some solutes, 1 mole of solute cannot dissolve to that extent in an organic solvent like ether. This is particularly true for an ionically bonded substance such as sodium chloride. Imagine if this much NaCl could dissolve in ether.

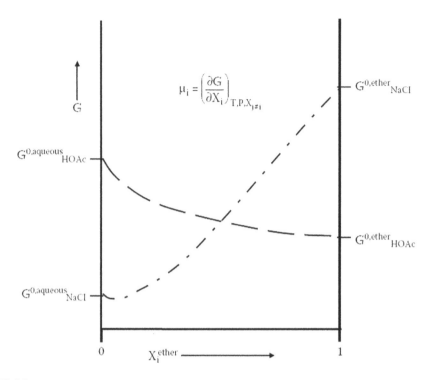

FIGURE 3.2

The free energy that would be required to dissolve as much as 1 mol NaCl in 1 L of ether would be expected to be extremely large indeed.

Such is not the case when considering the free energy required for the dissolution of 1 mole of HOAc in 1 L of ether. The mole fraction of solute partitioned into the ether at equilibrium is that point along the x-axis where G is at a minimum, or in other words, the slope of the tangent line (i.e., dG/dX_i) is zero. It becomes quite evident when viewing this graphical display that the magnitudes of standard-state Gibbs free energies are chiefly responsible for the position along the x-axis where G reaches a minimum. At this position, the mole fraction of each solute becomes fixed as defined by Equation (3.3). Figure 3.2 shows that the Gibbs free energy is minimized at equilibrium for NaCl at a much lower mole fraction when compared to the value of the mole fraction for HOAc, where its Gibbs free energy is minimized. In other words, the value of X_i where dG/dX_i is minimized at equilibrium depends entirely on the nature of the chemical compound. If a third solute is added to the original aqueous solution, as depicted in Figure 3.1, it too would exhibit its own G-as-a-function-of-X_i plot and reach a minimum at some other point along the x-axis. These concepts render Equation (3.3) a bit more meaningful when graphically represented.

3.5 CAN WE RELATE K^0 TO ANALYTICALLY MEASURABLE QUANTITIES?

It becomes important to TEQA to relate the thermodynamic distribution constant, K^0, to measurable concentration of dissolved solute in both phases. Because the chemical potential for a given solute must be the same in both immiscible phases that are in equilibrium, Equation (3.2) can be rewritten in terms of activity coefficients and concentration according to

$$\mu_{ether}^0 + RT\ln C_{ether} + RT\ln\gamma_{ether} = \mu_{aq}^0 + RT\ln C_{aq} + \ln\gamma_{aq}$$

Upon rearranging and simplifying, we get

$$RT \ln \frac{C_{ether}}{C_{aq}} + RT \ln \frac{\gamma_{ether}}{\gamma_{aq}} = -\Delta\mu^0$$

This equation can be solved for the ratio of measurable concentration of solute in the ether phase to that of the water phase; this is shown by

$$\frac{C_{ether}}{C_{aq}} = \frac{\gamma_{ether}}{\gamma_{aq}} e^{-\Delta\mu^0/RT} \qquad (3.6)$$

If we define a partition constant K_D as a ratio of measurable concentrations of solute in both phases, we get

$$K_D \equiv \frac{C_{ether}}{C_{aq}} \qquad (3.7)$$

Upon substituting Equation (3.6) into Equation (3.7), we obtained the relationship between the partition ratio and the thermodynamic distribution constant according to

$$K_D = \frac{\gamma_{ether}}{\gamma_{aq}} K^0 \qquad (3.8)$$

Equation (3.8) is the desired outcome. In many cases, with respect to TEQA, the activity coefficients of solutes in both phases are quite close to unity. The partition ratio and thermodynamic distribution constant can be used interchangeably.

For either NaCl or HOAc, or for any other solute distributed between immiscible liquids at a fixed temperature and pressure, provided that the concentration of solute is low (i.e., for the dilute solution case), K^0 can be set equal to the partition constant K_D because activity coefficients can be set equal to 1. The partition constant or Nernst distribution constant in our illustration for acetic acid partitioned between ether and water can be defined as

$$K_D = \frac{[HOAc]_{ether}}{[HOAc]_{aq}}$$

From the analytical results for measuring the concentration of HOAc in each phase introduced earlier, K_D can be calculated:

$$5\times 10^{-2}\,M\,/\,5\times 10^{-2}\,M = 1$$

Likewise, from the analytical results for measuring the concentration of NaCl in each phase introduced earlier, K_D can be calculated:

$$5\times 10^{-11}\,M\,/\,1\times 10^{-1}\,M = 1\times 10^{-10}$$

3.6   IS LLE A USEFUL CLEANUP TECHNIQUE?

Two examples of how LLE is used not only to isolate the analyte of interest from possible interferences from the sample matrix but also to provide an important cleanup are now discussed. Both procedures, which were then incorporated into respective methods, yield an extract that is ideally free of interferences that can be used in the determinative step to quantitate the presence of analyte that was originally in the sample.

In the first case, an environmental sample that contains a high concentration of dissolved inorganic salts such as NaCl is suspected to contain trifluoroacetic acid (TFA). TFA is a known by-product from the recently understood persistence of fluorocarbons in the environment.[9] The physical and chemical properties of TFA are well known. When dissolved in water, TFA is a moderately strong carboxylic acid with a pK_a lower than that of acetic acid. TFA also has an infinite solubility in water. TFA is not directly amenable to detection by GC because it cannot be sufficiently vaporized in the hot-injection port of the GC. It is not good practice to make a direct aqueous injection into a GC that possesses a column that contains the commonly used silicone polymer as a liquid phase. Hence, it is necessary to prepare an analytical reference standard in such a way that (1) TFA can be made amenable to analysis by GC, and (2) extracts that contain TFA must be nonaqueous. TFA could be determined by a direct aqueous injection if a different instrumental technique is used. The options here include either high-performance liquid chromatography (HPLC) in one of its several forms, ion chromatography (IC), or capillary electrophoresis (CE). There is a gain, however, if a sample preparation technique can be developed that concentrates the sample. Wujcik et al.'s group took the following approach to the determination of TFA in environmental waters.[10]

The highly salted aqueous sample that is expected to contain the targeted analyte TFA is initially acidified to suppress the ionization of the acid according to

$$CF_3COOH_{(aq)} \quad \overset{\leftarrow}{\leftrightarrow} \quad H^+_{(aq)} + CF_3COO^-_{(aq)}$$

where the subscript (aq) refers to the fact that each ionic species is dissolved in water and is surrounded by water dipoles. The triple-headed double-arrow denotes that when TFA is initially dissolved in water, a dynamic chemical equilibrium is quickly established whereby hydronium and trifluoroacetate ions exist in water along with undissociated TFA. Upon acidifying, the extent of this ionization is suppressed and a new equilibrium concentration of hydronium, trifluoroacetate, and TFA is reestablished with significantly higher concentration of TFA and hydronium ion and a much lower concentration of trifluoroacetate. Refer to any number of texts that elaborate on the principles of ionic equilibrium that govern the extent of acid dissociation.[11–15]*

The acidified aqueous environmental water sample is then extracted with a nonpolar solvent such as hexane, iso-octane, dichloromethane (methylene chloride), or some other common water-immiscible solvent. TFA is partitioned into the extractant to an appreciable extent owing to the fact that its ionization has been suppressed in the aqueous phase and the trifluoromethyl moiety gives a hydrophobic character to the molecule. The inorganic salts are left behind in the aqueous phase. Upon physically separating the phases and placing the organic phase in contact with a second aqueous phase that has been made alkaline or basic by the addition of NaOH or KOH, TFA molecules diffuse throughout the bulk of the extractant toward the interfacial surface area where they are ionized according to

$$CF_3COOH_{(aq)} \quad \overset{OH^-}{\underset{\leftrightarrow}{\rightarrow}} \quad H^+_{(aq)} + CF_3COO^-_{(aq)}$$

---

* In addition to the reference sources cited in reference 1–7, a number of texts on water chemistry discussing ionic equilibria and a recent book are listed in reference 8.

After the rate of TFA transport through to the interface from the bulk extractant and into the alkaline aqueous phase becomes equal to the rate of TFA from the bulk alkaline aqueous phase through to the extractant and equilibrium is reestablished, a new partitioning occurs, with most of the original TFA now in the alkaline aqueous phase. The cleanup has been accomplished because the aqueous phase contains TFA, as it conjugate base, without any dissolved inorganic salts. The alkaline aqueous matrix is then passed through a disk that contains anion exchange sites whereby trifluoroacetate can be retained by the ion exchange interaction. The disk is then placed into a 22 mL headspace vial containing 10% sulfuric acid in methanol and the vial is sealed tightly. Heating at 50°C for a finite period converts TFA to its methyl ester. The headspace, which now contains methyl trifluoroacetate, is sampled with a gas-tight GC syringe and injected into a GC. The headspace technique eliminates any solvent interference.

The second case, taken from the author's own work, uses LLE to initially clean up an aqueous sample taken from the environment that might contain, in addition to the analyte of interest, other organic compounds that may interfere in the determinative step.[16] The analytes of interest are the class of chlorophenoxy acid herbicides (CPHs) and include 2, 4-dichlorophenoxyacetic acid (2, 4-D), 2, 4, 5-trichlorophenoxyacetic acid (2, 4, 5-T), and 2, 4, 5-trichlorophenoxy propionic acid (Silvex). CPHs are used as herbicides in agricultural weed control, and because of this, CPHs are routinely monitored in drinking water supplies. CPHs are usually produced as their corresponding amine salts or as esters. An initial alkaline hydrolysis of the sample is needed to convert the more complex forms to their corresponding conjugate bases.

The sample is then extracted using a nonpolar solvent. This LLE step removes possible organic interferences while leaving the conjugate bases to the CPHs in the aqueous phase. Cleaned-up alkaline aqueous phase results can now be acidified and either re-extracted (LLE) or passed through a chemically bonded solid sorbent to isolate the CPHs, and possibly achieve a concentration of the CPHs from that in the original sample. As was true in the first case, the ionizable properties of these analytes can be exploited to yield a clean extract that can be quantitatively determined. Between 95 and 100% recoveries for the three CPHs cited were obtained from water spiked with each CPH. No influence of these high-percentage recoveries upon inserting an initial LLE step was observed.[16]

In contrast, the more conventional approach to trace CPH residue analysis serves to illustrate this difference in approaches to sample preparation. A water sample taken from the environment is initially acidified to preserve its chemical composition prior to sample preparation and analysis. At the onset of sample preparation, the water sample is made alkaline. To this alkaline aqueous phase, nonpolar solvent is added and the immiscible phases are shaken in a glass separatory funnel. Esters of CPHs, being nonpolar themselves, obey the universal principle that like dissolves like and partition into the organic phase. The free CPH acids remain in the aqueous phase. If only the formulated esters of CPHs are of interest, the extract can be cleaned up and analyzed. However, if it is desirable to convert the esters to acids, as is the case in most regulatory methods, a base hydrolysis is conducted on the organic phase that converts these CPH esters to their corresponding salts. The aqueous phase is reacidified and a second LLE is performed. The extracted CPHs are derivatized and converted to their corresponding methyl esters using any of the more common derivatization reagents. Following a cleanup step, the extract is ready for injection into a GC with a chlorine-selective detector such as an electron-capture detector (ECD) or an electrolytic conductivity detector (E1CD). This approach to sample preparation is a good example of the complexity involved in many of the methods of TEQA. If 1 L of an environmental water sample is taken through this method, it is likely that a concentration of 10 ppb 2, 4-D originally present in the sample can be separated from other CPHs, identified, detected, and quantified using all of the techniques available in TEQA.

These two examples clearly demonstrate the importance of secondary equilibria phenomena, particularly when the analyte of interest is ionizable in an environmental aqueous sample such as groundwater. Both examples exploit secondary equilibria in developing alternative methods that include LLE in extraction and in cleanup when applied to the complex sample matrices commonly encountered in TEQA. In the next section, the mathematical framework that underlies secondary equilibria will be presented.

3.7 HOW DO WE ACCOUNT FOR SECONDARY EQUILIBRIA IN LLE?

Let us return to the ether/aqueous-immiscible distribution equilibrium model introduced earlier (refer to Figure 3.1). What if the aqueous solution, prior to adding any ether, was made alkaline by the addition of NaOH? We know that the chloride ion concentration in the original aqueous solution would not change, but what about the HOAc? We also know that acetic acid is a weak acid and undergoes dissociation to hydronium ions and acetate ions. The extent of this dissociation is governed by the dissociation constant, K_a. The triple-head double-arrow notation is used in the following reaction to show that prior to the addition of hydroxide ion to an aqueous solution that contains dissolved acetic acid, the ionic equilibrium is already established:

$$CH_3COOH + H_2O \overset{OH^-}{\underset{\leftrightarrow}{\rightarrow}} H_3O^+ + CH_3COO^-$$

The effect of the added hydroxide ion is to shift the position of equilibrium to favor the product acetate, and thus to remove HOAc from the aqueous phase. HOAc molecules in the ether phase partition back to the aqueous phase until chemical potentials become equivalent and the magnitude of K_D is restored to the same value that the system had before the addition of the hydroxide ion.

Does this pH adjustment have any effect on the partitioning of HOAc between immiscible phases? By definition, only neutral HOAc can partition between phases. The value for the partition ratio must be preserved based on the thermodynamic arguments put forth earlier. This must mean that the concentrations of HOAc in the ether phase must be reduced due to the pH adjustment because the concentration of undissociated HOAc in the aqueous phase has also been reduced. This is illustrated for the HOAc only, in Figure 3.3. Our model assumes that the only chemical form of acetic acid in the ether phase is HOAc and that only acid dissociation of HOAc occurs in the aqueous phase. Because K_D accounts only for undissociated forms of acetic acid, a new constant is needed to completely account for the undissociated acetic acid and the acetate ion. This constant is called the distribution ratio, D, and is defined according to

$$D \equiv \frac{\sum_j [A]_o}{\sum_k [A]_{aq}} \tag{3.9}$$

where $[A]_o$ refers to the concentration or activity of the jth chemical species in the organic or extractant phase, and $[A]_{aq}$ refers to the concentration or activity of the kth chemical species in the aqueous phase.

The magnitude of D enables one to understand the extent to which all chemical forms of the analyte of interest are partitioned between two immiscible phases. D accounts for all secondary equilibrium effects that occur. Let us go back to the concept of acetic acid partitioning between diethyl ether and water while considering the influence of the secondary equilibrium, that of weak acid dissociation due to an adjustment of the pH of the aqueous phase. This discussion will help us enlarge the scope of LLE and set the stage for further insights into the role of secondary equilibrium.

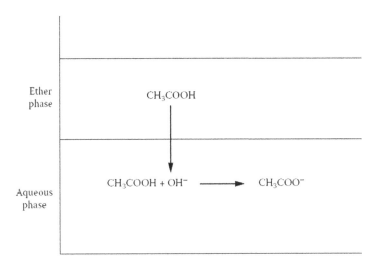

FIGURE 3.3

We start by using Equation (3.9) to define the different chemical species that are assumed to be present, and then we proceed to substitute secondary equilibrium expressions governed by acid–base dissociation constants or metal chelate formation constants. In the case of HOAc that is partitioned between ether and water, let us assume that only monomeric forms of HOAc exist in both phases and define the distribution ratio, D, according to

$$D = \frac{\left[HOAc\right]_{ether}}{\left[HOAc\right]_{aq} + \left[OAc^-\right]_{aq}} \tag{3.10}$$

Acetic acid dissociates in pure water to the extent determined by the magnitude of the acid dissociation constant, K_a. Based on the law of mass action, K_a is defined as

$$K_a = \frac{\left[H^+\right]\left[OAc^-\right]}{\left[HOAc\right]} \tag{3.11}$$

Let us solve Equation (3.11) for the acetate ion concentration that is in equilibrium with the hydronium ion, H^+, and undissociated HOAc:

$$\left[OAc^-\right] = K_a \frac{\left[HOAc\right]}{\left[H^+\right]}$$

Substituting for [OAc$^-$] in Equation (3.10) and simplifying yields a fruitful relationship:

$$D = \frac{\left[HOAc\right]_{ether}}{\left[HOAc\right]_{aq} + K_a \dfrac{\left[HOAc\right]_{aq}}{\left[H^+\right]}}$$

This expression can be further rearranged by factoring out the ratio of both molecular forms of HOAc:

$$D = \frac{[HOAc]_{ether}}{[HOAc]_{aq}} \left[\frac{1}{1 + K_a / [H^+]} \right]$$

This gives an expression for D in terms of a ratio of concentrations in both phases for the undissociated acid forms, which is exactly our definition of the distribution constant for the partitioning of HOAc between ether and water. Expressing D in terms of K_D yields an important relationship:

$$D = K_D \left[\frac{1}{1 + K_a / [H^+]} \right] \tag{3.12}$$

Equation (3.12) clearly shows the dependence of the distribution ratio on the secondary equilibrium (i.e., the weak acid dissociation) and on the extent of the primary equilibrium (i.e., the partitioning equilibrium of molecular HOAc between two immiscible phases). If Equation (3.12) is rearranged, we get

$$D = \frac{K_D [H^+]}{K_a + [H^+]} \tag{3.13}$$

A plot of D vs. $[H^+]$ is shown in Figure 3.4. The graph is hyperbolic, and upon careful examination, it would appear to resemble the Michaelis–Menten enzymes kinetics found in biochemistry.[17] The plot in Figure 3.4 as well as Equation (3.12) show that in the limit as the hydronium ion concentration gets very large, K_a becomes small in comparison to $[H^+]$, and in the limit of a very large hydronium ion concentration, the following can be stated: in the limit as

$$[H^+] \to \infty$$

it is evident that

$$D \to K_D$$

The partition constant, K_D, and the acid dissociation constant, K_a, for acetic acid can be found experimentally from a plot of D vs. $[H^+]$, as shown in Figure 3.4. Let $D = \frac{1}{2} K_D$ in Equation (3.13) so that

$$\frac{1}{2} K_D = K_D \left[\frac{[H^+]}{K_a + [H^+]} \right]$$

Eliminating K_D and solving this equation for K_a gives

$$K_a = [H^+]$$

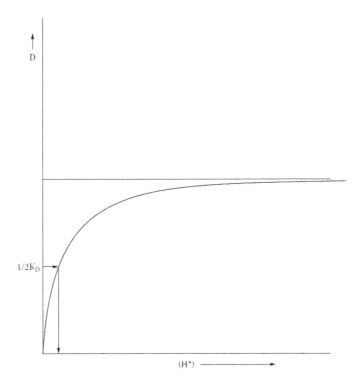

FIGURE 3.4

Hence, the acid dissociation constant for HOAc could be calculated. One would need to know experimentally exactly how D varies with the concentration of hydronium ion for this LLE in order to prepare a precise plot. It becomes difficult to estimate K_D from the hyperbolic curve shown in Figure 3.4. Equation (3.13) can be rearranged by taking reciprocals of both sides and rewriting Equation (3.13) in the form of an equation for a straight line of form $y = mx + b$, where m is the slope and b is the y intercept:

$$\frac{1}{D} = \left(\frac{K_a}{K_D}\right)\frac{1}{\left[H^+\right]} + \frac{1}{K_D}$$

A plot of $1/D$ vs. $1/[H^+]$ yields a straight line whose slope m is equal to the ratio K_a/K_D, and the y intercept b is equal to $1/K_D$. In this manner, both equilibrium constants can be determined with good precision and accuracy.[17]

Alternatively, Equation (3.12) can be viewed in terms of the primary equilibrium as represented by K_D and in terms of secondary equilibrium as represented by α_{HOAc}. Let us define α_{HOAc} as the fraction of neutral or undissociated HOAc present according to

$$\alpha_{HOAc} = \frac{\left[HOAc\right]_{aq}}{\left[HOAc\right]_{aq} + \left[OAc^-\right]_{aq}}$$

Upon simplifying, it can be shown that Equation (3.12) can be rewritten as

$$D = K_D \alpha_{HOAc}$$

Upon examination of this relationship among D, K_D, and α_{HOAc}, it becomes evident that the distribution ratio depends on the extent to which a solute (in our example, acetic acid) distributes itself between two immiscible phases (e.g., ether and water). At the same time, this solute is capable of exhibiting a secondary equilibrium (i.e., that of acid dissociation in the aqueous phase), as determined by the fraction of all acetic acid that remains neutral or undissociated. We will introduce this concept of fractional dissociation as just defined when we discuss LLE involving the chelation of transition metal ions from an aqueous phase to a water-immiscible organic phase.

3.8 WHAT IF THE CHEMICAL FORM OF HOAc CHANGES IN THE ORGANIC PHASE?

The above formalism assumed that only the monomeric form of HOAc exists in the ether phase. Carboxylic acids are known to dimerize in organic solvents that have a low dielectric constant. Let us assume we have acetic acid forming a dimer in the organic phase. This tendency may be more prominent if HOAc is dissolved in a nonpolar solvent like hexane, as compared to a moderately polar solvent like diethyl ether. The formation of a dimer can be depicted by

$$2\,\text{HOAc} \quad \xleftarrow{\quad K_{dim} \quad} \quad (\text{HOAc})_2$$

The extent to which the dimer is favored over that of the monomer is determined by the magnitude of K_{dim}. This added secondary equilibrium, this time appearing in the organic phase, is shown in Figure 3.5. The fundamental basis for the partitioning of HOAc between ether and water as introduced by the Nernst law is not violated and still is given by K_D. The measurable concentrations $[\text{HOAc}]_{ether}$ and $[\text{HOAc}]_{aq}$ will definitely differ with this added dimerization reaction. Let us define D for this distribution equilibrium involving weak acid dissociation of HOAc in the aqueous phase and, at the same time, dimerization of HOAc in the ether phase as follows:

$$D = \frac{[\text{HOAc}]_{ether} + 2\left[(\text{HOAc})_2\right]_{ether}}{[\text{HOAc}]_{aq} + [\text{OAc}^-]_{aq}} \tag{3.14}$$

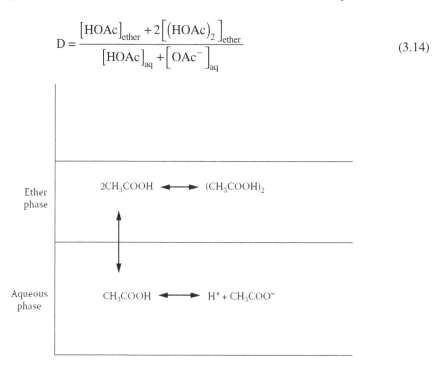

FIGURE 3.5

The following expressions are applicable to this distribution equilibrium and are defined as follows:

$$K_D = \frac{[HOAc]_{ether}}{[HOAc]_{aq}}$$

$$K_{dim} = \frac{\left[(HOAc)_2\right]_{ether}}{[HOAc]_{ether}^2}$$

$$K_a = \frac{[H^+]_{aq}[OAc^-]_{aq}}{[HOAc]_{aq}}$$

Substituting the above three definitions into Equation (3.14) and simplifying yields the following relationship:

$$D = \frac{K_D\left\{1 + 2K_{dim}[HOAc]_{ether}\right\}}{1 + K_a / [H^+]_{aq}} \tag{3.15}$$

Equation (3.15) shows that the value of the distribution ratio, D, depends not only on the equilibrium constants as indicated and the pH, but also on the concentration of HOAc in the ether phase.

It becomes instructive to compare Equations (3.12) and (3.15). The influence of dimerization in the organic phase results in an additional term in the numerator for the distribution ratio, D. This additional term depends on the magnitude of K_{dim} and the concentration of HOAc in this phase. In the case of HOAc, values for K_{dim} range from a high of 167 for benzene as the solvent to a low of 0.36 for diethyl ether as the solvent.[18] The larger the value for K_{dim}, the larger is the magnitude of D and, as we shall see in the next section, the higher is the percent recovery.

3.9   IF WE KNOW D, CAN WE FIND THE PERCENT RECOVERY?

The discussion so far has focused on first establishing the validity of the partition constant, K_D, for LLE and then extending this to the distribution ratio, D. We have shown that setting up expressions involving D becomes more useful when secondary equilibria exist. Before we consider other types of secondary equilibria, the importance of knowing how D relates to the percent recovery, %R, will be discussed. Percent recovery is an important QC parameter when LLE, SPE, and SFE techniques are used. Most EPA methods discussed in Chapter 1 require that the %R be measured for selected analytes in the same matrix as that for samples. This is particularly important as applied to the EPA methods for trace organics. In Chapter 2, we showed how %R is used in the statistical treatment of experimental data.

The determination of %R is paramount in importance toward establishing an alternative method in TEQA. A method that isolates phenol from wastewater samples using LLE and yields a consistently high %R is preferable to an alternative method that yields a low and inconsistent %R. As we showed in Chapter 2, a high %R leads to lower method detection limits (MDLs). However, if the alternative method significantly reduces sample preparation time, then a trade-off must be taken into account: lower MDLs vs. a long sample prep time. A practical question naturally arises here. What does the client want and what degree of trade-off is the client willing to accept?

Let C_0 represent the concentration of a particular analyte of interest after being extracted into an organic solvent whose volume is V_0 from an aqueous sample whose volume is V_{aq}. Assume also that the concentration of analyte that remains in the aqueous phase after extraction is C_{aq}. Let us define *the fraction of analyte extracted, E,* by

$$E \equiv \frac{amt_o}{amt\,(total)}$$

where amt_o refers to the amount of analyte extracted into the organic phase and amt(total) refers to the total amount of analyte originally present in the aqueous sample. The fraction extracted can be expressed as follows:

$$E = \frac{C_0 V_o}{C_0 V_o + C_{aq} V_{aq}} = \frac{D\beta}{1 + D\beta} \tag{3.16}$$

where β is defined as the ratio of the volume of the organic phase, V_o, to the volume of the aqueous phase, V_{aq}, according to

$$\beta = V_o \,/\, V_{aq}$$

The percent recovery is obtained from the fraction extracted, E, according to

$$\%\,\text{Recovery}\,(\%\,R) = E \times 100$$

Equation (3.16) shows that the fraction extracted and hence the percent recovery depend on two factors. The first is the magnitude of the distribution ratio, which is dependent on the physical/chemical nature of each analyte and the chemical nature of the extractant. The second factor is the phase ratio β. The magnitude is usually fixed if the extractant is not changed, whereas the phase ratio can be varied. If, instead of a single-batch LLE, a second and third successive LLE is carried out on the same aqueous solution by removing the extractant and adding fresh solvent, the %R can be maximized. After allowing time for partition equilibrium to be attained, while keeping the phase ratio constant, it can be shown that a second successive extraction will extract $E(1 - E)$ while a third successive extraction will extract $E(1 - E)^2$. The fraction remaining in the aqueous phase following n successive LLEs is $(1 - E)^{n-1}$. To achieve at least a 99% recovery, Equation (3.16) suggests that the product βD must be equal to or greater than 100. Even with a product βD = 10, two successive LLEs will remove 99% of the amount of analyte originally in an aqueous environmental sample. (pp. 255–257)[6]

3.10   ARE ORGANICS THE ONLY ANALYTES THAT WILL EXTRACT?

Our examples so far have focused on neutral organic molecules such as molecular acetic acid dissolved in water at a pH of < 7. At this lower pH, the ionization or weak acid dissociation of acetic acid, a secondary equilibrium process is negligible. The majority of priority pollutant organics of importance to TEQA are *acid–base neutral molecules* dissolved in *aqueous environmental samples*. Refer back to Figure 1.13. Note how all organic pollutants were initially classified as either VOCs, SVOCs or OCs/PCBs. *Semivolatile organic compounds* (SVOCs) are the principal class of pollutants amenable to LLE and LSE *sample prep* methods. Consider the significant difference in K_D for NaCl vs. HOAc partition constants discussed earlier. Ionic compounds have little to no tendency to partition into a moderate to nonpolar organic solvent. If, however, an ion can be converted to a neutral molecule via chemical change, this ion can exhibit a favorable K_D. This is accomplished in two ways: chelation of metal ions and formation of ion pairs. The mathematical development of a metal chelate is discussed in this section.

A number of organic chelating reagents exist that coordinate various metal ions, and the metal chelate that results consists of neutral molecules. This neutral or uncharged metal chelate will have a K_D much greater than 1. Metal ions initially dissolved in an aqueous phase such as a groundwater sample can be effectively removed by metal chelation LLE. Commonly used chelating reagents include four-membered bidentate organic compounds such as dialkyl dithiocarbamates, five-membered bidentates such as 8-hydroxyquinoline and diphenyl thiocarbazone, dithizone, and polydentates such as pyridylazonaphthol. 8-Hydroxyquinoline, commonly called oxine (HOx), is the chelating reagent used in this section to introduce the mathematical relationships for metal chelation LLE. Similar equations can be derived for other chelating reagents.

Figure 3.6 depicts the principal primary and secondary equilibria that would be present if oxine is initially dissolved in an appropriate organic solvent that happens to be less dense than water. If this solution is added to an aqueous solution that contains a metal ion such as copper (II) or Cu^{2+}, two immiscible liquid phases persist. The copper (II) oxinate that initially forms in the aqueous phase, oxine, itself is an amphiprotic weak acid and quickly partitions into the organic phase. Being amphiprotic means that oxine itself can accept a proton from an acid and can also donate one to a base. The degree to which oxine either accepts or donates a proton is governed by the pH of the aqueous solution. The acidic property is the only one considered in the development of the equations considered below. The formation of a Cu oxine chelate can proceed via 1:1 and 1:2 stoichiometry. The fact that it is the 1:2 chelate that is neutral, and therefore the dominant

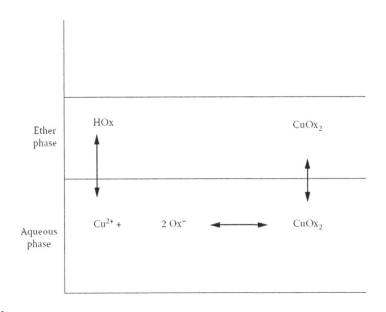

FIGURE 3.6

form that partitions into the nonpolar solvent, is important. All of the competing primary and secondary equilibria can be combined to yield a relationship that enables the distribution ratio to be defined in terms of measurable quantities.

The distribution ratio, D, for the immiscible phases and equilibria shown in Figure 3.6 is first defined as the ratio of chelated copper in the organic phase to the concentration of free and chelated copper in the aqueous phase. Expressed mathematically,

$$D \equiv \frac{\left[CuOx_2\right]_o}{\left[Cu^{2+}\right]_{aq} + \left[CuOx_2\right]_{aq}}$$

Similar to what was done earlier for HOAc, we can define α_{Cu} as the fraction of free Cu^{2+} in the aqueous phase: then,

$$\alpha_{Cu} = \frac{\left[Cu^{2+}\right]_{aq}}{\left[Cu^{2+}\right] + \left[CuOx_2\right]_{aq}}$$

so that

$$D = \frac{\left[CuOx_2\right]_o}{\left[Cu^{2+}\right] / \alpha_{cu}} \tag{3.17}$$

Use of α_{Cu} is a simple and convenient way to account for all of the many side reactions involving the metal ion. Substituting the equilibrium expressions into Equation (3.17) yields

$$D = \frac{K_D^{CuOx_2} \beta_2 K_a^2}{K_D^{HOx}} \frac{\left[HOx\right]_o^2}{\left[H^+\right]_{aq}} \tag{3.18}$$

We have assumed that the protonation of HOx as discussed earlier is negligible. Equation (3.18) states that the distribution ratio for the metal ion chelate LLE depends on the pH of the aqueous phase and on the ligand concentration. K_D^{HOx}, β_2, and α are dependent on the particular metal ion. This enables the pH of the aqueous phase to be adjusted such that a selected LLE can occur. One example of this selectivity is the adjustment of the pH to 5 and extraction as their dithizones to selectivity separate Cu^{2+} from Pb^{2+} and Zn^{2+} (pp. 615–617).[4]

The metal chelate LLE was much more common 25 years ago when it was the principal means to isolate and recover metal ions from aqueous samples of environmental interest. The complexes were quantitated using a visible spectrophotometer because most complexes were colored. A large literature exists on this subject.[19] The technological advances in both atomic absorption and inductively coupled plasma-atomic emission spectroscopy have significantly reduced the importance of metal chelate LLE to TEQA. However, metal chelate LLE becomes important in processes whereby selected metal ions can be easily removed from the aqueous phase.

3.11   CAN ORGANIC CATIONS OR ANIONS BE EXTRACTED?

We have discussed the partitioning of neutral organic molecules from an aqueous phase to a nonpolar organic solvent phase. We have discussed the partitioning of metal ions once they have been converted to neutral metal chelates. In this section, we discuss the partitioning of charged organic cations or charged organic anions. This type of LLE is termed ion pairing. Ion pair LLE is particularly relevant to TEQA, as will be shown below. We start by using equilibrium principles and assume that the only equilibria are the primary ones involving the partitioning of the ion pair between an aqueous phase and a lighter-than-water organic phase. The secondary equilibria consist of formation of the ion pair in the aqueous phase. Also, all cations and anions are assumed not to behave as weak acids or bases. For the formation of the ion pair in the aqueous phase, we have

$$C^+_{(aq)} + A^-_{(aq)} \overset{K_{IP}}{\leftrightarrow} CA_{(aq)}$$

The ion pair CA, once formed, is then partitioned into an organic solvent that is immiscible with water according to

$$CA_{(aq)} \overset{K_D}{\leftrightarrow} CA_{(org)}$$

The distribution ratio, D, with respect to the anion for IP-LLE, is defined as

$$D_{A^-} = \frac{[CA]_{org}}{[CA]_{aq} + [A^-]_{aq}}$$

In a manner similar to that developed earlier, D can be rewritten as

$$D_{A^-} = K_D \left\{ \frac{K_{IP}[C^+]}{1 + K_{IP}[C^+]} \right\} \tag{3.19}$$

The distribution ratio is seen to depend on the partition coefficient of the ion pair, K_D, to the extent to which the ion pair is formed, K_{IP} and on the concentration of the cation in the aqueous phase. Equation (3.19) shows some similarity to Equation (3.13).

3.12   IS THERE AN IMPORTANT APPLICATION OF IP-LLE TO TEQA?

Equation (3.19) suggests that if an ion pair that exhibits a high partition coefficient, K_D, forms the ion pair to a great extent (i.e., has a large value for K_{IP}), then a large value for D enables an almost complete transfer of a particular anion to the organic phase. Of all the possible ion pair complexes that could form from anions that are present in an environmental sample, the isolation and recovery of anionic surfactants using methylene blue is the most commonly employed IP-LLE technique used in environmental testing labs today. The molecular structure of this ion pair formed a large organic anion that is prevalent in wastewater such as an alkyl benzene sulfonate, a common synthetic detergent, using a large organic cation such as methylene blue, as follows:

Tetrapropylenebenzenesulfonate anion
an example of an alkylbenzenesulfonate (ABS)

6-dodecylbenzenesulfonate anion
an example of a linear alkylbenzene sulfonate (LAS)

Methylene blue cation

This ion pair absorbs visible light strongly at a wavelength of 652 nm. Because a method that might be developed around this ion pair and its high percent recovery into a nonpolar solvent (a commonly used one is chloroform) is nonselective, a cleanup step is usually introduced in addition to the initial LLE step. Of all possible anion surfactants, sodium salts of C_{10} to C_{20} do not form an ion pair with methylene blue, whereas anionic surfactants of the sulfonate and sulfate ester types do. Sulfonate type surfactants contain sulfur covalently bonded to carbon, whereas the sulfate ester type of surfactant contains sulfur covalently bonded to oxygen, which in turn is covalently bonded to sulfur. A good resource on the analysis of surfactants in all of its forms, including some good definitions, was published earlier.[20] Those surfactants that form an ion pair and give rise to a high percent recovery are termed *methylene blue active substance* (MBAS). A microscaled version to the conventional method[21] for the determination of MBAS in wastewater is introduced as one of the student experiments discussed in Chapter 5.

3.13 ARE THERE OTHER EXAMPLES OF NONSPECIFIC LLE PERTINENT TO TEQA?

In Chapter 1, the determination of total petroleum hydrocarbons (TPHs) was discussed in relation to EPA method classifications. This method is of widespread interest in environmental monitoring, particularly as this relates to the evaluation of groundwater or wastewater contamination. There are several specific determinations of individual chemical components related to either gasoline, fuel oil, jet fuel, or lubricant oil that involve an initial LLE followed by a GC determinative step. Methods for these require LLE, possible cleanup followed by GC separation, and detection usually via a flame ionization detector (FID). Specific methods are usually required when the type of petroleum hydrocarbon is of interest. There is almost equal interest among environmental contractors for a nonspecific, more universal determination of the petroleum content without regard to chemical specificity. A sample of groundwater is extracted using a nonpolar solvent. The extracted TPHs are then concentrated via evaporation either by use of a rotary evaporator, Kuderna–Danish evaporative concentrator, or via simple distillation to remove the extracting solvent. The residue that remains is usually a liquid, and the weight of this oily residue is obtained gravimetrically. An instrumental technique that represents an alternative to gravimetric analysis involves the use of quantitative infrared (IR) absorption. If the extracting solvent lacks carbon-to-hydrogen covalent bonds in its structure, then the carbon-to-hydrogen stretching vibration could be used to quantitate the presence of TPHs. The most common solvent that emerged was 1, 1, 2-trichlorotrifluoroethane (TCTFE). With the eventual total phasing out of Freon-based solvents, the EPA has reverted back to the gravimetric determinative approach. It is not possible to measure trace concentrations of TPHs via quantitative IR using a hydro-carbon solvent, due to the strong absorption caused by the presence of carbon-to-hydrogen cova-lent bonds. The author maintains that labs could recycle and reuse the spent TCTFE without any release of this Freon-type solvent to the environment while preserving the quantitative IR determinative method. Only time and politics will determine which method will dominate in the future. Nevertheless, the technique of LLE to isolate and recover TPHs from water contaminated with oil remains important.

3.14   CAN LLE BE DOWNSIZED?

Yes and we demonstrate this by referring to some earlier reported and interesting research that reinforces the basic concepts of LLE. Jeannot and Cantwell[22] have introduced the concept of a true LLE that has been downsized to a microextraction scale. In the past, the concept of a mini-LLE (mLLE), as introduced by the EPA and promulgated through their 500 series of methods, was designed to conduct TEQA on samples from sources of drinking water. Method 508 required that 35 mL of groundwater or tap water be placed in a 40-mL vial and extracted with exactly 2 mL of hexane. Organochlorine pesticides such as aldrin, alachlor, dieldrin, heptachlor, and so forth, are easily partitioned into the hexane. A 1-µL aliquot is then injected either manually or via autosampler into a GC-ECD to achieve the goal of TEQA. As long as emulsions are not produced, this downsized version of LLE works fine. Wastewater samples are prone to emulsion formation, and this factor limits the scope of samples that can be extracted by this mini-LLE technique.

Cantwell's group has taken this scale down by a factor of about 20 to the 1 mL and below sample volume levels. They refer to the technique as micro (µ) LLE. Some interesting mathem-atical relationships that serve to reinforce the principles discussed earlier are introduced here. It does not matter whether an analyst uses a liter of groundwater sample, a milliliter, or even a microliter. The principles of LLE remain the same.

The principle of mass balance requires that the amount of a solute that is present in an aqueous sample (e.g., groundwater), $amt_{initial}$, remain mathematically equivalent to the sum of solute in both immiscible phases. Matter cannot escape, theoretically, that is. If the initial amount of a solute is distributed between two immiscible phases, an organic phase, o, and an aqueous phase, aq, mass balance considerations require that

$$amt_{initial} = amt_{aq} + amt_o$$

We now seek to relate the concentration of solute that remains in the aqueous phase after µLLE to the original concentration of solute, $C_{initial}$. For example, a groundwater sample that contains dissolved organochlorine pesticides such as DDT can be mathematically related to the partitioned concentrations in both phases according to

$$C_{initial} = \frac{V_{aq}C_{aq} + V_oC_o}{V_{aq}}$$

We would like to express the concentration of analyte in the organic phase, C_o, in terms of the initial concentration of analyte that would be found in groundwater, $C_{initial}$. Dividing through by V_{aq} and eliminating C_{aq} for LLE,

$$C_{aq} = \frac{C_o}{K_D}$$

gives

$$C_{initial} = \frac{V_{aq}\left[\dfrac{C_o}{K_D}\right] + V_oC_o}{V_{aq}}$$

Substituting for the phase ratio, β, dividing the numerator and denominator by V_{aq}, and rearranging gives

$$C_{initial} = C_o\left[\frac{1}{K_D} + \frac{\beta}{1}\right]$$

Rearranging and solving for C_o gives

$$C_o = C_{initial}\left[\frac{K_D}{1 + \beta K_D}\right] \tag{3.20}$$

Hence, the concentration of solute present in the organic phase can be directly related to the concentration of solute initially present in the aqueous groundwater sample, $C_{initial}$, provided the partition coefficient and phase ratio, β, are known. The reader should see some similarity between Equations (3.20) and (3.16). Equation (3.20) was derived with the assumption that secondary equilibrium effects were absent. This assumption is valid only for nonionizable organic solutes.

With these mathematical relationships presented, we can now discuss the experimental details. The end of a Teflon® rod was bored to make a cavity. A volume of 8 μL of a typical organic solvent such as *n*-octane was introduced into the cavity, and a cap and rod were fitted to a 1 mL cylindrical vial with a conical bottom, into which a magnetic stirrer has been placed. After the solvent was placed on top of the aqueous sample, the sample was stirred for a fixed period at a fixed temperature, 25°C. This enables the solute to diffuse into the organic solvent. A 1 μL aliquot of this extract is taken and injected into a GC for quantitative analysis. This μLLE yields a β of 0.008. As values of β get smaller and smaller, the second term in the denominator of Equation (3.20) tends to zero. For a fixed K_D and $C_{initial}$, a low value for β results in a higher value for C_o, and hence a higher sensitivity for this μLLE technique.

Once a sample preparation method has been established, the analytical methodology, so important to achieving GLP in TEQA, can be sought. The analytical outcomes discussed in Chapter 2 can now be introduced for this μLLE technique.

An internal standard mode of calibration was used to conduct quantitative analysis using the minivial technique just described. The analyte studied was 4-methyl acetophenone and the internal standard was n-dodecane. The slope of the linear calibration was 4.88 L/mmol, with a y intercept of zero and a coefficient of determination of 0.998.[22]

3.15 DOES THE RATE OF MASS TRANSFER BECOME IMPORTANT IN μLLE?

It could. The kinetics of LLE can also be developed. Kinetics become a more important consideration when aqueous and organic phases cannot be afforded maximum contact, as is the case when a large *sep funnel* is used. It is worthwhile to consider kinetics in the context of the μLLE technique developed by Ma and Cantwell.[23] The general-rate equation for LLE can be written in terms of a differential equation that relates the rate of change of the concentration of analyte in the organic phase, C_o, to a difference in concentration between the aqueous phase, $C_{aq}(t)$, and the organic phase $C_o(t)$ according to the following:

$$\frac{\partial}{\partial t} C_o = \frac{A}{V_o} \Gamma_o \left(K_D C_{aq}(t) - C_o(t) \right)$$

where A is the interfacial area and Γ_o is the overall mass transfer coefficient with respect to the organic phase (in units of cm/sec). Thus, the time dependence of solute concentration in the organic phase can be seen as

$$C_o = C_{o,equil} \left(1 - e^{-kt} \right) \tag{3.21}$$

$C_{o,equil}$ represents the concentration of solute in the organic phase after equilibrium has been reached. k is the observed rate constant (in units of sec^{-1}) and is given by

$$k = \frac{A}{V_o} \Gamma_o \left[K_D + 1 \right]$$

Combining Equations (3.20) and (3.21) leads to an expression that is significant to TEQA:

$$C_{aq,initial} = C_o(t) \left[\frac{1 + K_D \beta}{K_D \left(1 - e^{-kt} \right)} \right] \tag{3.22}$$

The term in brackets in Equation (3.22) is usually held constant, and this term is evaluated by extracting a reference aqueous solution where the concentration is known. The concentration must, of course, be in the linear region of the distribution isotherm for both sample and standard.

Cantwell and coworkers have recently extended their μLLE technique to include a back-extraction using a modification of the minivial discussed earlier. An organic liquid membrane that consists of n-octane confined to within a Teflon ring sits on top of 0.5 or 1 mL of an aqueous sample whose pH is approximately 13 and contains an ionizable analyte. If an amine is dissolved in water and the pH adjusted to 13, the amine would remain unprotonated and therefore neutral. A large K_D would be expected, and the amine should partition favorably into the n-octane. A 100 or 200 μL acidic aqueous phase with a pH of approximately 2 is placed on top of the liquid membrane. The amine is then protonated and back-extracted into the acidic aqueous phase.[24] A further enhancement utilizes a microliter liquid-handling syringe to suspend a drop of acidic aqueous phase within the *n*-octane phase. The syringe that now contains the back-extracted analyte can be directly inserted into the injection loop of a high-performance liquid chromatograph (HPLC).

3.16   IS THERE ANY OTHER WAY TO PERFORM LLE?

Yes, indeed. There are several alternatives to separatory funnel LLE, mini-LLE, and μLLE (just described). *Sep funnels* are limited to ~1,000 mL or less, while mini-LLEs are limited to the size of ~40 mL (such as a typical screw-top cylindrical vial). For aqueous environmental samples whose volume exceeds 1,000 mL, continuous LLE (C-LLE) is often more appropriate and convenient within which to conduct LLE. To illustrate, if a 2 L wastewater effluent sample is to be extracted, C-LLE would be the technique of choice. C-LLE requires a relatively large glass apparatus whereby the receiving pot can vary in size. C-LLE can be performed using a lighter-than-water extractant or a heavier-than-water extractant. Typical lighter-than-water extractants include various lower-molecular-weight alkanes such as *n*-hexane, while typical heavier-than-water extractants include various chlorinated solvents such as methylene chloride (dichloromethane).

The operational procedure for lighter-than-water C-LLE has been described from a manufacturer of C-LLE glassware as follows:

The aqueous phase to be extracted and a stirring bar are placed in a 24/40 round-bottom flask. The flask size (up to and including the 5 L) is chosen so that it is not more than 2/3–4/5 full of aqueous phase. The flask is then filled with the lighter-than-water extracting solvent and gentle stirring is started. The extractor and an efficient condenser are put into place and a small flask containing an additional portion of the lighter-than-water extracting solvent is connected to the side-arm and the solvent in the small flask heated above its boiling point. The solvent vapors distill up the side-arm and condense at the condenser. The condensed solvent runs down the center tube where it is passed with stirring, through the aqueous phase. The extracting solvent removes a small amount of material and separates from the water. Since the density of the extracting solvent is less than that of water, the solvent rises past the joint at the top of the flask containing the aqueous phase and, when it reaches the side-arm, it flows back to the distilling flask though the side-arm. The extracted material then remains in the distilling flask while the solvent is distilled, condensed and is used to extract again. "Fresh" solvent is thus used over and over. In this way, by allowing the extractor to operate for long periods, materials only slightly soluble in the organic solvent can be removed from the aqueous phase in very high yields and only a relatively small amount of extracting solvent need be used.[25]

Heavier-than-water C-LLE designs are operated similarly:

Some heavier-than-water extracting solvent and a stirring bar are placed in the flask that contains the aqueous phase to be extracted. A good rule-of-thumb is that the flask should be about 1/5–1/6 full of heavier-than-water extracting solvent. The extractor is put in place and, with the aid of a funnel whose stem extends below the side-arm of the extractor, the aqueous phase to be extracted is added. The aqueous phase will fill the flask and may move up the vigreux column past the lower return tube. A small flask containing the heavier-than-water extracting solvent is then connected to the side-arm of the apparatus and the solvent therein heated above its boiling point.[25]

In *enviro-chemical* TEQA, the sample matrix determines whether LLE involving immiscible solvents is to be used. If the sample is a solid, such as a contaminated soil or sediment, C-LLE gives way to the Soxhlet extraction apparatus. There have also been attempts to modify the conventional Soxhlet via miniaturization or instrumentation that pressurizes and heats the extracting solvent. A recent technique promulgated by the EPA is called pressurized fluid extraction. EPA Method 3545 from Update III of SW-846 has been developed to enable priority pollutant semivolatile organics to be isolated and recovered from soils, clays, sediments, sludges, and other solid waste. The former Dionex Corporation (now part of Thermo Fisher Scientific) earlier developed *accelerated solvent extraction*, whereby a much smaller volume of extraction solvent is used. The vial containing the sample and extracting solvent is both heated and pressurized. These extreme temperature and pressure conditions supposedly accelerate the rate at which equilibrium is reached in LLE. The conventional technique for isolating and recovering semivolatile organics from solid matrices is called *Soxhlet extraction*. Soxhlet extraction as an analytical sample preparation technique has been around for over 100 years. The principle of Soxhlet extraction, abbreviated S-LSE (because it is a liquid–solid extraction technique), will be discussed in the following section.

3.17   WHAT IS SOXHLET EXTRACTION ANYWAY?

A solid matrix of environmental interest, such as soil that is suspected of containing any of the more than 100 priority pollutant semivolatiles, is placed into a cellulosic thimble. Vapors from heating a volatile organic solvent rise and condense above the thimble. This creates a steady-state condition called reflux. The refluxed solvent condenses into the thimble and fills until it overflows back into the distilling pot. Reflux is a common technique in organic chemistry and serves to bring the S-LSE process to a fixed temperature. Thus, solutes of interest are able to partition between a fixed weight of contaminated soil and the total extractant volume. Usually, a series of six vessels with six separate heaters are available as a single unit whereby the incoming and outgoing water lines for the reflux condensers are connected in series. A large phase ratio is obtained. S-LSE is usually conducted for one sample over a period of 12 to 24 h. An overnight continuous S-LSE is quite common. After the extraction time has ended, the glass S-LSE vessel is cooled to room temperature. The extractant in the thimble is combined with the refluxed extractant in the distilling pot. The pot is removed, and due to the large volume of solvent required, the analyst may have over 300 mL of extractant. Common solvents used in S-LSE are, in general, those solvents that are moderate to nonpolar and possess relatively low boiling points. Methylene chloride and petroleum ether are the two most commonly used to conduct S-LSE. Due to the low boiling points, both solvents can be easily removed in the next step after S-LSE.

The extractant from S-LSE must be concentrated. If a low boiling solvent such as methylene chloride is used to conduct S-LSE, it is straightforward to remove this solvent by use of either a Kuderna–Danish (K-D) evaporative concentrator or a rotary evaporator. More contemporary K-D designs provide a means to recover the spent solvent. Earlier designs did not include a means to recover the solvent, and because of this, most solvents, such as methylene chloride, were evaporated to the atmosphere usually via a fume hood. A few boiling chips are usually added to the receiving tube so as to prevent "bumping" during the vigorous boiling step. Suspending the K-D vessel above a large boiling water bath where steam can be in contact with the glass surface serves to rapidly remove solvent. Solvent is removed until a volume between 1 and 5 mL is reached. An extractant is obtained whose concentration of the analytes of interest has been greatly increased. A K-D evaporative concentrator is shown in Figure 3.7. The concentrated extract is further cleaned up, depending on what matrix interferences might be present.

Numerous priority pollutant semivolatiles such as PAHs, various substituted phenols, substituted monoaromatics, and other hazardous chemicals are often extracted along with higher-molecular-weight aliphatic hydrocarbons. If these hydrocarbons are not removed prior to the determinative step (e.g., separation and quantitation by GC-MS), the MDL can be significantly increased. A peak in the GC for benzo(*a*)pyrene might be obscured because the peak might sit on top of an envelope of hydrocarbons!

Most research papers published during the past 10 to 15 years that seek to show the value of alternative approaches to S-LSE start by condemning S-LSE as too labor intensive and consuming too large a volume of extraction solvent. Outside of needing to use approximately 300 mL of volatile solvent, as mentioned earlier, this researcher does not share the view that merely

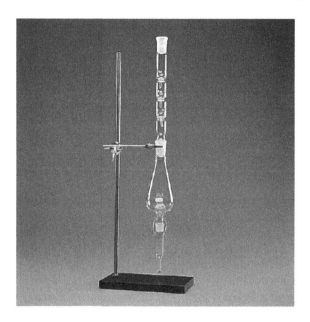

FIGURE 3.7

filling six thimbles with sample, reassembling the Soxhlet glassware, and turning on the heaters and coolant supply lines is really that time consuming. After all, once all of this is accomplished, the analyst is free to leave the laboratory and pursue other activities. All the while the sample is continuously refluxed with extractant. Nevertheless, a plethora of research has been done to supplant the classical low-cost Soxhlet, as is discussed next. Before we do that, however, let us cover automated Soxhlet extraction.

EPA Method 3540 uses conventional Soxhlet extraction (S-LSE) to isolate and recover various semivolatile and nonvolatile priority pollutant organic compounds from soil, sediment, sludges, and waste solids. EPA Method 3541 utilizes a unique, three-stage *automated Soxhlet extraction*, (AS-LSE). Arment has reviewed those aspects of AS-LSE relevant to TEQA.[26] The fundamental difference between S-LSE and AS-LSE is an improvement in design of the classical Soxhlet glass apparatus. Randall developed a Soxhlet apparatus in which the thimble containing the solid sample is immersed into the pot via a sliding rod that extends through the reflux condenser.[27] Samples can be rinsed of extractant by raising the thimble out of the pot and continuing to reflux as shown in Figure 3.8, the sketch from EPA Method 3541.[28]

Several SVOCs of environmental interest are shown in Figure 3.9, along with mean percent recoveries from spiked clay, using both S-LSE and AS-LSE techniques.[26]

1. Boiling—The thimble is immersed in the extracting solvent and refluxed for 60 min.
2. Rinsing—The thimble is raised out of the boiling solvent and suspended above it for another 60 min.
3. Evaporation-preconcentration—Condensed solvent is redirected away from the sample, and boiling solvent is collected in a condenser or reservoir. This reclaimed solvent could be reused. This step requires 10 to 20 min.

After the evaporation-preconcentration step, ~2 to 20 mL of preconcentrated extract remains in the cup. This extract can be further concentrated via nitrogen blowdown or mini-Kuderna–Danish

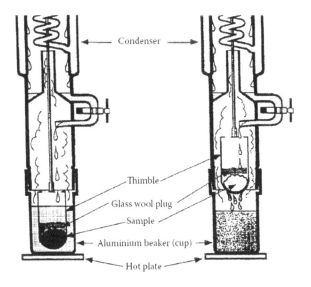

FIGURE 3.8

	Mean % Recovery (% RSD)	
Analyte	*S-LSE[a]*	*AS-LSE[b]*
δ-BHC (lindane)	*65.6 (27.1)*	*104 (9.7)*
Endrin	*81.0 (3.9)*	*112 (4.4)*
p,p'-DDT	*73.6 (38.5)*	*61.4 (6.5)*

[a] Extracted with 1:1 (v/v) hexane-acetone for 16 h; three replicate determinations.
[b] Extracted with 1:1 (v/v) hexane-acetone for 60 min of boiling and 60 min of rinsing; four replicate determinations. AS-LSE consists of three distinct steps:

FIGURE 3.9

evaporative concentration. The concentrated extract can be further cleaned up or solvent exchanged as dictated by the particular method or application.

AS-LSE seems ideally suited for the determination of total petroleum hydrocarbons (TPHs), in addition to targeted priority pollutants. To gravimetrically determine TPH content of a contaminated solid waste, the solvent is allowed to evaporate to dryness and the oily residue is weighed.

Several companies manufacture instrumentation to facilitate AS-LSE. These include Buchi, Foss-Tecator, and Gerhardt for the Randall type, while Labconco employs the Goldfisch type.[26]

The consumption of significantly less solvent and the smaller overall footprint and large sample throughout represent advantages for a laboratory to acquire an AS-LSE capability. Let us now consider alternative sample prep techniques to Soxhlet, classical or otherwise.

3.18   ARE THERE ALTERNATIVES TO S-LSE?

Yes, and the development of alternatives to conventional S-LSE has occurred during the past 25 years with an emphasis on reducing the extraction solvent volume while maintaining the high efficiency of S-LSE. These newer sample preparation *instrumental* methodologies applied to solid matrices exclusively are variations of S-LSE and include *ultrasonic probe liquid-solid extraction* (U-LSE), *microwave-assisted extraction* (MAE), and *accelerated solvent extraction* (ASE). These techniques assume that the analytical objective is to isolate and recover semivolatile to nonvolatile organic compounds from chemically contaminated soils, sediments, and sludges. These sample prep techniques fall into the *enviro-chemical TEQA* category. Beyond these variations to the classical Soxhlet, much progress has been made in performing extractions of solid matrices using supercritical fluids, a technique commonly called *supercritical fluid extraction* (SFE).

SFE can be conducted offline, whereby the solid sample is extracted and the extract is then transported to a solvent or adsorbent. In the case of an extract, the analytes of interest being dissolved in the solvent can be directly injected into a GC or, with a change of matrix, injected into an HPLC to complete the analytical steps that lead to quantification of the specific chemical contaminant. In the case of a sorbent that contains adsorbed analytes, the sorbent can be eluted with a solvent in much the same way that solid-phase extraction is conducted (refer to the extensive discussion of this technique in subsequent sections of this chapter). This eluent can then be directly injected into a GC or, with a change of matrix, directly injected into an HPLC in an effort to carry out the determinative step in TEQA. SFE can be interfaced to a chromatograph that uses supercritical fluids as chromatographic mobile phases. The technique is called online supercritical fluid extraction-supercritical fluid chromatography (SFE-SFC). This technique requires the availability of instrumentation to enable the extracted analytes to be directly injected into the chromatograph, where separation of such analytes takes place.

Supercritical fluid extraction as an alternative sample preparation method applied to solid sample matrices of interest to TEQA became quite popular during the late 1980s and early 1990s. However, SFE also requires that instrumentation be purchased. It has also been found that significant matrix dependence exists, and this matrix dependence contributes to differences in percent recoveries. The first generation of SFE instruments also suffered from problems with plugging of the low restrictors that are located after the extraction vessels. SFE will be discussed in more detail later in the chapter. We next discuss the three variations of S-LSE introduced earlier (i.e., U-LSE, MAE, and ASE).

3.19 WHAT IS ULTRASONIC LIQUID–SOLID EXTRACTION?

Ultrasonic LSE (U-LSE) is most applicable to the isolation of semivolatile and nonvolatile organic compounds from solid matrices, such as soil, sediment, clays, sand, coal tar, and other related solid wastes. U-LSE is also very useful for the disruption of biological material such as serum or tissue. U-LSE can be coupled with reversed-phase solid-phase extraction (RP-SPE) to give a very robust sample preparation method at relatively low cost in comparison to MAE and ASE approaches. The author has utilized U-LSE/RP-SPE to isolate and recover 9,10-dimethyl-1, 2-benzanthracene from animal bedding. An 89% recovery was obtained for bedding that was spiked with this polycyclic aromatic hydrocarbon (PAH) of interest to toxicologists.[29] An ultrasonic horn and tip are immersed into a mixture containing the liquid extractant and the solid sample. The mixture is sonicated at a specific percent of full power for a finite length of time, either continuously or pulsed.

Ultrasonication involves the conversion of a conventional 50/60 Hz alternating-current line power to 20 kHz electrical energy and transformation to mechanical vibration. A lead zirconate titanate electrostrictive (piezoelectric) crystal, when subjected to alternating voltage, expands and contracts. This transducer vibrates longitudinally and transmits this motion to the horn tip. The horn tip is immersed in the liquid slurry and cavitation results. Cavitation is the formation of microscopic vapor bubbles that form and implode, causing powerful shock waves to radiate throughout the sample from the face of the tip. Horns and microtips amplify the longitudinal vibration of the converter and lead to more intense cavitational action and greater disruption. U-LSE dissipates heat, and because of this, a sample should be placed in an ice-water bath. Proper care of the probe is essential. The intensity of cavitation will, after a prolonged period, cause the tip to erode and the power output to decrease without showing up on the power monitor.

Ultrasonic cell disruptors are manufactured by a half dozen or so companies. In the author's lab, the Model 450 Digital Sonifier (Branson Ultrasonics Corporation) has been in use. It becomes important to retune the generator when a new probe is changed. There are also additional tuning procedures to follow for microtips.

EPA Method 3550C (Revision 3, SW-846, February 2007) is a more recent procedure for extracting semivolatile and nonvolatile organic compounds from solids such as soils, sludges, and wastes using U-LSE. The method includes two approaches, depending on the expected concentration of contaminants in the sample. In the low-concentration method (<20 mg of each component/kg), a 30 g sample is mixed with anhydrous sodium sulfate to form a free-flowing powder and extracted three times with either 1:1 acetone/methylene chloride or 1:1 acetone/hexane. The use of a mixed solvent serves to adjust the extractant polarity, thereby enabling efficient extraction of some of the more polar analytes as well. In the high-concentration method (>20 mg of each component/kg), a 2 g sample is mixed with anhydrous sodium sulfate, again, to form a free-flowing powder. The extracts are concentrated by evaporating off the solvent using Kuderna–Danish evaporative concentrators. A solvent recovery system is recommended whereby the vaporized solvent can be recycled to prevent escape to the atmosphere. Surrogates and matrix spikes are added to the free-flowing powder once sodium sulfate has been added. Solvent evaporation can be combined with a solvent exchange, depending on which determinative technique is to be used. All solvents used should be pesticide-residue grade or equivalent. If further concentration of the extract is needed, either a micro-Snyder column technique or a nitrogen blowdown technique is used. Percent recoveries for 21 representative semivolatile priority pollutant organic compounds are listed in the method taken from the categories of base (B), neutral (N), and acid (A), the so-called BNAs from the over 100 compounds that are routinely monitored for in EPA contract type work.

Ultrasonic probe sonication utilizing a microtip effectively disrupts cell structures and liberates persistent organic pollutants from ~1 mL of rat plasma or ~0.5 g of rat liver homogenate when used with a water-miscible organic solvent such as acetonitrile in a test tube. This is an example of *enviro-health TEQA*. This author has demonstrated that specimens such as these that are known to contain PCBs such as those found in the commercial product, AR 1248, and can be isolated and recovered in high yield when combined with *reversed-phase solid-phase extraction* (RP-SPE).[30] Figure 3.10 (the first of numerous drawings devised by the author to illustrate the steps and logic of sample prep) is a flowchart that utilizes U-LSE, coupled with RP-SPE, for up-front analyte extraction and cleanup. The procedural details outlined in Figure 3.10 will be discussed in a subsequent section on RP-SPE techniques. A second alternative to S-LSE is microwave-accelerated extraction (MAE).

The realization that the physico-chemical conditions of the extract could be altered led to the development of microwave-accelerated extraction (MAE) and accelerated solvent extraction (ASE), or pressurized fluid extraction, whereby the same solvents used to conduct S-LSE are used. The elevation of the extract temperature and pressure serves to accelerate the mass transfer of analyte from solid matrix to extract, and hence reduce the time it takes to achieve a high percent recovery. The chemical nature of the extractant can also be changed, such as using carbon dioxide as a supercritical fluid. This is accomplished by elevating its temperature slightly above its critical temperature point while increasing its pressure to slightly above its critical pressure.

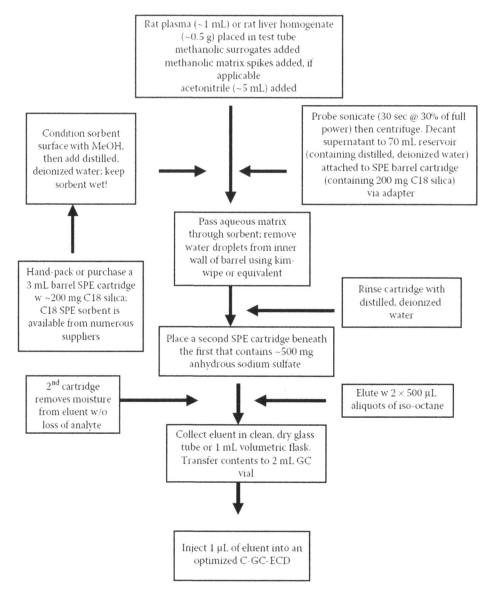

FIGURE 3.10

This sample prep technique, called supercritical fluid extraction (SFE), is another alternative to S-LSE, as applied to samples that are solids. We are going to digress a bit into phase diagrams before we discuss the underlying principles of MAE, ASE, and SFE.

3.20   CAN PHASE DIAGRAMS HELP TO EXPLAIN THESE ALTERNATIVE SAMPLE PREP TECHNIQUES?

The answer is yes, and we will digress a bit at this point to introduce these concepts, as we did earlier in the chapter. The temperature and pressure conditions that govern physico-chemical behavior of liquids are defined in terms of thermodynamics. The Gibbs phase rule is a direct outcome of the physical chemistry of changes in the state of matter. The phase rule helps to interpret the physico-chemical behavior of solids, liquids, and gases within the framework of the kinetic-molecular theory of phase equilibria.

If there are c distinct chemical species or components, the composition of any one phase is specified by $c - 1$ mole fractions.[31] The composition of the remaining component is fixed. For p phases, the number of composition variables that must be independently assigned is

$$\text{No. of composition variables} = p(c - 1)$$

In addition to the required composition variables, the two remaining parameters that change the thermodynamic state if varied are temperature and pressure. Hence,

$$\text{Total no. of variables} = p(c - 1) + 2$$

Not all of these variables are necessary to define a system; not all of these variables are independent. The fact that the chemical potential for a given component must be the same in every coexisting phase places restrictions on the number of independent variables necessary; hence, for a given component present in three phases (1, 2, and 3), stated mathematically,

$$\left(\frac{\partial G}{\partial n_1}\right)_{T,P} = \left(\frac{\partial G}{\partial n_2}\right)_{T,P} = \left(\frac{\partial G}{\partial n_3}\right)_{T,P}$$

Two equations are needed to satisfy the above condition for three phases, or $p - 1$ equations or restrictions per component. For c components, we have

$$\text{No. of restrictions} = c(p - 1)$$

The number of parameters that can be independently varied, f, is found from the difference between the total number of variables and the total number of restrictions. Stated mathematically,

$$\text{Total no. of independent variables} = p(c - 1) + 2 - \left[c(p - 1)\right]$$

The smallest number of independent variables that must be specified in order to describe completely the state of a system is known as the number of degrees of freedom, f. For a fixed mass of a gas, $f = 2$, because one can vary any two variables (e.g., pressure and temperature); the third variable is fixed by the equation of state (e.g., volume). Hence, only two properties of a fixed mass of gas are independently variable. Stated mathematically, for a system at equilibrium, the number of degrees of freedom, f, equals the difference between the number of chemical components, c, and the number of phases, p:[32]

$$f = c - p + 2$$

This is, in essence, the celebrated Gibbs phase rule. A generalized version of a phase diagram for a one-component system, $c = 1$, whereby the pressure exerted by the substance is plotted against the temperature of the substance, is shown in Figure 3.11. This means that different regions of the phase diagram yield different values for f. Regions shown in Figure 3.11 that correspond to a single phase and not a phase transition have $f = 2$. This means that both P and T can be varied independently. On one of the phase change lines, $f = 1$, and this means that if T were changed, P could not be changed independently of T if the two phases are to remain in equilibrium; rather, P must change in such a way as to keep the point on the line. At the triple point, $f =$

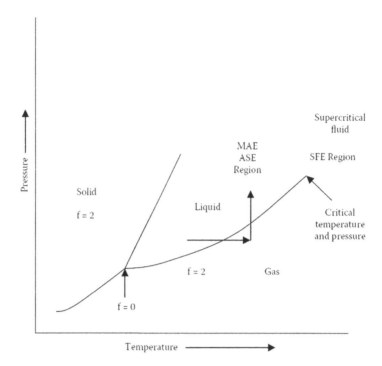

FIGURE 3.11

0, and there is a unique value for P and T. At this triple point, P and T can have only one value, and only at this point are solid, liquid, and gas allowed to exist in equilibrium. The phase diagram for most chemically pure substances exhibits a slight tilt to the right for their solid-to-liquid change of phase, the so-called fusion line. One notable exception is that of water, whose fusion line tilts slightly to the left. This enables a skater to apply a large enough pressure onto the surface of ice to enable melting to occur, and thus provide sufficient lubrication. Note the horizontal line that crosses the liquid–gas transition. This line indicates that the temperature of the extractant is being increased. To prevent vaporization from occurring, the vertical line shows that the pressure must be increased to again cross over the liquid–gas transition and keep the extractant as a liquid. This is exactly how MAE and ASE operate, and the phase diagram in Figure 3.11 is so noted. The region beyond the critical temperature and pressure, labeled as the supercritical fluid region, is where SFE is conducted. Phase diagrams for carbon dioxide, the most common substance for performing SFE, are well established. ASE involves conductive heating of extractant, while MAE requires microwave heating. MAE is introduced first, followed by ASE.

3.21   WHAT IS MICROWAVE-ACCELERATED LSE?

The key historical developments and the technical details of microwave-accelerated extraction (MAE) as applied to the extraction of solid matrices to isolate and recover priority pollutant organic compounds have been summarized.[33] Microwave heating is widely accepted as a replacement for hot-plate acid digestion of soil samples to determine trace metals. This use of microwave heating for inorganics sample prep will be discussed later in this chapter. Microwave heating of a sealed vessel that contains a solid matrix or slurry with an organic solvent is yet another and more contemporary alternative to S-LSE. Closed vessels used for MAE are designed to withstand temperatures as high as 200°C and pressures of 200 psi (14 bars). As Figure 3.11 indicates, an increase in the solvent temperature as well as its pressure in a closed vessel keeps the extractant

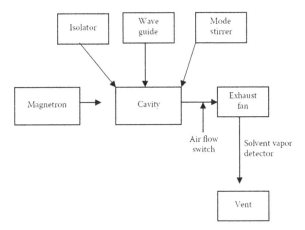

FIGURE 3.12

in the liquid phase while facilitating an increase in mass transfer of analyte from the matrix to the solvent in comparison with S-LSE. Microwave heating conjures up what many of us do every day in our kitchen (i.e., apply microwave energy to warm food). There is great danger in merely placing a soil sample to which an organic solvent has been added and heating in an open vessel such as a beaker. However, nonpolar solvents such as hexane do not heat when exposed to microwave radiation. In this case, an inert fluoro-polymer insert filled with carbon black, a strong microwave absorber, is placed into the solvent-sample mixture. Safety considerations demand a closed vessel and oven technology that is designed to prevent explosions and escape of toxic fumes.

Figure 3.12 outlines the essential features of a microwave oven designed for MAE. The sealed vessel is placed in the cavity. The magnetron generates microwave radiation that is propagated down the waveguide into the cavity. The mode stirrer distributes the energy in various directions, and the cavity serves to contain the energy until it is absorbed by the sample. The isolator protects the magnetron from radiation that would reflect back into the magnetron. The isolator acts as a one-way mirror. Microwave radiation generated by the magnetron goes to the cavity and is prevented from returning. A turntable is used to rotate the sample vessels within the cavity to evenly distribute the energy.[34] Figure 3.12 also shows important engineered safety features, such as an airflow switch, solvent vapor detector, and vent. A safe microwave oven for MAE should be designed to:

1. Eliminate possible ignition sources
2. Detect solvent leaks
3. Remove solvent vapors.

MAE joins ASE, SFE, and SPE in what the EPA has called the collection of "green sample prep techniques."

3.22 HAVE THERE BEEN STUDIES APPLYING MAE TO ENVIRONMENTAL SAMPLES?

Onuska and Terry are credited with publishing the first data on using MAE to extract pollutants from environmental samples.[35] Lopez-Avila and coworkers reported on their findings as part of an ongoing EPA evaluation of MAE.[36] The nature and types of organic compounds that were spiked into soil include:

- Semivolatiles (SVOCs), such as polycyclic aromatic hydrocarbons (PAHs), polychlorinated phenols, and phthalate esters
- Organochlorine pesticides (OCs)
- Polychlorinated biphenyls (PCBs)
- Organophosphorous pesticides (OPs).

Some of the soil was also aged before it was analyzed. Seventy-seven of the 95 SVOCs that were spiked into topsoil were isolated and recovered within the range of 80 to 120%. The recoveries of 14 compounds were below 80%. Upon spiking the topsoil and allowing 24-hour aging at 4°C, and adding water to the spiked soil to ensure good mixing of the target compounds with the matrix, only 47 compounds had recoveries between 80 and 120%.

EPA Method 3546 utilizes closed-vessel MAE to isolate and recover SVOCs, OPs, OCs, chlorinated herbicides, phenoxy acid, substituted phenols, PCBs, and PCDDs/PCDFs.[37] The extractant is a 1:1 mixture of hexane and acetone, and typical sample matrices include soil, glass fibers, and sand.

LeBlanc[33] compared MAE against Soxhlet. Table 3.1 shows several analyte/matrix environmental sample types with comparable results. The results of the two studies by Richter are shown in Tables 3.2 and 3.3. Two validation studies comparing ASE with S-LSE are summarized Table 3.4. Operating Conditions for conducting Accelerated Solvent Extraction (ASE) are shown in Table 3.5. Table 3.6 shows the percent recoveries (%R) for three PCB congeners extracted from fish meal.

The extractant from MAE remains in contact with the sample after the heating period is completed. This requires a filtration step. Instrumentation for MAE is quite expensive. Accelerated solvent extraction (ASE) is yet another alternative to S-LSE.

3.23 AUTOMATED LIQUID-SOLID EXTRACTION (A-LSE)

We next introduce a recently developed technology manufactured by CEM Corporation. The instrument is called the EDGE®. The EDGE is a fully automated solvent extraction system that performs solvent addition, *pressurized fluid extraction*, filtration, and system cleaning in minutes. The heart of the EDGE is a heated extraction chamber that allows for controlled heating of the sample and solvent and allows for diffusive flow and agitation of the sample, when necessary. Upon completion of the extraction the sample is filtered to 0.3 μm and deposited in a clean collection vial before the system removes the sample and performs a self-cleaning program.

The process for extraction begins with placing a Q-Disc® into the Q-Cup® and weighing the sample directly into the Q-Cup. The Q-Cup is then placed into a 12-place rack that allows the EDGE to perform 12 extractions sequentially without operator interaction. The operator needs only to select the pre-programmed method from the system software and press play. The EDGE then picks up the first sample and places it in the extraction chamber. After insertion into the chamber, a pressure cap is lowered to seal the chamber completely. Solvent is added through both the solvent spray nozzle on the top of the chamber and the solvent line at the bottom of the chamber. The lower solvent addition allows for a solvent gap to form between the Q-Cup and chamber walls. After solvent has been added the system heats the chamber to the programmed temperature. Heating the solvent in the solvent gap creates a pressure differential that forces the warmed solvent through the Q-Disc and up through the sample matrix and any sorbents or cleanup materials being used. Agitation can also be used to further increase the surface contact between the solvent and sample matrix to ensure optimal extraction. When the extraction is complete, the system opens a drain port and forces the extract through the Q-Disc filter, through a cooling coil, and deposits it in the collection vial associated with that sample. Since the EDGE performs extractions sequentially, it is necessary to perform a system clean after each extraction. The EDGE removes the Q-Cup from the extraction chamber and returns it to the 12-place rack. The pressure cap then closes on the empty chamber and the system washes itself with up to 5 customizable wash cycles to ensure complete cleanliness before it begins the cycle again with the next sample.

The biggest benefits of using an EDGE over other extraction techniques are speed and simplicity. What makes the EDGE so significant to the reduction in sample prep time is that the Q-Cup® sample holder is assembled with the Q-Disc® filter in seconds. This combined with the heated extraction chamber and dispersive mixing during extraction greatly increases extraction efficiency while cutting extraction time to a fraction of that of other techniques. The EDGE can be used for many types of environmental samples at the same time. Soils, PUF filters, and plant material can all be run in a single rack with up to 6 different solvents plumbed to the system.

Other extraction techniques and technologies can consume large quantities of solvent, take hours to perform, are very difficult to use, and/or require further handling of the sample prior to analysis. For instance, the extraction of SVOCs from soil can be performed by Soxhlet extraction in 6 hours and using 100 mL of solvent. The same extraction can be performed on the EDGE in only 7–10 minutes using 35 mL of solvent. The same extraction performed by microwave extraction will take only 25 mL of solvent but take over 30 minutes to perform the extraction then an additional hour or more to filter each sample. Sample cell setup of other pressurized fluid extraction systems can take up to 30 minutes per cell and if not done properly can lead to system leaking and inefficient extractions.[38] It would seem that A-LSE might be a smaller version of ASE which arrived in analytical laboratories much earlier and is discussed next.

3.24   HOW DOES ASE WORK?

A solid sample such as soil, whose priority pollutant content is of interest, is placed in a vessel that is capable of withstanding high pressures and temperatures. The vessel is filled with the extractant and extracted under elevated temperature (50 to 200°C) and pressure (500 to 3000 psi) for relatively short periods (5 to 10 min). The technique was first published by Richter et al. who also discuss the physico-chemical considerations that went into developing ASE.[39] The solubility of organic compounds in organic solvents increases with an increase in solvent temperature. Faster rates of diffusion also occur with an increase in solvent temperature. If fresh extracting solvent is introduced, the concentration gradient is increased, and this serves to increase the mass transfer rate or flux according to Fick's first law of diffusion. Thermal energy can overcome cohesive (solute–solute) and adhesive (solute–matrix) interactions by decreasing the energy of activation required for the desorption process. Higher extractant temperatures also increase the viscosity of liquid solvents, and thus allow better penetration of matrix particles and enhancing extraction. Increased pressure enables liquids to exist at the elevated temperatures. Increased pressure forces the solvent into areas of the matrix that would not normally be contacted by solvents using atmospheric conditions.

Instrumentation to set up and perform ASE on solid matrices of interest to TEQA has been commercially available for over a decade. The instrument was developed at the former Dionex Corporation which is now part of Thermo Fisher Scientific. The instrumentation to perform ASE is quite expensive, and for this reason, labs should have a large sample workload so that return on investment can be realized in as short a time frame as possible.

The ASE instrument works as follows. Solvent is pumped into the extraction cell after the cell is filled with the solid sample. The cell is heated and pressurized to enable the extractant to remain as a liquid well above its boiling point. A finite amount of time is taken for LSE to occur. While hot, the cell is flushed one or more times with solvent and the extractant is transferred to a collection vial. The remaining solvent is purged out of the cell with compressed nitrogen. These steps have been recently introduced and are outlined in Figure 3.13.[39]

3.25   WHAT SAMPLES HAS ASE BEEN APPLIED TO?

Two studies from the original paper by Richter and colleagues will now be discussed.[39] The first study involved isolating and recovering total petroleum hydrocarbons (TPHs) from soil using ASE. Four different extraction temperatures were used. TPHs were determined via infrared

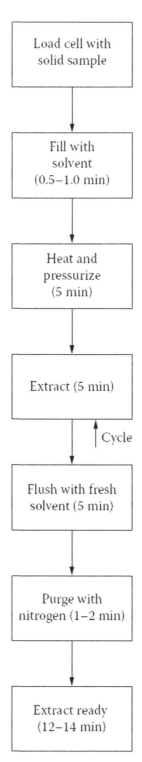

FIGURE 3.13

TABLE 3.1
Comparison of MAE vs. Soxhlet LSE for Various Semivolatile Priority Pollutants

	MAE		Soxhlet	
Sample	*V*ext(mL)/ *T*ext(min)	Concentration	*V*ext(mL)/ *T*ext(min)	Concentration
TPHs/soil	*30/7*	*943 mg/kg*	*300/60*	*773 mg/kg*
OCs/soil	*50/7*	*92.3% recovered*	*300/1080*	*83.4% recovered*
PCBs/soil	*25/7*	*47.7 µg/g*	*250/1080*	*44.0 µg/g*
CH3Hg/sediment	*10/6*	*80 µg/g*	*200/150*	*81 µg/g*
PCDDs/sediments	*30/8*	*565 µg/g*	*300/1440*	*542 µg/g*
Organotin/biomaterial	*20/5*	*1.28 µg/g*	*—*	*1.3 µg/g*

TABLE 3.2
Effect of Temperature on the Recovery of TPHs from Soil Using ASE

Extractant Temperature (°C)	**Amount Found (mg) ($n = 5$)**	**RSD (%)**
27	*974*	*6.0*
50	*1118*	*5.0*
75	*1190*	*2.0*
100	*1232*	*1.0*

Note: ASE conditions:
• Preheat method
• Perchloroethylene as extractant
• 2500 psi, 5 m in static extraction after equilibration
• 3 g of soil
• 3.5 mL of cell volume
• 4.5 to 5.0 mL of solvent used
• 1200 mg of TPH/kg of soil

absorption. The quantitative analysis of oil-laden soils has been introduced in Chapter 1, and the determinative technique for this will be discussed in Chapter 4. The results are shown in Table 3.2 and demonstrate that extractant temperature can significantly influence TPH recoveries. The precision over five replicate ASEs is seen to decrease as the extractant temperature is increased.

The second study involved isolating and recovering the selected polycyclic aromatic hydrocarbons (PAHs) from a certified reference sample of urban dust.

Table 3.3 shows the results and lists the ASE conditions for this study. Nearly 100% recoveries are reported together with good precision (%RSDs < 7% for six replicate ASEs) for these 10 priority pollutant PAHs.

Two validation studies conducted in 1994 compared ASE with S-LSE, automated Soxhlet, and wrist-shake or manual LLE for the determination of target analytes from solid wastes of interest to TEQA. These results are summarized in Table 3.4. ASE proved equivalent to the conventional procedures. The data generated in these two studies resulted in the generation of EPA Method 3545, which received final promulgation in June 1997 as part of SW-846 Method Update III, operational parameters for ASE are shown in Table 3.5.[41]

Dionex Corporation published a series of application notes that demonstrate the usefulness of ASE to effectively isolate and recover various persistent organic pollutants (POPs) from a fish tissue homogenate.[42] Fish tissue is a biological matrix that is of increasing interest as an indicator of the degree of surface-water contamination. Biomagnification of various semivolatile POPs in fish, combined with fish consumption by humans, lends an enviro-health importance to quantitatively determining the concentration of various POPs in fish. Consider using a Model 300 ASE (the model that allows for the largest sample size) whereby 30 g of fish tissue homogenate is weighed and spiked with 50 µL of a series of PCB congeners whose concentration in hexane is in the range of 50 to 250 ppm. After spiking, a concentration of PCB in the sample is in the 80 to 400 ng/g range. Twenty grams of Hydromatrix® (diatomaceous earth from Varian Sample Prep Products) is added and mixed in a mortar using a pestle. The solid mix is loaded into a 100 mL stainless-steel high-pressure extraction cell containing 10 g of alumina and a cellulose filter. ASE conditions as stated in the application note are listed as follows:

Extracts were dried using anhydrous sodium sulfate, then concentrated to 10 mL under nitrogen blowdown. The concentrated extract was analyzed using gas chromatography with electron-capture detection. According to Dionex, a mean %RSD of 6.1 ($n = 5$) was obtained. Alumina removes coextracted lipid from the extract as it passes from the cell. Alumina will retain ~75 mg of lipid per gram of sample under the ASE conditions described above. For fish tissue with appreciable lipid content, e.g., >5%, additional cleanup of the extract is recommended. Extract cleanup techniques are introduced later in the chapter. Models 200 and 300 are designed for high sample throughput and, as a consequence, are quite expensive. To address those labs that have low sample throughput, the ASE 100 has been developed.[43]

Fish tissue represents a sample matrix that is relatively high in lipid or fat content. LLE or any of the LSE techniques described earlier in this chapter when applied to fish tissue yield extracts rich in dissolved lipid. The ASE technique permits an adsorbent to be added to the extraction cell that removes significant amounts of lipid. Adsorbents that can be added are called *fat retainers.* A study, independent of Dionex Corporation, compared the degree to which the most common polar adsorbents, used in conventional sample cleanup, effectively removed lipid from ASE extracts.[44]

After proving that ASE could effectively extract PCB congeners from a certified reference material, in this case, spiked cod liver oil, various ratios of fat-to-fat retainer were prepared and used to conduct ASE. Plots of the percent retained fat vs. the *fat-to-fat retainer ratio,* using lard as the source of lipid, showed that the amount of coextracted fat could be reduced to only ~5% using a fat/fat retainer ratio of 0.05. The authors postulate that as much as 40× the amount of retainer is needed to achieve complete removal of lipid. Percent recoveries for three PCB congeners from a naturally contaminated fish meal sample using five different adsorbents as shown in Table 3.6. Two grams of fish meal were weighed using 10 g of fat retainer for each adsorbent below, except where 1 g of fish meal was weighed using 5 g of Florisil® (this adsorbent has a lower density than the others):

Sulfuric acid-impregnated silica gel emerged from this study as the most favorable fat retainer. A certified reference standard containing various PCB congeners dissolved in cod liver oil was extracted using this fat retainer at a fat-to-fat retainer ratio of 0.022. Percent recoveries for PCB congeners were near 100%, with %RSDs of 2 to 4% for triplicate ASEs performed per sample.

TABLE 3.3
Recovery of PAHs from Urban Dust (SRM 1649)

PAH	Mean% Recovery (n = 6)	RSD (%)
Phenathrene	*113*	*2.0*
Fluoranthene	*88.5*	*5.2*
Pyrene	*91.0*	*5.9*
Ben(a)anthracene	*97.2*	*5.7*
Chrysene	*101*	*4.2*
Benzo(b)fluoranthene	*115*	*6.2*
Benzo(k)fluoranthene	*112*	*6.7*
Benzo(a)pyrene	*125*	*6.2*
Benzo(g,h,i)perylene	*108*	*5.8*
Indeno(1, 2, 3-c,d)pyrene	*108*	*6.7*

Note: ASE conditions:
• Prefill method
• 100°C
• Methylene chloride/acetone 1:1 (v/v)
• 5 min of equilibration, 5 min of static
• 18 mL final volume
• 2000 psi

TABLE 3.4
Summary of Laboratory Studies Used for Validation of ASE for EPA Method 3545

Compound Class	Matrix	Comparison Technique	Relative Recovery (%)
Organochlorine pesticides	*Clay, loam, sand*	*Automated Soxhlet (Method 3541)*	*97.3*
Semivolatile compounds	*Clay, loam, sand*	*Automated Soxhlet (Method 3541)*	*99.2*
Organophosphorous pesticides	*Clay, loam, sand*	*Soxhlet (Method 3540)*	*98.6*
Chlorinated herbicides	*Clay, loam, sand*	*Manual LLE (Method 8150)*	*113*
PCBs	*Sewage sludge, river sediment, oyster tissue, soil*	*Various certified reference materials*	*98.2*
PAHs	*Urban dust, marine sediment, soil*	*Various certified reference materials*	*105*

TABLE 3.5
Typical Operational Parameters for ASE

Extracting solvent	*Methylene chloride*
Extraction temperature	*125°C*
Extraction pressure	*1500 psi (10 MPa)*
Heat-up time	*5 min*
Static time	*3 min*
Flush volume	*60%*
Purge time	*120 sec*
No. of static cycles	*3*
Total extraction time	*18 min/sample*
Total solvent consumed	*120–140 mL/sample*

TABLE 3.6
Percent Recoveries for Three Different PCBs Using ASE-LSE

PCB	A	B	C	D	E
22'455'PCBP	*117*	*110*	*112*	*116*	*116*
22'44'55'HCBP	*107*	*109*	*112*	*110*	*117*
22'344'55'HCBP	*90*	*95*	*91*	*97*	*96*

Note: A = sulfuric acid-impregnated silica gel; B = Florisil; C = basic alumina; D = neutral alumina; E = acidic alumina.

ASE extracts from Great Lakes fish were cleaned up using Florisil® combined with *activated* silica gel fractionation in the author's laboratory to quantitate selected Toxaphene parlar congeners whose analytical results including QC, IDLs and MDLs have been recently reported.[45]

We now turn our attention to sample preparation techniques to quantitatively determine *volatile organic compounds* (VOCs) in environmental samples (*enviro-chemical* TEQA). We then return to a discussion of the more recently developed sample prep techniques for SVOCs, namely, SFE, SPE, SPME, and QUECHERS. Because we have been discussing LLE and LSE techniques, it is appropriate for us to ask at this point the following question:

3.26   CAN VOLATILE ORGANICS BE ISOLATED FROM CONTAMINATED WATER USING LLE?

Yes. The original discovery of trihalomethanes (THMs) in the city of New Orleans drinking water was determined by conducting LLE using pentane as the extracting solvent.[46] If the sample contaminants are well known in a particular environmental sample, it is possible to use LLE as the principal sample prep technique. LLE is, however, unselective, and semivolatile and nonvolatile organics are equally likely to be extracted in samples that contain unknown contaminants. Consistent with the EPA protocols introduced in Chapter 1, methods have been developed to remove VOCs either by purging or by sampling the headspace in a sealed vial that contains the contaminated groundwater sample. The former technique will be discussed after we develop the principles behind the latter technique, commonly called static headspace sampling for the determination of trace volatile organics (VOCs), in drinking water, groundwater, surface water, wastewater, or soil.

3.27   ON WHAT BASIS CAN VOCs BE QUANTITATED USING STATIC HEADSPACE TECHNIQUES?

Liquid–liquid extraction has a counterpart for the determination of VOCs in the technique known as a static or equilibrium headspace sampling. The technique is most often combined with the determinative step and is often referred to as static headspace gas chromatography (HS-GC). The principles that underlie this technique will be outlined in this section. The decision to measure VOCs in the environment by either HS or purge-and-trap (P&T or dynamic headspace) represents one of the ongoing controversies in the field of TEQA. This author has worked in two different environmental laboratories in which one used P&T as the predominant technique to determine VOCs in the environment and the other used HS-GC.

Let us start by introducing the thermodynamic viewpoint for the distribution of a substance originally dissolved in the sample. The substance exists at room temperature in a closed vessel partitioned between an aqueous phase and a gaseous phase. The gaseous phase in the HS usually consists of air saturated with water vapor. Dissolved gases and volatile organic compounds are the two most common chemical substances that are likely to be found in the headspace. Dissolved gases equilibrate into the HS because they follow Henry's law. Volatile organic compounds, many of which are classified as priority pollutants by the EPA, behave similarly to fixed gases, and each different VOC has its own Henry's law constant, K_H. We will consider sulfur dioxide as being illustrative of a dissolved gas and proceed to develop the basis for the equilibrium. We will next use trichloroethylene, as illustrative of a dissolved VOC, and develop the more applied equations that must be considered in order to relate the HS technique to TEQA.

3.28   WHAT HAPPENS WHEN GASEOUS SO2 IS DISSOLVED IN WATER, THEN PLACED IN A VESSEL AND SEALED?

Consider a water sample that contains dissolved SO_2. The sample is placed in a cylindrical glass vessel and then sealed. Glass vials with a crimped-top seal and a volume of 22 mL are common-place. If the pH of the aqueous phase could be measured, the value would indicate that the effect of the dissolved gas acidified the water. It is well known that one molecule of SO_2 combines with one molecule of water to produce sulfurous acid according to

$$SO_{2(g)} + H_2O \leftrightarrow H_2SO_{3(aq)}$$

Sulfurous acid, in turn, is known to ionize according to

$$H_2SO_{3(aq)} \leftrightarrow H^+_{(aq)} + HSO^-_{3(aq)}$$

Both equations describe what we have been calling secondary equilibrium effects to the primary distribution equilibria, which, in this case, is the distribution of neutral SO_2 molecules between the aqueous phase and the HS, described as follows:

$$SO_{2(aq)} \xoverset{K_H^{SO_2}}{\longleftrightarrow} SO_{2(g)}$$

The extent to which SO_2 partitions into the HS is determined by the Henry's law constant, K_H.

In a manner similar to that developed for LLE [refer to Equations (3.1) and (3.2)], it becomes apparent that in a closed system between two phases, the chemical potentials for SO_2 in each phase become equal once dynamic equilibrium is reached. If this is so, then similar to what was stated earlier, we can mathematically state

$$\mu_{HS,SO_2} = \mu_{aq,SO_2}$$

The chemical potential for SO_2 in either phase is related to a standard-state chemical potential, μ^0, for SO_2 in both the headspace (HS) and the aqueous phase, and a term related to either the pressure exerted by the gas in the HS or the activity in the dissolved state according to

$$\mu_{HS,SO_2}^0 + RT \ln p^{SO_2} = \mu_{aq,SO_2}^0 + RT \ln a_{aq}^{SO_2}$$

We seek to express the ratio of activity to partial pressure so as to be consistent with a subsequent definition of the partition coefficient. Upon rearranging, we obtain the following relationship:

$$RT \ln \left(\frac{a_{aq,SO_2}}{p_{HS,SO_2}} \right) = - \left(\mu_{aq,SO_2}^0 - \mu_{HS,SO_2}^0 \right)$$

Upon further rearranging and proceeding to define the Henry's law constant, we obtain

$$\frac{a_{aq,SO_3}}{p_{HS,SO_2}} = e^{-\Delta\mu^0/RT} \equiv K_H^{SO_2} \tag{3.23}$$

Secondary equilibrium effects are handled by defining the degree of ionization by α, as done previously [refer to Equations (3.17) and (3.18)], and expressing the activity of neutral SO_2 in terms of α and the total activity, a_T, according to

$$a_{SO_2} = a_T (1 - \alpha)$$

For example, if 0.7643 mol SO_2 is dissolved in 1 kg of water in a closed system, a 0.7643 M solution results, and with $\alpha = 0.1535$, the vapor pressure exerted by SO_2 is such that K_H is 0.813. When 1.273 g of SO_2 is dissolved in the same weight of water, with α, this time being 0.1204, the vapor pressure exerted is such that K_H is again 0.813 (pp. 1083–1085).[47]

3.29 ON WHAT BASIS CAN WE QUANTITATE TCE IN GROUNDWATER BY HS-GC?

Consider how the priority pollutant VOC trichloroethylene or trichloroethene, abbreviated TCE, is distributed between a sample of groundwater and the headspace in a sealed HS vial, as shown schematically in Figure 3.14. Let us assume that the temperature is fixed such as would be found in a thermostated automated HS analyzer. A consideration of mass balance requires that the original amount of TCE be found in either the headspace or the aqueous sample. Mathematically stated,

$$amt_o^{TCE} = amt_S^{TCE} + amt_{HS}^{TCE} \tag{3.24}$$

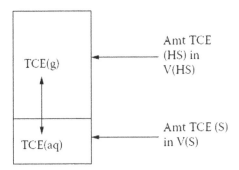

FIGURE 3.14

Using principles encompassed by Equation (3.23), we can proceed to define the *partition coefficient* as

$$K^{TCE} \equiv \frac{C_S}{C_{HS}} \tag{3.25}$$

It is helpful at this point to define the abbreviations used to develop the equations that lead to Equation (3.26), derived below. Consider the following definitions:

o = original concentration of TCE in the groundwater sample
S = sample
HS = headspace
amt = amount of TCE
V_S = volume of sample placed in the HS vial
V_{HS} = volume of headspace in the HS vial after V_S mL have been placed in the HS vial
C = concentration
β = phase ratio for an HS vial, equals V_{HS}/V_S.

We can proceed from both the mass balance principle and the definition of K^{TCE}. Let us divide Equation (3.24) by V_S:

$$\frac{amt_o}{V_S} = \frac{amt_S}{V_S} + \frac{amt_{HS}}{V_S}$$

$$C_0 = C_S + \frac{V_{HS} C_{HS}}{V_S}$$

$$= K_H^{TCE} C_{HS} + \frac{V_{HS}}{V_S} C_{HS}$$

$$= C_{HS} \left[K_H^{TCE} + \frac{V_{HS}}{V_S} \right]$$

$$= C_{HS} \left[K_H^{TCE} + \frac{V_{HS}}{V_S} \right] \tag{3.26}$$

Solving Equation (3.26) for C_{HS} gives a useful relationship:

$$C_{HS}^{TCE} = C_o \left[\frac{1}{K_H^{TCE} + \beta} \right] \tag{3.27}$$

Equation (3.27) suggests that for a given concentration of TCE in the original sample of groundwater C_o, the concentration of TCE expected to be found in the HS, C_{HS}^{TCE}, depends not only on C_o, but also on two factors: K^{TCE}, which is due to the physico-chemical nature of this particular VOC, and β, which is due to the volumes occupied by HS and sample. β might be thought of in terms of the physical characteristics of the HS vial. In practice, however, K^{TCE} may exhibit some dependence on temperature and sample ionic strength, whereas the range of values that β can take on is limited. As we shall see, both factors play off of one another in the consideration of analyte sensitivity in HS techniques.

Equation (3.27) defines the mathematical basis for static HS and is applicable to both *enviro-chemical* and *enviro-health* TEQA. This is true because for a given VOC and fixed sample volume in a headspace vial (i.e., K_{HS}^{VOC} and β), C_{HS}^{VOC} is directly proportional to the original concentration of VOC in the aqueous sample, C_o. The volume of headspace sampled and injected into a GC is usually held fixed so that the area under the curve of a chromatographically resolved GC peak, A^{VOC}, is directly proportional to C_{HS}^{VOC}. As discussed in Chapter 2, Equations (2.1) and (2.2), with respect to the external mode of instrument calibration, the *sensitivity* of the HS-GC technique is related to the magnitude of the response factor, R_F^{HS}, according to

$$R_F^{HS} = \frac{A_{HS}^{VOC}}{C_{HS}^{VOC}}$$

A fixed aliquot of the HS whose concentration of TCE, C_{HS}^{TCH}, is injected into a GC that incorporates a halogen-specific detector, such as an electrolytic conductivity detector (ElCD) (refer to EPA Methods 601 and 602, even though these methods employ P&T techniques). Alternatively, this aliquot can be injected into a gas chromatograph-mass spectrometer (GC-MS). The mass spectrometer might be operated in the selective ion monitoring (SIM) mode. The most abundant fragment ions of TCE would then be detected only (refer to EPA Method 624, even though this method employs P&T techniques). Determinative techniques such as GC-ElCD and GC-MS are introduced in Chapter 4. Nevertheless, it should be apparent to the reader at this point that the *partition coefficient* plays an important role in achieving a high sensitivity in HS techniques. We digress from here to discuss this a bit more.

3.30   ON WHAT BASIS DOES K^{TCE} DEPEND?

Let us approach an equivalent to Equation (3.25) from the perspective of applying the three great laws of phase equilibrium found in most physical chemistry texts: Dalton's law of partial pressures, Raoult's law of ideal solutions, and Henry's law for dissolved gases (pp. 16–18).[48] Applying Dalton's law enables one to state that the concentration of analyte in the HS is proportional to its partial pressure. The partial pressure exerted by TCE in the HS is independent of all other gases in the HS mixture and is related to the total pressure in the HS as follows:

$$p_i^{HS} = p_T X_i^{HS}$$

This partial pressure exerted by component i is, in turn, related to the vapor pressure exerted as if i were pure, p_i^o, and the mole fraction of component i dissolved in the sample, X_I^S, according to Raoult's law. An activity coefficient γ_i is included to account for nonideality so that

$$p_i^{HS} = p_i^o \gamma_i X_i^S$$

Also, the partial pressure exerted by component i can be related to the mole fraction of i dissolved in the sample, S, according to Henry's law, as follows:

$$p_i^{HS} = K_H X_i^S$$

The partial pressure of component i in the HS can be eliminated so that

$$p_T X_i^{HS} = p_i^o \gamma_i X_i^S$$

Rearranging the equation to a ratio of mole fractions yields

$$\frac{X_i^S}{X_i^{HS}} = \frac{p_T}{p_i^o \gamma_i} \approx \frac{C_i^S}{C_i^{HS}} = K$$

This equation yields an alternative relationship for the partition constant in HS, in which K is found to be inversely proportional to a product of the partial pressure of component i, if it is pure, and its activity coefficient γ_i according to

$$K \propto \frac{1}{p_i^o \gamma_i} \tag{3.28}$$

Because Equation (3.25) relates K to a ratio of concentrations, combining Equations (3.25) and (3.28) leads to the following:

$$\frac{C_i^S}{C_i^{HS}} \propto \frac{1}{p_i^o \gamma_i} \tag{3.29}$$

p_{TCE}^o refers to the intrinsic volatility exhibited by a chemically pure substance TCE. TCE is a liquid at room temperature, and if one opens a bottle that contains the pure liquid, one is immediately struck by virtue of the sense of smell with the concept of the vapor pressure of TCE. It is likely that P_{TCE}^o is greater than P_{PCE}^o, where PCE refers to perchlorethylene (tetrachloroethene). PCE is also on the EPA's priority pollutant list. One would expect to find that C_{TCE}^{HS} is greater than C_{PCE}^{HS}, assuming all other factors equal. This is largely due to the inverse relationship between K and p^o, as shown in Equation (3.28). This equation also suggests that if a matrix effect exists, K is also influenced by differences in the activity coefficient, γ. Changes in γ might be due to changes in the ionic strength due to the sample matrix and, in turn, influence K, as shown in Equation (3.28).

3.31   MUST WE ALWAYS HEAT A SAMPLE IN HS-GC TO CONDUCT TEQA?

Equations (3.27) to (3.29) lay the foundation for a theoretical understanding of what factors are involved in obtaining a sufficiently large value for the concentration of any VOC in the HS (i.e., C_{HS}^{VOC}). In this section, we look at the effect of increasing temperature of a headspace vial that contains priority pollutant VOCs dissolved in water. An example of this might be a sample of groundwater that has been contaminated with priority pollutant VOCs. The static equilibrium HS technique just described would be used to satisfy the criteria for TEQA.

The Clausius–Clapeyron equation, one of the most famous in physical chemistry, is most applicable for this discussion. The equation states that the partial differential with respect to absolute temperature of the logarithm of a pure liquid's vapor pressure is inversely related to the liquid's absolute temperature. We again consider the liquid TCE and state the Clausius–Clapeyron equation mathematically (pp. 275–279):[47]

$$\frac{\partial}{\partial T}\left(\ln p^{o}\right) = \frac{\Delta H_{v}}{RT^{2}}$$

Upon rearranging and expressing this relationship with respect to the VOC that we are considering, TCE, we obtain

$$\ln p_{TCE}^{o} = \frac{\Delta H_{v}^{TCE}}{R}\int\frac{dT}{T^{2}}$$

The indefinite integral is then evaluated as

$$-\frac{\Delta H_{v}^{TCE}}{RT} + C$$

The vapor pressure exerted by pure TCE can then be expressed as

$$p_{TCE}^{o} = e^{\left[-\left(\frac{\Delta H_{v}^{TCE}}{RT}\right)+C\right]} \tag{3.30}$$

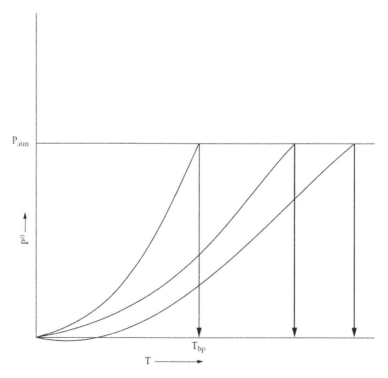

FIGURE 3.15

where ΔH_v^{TCE} is the molar heat of vaporization for pure TCE and R is the ideal gas constant. Equation (3.30) suggests that a pure liquid's vapor pressure increases exponentially as the temperature of the liquid is increased, until its vapor pressure reaches that exerted by the atmosphere. A plot of Equation (3.30) is shown in Figure 3.15 for three different liquids. For example, if three ClVOCs are plotted, such as for chloroform, TCE, and PCE, their respective vapor pressure/temperature curves would resemble the three shown in Figure 3.15. Because the partition coefficient and the partial pressure of a pure liquid such as TCE are inversely related according to Equation (3.28), we can relate K and T in the following manner:

$$\ln K_{HS}^{TCE} = \frac{A}{T} - B \tag{3.31}$$

where A and B are constants related to TCE. Thus, an increase in T serves to decrease K_{HS}^{TCE} and, according to Equation (3.29), partitions a greater percent of more TCE molecules into the HS vs. the condensed phase. In other words, the ratio C_S^{TCE} / C_{HS}^{TCE} is seen to decrease.

3.32   IS THERE AN EXAMPLE THAT ILLUSTRATES THE CONCEPT IN EQUATION (3.31)?

One application of these principles is evident in the hypothetical preparation of a mixture of C_5 through C_{20} alkanes; each alkane is then placed in an HS vial in equal amounts. Prior to sealing the vial, a 10 μL aliquot of the liquid is taken and injected into a gas chromatograph with a carbon-selective detector. A chromatogram is generated that shows a separate peak for each of the 16 alkanes. The HS vial is then sealed with a crimped-top cap and septum. The vial is placed inside a cylindrical heater block whose temperature has previously been set to 80°C. A 1.0 cc gas-tight syringe is used to withdraw an aliquot of the headspace, which is directly injected into the same GC. A chromatogram is generated that shows a distribution of the 16 peaks that is different from the first. The lower-carbon-number or lower-molecular-weight alkanes seem to have been enriched in the HS, compared to the direct liquid injection. These results are readily explained by considering Equation (3.26) and the subsequent reexpression of Equation (3.30) in terms of the HS partition coefficient [i.e., Equation (3.31)].

3.33   WHAT HAPPENS WHEN K_{HS}^{VOC} VALUES VARY SIGNIFICANTLY?

Kolb and Ettre recently discussed their work on determining values for K_{HS}^{VOC} and report some interesting findings. (pp. 22–27)[48] They observed that over a range of some 15 VOCs that vary in polarity from dioxane to lower-molecular-weight alcohols, through to ketones, and then to acetate type esters, monoaromatics, and polychlorinated ethenes, and finally to C_6 alkanes that K_{HS}^{VOC} values vary by over four orders of magnitude. Those VOCs that exhibit low K_{HS}^{VOC} values (i.e., favor a high concentration in the HS) (C_{HS}^{VOC} is high at equilibrium) are not influenced by an increase in temperature. Those VOCs that exhibit high K_{HS}^{VOC} (i.e., favor a low concentration in the HS) (C_{HS}^{VOC} is low at equilibrium) are strongly influenced by an increase in temperature. Kolb and Ettre chose five solutes as representative of the large number of VOCs that have a wide range of partition coefficients. These are listed below, along with their partition coefficients:

Solute	K_{HS}^{VOC} at 40°C
Ethanol	1355
Methyl ethyl ketone	139.5 at 45°C
Toluene	2.82

Solute	K_{HS}^{VOC} at 40°C
Tetrachloroethylene	1.48
n-Hexane	0.14

As the temperature is increased from 40 to 80°C while β is held fixed at 3.46, the partition coefficient for ethanol is observed to increase to the greatest degree, whereas that for methyl ethyl ketone is also increased, but to a lesser degree. Partition coefficients for toluene, tetrachloroethylene, and n-hexane do not increase to any extent as the temperature is increased from 40 to 80°C.

This phenomenon is explained by these authors in the following manner: VOCs whose partition coefficients are already low are enriched in the HS at the lower temperature, 40°C, and the percent of remaining dissolved VOCs in the condensed phase that partition into the HS as the temperature is raised to 80°C is not appreciable. Polar solutes, on the other hand, are predominantly dissolved in the condensed phase at 40°C, and the percent that partition into the HS as the temperature is raised to 80°C is much more appreciable.

Let us return to Equation (3.27) and briefly consider the *second factor*, the phase ratio β, in the headspace sampling of groundwater to determine trace concentrations of VOCs in the environment. Kolb and Ettre have studied the influence of β on HS *sensitivity*. For a fixed temperature and fixed original concentration of VOC in the aqueous phase, C_o, the influence of changing β from, for example, 4.00 (only 20% of the total volume of the HS vial contains the sample), to β = 0.250 (80% of the total volume of the HS vial contains the sample) depends on the magnitude of K_{HS}^{VOC}. For nonpolar aliphatic hydrocarbons like n-hexane or for chlorinated C_1 or C_2 aliphatics such as TCE, a change in β from 4.00 to 0.250 increases the HS sensitivity by almost a factor of 10. For monoaromatics, the same change in β gives an increase in HS sensitivity of about 4. For acetate type esters, ketones, lower-molecular-weight alcohols, and ethers, a change in β from 4.00 to 0.250 gives an insignificant increase in HS sensitivity (pp. 22–27).[48]

The influence of the sample matrix activity coefficient, γ, represents the *third factor* that serves to influence HS sensitivity. The theoretical basis for this is encompassed in Equation (3.28). The influence of increasing γ by adding salt to an aqueous sample, a well-known technique called *salting out,* has been shown to have a negligible effect on nonpolar VOCs such as TCE dissolved in water. Polar solutes, however, are more strongly influenced by changes in the sample matrix activity coefficient. It has also been observed that a high concentration of salt must be dissolved in the aqueous sample matrix to have any effect at all. The addition of a high amount of salt increases the volume of the liquid sample and thus serves to decrease β. There are a number of drawbacks to adding salt in static HS, including increases in sample viscosity and the addition of volatile impurities.

Few comprehensive studies have been published on the effect of the sample matrix on the partitioning of VOCs at trace concentration levels. Friant and Suffet chose four model compounds that are polar and representative of the intramolecular forces of dispersion, dipole orientation, proton-donor capability, and proton-acceptor capability.[49] These were methyl ethyl ketone, nitroethane, n-butanol, and p-dioxane. The pH of the aqueous sample matrix had no effect at all except for nitroethane. An optimum salt concentration was 3.35 M in sodium sulfate, and an optimum HS sampling temperature of 50°C. For example, the partition coefficient of methyl ethyl ketone increased from 3.90 to 260 when the temperature of the aqueous sample that contained the dissolved ketone was increased from 30 to 50°C, and at the same time, the concentration of salt increased from zero to 3.35 M. Otson et al. compared static HS, P&T, and LLE techniques for the determination of THMs in drinking water and found that static HS showed the poorest precision and sensitivity. However, a manual HS sampling technique was used back in this era and might possibly have contributed to this loss of precision. LLE was found to be quite comparable to P&T in this study.[50]

3.34 WHY IS STATIC HEADSPACE SAMPLING NOT MORE ACCEPTABLE TO THE EPA?

This might be close to the $64,000 question! The answer lies somewhere between "MDLs are not low enough in comparison to those obtained using P&T" and "HS techniques were not developed in EPA labs, while purge-and-trap (P&T) techniques were." Static HS, whose principles have already been introduced in this chapter, requires that analytes be classified as volatile (i.e., having boiling points <125°C). The fundamental principles behind the static HS technique introduced earlier are becoming better understood. Prior to the development of the PAL dual-rail autosampler (CTC Analytics AG) most static HS techniques are performed in automated analyzers. These analyzers were directly interfaced to the injection port of a gas chromatograph via a transfer line. Static HS being an equilibrium process as discussed earlier, cannot reach the much lower MDLs when compared to P&T techniques (to be discussed). Thus, the nature of the GC detector that is available is crucial to the notion of static HS sensitivity. If a less sensitive GC detector is used, a given VOC can be highly favored to partition into the headspace, yet have an MDL that is less than desirable.

In the author's laboratory, the technique of manual HS sampling is popular among environmental microbiologists who need analytical results almost immediately after they sample the headspace of their bioreactors. They are less interested in achieving the lowest MDLs. These researchers also cannot wait for the long equilibrium times of commercial automated HS analyzers. P&T techniques would likewise be out of the question as a means to measure trace VOCs, due to the long purging, trapping, and thermal desorbing times involved.

The author utilized the Multipurpose Sampler® (Gerstel GmbH & Co. KG) based on the CTC PAL design atop a 6890/5973N GC-MS (Agilent Technologies) to meet the Centers of Disease Control and Prevention's (CDC's) demand for quantitating trace cyanide (CN⁻) in human blood, an example of *enviro-health TEQA*.[51] Figure 3.16 depicts a schematic drawing of HCN partitioning

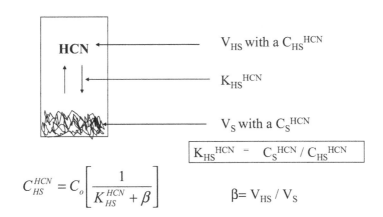

V_{HS} – volume of headspace

C_{HS}^{HCN} – concentration of HCN in ppb in headspace

K_{HS}^{HCN} – ratio of concentration of HCN in sample to concentration of HCN in headspace

V_S – volume of sample

C_S^{HCN} – concentration of HCN in sample

β – phase ratio, ratio of the volume of headspace to the volume of sample

C_o – original concentration of HCN in sample

FIGURE 3.16

between the HS and blood sample. Also shown in Figure 3.16 are the earlier definitions and the mathematics of static HS as applied to the partitioning of HCN between two phases, i.e. the HS (air) and the blood (liquid). Readers should review this figure and attempt to relate it to the general case to help reinforce the principles of static HS sample prep as shown earlier.

3.35 WHAT IF WE COULD CONCENTRATE THE ANALYTE IN THE HEADSPACE ONTO A NONPOLAR COATED FIBER AND THEN THERMALLY DESORB THE FIBER INTO THE HOT GC INLET?

This is precisely the question asked by Professor Pawliszyn as he and his research group at the University of Waterloo began to apply their polymer-coated fiber to the HS above an aqueous that contained dissolved VOCs.[52] VOCs partition into the fiber coating. The fiber is inserted into a hot GC inlet where VOCs are thermally desorbed inside the GC inlet and swept by the carrier gas onto the open tubular GC column. HS sampling and quantitative analysis could then be combined and automated! Such a technique is called *headspace solid-phase microextraction* (HS-SPME). The thermally desorbed fiber could then be inserted into the next HS sample vial! The HS vial must be agitated and kept at a constant temperature before sampling. The author recently applied hypothesis testing based on Student *t* statistics for calf serum samples spiked with ten priority pollutant VOCs. The concentration of VOCs from two batches of 20 QC serum samples per batch were compared. The first batch was repeatedly frozen then rethawed while the second batch remained frozen until analysis. A summary of a systematic study using null hypothesis testing and Student's *t* statistics was published recently.[53] A supplement to that study was also published and appears in this book as Appendix F. Appendix F shows how to conduct a null hypothesis study using *t* statistics (discussed in Chapter 2) using the VOCs data generated in the author's laboratory.

The working equation for HS-SPME was articulated earlier.[54] The basic mathematics for HS-SPME is briefly presented here. Mass balance requires that the *#mmoles n* (where $n = C_0 V_s$) present in a sample must be distributed after equilibriation and denoted by the infinity symbol ∞, can only be distributed within the headspace vessel, i.e among the sample (*s*), headspace (*h*) and fiber coating (*f*). Mass balance requires that:

$$C_0 V_S = C_f^\infty V_f + C_h^\infty V_h + C_s^\infty V_s$$

And the *#mmoles n* of analyte absorbed by the coating is given by:

$$n = \frac{K_{fs} V_f C_0 V_s}{K_{fs} V_f + K_{hs} V_h + V_s}$$

where n—#mmoles or mass of analyte absorbed by the fiber coating

    C_0—initial concentration of analyte in the sample or matrix
    K_{fs}—partition or distribution constant between the fiber and sample
    V_f—volume of the fiber coating
    V_s—volume of the sample
    K_{hs}—partition or distribution constant between the headspace and sample
    V_h—volume of the headspace

The coated fiber can also be directly inserted into an aqueous sample to extract SVOCs. This technique is called *solid-phase microextraction*. We will address SPME in a subsequent section when we introduce more contemporary *sample prep* approaches to isolate and recover priority pollutant SVOCs.

3.36   IS THERE A RECENTLY UPDATED EPA METHOD FOR VOCs USING STATIC HS TECHNIQUES?

Yes, there is, and it is Method 5021A: *VOCs in various Sample Matrices using Equilibrium Headspace Analysis.* The method was revised in July 2014 and is part of the SW-846 series of methods published by the Office of Solid Waste at EPA. The method uses the static HS technique to determine VOCs from soil or another solids matrix. This method also introduced some experimental considerations with respect to trace VOC analyses of soil samples.[55] Method 5021A is applicable to a wide range of organic compounds that have sufficiently high volatility to be effectively removed from soil samples using the static HS technique. The method is used in combination with a determinative technique such as described in EPA Methods 8015, 8021 or 8260. Determinative techniques are introduced in Chapter 4. The method cautions the user to the fact that solid samples whose organic matter content exceeds 1%, or compounds with high octanol–water partition coefficients, may yield a lower result for the determination of VOCs by static HS than dynamic headspace (P&T). It is recommended to add surrogates to each and every sample to evaluate the so-called matrix effect. Today, in contrast to a previous era, EPA methods are readily accessible online. Study of Method 5021 would be an excellent introduction to an EPA method for someone just entering the field of trace VOCs analysis.

3.37   HOW SHOULD I PROCEED TO QUANTITATE VOCs?

The EPA's approach to sample preparation to determine trace concentration levels of various VOCs is discussed in this section. We have already introduced the static HS technique from a theoretical point of view and shared the author's approach to quantitating trace cyanide in blood. Purge-and-trap (P&T) techniques, in contrast to static HS, involve passing an inert gas through an aqueous sample such as drinking water or groundwater. The purge gas is directed to a *trap,* a term commonly used to describe a packed column that contains an adsorbent that exhibits a high efficiency for VOCs. After sufficient time for purging and trapping has been allowed, the trap is rapidly heated to thermally desorb the VOCs off of the adsorbent and directly into the GC. In this manner, VOCs are removed from the environmental sample without a matrix effect, and the objectives of TEQA can be met.

Purge-and-trap is dynamic; the analyte is merely transferred from the sample to an organic polymeric matrix that exhibits a large surface area. One point of view assumes that static HS is merely a screen for dynamic HS (P&T) because MDLs for P&T are significantly lower than those for static HS. A second approach to screening for particular priority pollutant VOCs is called *hexadecane screening*. The logic as to what technique to use is depicted in the flowchart in Figure 3.17. The sample matrix plays a significant role as to whether one chooses the static or dynamic HS technique for the isolation and recovery of VOCs from environmental samples.

When analyzing environmental samples that are considered grossly contaminated, *screening techniques* are essential. Screening techniques serve to inform the analyst as to whether a dilution of the original sample is warranted. Laboratories that have both automatic static HS and automated P&T systems are more likely to use static HS for the initial screen, followed by P&T for the quantitative analysis. This is particularly true for labs that engage in EPA contract work. Labs that have either static HS or P&T, but not both, often elect to use hexadecane screening prior

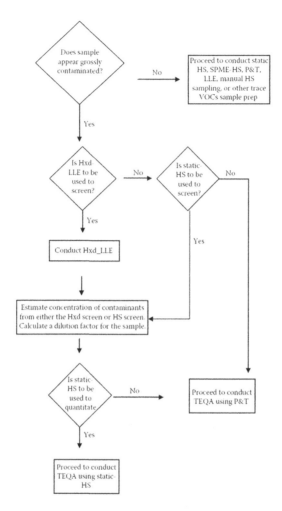

FIGURE 3.17

to the quantitative determination by either static HS or P&T. Labs that analyze predominantly drinking water samples most likely do not have a need to screen samples and usually proceed directly to the available determinative technique.

3.38 WHAT IS HEXADECANE SCREENING?

Hexadecane screening uses hexadecane (Hxd) as the extractant to perform an initial LLE of a water or wastewater sample that is suspected of containing appreciable levels of VOCs in a complex sample matrix. We thus refer to this technique with the abbreviation Hxd-LLE. Figure 3.18 is a *gas chromatogram* or *GC-gram*, of a groundwater sample that has been in contact with gasoline. This is a common occurrence in the environment because gasoline storage tanks are known to corrode with age and leak into the groundwater supply, thus contaminating sources of drinking water. The VOCs, commonly referred to as BTEX components, are clearly evident in the GC-gram. BTEX is an abbreviation of the aromatic VOCs benzene, toluene, ethyl benzene, and one or more of the three isomeric xylenes, *ortho, para,* and *meta.* A fourth

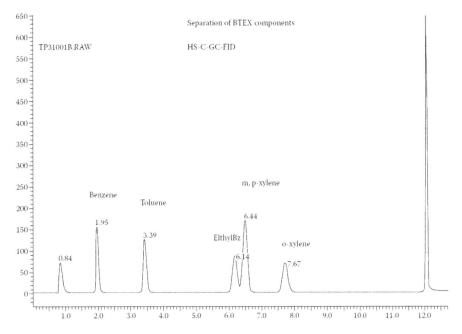

FIGURE 3.18

component is more frequently showing up in groundwater samples. This compound is methyl-*t*-butyl ether (MTBE). MTBE was added to gasoline over 20 years ago as an oxygenated hydrocarbon until it began showing up in groundwater. A groundwater sample that exhibits a significant concentration of BTEX should be diluted with high-purity water. This diluted sample should be subsequently analyzed to quantitatively determine the concentrations of BTEX and MTBE using either static HS or P&T. If a hydrocarbon of much lower molecular weight, such as *n*-hexane, were used to conduct the preliminary assessment of the degree of sample contamination, the instrument response due to the elution of *n*-hexane from the GC would interfere with one or more of the peaks due to the presence of BTEX. It is illustrative to compare the physical properties of *n*-hexane vs. *n*-hexadecane. The physical properties of both hydrocarbons are compared below:

Compound	MW	Tb (°C)	$\rho 20$°C (g/mL)	P'	Solubility in Water (%)
n-Hexane	86	68.7	0.659	0.1	0.014
Hexadecane	226	287	0.773	0.5	—

It is important to use as high a purity of Hxd as can be obtained; otherwise, impurities in this extracting solvent might cause interference with some VOCs. The absence of lower-molecular-weight impurities in Hxd early in the GC-gram greatly contributes to lowering MDLs. A GC that incorporates a flame ionization detector is usually designated as the Hxd-LLE screening instrument. A mini-LLE scale using a phase ratio, $\beta = V_{Hxd}/V_{aq} = 0.057$, such that approximately 35 mL of aqueous sample is extracted once into 2 mL of Hxd is a common technique. The limitations to the effective use of Hxd-LLE lie in the nature of the environmental sample matrix.

Samples such as wastewaters that are prone to cause emulsions prevent analysts from quickly injecting a 1 μL aliquot of the extract. Common techniques for breaking emulsions include the following:[56]

1. Adding salt to the aqueous phase
2. Using a heating-cooling extraction vessel
3. Filtering the emulsion through a glass wool plug
4. Filtering the emulsion through a phase separation filter paper
5. Using a centrifuge
6. Adding a small amount of a different organic solvent.

If this author is asked to estimate the concentration level of VOCs and static HS or P&T techniques are not available, he would employ Hxd-LLE. EPA Method 3820 from the SW-846 series is a useful starting point for this LLE technique. Seven different approaches to prepare environmental samples to determine VOCs are presented in SW-846. We will discuss each of these with an emphasis on the operational aspects.

3.39   WHAT ARE EPA'S SAMPLE PREP APPROACHES TO TRACE VOCs?

Method 5000 in the SW-846 lists seven methods to prepare samples for the quantitative determination of VOCs. These methods reflect the different sample matrices involved. Water, soil/sediment, sludge, and waste samples that require analysis for VOCs are extracted or introduced into a GC or GC-MS system by the various methods, as discussed below. The following methods are briefly introduced from an EPA point of view, and the advantages and limitations from that viewpoint are discussed.

Method 3585 describes a solvent dilution technique using hexadecane followed by direct injection into a GC-MS instrument for the determination of VOCs in waste oils. Direct injection of an oil waste could lead to instrument contamination problems. However, the method provides for a quick turnaround.

Method 5021 describes the automated static HS technique. A soil sample is placed in a tared septum-sealed vial at the time of sampling. A matrix modifier containing internal or surrogate standards is added. The sample vial is placed into an automated headspace sampler vial, which is placed into an automated equilibrium headspace sample. The vial's temperature is elevated to a fixed value that does not change over time, and the contents of the vial are mixed by mechanical agitation. A measured volume of headspace is automatically introduced into a GC or GC-MS. The method is automated and downtime is minimal. However, the cost of the automated system is appreciable. Contamination of the instrument is minimal. The MDL for this technique is significantly influenced by the choice of GC detector.

Method 5030 describes the technique of purge-and-trap for the introduction of purgeable organics into a GC or GC-MS. The method is applicable to aqueous samples such as groundwater, surface water, drinking water, wastewater, and water-miscible extracts prepared by Method 5035. An inert gas is bubbled through the sample and the VOCs are efficiently transferred from the aqueous phase to the vapor phase. The vapor phase is swept through a sorbent trap where the purgeables are trapped. After purging is completed, the trap is heated and backflushed with the inert gas to desorb the purgeables onto a GC column. P&T is easily automated and provides good precision and accuracy. However, the method is easily contaminated by samples that contain compounds that are present in the sample at the ppm level. Since P&T is an exhaustive removal of VOCs from the environmental sample, some argue that this technique offers the lowest MDLs. Again, the MDL would be strongly influenced by the choice of GC detector. Because this is the most commonly used method to determine VOCs, the method will be elaborated upon later.

Method 5031 describes an azeotropic distillation technique for the determination of nonpurgeable, water-soluble VOCs that are present in aqueous environmental samples. The sample is distilled in an azeotropic distillation apparatus, followed by direct aqueous injection of the distillate into a GC or GC-MS system. The method is not amenable to automation. The distillation is time consuming and is limited to a small number of samples.

Method 5032 describes a closed-system vacuum distillation technique for the determination of VOCs that include nonpurgeable, water-soluble, volatile organics in aqueous samples, solids, and oily waste. The sample is introduced into a sample flask that is in turn attached to the vacuum distillation apparatus. The sample chamber pressure is reduced and remains at approximately 10 torr (the vapor pressure of water) as water is removed from the sample. The vapor is passed over a condenser coil chilled to a temperature of 10°C or less. This results in the condensation of water vapor. The uncondensed distillate is cryogenically trapped on a section of 1/8 in. stainless-steel tubing chilled to the temperature of liquid nitrogen (−196°C). After an appropriate distillation period, the condensate contained in the cryogenic trap is thermally desorbed and transferred to the GC or GC-MS using helium carrier gas. The method very efficiently extracts organics from a variety of matrices. The method requires a vacuum system along with cryogenic cooling and is not readily automated.

Method 5035 describes a closed-system P&T for the determination of VOCs that are purgeable from a solid matrix at 40°C. The method is amenable to soil/sediment and any solid waste sample of a consistency similar to soil. It differs from the original soil method (Method 5030) in that a sample, usually 5 g, is placed into the sample vial at the time of sampling along with a matrix-modifying solution. The sample remains hermetically sealed from sampling through analysis as the closed-system P&T device automatically adds standards and then performs the purge-and-trap. The method is more accurate than Method 5030 because the sample container is never opened. This minimizes the loss of VOCs through sample handling. However, it does require a special P&T device. Oil wastes can also be examined using this method.

Method 5041 describes a method that is applicable to the analysis of sorbent cartridges from a volatile organic sampling train. The sorbent cartridges are placed in a thermal desorber that is in turn connected to a P&T device.

3.40   WHAT IS THE P&T TECHNIQUE?

The P&T method to isolate, recover, and quantitate VOCs in various environmental sources of water has and continues to remain the *premier technique* for this class of environmental contaminants. The technique was developed at the EPA in the early 1970s and remains the method of choice, particularly for environmental testing labs that are regulatory driven.[57] P&T has had the most success with drinking water samples when combined with gas chromatography and element-specific detectors. GC detectors will be discussed in Chapter 4. In this section, we will discuss the EPA methods that use the P&T technique to achieve the goals of TEQA. EPA Method 502.2 summarizes the method as follows:[58]

Highly volatile organic compounds with low water solubility are extracted (purged) from the sample matrix by bubbling an inert gas through a 5 mL aqueous sample. Purged sample components are trapped in a tube containing suitable sorbent materials. When purging is complete, the sorbent tube is heated and back-flushed with helium to thermally desorb trapped sample components onto a capillary gas chromatography (GC) column. The column is temperature programmed to separate the method analytes which are then detected with a photo-ionization detector (PID) and an electrolytic conductivity detector (ElCD) placed in series. Analytes are quantitated by procedural standard calibration ... Identifications are made by comparison of the retention times of unknown peaks to the retention times of standards analyzed under the same conditions used for samples. Additional confirmatory information can be gained by comparing the relative response from the two detectors. For absolute confirmation, a gas chromatography-mass spectrometry (GC-MS) determination according to EPA Method 524.2 is recommended.

The classical purge vessel is shown in Figure 3.19. Usually, 5 mL of an environmental aqueous sample is placed in the vessel. The sample inlet utilizes a two-way valve. A liquid-handling syringe that can deliver at least 5 mL of sample is connected to this sample inlet, and the sample is transferred to the purge vessel in a way that minimizes the sample's exposure to the atmosphere. Note that the incoming inert purge gas is passed through a molecular sieve to remove moisture. A hydrocarbon trap can also be inserted prior to the purge vessel to remove traces of organic impurities as well. The vessel contains a fritted gas sparge tube that serves to finely divide and disperse the incoming purge gas. These inert gas bubbles from the purging provide numerous opportunities for dissolved organic solutes to escape to the gas phase. The acceptable dimensions for the purge device are such, according to EPA Method 502.2, that it must accommodate a 5 mL sample with a water column at least 5 cm deep. The headspace above the sample must be kept to a minimum of <15 mL to eliminate dead-volume effects. The glass frit should be installed at the base of the sample chamber, with dispersed bubbles having a diameter of <3 mm at the surface of the frit.

Figure 3.19 also depicts a typical trap to be used in conjunction with the purge vessel. The trap must be at least 25 cm long and have an inside diameter of at least 0.105 in., according to Method 502.2. The trap must contain the following amounts of adsorbents:

* One third of the trap is to be filled with 2, 6-diphenylene oxide polymer, commonly called Tenax.
* One third of the trap is to be filled with silica gel.
* One third of the trap is to be filled with coconut charcoal.

It is recommended that 1 cm of methyl silicone-coated packing be inserted at the inlet to extend the life of the trap. Method 5030B from the SW-846 series is more specific than Method 502.2 and recommends a 3% OV-1 on Chromosorb-W, 60/80 mesh, or equivalent. Analysts who do not need to quantitate dichlorodifluoromethane do not need to use the charcoal and can replace this charcoal with more Tenax. If only analytes whose boiling points are above 35°C are to be determined, both the charcoal and the silica gel can be eliminated and replaced with Tenax. The trap needs to be conditioned at 180°C prior to use and vented to the atmosphere instead of the analytical GC column. It is also recommended that the trap be reconditioned on a daily basis at the same temperature. Tenax is a unique polymer and offers the advantage that water is not trapped to any great extent.

Commercially available P&T units are fully automated and consist of a bank of purge vessels with switching valves that enable one vessel to be purged after another. Figure 3.20 is a schematic diagram that clearly depicts the purge, trap, and thermal desorption steps involved in this technique. A six-port valve, placed after the trap and interfaced to the injection port of a GC, provides the needed connection. Referring to Figure 3.20a, the P&T step is shown. Inert gas, usually helium, enters the purge vessel while the trap outlet is vented to the atmosphere. Meanwhile, GC carrier gas flows through one side of the six-port valve directly to the GC. The purge vessel and

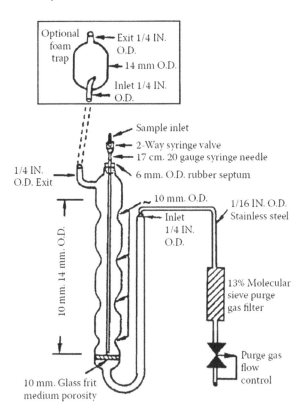

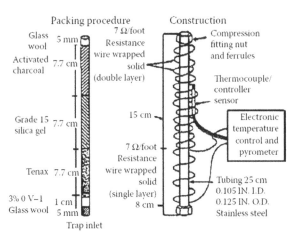

FIGURE 3.19

the trap are generally kept at ambient temperature. Some commercially available units provide for a heated purge vessel. Referring to Figure 3.20b, the direction of gas flow during the desorb step is illustrated. The valve is turned and inert gas enters and passes over the trap in a direction opposite to the P&T step. The trap is rapidly heated to the required final temperature and the trap outlet is directed to the injection port of the GC. GCs that are equipped with cryogenic cooling can deposit the VOCs from the trap to the GC column inlet as a plug. The GC can then

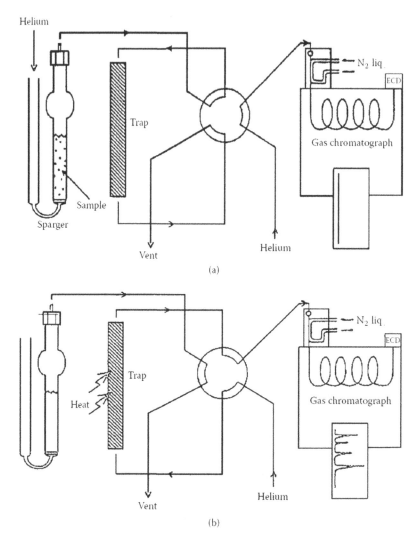

FIGURE 3.20

be temperature programmed from the cryogenic temperature to the final temperature, and hence complete the gas chromatographic separation of all VOCs.

Operating conditions drawn from EPA Method 5030B for the P&T system differ slightly based on which determinative method is used. Analytical data for VOCs from applying the P & T technique is presented in Table 3.7. Gas chromatographic results for selected VOCs using the P&T technique are shown in Table 3.8[50].

3.41 HOW DOES ONE GO ABOUT CONDUCTING THE P&T TECHNIQUE?

Let us assume that the decisions of whether to screen the sample prior to conducting P&T have been made and that we are ready to actually perform the technique. We have available in our laboratory organic-free water, the determinative method has been defined, and we have reviewed EPA Method 5000 for guidance with respect to internal and surrogate standards. Environmental samples such as groundwater, drinking water, wastewater, and so forth, have already been

TABLE 3.7
Operating Conditions Drawn from EPA Method 5030B

Analysis Method	8015	8021 or 8260
Purge gas	N_2 or He	N_2 or He
Purge gas flow rate (mL/min)	20	40
Purge time (min)	15.0	11.0
Purge temperature (°C)	85	Ambient
Desorb temperature (°C)	180	180
Back-flush inert gas (mL/min)	20–60	20–60
Desorb time (min)	1.5	4

TABLE 3.8
Retention Times, Purge Ratios and Coefficient of Variation for Various VOCs Taken from Reference 50

VOC	Tr (Min)A	Pi	% RSD
Chloromethane	1.82	339	36
1,2-Dichloroethane	9.99	2.21	6
1, 1, 1, 2-Tetrachloroethane	16.98	6.06	9
1, 2, 4-Trichlorobenzene	26.30	3.64	8
Acetone	4.30	1.05	2
2-Butanone	8.34	1.03	5

[a] 30 m × 0.53 mm DB-624 (J&W Scientific); refer to EPA Method 524.2 for GC column temperature program and other GC conditions.

collected, stored in capped bottles with minimum headspace, free of solvent fumes, and stored at 4°C. If the sample was found to be improperly sealed, it should be discarded. All samples should be analyzed within 14 days of receipt at the laboratory. Samples not analyzed within this period must be noted and data is considered to represent a minimum value. This is the essence of the so-called *holding times*. Holdings times, like detection limits that were discussed in Chapter 2, are a controversial topic between the regulatory agency and contracting laboratory. Using Method 5030B for guidance, the sequence of steps needed to carry out the method are as follows:

- Initial calibration: The P&T apparatus is conditioned overnight with an inert gas flow rate of at least 20 mL/min and the Tenax trap held at a temperature of 180°C. The P&T apparatus is connected to the GC, and each purge vessel is filled to its total volume with organic-free water. Methanolic solutions containing VOCs at high enough concentration such that only microliter aliquots need to be added to the water in each purge vessel are added to a series of purge vessels. A blank and set of working calibration standards are prepared right in the purge vessels. Internal standards and surrogates are also added as their methanolic solutions as standard references in a manner similar to that of the analytes to be calibrated. Refer to Chapter 2 for a discussion of the various modes of instrumental calibration.

- Conduct the P&T for the set of blanks and working calibration standards.
- Initial calibration verification (ICV): Prepare one or more purge vessels that contain the ICV. Sometimes, it is advantageous to prepare ICVs in triplicate to enable a preliminary evaluation of the precision and accuracy of the calibration to be made. The ICV criteria are usually given in a determinative method such as in the 8000 series of SW-846. These criteria must be met before real samples can be run.
- Adjust the purge gas flow rate referring to the guidance given in the above table.
- Sample delivery to the purge vessel: Remove the plunger from a 5 mL liquid-handling syringe and attach a closed syringe valve. Open the sample or standard bottle, which has been allowed to come to ambient temperature. Carefully pour the sample into the syringe barrel to just short of overflowing. Replace the syringe plunger and compress the sample. Open the syringe valve and vent any residual air while adjusting the sample volume to 5.0 mL. This process of taking an aliquot destroys the validity of the liquid sample for future analysis. If there is only one sample vial, the analyst should fill a second syringe at this time to protect against possible loss of sample integrity. Alternatively, carefully transfer the remaining sample into a 20 mL vial and seal with zero headspace. The second sample is maintained only until such time as when the analyst has determined that the first sample has been analyzed properly. In this way, the VOC content of the environmental sample is preserved during the transfer to the purge vessel.

A GC capillary or open tubular column chromatogram from EPA Method 502.2 that depicts the separated peaks from both the photo-ionization detector (PID) and the electrolytic conductivity detector (ElCD) is shown in Figure 3.21. Both chromatograms are overlaid and show the different response for each chromatographically resolved peak between detectors.

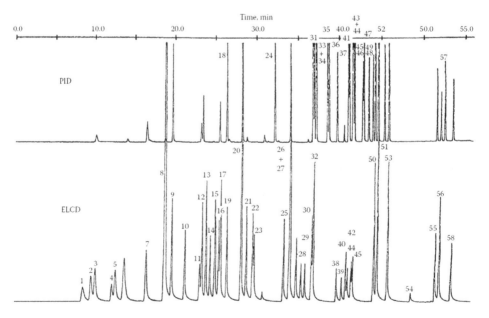

FIGURE 3.21

3.42 DO ALL VOCs PURGE WITH THE SAME RATE?

No, they do not, and until recently, little was known about the kinetics of purging for priority pollutant VOCs using conventional P&T techniques. We describe the findings of Lin et al., who demonstrated that first-order kinetics are followed for the removal of VOCs from P&T vessels.[59] The authors make the point that because EPA methods such as 524.2 and 624 require the use of internal standards to obtain relative response factors, the percent purge efficiency is never considered. Some 28 priority pollutant VOCs were studied. Each VOC, dissolved in methanol, was spiked into high-purity water at concentration levels of 1 and 10 ppb, with a 10 ppb internal standard. Spiked samples were purged for 11 min. A GC-MS was used, and the absolute peak area for the characteristic primary ion was used to quantitate. Experimentally, samples were purged with helium for 11 min and GC-MS data were obtained. These purged samples were then purged a second time for the same 11 min duration under identical conditions. Peak areas were obtained in the same manner as for the first purge.

Let us begin to construct a mathematical view of the kinetics of P&T by first defining the purge ratio, P_i, as being the ratio of the mean GC-MS area count for the first purge to the mean GC-MS area count for the second purge according to

$$P_i = \frac{A_i^1\left(\text{ave}\right)}{A_i^2\left(\text{ave}\right)} = \frac{\sum_j^5 A_{ij}^1 / 5}{\sum_j^5 A_{ij}^2 / 5}$$

Each peak area for the ith analyte was obtained by averaging peak areas over j replicate purges, with j taken from 1 to 5. A selected list of VOCs, along with each VOC retention time, purge ratio, and coefficient of variation (refer to Chapter 2) from their work, is given in the following table:

The low percent RSDs, except for the very volatile chloromethane, indicate that P_i is constant over the range of concentrations studied. According to these authors, consistent values of P_i would indicate that the purging VOCs from an aqueous solution can be mathematically described by first-order chemical kinetics. How fast any of the 28 VOCs studies are removed from a fixed volume of an aqueous sample such as groundwater in a conventional purge vessel is expressed as $-dC/dt$, where the negative sign indicates that the concentration of a VOC decreases with time. First-order kinetics suggests that the rate is directly proportional to the concentration of each VOC. (pp. 1115, 1298, Appendix 9)[47] Expressed mathematically,

$$\frac{dC}{dt} = -kC \tag{3.32}$$

where C is the analyte concentration at any time t and k is the first-order rate constant.

Differences in the magnitude of k depend on the chemical nature of the particular VOC and the degree of intermolecular interaction with the sample matrix—in this case, a relatively clean water sample. Equation (3.32) is first rearranged as

$$\frac{dC}{C} = -k\,dt$$

Integrating between the initial concentration C_1 and the concentration after the first purge, C_2, on the left side, while integrating between time $t = t_0$ and time $t = t_0 + \Delta t$ on the right side is shown below. Identical integration is also appropriate between C_2 and C_3. Δt is the time it takes to purge the sample. We then have

$$\int_{C_1}^{C_2} \frac{dC}{C} = -k \int_{t_0}^{t_0 + \Delta t} dt$$

Evaluating the definite integrals for both sides of the equation yields

$$\ln\left(C_2 - C_1\right) = -k\Delta t$$

Utilizing a property of logarithms gives

$$\ln \frac{C_2}{C_1} = -k\Delta t$$

Expressed in terms of exponents,

$$\frac{C_2}{C_1} = e^{-k\Delta t}$$

$$\therefore C_2 = C_1 e^{-k\Delta t} \tag{3.33}$$

Equation (3.33) implies that C_2 can be expressed in terms of C_1, k, and Δt.

The experiment starts with each VOC at a concentration C_1. After purge 1, each analyte is at concentration C_2. After purge 2, each analyte is at concentration C_3. This is depicted by

$$C_1 \xrightarrow{\text{Purge1}} C_2 \qquad C_2 \xrightarrow{\text{Purge2}} C_3$$

Because VOCs are continuously removed from a fixed volume of aqueous sample, it is also true that

$$C_3 < C_2 < C_1$$

The peak area obtained after the first purge A^1 is proportional to the difference in concentration between C_1 and C_2; that is,

$$A^1 \propto C_1 - C_2$$

Substituting for C_2 using Equation (3.33) gives

$$A^1 \propto C_1 \left(1 - e^{-k\Delta t}\right)$$

Likewise, the peak area obtained for each analyte after the second purge is proportional to the difference in concentration between C_2 and C_3 according to

$$A^2 \propto C_2 - C_3$$

Substituting for C_3 using Equation (3.33) gives

$$A^2 \propto C_2 \left(1 - e^{-k\Delta t}\right)$$

The ratio of peak areas between the first and second purges for a given analyte, P, can be related to these differences in concentration as follows:

$$P = \frac{C_1 - C_2}{C_2 - C_3} = \frac{C_1}{C_2}$$

Substituting for Equation (3.33) yields

$$P = \frac{C_1}{C_1 e^{-k\Delta t}}$$

This simplifies to

$$P = e^{k\Delta t} \tag{3.34}$$

Equation (3.34) suggests that P is always greater than 1 and is independent of the initial analyte concentration. The development of Equation (3.34) has assumed first-order kinetics. The authors observed good precision over the five replicate experiments for all 28 VOCs studied, with the exception of the very volatile chloromethane.

If P is determined experimentally, by first obtaining the GC-MS peak area, A_1, and then obtaining A_2, Equation (3.34) can be solved for the first-order rate constant, k.

In addition, when $C_2 = \frac{1}{2}C_1$, the time it takes for half of the analyte to be removed can be found according to

$$\frac{C_1}{C_2} = \frac{C_1}{\frac{1}{2}}C_1 = 2\frac{C_1}{C_1} = e^{kt^{1/2}}$$

$$2 = e^{kt]^{1/2}}$$

$$\ln 2 = kt_{1/2}$$

$$\therefore t_{1/2} = \frac{1}{k}\ln 2 \tag{3.35}$$

Using Equation (3.35), the $t_{1/2}$ values given for the three VOCs from the author's work are shown below:

VOC	$t_{1/2}$ (min)
Chloromethane	1.3
1, 1, 1, 2-Tetyrachloroethane	4.2
Acetone	156

A purge time, $\Delta t = 11$ min, is used in indicated EPA methods. For hydrophobic VOCs, the $t_{1/2}$ seems to be too low, and for hydrophilic VOCs, the $t_{1/2}$ seems to be too high. Other generalizations emerge with respect to the chemical nature of the VOCs studied. The effect of deuterating 1,2-dichlorobenzene yields values for $t_{1/2}$ that are quite close to each other, 5.6 and 5.8 (1,2-dichlorobenzene-d$_4$), respectively. Differences in structural isomers such as 1, 1, 2, 2- vs. 1, 1, 1, 2-tetrachloroethanes are reflected not only in their volatility in water, but also in their gas-phase stability of both neutral and ionic states. It would also appear that the choice of internal standard should reflect the kinetics of purging in addition to GC relative retention time. Choosing internal standards based on relative retention times has been the hallmark of most EPA methods. This work would suggest that $t_{1/2}$ also be taken into consideration when deciding on which internal standard to use when using P&T techniques.

We leave sample prep for VOCs and return to *semivolatile organic compounds* (SVOCs). Conventional LLE sample extracts containing SVOCs undergoing cleanup techniques are first introduced followed by an introduction to supercritical fluid extraction (SFE). We then introduce the more recent alternatives to LLE: reversed-phase (RP-SPE), normal-phase (NP-SPE), and ion-exchange solid-phase extraction (IE-SPE). We then introduce solid-phase microextraction as applied to SVOCs (SPME). We introduce a relatively newer sample prep technique: microextraction by packed sorbent (MEPS). We also introduce stir-bar sorptive extraction (SBSE), solid-phase matrix dispersion (SPMD), and QUECHERS. These are relatively newer sample prep techniques as well.

We already discussed how LLE is used to remove unwanted neutral interferences when the target analyte of interest is a weak acid and can therefore be ionized. *Cleanup* introduced here refers to the need to remove neutral polar interferences from neutral targeted SVOC analytes.

3.43   WHAT IS CLEANUP?

It is not the use of detergent to remove dirt! It is the removal of chemical interferences from the extract following any of the extraction techniques discussed earlier. The dirtier the sample matrix, the greater is the need for extract cleanup. Cleanup plays an important role in both *enviro-chem*ical and in *enviro-health* TEQA. For the *enviro-chemical* sample, drinking and surface water may not need cleanup while wastewater, sludge and petroleum-contaminated soil need cleanup. For the *enviro-health* sample, urine may not need cleanup while blood may need cleanup. Chromatographers know very quickly when a sample extract has not been sufficiently cleaned up! Separated analyte peaks in the chromatogram are obscured in contrast to those analyte peaks in the reference standards chromatogram! Chromatographers have been known to shout out in the laboratory something like "looks like a matrix effect"!

Cleanup is accomplished by using low pressure or gravity-fed column liquid chromatography (LPLC). LPLC is a form of column liquid chromatography. Tswett was performing LPLC when he "discovered chromatography" in the year 1906!|

Meloan lists 16 adsorbents or stationary phases that have been used historically to conduct LPLC.[60] These adsorbents are listed in Table 3.9 from most active (polar interferences most strongly retained) to least active (polar interferences weakly retained).

EPA Method 3600C from the SW-846 series of solid waste methods provides an overview of sample cleanup techniques. Grossly contaminated solid wastes represent a significant challenge to cleanup techniques. The 3000 series includes eight cleanup methods listed in Table 3.10. To quote from Method 3600C:[61]

The purpose of applying a cleanup method to an extract is to remove interferences and high boiling material that may result in ... errors in quantitation ... false positives ... false negatives ... rapid deterioration of capillary GC columns ... instrument downtime due to contamination of detectors, inlets and mass spectrometer ion sources.

These techniques may be used individually or in various combinations, depending on the extent and nature of the coextractives.

TABLE 3.9
Adsorbents for Extract Cleanup

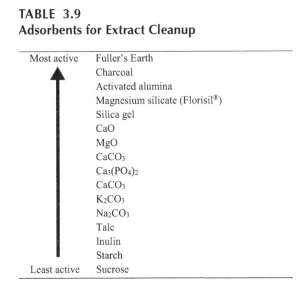

Most active	Fuller's Earth
	Charcoal
	Activated alumina
	Magnesium silicate (Florisil®)
	Silica gel
	CaO
	MgO
	$CaCO_3$
	$Ca_3(PO_4)_2$
	$CaCO_3$
	K_2CO_3
	Na_2CO_3
	Talc
	Inulin
	Starch
Least active	Sucrose

TABLE 3.10
List of EPA Cleanup Methods in SW 846

Method	Method Name	Cleanup Type
3610	Alumina Cleanup	Adsorption
3611	Alumina Cleanup and Separation of Petroleum Wastes	Adsorption
3620	Florisil Cleanup	Adsorption
3630	Silica Gel Cleanup	Adsorption
3640	Gel-Permeation Cleanup (GPC)	Size separation
3650	Acid–Base Cleanup	Acid–base partitioning
3660	Sulfur Cleanup	Oxidation/reduction
3665	Sulfuric Acid/Permanganate Cleanup	Oxidation/reduction

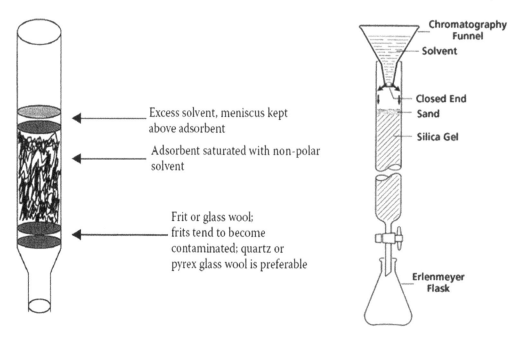

Excess solvent, meniscus kept
above adsorbent

Adsorbent saturated with non-polar
solvent

Frit or glass wool;
frits tend to become
contaminated; quartz or
pyrex glass wool is preferable

Chromatography
Funnel
Solvent

Closed End
Sand

Silica Gel

Erlenmeyer
Flask

FIGURE 3.22

Stroll into any pesticide residue analysis laboratory that utilizes LLE and S-LSE sample prep
techniques and you will see cylindrical glass columns of dimensions: from 1 cm to 4 cm i.d. wide
× 15 cm to 60 cm in length. These glass columns contain a retaining frit or glass wool plug at the
bottom. The column is packed with one or more of the commercially available and highly purified adsorbents listed above. A sketch of a simple glass column used to conduct LPLC is shown in
Figure 3.22.

Adsorption used to clean up nonpolar extracts is a normal-phase chromatographic phenomenon. Polar interferences are separated from nonpolar targeted analytes of interest to both
enviro-chemical and enviro-health analysis. For example, an extract that contains both PAHs
and fatty acids upon being passed through a properly deactivated alumina column will separate both compound classes. *Deactivation* is a term used in adsorption chromatography to
mean that the active sites have been "tied up" by the addition of water. PAHs would be weakly
retained while fatty acids would be strongly retained owing to the terminal carboxylic acid
functional group. The strongly retained and more polar fatty acids will not elute if the mobile
phase is of low polarity. Common low-polarity solvents include hexane, heptane, and cyclohexane. The underlying principles of column chromatography are introduced in Chapter 4.
Figure 3.23 places each of the 3000 series cleanup methods in a logical tree structure while
adding acid–base hydrolysis. Acid–base hydrolysis is not listed as a method in SW-846. In
addition to adsorption, each cleanup method from SW-846 is briefly described in the next
paragraph.

"Acid–Base Partitioning," Method 3650, is an LLE technique that separates organic analytes
and interferences from neutral organic compounds of enviro-chemical or enviro-health interest.
For example, neutral PAHs are separated from acidic phenols when analyzing a hazardous
waste site contaminated with creosote and pentachlorophenol. (pp. 1115, 1298, Appendix 9)[61]
"Sulfur Cleanup," Method 3660, is used to eliminate elemental sulfur from sample extracts that
could interfere with chromatographic determinative techniques. "Sulfuric Acid / Permanganate
Cleanup," Method 3665, provides rigorous cleanup of sample extracts prior to the analysis of
samples that might contain PCBs. Most OCs, such as Dieldrin and other cyclodiene insecticides,

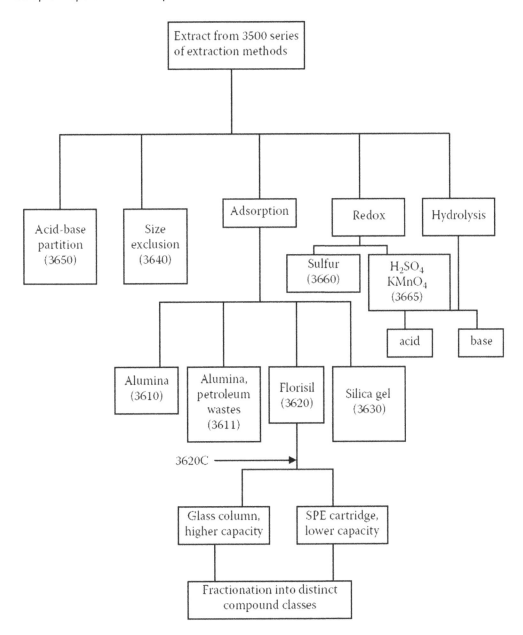

FIGURE 3.23

are destroyed chemically by this vigorous oxidation. Size exclusion or "Gel Permeation Cleanup (GPC)," Method 3640, is the most universal of the cleanup techniques and is based on the principle of size exclusion liquid chromatography. GPC can separate broad classes of SVOCs and pesticides. GPC effectively removes lipids and other high-molecular-weight interferences from nonpolar sample extracts. GPC has most often been applied to clean up sample extracts with appreciable oil or lipid (fat) content, such as petroleum-contaminated soil or fish tissue and, as we shall see, animal feed. Application of GPC to cleanup will be discussed in more detail shortly. Adsorption chromatography enables another important facet of sample prep to be considered: the separation of compound classes, commonly referred to as *fractionation.*

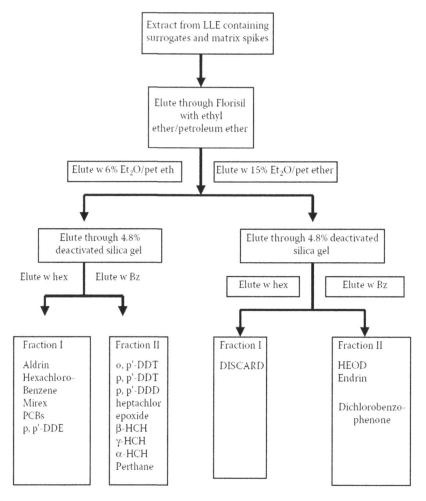

FIGURE 3.24

3.44 CAN ADSORPTION CHROMATOGRAPHY BE USED TO ACHIEVE COMPOUND CLASS FRACTIONATION IN ADDITION TO CLEANUP?

Yes, indeed. Both Florisil and silica gel adsorption chromatography, under LPLC conditions, provide opportunities to achieve compound class fractionation. Najam and coworkers at the CDC earlier published their fractionation scheme that enabled them to resolve a mixture of OCs and all 209 PCB congeners from human serum.[62] Figure 3.24 is adapted from this paper and it shows the logic of how both adsorbents were used in sequence to achieve a partial separation of polychlorinated biphenyls (PCBs) from organochlorine pesticides (OCs). This flowchart is a good example of *enviro-health* TEQA sample prep. EPA Method 3620C extends adsorption chromatography using larger glass column LPLC *and* smaller NP-SPE cartridges to achieve a compound class fractionation. Figure 3.25 is a flowchart from 3620C that outlines how a mixture of OCs and PCBs can be class separated by the use of *binary eluents* that are miscible yet of sufficient solvent polarity (P') differences and represents a good example of *enviro-chemical* TEQA sample prep.[63] Principles that underlie P' will be discussed later in this chapter. Appendix E includes other innovative sample prep flowcharts that caught the author's attention over the past

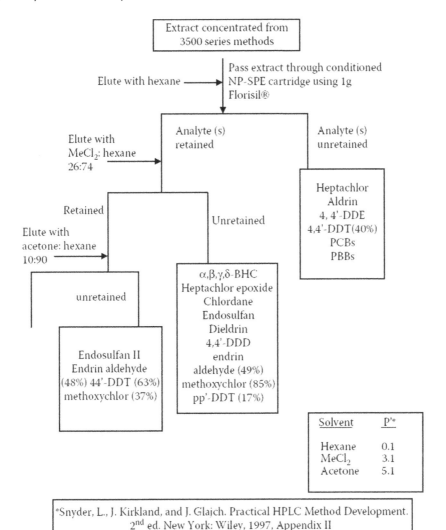

FIGURE 3.25

decade. When oil and grease contaminate an environmental sample or when the lipid content of a biological sample is appreciable, *gel permeation chromatography* (GPC) is a good first cleanup approach to the concentrated sample extract.

3.45   HOW DOES GPC CLEAN UP OILY SAMPLE EXTRACTS?

Gel permeation chromatography (GPC) involves the passage of a mobile phase of low polarity across a polystyrene-divinyl benzene (PS-DVB) or silica stationary phase whose average pore size covers a range from 100 to 1,000,000 Å (silica). Preparative-scale GPC in contrast to analytical-scale GPC is most likely found in laboratories that conduct TEQA. EPA Method 3640 established GPC as the principal means to remove lipid and other higher-molecular-weight coextractives.[64] The method requires a preparative-scale liquid chromatograph with or without an autosampler injector. An ultraviolet absorbance detector equipped with a preparative flow-through cuvette assembly is required for real-time readout. Of course, this analog detector can

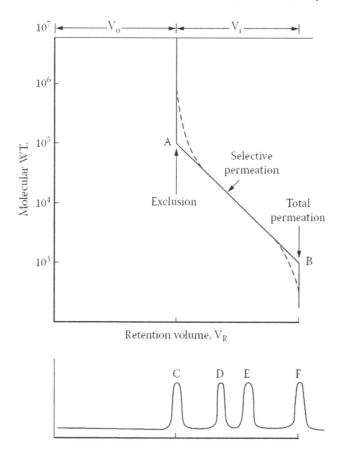

FIGURE 3.26

be interfaced through an A-to-D converter to a PC for complete automated data acquisition and control of the chromatograph. This author first set up a GPC for cleanup of oil-laden soil from Superfund hazardous waste sites back in the mid-1980s. The sole supplier back then of preparative GPC instruments for environmental testing labs to implement EPA Method 3640 was ABC Laboratories, Inc. Several companies today offer preparative GPC instruments configured to provide cleanup of oily or lipid-enriched sample extracts.

The principle of separation via GPC is briefly described as follows. Separation in GPC depends on molecular size, which in turn depends on molecular weight (MW). Figure 3.26, a hypothetical plot of MW against GPC retention volume, V_R, is a classic drawing that explains chromatographic separation by size exclusion very nicely. To illustrate, for all compounds with MWs of <1000, total permeation occurs. This means that the volume of mobile phase required for low-MW compounds to elute from the column is very large. On the other extreme, compounds with MWs of >100,000 do not penetrate the gel and elute with the mobile phase or eluent. This is called size exclusion, and the large MW compounds are said to be size excluded from the gel. Molecules represented by those from peak F are much smaller than molecules represented by those from peak C. Compounds represented by bands D, E, and F are separated in the selective permeation region. This region is often referred to as the fractionation region of the packing.

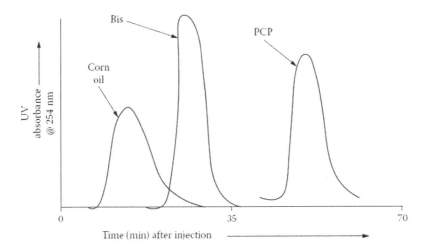

FIGURE 3.27

The calibration procedure in Method 3640 recommends that corn oil, bis(2-ethyl hexyl) phthalate (Bis), and pentachlorophenol (PCP) be separated and collected in the fractionation region. A typical GPC chromatogram obtained is shown in Figure 3.27. Corn oil consists of higher-MW triglycerides and could represent band D in Figure 3.26. Bis is typical of many polycyclic aromatics and could represent band E in Figure 3.26 while PCP is typical of many substituted phenols and could represent band F in Figure 3.26. It is important to know exactly how many milliliters of eluent are required to dump oils while desiring to collect analytes of interest, such as priority pollutant organic compounds of enviro-chemical or enviro-health interest.

Once it is known how many milliliters of mobile phase is required to elute the analytes of interest while discarding the high-MW fats and oils, extracts from Kuderna–Danish or rotary evaporative concentrative techniques can be injected directly into the 5 mL sample loop of the GPC chromatograph either manually or via autosampler. EPA Method 3640A is a more recently modified method that eliminates PCP in the recommended calibration mixture. Methoxychlor (representative of OCs), perylene (representative of PAHs), and sulfur (common interference in petroleum-contaminated soil) are added to the calibration mixture. The elution order then becomes:

corn oil peak 1 < corn oil peak 2 < bis < methoxychlor < perylene < sulfur

Bringing contemporary GPC capability to a laboratory requires an investment in instrumentation. Hydrolysis as a cleanup technique, shown in Figure 3.23, although not included in SW-846, becomes useful when the lipid content is appreciable and offers an alternative to GPC. Acid and base hydrolysis can be applied; however, lipids are easily decomposed in the presence of alcoholic potassium hydroxide. The author's work in isolating PCBs from animal feed nicely illustrates the use of saponification as yet another cleanup tool in the arsenal.

3.46   HOW IS SAPONIFICATION USED TO CLEAN UP RAT FEED TO ISOLATE PCBs?

This question is answered by focusing on a specific laboratory procedure that this author has used to chemically decompose lipid and release lipophilic PCBs from animal feed (rat ration reported with ~7% lipid as measured gravimetrically). This procedure is adapted from that published earlier by Erickson and is given via a series of steps below:[65]

- To prepare unspiked and spiked control feed samples (a control feed sample is a sample of feed unadulterated by PCBs), place 10 g of control feed into a 40 mL vial, seal and place on a shaker, and label as "unspiked control feed."
- Place 10 g of control feed into a 40 mL vial and add 500 μL of 10 ppm TCMX (dissolved in MeOH). Place on a shaker and label as "spiked control feed."
- Shake for 1 h, transfer contents of 40 mL vial to Soxhlet thimble, and place thimble in *Soxhlet apparatus*. Weigh a 250 mL round-bottom flask, add 230 mL of methylene chloride, and conduct S-LSE for anywhere from 12 to 24 h.
- Remove solvent from the extract via rotary evaporation. This should leave a mass of oily residue at the bottom of the round-bottom flask. Weigh the flask; this gives a gravimetric determination of the percent lipid in the 10 g of control feed.
- Add 25 mL of 2.5% KOH (dissolved in ethanol), add boiling chips to prevent *bumping*, and connect the flask to the water condenser and reflux for 1 h or until the oily phase disappears.
- Cool, then transfer the solution to a 250 mL *separatory funnel* that already contains 25 mL of distilled deionized water. Rinse the extraction flask with 25 mL of hexane and add this washing to the *separatory funnel*.
- Shake the contents of the *separatory funnel* for 1 min. This allows the PCBs to partition into the hexane phase.
- Allow phases to separate and withdraw the lower layer (aqueous) into a second *separatory funnel*.
- Extract the saponification solution with a second 25 mL aliquot of hexane. After the phases have separated, withdraw the aqueous phase. Add 25 mL of hexane to the aqueous phase and perform a third LLE.
- Combine all three hexane extracts and discard the saponified aqueous phase.
- Concentrate the hexane extract to 5 mL or less using either a *rotary evaporator* or a *Kuderna–Danish* evaporative concentrator.
- Transfer the concentrated hexane extract to either a 1, 2, or 5 mL volumetric flask, and adjust to the calibration mark of the flask with hexane.
- Transfer to a 2 mL GC vial and inject 1 μL via autosampler into a C-GC-ECD or equivalent.

A 99.0% recovery of TCMX and a 107% recovery of AR 1242 (a commercially made mixture of 20 to 30 PCB congeners) was obtained using the procedure as outlined above.[66]

Let us return to a consideration of the phase diagram shown in Figure 3.11 and consider a move both horizontal and vertical to the so-called supercritical region. We can accomplish similar analytical objectives to those addressed by ASE and those of interest to the practice of TEQA by choosing a fluid whose supercritical pressures and temperatures can be reached in the laboratory. With the availability of carbon dioxide, an inexpensive fluid with relatively low supercritical values for temperature and pressure, SFE, as a laboratory sample prep technique, advanced.

3.47 HOW DO I GET STARTED DOING SFE?

In principle, all one needs is a means to introduce CO_2 into a high-pressure extraction cell and to collect the extract. You would need a means to both pressurize and elevate the temperature of carbon dioxide. A critical temperature of 31.4°C with a critical pressure of 73 atm is needed to maintain CO_2 in a supercritical state. Consider again the phase diagram shown in Figure 3.11. Unlike liquids, if CO_2 is kept above 31.4°C, it cannot be liquefied no matter how high the pressure. In this supercritical fluid state, the physical properties, such as liquid-like density, intermediate diffusivity, gas-like viscosity, and gas-like surface tension, provide a medium for rapid and selective extraction. SFE as an alternative to S-LSE was, in essence, oversold during the late 1980s and early 1990s. SFE does exhibit a matrix dependence. The use of organic modifiers has helped, however. Like ASE, SFE instrumentation is very expensive, and if only a few samples are to be prepared and analyzed for trace organics in environmental matrices, is it really worth the cost? In this section, we will discuss how SFE works and then, as we did for ASE, we will delve into some recovery studies.

3.48   WHAT LIQUIDS AND GASES CAN BE USED TO CONDUCT SFE?

Table 3.11 lists nine fluids together with their respective boiling points, critical temperatures, critical pressures, and critical densities. From this list of nine, several can be eliminated as being either too hazardous or too corrosive to work with, such as ammonia, or as requiring too high a temperature, such as water and methanol. Carbon dioxide, nitrous oxide, xenon, ethane, ethylene, sulfur hexafluoride, and very recently fluoroform, CHF_3, have emerged as the likely candidates. Of these most suitable supercritical fluids (SFs), carbon dioxide emerges as the supercritical fluid of choice because CO_2 is nontoxic and inexpensive and exhibits a relatively low critical temperature and pressure. Its critical density is also appropriate. CO_2 is considered under SF conditions as having a solvent polarity about equal to hexane—in other words, nonpolar. The need to add polar, so-called matrix modifiers has led to increases in the percent recoveries of some of the more polar priority pollutants. The poor recovery of polar analytes from relatively nonpolar sample matrices has led to the notion of inverse SFE, whereby the sample matrix can be completely removed by SFE while the more polar analyte remains. This has much more relevance to pharmaceutical analysis than to TEQA because sample matrices tend to be more polar than the analyte of interest. In the determination of polar ingredient in an ointment, the sample matrix is extracted by the relatively nonpolar CO_2, leaving the polar-active ingredient in the SFE extraction vessel. Earlier, Chester reviewed the eight most likely candidates for use as supercritical fluid chromatography (SFC); this information is also pertinent to SFE.[67]

TABLE 3.11
Physico-Chemical Properties for the Common Supercritical Fluids

Fluid	Tbp (°C)	Tc (°C)	Pc (atm)	$P\gamma$ (g/mL)
CO_2	−78.5	31.3	72.9	0.448
NH_3	−33.3	132.4	112.5	0.235
H_2O	100.0	374.1	218.3	0.315
N_2O	−88.6	36.5	71.7	0.450
C_2H_6	−88.6	32.3	48.1	0.203
Xe	−108.1	16.6	58.4	1.10
CH_3OH	64.7	240.5	78.9	0.272
SF_6	63.8 (sublimes)	45.5	7.1	0.740
C_2H_4	−103.7	9.2	49.7	0.218

3.49   WHAT OTHER PHYSICO-CHEMICAL PROPERTIES ARE OF INTEREST IN SFE?

Supercritical fluids begin to exhibit significant solvent strength when they are compressed to liquid-like densities. Figure 3.28 shows how the density of a pure SF changes as the SF is compressed. The greatest change in SF density, ρ, is seen to occur in the region near to its critical point. An increase in the density of a supercritical fluid has been shown to increase the amount of a chemical compound that will dissolve in the SF.[68] Certain other physico-chemical properties give SFs advantages with respect to SFE as well. The more an SF is compressed, the more liquid-like the fluid becomes. Very high density SFs (i.e., $P \gg Pc$) are very liquid-like fluids. Very low density SFs, where $T > Tc$ and $P \sim Pc$, are very gaseous-like. The three most useful physico-chemical properties of fluids, density, viscosity, and diffusion coefficient, are listed for gases, SFs, and liquids in Table 3.12. Interpretation of Table 3.12 clearly shows how SFs are intermediate between gases and liquids with respect to all three properties. It is not surprising that SFs are denser than gases and, in turn, less dense with respect to liquids. SFs are more viscous than gases, yet less viscous by one order of magnitude than liquids. Diffusion coefficients are

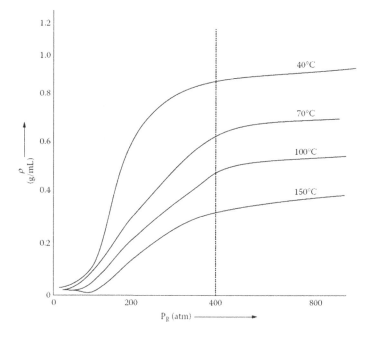

FIGURE 3.28

TABLE 3.12
Typical Ranges of Densities (ρ), Viscosities (η), and Diffusion Coefficients (D) for Gases, Supercritical Fluids (SF), and Liquids

Fluid	ρ (g/mL)	η (poise)	D (cm²/sec)
Gas	0.001	0.5–3.5 ($\times 10^{-4}$)	0.01–1.0
SF	0.2–0.9	0.2–1.0 ($\times 10^{-2}$)	0.3–1.0 ($\times 10^{-5}$)
Liquid	0.8–1.5	0.3–2.4 ($\times 10^{-2}$)	0.5–2.0 ($\times 10^{-5}$)

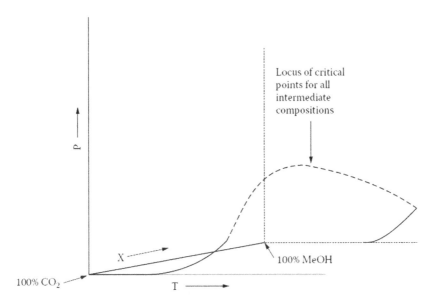

FIGURE 3.29

more important in consideration of SFC; however, it is interesting to compare the range of values of D for SFs vs. liquids. SFs diffuse over 100 times faster than liquids.

The need to add organic modifiers to the supercritical CO_2 implies that a second component be considered from the perspective of the Gibbs phase rule. For two components, the number of degrees of freedom, f, is found according to $f = 4 - p$. Hence, for a one-phase region in the phase diagram where $p = 1$, three degrees of freedom are needed. In general, in addition to pressure and temperature, composition becomes this third degree of freedom. Therefore, phase diagrams must be viewed in three dimensions. To facilitate a two-dimensional view, in which regions of identical temperature can be plotted, so-called isotherms can be shown. Figure 3.29 is a three-dimensional representation of a pressure–temperature–composition (P-T-X) plot for a binary mixture. Figure 3.29 has been labeled to show the extremes of the composition axis where, using the example of a CO_2/MeOH binary mixture, we start with pure CO_2 on one end and pure methanol on the other. The liquid–vapor equilibrium line is drawn for 0% MeOH (100% CO_2) and for 100% MeOH (0% CO_2). The dashed line connects the locus of mixture critical points. The significance of P-T-X plots for binary SF mixtures is that for values of P, T, and X values that reside below the curve, two phases are possible depending on the composition. If 10% MeOH is added as an organic modifier when using carbon dioxide as the SF, this plot is useful in knowing what conditions need to be reached to keep the SFE in one phase.

3.50   WHAT DOES AN SFE INSTRUMENT LOOK LIKE?

The principal components of an SFE instrument are shown in Figure 3.30. It must be recognized that SFE-SFC-grade CO_2 requires a purity greater than 99.999%. The SF delivery system can be either a high-pressure reciprocating pump or a high-pressure syringe pump. One's preference as to which type of pump to use depends on choice and cost considerations. It is necessary to cool the pump head when using a reciprocating pump. This need for cooling is unnecessary when using a syringe pump because CO_2 is liquefied by pressure, not temperature. A syringe pump also eliminates the inconsistencies in repressurization that might be present in a reciprocating pump; in other words, it provides a pulseless flow of CO_2. A syringe pump, however, is limited in capacity and must be refilled while a reciprocating pump draws continuously from a large reservoir.

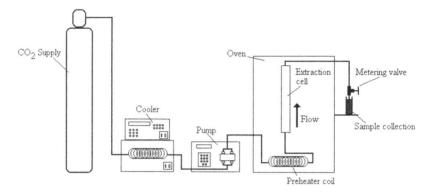

FIGURE 3.30

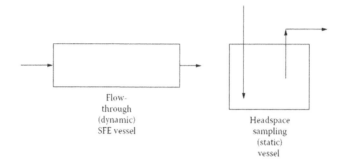

FIGURE 3.31

A second reciprocating pump is needed if the addition of an organic modifier is needed. This second pump does not need to be cooled. There have been other designs discussed in the literature as well, necessitated by the high cost of having to cool reciprocating pumps.[69]

The sample cell or vessel is usually made of stainless steel of some other inert material and ranges in size from 150 μL to 50 mL. It is most important that the sample vessels are rated for the maximum working pressure of the system. Cell designs are either of a flow-through (dynamic) or headspace sampling (static) type. Flow-through cells enable the SF to continuously pass through the sample. Static headspace sample cells enable the SF to surround the sample while the headspace above the sample is withdrawn. Headspace sampling tends to be used for large samples, typically 40 mL. Figure 3.31 illustrates the fundamental difference between these two cell designs.[69]

The restrictor or depressurization step is very important and, over the years, has been the Achilles' heel of SFE. Plugging of restrictors was a major problem when this device consisted of an unheated fused-silica capillary tube. Restrictor problems have been minimized or even eliminated with the development of automated, variable restrictors.[71] Variable restrictors use position sensors, motorized gears, and feedback from the CO_2 pump to control the flow set point and make flow adjustments. The restrictor is responsible for maintaining the pressure of the SF above its Pc. The restrictor induces a back pressure in the sample cell. For a CO_2-based SFE, the pump delivers the CO_2 in liquid form at a constant flow rate and SF conditions are established in the heated extraction chamber. The smaller the restrictor orifice, the higher the back pressure in the line and the greater the mass flow rate. The final part of SFE instruments is the collection device. Two approaches have been developed for this aspect of SFE instruments; online and offline. Online SFE refers to the coupling of the analyte-containing SF to a chromatographic separation

TABLE 3.13

Comparison of Soxhlet Extraction and Different SFE Fluids for the Recovery of PCB Congeners from River Sediment

PCB Congener	Soxhlet	CHClF2	N2O	CO2	CO2/MeOH
2, 2′, 3, 5′	1070	1158	497	470	855
2, 3′, 4, 4′, 5′	510	631	319	382	482
2, 2′, 3, 4′, 5, 5′	160	166	145	144	168

Source: Friant S. and Suffet I., *Anal. Chem.*, 51, 2167–2172, 1979.

system, and abbreviations are used to indicate which determinative technique is interfaced. For example, SFE-GC implies that supercritical fluid extraction has been interfaced to a gas chromatograph. Offline SFE refers to the stand-alone instrument whose block diagram is shown in Figure 3.30.

The collection device stores the extracted analyte in one of three ways: (1) via thermal trapping, (2) by sorbent trapping, or (3) by solvent trapping. Thermal trapping is the simplest experimentally and merely involves directing the depressurized SFE effluent into a cooled vessel. Moderately volatile analytes are known to be lost using this type of SFE trapping; therefore, use of thermal trapping is limited to semivolatile and nonvolatile organic analytes. Sorbent trapping incorporates a solid adsorbent whose surface is capable of removing the extracted analytes from the SF effluent. The sorbed analytes are removed from the adsorbent by elution with a solvent. Sorbent trapping affords the opportunity to incorporate a degree of analyte selectivity either by selection of an appropriate sorbent or by choice of elution solvent. Commercially available chemically bonded silicas have been used with success in sorbent trapping. We will have much more to say about the use of these materials with respect to the sample prep technique called solid-phase extraction later. Solvent trapping is the most widely used method of trapping analytes from the extracted SF effluent and merely involves depressurizing the SF directly into a small vial containing a few milliliters of liquid solvent. Hawthorne reports a tendency toward higher percent recoveries for sand that has been spiked with selected semivolatile organics such as phenol, nitrobenzene, and benzo(a)pyrene with a change in the chemical nature of the collection solvent. The following solvents were placed in a collection vial, and the order in terms of a percent collection efficiency was hexane < methanol < methylene chloride < methylene chloride kept at 5°C (by using a heating block).[72]

It has been well established that SF CO_2 is not a good extractant for polar analytes. In addition, some analyte–matrix interactions are sufficiently strong to yield low percent recoveries. The recovery of three PCB congeners was compared against Soxhlet extraction while using four different SFs, including SF CO_2 modified with methanol, shown in Table 3.13.[72] The results show that the more polar SFE (CO_2/MeOH) yields as good a recovery as $CHClF_2$, whose dipole moment is 1.4 Debyes. Pure CO_2 has a dipole moment of 0.0 Debyes, and the addition of a polar modifier like MeOH serves to increase the dipole moment of the SF fluid. This author is not aware if there are any published dipole moment values for polar-modified SF CO_2.

3.51   IS THERE AN EPA METHOD THAT REQUIRES SFE? IF SO, PLEASE DESCRIBE

Yes, it is Method 3562 from the SW 846 series and it titled: "Supercritical Fluid Extraction of polychlorinated biphenyls (PCBs) and organochlorine pesticides"[73] Twelve specific PCB congeners of either the tri-, tetra-, penta-, hexa-, or heptachlorobiphenyl variety have been studied in this method. Fourteen of the priority pollutant OCs from aldrin through to heptachlor epoxide have also been studied in this method. SFE is used to extract these analytes from soils, sediments, fly ash, solid-phase extraction media, and other solids materials that are amenable to extraction with conventional solvents. The uniqueness of this method is that it is possible to achieve a class separation (i.e., PCBs from OCs) by differences in SFE conditions. Samples to be analyzed for PCBs are subjected to a 10 min static extraction, followed by a 40 min dynamic extraction. Samples for OCs are subjected to a 20 min static extraction, followed by a 30 min dynamic extraction. The analyst must demonstrate that the SFE instrument is free from interferences. This is accomplished by extracting method blanks and determining the background contamination or lack thereof. The method cautions the user about carryover. Reagent blanks should be extracted immediately after analyzing a sample that exhibits a high concentration of analytes of interest. This method does not use a polar organic modifier. Referring again to the schematic of SFE instrumentation shown in Figure 3.30, Method 3562 specifies the nature of the extraction vessel, restrictor, and collection device. Vessels must be stainless steel with end fittings, with 2 μm first. Fittings must be able to withstand pressures of 400 atm (5,878 psi). The method was developed using a continuously variable nozzle restrictor. Sorbent trapping is the collection device used in the method. Florisil with a 30 to 40 μm particle diameter is recommended for PCBs, whereas octadecyl silane is suggested to trap OCs. Solvent trapping is also suggested, with a cautionary remark concerning the loss of volatile analytes. The method requires the use of internal standards and surrogate standards. For sorbent trapping, *n*-heptane, methylene chloride, and acetone are recommended. The method also requires that a percent dry weight be obtained for each solid sample. The weight loss that accompanies keeping 5 to 10 g of the sample in a drying oven overnight at 105°C is measured. The sample may need to be ground to ensure efficient extraction. To homogenize the sample, at least 100 g of ground sample is mixed with an equal volume of CO_2 solid "snow" prepared from the SFE CO_2 solid. This mixture is to be placed in a food type chopper and ground for 2 min. The chopped sample is then placed on a clean surface, and the CO_2 solid snow is allowed to sublime away. Then, 1 to 5 g of the homogenized sample is weighed and placed in the SFE vessel. If the samples are known to contain elemental sulfur, elemental copper (Cu) powder is added to this homogenized sample in the SFE vessel. No adverse effect from the addition of Cu was observed, and EPA believes that finely divided Cu may enhance the dispersion of CO_2. To selectively extract PCBs, the following SFE conditions using SF CO_2 are listed below:

Pressure: 4417 psi
Temperature: 80°C
Density: 0.75 g/mL
Static equilibration time: 10 min
Dynamic extraction time: 40 min
SFE fluid flow rate: 2.5 mL/min.

To extract OCs, the recommended SFE conditions using SF CO_2 are listed below:

Pressure: 4330 psi
Temperature: 50°C
Density: 0.87 g/mL
Static equilibration time: 20 min
Dynamic extraction time: 30 min
SFE fluid flow rate: 1.0 mL/min.

Study of Method 3562 is a good starting point for those analytical chemists and/or laboratory technicians who anticipate implementing an SFE *sample prep* method.

3.52   HAVE COMPARATIVE STUDIES FOR SFE BEEN DONE?

Yes, there have been comparative studies in which the percent recovery has been measured using not only SFE, but also ASE, as well as comparing these results to percent recoveries from the more conventional Soxhlet extraction (S-LSE). Generally, these studies are done on certified reference samples or grossly contaminated samples. We will discuss two studies from the analytical chemistry literature. The first study compared the efficiencies of SFE, high-pressure solvent extraction (HPSE), to S-LSE for removal of nonpesticidal organophosphates from soil.[74] HPSE is very similar to accelerated solvent extraction; however, we will reserve the acronym ASE for use with the commercial instrument developed by Dionex Corporation (now a part of Thermo Fisher Scientific, Inc.). HPSE as developed by these authors used parts available in their laboratory. The authors compared S-LSE with SFE and HPSE for extraction of tricresyl phosphate (TCP) and triphenyl phosphate (TPP) from soil. Molecular structures for these two substances are as follows:

Triphenyl phosphate   Tricresyl phosphate

Ortho-, *meta-*, and *para*-substitution within the phenyl rings can lead to isomeric TCPs. Aryl phosphate esters are of environmental concern due to their widespread use and release. They have in the past been used as fuel and lubricant additives and as flame-retardant hydraulic fluids. Widespread use of TCP associated with military aircraft have led to contamination of soil at U.S. Air Force (USAF) bases. TCP-contaminated soil samples were obtained from a USAF site. The percent water was determined. Spiked oil samples were prepared by adding an ethyl acetate solution containing TPP and TCP to 500 g of a locally obtained soil. This soil was placed in a rotary evaporator whereby the analytes of interest could be uniformly distributed throughout the soil. Methanol was added as the polar modifier directly into the soil in the SFE extraction chamber. It was deemed important to add methanol because TCP and TPP are somewhat polar analytes. The actual SFE consisted of 10 min of static SFE followed by 20 min of dynamic SFE. The sample, which consisted of 1.5 g of soil with 1.0 mL of sand, was placed in the bottom of the SFE vessel. A sorbent trapping technique using glass beads with subsequent washing with ethyl acetate was used to recover the analytes. The equivalent of ASE was conducted in this study, not with a commercial instrument as discussed earlier, but with a combination of a commercially available SFC syringe pump (Model SFC-500 from Isco Corp.) and a gas chromatographic oven (Model 1700 GC from Varian Associates). We will use the author's abbreviation for high-pressure solvent extraction

(HPSE). A series of SFE-related method development studies were first undertaken to optimize the effect of the volume of methanol and the effects of temperature and pressure on percent recoveries of TCP from native soil. In a similar manner, extraction conditions for HPSE were optimized by varying temperature and pressure. Optimized conditions arrived at were the following:

SFE: 80°C, 510 atm, and 1250 μL of MeOH
HPSE: 100°C and 136 atm

The determinative technique was capillary gas chromatography equipped with a nitrogen–phosphorous detector. The authors found the percent recoveries to be quite close to S-LSE while reporting that considerable time and solvent consumption could be saved. Next, we consider a second comparative study.

Heemken and coworkers in Germany reported on percent recovery results for the isolation and recovery of PAHs and aliphatic hydrocarbons from marine samples. [75] Results were compared using accelerated solvent extraction (ASE), SFE, ultrasonication (U-LSE), methanolic saponification extraction (MSE), and classical Soxhlet (S-LSE). Both ASE and SFE compared favorably to the more conventional methods in terms of a relative percent recovery against the conventional methods, so that extracted analytes from ASE and SFE were compared to the same extracted analytes from S-LSE and U-LSE. Relative percent recoveries ranged from 96 to 105% for the 23 two through seven (coronene) fused-ring PAHs and methyl-substituted PAHs. To evaluate the percent recoveries, the authors defined a bias, D_{rel}, according to

$$D_{rel} = \frac{X_1 - X_2}{X_1} \times 100\%$$

where X_1 and X_2 are the extracted yields for one method vs. another. For example, a summation of the amount of nanograms extracted for all 23 PAHs using SFE was 33,346ng, whereas for S-LSC, it was 33,331ng. Using the above equation gives a bias of 3.0%. Table 3.14 gives values for D_{rel} for PAHs for (1) SFE vs. either S-LSE or U-LSE and (2) ASE vs. either S-LSE or U-LSE. Three different environmental solid samples were obtained. The first was a certified reference marine sediment sample obtained from the National Research Council of Canada. The second sample was a suspended sediment obtained from the Elbe River in Germany using a sedimentation trap. This sample was freeze-dried and homogenized. The third sample was a suspension obtained from the Weser River and collected via a flow-through centrifuge. This sample was air-dried and had a water content of 5.3%. Values of D_{rel} were all within 10% among the four methods shown in Table 3.6, and it is safe to conclude that it makes no

TABLE 3.14
Values of Bias D_{rel} for PAHs for Various Sample Preparation Methods

Sample	SFE vs. S-LSC	SFE vs. U-LSC	ASE vs. S-LSC	ASE vs. U-LSC
Certified reference marine sediment	3.0	4.7	0.3	8.3
Freeze-dried homogenized suspended particulate matter	5.0	N	2.4	N
Air-dried suspended particulate matter	−3.8	N	N	N

Note: N = not investigated.

difference whether SFE or ASE, or S-LSE or U-LSE is used to extract PAHs from the three
marine sediment samples. The other criterion used to compare different sample preparation
methods is precision. Are SFE and ASE as reproducible as S-LSE and U-LSE? For the certified
reference standard, an overall relative standard deviation for PAHs was found to be 11.5%. For
the first sediment, an RSD was found to range from 3.4 to 5.0%, and for the second sediment,
RSD ranged from 2.7 to 7.5%. Let us now consider the isolation and recovery of one priority
pollutant PAH from this work.

Benzo(*g*, *h*, *i*)perylene (BghiP) is a five-membered fused-phenyl-ring PAH of molecular
weight 277 and could be considered quite nonpolar. Its molecular structure is as follows:

Benzo (g, h, i) perylene

Shown in Table 3.15 are the yields of analyte in amount of nanograms of BghiP per
gram of dry sediment. Also included are yields from the MSE approach. MSE as defined
in the Heemken paper is essentially a batch LLE in which the sediment is refluxed in alkaline methanol, followed by dilution with water, and then extracted into hexane. The yield of
BghiP is similar for all five sample prep methods, as reported by the authors. The other significant outcome was to show that for a nondried marine sediment, the addition of anhydrous
sodium sulfate increased the yield of BghiP from 35 to 96 ng, and this result came close to that
obtained by the MSE approach.

TABLE 3.15
**Recovery of BghiP in Amount of ng of BghiP/g of Sediment Using Five Sample
Preparation Methods**

Sample	SFE	ASE	S-LSE	U-LSE	MSE
Certified reference marine sediment 1726 ± 720	1483 (5.6), $n = 6$	1488 (7.4), $n = 6$	1397 (3.9), $n = 3$	1349 (3.8), $n = 3$	1419 (3.4), $n = 3$
Free-dried homogenized suspended particulate matter, I	147 (2.3), $n = 6$	156 (1.6), $n = 3$	207, $n = 1$		180 (1.6), $n = 3$
Air-dried suspended particulate matter, II	122.5 (1.2), $n = 3$		129.1 (0.4), $n = 3$		127.2 (0.1), $n = 6$
II, nondried 56% water	35 (9.0), $n = 3$	32 (14.1), $n = 3$			98 (0.6), $n = 3$
II, nondried, anhydrous sodium sulfate	96 (3.7), $n = 6$				

Note: n is the number of replicate extractions. The number in parentheses is the relative standard deviation expressed as
a percent.

Before we leave the topic of SFE, it is interesting to note that SFE on a larger scale is used to decaffeinate coffee. SFE using CO_2 under subcritical conditions, can extract mono- and di-terpenes, while SFE using CO_2 under supercritical conditions can extract cannabinoids, fats, lipids, chlorophyll and flavonoids from cannabis. SFE using CO_2 leaves no chemical residue and therefore is of great importance to the ever-growing cannabis industry whose principal chemical products are tetrahydrocannabinol (THC) and cannabidiol (CBD).[76] Molecular structures for all three organic compounds just discussed are shown below:

Caffeine   THC   CBD

Water maintained at a subcritical temperature and pressure, i.e. below 374°C and below 3,210 pounds/in^2 (recall that 1 atmosphere = 14.7 pounds/in^2 = 1.013 x 10^5 newtons/m^2 = 1.013 x 10^5 Pa) exhibits physiochemical properties that are similar to that of a nonpolar extraction solvent and has been shown by the research of the Hawthorne group at the University of North Dakota to isolate and recover a variety of classes of priority pollutants.[77]

This completes our digression into alternative *sample prep techniques* to extract priority pollutants from contaminated solid matrices—an important aspect of *enviro-chemical* TEQA. We now return to aqueous samples and entertain a somewhat detailed discussion of an important alternative to LLE used to isolate and recover SVOCs that emerged over 30 years ago. This alternative sample prep technique uses various chemically bonded silica gels and in selected applications, ion-exchange resins to isolate and recover trace SVOCs from aqueous samples. Today we refer to this sample prep technique as *reversed-phase solid-phase extraction* (RP-SPE). RP-SPE is used in both realms of TEQA, *enviro-health* as well as *enviro-chemical*.

3.53   WHAT IS SOLID-PHASE EXTRACTION?

Solid-phase extraction (SPE) is today a viable *sample prep* alternative to LLE. SPE originated during the late 1970s and early 1980s as a means to preconcentrate aqueous samples that might contain dissolved SVOCs that are amenable to analysis by gas chromatographic determinative techniques. SPE was first applied to sample preparation problems involving clinical or pharma-ceutical types of samples and only much later evolved as a viable sample preparation method in TEQA. Today, SPE techniques are used to extract chemical analytes from environmental samples such as drinking, ground and wastewater and even from contaminated soil as well as from blood, serum, urine, and other biological fluids.

The concept of column liquid chromatography as a means to perform environmental sample preparation was not immediately evident after the development of high-performance (once called high-pressure) liquid chromatography (HPLC) using reversed-phase silica as the stationary phase in the late 1960s. This technique is termed *reversed-phase HPLC* (RP-HPLC). It is also referred to as bonded-phase HPLC, and it is estimated that over 75% of HPLC being done today is RP-HPLC. As early as the 1930s, silica, alumina, Florisil, and Kieselguhr or diatomaceous earth were used as solid sorbents primarily for cleanup of nonpolar extractants, and the mechanism of retention was based on adsorption. The early realization that hydrophobic surfaces could isolate polar and nonpolar analytes from environmental aqueous samples such as groundwater came about with the successful use of XAD resins, whereby as much as 10 L of sample could be passed through the resin.[78] By this time, bonded silicas that contained a hydrocarbon moiety covalently bonded were increasingly predominant in HPLC. One way that this rise in prominence for RP-HPLC can be seen is by perusing the earlier and pioneering text on HPLC and then comparing the content in this text to a text published later by the same authors.[79,80]

It becomes clear that during this 20+-year period, RP-HPLC dominated the practice of column liquid chromatography. It will also become evident that the demands being made by EPA and state regulatory agencies on environmental testing labs required that sample extracts be made as free of background interferences as possible. Terms began to be used such as "the quality of the chromatogram is only as good as the extract." A well-used cliché also applied: "Garbage into the GC-MS, garbage out." All EPA methods during the 1980s that considered aqueous samples required LLE as the sole means to prepare environmental samples for TEQA. LLE was done on a relatively large scale whereby 1,000 mL of groundwater or other environmental sample was extracted three times using 60 mL of extracting solvent each time. Because only 1 μL of extract was needed to be injected into a GC, this left almost all of the extractant unused! As MDLs began to go to lower and lower values due in large part to regulatory pressures, it became evident that the 180 mL of extract could be reduced to 5 mL or less via some sort of vaporization of the solvent. This led to the use of lower-boiling solvents such as methylene chloride (dichloromethane), whose boiling point is 34°C. After all of this, the reduced volume of extract still had to be cleaned up so as to obtain a good signal-to-noise ratio, and hence to satisfy the now stringent MDL requirements.

The evolution of RP-SPE came about after the development of RP-HPLC. The concept that the hydrophilic silica gel, as a stationary phase, that had been used to pass a nonpolar mobile phase across it could be transformed to a hydrophobic stationary phase was a significant development. Organic compounds of low to moderate polarity could be retained on these hydrophobic sorbents provided that the mobile phase was significantly more polar than that of the stationary phase. If the particle size could be decreased down to the smallest mesh size, a sorbent material with a relatively large surface area would provide plenty of active sites. A surface that was also significantly hydrophobic as well would facilitate removal of moderately polar to nonpolar analytes from water. The stage was set then to bring RP-SPE into the domain of TEQA. This was accomplished throughout the 1980s and led to a plethora of new methods, applications, and SPE suppliers. EPA, however, could not at first envision a role for RP-SPE within its arsenal of methods, and hence largely ignored these developments. It did, however, offer to fund some researchers who were interested in demonstrating the feasibility of applying RP-SPE to environmental samples. It became necessary then for analytical chemists engaged in method development to conduct fundamental studies to determine the percent recovery of numerous priority pollutant semivolatile organics from simulated or real environmental samples. This was all in an attempt to validate methods that would incorporate the RP-SPE technique. This author was one such researcher who, while employed in a contract lab in the late 1980s, received a Phase I Small Business Innovation Research (SBIR) grant from the U.S. EPA to conduct just such studies. This grant was subsequently matched by the New York State Science and Technology Foundation.[81,82]

As is true for any emerging and evolving technology, SPE began with a single commercially available product, the Sep-Pak cartridge, developed in 1978 at Waters Associates. These devices, designed to fit to a standard liquid-handling syringe, were at first marketed to the pharmaceutical industry. The focus was on a silica cartridge for what would be considered today the practice of normal-phase SPE (NP-SPE). The idea that a large volume of sample could exceed 10 mL led to the development in 1979 of the so-called barrel design. Analytichem International, then Varian Sample Preparations and now Agilent Technologies, Inc, was first to introduce the SPE barrel, which in turn could be fitted to a vacuum manifold. This design consisted of a cylindrical geometry that enabled a larger reservoir to be fitted on top while the Luer tip on the bottom of the barrel was tightly fitted to the inlet port of a vacuum manifold. The vacuum manifold resembled the three-dimensional rectangular-shaped thin-layer chromatographic development tank. The sorbent bed consisted of chemically bonded silica of irregular particle size with a 40 μm average and was packed between a top and a bottom 20 μm polypropylene frit. This configuration opened the door to the application of SPE techniques to environmental aqueous samples such as drinking water. It was left to researchers to demonstrate that priority pollutant organics dissolved in drinking water could be isolated and recovered to at least the same degree as was well established using LLE techniques. J.T. Baker followed and began to manufacture SPE cartridges and syringe formats in 1982. It is credited with coining the term *SPE*, as opposed to the EPA's term *liquid-solid extraction*. Further evolution of the form that SPE would take was sure to follow.

In 1989, the bulk sorbent gave way to a disk format in which 8 to 12 μm C_{18} silica was impregnated within an inert matrix of polytetrafluoroethylene (PTFE) fibrils in an attempt to significantly increase the volumetric flow rate of water sample that could be passed through the disk. These developments were made by the 3M Corporation. It coined the term *Empore Disk*. The next major development in SPE design occurred in 1992 as Supelco introduced a device called solid-phase microextraction (SPME). This followed the pioneer developments by Pawliszyn and coworkers as introduced earlier.[54] The GC capillary column, a topic to be discussed in Chapter 4, was simply inverted. The polydimethyl siloxane polymer coating was deposited on the outer surface of a fused-silica fiber. This fiber is attached to a movable stainless-steel "needle in a barrel" syringe. This design is similar to the 7000 series manufactured by the Hamilton Company for liquid-handling microsyringes. SPME is a solventless variation of SPE in that the coated fiber is immersed within an aqueous phase or in the headspace above the aqueous phase. After a finite period, the fiber is removed from the sample and immediately inserted directly into the hot-injection port of a GC. The analytes are then removed from the fiber by thermal desorption directly into a GC column. We will discuss the principles and practice of SPE and then focus on RP-SPE as applied to TEQA. Some of the work of the author will also be included. A very readable and comprehensive discussion of all aspects of the science that underlies SPE is found in an earlier book by Thurman and Mills.[83]

3.54 HOW IS SPE DONE?

SPE is performed using a variety of consumable items. These items include encapsulated Sep-Pak SPE devices that fit on the end of a handheld plastic disposable syringe; 47-mm octadecyl or octyl-bonded silica-impregnated PFTE circular disks; and barrel-type cartridges packed with bulk sorbent or fitted with bonded silica-impregnated PFTE disks with Luer® adapters. Volumes ranging from 1 to 60 cc present the most contemporary barrel cartridge design to conduct up-front reversed-phase (RP-SPE) or normal-phase (NP-SPE) solid-phase extraction. Perusal of online websited (formerly paper) catalogs from current SPE suppliers such as Millipore Sigma (who offer the former Supelco SPE consumables), Agilent Technologies (who offer the former Analytichem International, AnSys/Toxi-Lab and Varian SPE consumables), Waters, UCT, and Phenomenex SPE consumables among others, is the quickest way to become knowledgeable as to what is available. A sketch (not drawn to scale) that depicts the barrel-type design for passing an aqueous sample (sample loading) through a conditioned RP-SPE sorbent is shown in Figure 3.32 while more realistic drawings that depict various types of SPE laboratory devices to include the vacuum manifold are shown in Figure 3.33.

Barrels and reservoirs have remained the same since their inception; however, some innovation is evident in making the Luer adapter fit more than one barrel mouth size. If an adapter is placed atop the 70 mL polypropylene reservoir and connected to a much larger reservoir, such as an HPLC type container, the vacuum manifold can be modified to pass much larger water samples through the sorbent. SPE as a system has been completely automated. Companies such as Caliper Life Sciences, formerly Zymark, Hamilton, TomTec, and Gilson, among others, have automated SPE systems commercially available. Automated systems that incorporate a 96-well plate have become a popular adaptation of SPE to pharmaceutical industry interest in recent years and today have made inroads into environmental testing laboratories.

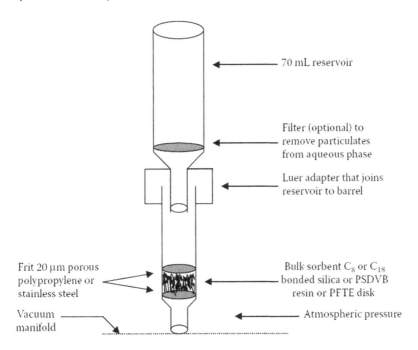

70 mL reservoir

Filter (optional) to
remove particulates
from aqueous phase

Luer adapter that joins
reservoir to barrel

Frit 20 µm porous
polypropylene or
stainless steel

Bulk sorbent C_8 or C_{18}
bonded silica or PSDVB
resin or PFTE disk

Vacuum
manifold

Atmospheric pressure

FIGURE 3.32

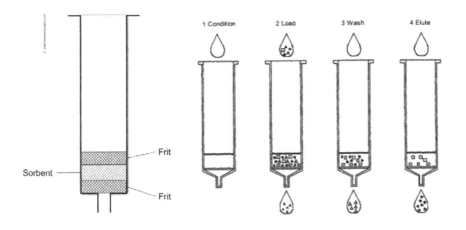

Frit

Sorbent

Frit

1 Condition   2 Load   3 Wash   4 Elute

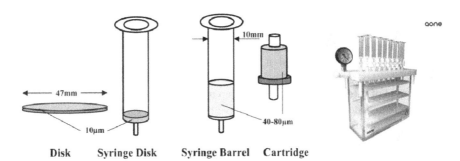

47mm

10µm

10mm

40-80µm

Disk   Syringe Disk   Syringe Barrel   Cartridge

FIGURE 3.33

All forms of SPE generally follow the same four-step procedure. This process comprises (1) sorbent conditioning, (2) passage of the sample through the sorbent, commonly called sample loading, (3) removal of interferences, and (4) elution of the analyte of interest into a receiving vessel. The four steps that are involved in performing methods that require SPE as the principal means of sample preparation are now discussed:

1. **Sorbent conditioning:** This initial step is particularly important in order to maximize the percent recovery when using RP-SPE with an aqueous sample. Also, once conditioned, the sorbent should not be allowed to dry. If dried, the sorbent should be reconditioned. For RP-SPE, methanol (MeOH) is usually used to condition, whereas for NP-SPE, n-hexane is common. The need for conditioning in RP-SPE has been discussed in the literature.[84] The brush-like nature of the organo-bonded hydrophobic surface is somewhat impermeable. It is as if a thin hydrophobic membrane has been placed on top of the surface. Organics dissolved in water upon passing through this sorbent only partially penetrate this semiper-meable membrane. If the sorbent is not conditioned, low and variable percent recoveries from RP-SPE will result. Upon passing MeOH or other polar solvent across the sorbent, the brushes line up and, by interacting with MeOH, become more "wetted." This hydro-phobic membrane becomes permeable, and hence dissolved organics can more effectively penetrate this hypothetical membrane barrier. Conditioning has been thought of in terms of being brush-like. This author has found that by careful adjustment of the level of MeOH by removing trapped air between the reservoir and cartridge in the barrel-type SPE column, MeOH can be passed through the sorbent, followed by the water sample, without exposing the sorbent to air. Some analysts, after conditioning, will pass the eluting solvent through the wetted sorbent. The eluting solvent in many cases of RP-SPE is much less polar than MeOH. In this way, the sorbent can be cleaned of organic impurities. This is also a useful technique to recondition and reuse SPE cartridges. This author has observed even slightly higher percent recoveries when conducting RP-SPE on a previously used cartridge!

2. **Sample loading:** The second step in SPE involves the passage of the sample through the previously conditioned sorbent. The sorbent might consist of between 100 mg and 1 g of chemically bonded silica. The two most common hydrophobic chemically bonded silicas are those that contain either a C_8 or a C_{18} hydrocarbon moiety chemically bonded to 40-μm-particle-size silica gel. Alternatively, a disk consisting of impregnated C_8 or C_{18} silica in either a Teflon or glass–fiber matrix is used. For RP-SPE, an aqueous water sample is usu-ally passed through the sorbent. Samples that contain particulates or suspended solids are more difficult to pass through the sorbent. Eventually, the top retaining frits will plug. This drastically slows the flow rate, and if the top frit gets completely plugged, there is no more SPE to be done on that particular cartridge! Suspended solids in water samples present a severe limitation to sample loading. A filtration of the sample prior to SPE sample loading will usually correct this problem. Adding the 70 mL reservoir on top of the 3 cc SPE barrel-type cartridge enables a relatively large sample volume to be loaded. Upon frequent refilling of the reservoir, a groundwater sample of volume greater than the 70 mL capacity, up to perhaps 1,000 mL, is feasible. Alternatively, the top of the 70 mL reservoir can be fitted with an adapter and a plastic tube can be connected to a large beaker containing the groundwater sample. The design of the multiport SPE vacuum manifold enables 5 or 10 or more samples, depending on the number of ports, to be simultaneously loaded.

3. **Removal of interferences:** The third step in SPE involves passing a wash solution through the sorbent. The analytes of interest have been retained on the sorbent and should not be removed in this step. In RP-SPE, this wash solution is commonly found to be distilled deionized water. In other cases, particularly when ionizable analytes are of interest, water that has been buffered so that the pH is fixed and known is used as the wash solution. The wash solution should have a solvent strength not much different from that of the sample, so

that retained analytes are not prematurely removed from the sorbent. In addition, air can be passed through the sorbent to facilitate moisture removal. It is common to find droplets of water clinging to the inner wall of the barrel-type cartridge. Passing air through during this step is beneficial. Also, a Kim-Wipe® or other clean tissue can be used to remove surface moisture prior to the elution step. Removal of surface water droplets requires disassembly of the reservoir-adapter-barrel. If a full set of SPE cartridges is used, it is a bit time consuming to complete. At this point in the process, the sorbent could be stored or transported. Too few studies have been done to verify or refute the issue of whether analytes are stable enough to be sampled in the field and then transported to the laboratory.

4. **Elution of retained analyte:** The fourth and last step in SPE involves the actual removal of the analyte of interest; hopefully, the analyte is free of interferences. This is accomplished by passing a relatively small volume of a solvent, called an eluent, whose solvent strength is sufficient so that the analyte is removed with as small a volume of eluent as possible. This author has been quite successful, particularly when using barrel-type SPE cartridges, in using less than 1 mL of eluent in most cases. A common practice is to make two or three successive elutions of sorbent such that the first elution removes >90% of the retained analyte, and the second or third removes the remaining 10%. If there is evidence of water in the receiving vessel, the analyst can add anhydrous sodium sulfate to the receiving vessel. This is particularly relevant for RP-SPE after passage of a drinking water or groundwater sample. Alternatively, a second SPE cartridge can be placed beneath the first in a so-called piggyback configuration to remove water, remove lipid interferences, and fractionate. As a nonpolar to moderately polar elution solvent or binary solvent is passed through the SPE cartridge while eluting the retained analytes, the eluent is also passed through this second SPE cartridge that contains anhydrous sodium sulfate. Figure 3.34 is a sketch (not drawn to scale) that depicts the piggyback style for the common 3 cc SPE barrel size.

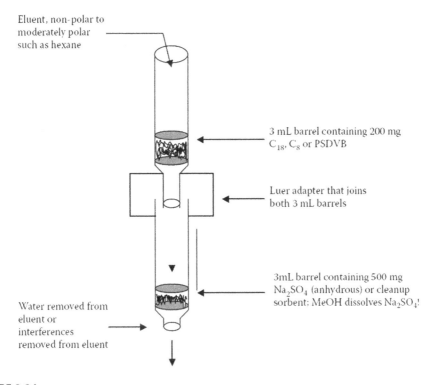

Eluent, non-polar to moderately polar such as hexane

3 mL barrel containing 200 mg C_{18}, C_8 or PSDVB

Luer adapter that joins both 3 mL barrels

3mL barrel containing 500 mg Na_2SO_4 (anhydrous) or cleanup sorbent; MeOH dissolves Na_2SO_4!

Water removed from eluent or interferences removed from eluent

FIGURE 3.34

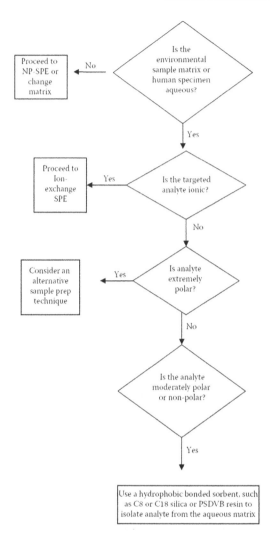

FIGURE 3.35

These four steps, which comprise the contemporary practice of SPE, serve to transform an environmental sample that is, by itself, incompatible with the necessary determinative technique, most commonly gas chromatography. Figure 3.35 is a flowchart to assist the environmental analytical chemist in deciding how to approach analytical method development utilizing SPE. The logic used strongly depends on the chemical nature of the semivolatile analyte of interest and on the sample matrix in which the analyte of interest is dissolved. SPE is applicable only to *semivolatile* to *nonvolatile* organic analytes. The use of a vacuum to drive the aqueous sample through the sorbent eliminates any applicability that SPE might have in isolating volatile organics.

3.55   HOW CAN I LEARN MORE ABOUT SPE TECHNIQUES?

By attempting to keep abreast of the analytical chemistry literature! Not an easy task! Journey through the "maze of verboseness" to find out how SPE is being used in laboratories. Ask yourself "How has this author used SPE creatively?" Novel strategies for incorporating SPE techniques as part of an overall *sample prep* strategy are likely to be found in the primary literature. A list of relevant journals is given below:

American Laboratory (print issues terminated in 2019, archive issue are still available online)
Analyst
Analytica Chimica Acta
Analytical Biochemistry
Analytical Chemistry
Chemosphere
Chromatographia
CRC Critical Reviews of Analytical Chemistry
Environmental Science and Technology
International Journal of Environmental Analytical Chemistry
Journal of Analytical Toxicology
Journal of the Association of Official Analytical Chemists International
Journal of Chromatographic Science
Journal of Chromatography
Journal of Liquid Chromatography and Related Technologies
LC-GC: Solution for Separation Scientists (this is a must read and complimentary)
Separation Science and Technology
Trends in Analytical Chemistry.

Book chapters as well as books on SPE have been published.[83,85–87] LC-GC has published three supplements on sample prep, and this series includes discussions of sample prep techniques to determine VOCs.[88–90]

3.56   HOW DOES SPE WORK?

The chemically bonded silica removes moderately polar to nonpolar solutes from an aqueous matrix based on thermodynamic principles. These principles were introduced early in this chapter as a means to discuss the underlying physical chemistry of LLE. Equation (3.4) can be restated pertinent to SPE as follows:

$$K^o = \frac{a_{\text{surface}}}{a_{\text{aq}}}$$

This relationship resulted from the spontaneous tendency for the analyte, initially solubilized in an aqueous matrix, to partition into the surface monolayer of hydrophobic octadecyl- or octyl-bonded silica. Organic compounds as analytes will have a unique value for the thermodynamic distribution constant. One fundamental difference between LLE and SPE stands out. Partition or adsorption of solute molecules onto a solid surface follows the principles of the Langmuir adsorption isotherm, whereas LLE does not. We will briefly develop the principles below. This model assumes that an analyte, A, combines with a site of adsorption, S, in which there is a finite number of such sites according to

$$A + S \overset{K^L}{\leftrightarrow} A - S$$

An equilibrium constant, K^L, can be defined in terms of the mole fraction of adsorbed sites, X_{A-S}, the mole fraction of unadsorbed sites, X_S, and the mole fraction of analyte in equilibrium according to X_A:

$$K^L = \frac{X_{A-S}}{X_S X_A}$$

A fractional coverage, θ, can be defined whereby

$$\theta = \frac{K^L X_A}{1 + K^L X_A}$$

When the mole fraction of analyte is low, the above relationship can be simplified to

$$\theta = K^L X_A$$

When the mole fraction of analyte is high, θ approaches 1. These equations can be visualized graphically in Figure 3.36.

An organic compound such as tetrachloro-m-xylene (TCMX), a popular organic compound used as a surrogate whose structure is shown below, and initially dissolved in water (to the extent that a nonpolar organic compound can be dissolved), is thermodynamically unstable.

Upon passage of a water sample containing TCMX through a chemically bonded silica sorbent, the degree of interaction between the octyldecylsiloxane or octysiloxane moieties chemically bonded to a silica matrix (SPE sorbent) surface and TCMX exceeds the degree of interaction between the aqueous matrix and TCMX. Hence, the analyte–sorbent (A–S) interaction is lower than that for the analyte–matrix (A–M) interaction. In other words, the minimum in the free-energy curve, as shown in Figure 3.2, occurs well displaced toward the side of the analyte–sorbent interaction. This concept can be viewed as an "eternal triangle," whereby both the sorbent and matrix compete for the analyte as shown below:

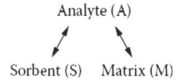

Efficient removal of analyte from the matrix requires that the strength of the A–S interaction be increased while that of the A–M interaction be decreased. This is accomplished chemically by matching the polarity of the sorbent to that of the analyte while mismatching the polarity of the matrix to that of the analyte. Consider TCMX as the analyte and octysiloxane (C_8SiO_2)

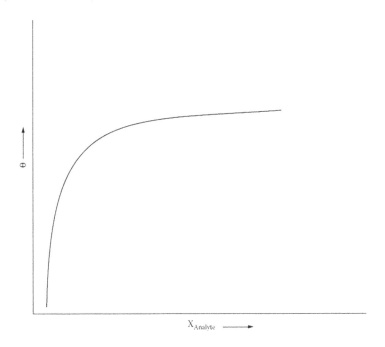

FIGURE 3.36

bonded silica as the sorbent. If TMCX could be dissolved in water (a very small amount of TCMX can actually dissolve in water until a saturated solution is reached), an aqueous sample would be created, and if ultrapure water was used, this sample would be known as a spiked sample. One way to increase the amount of TCMX that can be dissolved in a given volume of water is to also spike the water with a methanol. Methanol serves as a molecular interface by enabling TCMX to permeate the hydrogen-bonded matrix. The solubility of TCMX is now significantly increased, and a series of spiked samples can be prepared that cover a wide range of solute concentration.

3.57   IS SPE A FORM OF COLUMN CHROMATOGRAPHY?

Yes, SPE, as a form of liquid column chromatography, is one of three broad classifications of column chromatography. Historically, SPE would have been called frontal development. The other two categories are elution and displacement chromatography. Elution chromatography is the principal means to conduct GC and HPLC and is of utmost importance to TEQA and as an instrumental technique; it will be discussed in sufficient detail in Chapter 4. In elution chromatography, a mobile phase is continuously passed across a stationary phase and a small plug of sample is injected into this flowing stream. Displacement chromatography is similar to elution, except that the mobile phase is much more strongly retained by the stationary phase than is the sample (p. 129).[6] Displacement chromatography is a rarely used form and will not be considered here.

   The principles underlying SPE as an example of frontal chromatography can be more easily understood if the following experiment is designed, the instrument configured, and the results interpreted (pp. 88–92).[83] Figure 3.37 is a schematic of a frontal chromatographic instrumental configuration in which the bonded sorbent is placed into an HPLC column and installed in such a way that the sample is passed through the SPE sorbent and the effluent is also passed through an ultraviolet (UV) absorbance detector via a flow-through microcell. Provided that the total extra

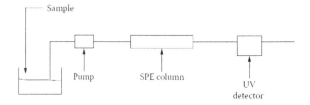

FIGURE 3.37

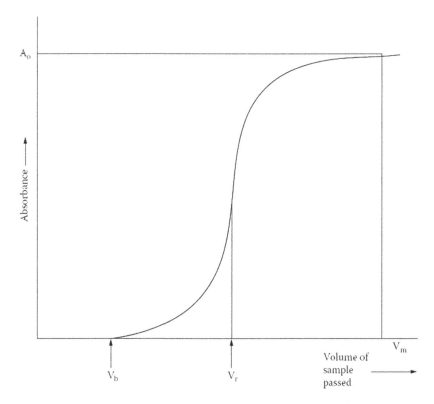

FIGURE 3.38

column void volume is not large in comparison to the size of the SPE column, a frontal chromatogram can be obtained. The reservoir contains the analyte dissolved in a matrix such as groundwater, where the analyte–matrix (A–M) interaction exists and the analyte–sorbent (A–S) interaction takes over in the SPE column. Figure 3.38 is the frontal chromatogram that would be typical. This assumes that the analyte is a strong absorber in the ultraviolet region of the electromagnetic spectrum. Referring to Figure 3.38, it should be noted that there is a certain volume of sample that can be passed through the sorbent without observing any rise in the absorbance. At Vb, a volume that represents the volume of sample that gives an absorbance that is 1% of Ao, the first signs of breakthrough are seen to emerge. The inflection point in the curve, denoted by a volume of sample passed through the column, Vr, is unique to a given solute and is often called the *analyte breakthrough volume*. The absorbance is seen to increase sigmoidally as shown in Figure 3.39 and plateaus at Ao. This plateau, in essence, is the same one shown by the Langmuir isotherm in Figure 3.36.

Let us develop this concept a bit more by assuming that approximately 200 mL of water has been spiked with traces of dimethyl phthalate (DMP). DMP strongly absorbs in the UV, and

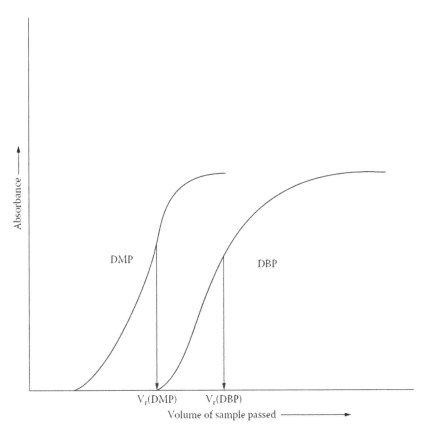

FIGURE 3.39

one can expect a sigmodal curve, as shown in Figure 3.38, to result. If, instead of DMP, dibutyl phthalate (DBP) is added to water, a similar curve is obtained. In this case, a significant increase in Vb and Vr is observed. This is shown in Figure 3.39. If both DMP and DBP were spiked together at the onset to 200 mL of high-purity water and the breakthrough experiment performed, the curve would resemble that shown in Figure 3.40. Much before the commercial development of SPE, a curve such as that shown in Figure 3.40 was already established and the following predictions made (pp. 128–129):[6]

> Separation results in the formation of fronts ... the least retained component emerges first from the bed followed by a mixture of the least and next most strongly retained component ... This technique is useful for concentrating trace impurities.

This author doubts whether these authors could have foreseen the impact of their words on the eventual development of commercial products that rely on frontal development, namely, SPE.

Let us now set aside this spiked aqueous sample that consists of DMP and DBP dissolved in water. This sample easily simulates an environmental sample that contains phthalate esters, since this class of organic compound is commonly found in landfill leachates due to the use of phthalate esters as plasticizers. Let us reconfigure our HPLC to that shown in Figure 3.41, so that ultrapure water is pumped through our SPE column. The results, shown in Figure 3.42, corresponding to elution chromatography, give, instead of the sigmoidal shape, the Gaussian peak profile, with the apex of the peak located at Vr. Because Vr differs between DMP and DBP, we have a basis of separation. In Chapter 4 we will discuss the theory underlying column elution liquid chromatography as a means to separate and quantitate the presence of organic compounds.

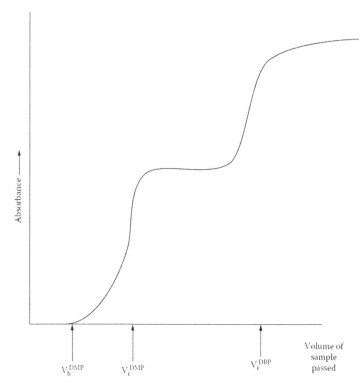

FIGURE 3.40

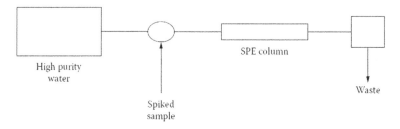

FIGURE 3.41

Solid-phase extraction has been called digital chromatography because its chief use is to retain and then remove analytes of interest in an on–off fashion. Outside of conducting breakthrough studies, the purely frontal chromatographic form has little practical value in TEQA. The elution step, in which an eluent's polarity is near to that of the analyte, removes all analytes and can be viewed in a manner similar to that done for the sorption step.

The analyte–matrix interaction becomes so strong that analytes are easily and quickly eluted from the sorbent. The polarity of the matrix now matches the polarity of the analyte.

3.58   CAN SPE BREAKTHROUGH BE PREDICTED?

We saw earlier just how Vb and Vr are obtained experimentally by reconfiguring a high-pressure liquid chromatograph (refer to Figure 3.37). An HPLC column packed with SPE sorbent replaces the conventional HPLC column. A spiked sample is pumped through the SPE column to develop

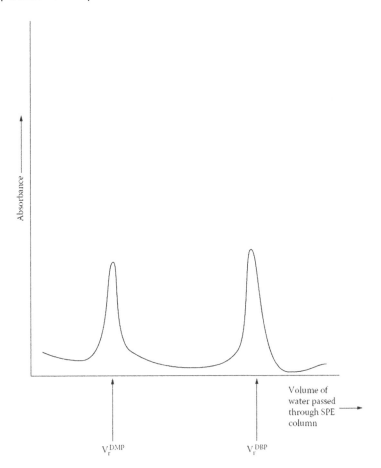

FIGURE 3.42

the frontal chromatograms, as shown in Figure 3.38. This configuration is all it takes to conduct the breakthrough study. Is there sufficient theory developed to predict Vb or Vr? Yes, there is, and we proceed to consider this below. Before we do, though, we need to introduce certain concepts and definitions that are fundamental to achieving a somewhat deeper understanding of liquid chromatography. We must introduce the important elements from the more contemporary approaches taken to understand exactly what one means by the term *solvent polarity*. We must also discuss what one means by the term *aqueous solubility* for an organic compound acting as a solute, and we need to discuss how this important parameter relates to the solute's octanol–water partition coefficient. We admit to digressing somewhat into the physical chemistry of solution behavior insofar as it aids the reader in understanding the theory that underpins RP-SPE. Introducing these concepts in this chapter on sample preparation will facilitate the discussion of HPLC as a determinative technique in the next chapter.

3.59   WHAT IS SOLVENT POLARITY?

Solvent polarity is a numerical value that can be assigned to each of the common organic solvents used in TEQA. A solvent polarity index, denoted by P', is defined based on experimental solubility data using test solutes and was originally developed by Rohrschneider[91] and further discussed by Snyder.[92] Table 3.16 lists a number of representative solvents in order of

TABLE 3.16
Polarity of Selected Organic Solvents

Solvent	P'	Solvent	P'
Pentane	0.0	Tetrahydrofuran	4.0
1, 1, 2-TCTFE	0.0	Chloroform	4.1
Hexane	0.1	Ethyl acetate	4.1
Iso-octane	0.1	MEK	4.7
Petroleum ether	0.1	1, 4-dioxane	4.8
Cyclohexane	0.2	Acetone	5.1
Hexadecane	0.5	MeOH	5.1
Toluene	2.4	Acetonitrile	5.8
MTBE	2.5	N, N-DMF	6.4
Ethyl ether	2.8	DMSO	7.2
Methylene chloride	3.1	Water	10.2
Isopropyl alcohol	3.9		

increasing P'. (p. 273)[3] The introduction of heteroatoms such as oxygen, sulfur, and nitrogen significantly contributes to solvent polarity. Consider what happens when a test solute such as methanol is added to the very polar solvent water vs. what happens when methanol is added to the very nonpolar solvent hexane. The classic cliché "like dissolves like" certainly applies, whereby methanol, whose P' is 5.1, is infinitely soluble or completely miscible in all proportions with water, whose P' is 10.2. However, methanol, with a density of 0.7915 g/mL at 20°C, will tend to sink to the bottom when added to hexane, whose density is 0.660 g/mL at the same temperature. The presence of hydroxyl groups in both methanol and water creates an infinite number of hydrogen bonds and is responsible for defining both solvents as associated liquids. The degree to which methanol molecules interact through intramolecular hydrogen bonding is of a similar attractive potential energy, as is the degree to which water molecules interact through intramolecular hydrogen bonding. The energy of interaction is due to hydrogen bonding between methanol and water, the so-called intermolecular hydrogen bonding; this is shown in Figure 3.43. Because there is little difference in the potential energy for the interaction intramolecularly and intermolecularly, the tendency for matter to spread (entropy) becomes the driving force between methanol and water.

Consider what happens when the two immiscible liquids, methanol whose P' is 5.1 and hexane whose P' is 0.1, are in contact with each other. The weak dispersion forces are present throughout hexane, whereas the much stronger interaction due to hydrogen bonding among methanol molecules exists; this is shown in Figure 3.44. It is impossible to overcome the high intermolecular potential energy that exists to afford a true dissolution of hexane into methanol and vice versa.

These intramolecular and intermolecular forces that exist among molecules in liquids, as in Figures 3.43 and 3.44 not only serve to explain why hexane and methanol do not mix and remain as two immiscible liquids, but also help in understanding how to go about developing a method that incorporates SPE as the principal sample preparation technique. Hexane is a suitable elution solvent for relatively nonpolar analytes that are sorbed onto and into the octyldecyl

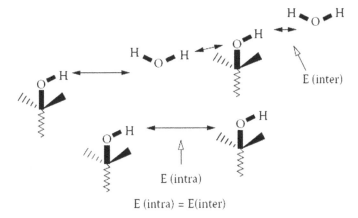

FIGURE 3.43

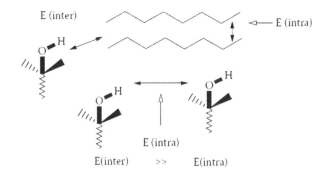

FIGURE 3.44

or octyl siloxane moiety covalently bonded to the silica sorbent (so-called reversed-phase sorbent), whereas methanol is far too polar a solvent to be used as an elution solvent for relatively nonpolar analytes.

3.60 CAN A SOLUTE'S AQUEOUS SOLUBILITY OR OCTANOL–WATER PARTITION COEFFICIENT BE USED TO PREDICT RP-SPE PERCENT RECOVERIES?

Yes, both solute physico-chemical properties give important clues about the possibility of whether a given organic compound originally dissolved in an aqueous matrix will be effectively recovered by RP-SPE. Again, because SPE is a two-step transfer of solute (sorption from the matrix followed by elution from the sorbent), both factors serve to limit the percent recovery. It is only percent recoveries that can be measured by determinative techniques. There is no direct way to measure the amount of analyte sorbed except through the breakthrough experiments previously discussed. Let us assume then that the analyte, once sorbed, is efficiently removed from the sorbent. The differences in our measured percent recoveries then reflect only the mass transfer of analyte from an aqueous matrix to the chemically bonded sorbent. Solutes that in general exhibit low aqueous solubilities, S_{aq}, and have relatively large octanol–water partition coefficients, K_{ow}, have a strong tendency to sorb on hydrophobic surfaces from an aqueous matrix. The mathematical relationship between S_{aq} and K_{OW} has been the focus of extensive study among physical chemists. A nearly inverse and linear relationship exists between two solute parameters. This is stated from one resource as follows:[93]

$$\log K_{OW} = -\log S_{aq} - \log \gamma_0 - \log V_0 \qquad (3.36)$$

where γ_0 represents the activity coefficient for a water-saturated octanol phase and V_0 is the molar volume for the water-saturated octanol phase, which is 0.12 L/mol. If $\log K_{OW}$ is plotted against the values for the aqueous solubility, S_{aq}, for a variety of organic compounds of environmental interest, a linear relationship with a negative slope emerges. The majority of priority pollutant organic compounds have activity coefficients in water-saturated octanol, γ_0, that fall between 1 and 10. This suggests that these solutes exhibit ideal solution behavior and reflects a strongly hydrophobic character.[93]

In general, polar solutes have $\log K_{OW}$ values less than 1; moderately polar solutes have $\log K_{OW}$ values between 1 and 3; nonpolar solutes have $\log K_{OW}$ values greater than 3. To illustrate the influence of various substituents on the $\log K_{OW}$ for a parent compound such as phenol, consider the following list of phenols as shown in Table 3.17.[94]

Clearly, the effect of either chloro or methyl substitution serves to significantly increase the octanol–water partition coefficient. This increase in hydrophobicity is an important consideration in developing an RP-SPE method that is designed to isolate one or more phenols from a groundwater sample. Those substituted phenols with the greater values for K_{OW} would be expected to yield the higher percent recoveries in RP-SPE. The substitution of a trimethyl silyl moiety in place of hydrogen significantly increases the hydrophobicity of the derivatized molecules. This use of a chemical derivative to convert a polar molecule to a relatively nonpolar one has important implications for TEQA and will be discussed in Chapter 4.

3.61 WHAT IS AQUEOUS SOLUBILITY ANYWAY?

The amount of solute that will dissolve in a fixed volume of pure water is defined as that solute's aqueous solubility, S_{aq}. It is a most important concept with respect to TEQA because there is such a wide range of values for S_{aq} among organic compounds. Organic compounds possessing low aqueous solubilities generally have high octanol–water partition coefficients. This leads to a tendency for these compounds to bioaccumulate. The degree to which an organic compound will dissolve in water depends not only on the degree of intramolecular vs. intermolecular

TABLE 3.17
Log K_{ow} Values for Selected Phenols

Phenol	log K_{ow}
Phenol	1.46
o-Methyl phenol	1.95
o-Chlorophenol	2.15
2, 6-Dimethyl phenol	2.36
p-Chlorophenol	2.39
2, 4-Dichlorophenol	3.08
p-Trimethyl silyl phenol	3.84

interactions, as we saw earlier, but also on the physical state of the substance (i.e., solid, liquid, or gas). The solubility of gases in water is covered by Henry's law principles, as discussed earlier.

A solute dissolved in a solvent has an activity a. This activity is defined as the ratio of the solute's fugacity (or tendency to escape) in the dissolved solution to that in its pure state. The activity coefficient for the ith solute, γ_w^i, relates solute activity in pure water to the concentration of the solute in water, where the concentration is defined in units of mole fraction, x_w, according to

$$x_w^i = \frac{\text{moles}^i}{\displaystyle\sum_i \text{moles}^i} \qquad \gamma_w^j \equiv \frac{a_w^i}{x_w^i}$$

We usually think of aqueous solubilities in terms of the number of moles of the ith solute per liter of solution. Alternative units commonly used to express aqueous solubility include milligrams per liter or parts per million and also micromoles per liter. The concentration in moles per liter can be found by first considering the following definition:

$$C_i = \frac{x_i \left(\text{moles} / \text{mole total}\right)}{V_{mix} \left(L / \text{mole total}\right)}$$

Vw is the molar volume of water (0.018 L/mol) and Vi is the molar volume of a typical solute dissolved. The molar volume of the mixture, V_{mix}, is found by adding the molar volume contributions of the solute and solvent, water, according to

$$V_{mix,w} = 0.2 x_{i,w} + 0.018 x_w$$

This expression can be further simplified to

$$V_{mix,w} = 0.182 x_{i,w} + 0.018$$

Because the mole fraction for solutes dissolved in water is very low, $x_{i,w} < 0.002$, which corresponds to about 0.1 M, the molar volume for the mix can be approximated at 0.018, and hence the above equation for the concentration in moles per liter can be expressed as

$$C_w^{sat} = \frac{x_w^{sat}}{V_{mix,w}} \cong \frac{x_w^{sat}}{0.018} \left(\text{moles/L}\right)$$

The above equation converts the mole fraction of a solute dissolved in water, up to saturation conditions, x_w^{sat}, to the corresponding concentration of that solute in units of moles per liter. By knowing the molecular weight of a particular solute, it is straightforward to obtain the corresponding concentration in milligrams per liter.

For hydrophobic solutes, γ_w is greater than 1. When enough solute has been added to a fixed volume of water until no more can be dissolved, the solution is said to be saturated and a two-phase system results. This is a condition of dynamic equilibrium whereby the solute activity in both phases is equal. Denoting o as the organic phase, we have

$$a_w = a_o$$

so that

$$x_w \gamma_w = x_o \gamma_o$$

For immiscible liquids in water, the mole fraction of solute in itself and its activity coefficient approximate 1, so that

$$x_w \gamma_w = 1$$

and the mole fraction, x_w, is related to the aqueous activity coefficient by

$$x_w = \frac{1}{\gamma_w}$$

and in terms of logarithms,

$$\log x_w = -\log \gamma_w$$

The purpose of deriving this is to show that attempts to measure the aqueous solubility, xw, are, in essence, attempts to estimate the activity coefficient, γw. This relationship applies only to liquid solutes. Solid or crystalline solutes require an additional term where

$$\log x_w = \log \frac{x^c}{x^{\text{sel}}} - \log \gamma_w$$

where xc/x_{sel} represents the ratio of the solubility of the crystal to that of the hypothetical supercooled liquid. (pp. 60–65)[95] The additional term above is approximated by the term $-0.00989 \, (T_{mp} - 25)$ for rigid molecules at 25°C and $-[0.01 + 0.002(n - 5)]T_{mp}$ for long-chain molecules, with n monomers per polymer and T_{mp} the melting point of the solid. Octanol–water partition coefficients can also be estimated from solute retention times using, for example, reversed-phase HPLC.

Aqueous solubility for various solutes can be estimated by applying any of the techniques in the following four categories (pp. 41–42):[95]

1. Methods based on experimental physico-chemical properties, such as partition coefficient, chromatographic retention, boiling point, and molecular volume.
2. Methods based directly on group contributions to measured activity coefficients.
3. Methods based on theoretical calculations from molecular structures, including molecular surface area, molecular connectivity, and parachor.
4. Methods based on combinations of two or more parameters that can be experimentally measured, calculated, or generated empirically. These include the solubility parameter method and the UNIFAC technique (a method based on linear solvation energy relationships and on the use of multivariate statistical methods).

However, for many of the very hydrophobic solutes of interest to TEQA, the aqueous solubility is extremely low. For example, iso-octane, a nonpolar hydrocarbon, will dissolve in water up to 0.0002% at 25°C (p. 193).[3] Higher-order aliphatic hydrocarbons have such low values for their aqueous solubilities that it is impossible to directly measure this physico-chemical property.

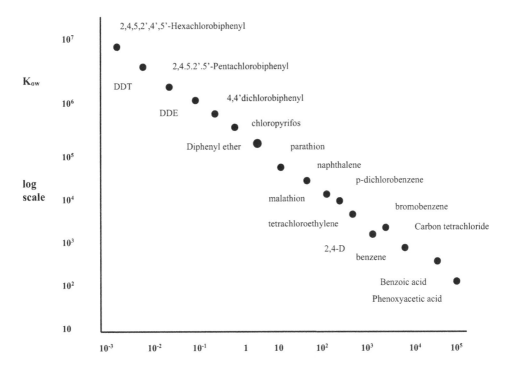

Aqueous Solubility, S_{aq}, in μmoles/L (logarithmic scale)

FIGURE 3.45

3.62   WHAT DOES A PLOT OF K_{OW} VS. AQUEOUS SOLUBILITY TELL US?

Figure 3.45 is a qualitative plot of the logarithm of the octanol–water partition coefficient, K_{OW}, against the logarithm of the aqueous solubility, S_{aq}, of a selected number of organic solutes of environmental interest. This qualitative plot makes it very evident that polar solutes with appreciable aqueous solubility exhibit low values of K_{OW}. Hydrophobic solutes that are not soluble in water to any great degree exhibit large values of K_{OW}. The importance of knowing K_{OW} values as this pertains to the practice of TEQA and to environmental science in general has been stated earlier:

> In recent years the octanol/water partition coefficient has become a key parameter in studies of the environmental fate of organic chemicals. It has been found to be related to water solubility, soil/sediment adsorption coefficients, and bioconcentration factors for aquatic life. Because of its increasing use in the estimation of these other properties, K_{OW} is considered a required property in studies of new or problematic chemicals.[96]

Much effort has been expended in finding a linear equation that relates octanol–water partition coefficients for any and all solutes to the aqueous solubility in octanol-saturated water. Two useful equations have emerged in which the logarithm of the aqueous solubility, log S_{aq}, is, in general, linearly related to log K_{OW} (pp. 60–65)[95]

For liquid solutes:

$$\log K_{OW} = 0.8 - \log S_{aq} \qquad (3.37)$$

TABLE 3.18

Degree of Correlation between the Octanol–Water Partition Coefficient and Aqueous Solubility for Four Classes of Organic Compounds

Class	$\log K_{OW}$	Correlation Coefficient
Aromatic hydrocarbons	$0.786{-}1.056 \log S_{aq}$	0.995
Unsaturated hydrocarbons	$0.250{-}0.908 \log S_{aq}$	0.993
Halogenated hydrocarbons	$0.323{-}0.907 \log S_{aq}$	0.993
Normal hydrocarbons	$0.467{-}0.972 \log S_{aq}$	0.999

For crystalline solutes:

$$\log K_{OW} = 0.8 - \log S_{aq} - 0.01\left(T_{mp} - 25\right) \tag{3.38}$$

Yalkowsky and Banerjee have reviewed specific studies and summarized the relationship in Equations (3.37) and (3.38) as applied to liquid organic compounds and crystalline organic solids. [95] The results are shown in Table 3.18 for four classes of liquid hydrocarbons. A plot of $\log K_{OW}$ against $\log S_{aq}$ reveals a slope close to the value of -1 for each of these four classes of hydrocarbons. The differences in the y intercept reflect differences in the activity coefficients as defined in Equation (3.36).

3.63   CAN WE PREDICT VALUES FOR K_{OW} FROM ONLY A KNOWLEDGE OF MOLECULAR STRUCTURE?

Yes, and the predictions are quite good. Two different methods emerge from a host of others and are most commonly used to predict octanol–water partition coefficients for the many organic compounds that exist. One approach is to calculate K_{OW} from a knowledge of structural constants, whereas the second approach requires that a chemical's partition coefficient be measured between a solvent other than octanol and water, K_{SW}. K_{OW} can then be calculated from linear regression equations that relate $\log K_{SW}$ (for a particular solvent) and $\log K_{OW}$. Two forms of the structural constant approach are most popular: (1) the Hansch π hydrophobic character of substituents approach and (2) the Leo fragment constant approach. The Hansch π approach is based on the assignment of a value for π_X as the difference between octanol–water partition coefficients for a substituted vs. unsubstituted or parent compound. Mathematically, this difference can be stated as follows:

$$\pi_X = \log K_{OW}^X - \log K_{OW}^H$$

For example, the $\log K_{OW}$ for chlorobenzene is 2.84, whereas that for benzene is 2.13, and thus

$$\pi_X = 2.84 - 2.13 = 0.71$$

and the πX for the substituent chlorine on the monoaromatic ring is 0.71. The Hansch π approach has been recently criticized because it ignores hydrogens attached to carbon, and this led to some erroneous values for a few aliphatic hydrocarbons.

The Leo fragment method has emerged as a more powerful way to predict $\log K_{OW}$. The working mathematical relationship is

$$\log K_{OW} = \sum_i f_i + \sum_j F_j \tag{3.39}$$

where the logarithm of the octanol–water coefficient for a particular organic compound is found by algebraically summing the contributions due to the ith structural component (f) (building block or functional group), overall components, and the algebraic sum of the jth factor (F) due to intramolecular interactions caused largely by geometric or electronic effects. For complex molecules, it is desirable to have a known log K_{OW} value for a given compound and use the fragment approach to add and subtract structural and geometric/electrical contributions such that

$$\log K_{OW}\,(\text{unknown}) = \log K_{OW}\,(\text{known}) - \underset{\text{removed}}{\sum f} + \underset{\text{added}}{\sum f} - \underset{\text{removed}}{\sum F} + \underset{\text{added}}{\sum F} \tag{3.40}$$

Thus, two ways emerge to calculate log K_{OW} using the Leo fragment method. One is to build the entire molecule from fragments and factors using Equation (3.40), and the other is to calculate log K_{OW} from structurally related compounds. To illustrate how one goes about calculating the log K_{OW} for a specific organic compound, we return to the organic compound: tetrachloro-m-xylene (TCMX). We begin by first using Equation (3.39) to establish the log K_{OW} for benzene by using previously published fragment factors. For carbon that is aromatic, a fragment factor of 0.23 has been established. Hence, because benzene consists of six aromatic carbons that in turn are bonded to six hydrogens,

$$\log K_{OW}\,(C_6H_6) = 6 f_{c(aromatic)} + 6 f_h^{\phi} = 6(0.13) + 6(0.23) = 2.16 \tag{3.41}$$

This prediction is quite close to the well-established and observed value for benzene, which is 2.13. Using this established value for benzene, we begin to view our molecule of interest, TCMX, as a substituted benzene. Theoretically, we can arrive at TCMX by removing all six aromatic hydrogens and then adding two methyl groups and four chlorine groups. Stated mathematically,

$$\log K_{OW(TCMX)} = \log K_{OW(C_6H_6)} - 6 f_H^{\phi} + 2 f_{CH_3}^{\phi} + 4 f_{Cl}^{\phi} \tag{3.42}$$

The methyl groups are viewed as being composed of an aromatic carbon and three aliphatic bonded hydrogens and a fragment constant. Equation (3.42) is modified as follows:

$$\log K_{OW(TCMX)} = \log K_{OW(C_6H_6)} - 6 f_H^{\phi} + 2 f_C^{\phi} + 3 f_H + 4 f_{Cl}^{\phi} \tag{3.43}$$

Upon substituting the values for the various fragment constants into Equation (3.43), we get

$$\log K_{OW(TCMX)} = 2.13 - 6(0.23) + 2\big[(0.13) + 3(0.23)\big] + 4(0.19) = 6.15$$

Such a large value for the octanol–water partition coefficient for TCMX suggests that the compound exhibits a very low aqueous solubility. Note that the substitution by hydrogen by methyl and chloro groups significantly increases the hydrophobicity of the molecule. The addition of a methyl group is seen to add about 0.5 of a log unit, irrespective of the compound involved. For example, substituting a methyl for hydrogen in benzene contributes the same, as if a methyl were substituted for hydrogen in cyclohexane.

As long as the building blocks lead to a complete molecular structure, only structural fragment constants are necessary, as in the TCMX example. However, for molecules that are more complex, the molecular components begin to exert significant influences on one another through steric, electronic, and resonance intramolecular interactions. These interactions affect both the aqueous and octanol activity coefficients, as ability of groups to rotate about a carbon–carbon single bond, chair/boat confirmation in cyclohexane, and carbon-chain branching in aliphatic structures. Geometric effects increase aqueous solubility, decrease K_{OW}, and hence their F factor contributes negativity to Equation (3.40). Electronic effects due to electronegativity differences between the elements of both atoms that comprise a polar covalent bond, such as a carbon–chlorine bond, decrease aqueous solubility, increasing K_{OW}, and hence their F factor contributes positively to Equation (3.40). Other electronic effects include nearby polyhalogenation on carbon, nearby polar groups, and intramolecular hydrogen bonding. These other factors also contribute in a positive manner to Equation (3.40).

3.64   ARE VALUES FOR K_{OW} USEFUL TO PREDICT BREAKTHROUGH IN RP-SPE?

The capacity factor for an RP-SPE cartridge when pure water is used, k_W', discussed earlier, has been found to be closely related to K_{OW} and is related by (p. 52)[95] and by (p. 95)[83]

$$k_W' = 0.988 \log K_{OW} + 0.02$$

Hence, merely obtaining octanol–water partition coefficient values from various tabulations in the literature can yield a predictive value for k_W' because we have shown that k_W' is directly related to the breakthrough volume, Vr.

3.65   WHERE CAN ONE OBTAIN K_{OW} VALUES?

The resources cited are available in most university libraries in order to find these values, as well as to obtain good introductions. Leo and colleagues published a comprehensive listing and followed this with a more systematic presentation.[94,97] Even earlier, Hansch and colleagues published their findings.[98] Lyman and colleagues have published a handbook on the broad area of physico-chemical property estimations.[96] These reference sources also provide detailed procedures for calculating and estimating K_{OW}.

3.66   CAN BREAKTHROUGH VOLUMES BE DETERMINED MORE PRECISELY?

A more precise way to determine the *breakthrough volume* for a particular *analyte* on a given SPE sorbent, Vr, is to use reversed-phase HPLC. The SPE sorbent is efficiently packed into a conventional HPLC column. The analyte of interest is injected into the instrument, while the mobile-phase composition is varied. This is accomplished by varying the percent organic modifier, such as acetonitrile or methanol, and measuring the differences in analyte retention time. Extrapolation of a plot of k' vs. percent organic modifier to zero percent modifier yields a value for the capacity factor for that analyte at 100% water. This capacity factor is represented by $k_{W,SPE,HPLC}'$, where the subscript refers to 100% water. The breakthrough volume is related to the capacity factor in 100% water according to

$$V_r = V_0 \left(1 + k_{W,HPLC}' \right)$$

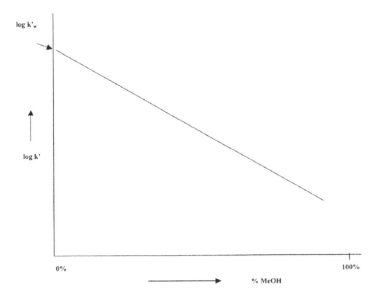

FIGURE 3.46

V_0 is the void volume of the SPE column and k'_W (SPE, HPLC) is the capacity factor of the analyte eluted by water. This capacity factor should theoretically be the same for any technique used, as V_0 is found by knowing the porosity of the sorbent and the geometry of the SPE column or sorbent bed in the cartridge. The capacity factor of a given analyte, k' (SPE, HPLC), is generally obtained from HPLC. From an HPLC chromatogram, k' (SPE, HPLC) is obtained by taking the ratio of the difference in the retention time between an analyte and its void retention time to the void retention time. Expressed mathematically,

$$k'_{\text{HPLC}} = \frac{t'_R - t'_0}{t'_0}$$

where t'_R represents the *retention time for the analyte of interest* under the HPLC chromatographic conditions of a fixed mobile-phase composition. t'_0 refers to the retention time for an analyte that is *not retained* on the stationary phase. t'_0 also relates to the void volume in the column.

Log $k'_{W,SPE,HPLC}$ is obtained by extrapolation of a plot of log k' vs. percent MeOH to zero. This plot is shown in Figure 3.46. A linear relationship exists between the logarithm of the capacity factor, k' (SPE, HPLC), and the percent MeOH concentration for a specific organic compound in HPLC. Because $k'_{W,SPE,HPLC}$ has also been shown to be related to the octanol–water partition coefficient, K_{ow}, $k'_{W,SPE,HPLC}$, and hence the *breakthrough volume Vr*, can be obtained for a given RP-SPE sorbent by this approach.

3.67 DOES AN SPE CARTRIDGE OR DISK HAVE A CAPACITY FACTOR AND IF SO, HOW DO WE CALCULATE IT?

Yes and it can be shown (see below) that the extent to which any analyte, initially dissolved in water, can be isolated on an octydecyl- or an octyl siloxane chemically bonded silica sorbent (either in a 3 cc sorbent barrel cartridge or a disk format) can be mathematically related to a ratio of chromatographic capacity factors provided secondary equilibrium effects are ignored.

A capacity factor k' can be related to the ratio of analyte sorbed to that of the total amount of analyte present in a given sample according to:

$$\frac{n_S}{n_0} = \frac{k'}{1+k'}$$

where

n_S—#millimoles of analyte adsorbed/partitioned onto and/or into the sorbent or disk
n_0—#millimoles of analyte in the volume of water or sample passed through
k'—capacity factor for the cartridge or disk.

The capacity factor k' for a specific RP-SPE cartridge or disk can in turn be related to the partition or distribution constant for a specific analyte according to:

$$k' = \frac{V_S}{V_m} K_D = \beta\, K_D \tag{3.44}$$

where

V_S—volume of reversed-phase of sorbent or disk
V_m—void volume of sorbent or disk
β—phase ratio, V_S/V_m
K_D—distribution constant.

To achieve a 90% recovery ($n_S/n_0 \times 100$) chemical analyte from, for example, an environmental drinking water sample, requires a capacity factor k' = 9. Table 3.19 uses Equation (3.44) to calculate what value β must have given a knowledge of K_D. For example: A 40 μm, 60 Å (angstrom) length pure octyldecylsiloxane chemically bonded silica of mass 250 mg has a V_m = 300 μL. Assuming a 20 Å length for a C_{18} ligate and a 350 m^2/g surface area, this same 250 mg sorbent can be estimated to have a V_S = 175 μL. We can roughly estimate a phase ratio β (175/300) = 0.6.[99]

3.68   HOW DID RP-SPE ONCE SOLVE AN ANALYTICAL LABORATORY DILEMMA IN *ENVIRO-CHEMICAL* TEQA?

The author, working in an EPA contract lab during the late 1980s, was once confronted with the following problem. A brief review of EPA Method 8151 (Chlorinated Herbicides by GC

TABLE 3.19
Application of Equation 3.44

If KD =	Then β Must Be
0.001	≥9000
0.01	≥900
0.1	≥90
1	≥9
10	≥0.9
100	≥0.09
1000	≥0.009

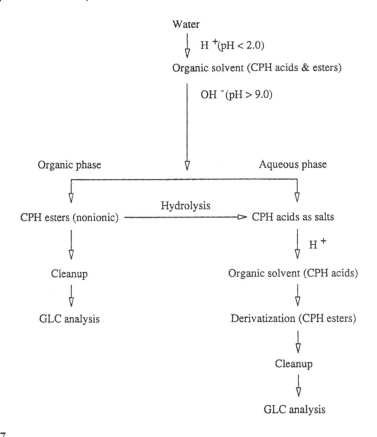

Water

H$^+$(pH < 2.0)

Organic solvent (CPH acids & esters)

OH$^-$(pH > 9.0)

Organic phase   Aqueous phase

CPH esters (nonionic) ———— Hydrolysis ————▷ CPH acids as salts

Cleanup   H$^+$

GLC analysis   Organic solvent (CPH acids)

Derivatization (CPH esters)

Cleanup

GLC analysis

FIGURE 3.47

TABLE 3.20
RP-HPLC Elution Order, pK$_a$ and Aqueous Solubility of the Five Weak Acid Herbicides Studied

Acid Herbicide	RP-HPLC Elution Order	pKa	Aqueous Solubility (#mg/L)
Dicamba	1	1.9	4500,7900
2,4-D	2	2.8, 3,3	650,724
2,4,5-T	3	2.8, 3.3	250,270
Silvex	4	3.0	140
Dinoseb	5	1.5, 4.4	50

Acid Herbicide RP-HPLC Elution Order pKa Aqueous Solubility (#mg/L)

using methylation or penafluorobenzylation derivatization) Revision 1 published in 1996 reveals that unlike an earlier version of the method, the analyst now has a choice of reagents in order to chemically derivatize the chlorophenoxy and related weak acid herbicides (thus converting them to their corresponding methyl esters) prior to analysis. Only the diazomethane derivative was approved back then. My colleagues while attempting to implement EPA method at that time reported 0% recoveries! Figure 3.47 shows a flowchart that describes a seemingly complex multi-step procedure to isolate and recover chlorophenoxy and related acid herbicides from drinking or ground water using conventional LLE and cleanup prior to chemical derivatization. I was asked to develop an alternative approach.

Five weak acid herbicides were placed in distilled, deionized water. These analytes were isolated and recovered using RP-SPE *sample prep* techniques followed by analysis via direct injection into a high-performance liquid chromatograph (HPLC) under reversed-phase HPLC conditions. The principles and practice of HPLC will be discussed in Chapter 4. It is instructive to list the analyte of interest's RP-HPLC elution order along with its pKa values and its known aqueous solubility. These parameters are shown in Table 3.20. RP-HPLC elution order under isocratic and reversed-phase conditions for these five analytes is inversely proportional to their aqueous solubilities. Dinoseb, being a phenol, might be acidic enough due to the two nitro groups to possess a pKa of 1.5. Maximum contaminant levels in drinking water (MCLs, refer to Chapter 1) for 2,4-D and 2,4,5-T are 70 and 50 ppb respectively).[100] Molecular structures for the five weak acid herbicides (some are still used) for weed control in agriculture are shown below:

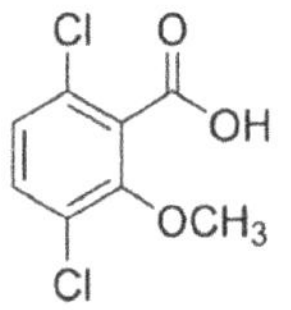

Dicamba 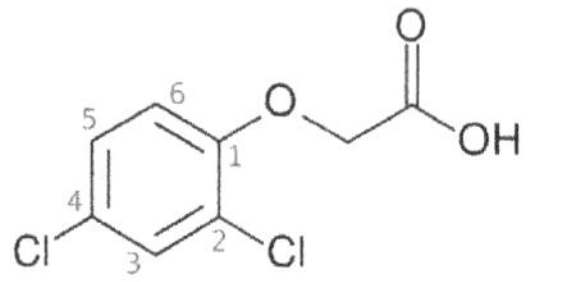   2, 4-Dichlorophenoxyacetic acid 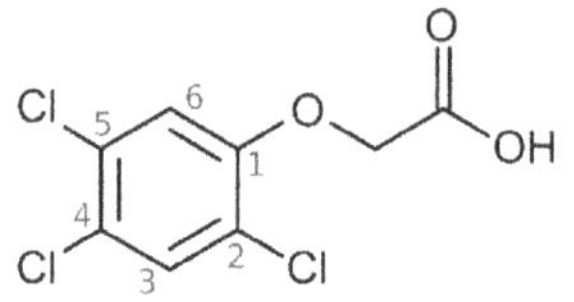   2,4, 5-Trichlorochlorophenoxyacetic acid

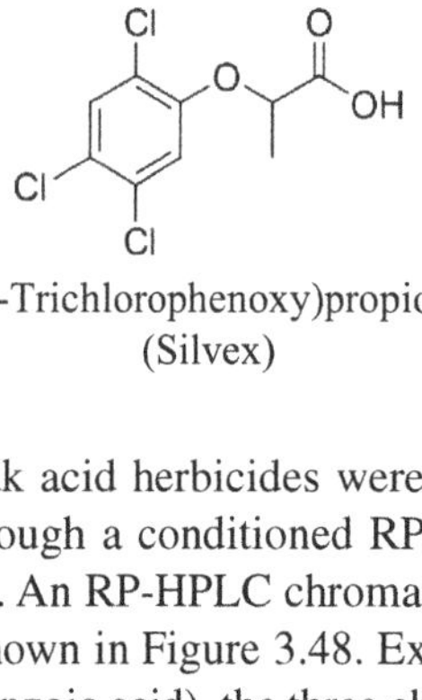

2-(2,4, 5-Trichlorophenoxy)propionic acid
(Silvex)

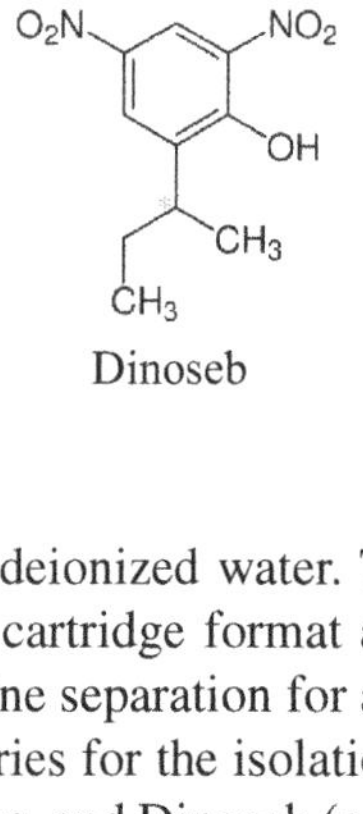

Dinoseb

These five weak acid herbicides were spiked into distilled, deionized water. The spiked water was passed through a conditioned RP-SPE in a 3 cc barrel cartridge format and % recoveries were measured. An RP-HPLC chromatogram showing baseline separation for all five weak acid herbicides is shown in Figure 3.48. Excellent percent recoveries for the isolation of Dicamba (a derivative of benzoic acid), the three chlorophenoxy herbicides, and Dinoseb (a multi-substituted phenol) using reversed-phase SPE quantitated using HPLC (abbreviated RP-SPE-RP-HPLC-PDA) were obtained and are shown in Table 3.21A. Thus demonstrating early on that toxic analytes with known MCLs (maximum contaminant levels), not directly amenable to analysis by GC instrumental techniques, can be effectively quantitated using alternative instrumental techniques such as high-performance liquid chromatography (HPLC)!

3.69   WHAT ROLE DOES RP-SPE HAVE IN TEQA?

Because *enviro-chemical* matrices are largely air, soil, and water of some sort, the objective in TEQA is to isolate and recover a hydrophobic organic substance from a more polar sample matrix. RP-SPE can directly remove organic contaminants and represents the dominant mode of SPE most relevant to TEQA. Normal-phase SPE (NP-SPE) also has a role to play in TEQA. NP-SPE has served as an important cleanup step following LLE using a nonpolar solvent to remove polar organic interferences. This was accomplished in large-diameter glass columns instead of the familiar barrel type of SPE cartridge as discussed earlier in the section on cleanup.

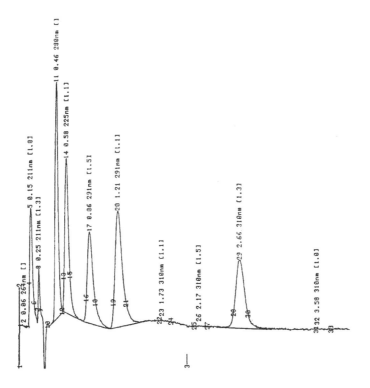

FIGURE 3.48

TABLE 3.21A
Percent Recoveries for the Isolation and Recovery of Dicamba,
Chlorophenoxy Herbicides and Dinoseb Using RP-SPE and
RP-HPLC-PDA

Analyte	Mean% Recovery	RSD (%)*
Dicamba	62	9.1
2,4-D	101	4.0
2,4,5-T	95.5	7.1
Silvex	91.4	3.4
Dinoseb	65.0	3.4

* RSDs reflect replicate SPEs performed and do not propagate error between sample and control

Hydrophobic organic contaminants found in the environment can be easily transferred from the native matrix to the RP-SPE sorbent using appropriate *sample prep* procedures.

Since 1985, the analytical literature has been replete with articles that purport good to excellent recoveries of various organic compounds that have been isolated and recovered by applying RP-SPE techniques. The rise in popularity of RP-SPE between 1985 and the present has served to limit the growth of both LLE and supercritical fluid extraction (SFE). Hydrophobic organic contaminants are partitioned from the environmental matrix to the surface layer of the octyorgano moiety. The contaminants are then eluted off of the sorbent using a solvent with the appropriate polarity. This eluent is then introduced into the appropriate determinative technique, such as a gas or liquid chromatograph.

TABLE 3.21B
Comparison of the Percent Recoveries (Triplicate Replicate SPEs) of Various Organophosphorous Pesticides from Spiked Water between a C$_8$ and a C$_{18}$ Chemically Bonded Silica Sorbent

OP	C$_8$ % Recovery (Coefficient of Variation)	C$_{18}$ % Recovery (Coefficient of Variation)
TEPT	77.6 (3.9)	94.0 (3.1)
Dichlorovos	42.0 (3.0)	56.9 (2.6)
Ethoprop	85.0 (2.1)	92.0 (1.9)
Phorate	80.1 (2.5)	54.6 (2.2)
Diazinon	83.5 (2.3)	89.3 (2.4)
Dimethoate	4.1	5.0
Methyl parathion	89.0 (3.0)	96.3 (2.7)
Parathion	83.6 (2.5)	94.3 (2.9)
Tokuthion	41.1 (2.6)	70.4 (3.0)
Famphur	90.0 (2.3)	99.9 (3.3)
EPN	61.8 (2.7)	86.3 (3.3)
Azinphos methyl	92.1 (3.4)	97.2 (3.8)

Historically, RP-SPE was applied principally in two areas of TEQA. The first application is in the determination of organochlorine pesticides (OCs) in connection with the determinative technique of GC using a chlorine-selective detector such as the ECD (electron-capture detector). The second application is in the determination of priority pollutant semivolatile organics that are recovered from drinking water. Offline RP-SPE coupled to a determinative technique such as GC using a low-resolution mass spectrometer is a very powerful combination with which to achieve the objectives of TEQA.

Organophosphorous (OPs) pesticides are used primarily as insecticides, and in mammals, they act as cholinesterase inhibitors. A number of OPs are listed as priority pollutants. EPA Method 8141B from SW-846, Revision 2, February 2007, is an analytical method that uses LLE coupled to GC with nitrogen/phosphorous element selective detection. The method can separate and quantitate OPs as well as triazine herbicides and selected carbamates. OPs were quantitated using a GC with a dual capillary column / dual [NPD (nitrogen/phosphorous) and FPD (flame photometric)] element-specific detection. The author also studied the isolation and recovery of various organophosphorous pesticides (OPs) from spiked water using RP-SPE techniques comparing octyl and octyldecyl siloxane silica in the barrel format.[101] Table 3.21B summarizes % recoveries from this comprehensive RP-SPE study. Equations (2.53) (calculation of percent recovery) (2.54) (propagating error between sample and control, and (2.55) (RSD or coefficient of variation) were used to calculate the % recoveries and coefficients of variations (relative standard deviations) in the % recovery. Three molecular structures of OPs drawn from Table 3.21B are shown below:

Dichlorvos   Diazinon   Parathion

We now discuss the author's earlier experience with RP-SPE using selected organochlorine pesticides (OCs).

3.70   SHOULD I USE RP-SPE TO MEASURE OCS IN DRINKING WATER? HOW COULD I USE RP-SPE TO ISOLATE AND RECOVER THESE SELECTED OCS FROM SOIL?

This author's experience in applying RP-SPE to spiked water answers the question in the affirmative. OCs that exhibit insecticidal properties and were in widespread use early in the 20th century are easily isolated from water and recovered by solvent elution. The majority of OCs that were used as insecticides are relatively high-molecular-weight semivolatile organic molecules having a complex structure. The original approach adopted by the EPA was encompassed in Method 608 as applied to wastewater and required LLE, followed by column cleanup using Florisil. OCs were quantitated using packed-column GC-ECD (electron-capture detector). This method could monitor 17 OCs. With selective use of elution solvents off of the Florisil cleanup column, some fractionation, hence separation, of OCs from multicomponent OCs such as toxaphene and Aroclors (mixtures of polychlorinated biphenyl congeners commercially manufactured) could be achieved. The replacement of conventional LLE with Florisil cleanup with a direct isolation of OCs from samples of environmental interest using RP-SPE represented a clear alternative to EPA Method 608. When RP-SPE is used in combination with high-resolution capillary GC-ECD, a much more robust and powerful analytical method emerges. Samples can be prepared and analyzed much faster, and the increase in GC resolution afforded by capillary columns enables a greater degree of delineation among all OCs of interest.

A soil sample cannot be directly placed atop an SPE cartridge! An aliquot of a moderately polar solvent such as methanol or acetonitrile can be added to a given weight of soil. After the slurry is allowed to stand without agitation, the supernatant liquid can be transferred to a 70 mL SPE sample tube filled with distilled, deionized water. This sample is then passed through a conditioned SPE cartridge or disk, then subsequently eluted with a nonpolar solvent. This author has always been rather taken back by how efficiently these OCs can be removed from spiked water samples. Of the some dozen or so OCs that are routinely monitored in drinking water (EPA Method 508), wastewater (EPA Method 608), and solid waste (EPA Method 8081B (Revision 2, February 2007)), I chose three that are most representative of all of these OCs: lindane or γ-BHC, endrin, and methoxychlor.[102] Of these three, methoxychlor is still in use. Molecular structures for these three representative OCs are shown below:

Lindane (γ-BHC)   Endrin   Methoxychlor

3.71   CAN LINDANE, ENDRIN, AND METHOXYCHLOR BE ISOLATED AND RECOVERED USING RP-SPE?

Yes, and the data from one such study by this author is now discussed. Lindane, being a hexachlorocyclohexane, has a higher aqueous solubility than hexachlorobenzene. The gamma isomer, γ-BHC, is 10 times more soluble than the alpha or beta isomer. Lindane (γ-BHC) represents the class of chlorinated aliphatic hydrocarbons with an S_{aq} of between 7.5 and 10 ppm. Endrin is a member of the cyclodiene insecticides, possessing the characteristic endomethylene bridge structure. The S_{aq} for endrin is 0.23 ppm, which is similar to that of aldrin, dieldrin, and heptachlor epoxide, but greater than that of heptachlor and chlordane. Endrin is known to partially break down to endrin aldehyde and endrin ketone in packed GC columns. It is not much of a problem when capillary columns are used. Methoxychlor, a substituted diphenyl trichloroethane similar to DDT, has an S_{aq} of between 0.1 and 0.25 ppm, and it is about 100 times more soluble in water than DDT. Using ethyl acetate as an elution solvent and some of the early chemically bonded silica, of 60 Å pore size and 40 μm irregular particle size (Separalyte, Analytichem International), this author conducted a series of systematic studies of lindane, endrin, and methoxychlor (L, E, M). The work was focused on a sample matrix consisting of distilled deionized water (DDI), and the analysis was performed on a gas chromatograph that used a packed-column and electron-capture detector.

One such study is reported here. Nine cartridges were packed with an octyl-bonded silica, C_8 Separalyte. The mass of the sorbent packed in each cartridge varied from a low of 20 mg to a high of 320 mg. In each cartridge, after methanol conditioning, a spiked aqueous sample was passed through the packed bed under a reduced-pressure SPE manifold. Then 8.8 μL of a methanolic reference standard containing L, E, and M, each at a concentration of 100 ng/μL (ppm), was added to approximately 60 mL of DDI. This spiking gave a concentration level in the aqueous sample of 15 pg/μL (ppb). This concentration level is considered low enough to approximate the realm of TEQA. After the sample was passed through the nine replicate RP-SPE cartridges, two 500-μL aliquots of ethyl acetate were used to elute the sorbed analytes into a 1 mL receiver volumetric flask. The contents of the volumetric flask were quantitatively transferred to a 2 mL GC autosampler glass vial, and 2 μL of this eluent was injected into a GC that contained a packed chromatographic column. Only one injection per GC vial was made in this particular study, so that variation in the percent recovery reflects only the random error associated with the RP-SPE process only. Table 3.22 lists the percent recovery for all three OCs from each of the nine cartridges. The concentration of analyte that constituted a 100% recovery was calculated and not actually measured; hence, there is no contribution to the relative standard deviation from the control standard. It is evident from review of the percent recovery results shown in Table 3.22 that breakthrough was not reached even in the case where only 20 mg of sorbent was taken to prepare the packed cartridge.

An overall mean percent recovery can be found from these nine replicate % recoveries. The standard deviation in the mean percent recovery is then found using the fundamental equation for calculating standard deviations. Refer to mathematical equations presented in Chapter 2 such

TABLE 3.22
Percent Recoveries of OCs Using a C_8-Bonded Silica Sorbent from Spiked Water

Sorbent (mg)	Lindane	Endrin	Methoxychlor
115	94.3	111	93.2
291	97.1	87.3	80.8
225	106	133	107
241	100	111	100
98	100	122	100
20	100	106	93.2
52	77.1	87.3	86.3
130	97.1	113	93.2
320	100	95.2	93.2

Note: 8.8 μL of 100 ppm each of lindane, endrin, and methoxychlor, dissolved in methanol, was added to approximately 60 mL of distilled deionized water. The spiked samples were passed through the cartridge, previously conditioned with methanol, and eluted with two 500 μL aliquots of ethyl acetate. The volume of eluent was adjusted to 1.0 mL, and 2 μL of eluent was injected into a GC.

TABLE 3.23
Mean Percent Recoveries and Statistical Evaluation from the Data Shown in Table 3.22 for the Representative OCs: Lindane, Endrin and Methoxychlor

QC	Mean% Recovery (n = 9)	RSD (%)	Confidence Interval (95%)
Lindane	96.8	8.3	6.2
Endrin	107.3	14.3	11.8
Methoxychlor	94.1	8.2	5.8

as: Equations (2.53), (2.54), and (2.55). If replicate injections per GC vial were made, a pooled standard deviation relationship, as given in Equation (2.56), would be most appropriate. From the standard deviation, a relative standard deviation or coefficient of variation is obtained, followed by calculation of the corresponding confidence interval. For the percent recoveries using an octyl siloxane chemically bonded silica as shown in Table 3.22, a mean % recovery over 9 replicate SPEs, an RSD in the mean % recovery and a confidence interval can be calculated using relationships developed in Chapter 2. These results are shown in Table 3.23.

The mean percent recoveries in this replicate series of RP-SPEs are very high and represent a statement of accuracy in the measurement. A relative standard deviation of 8 or 14% among replicate SPEs represents the precision of the method. The confidence interval states that of the next 100 SPEs, 95 of these should fall within the interval specified. For example, if 100 additional percent recoveries using the SPE method could be performed for lindane, one could expect that 95 would fall within $96.8 \pm 6.2\%$.

This high and reproducible percent recovery for the isolation and recovery of lindane from water strongly suggests that RP-SPE is very appropriate as a sample preparation method for this analyte. Lindane or γ-BHC is of continued interest to environmental and toxicological scientists. One such study, discussed next, taken from the author's work, involves the isolation and recovery of this important OC pesticide from homogenized myometrial tissue suspended in an aqueous matrix.

3.72 WAS LINDANE ISOLATED AND RECOVERED FROM A BIOLOGICAL MATRIX USING RP-SPE? IF SO, HOW?

One hundred microliters of a methanolic solution containing 5 ng/μL (ppm) lindane was placed in a 1 mL volumetric flask half filled with iso-octane, while 100 μL of the same lindane reference standard was added to approximately 70 mL of distilled deionized water (DDI). One microliter of the former solution containing 500 ng of lindane was injected into a GC, and the resulting peak area served to define a *control* sample. This laboratory sample represents a 100% recovery of lindane. One microliter of the 1 mL eluent from performing the RP-SPE of the spiked DDI was also injected into the same instrument, and the resulting peak area served to define the *spiked recovery sample*. In this study, quadruplicate injections of the control yielded a mean concentration of 400 ppb lindane from interpolation of the least squares calibration curve. Seven replicates of the eluent from the spiked DDI sample were injected, and a mean concentration of 406 ppb lindane was obtained from interpolation of the same calibration curve. Equations (2.53) was used to find a 106% recovery of lindane. RSDs in the % recovery were found by propagating random error between spiked samples and control reference standards as shown in Equations (2.54) and (2.55). A complete result can be given by stating both the accuracy, 106% recovery, and the relative standard deviation of 15.6%. Myometrium samples were then prepared and lindane appeared as expected. This work showed that lindane could easily be isolated and recovered from an aqueous matrix and confirmed the earlier work on lindane isolation and recovery, discussed previously. One would be led to believe that this is a robust sample preparation method for lindane because it can be reproduced with confidence. Thus, a subsequent request for additional sample analyses does not require extensive QC and should merely report the analytical results.

3.73   WHAT DOES A SAMPLE ANALYSIS REPORT USING RP-SPE TECHNIQUES LOOK LIKE?

In addition to a tabular format for the determination of lindane in each of the myometrial tissue samples submitted, a method summary should be included. A one-paragraph method summary serves to inform the reader as to how the samples were handled once they arrived in this analyst's laboratory. The summary should also provide a brief overview of the sample preparation and a brief description of the determination technique used. The following report illustrates these concepts.

REPORT ON THE QUANTITATIVE DETERMINATION OF LINDANE IN MYOMETRIAL TISSUE SUSPENDED CELLS IN SALINE

SUMMARY OF METHOD

The entire contents (1 mL) of the sample that was received by the client were refrigerated upon receipt until the sample was prepared for analysis. Upon thawing, the entire contents of each sample were added to a reservoir that contained approximately 70 mL of distilled deionized water. This aqueous solution was passed across a previously conditioned octadecyl-bonded silica sorbent ($C_{18}RPSiO_2$). The retained analyte was eluted off of the sorbent with two 500 μL aliquots of pesticides-residue-grade iso-octane. The iso-octane eluent was then passed through a second SPE cartridge. This second cartridge was packed with approximately 0.5 g of anhydrous sodium sulfate. The volume of eluent that now contained the recovered analyte was adjusted to a final volume of 1.0 mL using a volumetric flask. Three microliters of this eluent was injected via autosampler into an Autosystem Gas Chromatograph (PerkinElmer) incorporating an electron-capture detector. A 30 m × 0.32 mm capillary GC column containing DB-5 (J&W Scientific) was used to separate the organics in the eluent. This instrument is abbreviated C-GC-ECD to distinguish it from other gas chromatographs in our laboratory. The column was temperature programmed following injection from 200 to 270°C at a rate of 10°C/min. The C-GC-ECD is connected to a 600 Link (PE-Nelson) interface module. This interface, in turn, is connected to a 386 personal computer. This PC uses Turbochrom® (PE-Nelson) software for data acquisition, processing, and control. A method specific for lindane was written, and one peak was identified within a 3% relative time interval (retention time window). A retention time, $t[R]$, of 1.8 min was consistently reproduced using autosampler injection.

CALIBRATION

A series of calibration or working standards were prepared from the methanolic stock solution containing lindane. This stock solution was prepared by carefully weighing out pure solid lindane on an analytical balance. These standards were injected into the C-GC-ECD from lowest to highest lindane concentration. The peak at 1.8 min was identified as a reference peak in the Turbochrom software, and a narrow $t[R]$ window was defined around the $t[R]$ of the apex of the peak. A calibration curve was constructed using a least squares regression algorithm in the software. A correlation coefficient of 0.9990 was obtained.

SAMPLE ANALYSIS

The sequence within which the samples were run after calibration was created by using the Sequence File Editor within Turbochrom. Samples are injected via autosampler according to the instructions in the Sequence File. After the sequence is completed, a Summary File is created and the analytical results are reported within this summary format. In the Summary File, for each sample analyzed, the following is given:

(a) The sample number
(b) The concentration of lindane in ppb for 1.0 mL of eluent, interpolated from the external standard mode of instrument calibration
(c) The retention time for lindane, $t[R]$, in minutes
(d) The integrated peak area, in microvolts-seconds.

The reported concentration is that for each eluent and should be multiplied by the eluent volume (in this case, 1.0 mL) to obtain the number of nanograms of lindane found. The number of nanograms found divided by the volume of the aqueous sample used in RP-SPE gives the reported concentration in ppb for lindane. The reported concentration should be divided by the volume of the sample to obtain the reported concentration in ppb for lindane. The reported concentration should also be divided by the percent recovery, expressed as a decimal, in order to find the true and final concentration of lindane in the original sample.

METHOD EVALUATION

Lindane is efficiently recovered from aqueous matrices with percent recoveries between 75 and 100% using RP-SPE. The biological sample closely approximates an aqueous matrix.

3.74 DOES IT MATTER WHICH ELUTION SOLVENT IS USED IN RP-SPE?

Answers to this can be given as yes and no and a qualifying statement that it might depend on the polarity of the retained analyte. With a series of analytes that have similar octanol–water partition coefficients, differences in eluting solvent polarity may not significantly influence analyte recovery. If breakthrough studies are not performed, as discussed earlier, one never knows whether low percent recoveries are due to poor sorption or to inefficient elution or solvent removal using the RP-SPE technique. Earlier, this author had the opportunity to quantitatively determine trace concentrations of 2, 4, 6-trichlorobiphenyl (246-TCBP) and 4-hydroxy-2′, 4′, 6′-trichloro-biphenyl (4HO-2′4′6′TCBP) in biological matrices. Molecules from these two organic compounds of *enviro-chemical* interest differ in that a hydroxyl group is substituted for a hydrogen on the phenyl ring *para* to the second phenyl group in the molecule. This difference in molecular structure is evident as shown below:

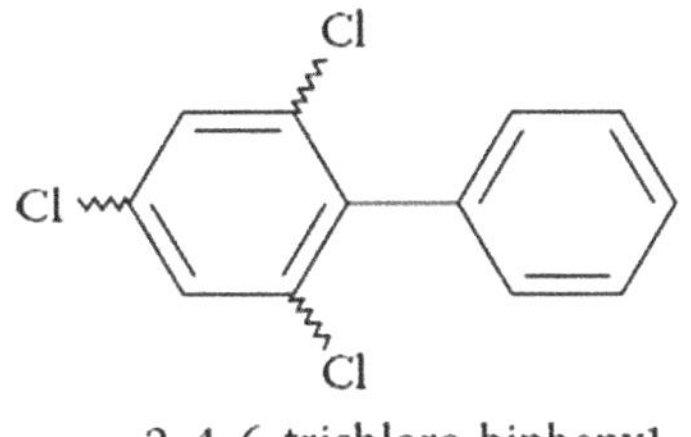

2, 4, 6-trichloro-biphenyl

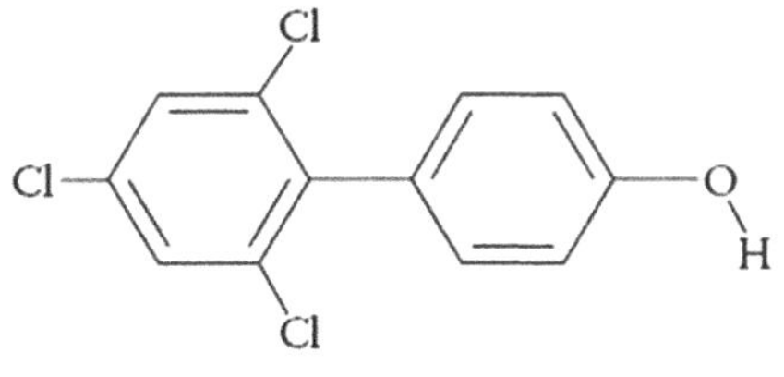

4-hydroxy-2, 4, 6-trichloro-biphenyl

We were quite surprised to measure a zero percent recovery when we attempted to apply our method for OCs to isolate and recover 4HO-2'4'6'TCBP from spiked water. Our recoveries of 246-TCBP using our OC method were very high. Replacing iso-octane with methanol as the eluting solvent resulted in a significant increase in percent recoveries for 4HO-2'4'6'TCBP. This surprising finding led us to conduct a series of systematic RP-SPE recovery studies. An aqueous spiked sample was prepared that contained both PCB congeners, and two elution schemes were used. Refer to Table 3.24: in Scheme A, the retained analytes were eluted with methanol followed by a separate, second elution with iso-octane. Two receivers were used, and their final volumes were both adjusted to a 1.0 mL total eluent volume. In Scheme B, the retained analytes were eluted with iso-octane followed by methanol. The integrated peak areas (in units of μV-sec) for the 1 mL eluents reflect the extent to which each analyte was isolated and recovered. It is a good assumption that the response factors for both congeners are near one another because the substitution of a hydroxyl for hydrogen in the phenyl ring contributes little or nothing to the ECD response. Eluting first with two 500 μL aliquots of methanol recovers more 4HO2'4'6'TCBP than 246TCBP, whereas the second elution with iso-octane recovers nearly equal amounts of both congeners. Thus, Scheme A reflects incomplete recovery of the congeners with two 500 μL aliquots of methanol. Eluting first with two 500 μL aliquots of iso-octane recovered even more 246TCBP than MeOH, yet recovered no 4HO2'4'6'TCBP! This observation was consistent with our preliminary finding, discussed earlier. Upon eluting with MeOH via Scheme B, a relatively small amount of 246TCBP was recovered, whereas a relatively large amount of 4HO2'4'6'TCBP was recovered. The percent recoveries for 4HO2'4'6'TCBP were similar between Schemes A and B when MeOH was used to elute off the retained analyte.

TABLE 3.24
Isolation and Recovery of an Aqueous Sample Spiked With Both 2, 4, 6-Trichlorobiphenyl and 4- Hydroxy-2', 4', 6'-Trichlorobiphenyl Using RP-SPE. Demonstrates the Effect of Using Two Elution Solvents That Differ in Polarity

	246TCBP	4HO2'4'6'TCBP
Scheme A		
Elute with MeOH, mean of 4 SPEs	137,026	218,526
Elute with iso-octane, mean of 4 SPEs	40,413	58,259
Scheme B		
Elute with iso-octane, mean of 4 SPEs	210,234	0
Elute with MeOH, mean of 4 SPEs	24,825	194,002

TABLE 3.25
Effect of Eluting Solvent Polarity on the % Recovery of the Organophosphorus Pesticides Diazinon and Malathion Using RP-SPE: Comparing Hexane vs. MTBE

Matrix: 10% NaCl (aq), 0.2M KH_2PO_4(aq)	% R: Diazinon	%R: Malathion
First elute with hexane (P'=0.1)	20	0
Then elute with MTBE (P'=2.5)	112	98

Another illustration from the author's experience is the differences in percent recovery when hexane is used vs. methyl-*tert*-butyl ether (MTBE) to elute off of a C_{18} sorbent. An aqueous matrix containing 10% NaCl and 0.2N potassium dihydrogen phosphate KH_2PO_4 was spiked with the OPs diazinon and malathion whose molecular structures are shown below. Two 500-μL aliquots of hexane were used to elute off of the cartridge, followed by two 500-μL aliquots of MTBE eluted off of the same cartridge. Results from this % recovery study are shown in Table 3.25. It is clearly evident that the somewhat more polar MTBE more effectively removes both OPs from the hydrophobic sorbent than the less polar hexane. This fact has perhaps more to do with the polarity differences between these two OPs and the nature of the *sorbent–analyte–solvent triangle*, as discussed earlier in this chapter. Based on the results shown in Table 3.25, which OP pesticide is somewhat less polar of the two and why?

Diazinon   Malathion

3.75   HOW CAN PCBs BE ISOLATED AND RECOVERED FROM SERUM, PLASMA, OR ORGAN FOR TRACE ENVIRO-HEALTH QA?

Figure 3.10 (introduced earlier) is a flowchart that summarizes this author's attempt to use RP-SPE as part of a combination of sample prep techniques that isolate and recover AR 1248 with nearly 100% recoveries.[103] PCBs are easily released from a serum, plasma, or tissue homogenate via probe sonication (PC), coagulation using acetonitrile, or other water-miscible organic solvent or salt. The supernatant aqueous phase is easily separated from the coagulated protein via mini-centrifugation. RP-SPE serves in this case, similar to the situation when an environmental water sample is passed through, as an on–off or *digital* extraction step. The dilution of the aqueous supernatant serves to decrease the analyte–matrix interaction, as discussed earlier, which in turn strengthens the analyte–sorbent interaction. In this case, the combination of PS-RP-SPE with C-GC-ECD as the determinative technique eliminated the need for further cleanup. Such might not be the case if C-GC-MS or another universal determinative technique is used. Cleanup techniques might then need to be considered. Table 3.26 lists the excellent % recoveries obtained from the author's method development study.

TABLE 3.26

% Recoveries for the Isolation and Recovery of AR 1248 Comparing Spiked Acetonitrile to Spiked Rat Liver Homogenate

Sample	Recovery as Total PCB (%)
Spiked acetonitrile 1	102
Spiked acetonitrile 2	102
Spiked rat liver 1	115
Spiked rat liver 2	88.2
Spiked rat liver 3	77.7

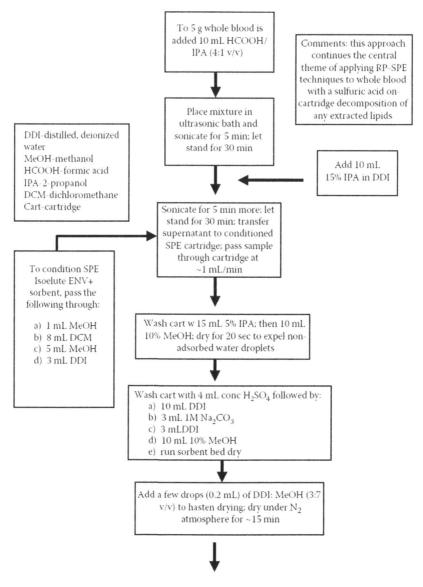

FIGURE 3.49A

Janák and coworkers have discussed a similar approach to the isolation and recovery of PCBs from whole blood.[104] Details of their approach are outlined via a flowchart in Figure 3.49A and Figure 3.49B. Referring to Figure 3.49A, formic acid dissolved in isopropyl alcohol (IPA) was used to coagulate protein from 5 g of whole blood prior to bath sonication. After passing the supernatant through a conditioned RP-SPE cartridge, the sorbent is washed with 5% IPA dissolved in water, followed by 10% methanol in water. The sorbent is then washed again with a series of solutions as indicated and dried. Referring to Figure 3.49B, the sorbent is eluted with methylene chloride (dichloromethane), then evaporated down to 10 μL, and finally reconstituted with enough heptane to yield ~100 μL. This heptane eluent is transferred to a Pasteur pipette filled with mixed adsorbents and previously rinsed with heptane. The cleaned-up PCBs are subsequently eluted with a binary solvent eluent consisting of 4:1 heptane to methylene chloride. Mean percent recoveries reported from spiked whole-blood specimens were 78 ± 8%.[104]

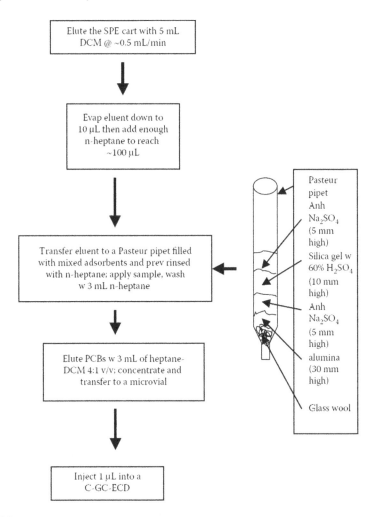

FIGURE 3.49B

Readers can refer to the numerous logical flow chart diagrams shown in Appendix D that describe many innovative approaches to conduct sample prep for both *enviro-chemical* and *enviro-health* TEQA.

3.76 PBDEs, USED AS FLAME RETARDANTS, ARE MORE CONTEMPORARY POLLUTANTS. HAVE THESE ORGANIC COMPOUNDS BEEN STUDIED BY RP-SPE SAMPLE PREP TECHNIQUES?

Within the first decade of the 21st century, a class of organic compounds very similar in its chemistry was making headlines. *Polybrominated diphenyl ethers*, PBDEs, whose molecular structures are closely related to the polybrominated biphenyls, PBBs. There exists 209 distinct PBDE congeners as is found for both PBBs and PCBs. Scheter and coworkers reported that dioxin, dibenzofuran, and PCB levels in human blood are much lower in 2003 when compared to 1973 levels; however, the opposite is true for PBDEs. Unlike dioxin, dibenzofuran, and PCBs, they found no significant correlation between PBDE levels and age.[105] A comprehensive study of the most common PBDE congeners: BDEs 28, 47, 77,100, 99, 85, 154, 153, 183 that compared

% recoveries from human and sheep serum using both LLE and reversed-phase solid-phase disk extraction (RP-SPDE) *sample prep* techniques brought unique insights.[106]

RP-SPE as a sample prep technique is low cost, effective, and continues to gain popularity as regulatory restrictions lift over time. For *enviro-chemical* QA considerations, Font et al. have reviewed developments in water pollution analysis for PCBs and earlier published a review applicable to multiresidue pesticide analysis of water.[107,108] For *enviro-health* QA considerations, a comprehensive RP-SPE study involving bovine and human serum conducted by Brock and colleagues at the National Center for Environmental Health (NECH), Centers for Disease Control and Prevention (CDC) is noteworthy.[109] A plethora of analytical papers utilizing RP-SPE techniques as a viable alternative to LLE continue to be seen in the primary analytical chemistry literature. Over the last decade, advances in the technology for SPE have added the so-called 96-well plate with accompanying *sample prep automation* to the analytical chemist's arsenal. We now consider a sample prep technique developed during the late 1980s that relates to SPE, yet is applicable to solid matrices, in particular biological tissue specimens. It is called *matrix solid-phase dispersion.*

3.77 WHAT IS MATRIX SOLID-PHASE DISPERSION AND IS IT APPLICABLE TO BIOLOGICAL TISSUE?

In the same manner that S-LSE complements LLE, matrix solid-phase dispersion (MSPD) has emerged as a complement to RP-SPE. MSPD is most applicable to biological tissue. Biological tissue is easily "ground" into a chemically bonded silica sorbent. Soil may prove to be a more difficult sample matrix to achieve MSPD with. MSPD came about when it was realized that the octadecyl siloxane–silica gel sorbent used in RP-SPE could have abrasive properties, while the octadecyl ligates could extract analytes from the biological tissue. Barker, along with coworkers, who published the benchmark paper on MSPD puts it this way.[110,111] MSPD combines aspects of several techniques for sample disruption while also generating a material that possesses unique chromatographic character for the extraction of compounds from a given sample.

A biological sample matrix such as fish tissue is ground into conventional RP-SPE sorbents such as C_{18} silica using a mortar and pestle. The cell structure is disrupted, and organic compounds are released and partitioned into the C_{18} silica based on their respective partition coefficients along the same lines as already discussed for RP-SPE. Barker has compared scanning electron micrographs for bovine liver tissue ground with underivatized silica to those of the same tissue ground with C_{18} silica.[112] Although dispersed with both silicas, the underivatized silica shows an intact cell structure evenly dispersed over the material.

MSPD is straightforward to perform in the laboratory. A schematic representation of an eight-step process to perform MSPD is shown in Figure 3.50A, which depicts the first four steps of blending the tissue with the bonded silica, while Figure 3.50B depicts the last four steps of solvent elution. A glass or agar mortar and pestle is preferable to the more porous porcelain type to minimize analyte loss. Only gentle blending is recommended. A ratio of 4:1, i.e., 4 g of C_{18} silica sorbent material to 1 g of biological tissue, has evolved as the optimum amounts to blend.[113]

Chemically bonded silicas serve several functions in MSPD. Bonded silicas serve as:[113]

- An abrasive that promotes cell disruption
- A lipophilic, bound solvent that disrupts and lyses cell membranes
- A sorbent capable of being packed into a column and can be eluted with solvents of differing polarity
- A bonded-phase support that enables sample fractionation.

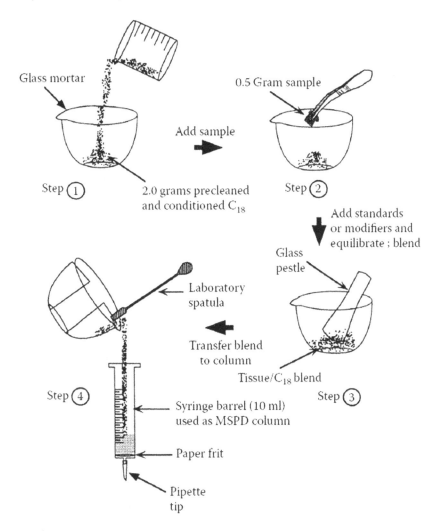

Glass mortar

Add sample

Step ① 2.0 grams precleaned and conditioned C_{18}

0.5 Gram sample

Step ②

Add standards or modifiers and equilibrate ; blend

Glass pestle

Laboratory spatula

Transfer blend to column

Tissue/C_{18} blend

Step ④

Step ③

Syringe barrel (10 ml) used as MSPD column

Paper frit

Pipette tip

FIGURE 3.50A

It is all too easy to compare RP-SPE with MSPD. However, after the biological tissue has been blended with the bonded silica sorbent, a new sorbent or stationary-phase results. In RP-SPE, after a sample has passed through a column, most components of the sample accumulate at the top of the column bed. In MSPD, the blended sample suggests that the components are more uniformly distributed throughout the entire sorbent. This creates what Barker calls a "unique chromatographic phase" whose dynamic interactions are "not completely understood."[113]

3.78   WHAT FACTORS INFLUENCE PERCENT RECOVERIES IN MSPD?

A robust sample prep technique ought to yield a decent percent recovery. What have researchers who have sought answers to this question found?[111]

- General principles of SPE apply.
- A unique chromatographic stationary phase is created when biological tissue is blended with chemically bonded silica.

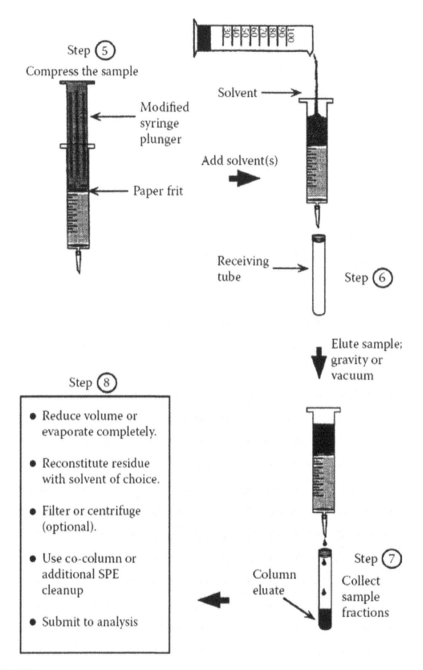

FIGURE 3.50B

- Underivatized silanols on the surface and in the pores of the support apparently remove water from the blend, thus yielding a drier support.
- Pore size does not seem to be significant.
- Particle size is relatively important since particles of 3 to 20 μm do not permit flow of elution solvent via gravity, whereas the conventional 40 to 100 μm-diameter particles do permit flow.

- A lipophilic bonded phase is believed to lead to the formation of a new phase that resembles a cell membrane bilayer assembly, giving the MSPD material unique chromatographic properties.
- The percent carbon load does not appear to have an appreciable effect.
- Conditioning of the bonded silica is as essential to MSPD as it is to SPE.
- The matrix has a profound impact on percent recoveries, unlike SPE, due to the fact that the matrix becomes part of the chromatographic phase.
- Elution yields certain coeluted matrix components with certain analytes of interest that are not well predicted based on SPE principles.

Huang and coworkers found that percent recoveries of two different PCB (presumably Aroclors)-fortified fish (grass carp) samples utilizing MSPD compared favorably with the more conventional approach of saponification, followed by LLE.[114] Acidified silica gel was added to the elution syringe and provided cleanup of coextractants that resulted in the appearance of a distinct yellow color for the hexane eluent when silica gel was not used. Ling and Huang followed up their benchmark paper on isolating and recovering PCBs from fish tissue using MSPD with a second paper related to OCs as well as PCBs.[115] These authors found an optimum cleanup and elution solvent combination following the blending of fish tissue with octadecyl silyl-derivatized silica (ODS), which maximized percent recoveries while minimizing coextractive interferences. For a given ODS, the following adsorbents were tried: Florisil, acidic silica gel (44% H_2SO_4), and neutral alumina. The mass of fish muscle tissue was obtained by subtracting the mass of the adsorbent-loaded-only column from the mass of the sample-loaded column. Percent recoveries of >90% were realized for most priority pollutant OCs using the Florisil/hexane-acetone (9:1) combination. The authors applied MSPD to a number of fish samples from fish caught from a river that passes through an incineration facility in Taiwan. PCB levels were significantly higher than OC levels. One fish, caught upstream from the facility, showed much lower concentration levels of PCBs.

Rodriguez and colleagues using only Florisil and silica gel MSPD developed a screening method that succeeded in accomplishing a compound class separation between PCBs and PBDEs from a biota sample.[116] A flowchart that describes how this compound class fractionation was accomplished is shown in Figure 3.51. The authors note that some lipid was retained in the Florisil. Lipids were oxidized on the acidic silica gel. PBDEs demonstrated that they elute from the acidified silica gel with hexane. Generalized molecular structures for the 209 congeners for each of these two important classes of organic compounds that have contaminated the environment are shown below:

Rodriguez and colleagues also reported on their success using MSPD to isolate and recover four parabens (esters of p-hydroxybenzoic acid) and triclosan (2-(2,4-dichlorophenoxy)-5-chlorophenol) from indoor dust. These analytes were dispersed into octydecyl siloxane chemically bonded silica, cleaned up using Florisil, subsequently extracted to remove less polar species, chemically derivatized and then quantitated by GC-MS-MS.[117]

MSPD also uses chemically bonded silica and provides an alternative to conventional LLE, cleanup, and fractionation sample prep methods as applied to biological tissue. Of a more recent vintage is a miniaturized RP-SPE sample prep technique called *microextraction by packed sorbent* (MEP).

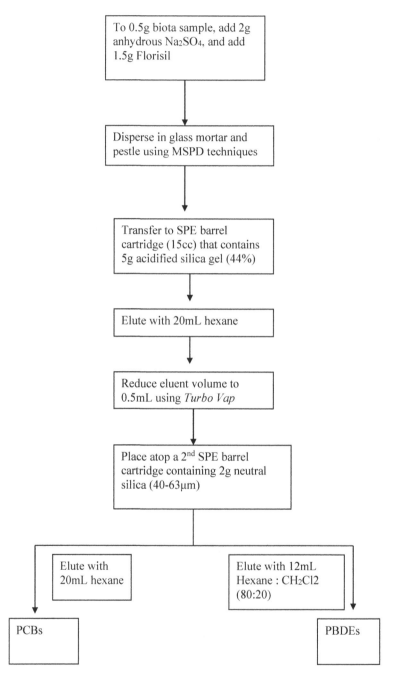

To 0.5g biota sample, add 2g anhydrous Na$_2$SO$_4$, and add 1.5g Florisil

↓

Disperse in glass mortar and pestle using MSPD techniques

↓

Transfer to SPE barrel cartridge (15cc) that contains 5g acidified silica gel (44%)

↓

Elute with 20mL hexane

↓

Reduce eluent volume to 0.5mL using *Turbo Vap*

↓

Place atop a 2nd SPE barrel cartridge containing 2g neutral silica (40-63μm)

Elute with 20mL hexane → PCBs

Elute with 12mL Hexane : CH$_2$Cl2 (80:20) → PBDEs

FIGURE 3.51

3.79   WHAT IS MICROEXTRACTION BY PACKED SORBENT (MEPS)?

It is a miniaturized RP-SPE and IE-SPE *sample prep* device! It was developed relatively recently. MEPS is most applicable to sample volumes as small as 10 μL. It was developed by the pharmaceutical industry where the application of MEPS in clinical and pre-clinical studies for the quantification of drugs and metabolites in blood, plasma and urine is important. Both reversed-phase (RP) and ion-exchange (IE) modes of solid-phase extraction *sample prep* techniques are

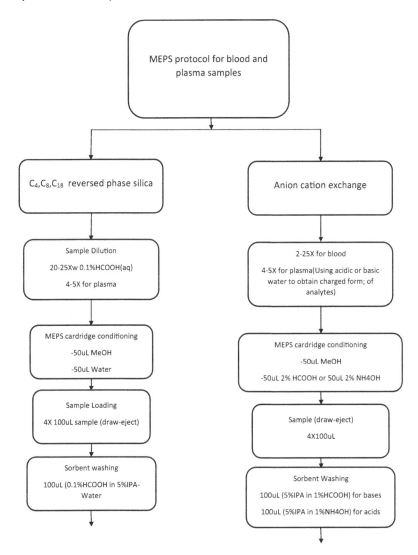

FIGURE 3.52A

incorporated in the MEPS technique. The MEPS bed is integrated into a liquid-handling syringe that enables a low void volume V_m (as introduced earlier for SPE). The MEPS syringe can be used manually or integrated in combination with laboratory robotics. The sorbent(1–4 mg) can be inserted into the (100–250 μL) syringe barrel or placed between the needle and the barrel as a cartridge. Chemically bonded silica with ethyl (C_2), octyl (C_8) and octadecyl (C_{18}) siloxane as well as strong cation exchanger (SCX) using sulfonic acid bonded silicas, restricted access material (RAM), hilic, carbon, and polystyrene. In addition, the ability to fully elute the sorbent with volumes of less than 10μL through a syringe needle serves to minimize solvent, reagent, and most importantly sample volume consumption when compared to conventional LLE and SPE techniques. Figure 3.52A and 3.52B show a logical flowchart for the isolation and recovery of various biomolecules including potentially toxic contaminants from blood and plasma samples using the MEPS *sample prep* technique.[118]

This sample prep technique is clearly applicable to enviro-health TEQA in direct and indirect ways. Consider using a MEPS syringe to effect a solvent change. Suppose you are evaluating the

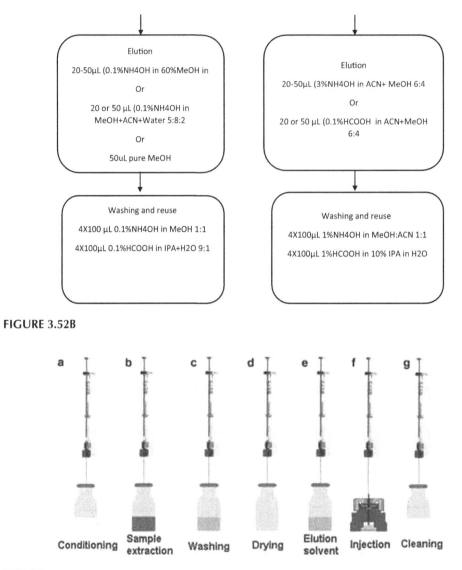

FIGURE 3.52B

FIGURE 3.53

most suitable determinative technique to use to quantitate analytes retained on the MEPS cartridge. A lab chemist or technician might elute the MEPS cartridge with a nonpolar solvent and upon subsequent injection into a GC, discover that the analytes give no peaks in the GC chromatogram! He or she could try eluting the MEPs cartridge with a more polar solvent with subsequent injection into an HPLC and observes peaks in the HPLC chromatogram! Figure 3.53 illustrates the procedural steps involved when implementing the MEPs sample preparation technique.

3.80   CAN AN SPME FIBER EXTRACT SVOCs FROM AQUEOUS SAMPLES?

Yes, SPME is applicable to SVOCs as well as applicable to VOCs as introduced earlier. Another way to answer this question in the affirmative is to state that solid-phase microextraction (SPME) *is* to the capillary GC column what SPE *is* to the HPLC column! For SPE, consider taking a portion of the HPLC column packing, increasing the particle size, and allowing for a wider distribution of the particle size; then pack this material into a cartridge and use this for sample prep. For SPME, on the other hand, coat a fused-silica capillary column on the outside, instead of the inside, and use this for sample prep. Janusz Pawliszyn at the University of Waterloo, who understood the limitations of SFE and SPE, was the first to modify a 7000 Series (Hamilton) liquid-handling microsyringe by coating polydimethyl siloxane on a fine rod such as fused-silica fiber. Refer to two of the pertinent articles published by Pawliszyn and coworkers.[119,120] SPME is also *solventless* in that a sorbed analyte is easily thermally desorbed off of the coated fiber in the injection port of a gas chromatograph. The modified microsyringe of Pawliszyn has given way to the commercial SPME device that incorporates a retractable fiber and is manufactured and marketed by the Supelco division of Millipore Sigma. Figure 3.54 illustrates an SPME fiber incorporated in a sampling device is immersed into a stirred aqueous sample as well as sampling the headspace above the liquid. The SPME sampling device can be immersed directly into an aqueous sample, such as groundwater, for a finite period, then withdrawn, the fiber and rod retracted back into the needle, brought to the hot-injection port of a gas chromatograph, and then inserted into the septum and the fiber and rod extended into the injection port for a finite period, in which thermal desorption is accomplished. The rate of SPME depends on the mass transport from a matrix to the coating. The effectiveness of mass transport depends on the following:

1. Convective transport in air or liquid
2. Desorption rate from particulates that might be present in the sample
3. Diffusion of analytes in the coating itself.

For direct SPME sampling with agitation, convective effects can be minimized. In the absence of particulates, the mass transport rate is determined by diffusion of analytes into the coating. For gaseous samples, the rate of mass transport is determined by diffusion of the analyte into the coating, and equilibrium can be achieved in less than 1 min. For aqueous samples, vigorous agitation of the sample is necessary. A common technique is to simply stir the sample with a magnetic stirrer. In this case, it takes much longer for the analyte to diffuse through the static layer of water that surrounds the fiber. Zhang et al. articulated the challenge for SPME as follows:[119]

For volatile compounds, the release of analytes into the headspace is relatively easy because analytes tend to vaporize once they are dissociated from their matrix. For semivolatile compounds, the low volatility and relatively large molecular size may slow the mass transfer from the matrix to the headspace and, in some cases, the kinetically controlled desorption or swelling process can also limit the speed of extraction, resulting in a long extraction time. When the matrix adsorbs analytes more strongly than the extracting medium does, the analytes partition poorly into the extraction phase. Because of the limited amount of the extraction phase in SPME (as in SPE), the extraction will have a thermodynamic limitation. In other words, the partition coefficient, K, is too small, resulting in poor sensitivity. If the coating has a stronger ability to adsorb analytes than the matrix does, it is only a matter of time for substantial amount of analytes to be extracted by the fiber coating and only kinetics plays an important role during extraction. One of the most efficient ways to overcome the kinetic limitation is to heat the sample to higher temperatures, which increases the vapor pressure of analytes, provides the energy necessary for analytes to be dissociated from the matrix, and at the same time speeds up the mass transport of analytes.

As was developed for other distribution equilibria, such as LLE [Equation (3.4)] and static HS [Equation (3.25)], we start to discuss the principles that underlie SPME by considering the equilibrium for the ith analyte between a sample S and the coated fiber, f, once dynamic equilibrium has been reached.

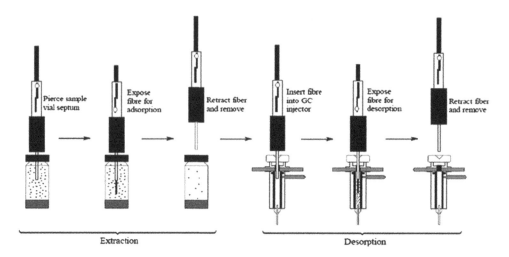

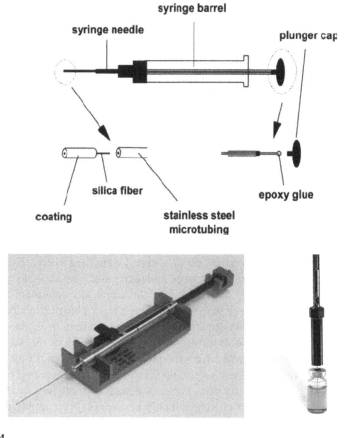

FIGURE 3.54

Let us consider this in more detail. We start by defining the partition coefficient, $k^i_{fs(SPME)}$, for analyte i between the fiber, denoted by f, and a sample, denoted by s, as follows:

$$k^i_{fs(\text{SPME})} = \frac{C_f}{C_s} = \frac{n/V_f}{C_0 - (n/V_s)} \tag{3.44}$$

where

n = amount of analyte i adsorbed on the SPME polymer film (in moles)
Vf = volume of coating on SPME fiber
VS = volume of sample
C_0 = initial concentration of the ith analyte in the sample
Cf, CS = concentration of the ith analyte in the fiber and sample, respectively, once equilibrium
is reached

Solving Equation (3.44) for n yields

$$n = \left[\frac{K^i_{fs(\text{SPME})} V_f V_s}{V_s + K^i_{fs(\text{SPME})} V_f} \right] C_0 \tag{3.45}$$

Equation (3.45) presents an opportunity to make two simplifying assumptions. If

$$K^i_{fs} \gg V_s$$

then

$$n_0 \approx V_s C_0.$$

The amount of analyte sorbed on the fiber is directly proportional to the original concentration of that analyte in the sample. If, however, a large sample volume is used,

$$K^i_{fs} V_f \ll V_s$$

then

$$n_0 \approx K^i_{fs} V_f C_0.$$

The amount of analyte sorbed on the fiber is also directly proportional to the original concentration of that analyte in the sample. Both simplifying assumptions lead to the conclusion that the amount of analyte sorbed, after equilibrium is attained, can be related to C_0. The term used to describe this is *exhaustive SPME*, and the maximum number of moles of the ith analyte that can be adsorbed or partitioned into the fiber is denoted by n_0.

3.81   CAN WE QUANTITATE BEFORE WE REACH EQUILIBRIUM IN SPME?

We answer the question by considering the derivation first proposed by Ai.[121] The assumption is that analyte molecules diffuse from (1) the sample matrix to the surface of the polymer and (2) through the polymer surface to the inner layers. For a steady state, with diffusion as the rate-controlling factor, the mass flow rate of analyte molecules from the sample matrix to the SPME polymer surface would equal the flow rate from the polymer surface to its inner layers. With D_1 as the diffusion coefficient of analyte molecules in the sample matrix and D_2 as the diffusion coefficient for molecules in the polymer phase whose surface area is denoted by A, and Cs and Cf are concentrations of analyte in the sample matrix and polymer film, respectively, Fick's first law can be used to give the rate-determining steps governed by

$$\frac{1}{A} \frac{\partial n}{\partial t} = -D_1 \frac{\partial C_s}{\partial x} = -D_2 \frac{\partial C_f}{\partial x}$$

By assuming that a steady-state mass transfer occurs when agitation is effectively applied, that the diffusion layer is a thin film, and that steady-state diffusion in this thin film is in effect, a simplified normal differential equation can be considered:

$$\frac{1}{A}\frac{dn}{dt} = \frac{D_1}{\delta_1}\left(C_s - C_s'\right)$$

$$= \frac{D_2}{\delta_2}\left(C_f - C_f'\right) \tag{3.46}$$

The terms used in Equation (3.46) are defined as follows:

C_s = concentration of the analyte in the bulk sample matrix
C_s' = surface concentration of the analyte in bulk sample matrix
δ_1 = diffusion layer thickness in the sample matrix
δ_2 = thickness of polymer film
C_f = analyte concentration in the polymer film at the surface
C_f' = analyte concentration in the polymer film in contact with the silica fiber.

Figure 3.55 is a schematic of the interface of the polymer-coated silica fiber in contact with an aqueous solution. The illustration shows the diffusion layer thickness in the sample matrix and the thickness of the polymer film. The slopes of the concentration gradients are shown at the interface between the bulk sample and polymer film surface. A steady-state diffusion is assumed when the aqueous solution is effectively agitated. The concentration gradient in the SPME film is assumed to be linear. Equation (3.46) can be rearranged and integrated, and within a set of boundary conditions this ordinary differential equation can be solved for n to yield[123]

$$n = \left[1 - e^{-A\{B\}t}\right]\left[\frac{KV_fV_s}{KV_f + V_s}\right]C_0 \tag{3.47}$$

Equation (3.47) suggests that the number of moles of analyte sorbed depends only on the original concentration of analyte in the sample, C_0, provided that the sampling time, t, and the rates of diffusion remain constant between samples. This diffusion rate is reflected in the slope of the lines between C_s and C_s' and C_f and C_f', as shown above. As the sampling time gets very

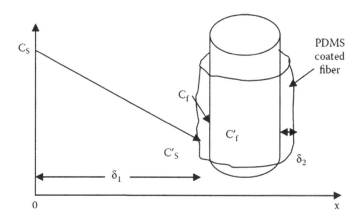

FIGURE 3.55

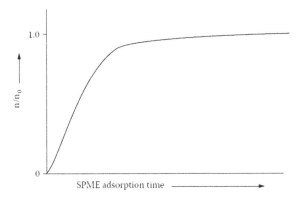

FIGURE 3.56

large (i.e., n approaches infinity), Equation (3.47) suggests that n approaches n_0, where n_0 is the number of moles adsorbed by the coating at equilibrium, so that Equation (3.47) reduces to Equation (3.45). Hence, Equation (3.47) can be rewritten in terms of n as a function of sampling time according to

$$\frac{n}{n_0} = 1 - e^{-at} \tag{3.48}$$

where a is a constant that is composed of various mass transfer coefficients, $K_{fs(SPME)}$, Vf, and Vs. If plots are made of n/n_0 vs. SPME adsorption or sampling time, curves such as that shown in Figure 3.56 result.

In a second paper, Ai developed a theoretical treatment for HS-SPME.[122] Lakso and Ng used SPME to isolate nerve gas agents such as Sarin from drinking water and observed similar uptake vs. adsorption time for direct aqueous sampling.[123] A plethora of SPME applications continue to appear in the literature. This author began to use SPME in pursuit of alternative sample prep methods designed to isolate and recover trace concentration levels of specific polychlorinated biphenyl (PCB) congeners and various Aroclors, along with the necessary surrogates, as utilized in EPA Method 8081A.

3.82   WHAT FACTORS NEED TO BE CONSIDERED WHEN USING SPME TO ISOLATE PCBs FROM CONTAMINATED GROUNDWATER?

The sample prep considerations are different for SPME from anything we have encountered because this is a solventless technique and requires that we consider the kinetics of sampling. Some factors are as follows:

- Sorption time for the fiber, as discussed earlier
- Thermal desorption temperature of the GC injector port
- Isothermal column temperature and time during thermal desorption
- Chromatographic temperature program to adequately resolve peaks in minimum run time.

The author studied *tetrachloro-m-xylene* and three of the most important of the 209 congeners of the polychlorinated biphenyls, *3,3',4,4'-tetrachlorobiphenyl, 2,2',4,4'-tetrachlorobiphenyl,* and *decachlorobiphenyl* whose molecular structures are shown below:

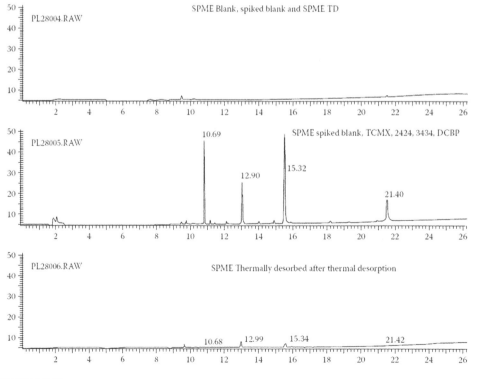

tetrachloro-m-xylene   decachlorobiphenyl

2,2',4,4'-tetrachlorobiphenyl   3,3',4,4'-tetrachlorobiphenyl

Three chromatograms that were obtained in the author's laboratory and are overlaid for comparative purposes are shown in Figure 3.57A.[124] The SPME device was used to extract the four polychlorinated PCBs shown earlier and then thermally desorbed these analytes of *enviro-chemical* importance into the injection port of an Autosystem® (PerkinElmer) gas chromatograph that incorporated a capillary or open tubular column chlorine-selective electron-capture detector

FIGURE 3.57A

TABLE 3.27
C-GC-ECD Retention Times t[R] for Selected PCBs Studied by SPME

		$t[R]$
PCB Congener	Abbreviation	Minutes after Injection
Tetrachloro-*m*-xylene	TCMX	10.29
2,2′, 4,4′-Tetrachlorobiphenyl	2244-TCBP	12.90
3,3′4, 4′-Tetrachlorobiphenyl	3344-TCBP	15.32
Decachlorobiphenyl	DCBP	21.40

TABLE 3.28
Results from Fitting the Sum of Peaks for AR 1242 versus #ppb AR 1242 to a Third-Order Polynomial

Concentration (ppb)	Sum of Peaks from AR 1242
11.7	217,483
23.3	464,987
46.6	718,531
93.2	777,197
117	976,544
233	1,510,288
466	2,243,387
932	4,666,817

(C-GC-ECD). The top chromatogram shown in Figure 3.57A was obtained by immersing the fiber into distilled deionized water and thermally desorbing the fiber into the GC inlet. This sample constitutes a method blank. The method blank is a useful quality control practice in that a clean blank, as this one shows, demonstrates that the lab and analyst are free of contamination. The absence of any lab contamination suggests that the method used is free of interferences. The center chromatogram in Figure 3.57A was obtained by exposing the fiber to water that had been spiked with the four PCBs. All four analytes had been extracted out of the sample and thermally desorbed into the C-GC-ECD. The separated analytes are given with their abbreviated names and GC retention times ($t[R]$) in Table 3.27.

The bottom chromatogram was obtained from merely injecting the already spent fiber into the hot-injection port of the C-GC-ECD a second time. Evidence of a tiny trace of each analyte is observed.

Two chromatograms are overlaid in Figure 3.57B. A comparison is shown between a spiked and unspiked aqueous sample using AR 1242. AR 1242 includes numerous PCB congeners. The ability of SPME to preconcentrate analytes is shown in Figure 3.57C. The top chromatogram in Figure 3.57C was obtained by injection of 1 µL of a methanolic 10 ppm AR 1242 reference standard, and the bottom chromatogram was obtained by applying the SPME technique to a sample that consisted of 50 µL of methanolic 10 ppm AR 1242 added to 10 mL of distilled deionized water. The sum of the numerous peaks in the chromatogram for AR 1242 provides a quantitative analysis for total Aroclor. A plot of this sum vs. the SPME sorption time is shown in Figure 3.57D for four sampling times. This plot roughly follows the theoretical profile shown

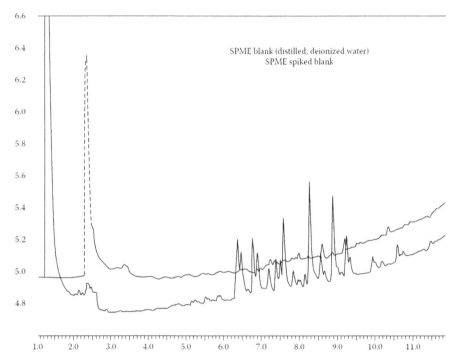

FIGURE 3.57B

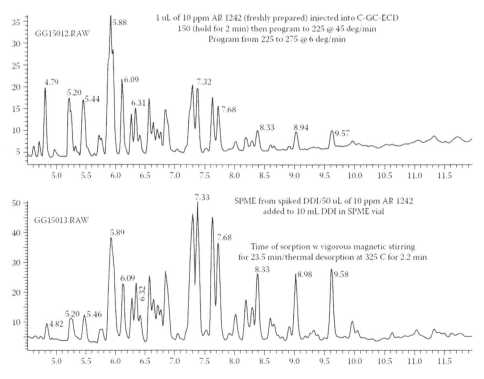

FIGURE 3.57C

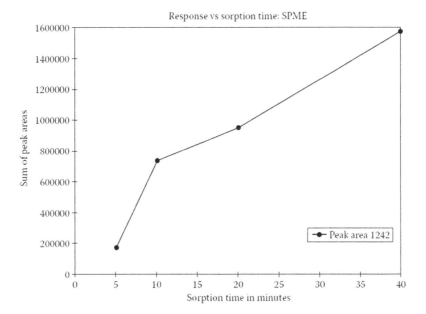

FIGURE 3.57D

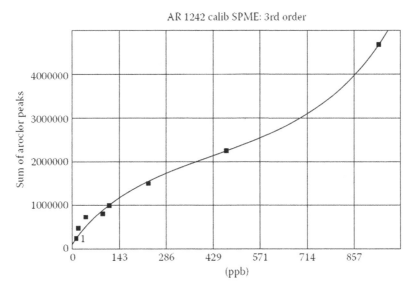

FIGURE 3.57E

earlier (Figure 3.56) for SPME. For a fixed sorption time and magnetic stir rate, a series of calibration standards were prepared and "SPME injected" into our C-GC-ECD. The calibration curve is shown in Figure 3.57E for eight standards. The raw data points are fitted to a third-order polynomial, as shown using Turbochrom (PE-Nelson) software. This author attempted to fit the experimental data to a least squares regressed calibration curve. The results of applying a linear regression to the x,y data shown for AR 1242 (Table 3.28) is given in Table 3.29. Refer to Chapter 2 for a review of parameter definitions shown in Table 3.29.

TABLE 3.29
Results of Applying a Simple (1st Order) Least Squares Regression Fit to the Experimental Data as Shown in Figure 3.57E

Parameter	Value
Slope of least squares regression line (linear fit)	4548 µV-sec/ppb AR 1242
y intercept of least squares line	353,887
Correlation coefficient	0.9954
Y_C	674,348 µV-sec
X_C	70.4 ppb
X_D	139.7 ppb
Y_D	985.686, x_{DD} = 138.9ppb
ICV (low) ± confidence interval at 95%	80.2 ± 87 ppb
ICV (high) ± confidence interval at 95%	415.4 ± 87 ppb

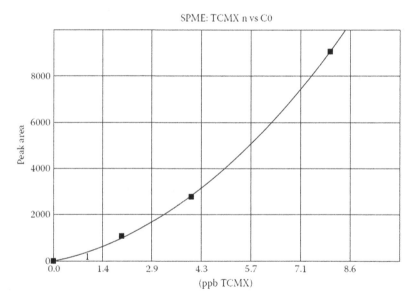

FIGURE 3.57F

It is clearly evident that, at least for this first attempt to simultaneously conduct sample prep and calibrate the instrument, good linearity was not achieved. Note that the initial calibration verification (ICV) represented by x_0 is much more accurate at the higher concentration level than at the lower level. In addition, the precision determined by the confidence interval at the 95% significance level at the lower level for the ICV is much larger than is the precision for the ICV at the higher concentration level.

A calibration plot for combined SPME/calibration for the surrogate TCMX is shown in Figure 3.57F. Again, there is a tendency for a nonlinear relationship as the concentration of TCMX increases, as shown. A second-order least squares fit gave a correlation coefficient of 0.9999.

3.83   MIGHT THE SAMPLE MATRIX INFLUENCE SPME EFFICIENCY?

Yes, the sample matrix may significantly influence the analyte percent recovery. This became apparent to the author when a request to isolate and recover AR 1242 from adulterated rat feed was made. It became necessary to remove the lipid content (at approximately 7 to 8%), and this was accomplished via base saponification. We attempted to isolate the AR 1242 from this base-saponified matrix, and the three chromatograms shown in Figure 3.58 reveal what success we had using SPME and LLE sample prep techniques. The top chromatogram reveals no recovery of AR 1242 in this base-saponified matrix

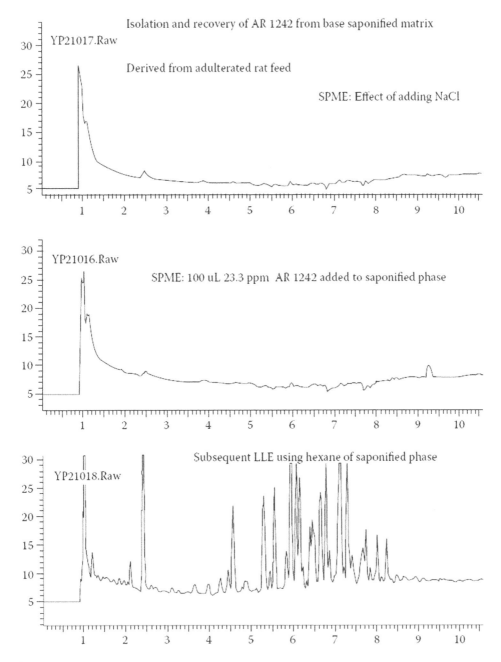

FIGURE 3.58

despite adding salt. The middle chromatogram reveals again no recovery of AR 1242 from a portion of the matrix that was spiked with AR 1242, as shown. However, an LLE of the base-saponified matrix recovered AR 1242, as shown. This observation is difficult to explain because PCBs are neutral and an alkaline matrix should not play such a significant role. What role did pH play in yielding almost a zero % recovery of AR 1242? Perhaps this is a worthwhile SPME research problem.

3.84   WHAT IS THIS SOLID-PHASE EXTRACTION THAT USES A STIR BAR?

This author has and continues to be fascinated with the magnetic stirrer/stir-bar device that has been available to laboratories for over 30 years. A Teflon-coated stir bar is dropped into a beaker that is approximately half filled or less with a liquid, the power is turned on, and a vortex is created within the liquid. This swirling vortex greatly facilitates mixing.

Pat Sandra's research group at the University of Ghent in Belgium coated a small stir bar with polydimethyl siloxane (PDMS).[125] They found identical sorptive extraction properties to SPME, rendering this new sample prep extraction technique quite suitable for accomplishing the goals of TEQA; they called this new approach, stir-bar sorptive extraction (SBSE).[125] In a manner similar to that developed for LLE and SPME, we can first define a molecular partition constant K_{SBSE} such that

$$K_{\text{SBSE}}^i = \frac{C_{\text{PDMS}}^i}{C_{\text{aq}}^i}$$

where C_{PDMS}^i is the concentration for the ith analyte sorbed on the stir bar with a PDMS coating whose volume is V_{PDMS}. C_{aq}^i is the concentration for the ith analyte after equilibrium has been attained for the aqueous or sample phase whose volume is V_{aq}. A percent recovery, $\%E_{SBSE}^i$, for the ith analyte initially dissolved in the aqueous phase is given by

$$\%E_{SBSE}^i = \frac{K_{SBSE}^i \beta}{1 + K_{SBSE}^i \beta}$$

where $\beta = V_{\text{PDMS}}/V_{\text{aq}}$ and $K_{\text{SPME}} \sim K_{\text{SBSE}}$.

Let us consider the actual volumes used in SBSE while comparing these to SPME:

$$V_{\text{aq}} = 10 \text{ml}$$

$$V_{\text{PDMS}}^{\text{SPME}} \sim 0.5 \mu l \text{ and } V_{\text{PDMS}}^{\text{SBSE}} \sim 25 - 125 \mu l$$

$$\beta_{\text{SPME}} \ll \beta_{\text{SBSE}}$$

Octanol–water partition coefficients, K_{OW}, are good approximations to either K_{SPME} or K_{SBSE}. K_{OW} (as introduced earlier) can be used to make predictions about $\%Ei$. A low phase ratio, β_{SPME} for SPME, coupled to low values for K_{OW} (i.e., more polar analytes) has resulted in low $\%E$ values. Contrast these low percent recoveries with the much higher $\%E$ values using SBSE, due to the significant increase in the phase ratio, β_{SBSE}. To illustrate this difference, a plot of $\%Ei$ against K_{OW} is shown in Figure 3.59 and suggests that SPME is more applicable to nonpolar analytes, while SBSE extends down to more polar analytes as a consequence of an increased phase ratio β.

3.85   WHAT INSTRUMENT ACCESSORIES ARE NEEDED TO CONDUCT SBSE?

Unlike SPME, a gas chromatograph must be outfitted with SBSE-related accessories to successfully conduct the technique. These accessories enable complete automation. This author's experience with the SBSE sample prep technology is from that offered by Gerstel GmbH. These items include:

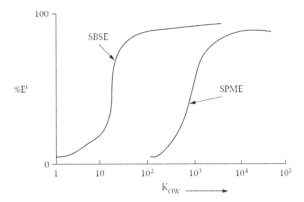

FIGURE 3.59

- A coated stir bar (Twister®) that consists of a magnetic rod covered by a glass jacket and coated with PDMS
- A thermal desorption tube to place the stir bar in
- A means to thermally desorb analytes from the stir bar
- A programmed-temperature vaporization inlet such as the cooled inlet system (CIS-4) from Gerstel. This inlet enables cryofocusing of the thermally desorbed analytes to become trapped at the head of the wall-coated open tubular (WCOT).

Two systems are commercially available that provide the items just listed: a TDS-A® classic thermal desorption system and a specifically designed Twister Desorption abbreviated TDU (both from Gerstel). Both systems can be mounted on GCs equipped with a CIS-4® programmed-temperature vaporizing inlet (Gerstel).[125]

An example of the specific analytical instrumentation (determinative techniques are introduced in Chapter 4) that the laboratory analytical chemist or technician needs to extract and quantitate SVOC analytes of *enviro-health* TEQA is best described (see below) by drawing on the author's own work with SBSE:[126]

The system consisted of a dual-rail multipurpose robotic sampler MPS2® installed atop a 6800®gas chromatograph, a 5973N® Mass Selective Detector (MSD) utilizing Chemstation® software, Version D.01.02 (Agilent Technologies) operated within Windows XP® Version 5.1(Microsoft Corporation). A Gerstel MPS2/TDU bundle consisting of the following was installed: MPS2 autosampler; CIS4 (a programmed-temperature vaporizer inlet with a septumless sampling head) and a C505 Controller; TDU (thermal desorption unit); TDU Automation Kit for the MPS2;Universal Peltier Cooling Unit for the TDU®; Maestro® software for MPS2/TDU/CIS-4 control.

3.86 HOW IS SBSE PERFORMED?

The coated stir bar is added to an aqueous or similar liquid sample that has been previously placed in a cylindrical vial, such as a headspace vial or equivalent. To extract solutes from serum or plasma requires dilution with water or buffer, treated similarly to that discussed for RP-SPE. The sample is stirred for anywhere from 30 to 240 min, depending upon sample volume, stir speed, and stir-bar dimensions, and must be optimized for each application. This is accomplished by measuring percent recoveries against SBSE extraction time.

After extraction, the stir bar is removed, dipped on a clean paper tissue to remove water droplets, and placed into an empty glass thermal desorption tube. The stir bar can be rinsed with distilled deionized water to remove adsorbed sugars, proteins, or other interferents. Analytes of *enviro-chemical* or *enviro-health* interest are then thermally desorbed, cryofocused, and temperature programmed to generate a chromatogram. Analytes of interest can also be eluted off of the surface of the stir bar with organic solvents in a manner similar to that in RP-SPE. The much larger β for SBSE requires a longer desorption time than SPME.

Sandra and coworkers report successful *in situ* chemical derivatization of polar analytes conducted just prior to SBSE.[125] These analytes range from biological markers such as phenols, hormones, and fatty acids to artificial contaminants such as drugs and plasticizers. Ethyl chloroformate and acetic acid anhydride in an aqueous medium have yielded useful derivatives that can be subsequently extracted via SBSE. To illustrate, the bacteriostate to some dentifrices and mouth rinse products, triclosan, was found in the urine of a male volunteer using SBSE. Dissolving 1 mg of toothpaste containing triclosan at the 0.1% level in 10 mL of water, an analysis of the mixture by SBSE, combined with thermal desorption and GC-MSD, verified its relationship with the dentifrice source.[127]

A summary of the different *sample prep* techniques that have evolved into methods to isolate and recover SVOCs from drinking, ground and wastewater drawn from the environment which are examples of *enviro-chemical* TEQA, is found including important procedural comments in Table 3.30.

3.87 CAN SPE, SPME, AND SBSE BE AUTOMATED?

Yes, all three sample prep techniques have been automated. The Zymark Corporation (now Caliper Life Sciences, Inc.) was first to automate the barrel-type SPE cartridge. Gilson, Hamilton,

---

TABLE 3.30
Review of Sample Prep Approach to Isolate Persistent Organic Pollutants (POPs) or Semivolatile Organic Compounds (SVOCs) From an Aqueous Environmental Sample (*Enviro-Chemical* TEQA). Analytes are Amenable to Analysis by GC

Extraction Technique	Volume of aqueous sample	Volume of extractant or eluent (# of successive extractions)	Post extraction and prior to GC injection
Macro-LLE[1]	100–1,000 mL	30–60 mL (3 X)	Evaporate extractant to < 5 mL, 1 µL injection into GC
Mini-LLE[2]	10–40 mL	1-2 mL (1X)	1 µL injection into GC
Micro-LLE[3]	< 10 mL	1 drop (1X)	1 µL injection into GC
RP-SPE[4]	~70 mL or greater	1-5 mL 1X	1 µL injection into GC
RP-SPDE[4]	~70 mL or greater	1-5 mL (1X)	1 µL injection into GC
RP-SPME[5]	10–50 mL	None	Thermally desorb into GC inlet
RP-SBSE[5]	10–50 mL	None	Thermally desorb into GC inlet

Notes: [1]Use a 1,000 mL separatory funnel. Make 3 successive LLE extractions. Use an evaporative concentrator to reduce extract volume. This yields the lowest MDLs; however, it is the most time-consuming *sample prep* technique.[2] It is best to perform this mini-LLE in a 40 mL glass VOA bottle with a narrow mouth. The nonpolar extractant will rise into the neck of the bottle. A 1 µL aliquot can be sampled with a 10 µL liquid-handling syringe and then directly injected into a GC. An evaporative concentrator could also be used to reduce extract volume and therefore to lower MDLs.[3] One drop of nonpolar extractant is suspended in a test tube that contains < 10 mL of aqueous sample. The GC liquid-handling syringe used to suspend the drop is retracted and subsequently injected into the GC inlet.[4, 5] The nonpolar eluent can be transferred to a 1.0 mL volumetric flask and adjusted to a precise final volume prior to injecting 1 µL into a GC inlet. The same is true for the RP-SPDE technique.[5] Both the SPME polymer-coated fiber and the polymer-coated stir bar are then directly injected into the GC inlet. Most suitable to ultratrace quantitative analysis.

TomTec and other companies also have developed automated SPE, with an emphasis on the *96-well plate* SPE design. This author has used the 96-well plate SPE design manufactured by Phenomenex Inc. Gerstel GmbH & Co. and Leap Technologies, Inc., have automated both SPME and SBSE for trace enviro-chemical and enviro-health quantitative analysis within the dual-rail multipurpose robotic technology originally developed by CTC Analytics AG. Thurman and Mills devote an entire chapter to SPE automation up through the late 1990s. This book provides a good starting point on this topic (pp. 243–279).[83] The future looks very bright for continued advancements in automating SPE, SPME and SBSE by interfacing with determinative techniques such as gas chromatography (GC) and high-performance liquid chromatography (HPLC).

3.88 ARE THERE EXAMPLES OF AUTOMATED SPE OUT THERE?

Yes, but not as many published reports as you might think. Of enviro-health interest is an earlier paper from the Centers for Disease Control and Prevention (CDC) on incorporating the Rapid Trace SPE Workstation as part of a faster *sample prep* approach to isolating and recovering persistent organic pollutants (POPs) from archived plasma samples. The method consisted of upfront RP-SPE of selected OCs, followed by NP-SPE cleanup using silica gel, with subsequent injection into an analytical HPLC column incorporating an analytical gel permeation column. A fraction is obtained from the GPC column, which is subsequently injected into a GC or GC-MS. Mean recoveries of the ^{13}C-labeled internal quantification standards ranged from 64 to 123% for the 11 monitored OCs.[128] A semiautomatic high-throughput extraction and cleanup method developed around the use of the Rapid Trace SPE Workstation was also developed at the CDC and was reported earlier and was at that time a substantial extension of the work just cited. This paper shows how automated spiking of samples and automated RP-SPE and NP-SPE can be coupled together while extending the method to poly-brominated diphenyl ethers (PBDEs), PBBs, and PCBs in human serum.[129]

3.89 WHAT IS THIS SAMPLE PREP TECHNIQUE CALLED QUECHERS?

QUECHERS is an abbreviation for the phrase: *Quick, Easy, Cheap, Rugged, Effective, Safe* whose origins go back to pioneer work done at a United States Department of Agriculture (USDA) laboratory.[130] This author unknowingly applied what came to be known as QUECHERS when he developed a LLE approach while attempting to find a way to extract three polar organic compounds: 2-aminoethanol, N-methyl-2-aminoethanol and N,N'-dimethyl-2-aminoethanol from an aqueous solution. He was advised to add enough Na_2SO_4(s) to 4 mL of a deep-blue liquid (copper (II) hydroxide complexed 2-aminoethanol) to saturate the solution followed by adding 2 mL of 2-propanol (isopropyl alcohol). After shaking this mixture and allowing it to settle, an upper phase developed![131] The schematic shown in Figure 3.60 which describes the first step in implementing the QUECHERS *sample prep* technique. When the ionic strength of an aqueous solution is significantly increased, the degree to which a polar molecule such as 2-propanol interacts intermolecularly with water is significantly reduced. The reader would do well at this point to go back and review Figures 3.43 and 3.44.

During the early development of reversed-phase HPLC (RP-HPLC) when a typical mobile phase consisted of acetonitrile (MeCN) and an aqueous phosphate buffer such as (Na_2HPO_4 – Na_3PO_4), it was realized that a limit to the ionic strength of the phosphate buffer was important. When the concentration of phosphate buffer exceeded a certain molarity, the HPLC pump pressure would start to increase significantly! The acetonitrile phase became immiscible and thus separated out! Therefore, it was "discovered" that one way to extract polar solutes from a polar matrix is to saturate the polar matrix (e.g. mashed fruits with added water) with inorganic salts and thus separate the polar phase which floats atop the aqueous phase. This enables polar analytes to be effectively partitioned from the aqueous phase into the polar phase!

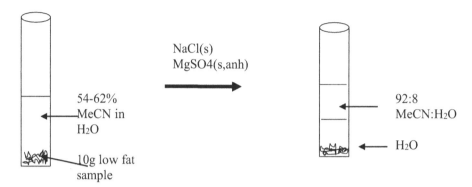

FIGURE 3.60

Acetonitrile (MeCN) emerged as the ideal extraction solvent for the QUECHERS *sample prep* technique. Referring back to Table 3.16 where a list of solvent polarities is given, it is noted that MeCN with a P'= 5.8 is one of the most polar solvents used in analytical chemistry. Acetonitrile is miscible with water, yet immiscible in an inorganic salt saturated aqueous solution.

This first step in the QUECHERS procedure is the only step needed provided the solid, semi-solid or aqueous matrix is low in fat content. Fat content in a sample is often expressed in terms of the % lipid. This explains the success of the early QUECHERS *sample prep* technique when applied to isolating and recovery semi-polar SVOCs such as persistent organic pesticides (also called POPs) from a fruit or vegetable sample/aqueous matrix with a low % lipid content. Lehotay and colleagues at the USDA extended the QUECHERS sample prep technique to fatty food matrices by studying dispersive SPE along with MSPD.[132]

Two analytical methods of sample preparation with quantitative analysis were evaluated and compared for 32 pesticide residues that represent a wide range of physico-chemical properties. The *first method* studied began by extracting 15 g of either milk, eggs, or avocado (foods with appreciable fat content) with MeCN containing 1% acetic acid followed by addition of 6 g anhydrous $MgSO_4$ and 1.5 g sodium acetate (NaOAc). After centrifugation, 1 mL of the buffered MeCN extract underwent *two different cleanup steps* (dispersive SPE and column SPE, respectively) using 50 mg each [C_{18} chemically bonded silica and primary secondary amine (PSA) sorbents] plus $MgSO_4$ for dispersive SPE and 500 mg each (C_{18} chemically bonded silica and PSA) for column SPE. The *second method* incorporated a form of MSPD in which 0.5 g sample plus 2 g C_{18} chemically bonded silica and 2 g anhydrous Na_2SO_4 was mixed in a mortar and pestle. The sample was transferred to an empty SPE barrel-type cartridge. This cartridge was stacked on the SPE above a second barrel-type cartridge containing 2 g Florisil column on a vacuum manifold. Then 5 x 2 mL MeCN was used to elute the pesticide analytes from the sample into a collection tube, and the extract was concentrated to 0.5 mL by evaporation. Extracts in both methods were analyzed concurrently by gas chromatography-mass spectrometry (GC-MS) and by liquid chromatography tandem mass spectrometry (LC-MS-MS). Percent recoveries of polar and semi-polar pesticides were typically 100% in both methods (except for some so-called basic pesticides). However, % recoveries for nonpolar pesticides decreased as fat content of the sample increased. This trend was more pronounced in the QUECHERS method, in which the most lipophilic analyte tested, hexachlorobenzene, yielded a 27±1% recovery of six replicate samples in avocado which is 15% fat. A limit of quantitation was reported as <10 ng/g. Table 3.31 lists mean % recoveries with corresponding %RSDs in the mean % recoveries at one spiking concentration

TABLE 3.31
Mean % Recoveries of Fortified Pesticides in Milk and Egg from a Buffered QUECHERS Method with Dispersive SPE and Column SPE, and the MSPD Method Quantitated Using GC-MS and LC-MS-MS Determinative Techniques

POP	Dispersive SPE (Milk) 500ng/g*	Dispersive SPE (Egg) 500ng/g*	Column SPE(Milk) 500ng/g*	Column SPE (Egg) 500ng/g*	MSPD (Milk) 500ng/g*	MSPD (Egg) 500ng/g*
Chlorpyrifos	94(4)	90(3)	93(2)	90(2)	100(5)	104(3)
DDE	75(5)	63(5)	70(6)	56(9)	102(9)	102(6)
Dichlorvos	115(5)	106(3)	107(3)	115(5)	96(10)	86(20)
Dieldrin	95(5)	79(2)	86(5)	76(5)	100(10)	103(5)
Lindane	106(4)	95(0)	97(4)	94(1)	97(5)	96(3)

*Results reported at the indicated pesticide spike level; a second spike level of 50 ng/g was also studied.

Adapted from S. Lehotay, K. Mastovska and S. Yun, *J AOAC International* 88(2): 630–638, 2005.

for five of the numerous POPs studied.[132] Molecular structures for five representative POPs from this study are given below:

Chlorpyrifos   DDE   Dichlorvos

Dieldrin   Lindane

A recent application note presented both % recovery and %RSD in the mean recovery for some commonly used pesticides such as carbaryl, atrazine, and methyl parathion among others from brewed coffee using products designed by industry to implement the QUECHERS *sample prep* technique.[133]

3.90 HAVE BLOODSPOTS BEEN ISOLATED AND RECOVERED USING THESE MINI-EXTRACTION TECHNIQUES?

Yes. We will discuss other contributions and finish with the author's work on this important topic in *enviro-health* TEQA. One advantage of the lower instrument detection limits (IDLs) due to advances in instrumentation and computer technology has been the significant decrease in sample size necessary to reach low part-per-billion (ppb) and moderate to low part-per-trillion (ppt) concentration levels. These advantages enable a 20 µL volume of human or animal blood to be "spotted" on filter paper, dried, and subsequently extracted. The chemical analytes in the extract are subjected to the determinative techniques (introduced in Chapter 4) that enable analytical chemists to furnish the mathematical tools (introduced in Chapter 2) to satisfy the ultimate goals (introduced in Chapter 1) of providing trace *enviro-health* quantitative analysis.

Earlier, it was recognized that the use of dried-blood spot specimens to assess inborn errors of metabolism (e.g. galactose, leucine, methionine, phenylalanine, etc.) in newborns could also be used to conduct biomonitoring. Blood spots on filter paper obtained from newborn screening were soaked in a phosphate buffer, extracted with an organic solvent (pesticide residue grade hexane) and eluted through silica gel. The concentrated eluates were analyzed by capillary column gas chromatography with electron-capture detection (C-GC-ECD). Blood collected from 10 newborns was analyzed and found to contain DDE concentrations ranging from 0.1 ppb to 1.87 ppb. One newborn had a whole-blood DDE concentration of 1.87 ppb which was greater than the concentration of 1.34 ppb from an adult donor whose blood serum was shown to contain DDE![134]

Today, most of the 50 states in the United States conduct routine newborn screening programs and collect blood as dried spots for temporal biomonitoring. With funding from the CDC, the New York State Department of Health in collaboration with SUNY Albany report on a study designed to quantitate five perfluoroalkyl substances (PFAS) whose molecular structures are shown below in newborn blood spots. Between 1997 and 2007 blood spot specimens were collected that represent a total of 2,640 infants. All five analytes were detected in ≥ 90% *of the specimens*! Concentrations of PFOS, PFOSA, PFHxS, and PFOA exhibited significant exponential declines after the year 2000. This coincided with the phase-out in PFOS production in the U.S. The utility of using newborn screening program for assessment of temporal trends in exposure was demonstrated.[135]

The blood spots were extracted using Ion Pair LLE while extracts were injected and quantitated using an HPLC interfaced to a triple quadrupole mass spectrometer abbreviated LC-MS-MS. Molecular structures are shown below for the PFAS organic compounds studied:

Perfluorooctanesulfonic acid (PFOS)
SRM 499 ⟶ 99

Perfluorooctanesulfonamide (PFOSA)
SRM 498 ⟶ 78

Perfluorooctanoic acid (PFOA)
SRM 413 ⟶ 369

Perfluorononanoic acid (PFNA)
SRM 463 ⟶ 219

Perfluorohexanesulfoic acid (PFHxS)
SRM 399 ⟶ 80

Perfluorobutanesulfonic acid
SRM 297 ⟶ 80

^{13}C Perfluorooctanesulfonic acid
SRM 503 ⟶ 99

^{13}C Perfluorooctanoic acid
SRM 417 ⟶ 372

Note the use of a heavy black spot in the molecular structures of the two internal standards to *denote the* 13C *isotope of carbon*. The placement of the ^{13}C atoms was arbitrary and does not reflect the true position of the isotopically labeled molecules. Because a tandem mass spectrometer was used, each PFAS compound was quantitated based on selective reaction monitoring (SRM). The principles of SRM in tandem mass spectrometry will be introduced in Chapter 4.

Given the success of extracting toxic chemicals from blood spots, this author coupled his prior work with extracting PBDE/PBBs using SBSE[126] and investigated the isolation and recovery of these toxic compounds from blood spots while replacing the mini-LLE technique with the SBSE *sample prep* technique. The experimental procedure is outlined as follows: ~0.5 mL of blood is placed in a glass cylindrical headspace vial and spiked with 250 pg each BDE congener.

The generalized molecular structure of polybrominated diphenyl ethers and also that for polybrominated biphenyls are shown below:

Recall, similar to PBBs and PCBs, there are 209 possible BDE congeners. A mix of 10 BDE congeners and two BB congeners comprised the spike reference. Each circle on the Whatman Card® is spotted with ~20 µL which corresponds to 10 pg each BDE congener (20 µL x 500 pg/ mL x 1 mL/1000 µL). One spotted circle is placed in a 5 mL headspace vial using a glass stir rod. 1 mL distilled, deionized water (DDI) is added to the headspace vial followed by 1 mL 5% acetonitrile (MeCN) in concentrated formic acid (HCOOH) added to the headspace vial. A clean Twister® is placed into the headspace vial and the contents are stirred for ~ 2 hours. The Twister is then removed and placed in a 50 mL glass beaker that contains DDI to remove debris. Remove the Twister and place it in a glass TDU adapter glass tube. All five twisters were sequentially thermally desorbed and gas chromatographed under TD-GC-ECNI-MS (thermal desorptive capillary gas chromatography interfaced to electron capture negative ion mass spectrometry). Results for the five replicate samples are shown in Table 3.32. Some surprising results are shown in Table 3.32! BDEs 47 and 99 yielded much higher concentrations when compared to the other eight BDE congeners! Might the unspiked blood spot contain a background of these two BDE congeners?

To test this hypothesis, this author conducted an experiment that compared the isolation and recovery from a blank blood spot vs. a spiked blood spot. Figure 3.61 compares the concentration in terms of the #ppb found for each BDE congener from the spiked blood spot against the #ppb found from the unspiked blood spot. It would appear that BDE congeners 47 and 99 are present in the blank sample. This finding helps to explain some of the results from Table 3.32! BDE congeners 47 and 99 are known to tenaciously cling to glass surfaces! Referring again to Figure 3.61, might other BDE congeners show a similar tendency? What about BB congeners 155 and 153? A typical TD-C-GC-ECNI-MS chromatogram from this bloodspot/SBSE work is shown in Figure 3.62. Question marks denote unidentifiable chromatographically resolved peaks.

TABLE 3.32
Results from Isolating and Recovering Five Replicate Bloodspots Spiked With PBDEs/PBBs

BDE#	1	2	3	4	5	$\bar{x}$	RSD(%)
28	130	108	80	70	99	97.4	24.2
47	476	492	349	422	402	428.2	13.4
66	109	90	75	70	94	87.6	17.8
77	62	74	54	42	42	54.8	25.0
BB-155	171	271	186	191	226	209	19.2
100	162	191	129	157	158	159.4	13.8
99	419	434	340	519	408	424	15.1
85	104	100	82	82	86	90.8	11.5
154	144	153	133	131	139	140	6.3
BB-153	243	289	237	209	253	246.2	11.8
153	197	215	195	185	153	189	12.1
183	3.1	2.8	0.1	ND	ND	NA	NA

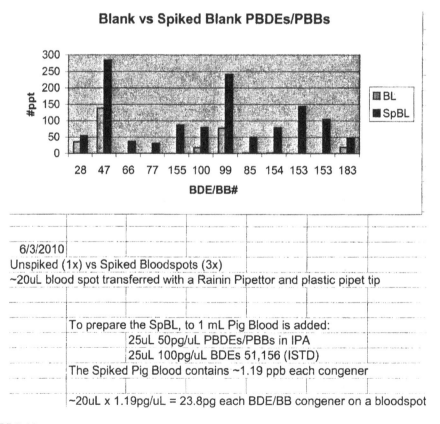

6/3/2010
Unspiked (1x) vs Spiked Bloodspots (3x)
~20uL blood spot transferred with a Rainin Pipettor and plastic pipet tip

To prepare the SpBL, to 1 mL Pig Blood is added:
   25uL 50pg/uL PBDEs/PBBs in IPA
   25uL 100pg/uL BDEs 51,156 (ISTD)
The Spiked Pig Blood contains ~1.19 ppb each congener

~20uL x 1.19pg/uL = 23.8pg each BDE/BB congener on a bloodspot

FIGURE 3.61

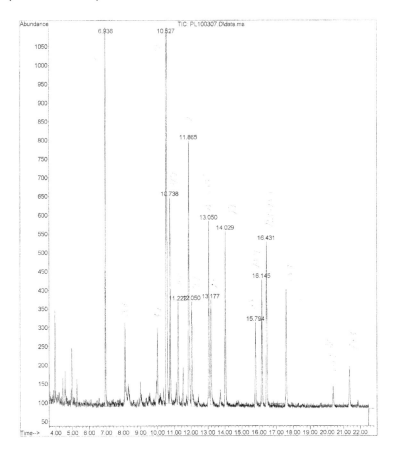

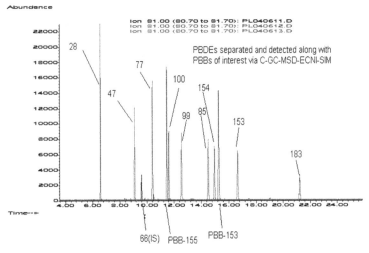

FIGURE 3.62

3.91 HOW ARE HETEROGENEOUS MAGNETIC MATERIALS BASED ON METAL-ORGANIC FRAMEWORKS (MOFs) USED TO ISOLATE AND RECOVER PRIORITY POLLUTANT ORGANICS FROM CONTAMINATED WATERS DRAWN FROM THE ENVIRONMENT?

By incorporating magnetic nano particles (MNPs) into metal-organic frameworks (MOFs) and thus creating a relatively new miniaturized version of the MSPD *sample prep* technique known as magnetic dispersive micro solid-phase extraction (M-D-μSPE).[136-138] MNPs composed of nanoparticles in the magnetic phase Fe_3O_4 protected by a silica-protected coating phase $Fe_3O_4@SiO_2$ and with post functionalization generating $Fe_3O_4@SiO_2-NH_2$ or $Fe_3O_4@SiO_2-NH_2-COOH$ materials. Four heterogeneous magnetic composites based on MOFs are selected as magnetic sorbents in M-D-μSPE procedures used to isolate and recover the sixteen common polycyclic aromatic hydrocarbons (PAHs) from environmental waters. The 16 PAHs were quantitated using ultra high pressure liquid chromatography and fluorescence detection (UHPLC-FL).[139] Molecular structures for representative PAHs studied showing a consecutive increase in ring size are shown below:

Naphthalene   Phenanthrene   Chrysene   Benzo(a)pyrene

Experimentally, the optimum M-D-μSPE procedure used $Fe_3O_4@SiO_2-NH_2-COOH/ZIF-8$ as the sorbent, 100 mg of this magnetic composite was added to 10 mL of aqueous standard or water sample containing PAHs. The mixture was subjected to 5 min of stirring using ultrasound. Afterwards, a magnet was placed outside the extraction tube for a few seconds, and the aqueous phase was simply removed. The magnetic composite containing the extracted PAHs was subjected to elution using 100 μL of acetonitrile and 5 min of ultrasound. The magnet was again placed outside the extraction tube for a few seconds, and the eluate was easily collected with a syringe and injected into the UHPLC-FL system.[139] The method was validated: the calibration yielded a correlation coefficient r > 0.9940; limit of detections and quantitation using the 3σ model varied from a low of 0.3 ppt for Chrysene to 13 ppt for Indeno(1,2,3-c,d) pyrene.[139]

Our journey through the variety of traditional, novel, green, and miniaturized *sample prep* techniques available to isolate and recover most toxic semivolatile organic compounds (SVOCs) from *enviro-health* and *enviro-chemical* samples has come to an end! These important tools are contained in the analytical chemist's arsenal. *Sample prep* techniques that isolate and recover metals and several specialized inorganic analytes are considered next.

3.92 HOW ARE THE METHODS CATEGORIZED FOR TRACE INORGANICS ANALYSIS?

An earlier compilation of EPA methods organizes the numerous analytical methods for inorganics analysis according to the following six major categories:[140]

1. *Trace metals* identified by flame (FlAA) and by graphite furnace (GFAA) atomic absorption spectroscopy.

2. *Trace metals* identified by inductively coupled plasma-atomic emission spectroscopy (ICP-AES).

3. *Trace metals* identified by inductively coupled plasma-mass spectrometry (ICP-MS).

4. *Mercury* identified by cold-vapor atomic absorption spectroscopy.

5. *Cyanide* (total and amenable).

6. *Inorganic carbon* (total carbon less organic carbon).

Not listed in these categories are analytical methods for the principal *inorganic anions* derived from strong acids that are prevalent in groundwater: chloride, bromide, nitrite, nitrate, phosphate, and sulfate. These analytes are currently measured routinely by the application of either ion chromatography (IC) or specific colorimetric procedures following the conversion of the anion to a colored complex. Oxyhalides such as the bromate ion have been found in chlorinated drinking water. IC methods have been developed for various oxyhalide ions in recent years. Water that is free of dissolved organics and heavy metal can be directly injected into ion chromatographs or filtered if particulates are present. Aqueous samples that contained dissolved biomatter or heavy metal ions pose severe challenges to IC because the columns employed in the technique are susceptible to column fouling.

In this section, we will discuss the basis of sample preparation for five of the six categories listed above and focus on the determinative techniques for all six in Chapter 4. Total organic carbon (TOC) is a combustion technology in which aqueous samples can be injected directly without the need for sample preparation; therefore, we will not discuss it any further in this chapter. Let us start with a discussion of the principles of sample prep with respect to trace metals.

3.93 HOW DO YOU PREPARE AN ENVIRONMENTAL SAMPLE TO MEASURE TRACE METALS?

Sample preparation for the determination of trace concentration levels of the many priority pollutant metals is strongly connected to the nature of the determinative technique. Historically, FlAA was first used to measure metals. The more sensitive GFAA technique followed. About the same time as GFAA was being developed, ICP-AES came along. ICP-AES afforded the opportunity to measure more than one metal in a sample at a time, the so-called multielement approach. In recent years, the development of ICP-MS has carried trace metal analysis to significantly lower IDLs and introduced the opportunity to identify and quantitate the various elemental isotopes.

The metals of greatest interest are those listed earlier in Table 1.6 and have the lowest maximum contaminant levels (MCLs). A comparison of the sheer number of oxidation states among metallic chemical elements is presented. A metallic chemical element is most likely to be found in an oxidized state as indicated for each metallic chemical element. The design of instrumentation for either FlAA, GFAA, ICP-AES, or ICP-MS requires that an aqueous matrix containing the oxidized form of a given metal be introduced into the instrument. This feature significantly differs from gas or liquid chromatographs (GCs or LCs). For example, GCs require that a sample matrix must be an organic solvent. Aqueous samples that contain inorganics, in contrast to aqueous samples that contain organics, can be introduced directly into an FlAA, GFAA, ICP-AES or ICP-MS without removal of the analyte from its sample matrix. Chemically, metals might exist in the environment as ions in one or more oxidation states or partially or wholly chelated to ligands of various sorts. They may be bound or complexed to soil/sediment particles and therefore cannot be easily released via a liquid–solid extraction or leaching. Some metals form the structural composition of solid matrices derived from the environment, such as aluminosilicates in clay. In these cases, decomposition of the sample removes the organic portion of the matrix. Before technological advancements in instrumentation, specifically in ICP-MS, to quantitate trace metals listed in Table 3.32, it was required in earlier EPA methods that the five of the more toxic metals namely: As, Cd, Hg, Pb, and Tl be quantitated by GFAA.

Decomposition methods include: 1) combustion with oxygen with and without fluxes and 2) digestion with strong acids. Both methods for decomposing the sample matrix require heat, and both have benefited from replacement of resistive heating (e.g., laboratory hot-plate equipment) with microwave heating (pp. 60–61).[140]

The following procedure has been used by this author to prepare a sample of fly ash for determination of priority pollutant metals.[142] Assume that the ash has been obtained from a previous combustion procedure:

1. Weigh 0.2 g of sample into a 100 mL beaker. Record the weight to the nearest 0.001 g using an analytical balance.
2. Add 5 mL of concentrated nitric acid (HNO_3) and 5 mL of concentrated hydrochloric acid (HCl).
3. Place a watch glass over the beaker and digest at medium heat for 60 min.
4. Evaporate to dryness.
5. Add 5 mL of concentrated HNO_3 and evaporate to dryness.
6. Add 1 mL of concentrated HNO_3 and warm.
7. Add 1 mL of distilled deionized water and warm. Filter into a 25 mL volumetric flask.
8. Cool and dilute to the mark using 1% HNO_3. This gives exactly 25 mL of an aqueous sample containing the solubilized metal ion.
9. Aspirate into a previously calibrated FlAA and record the absorbance.

Notes:

1. Be sure to choose the appropriate wavelength and oxidizer gas for each metal of interest. The aqueous solutions used as calibration standards should be prepared in 1% HNO_3.
2. For every batch of samples that has been digested, one blank, one matrix spike, and one matrix spike duplicate should also be prepared. For the matrix spike and matrix duplicate, designate one sample from the batch for these.

Sample matrices from a biological origin such as blood, urine, and serum can be merely diluted with water, dilute nitric acid, or a dilute surfactant such as Triton X-100 and aspirated directly into the flame for FlAA or placed directly into the graphite tube for GFAA. With respect to GFAA, a plethora of matrix modifiers have been developed over the years to deal with spectral and chemical types of interference (pp. 109–114, Table C.1).[143] Spectral interferences arise when the absorption on emission of an interfering species either overlaps or lies so close to the analyte absorption or emission that resolution by the monochromator becomes impossible. Chemical interferences result from various chemical processes occurring during atomization that alter the absorption characteristics of the analyte.[144] Spectral interferences refer to the presence of concomitants that affect the quantity of the source light that reaches the detection system, whereas chemical interferences reduce the analyte absorbance signal by interactions with one or more concomitants compared to standards without concomitants (pp. 228, 237).

3.94   WHAT IS MATRIX MODIFICATION IN GFAA?

Because matrix modification is one aspect of sample preparation, even though it is most often accomplished automatically in GFAA, we will discuss it here. For example, National Institute for Occupational Safety and Health (NIOSH) Method 7105 recommends a matrix modifier that consists of a mixture of ammonium dihydrogen phosphate: $(NH_4)H_2PO_4$, magnesium nitrate: $Mg(NO_3)_2$, and nitric acid: HNO_3 for the determination of airborne Pb. Because the graphite furnace can be viewed as a chemical reactor whereby the sample with its matrix is placed on a graphite platform (the L'vov platform, discussed in Chapter 4) and heated to a very high temperature, reactions can take place that involve both the metal analyte of interest and sample matrix components.

The concept of matrix modification, from the sample prep perspective, is to add to the sample a chemical reagent that will cause a desirable chemical reaction or inhibit an undesirable reaction.[143] For metals that tend to volatilize, one can add a modifier that reduces analyte volatility by increasing the volatility of the matrix. Consider the determination of Pb in highly salted aqueous samples such as seawater. Seawater contains appreciably elevated levels of chloride (Cl^-) salts. Adding an ammonium ion (NH_4^+) to the seawater, followed by heating the sample to a high temperature, causes the following reaction to occur:

$$Cl^-_{(aq)} + NH^+_{4(aq)} \xrightarrow{\text{heat}} NH_4Cl_{(g)}$$

This reaction removes chloride ions from the sample while minimizing the loss of Pb as the more volatile $PbCl_2$.

Harris discusses the findings of Styris and Redfield, who studied the effect of magnesium nitrate on the determination of Al.cc[145] At high temperature, MgO is formed and steadily evaporates. This maintains a steady vapor pressure of MgO in the GFAA tube. The presence of MgO serves to keep Al as the oxide by establishing the following equilibrium:

$$3MgO + 2Al_{(S)} \rightleftharpoons 3Mg_{(g)} + Al_2O_{3(s)}$$

When most of the MgO has evaporated, the equilibrium begins to shift to the left as Al_2O_3 is converted to elemental Al. This reaction serves to delay the Al from evaporating until a higher temperature is reached.

Butcher and Sneddon describe the work of Schlemmer and Welz, who investigated the use of Pd and Mg nitrates as matrix modifiers in the determination of nine metallic elements; they showed that higher pyrolysis temperatures could be used, compared to no modifier or using other common modifiers. (pp. 111–113)[141,146] Thus, we have shown how additions to the sample matrix led to improved performance in GFAA.

With respect to FlAA and ICP, the addition of so-called matrix modifiers developed for GFAA serves no useful purpose. FlAA and ICP use nebulization into an oxidizing-reducing high-temperature source to introduce a liquid into the flame and plasma, respectively, and because of this, both techniques require that samples have a low dissolved solids content so as to prevent clogging. ICP is essentially free from most spectral and chemical interferences due to the extremely high temperature of the plasma (8,000 to 10,000°C), whereas these interferences are prevalent in FlAA and serve to influence the IDL for a given metal.

3.95   HOW DO I PREPARE A SOLID WASTE, SLUDGE, SEDIMENT, BIOLOGICAL TISSUE, OR SOIL SAMPLE?

EPA Method 3050B Revision 2, 1996 from the SW-846 series of methods involves solubilizing a solid sample with acids/peroxide and removing the insoluble residue by filtration. EPA Method 3051 is a microwave-assisted acid digestion procedure. EPA Method 200.3 is applicable to the preparation of biological tissue samples prior to using atomic spectrometry for quantifications of Al, Sb, As, Ba, Be, Cd, Ca, Cr, Co, Cu, Fe, Pb, Li, Mg, Mn, Hg, Mo, Ni, P, K, Se, Ag, Na, Sr, Tl, Th, U, V, and Zn, and an outline of the sample prep procedure is as follows: (p. 74)[141]

- Place up to a 5 g subsample of frozen tissue into a 125 mL Erlenmeyer flask. Any sample-spiking solutions should be added at this time and allowed to be in contact with the sample prior to the addition of acid.
- Add 10 mL of concentrated nitric acid and warm on a hot plate until the tissue is solubilized. Gentle swirling of the sample will aid in this process.
- Increase the temperature to near boiling until the solution begins to turn brown. Cool sample, add an additional 5 mL of concentrated nitric acid, and return to the hot plate until the solution once again begins to turn brown.
- Cool sample, add an additional 2 mL of concentrated nitric acid, return to the hot plate, and reduce the volume to 5 to 10 mL. Cool sample, add 2 mL of 30% hydrogen peroxide, return sample to the hot plate, and reduce the volume to 5 to 10 mL.
- Repeat the previous step until the solution is clear or until a total of 10 mL of peroxide has been added.
- Cool the sample, add 2 mL of concentrated hydrochloric acid, return to the hot plate, and reduce the volume to 5 mL.
- Allow the sample to cool and quantitatively transfer to a 100 mL volumetric flask. Dilute with DDI, mix, and allow any insoluble material to separate. The sample is now ready for either ICP-AES, ICP-MS, or GFAA.

3.96   WHAT ARE THE EPA'S MICROWAVE DIGESTION METHODS?

EPA Method 3015A Revision 1 2007 (applicable to an aqueous sample such as groundwater) and EPA Method 3051A, Revision 1 2007 (applicable to soils, sediments, sludges, and oils) utilize advances in microwave heating technology.[148] Microwave heating significantly reduces the more labor-intensive hot-plate heating techniques described earlier in this chapter. These methods developed earlier enable environmental samples to be digested so that a quantitative determination of up to 26 metals can be made. These metals are listed as their chemical symbols as follows:

Al	B	Cu	Hg	Ag	Zn
Sb	Cd	Fe	Mo	Na	
As	Ca	Pb	Ni	Sr	
Ba	Cr	Mg	K	Tl	
Be	Co	Mn	Se	V	

Figures 3.64A and 3.64B, adapted for Method 3051A, outline the microwave digestion procedure using logical flow charts to prepare soil, sediments, sludges, and oils for trace metals analysis. This sample prep technique is an example of trace *enviro-chemical* TEQA.
To summarize Method 3051A:

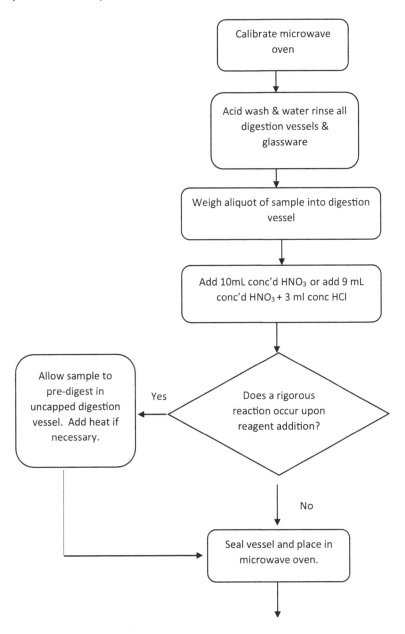

FIGURE 3.63A

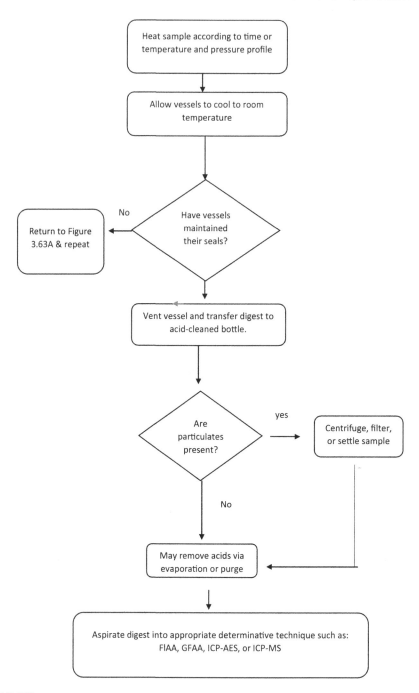

FIGURE 3.63B

A representative sample of up to 0.5 g is extracted and/or dissolved in 10 mL of concentrated nitric acid or 9 mL of concentrated nitric acid and 3 mL of concentrated hydrochloric acid for 10 min using microwave heating with a suitable laboratory unit. The sample and acid(s) are placed in a fluorocarbon polymer or quartz vessel or vessel liner. The vessel is sealed and heated in a microwave unit. After cooling, the vessel contents are filtered, centrifuged or allowed to settle and then diluted to volume and analyzed by the appropriate determinative technique.

Safety considerations are paramount in digestion involving microwave heating technology. It cannot be emphasized enough that a suitable laboratory microwave unit must be used. There is a mind-set that a microwave oven designed for ordinary kitchen use is a suitable low-cost substitute. The microwave unit cavity must be corrosion resistant and well ventilated. All electronics must be protected against corrosion for safe operation. The oven exterior, including the door, must be explosion-proof as well as equipped with temperature and pressure sensors and alarms.

3.97   WHAT IS DONE TO PREPARE A BLOOD, SERUM, OR URINE SPECIMEN FOR TRACE METALS ANALYSIS?

Public health laboratories have a long history of quantitatively determining Pb in human blood. Some states are required by law to measure blood Pb in children from at-risk environments. Flame atomic absorption and graphite furnace atomic absorption (whose principles will be introduced in Chapter 4) have been the principal instrumental analysis approach to quantitatively determine blood Pb. ICP-MS is rapidly emerging as an instrument of choice for multielement quantitative analysis for *enviro-health* TEQA since whole blood and urine can be (after an appropriate dilution) directly aspirated into the inductively coupled plasma without concomitant interferences.w[149]

So far, we have considered destroying or transforming the sample matrix in some way while leaving the metal ion intact. We now consider ways to transfer the metal ion from the sample matrix, and this leads to sample prep techniques that preconcentrate the metal in the sample.

3.98   WHAT CAN I DO TO PRECONCENTRATE A SAMPLE FOR TRACE METAL ANALYSIS?

There is a need for methods that can detect ultratrace concentration levels of the priority pollutant metal ions that are present in the environment. It may be desirable to detect and measure the concentration of one or more metals from a matrix that is highly salted, and it is expected that the concentrations of the metals of interest are extremely low (e.g., at parts per trillion (ppt) levels). An aqueous sample can be preconcentrated by evaporating off the water, precipitating the analyte, followed by redissolution of the precipitate, extracting the analyte via cation exchange or after forming an anionic species from the metal ion, via anion exchange, and isolating the metal ion by first forming the neutral metal chelate and sorbing the chelate on a hydrophobic surface, such as a chemically bonded silica, or extracting the metal chelate into a nonpolar solvent via LLE. We will focus on three aspects of preconcentration by considering the following examples:

1. The coprecipitation of Cr (VI) using lead sulfate.
2. The mathematics for the general case of LLE involving metal ions and the neutral metal chelates that can be formed.
3. The mathematics for isolating and recovering a neutral metal chelate, cadmium(II) oxinate, from a spiked aqueous sample.

3.99   HOW DO I PRECONCENTRATE Cr(VI) FROM A LEACHATE USING COPRECIPITATION?

In the environment, Cr (VI) is in either the chromate $\left(CrO_4^{2-}\right)$ or dichromate $\left(Cr_2O_7^{2-}\right)$ form. Whether Cr(VI) is predominantly one or the other strongly depends on the pH of the aqueous solution within which it is dissolved. The following ionic equilibrium is established between the two:

$$CrO_4^{2-} + 2H^+ \rightleftharpoons Cr_2O_7^{2-} + H_2O$$

Thus, if the aqueous solution that contains CrO_4^{2-} is made acidic by adding a source of the hydronium ion, the equilibrium adjusts to produce more $Cr_2O_7^{2-}$. If the hydronium ion is removed by reaction with the hydroxide ion or other base, the equilibrium adjusts to produce more CrO_4^{2-}. Let us assume that we have an aqueous solution containing Cr (VI) as either chromate or dichromate and chromium (III) as the inorganic compound $CrCl_3$. The addition of a source of the Pb^{2+} ion, such as lead (II) sulfate will precipitate Cr (VI) as the insoluble $PbCrO_4$ while leaving Cr(III) in the supernatant as Cr^{3+}. Thus, a speciation of chromium via coprecipitation can be realized. Hence,

$$CrO_{4(aq)}^{2-} \xrightarrow{\ Pb^{2+}\ } PbCrO_{4(s)}$$

The $PbCrO_4$ precipitate is washed clean of occluded Cr^{3+} and chemically reduced by the addition of hydrogen peroxide and nitric acid according to

$$PbCrO_{4(s)} \xrightarrow{HNO_3, H_2O_2} Cr_{(aq)}^{3+} + Pb_{(aq)}^{2+}$$

If it is expected that the concentration is within the range of 1 to 10 mg/L Cr, the dissolved Cr(III) can now be analyzed after adjustment to a precise volume of either FlAA or ICP-AES. If it is expected that the concentration is within the range of 5 to 100 µg/L Cr, the Cr(III) could be injected into the GFAA. A procedure for the possible speciation of chromium that might be found in the environment between Cr(III) and Cr(VI), along with the appropriate sample prep, is found in EPA Method 7195 from the SW-846 series.

Chromium (VI) can be quantitated without coprecipitation by forming a metal chelate. Method 7196A provides a procedure to prepare the diphenyl carbazone complex with Cr(VI) in an aqueous matrix. The method is not sensitive in that it is useful for a range of concentrations between 0.5 and 50 mg/L Cr. A more sensitive colorimetric method converts Cr(VI) to a Cr(VI) chelate with ammonium pyrrolidine dithiocarbamate (APDC), followed by LLE into methyl isobutyl ketone (MIBK). The molecular structure for APDC with the negative charge density surrounding the sulfur moieties are shown below:

Ammonium pyrrolidine dithiocarbamate

The APDC forms chelates with some two dozen metal ions (p. 79).[141] The extent of formation of the metal chelate is determined by the magnitude of the formation constant β. The efficiency of LLE as defined by a metal chelate's percent recovery depends on the distribution ratio, D, of the metal chelate in the two-phase LLE. We use the fundamental definition of D to develop a useful relationship for metal chelate LLE.

3.100 TO WHAT EXTENT CAN A GIVEN METAL CHELATE BE RECOVERED BY LLE?

Recall the definition of a distribution ratio for a specific chemical species as defined by Equation (3.9). If a chelate itself is a weak acid, secondary equilibrium plays a dominant role. We start by writing down a definition for the distribution ratio that accounts for all chemical species involving a metal ion M:

$$D = \frac{\sum C_M \left(\text{organic}\right)}{\sum C_M \left(\text{aqueous}\right)}$$

This generalization can be reduced to

$$D = \frac{\left[ML_n\right]_{\text{organic}}}{\left[ML_n\right]_{\text{aqueous}} + \left[M^{n+}\right]_{\text{aqueous}}} \tag{3.49}$$

This definition assumes that the only metal-containing species are the neutral chelate, MLn, and the free metal ion, Mn^+. This assumption simplifies the mathematics. Other chemical species that might contain the metal are not in appreciable enough concentrations to be considered. The degree to which a metal remains as the uncomplexed metal ion, Mn^+, is given by α_M, whereby α_M is the fraction of all of the metal-containing species in the aqueous phase that is in the Mn^+ form:[141]

$$\alpha_M = \frac{\left[M^{n+}\right]_{\text{aqueous}}}{\left[ML_n\right]_{\text{aqueous}} + \left[M^{n+}\right]_{\text{aqueous}}} \tag{3.50}$$

Equation (3.50) is solved for the term $\{[ML_n]+ [M^{n+}]\}$ and then substituted into Equation (3.49) to give the following relationship:

$$D = \frac{\left[ML_n\right]_{\text{organic}}}{\left[M^{n+}\right]} \alpha_M \tag{3.51}$$

Equation (3.51) states that the degree to which a given metal chelate, MLn, partitions into the organic phase depends on the ratio of the concentration of extracted MLn to the concentration of free metal ion, and on the degree to which metal ion remains uncomplexed in the aqueous phase. Equation (3.51) is quite complex, and as it stands, this equation is not too useful in being able to predict the extraction efficiency. Figure 3.64 is a diagrammatic representation of what happens when a metal ion, Mn^+, forms a metal chelate with a weak acid-chelating reagent, HL. The metal chelate is formed where one metal ion complexes to n singly charged anionic ligands, L, to form the metal chelate, MLn. The several equilibria shown set the stage for secondary equilibrium effects in which the concentrations of HL and

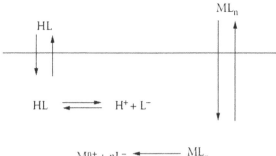

FIGURE 3.64

MLn, both present in the organic phase, can be changed from a consideration of the effect of pH in the aqueous phase. We now proceed to discuss and derive a much more useful relationship that, once derived, enables some predictions to be made about the LLE extraction efficiency of metal chelates.

3.101   HOW DO YOU DERIVE A MORE USEFUL RELATIONSHIP FOR METAL CHELATES?

We derive a more useful relationship for the LLE of metal chelates by considering the well-known secondary ionic equilibria described in Figure 3.64. Let us start by assuming that the chelating reagent to be used to complex with our metal ion of environmental interest is, in general, a weak acid. This monoprotic (our assumption) weak acid can ionize only in the aqueous phase and does so according to

$$HL_{(aq)} \rightleftharpoons H^+ + L^-$$

The extent of this dissociation is governed by its acid dissociation constant, Ka, and is defined in the case of a monoprotic weak acid, HL, as follows:

$$K_a = \frac{\left[H^+\right]\left[L^-\right]}{[HL]}$$

The neutral chelate can also partition into the organic phase according to

$$HL_{(aq)} \xrightarrow{K_D^{HL}} HL_{(organic)}$$

The extent to which HL partitions into the organic phase is governed by its partition coefficient, K_D^{HL}, and is defined as

$$K_D^{HL} = \frac{[HL]_{organic}}{[HL]_{aqueous}} \tag{3.52}$$

Chelating reagents that are amphiprotic, such as 8-hydroxyquinoline, HOx, have a more limited pH range within which the distribution ratio for HOx approximates K_D^{HL}.

Free metal ions, M^{n+}, and the conjugate base to the weak acid chelate, L^-, that are present in the aqueous phase will form the metal chelate by reaction of n ligands coordinating around a central metal ion. The extent of complexation is governed by the formation complex, β, according to

$$M^{n+} + nL^- \underset{\longleftarrow}{\overset{\beta}{\longrightarrow}} ML_n$$

The formation constant of the metal chelate in aqueous solution is defined as

$$\beta = \frac{\left[ML_n\right]_{\text{aqueous}}}{\left[M^{n+}\right]_{\text{aqueous}}\left[L^-\right]^n_{\text{aqueous}}} \tag{3.53}$$

The partition coefficient for the neutral metal chelate, MLn, where a relatively nonpolar and water-immiscible solvent is added to an aqueous solution containing the dissolved metal chelate, is given as follows:

$$K_D^{ML_n} = \frac{\left[ML_n\right]_{\text{organic}}}{\left[ML_n\right]_{\text{aqueous}}} \tag{3.54}$$

The acid dissociation constant expression, the formation constant expression, and the two expressions for the partition coefficients can be substituted into the defining equation for D ([Equation (3.49)], rearranged, and simplified.

3.102 CAN WE DERIVE A WORKING EXPRESSION FOR THE DISTRIBUTION RATIO?

Yes, we can, and we utilize all of the above equations to do so. This is an instructive exercise that can be found in a number of analytical chemistry texts that introduce the topic of metal chelate extraction. Usually the derivation itself is not included and only the final equation is given and interpreted. In the derivation that follows, we find that we do not need to add any more simplifying assumptions to those already given to reach the final working equations.

Let us consider eliminating the concentration of free metal ion by solving Equation (3.53) for $[Mn^+]$ and substituting this expression into Equation (3.51). This gives

$$D = \frac{K_D^{ML_N}\left[ML_n\right]}{\left[ML_n\right]\beta\left[L^-\right]^n}\alpha_M$$

The concentration of metal chelate in the aqueous phase cancels, and we obtain the following expression for D:

$$D = K_D^{ML_n}\beta\left[L^-\right]^n\alpha_M$$

This equation can be further simplified by eliminating the ligand concentration term by solving the equation for Ka given earlier for $[L-]$ and substituting this expression. We get

$$D = K_D^{ML_n} \beta \left\{ \frac{K_a[HL]}{[H^+]} \right\}^n \alpha_M$$

Rearranging this equation gives

$$D = \frac{K_D^{ML_n} \beta K_a^n}{[H^+]^n} [HL]_{aqueous}^n \, \alpha_M$$

This equation can be further simplified by eliminating the concentration of the undissociated weak acid chelate in the aqueous phase by substituting for $[HL]_{aqueous}$ using Equation (3.52). Upon rearrangement, we have the final working relationship for the distribution ratio of the metal chelate MLn when the chelate itself is a monoprotic weak acid HL:

$$D = \frac{K_D^{ML_n} \beta K_a^n [HL]_{organic}^n}{\left(K_D^{HL}\right)^n [H^+]^n} \alpha_M \tag{3.55}$$

Equation (3.55) is the generalized relationship. This relationship was developed earlier for a specific metal chelate, as was shown in Equation (3.18). Equation (3.55) shows that the magnitude of the distribution ratio depends on the magnitude of the four equilibrium constants. These constants depend on the particular metal chelate. The distribution ratio can be varied by changing either the concentration of chelate in the organic phase or the pH of the aqueous phase. The number of ligands that bond to the central metal ion, n, is also an important parameter. As shown earlier [Equation (3.16)], once we know D we can calculate E if we know or can measure the phase ratio for LLE. Knowing E enables us to determine the percent recovery and hence to quantitatively estimate the extraction efficiency. Because $100 \times E$ equals the percent recovery for a given metal chelate, a plot of the percent recovery of a given metal chelate vs. pH reveals a sigmoid-shaped curve. The inflection point in the curve yields the $pH_{1/2}$ for the specific metal. This is the pH at which 50% of a metal is extracted.

Figure 3.65 is a plot of the percent extracted against the solution pH for four metals, Cu(II), Sn(II), Pb(II), and Zn(II), as their respective dithizones. The exact values for each metal's $pH_{1/2}$ are as follows:

Metal Dithizone	$pH_{1/2}$
Cu(II)	1.9
Sn(II)	4.7
Pb(II)	7.4
Zn(II)	8.5

It should become clear that pH is a powerful secondary equilibrium effect that can be used to selectively extract a particular metal from a sample that may contain more than one metal.

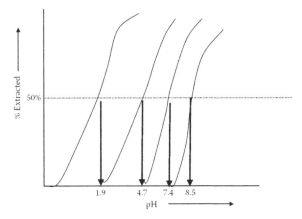

FIGURE 3.65

Day and Underwood have shown that if the logarithm is taken on both sides of Equation (3.55), we obtain a more useful form.[151] Let us first rewrite Equation (3.55) combining the various equilibrium constants as follows:

$$D = \frac{K_{ex}\left[HL\right]_{organic}^{n}}{\left[H^{+}\right]_{aqueous}^{n}}$$

(3.56)

where K_{ex} is substituted for all of the equilibrium constants in Equation (3.55). Upon taking the logarithm of both sides of Equation (3.56), we obtain

$$\log D = \log K_{ex} + n\log\left[HL\right] - n\log\left[H^{+}\right]$$

This equation can be rewritten in terms of pH as follows:

$$\log D = \log K_{ex} + n\log\left[HL\right] + n\text{pH}$$

A plot of $\log D$ vs. pH should in theory be a straight line whose slope is n and whose intercept on the $\log D$ axis (i.e., when pH = 0) is {$\log K_{ex} + n\log$ [HL]}. Figure 3.66 shows such a plot in general terms. Note that the two lines drawn correspond to two different values for $[HL]_{organic}$. These straight lines eventually curve and plateau as the pH of the aqueous phase becomes very high. In this case, the $[H^{+}]$ becomes so low that abundant L^{-} is made available, which in turn drives the formation of the metal chelate equilibrium to the right. As more aqueous metal chelate becomes available, the partitioning of the metal chelate shifts in favor of the organic metal chelate. Hence, in the absence of any hydroxide, the value for D approaches the value for the molecular partition coefficient for the metal chelate, $K_{D}^{ML_{n}}$. Metal hydroxide precipitation at a high pH competes for the free metal ion, M^{n+}, a factor not taken into account during the development of Equation (3.55).

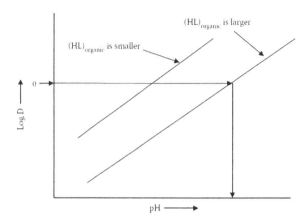

FIGURE 3.66

3.103 ARE THERE OTHER WAYS TO PRECONCENTRATE METAL IONS FROM ENVIRONMENTAL SAMPLES?

Yes, there are, and these are in essence SPE type methods. Both cation and anion exchange resins have been used to preconcentrate inorganic metal cations while removing anionic interferents. A large body of work has been related to the formation of the polychloro-anionic complex such as $FeCl_4^-$ and its isolation using anion exchange resins.

Chelating resins, styrene–divinyl benzene copolymer, containing iminodiacetrate functional groups have been successfully used to preconcentrate transition metal ions from solutions of high salt concentrations. Selective neutral metal chelates have been found to be isolated and recovered using C_{18} chemically bonded silica gel.[152] We now digress to some of the author's findings in this area.

3.104 CAN WE ISOLATE AND RECOVER A NEUTRAL METAL CHELATE FROM AN ENVIRONMENTAL SAMPLE USING BONDED SILICAS?

Neutral metal chelates should behave no differently with respect to the adsorption/partitioning of the species from water to an octadecylsiloxane bonded silica than neutral organics, as discussed earlier. This author has developed mathematical equations for the reaction of cadmium ion, Cd^{2+}, with 8-hydroxyquinoline (HOx), also referred to as oxine, to form a series of complexes. If it is assumed that only the *neutral* 1:2 complex, $CdOx_2$, will partition into the bonded sorbent, equations can be derived that relate the distribution ratio to measurable quantities. We now proceed through this derivation and start with a consideration of the secondary equilibria involved. We first need to consider a more expanded concept, enlarging upon that shown for metal chelate LLE (refer back to Figures 3.6 or 3.64). The schematic in Figure 3.67 depicts the bonded silica–aqueous interface, in which only the neutral 1:2 engages in the primary equilibrium, that of partitioning onto or into the monolayer of wetted C_{18} ligates.

The extent to which the free cadmium ion complexes with the oxinate ion to form the 1:1 cation is governed by

$$Cd^{2+} + Ox^- \underset{}{\overset{\beta_1}{\rightleftharpoons}} CdOx^+$$

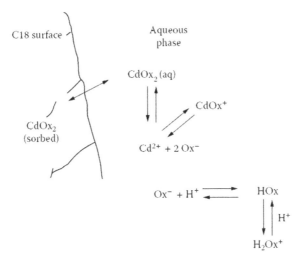

FIGURE 3.67

The extent to which the cadmium ion complexes with two oxinate anions to form the 1:2 complex is governed by

$$Cd^{2+} + 2Ox^- \xrightleftharpoons{\beta_2} CdOx_2$$

Oxine, itself being amphiprotic, can exist as either the neutral, weak acid or a protonated species. A molecular structure for oxine is as follows:

The hydroxyl group behaves as a weak acid, whereas the aromatic nitrogen can accept a proton and behave as a weak base. The degree to which oxine is protonated, H_2Ox^+, remains neutral, HOx, or dissociates to H^+, and oxinate, Ox^-, is governed by the pH of the aqueous phase.

For the protonated species, we can write an equilibrium for the acid dissociation:

$$H_2Ox^+ \xrightleftharpoons{K_{a1}} H^+ + HOx$$

The extent of dissociation of this weak acid is governed by the first acid dissociation constant, Ka_1.

For the neutral weak acid, it too dissociates in aqueous media according to:

$$HOx \xrightleftharpoons{K_{a2}} H^+ + Ox^-$$

The extent to which the neutral form of oxine dissociates is governed by the magnitude of the second acid dissociation constant, Ka_2.

If Cd^{2+} and oxine are present in the aqueous phase, these four ionic equilibria would be present to the extent determined by their formation constants. Which complex predominates—1:1 or 1:2? We need to define what we mean by a formation constant for metal complexes, and these definitions we take from well-established chemical concepts. The formation constant for the 1:1 complex is defined as

$$\beta_1 = \frac{\left[CdOx^+\right]}{\left[Cd^{2+}\right]\left[Ox^-\right]} \tag{3.57}$$

Likewise, the equilibrium expression for the formation of the 1:2 complex is given by

$$\beta_2 = \frac{\left[CdOx_2\right]}{\left[Cd^{2+}\right]\left[Ox^-\right]^2} \tag{3.58}$$

The acid dissociation constants for the amphiprotic oxine are

$$K_{a1} = \frac{\left[H^+\right][HOx]}{\left[H_2Ox^+\right]} \tag{3.59}$$

$$K_{a2} = \frac{\left[H^+\right]\left[Ox^-\right]}{[HOx]} \tag{3.60}$$

Because our discussion centers around trace Cd analysis, let us consider the fraction of all forms of this metal existing as the free, uncomplexed ion δ_0, where this fraction is defined as

$$\delta_0 = \frac{\left[Cd^{2+}\right]}{C_M} \tag{3.61}$$

Likewise, the fraction of cadmium complexed 1:1 is given by

$$\delta_1 = \frac{\left[CdOx^+\right]}{C_M} \tag{3.62}$$

Furthermore, the fraction of cadmium complexed 1:2 is given by

$$\delta_2 = \frac{\left[CdOx_2\right]}{C_M} \tag{3.63}$$

We also need to distinguish between the concentration of oxine in the aqueous phase, [HOx], and the total concentration of oxine, C_{HOx}. This total oxine concentration results from the actual addition of a given amount of the substance to water, with adjustment of a final solution volume. With all of these definitions in mind, we can now proceed to substitute and manipulate using simple algebra to help us arrive at more useful relationships than those given by just the definitions.

We start by considering the fraction of cadmium complexed as the 1:2 complex. Solving Equation (3.58) for $[CdOx_2]$ yields

$$\left[CdOx_2\right] = \beta_2\left[Cd^{2+}\right]\left[Ox\right]^2 \tag{3.64}$$

Equation (3.57) can also be solved for the concentration of the 1:1 complex:

$$\left[CdOx^+\right] = \beta_1 Cd^{2+}Ox^- \tag{3.65}$$

Substituting Equation (3.64) into Equation (3.62) gives

$$\delta_2 = \frac{\beta_2\left[Cd^{2+}\right]\left[Ox\right]^2}{\left[Cd^{2+}\right]+\left[CdOx^+\right]+\left[CdOx_2\right]}$$

$$= \frac{\beta_2\left[Ox^-\right]^2}{1+\left[CdOX^+\right]+\left[CdOx_2\right]\left[Cd^{2+}\right]}$$

Upon substituting Equations (3.64) and (3.65) into the denominator in the above expression, we get the following simplified result:

$$\delta_2 = \frac{\beta_2\left[Ox^-\right]^2}{1+\beta_1\left[Ox^-\right]+\beta_2\left[Ox\right]^2} \tag{3.66}$$

Equation (3.66) is important, but this equation is not the ultimate objective of this derivation. Equation (3.66) states that the fraction of all cadmium can be found in the aqueous phase from knowledge of only the two formation constants and the free oxinate ion concentration. It is difficult analytically to measure this free $[Ox^-]$. There is a solution to this dilemma. Let us consider how the chelate, oxine, is distributed in the aqueous phase. We start by writing a mass balance expression for the total concentration of oxine, C_{HOx}, as follows:

$$C_{HOx} = H_2Ox^+ + \left[HOx\right] + Ox^-$$

We have thus accounted for all forms that the chelate can take. Equations (3.59) and (3.60) can be solved for the undissociated forms and substituted into the mass balance expression to yield

$$\frac{\left[H^+\right]\left[HOx\right]}{K_{a1}} + \frac{\left[H^+\right]\left[Ox\right]\left[Ox^-\right]}{K_{a2}} + \left[Ox^-\right]$$

Upon further rearrangement and simplification,

$$\frac{\left[H^+\right]}{K_{a1}}\left\{\frac{\left[Ox^-\right]\left[H^+\right]}{K_{a2}}\right\} + \frac{\left[H^+\right]\left[Ox^-\right]}{K_{a2}} + \left[Ox^-\right]$$

Factoring out [Ox] and further rearranging leads to a useful expression:

$$\left[Ox^-\right]\left\{1 + \frac{\left[H^+\right]}{K_{a2}} + \frac{\left[H^+\right]^2}{K_{a1}K_{a2}}\right\}$$

As a result, we now have the total concentration of oxine in the aqueous phase as being equal to the product of the free oxinate ion concentration and a term that is entirely dependent on the magnitude of the two acid dissociation constants for oxine and the hydrogen ion concentration or pH. Let us define this term collectively as α and define α as

$$\alpha \equiv \left\{1 + \frac{\left[H^+\right]}{K_{a2}} + \frac{\left[H^+\right]^2}{K_{a1}K_{a2}}\right\}$$

The total oxine concentration is thus seen as the product of two terms, so that

$$C_{HOx} = \left[Ox^-\right]\alpha \tag{3.67}$$

Equation (3.67) can be solved for [Ox$^-$] and substituted into Equation (3.66) to yield the important outcome whereby the fraction of cadmium as the 1:2 complex can be expressed entirely in terms of known equilibrium constants and the measurable quantities C_{HOx} and α:

$$\delta_2 = \frac{\beta_2 C_{Hox}^2}{\alpha^2 + \beta_1 C_{HOx}\alpha + \beta_2 C_{Hox}^2} \tag{3.68}$$

The fraction of cadmium as the 1:1 complex, δ_1, is defined as

$$\delta_1 = \frac{\left[CdOx^+\right]}{C_M}$$

$$\delta_1 = \frac{\beta_1\left[Ox^-\right]_1}{1 + \beta_1\left[Ox^-\right] + \beta_2\left[Ox^-\right]^2}$$

δ_1 can be expressed in terms of the concentration of free oxinate and is shown as follows. As we did for the 1:2 complex, δ_1 can be expressed in terms of measurable quantities as follows:

$$\delta_1 = \frac{\beta_1 C_{HOx}}{1 + \beta_1 C_{HOx} + \beta_2 C_{HOx}^2 / \alpha} \tag{3.69}$$

Finally, the fraction of cadmium as the free ion, δ_0, is

$$\delta_0 = \frac{\left[Cd^{2+}\right]}{C_M}$$

δ_0 can be expressed as well in terms of the free oxinate concentration:

$$\delta_0 = \frac{1}{1+\beta_1\left[Ox\right]+\beta_2\left[Ox^-\right]^2}$$

δ_0 can be expressed in measurable quantities and is shown as

$$\delta_0 = \frac{1}{1+\beta_1\left\{\dfrac{C_{HOx}}{\alpha}\right\}+\beta_2\left\{\dfrac{C_{HOx}}{\alpha^2}\right\}^2} \tag{3.70}$$

3.105 HOW DOES THE FRACTION OF CADMIUM FREE OR COMPLEXED VARY?

Equations (3.68) to (3.70) show that the fraction of cadmium as the 1:2 complex, 1:1 complex, and the free cation depends entirely on the magnitude of β_1, β_2, the concentration of total oxine, C_{HOx}, and pH. This is a significant finding. Using the following values for the respective equilibrium constants, the fraction δ_0 is plotted against pH for a given value for C_{HOx}. Likewise, δ_1 and δ_2 can be plotted against pH on the same graph to yield three distribution curves:[153]

$$K_{a1} = 9.84\times10^{-6}, \quad \beta_1 = 1.59\times10^7$$

$$K_{a2} = 1.23\times10^{-10}, \quad \beta_2 = 12.56\times10^{13}$$

Figure 3.68A shows how the fraction of free cadmium ion, δ_0, the fraction of 1:1 cadmium oxinate, δ_1, and the fraction of 1:2 cadmium oxinate, δ_2, vary with a change in aqueous phase pH for a C_{HOx} of 0.5 ppm. As the pH of the aqueous phase is increased, most of the cadmium starts out as Cd^{2+} but gradually is complexed as the 1:1 and 1:2 complexes. The fraction of total cadmium existing as the 1:1 complex reaches a maximum at around 8.8, then decreases, whereas the fraction as the 1:2 complex continuously increases until all of the metal is complexed as the 1:2 complex at around pH = 10. Figure 3.68B is a similar plot, but this time for a C_{HOx} of 5.0 ppm. Note that the pH in which the 1:1 and 1:2 complexes are formed is lower than that for the lower value of C_{HOx}. Figure 3.68C is a similar plot except that C_{HOx} is 50 ppm. The pH at which the complexes are formed is further shifted to lower values. Figure 3.68D is a similar plot except that C_{HOx} is 400 ppm. At this higher total oxine concentration, the pH at which the complexes form is shifted over one pH unit lower vs. the case for a C_{HOx} of 50 ppm.[145] A knowledge of the various δ values enables a relationship to be developed for the distribution ratio.

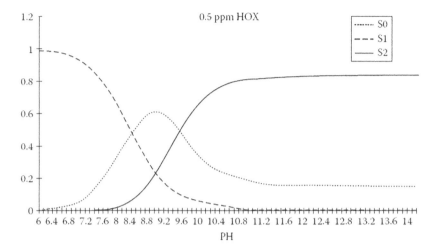

FIGURE 3.68A

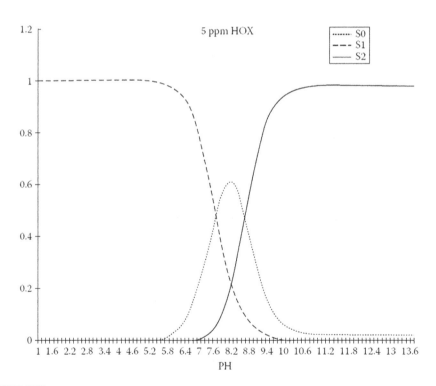

FIGURE 3.68B

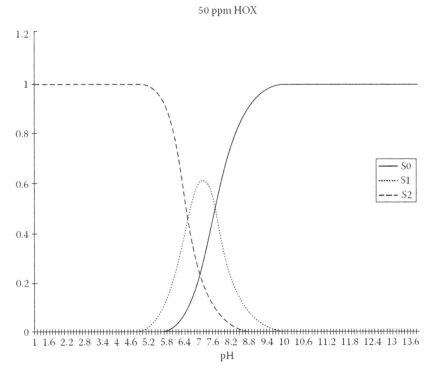

FIGURE 3.68C

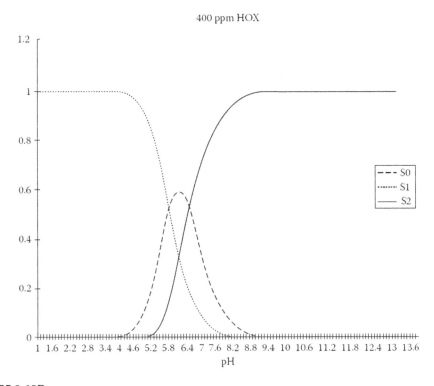

FIGURE 3.68D

3.106   CAN AN EQUATION BE DERIVED USING THESE δ VALUES TO FIND *D*?

The distribution ratio for the partitioning of the cadmium oxinate 1:2 complex from an aqueous phase to a C_{18}-bonded surface can be defined in terms of a ratio of the concentration of $CdOx_2$ that would be present on the surface to the sum of all of the forms of the metal in the aqueous phase. This is defined as

$$D_{Cdox_2} \equiv \frac{\left[CdOx_2 \right]_{surface}}{\left[Cd^{2+} \right] + \left[CdOx^+ \right] + \left[CdOx_2 \right]} \tag{3.71}$$

and the partitioning coefficient for the molecular form of the species can be defined as

$$K_D^{CdOx_2} = \frac{\left[CdOx_2 \right]_{surface}}{\left[CdOx_2 \right]_{aqueous}} \tag{3.72}$$

given

$$Cd^{2+} = \delta_0 C_M$$

$$CdOx^+ = \delta_1 C_M$$

$$CdOx_2 = \delta_2 C_M$$

Upon eliminating the bracketed concentrations in Equation (3.72), we get

$$D_{CdOx_2} = \frac{K_D^{CdOx_2} \delta_2 C_M}{\delta_0 C_M + \delta_1 C_M + \delta_2 C_M}$$

Upon simplifying and rearranging, we get an important result:

$$D^{CdOx_2} = K_D^{CdOx_2} \left\{ \frac{\delta_2}{\delta_0 + \delta_1 + \delta_2} \right\} \tag{3.73}$$

Equation (3.73) is an important outcome of our derivation efforts. Let us proceed to interpret this equation as it relates to the cadmium–oxine system. Equation (3.73) states that the distribution ratio for the 1:2 complex between an aqueous phase and a chemically bonded silica such as a C_{18} RP-SPE sorbent depends on the magnitude of the molecular partition coefficient and the degree to which cadmium is found as the 1:2 complex. Given that Equations (3.68) to (3.70) enable one to calculate δ_0, δ_1, and δ_2, respectively, and that we have already established that these fractions depend only on C_{Hox} and the pH, we need only to know the pH of the aqueous phase, the molecular partition coefficient from independent studies, and the total oxine concentration in the aqueous phase to calculate *D*. These parameters were incorporated into an Excel spreadsheet to calculate *D* for various values of pH at different C_{HOx} values. Figure 3.69A is a plot of log *D* against pH for C_{HOx} = 0.5 ppm; Figure 3.69B is similar, but for C_{HOx} = 5.0 ppm, and in a similar manner for Figure 3.69C, for C_{HOx} = 50 ppm. Referring to Figure 3.69A, we observe that the curve is linear between a pH of around 6.4 and approximately 8.8, and then the curve levels off, hence becoming independent of pH from about a pH of 9.2 all the way to 14. Figure 3.69B and Figure 3.69C look similar except where on the pH scale the curve crosses. This crossing corresponds to a log D = 0.

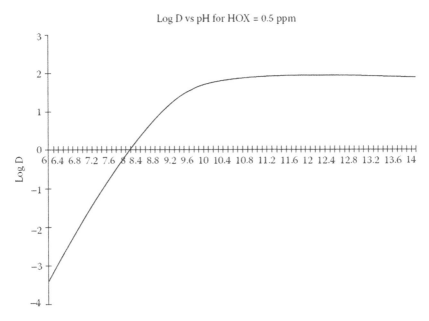

FIGURE 3.69A

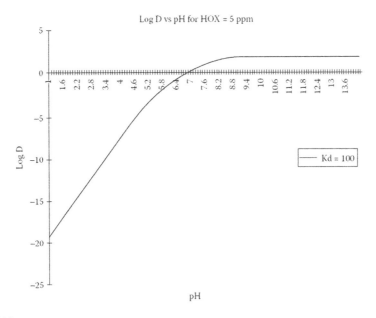

FIGURE 3.69B

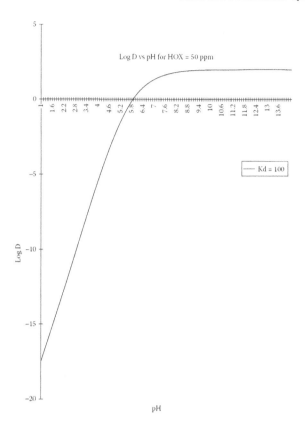

FIGURE 3.69C

Let us examine this a bit further. Taking the logarithm of both sides of Equation (3.73) gives the following:

$$\log D_{CdOx_2} = \log K_D^{CdOx_2} + \log \frac{\delta_2}{\delta_0 + \delta_1 + \delta_2}$$

Where the curve crosses the pH axis, $\log D = 0$ so that

$$0 = \log K_D^{CdOx_2} + \log \frac{\delta_2}{1}$$

Hence, at the pH where the curve crosses,

$$-\log K_D^{CdOx_2} = \log \delta_2$$

We have assumed a value for $K_D = 100$ so that $\delta_2 = 0.01$. The curve crosses the pH axis at which the fraction of the 1:2 complex formed is 0.01 and the percent of the 1:2 complex is 1%. The pH at which the δ_2 curve just begins to appear, for example, in Figure 3.68A, is about 7.8. This is the same pH value at which the curve in Figure 3.69A crosses the pH axis!

TABLE 3.33
Measured Absorbance for Recovered Cadmium as the Oxine Complex at an Acidic and an Alkaline pH

SPE Cartridge No.	Weight (mg)	A at pH 6.2	A at pH 8.8
1	199	0.003	0.217
2	202	0.035	0.237
3	272	0.021	0.272
4	310	0.032	0.236
5	415	0.007	0.301
6	415	0.006	—
7	508	0.009	0.290
8	517	0.006	0.290
9	648	0.006	0.348
10	785	0.004	0.215
Mean for n replicates		0.01	0.27
Confidence interval at 95% probability		0.009 ($n = 10$)	

3.107   HOW GOOD IS THE PREDICTION OF EQUATION (3.73)?

We now discuss the results of studies that show the strong dependence of D on the pH of the aqueous phase. However, we first introduce the experimental procedure used to generate the data. To about 50 mL of distilled deionized water (DDI) is added an aliquot of a 1% solution of oxine dissolved in methanol. An aliquot of a cadmium salt solution is then added. The pH is adjusted to the desired value with 1% ammonia. This spiked sample is passed through a conditioned C_{18}-bonded SPE sorbent, and a brightly colored band is seen near the top of the column. The SPE cartridge is subsequently eluted with 5 mL MeOH directly into a 10 mL volumetric flask used as a receiver. The contents of the volumetric flask are brought to the calibration mark with 1% nitric acid in DDI. The source of nitric acid (HNO_3) must be a *high-purity acid* deemed suitable for atomic absorption spectrophotometric analysis. The contents in the volumetric flask were aspirated into a Model 303 Flame AA (PerkinElmer). The metal, cadmium, is quantitated based on an external standard mode of instrument calibration in 50:50 MeOH:1% HNO_3. Ten replicate SPEs were conducted at pH values 6.2 and 8.8. The absorbance (A) was measured and the data is shown in Table 3.33.[153] It is evident that at a pH of 6.2, little to no cadmium was recovered, whereas at pH 8.8, almost 30 times as much cadmium was recovered! Two other studies on the isolation and recovery of Cd^{2+} by chelating the ion with oxine and isolating the complex on C_{18} and C_8 chemically bonded silicas gave similar results.

3.108   HOW ARE OTHER TOXIC METALS/METALLOIDS, CYANIDE, AND PHENOLS (OFTEN FOUND IN HAZARDOUS WASTE SAMPLES) PREPARED FOR QUANTITATIVE ANALYSIS?

By implementing well-established laboratory procedures. We discuss how to prepare wastewater samples in order to quantitate the following toxins: metals, metalloids, cyanide and phenols while accomplishing the goals of *enviro-chemical* TEQA. First, we step back and discuss how samples containing the toxic metal mercury (Hg) and the toxic metalloids selenium (Se) and Arsenic (As) were prepared much earlier when FlAA was the sole determinative technique available to quantitate these chemical elements. The development of GFAA, ICP-AES, and ICP-MS have pretty much eliminated the need for these sample prep techniques in order to quantitate Hg, Se and As. In contrast to the alkali, alkaline earth and transition metals, a direct injection of an *enviro-chemical* sample into a FlAA to quantitate Hg, Se and/or As does not directly lead to a quantitative result! In general, the presence of either Hg or cyanide renders a sample hazardous due to the acute toxicity of both species. Mercury exists either in elemental form, as the dimer mercurous ion Hg_2^{2+}, in which the oxidation state of the element is +1, or as the divalent mercuric ion Hg^{2+}. Its elemental form is the familiar silvery liquid metal, Hg^0, originally obtained by roasting cinnabar, HgS.[146] The cyanide ion (CN^-) is a moderately strong base derived from the weak acid hydrocyanic acid, HCN. HCN is a gas at room temperature and highly toxic. The fear is that hazardous wastes that contain CN^-, if acidified, could release HCN. CN^-, also forms complexes to many metal ions. We introduce the ingenious approaches developed much earlier when only the determinative technique FlAA was available for the general education of the reader.

3.109   HOW DO YOU PREPARE A SAMPLE FOR Hg DETERMINATION?

This author has always been fascinated by the oxidation and reduction reactions that Hg undergoes. Because mercury might be present in environmental samples in any of its three oxidation states, an initial and vigorous oxidation would convert all forms of the element to its highest oxidation state, i.e. to the mercuric ion, Hg^{2+}. In addition, organomercury compounds whereby Hg is in a further reduced state, such as dimethyl mercury, $(CH_3)_2Hg$, also would be oxidized to Hg^{2+}. The common oxidizing agent used is potassium permanganate, $KMnO_4$. This is diagrammed as follows:

$$(CH_3)_2Hg$$
$$Hg_2^{2+} \xrightarrow{\quad [O] \quad} Hg_{(aq)}^{2+}$$
$$Hg^0$$

Both EPA Method 7470, applicable to liquid waste, and Method 7471A, applicable to solid waste, use $KMnO_4$. The purple color of the permanganate ion is discharged as this oxidizing agent gets reduced. An excess of MnO_4^- that persists provides visual evidence that oxidation is complete. Further oxidation is afforded by adding potassium persulfate, $K_2S_2O_8$. Excess MnO_4^- is reduced using hydroxylamine according to:

$$MnO_4^- \xrightarrow{\quad [NH_2OH] \quad} Mn^{2+}$$

Tin(II) sulfate is used to reduce all Hg^{2+} to elemental Hg:

$$Hg_2^{2=} \xrightarrow{\quad Sn^{2+} \quad} Hg^0$$

Elemental Hg^0 is subsequently purged and swept out of the sample. Figure 3.70 illustrates what accessories are needed to prepare a sample for direct introduction of elemental mercury

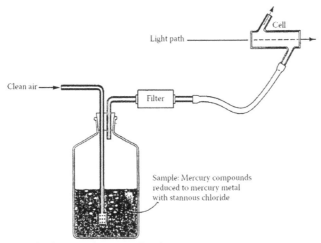

Flameless atomic absorption analyzer for mercury

FIGURE 3.70

vapor into a flow-through atomic absorption cell. This modification to the atomic absorption spectrophotometer is called a *cold vapor atomic absorption method*.

A typical procedure to determine the concentration of Hg for a drinking water sample is now summarized. This is a routine procedure used in environmental testing labs that perform trace Hg determinations. It reflects EPA Method 245.1 for the determination of Hg in aqueous environmental samples such as surface, saline, and wastewater. A 100 mL sample or equivalent spiked sample is placed into the reaction vessel of the *cold vapor generator*. A 300 mL biological oxygen demand (BOD) bottle is frequently used for this purpose. To this vessel are added 5 mL of 5% $KMnO_4$ and 50 mL of DDI. The contents of the container are mixed well and allowed to stand for 15 min. If the purple color is discharged during this time, aliquots of 5% $KMnO_4$ can be added until the purple coloration persists for 15 min. After this, 8 mL of 5% $K_2S_2O_8$ is added to the reaction vessel. The vessel is capped and heated in a 95°C heater block or water bath for 2 hours. The contents of the reaction vessel are then cooled and 6 mL of a 12% NaCl/12% hydroxylamine hydrogen sulfate (or alternatively hydroxylamine hydrochloride) solution is added. The reaction vessel is immediately connected to the cold vapor generator, and the headspace is purged for 30 sec to remove chlorine and other interferences. Next, 5 mL of a 10% tin(II) chloride solution in 0.5 N sulfuric acid is injected into the reaction vessel. The Hg vapor is swept into the absorption cell of the atomic absorption (AA) spectrometer, where the absorbance at 253.7 nm is measured. Labs generally dedicate one of the AA spectrometers to trace Hg determinations.

Cold vapor generators are but one of the four major atomization devices used in AA, the other three being flame (FlAA), electrothermal (GFAA), and hydride generators. The cold vapor generator shown in Figure 3.70 could also be used as a hydride generator. Hydride generators convert the following elements to their gaseous hydrides: Sb, As, Bi, Se, Te, and Sn. A comparison of method detection limits (MDLs) reveals the real advantage of hydride generators. For example, an MDL for direct aspiration of selenium into a flame with measurement as atomic Se gives 100 ppm, whereas measurement as the hydride gives an MDL of 0.03 ppm (p. 12).[141]

3.110 COULD HS-SPME COMBINED WITH AN ELEMENT-SPECIFIC GC BE USED TO SPECIATE HG?

Yes, indeed. The combination of solid-phase microextraction sampling of the headspace (SPME-HS) with capillary gas chromatography and atomic emission detection (C-GC-AED) enables a speciation of Hg to be achieved. Principles that underlie the GC-AED determinative technique will be introduced in Chapter 4. Organo-Hg compounds such as methyl-Hg and dimethyl-Hg are much more toxic than inorganic Hg. Carro and coworkers have recently demonstrated that methyl mercury, ethyl mercury, and phenyl mercury can be isolated, recovered, derivatized, then separated and detected down to a concentration of 100 ppt from seawater using SPME-HS–C-GC-AED.[155] A recent review published earlier discusses speciation of mercury, tin, and lead using C-GC-AED.[156]

3.111 CAN ARSENIC BE SPECIATED?

Yes, indeed. The combination of reversed-phase high-performance liquid chromatography (RP-HPLC) with inductively coupled plasma-mass spectrometry (ICP-MS) has enabled a speciation of As containing species both organic and inorganic.[157] Caruso and coworkers have reviewed the elemental speciation of As using ICP-MS.[158,159]

3.112 HOW IS A WASTEWATER SAMPLE PREPARED IN ORDER TO QUANTITATE TRACE CYANIDE? WHAT ABOUT SULFIDES? WHAT ABOUT PHENOLS?

The tendency of free or bound CN^- to become the toxic gas HCN and emanate from a hazardous wastewater sample as was discussed earlier is exploited in the *sample prep* approach

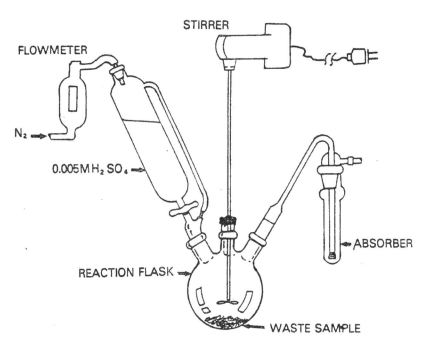

FIGURE 3.71

to quantitatively determining CN^-. A wastewater or hazardous solid waste sample is placed in (and preferably) a three-hole glass apparatus. Two-hole round-bottom flasks can also be used with a somewhat different configuration that involves refluxing. The scrubber contains a dilute NaOH solution that converts HCN to the cyanide ion, CN^-. The contents of the scrubber now consist of Na^+ CN^- in excess hydroxide (OH^-). This completes the sample prep. The sample is now subject to whatever determinative technique is used to quantitate CN^-. Figure 3.71 depicts a schematic for a three-holed round-bottom flask designed for this *sample prep* application. This apparatus can be used to isolate and recover toxic *cyanide, sulfide* and various organic *phenols* from wastewater samples. These three culprits are often found in hazardous waste samples. Referring again to Figure 3.71, note that the left hole on the flask enables a steady drip of acid under a controlled N_2 purge into the round-bottom flask while the center hole enables mechanical stirring of the hazardous waste sample. The right hole allows the HCN, H_2S, or the various priority pollutant organic phenols, to be purged out and sparged directly into the alkaline scrubber while excess N_2 is safely vented. *This entire apparatus should be operated in a fume hood.* Traditionally, to quantitate cyanide, the contents of the alkaline scrubber is chemically reacted with chloramineT at a pH < 8.0 to form the toxic cyanogen chloride, CNCl. CNCl forms a red-blue dye upon addition of a pyridine-barbituric acid reagent. The absorbance is read at 578 nm on a uv-vis spectrophotometer.[160] Molecular structures for these chemical reagents are shown below:

chloramineT   cyanogen chloride   pyridine   barbituric acid

Interested in how wastewater samples are prepared to quantitate CN^-? Please "google" EPA Method 335.4 and study! EPA Method 9010 Revision 0, 2010 introduces a microdiffusion device that consists of concentric circles whereby NaOH solution is placed in the center compartment while the wastewater sample is placed in the outer compartment. Sample pH is adjusted to 8 and CN^- simply diffuses into the central compartment! A sort of miniaturized scrubber! In a similar fashion, sulfide from the hazardous waste sample trapped in the scrubber in the form of Na_2S can be quantitated using the appropriate methods. Likewise, various priority pollutant organic phenols can be trapped by the scrubber. The contents of the scrubber for phenols consists of a dilute solution containing sodium phenolate. This solution is subject to whatever determinative technique is applicable to measure trace total phenols. Please peruse the numerous molecular structures shown in this book and find the various priority pollutant phenols.

3.113   WHAT IS CHEMICAL DERIVATIZATION AND WHY IS IT <u>STILL</u> IMPORTANT TO TEQA?

Many priority pollutants (enviro-chemical) or persistent organic pollutants (enviro-health) can be directly injected into a gas chromatograph owing to their physico-chemical properties of being relatively nonpolar, semivolatile, and thermally stable in the hot-injection port of the GC. However, those organic compounds with heteroatom functional groups are polar, nonvolatile, and sometimes thermally labile. Flip the pages ahead to Chapter 4 and find Figure 4.4 which shows the degree of analyte volatility plotted against the degree of analyte polarity. Polar, non-volatile analytes are converted to less polar ones, which become semivolatile derivatives. These derivatized organic compounds fall into the realm of GC and are said to be *amenable* to analysis by GC. Derivatives can also be prepared from analytes that yield a more sensitive means of detection for the GC determinative technique (to be introduced in Chapter 4). Advances in HPLC such as LC-MS-MS using the electrospray interface have eliminated much of the need for chemical derivatization. This author believes the topic of chemical derivatization is *still* important for the general education of the reader.

This author's first encounter with the need to make a chemical derivative, as discussed earlier, involved the *three chlorophenoxy acid herbicides* (CPHs)—2, 4-D, 2, 4, 5-T, and 2, 4, 5-TP (Silvex)—in drinking water. EPA Methods 515.1 (drinking water) and 8150 (solid waste) require that CPHs and other organic acids be converted to methyl esters. Earlier, boron trifluoride–methanol (BF_3-MeOH) was used to convert carboxylic acids to their corresponding methyl esters (with mixed results from this author's experience), while more recent methods favor the more vigorous *in situ* generation of *diazomethane* gas whose molecular structure is shown below:

EPA Method 8151A Revision 1 December 1996 also considers that pentafluorobenzyl (PFB) esters of CPHs and other "chlorinated acids of environmental interest" can be made and chromatographed using a GC-ECD. The PFB moiety in the derivatized ester of the CPH makes the ester extremely sensitive to detection via GC-ECD. The molecular structure for the chemical derivatizing reagent pentafluorobenzyl bromide is shown below:

Let us take a broad view of chemical derivatization in analytical chemistry. The flowchart shown in Figure 3.72 summarizes how most commercially available derivatization reagents are categorized. *Silylation* is the conversion of active hydrogen in a functional group to a trimethyl silyl (TMS) derivative. This was the first means to chemically convert carboxylic acids, alcohols, thiols, and primary and secondary amines to TMS esters. TMS esters are most appropriate where GC-MS is the principal determinative technique. *Acylation* is the conversion of active hydrogen, as is found in alcohols, phenols, thiols, and amines, into esters, thioesters, and amides by reacting organic compounds that contain these functional groups with fluorinated acid anhydrides. Heptafluorobutyrylimidazole and *N*-methyl-*N*-bis (trifluoroacetamide) are particularly effective in converting *primary amines* to *fluorinated amides*. Introduction of a perfluoroacyl moiety in the derivative leads to a significant increase in analyte sensitivity when using GC-ECD as the determinative technique. *Alkylation* is the conversion of active hydrogen by an alkyl or benzyl group to an ester or ether, depending upon whether the functional group in the organic compound is a carboxylic acid or alcohol or phenol, respectively. Diazomethane via *in situ* generation, BF_3-MeOH, dimethyl formamide-dialkyl acetals, and pentafluorobenzyl bromide are commonly used derivatizing reagents. *Enantiomeric purity analysis reagents* form diastereomers when reacted with optically active analytes. Diastereomers are easily separated by GC. Commercially available reagents include (−) methyl chloroformate that reacts with

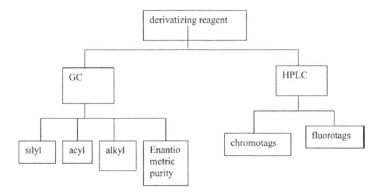

FIGURE 3.72

enantio-enriched alcohols and N-TFA-L-prolyl chloride that couples with amines to form diastereomers. *Chromotags* are derivatizing reagents that add an ultraviolet-absorbing chromophore to an aliphatic carboxylic acid that converts the aliphatic acid to a UV-absorbing derivative to enhance sensitivity in HPLC-UV. *Fluorotags* convert a minimally fluorescent analyte to a highly fluorescent derivative, and hence enhance sensitivity in HPLC-FL. The reaction of aliphatic carboxylic acids with *p*-bromophenacyl bromide in the presence of 18-crown-6 under alkaline conditions to form a strong ultraviolet-absorbing ester, and the conversion of aliphatic carboxylic acids to highly fluorescent 4-bromomethyl-7-methoxycoumarin represent common uses of chromotags and fluorotags.[161, 162, 163]

Analytes are usually isolated and recovered via any of the extraction and cleanup techniques described in this chapter. The extractant or eluent is evaporated to either dryness or close to dryness in order to concentrate the analyte. The derivatizing reagent, catalysts, acids, or bases, and any other reagents are then added. Heat can be applied if necessary to increase the reaction rate. The derivatized analyte is extracted from the product mix and further cleaned up, excess derivatizing reagent is removed if possible, and then the extract is injected into the appropriate chromatographic determinative technique. It is important that the excess derivatizing reagent be chromatographically separated from the derivative(s) to enable quantitative analysis. Let us digress a bit to some specific examples of the use of chemical derivatization to accomplish the goals of TEQA.

3.114 HOW DO YOU MAKE A PFB DERIVATIVE OF SOME BUTYRIC ACIDS?

This author once attempted to prepare PFB esters of *n*-butyric, *i*-butyric, and 2-methyl butyric acids whose molecular structures are shown below.[165,166]

n-butyric acid   i-butyric acid   2-methylbutyric acid

Here is what you need to do:

3.114.1 To Prepare the Reagents

30% potassium carbonate: Dissolve 7.5 g of K_2CO_3 (anhydrous) in ~20 mL of distilled deionized water (DDI). Transfer to a 25 mL volumetric flask and adjust to mark with DDI. Transfer to storage vial and label as "30% K_2CO_3(aq)."

1% PFBB: Dissolve 0.25 g of PFBB in ~20 mL of acetone. Transfer to a 25 mL volumetric flask and adjust to mark with acetone. Transfer contents to storage vial and label as "1% PFBB (acetone)."

1000 ppm each carboxylic acid: Weigh ~0.010 g of each acid into a 10 mL volumetric flask and already half filled with DDI. Label as "1000 ppm each acid."

3.114.2 To Synthesize and Extract the PFB Ester

Into a 22 mL headspace vial with crimp top, place 200 μL of the 100 ppm acid solution, 200 μL of 1% PFBB, 50 μL of 30% K_2CO_3, and 4 μL acetone. Shake vigorously and allow the contents of the vial to stand at room temperature for 3 hours. Add enough DDI to reach the neck of the headspace vial. Add 2 mL of *pesticide-grade iso-octane*. Transfer 1.0 mL of extract to a 2 mL GC vial and inject 1 μL of extract into a gas chromatograph incorporating an electron-capture detector (GC-ECD). For a 30 m × 0.32 mm DB-5 (formerly J&W Scientific) capillary column, the following temperature program adequately separates the PFB esters of C_3, C_4, and C_5 carboxylic acids. Start at 100°C and hold for 3 min, then raise the temperature at a rate of 8°C/min to 150°C, and then hold for 0.5 min. Under these conditions, we found that propionic acid elutes at 3.099 min, *n*-butyric at 3.65 min, and valeric at 6.09 min (principles of programmed-temperature GC will be considered in Chapter 4). Figure 3.73 shows two chromatograms in a stacked arrangement for the derivatization of *i*-butyric, *n*-butyric, and 2-methyl butyric as their PFB esters. A blank (lower chromatogram) and a spiked blank (upper chromatogram) reveal the presence of these PFB esters. Note that a 40 ppb concentration level for these PFB esters can easily be detected and therefore quantitated. After these butyric acids are converted to their respective PFB butyrates, not only are polar acids converted to nonpolar esters, but also significant increases in analyte sensitivity (using a GC-ECD as stated earlier) are realized. Let us consider a second illustration of chemical derivatization, this time for HPLC.

3.115 HOW DO YOU PREPARE A *P*-BROMOPHENACYL ESTER OF *N*-BUTYRIC ACID AS A *CHROMOTAG* AND CONDUCT A QUANTITATIVE ANALYSIS?

The following laboratory procedure answers this question:

3.115.1 Preparation of Mixed Alkylating (Alk Rgt) Reagent

Weigh 0.47 g of *p-bromophenacyl bromide* (2,4-dibromoacetophenone) and 0.045 g of 18-crown-6 and dissolve in enough acetone (<10 mL), then adjust to a final volume of 10 mL.

Molecular structures for the derivatizing reagent and for the crown ether (acting as a catalyst) are shown below:

p-bromophenacyl bromide or 2,4-dibromoacetophenone   18-crown

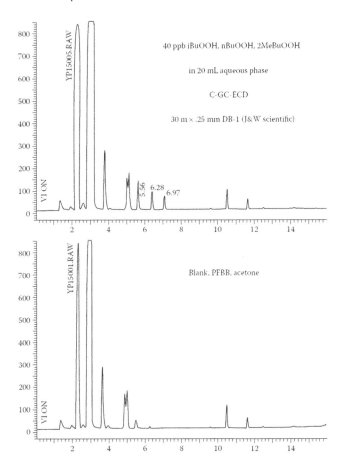

FIGURE 3.73

3.115.2 Preparation of Fatty Acid Stock Reference Standard

Prepare an approximately 10,000 ppm stock solution of *n-butyric acid* (*n*-BuOOH) in water by weighing out approximately 0.1 g of the acid and dissolving in a beaker filled with approximately 5 to 8 mL of water. Dissolve, then transfer to a 10 mL volumetric flask and adjust to the calibration mark with DDI.

3.115.3 Preparation of 1 *M* Aqueous KHCO₃

Prepare a 1 *M* solution containing potassium bicarbonate (KHCO$_3$) dissolved in DDI by dissolving approximately 10 g of KHCO$_3$ in enough to reach 100 mL. Transfer to storage bottle.

3.115.4 To Prepare the Potassium Salt of Butyric Acid (*n*-BuOOH)

To 5 mL of the stock fatty acid reference standard, in a 50 mL beaker, add enough 1 *M* KOH solution to adjust the pH to 7 to 8. This is best accomplished by filling a buret with the 1 *M* KHCO$_3$ solution and titrating to the desired pH. Adjust the acid solution to a precise final volume and record. Transfer to a storage vial and label with a new concentration for the fatty acid.

3.115.5 Preparation of Working Calibration Standards

Create a series of working calibration solutions with the same final volume according to the following table. Use a 22 mL headspace vial with crimp top:

Standard No.	Alk Rgt (mL)	Acetone (mL)	RCOOK (μL)	V(total) Adjusted with DDI
0 (blank)	1	3	0	5
1	1	3	10	5
2	1	3	50	5
3	1	3	100	5
4	1	3	500	5

3.115.6 Derivatization

Place the 22 mL headspace vial or equivalent into a heater block set at 80°C and heat for 30 min. Alternatively, the contents of the vial may be evaporated to dryness and the residue reconstituted in the HPLC-compatible solvent.

Determination of the ester via HPLC-UV: Inject 5 μL of the content of the 22 mL headspace vial into an HPLC with either a fixed-wavelength uv absorbance detector or a photo-diode array detector. Use a gradient elution reversed-phase approach as previously developed. Set the fixed wavelength detector at $\lambda = 254$ nm. If you have a photo-diode array (PDA) detector interfaced to an HPLC, try scanning the uv absorption spectrum and then choose the λ that gives the maximum absorbance.

For a third application of chemical derivatization, an important tool in the arsenal of *sample prep* techniques for TEQA, we consider the use of a fluorescent derivatizing reagent to convert a carboxylic acid to a highly fluorescent derivative.

3.116 WHAT IS THE SAMPLE PREP APPROACH TO PLACING A *FLUOROTAG* ON A CARBOXYLIC ACID?

Figure 3.74 is a flowchart that outlines the sample prep approach for isolating and recovering *perfluorocarboxylic acids* from liver homogenate, followed by the preparation of a highly fluorescent derivative using 3-bromoacetyl-7-methoxycoumarin.[166] The fact that methoxycoumarins can be used as fluorotags for carboxylic acids has been known for some time.[167,168] In this case, shown in Figure 3.74, the perfluorocarboxylate anion is *ion pair extracted* into 1:1 ethyl acetate:hexane using a tetrabutyl ammonium cation (TBA$^+$HSO$_4^-$) under alkaline conditions following bath sonication. The extract is evaporated to just dryness and acetonitrile (a polar solvent) is added, followed by the 3-bromoacetyl-7-methoxycoumarin (BrAMC) reagent. The derivatized perfluorocarboxylic acid is subsequently injected into an RP-HPLC-FL, as noted in Figure 3.74. A recent publication used the same chemical derivatizing reagent but isolated and recovered the C_2–C_{12} perfluorocarboxylics using C_{18} chemically bonded silica RP-SPE.[169]

HPLC-FL as a determinative technique will be introduced in Chapter 4. There are other derivatizing reagents that do not quite fit into the categories described earlier. We will encounter other derivatization concepts as we proceed through Chapter 4. A good starting point to better understand the scope and importance of *chemical derivatization of organic compounds* can be found elsewhere.[170]

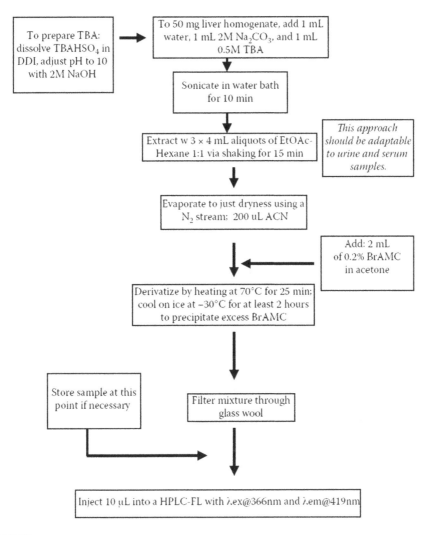

FIGURE 3.74

3.117   WHAT ABOUT SAMPLE PREP TECHNIQUES APPLICABLE TO AMBIENT AIR ANALYSIS?

They're important even though this book gives a much greater emphasis to condensed states of environmental sample matrices, i.e. liquids and solids! The reader should refer to Table 1.12 in Chapter 1 for a brief update on the numerous EPA air monitoring analytical methods available as of this writing. This author back in the 1990s once installed a porous layer open tubular (PLOT) column into a gas chromatograph with a thermal conductivity detector. He created a temperature program for the oven. He then injected air into the instrument using a gas-tight plunger-in-barrel syringe. Eventually two peaks emerged in the developing *gas chromatogram*! After visually adding the peak areas for both peaks, it appeared that the first peak included ~80% of the total peak area while the second peak included ~20% of the total peak area. In the earth's atmosphere, elemental nitrogen (N_2) is present at ~789,900 ppm while elemental oxygen (O_2) is present at ~209,500 ppm.[171]

The author once was asked to sample, using a gas-tight headspace glass syringe, the air ~18 inches above a laboratory benchtop. The objective was to measure the concentration at that height of methyl methacrylate monomer in the breathing space above the polymerization reaction (shown below) that converts the monomer to a polymer. This need for air sampling arose in an attempt to assess the exposure of this toxic volatile chemical. The reaction is depicted below:

methyl methacrylate monomer   polymethylmethacrylate

The air sample that was captured by the gas-tight headspace glass syringe was directly injected into a gas chromatograph equipped with a flame ionization detector, refer to Figure 3.75 for schematic drawings of air sampling devices. We return to the topic of air pollution by introducing a novel determinative technique near to the end of Chapter 4.

3.118   WHAT COMMMON SAMPLE PREP DEVICES ARE USED TO OBTAIN A SUITABLE AIR SAMPLE?

A quick perusal of Table 1.12 in Chapter 1 provides the answer. Three sample prep techniques for air sampling are applicable to most of the methods shown in Table 1.12:

- Adsorbent tubes that contain finely divided particulates with relatively large surface areas that are capable of adsorbing volatile organic pollutants from air that is forced through the tubes. The adsorbent tube is thermally desorbed into a gas chromatograph in the reverse direction to that used in taking the sample
- Glass impingers that are connected to a source of partial vacuum and contain a reagent dissolved in a solvent that instantly reacts with specific air pollutants that form a chemical derivative that can subsequently be measured by gas or liquid chromatography or directly via a spectrophotometer
- Partially evacuated Summa canisters that allow toxic air pollutants (VOCs) to be easily transferred from the atmosphere to the canister by merely opening the valve. The canister is subsequently taken to a laboratory where the VOCs can be directly introduced into a gas chromatograph

3.119   WHAT CAN WE CONCLUDE ABOUT SAMPLE PREP?

An attempt was made to introduce most of the recently developed sample prep techniques as well as provide for the underlying principles of established techniques. The link between true *enviro-chemical* quantitative analysis and true *enviro-health* quantitative analysis was attempted from the sample prep perspective. Hopefully, the reader comes away with a deeper appreciation of how samples and specimens are prepared so that these materials can be more properly introduced to the various determinative techniques introduced in the next chapter.

> One of the unique features of solvent extraction, particularly for metal ions, is the large variation in distribution ratios and separation factors made possible by controlling the chemical parameters of the system.
>
> **—Henry Freiser**

Air In →

Air Out →

Sketch of an adsorbent tube filled with either Tenax®, carbon molecular sieve, Thermosorb®, PUF, or XAD-2® for sampling VOC air pollutants

Air In

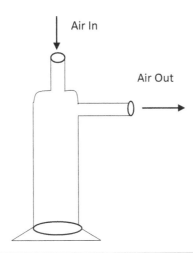

Air Out →

Sketch of a glass impinger filled with a liquid solution of a chemical derivatizing reagent used for example, to sample aldehydes and ketones from air

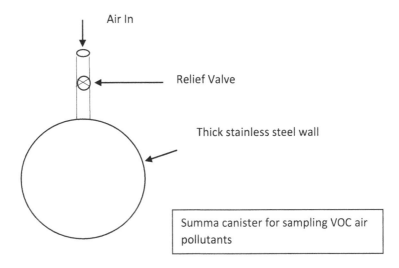

Air In

Relief Valve ←

Thick stainless steel wall ←

Summa canister for sampling VOC air pollutants

FIGURE 3.75

REFERENCES

1. Kolthoff I., E. Sandell, E. Meehan, S. Bruckenstein. *Quantitative Chemical Analysis,* 4th ed. New York: Macmillan, 1969, p. 355.
2. Swift T. *Principles of Chemistry: A Models Approach.* Boston: D.C. Heath, 1975, pp. 432–433.
3. Hoffman C. (Ed.). *Solvent Guide.* Muskegon, MI: Allied Signal Inc., Burdick and Jackson, 1997.
4. Harris D.C. *Quantitative Chemical Analysis,* 3rd ed. San Francisco: Freeman, 1991, chapter 6.
5. Laitinen H. *Chemical Analysis.* New York: McGraw-Hill, 1960, chapter 2.
6. Karger B, L Snyder, C Horvath. *An Introduction to Separation Science.* New York: Wiley Interscience, 1973, chapter 2.
7. Skoog D, D West. *Fundamentals of Analytical Chemistry,* 3rd ed. New York: Holt, Rinehart, Winston, 1976, chapter 5.
8. Loconto P.R., Y. Pan, D.P. Kamdem. *J Chromatog Sci* 36: 299–305, 1998.
9. Key B., R. Howell, C. Criddle. *Environ Sci Technol* 31: 2445–2454, 1997.
10. Wujcik C., T. Cahill, J. Seiber. *Anal Chem* 70: 4074–4080, 1998.
11. Stumm W., J. Morgan. *Aquatic Chemistry.* New York: Wiley, 1996.
12. Pankow J. *Aquatic Chemistry Concepts.* Lewis Publishers, Chelsea, MI, 1991.
13. Evangelow V. *Environmental Soil and Water Chemistry.* New York: Wiley, 1998.
14. Moral F., J. Hering. *Principles and Applications of Aquatic Chemistry.* New York: Wiley, 1993.
15. Snoeyink V., D. Jenkins. *Water Chemistry.* New York: Wiley, 1980.
16. Loconto P.R. *J Liq Chromatogr* 14: 1297–1314, 1991.
17. Day R., A. Underwood. *Quantitative Analysis,* 5th ed. Englewood Cliffs, NJ: Prentice Hall, 1986, pp. 507–510.
18. Sekine T., Y. Hasegawa. *Solvent Extraction Chemistry.* New York: Marcel Dekker, 1977, pp. 55–57.
19. Sandell E., J. Onishi. *Photometric Determination of Traces of Metals.* New York: Wiley, 1978.
20. Schmitt T. *Analysis of Surfactants,* 2nd ed. New York: Marcel Dekker, 2001.
21. *Standard Methods for the Examination of Water and Wastewater,* 19th ed. 1995, pp. 5–42–5–44.
22. Jeannot M., F. Cantwell. *Anal Chem* 68: 2236–2240, 1996.
23. Ma M., F. Cantwell. *Anal Chem* 70: 3912–3919, 1998.
24. Ma M., F. Cantwell. *Anal Chem* 71: 388–399, 1999.
25. *The Complete Laboratory Glassware Catalog.* Vineland, NJ: Kimble/Kontes, 2002, p. 276.
26. Arment S. *LC-GC Curr Trends Dev Sample Prep* 17: S38–S42, 1999.
27. Randall E. *J AOAC* 57: 1165–1198, 1974.
28. Method 3541, Automated Soxhlet Extraction. *Test Methods for Evaluating Solid Wastes, Physical/Chemical Methods,* SW-846, 3rd ed., Revision 0, September 1994.
29. Loconto P.R. *J Chromatogr A* 774: 223–227, 1997.
30. Loconto P.R. *LC-GC* 11:1062–1068, 2002.
31. Ander P., A. Sonessa. *Principles of Chemistry.* New York: Macmillan, 1965, pp. 698–703.
32. Cole R., J. Coles. *Physical Principles of Chemistry.* San Francisco: Freeman, 1964, pp. 644–649.
33. LeBlanc G. *LC-GC: Curr Trends Dev Sample Prep* 17: S30–S37, 1999.
34. Lesnik B. *Environ Test Anal* 4: 20, 1998.
35. Onuska F., K. Terry. *Chromatographia* 36: 191–194, 1993.
36. Lopez-Avila V. et al. *Anal Chem* 67: 2096–2102, 1995.
37. Method 3546, Microwave Extraction. *Test Methods for Evaluating Solid Wastes, Physical/Chemical Methods,* SW-846, 3rd ed., Update III, Revision 0, November 2000.
38. This description of the EDGE® automated LSE extractor was kindly provided by LeAnne Anderson from CEM Corporation.
39. Richter B., et al. *Anal Chem* 68: 1033–1039, 1996.
40. Richter B. *LC-GC* 17: LS22–LS28, 1999.
41. U.S. EPA. SW-846, Update III, Test Methods for Evaluating Solid Waste, Method 3545. *Fed Reg* 62: 32, 451, 1997.
42. Application Notes 322 (for Model 200, 1996) and 342 (for Model 300, 2000). Dionex Corporation.
43. Richter B., et al. Abstract 1131. 2002 Pittsburgh Conference on Analytical Chemistry and Applied Spectroscopy, New Orleans.

44. Bjrklund E., et al. *Anal Chem* 73: 4050–4053, 2001.

45. Loconto, P.R., Quantitating Toxaphene Parlar congeners in fish using large volume injection isotope dilution GC with electron-capture negative ion MS, *LC-GC NA* 36(5): 320–328 (2018)

46. Rook J. *J Water Treat Exam* 23: 223–243, 1974.

47. Moelwyn-Hughes E. *Physical Chemistry,* 2nd ed. Elmsford, NY: Pergamon, 1961.

48. Kolb B., L. Ettre. *Static Headspace-Gas Chromatography: Theory and Practice.* New York: Wiley VCH, 1997.

49. Friant S., I. Suffet. *Anal Chem* 51: 2167–2172, 1979.

50. Otson R., et al. *Environ Sci Technol* 13: 936–939, 1979.

51. Loconto, P.R., A decade of quantitating cyanide in aqueous and blood matrices using automated cryo-trapping isotopic dilution static headspace GC-MS, *LC-GC NA* 33(7): 490–505 (2015).

52. Zhang, Z., J. Pawliszyn. *Anal Chem* 65: 1843–1852, 1993.

53. Loconto, P.R., Quantitating VOCs in serum using automated headspace-SPME/cryofocusing/isotope dilution/capillary GC-MS, *American Laboratory*, 49(4): 28–29 (2017).

54. Pawliszyn, J., *Solid Phase Microextraction: Theory and Practice*. New York: Wiley-VCH. 1997.

55. Method 5021: Volatile Organic Compounds in Soils and Other Solid Matrices Using Equilibrium Headspace Analysis. *Test Methods for Evaluating Solid Wastes, Physical/Chemical Methods,* SW-846, Update V, Revision 2, July 2014.

56. Majors R. *LC-GC* 14: 936–943, 1996.

57. Bellar T., J. Lichtenberg. *J Am Water Works Assoc* 66: 739, 1974.

58. Method 502.2, Volatile Organic Compounds in Water by Purge and Trap Capillary Column Gas Chromatography with Photoionization and Electrolytic Conductivity Detectors in Series. *Methods for the Determination of Organic Compounds in Drinking Water,* Supplement III, EPA/600/R-95/131, 50.

59. Lin D., et al. *Anal Chem* 65: 999–1002, 1993.

60. Meloan, C. *Chemical Separations: Principles, Techniques, and Experiments.* New York: Wiley, 1999, p. 155.

61. Method 3600C, Cleanup. *Test Methods for Evaluating Solid Wastes, Physical/Chemical Methods,* SW-846, 3rd ed., Revision 3, December 1996.

62. Najam A., M. Korver, C. Williams, V. Burse, L. Needham. *J AOAC Int* 82: 177–185, 1999.

63. Method 3620C, Florisil Cleanup. *Test Methods for Evaluating Solid Wastes, Physical/Chemical Methods*, SW-846, 3rd ed., Revision 3, November 2000.

64. Method 3640A, Gel-Permeation Cleanup. *Test Methods for Evaluating Solid Wastes, Physical/ Chemical Methods,* SW-846, 3rd ed., Revision 1, September 1994.

65. Erickson M. Analytical Chemistry of PCBs, 2nd ed. Boca Raton, FL: CRC/Lewis Publishers, 1997, pp. 225–226.

66. Loconto P.R. Meeting the Analytical Challenge: Quantitative Determination of AR 1242 in Adulterated Rat Feed. Paper presented at the 30th Central Regional Meeting of the American Chemical Society, Cleveland, OH, May 27–29, 1998.

67. Chester T. In G. Ewing, Ed. *Analytical Instrumentation Handbook,* 2nd ed. New York: Marcel Dekker, 1997.

68. McHugh M., V. Krukonis. *Supercritical Fluid Extraction: Principles and Practice,* 2nd ed. Boston: Butterworth-Heinemann, 1994, pp. 11–13.

69. Dean J., Ed. *Applications of Supercritical Fluids in Industrial Analysis.* New York: Chapman & Hall, 1993, pp. 50–53.

70. U.S. EPA SW-846, Update III, *Test Methods for Evaluating Solid Waste,* Method 3562.

71. Levy J. *LC-GC* 17: S14–S21, 1999.

72. Hawthorne S. In S. Westwood, Ed. *Supercritical Fluid Extraction and Its Use in Chromatographic Sample Preparation.* New York: Chapman & Hall, 1993, pp. 50–51.

73. Method 3562: *Supercritical Fluid Extraction of polychlorinated biphenyls (PCBs) and organochlorine pesticides*, Revision 0, February 2007.

74. David M., J. Seiber. *Anal Chem* 68: 3038–3044, 1996.

75. Heemken O., et al. *Anal Chem* 69: 2171–2180, 1997.

76. Reed R., A. Wise. *Cannabis Science and Technology* 2(5): 14–19, 2019.

77. Hawthorne S. et al. Comparison of Soxhlet extraction, pressurized liquid extraction, supercritical fluid extraction, and subcritical water extraction for environmental solids: recovery, selectivity, and effects on sample matrix. *JChrom A* 892 (1–2): 421–433, 2000

78. Junk G., J. Richard, M. Grieser, D. Witiak, J. Witiak, M. Arguello, R. Vick, H. Svec, J. Fritz, and G. Calder. *J Chromatogr* 99: 745–762, 1974.

79. Snyder L., J. Kirkland. *Introduction to Modern Liquid Chromatography*, 2nd ed. New York: Wiley Interscience, 1979.

80. Snyder L., J. Kirkland, J. Glajch. *Practical HPLC Method Development.* 2nd ed. New York: Wiley Interscience, 1997.

81. Loconto P.R. Contract SBIR Study 68-02-4481, awarded in 1987 from the U.S. Environmental Protection Agency.

82. Loconto, P.R. Contract SBIR # (88)-172, awarded in 1988 from the New York State Science and Technology Foundation.

83. Thurman, E., M. Mills. *Solid-Phase Extraction: Principles and Practice.* Volume 147 in the Chemical Analysis series of monographs. New York: Wiley-Interscience, 1998, pp. 198–200.

84. Yonker C., et al. *J Chromatogr* 241: 269–280, 1982.

85. Snow N., G. Slack in *Modern Practice of Gas Chromatography,* 4th ed., R. Grob and E. Barry (Eds.). Hoboken, NJ: Wiley-Interscience, 2004, Chapter 11.

86. Majors R., G. Slack in *Practical HPLC Method Development,* 2nd ed., L. Snyder, J. Kirkland, J. Glajch (Eds.). New York: Wiley-Interscience, 1997, Chapter 4.

87. Simpson N. *Solid-Phase Extraction: Principles, Strategies and Applications.* New York: Marcel Dekker, 1997.

88. LC-GC: Current Developments in Sample Preparation Vol 16, May 1998.

89. LC-GC: Current Developments in Sample Preparation: Vol. 17, June 1999.

90. LC-GC: Current Developments in Sample Preparation: Vol. 17, September 1999.

91. Rohrschneider L. *Anal Chem* 45: 1241, 1973.

92. Snyder L. *J Chromatogr Sci* 16: 223, 1978.

93. Schwarzenback R., P. Gschwend, D. Imboden. *Environmental Organic Chemistry.* New York: Wiley Interscience, 1993, pp. 132–134.

94. Leo A., C. Hansch, D. Elkins. *Chem Rev* 71: 525–616, 1971.

95. Yalkowsky S., S. Banerjee. *Aqueous Solubility: Methods of Estimation for Organic Compounds.* New York: Marcel Dekker, 1992.

96. Layman W., W. Reehl, D. Rosenblatt. *Handbook of Chemical Property Estimation Methods: Environmental Behavior of Organic Compounds.* New York: McGraw-Hill, 1982, pp. 1–2.

97. Hansch C., A. Leo. *Substituent Constants for Correlation Analysis in Chemistry and Biology.* New York: Wiley Interscience, 1979.

98. Hansch C., J. Quinlan, G. Lawrence, *J Org Chem* 33: 347–350, 1968.

99. Loconto, P.R., Isolation and recovery from water of selected chlorophenoxy acid herbicides and similar weak acid herbicides by solid-phase extraction HPLC and photodiode array detection, *J Liq Chrom* 14(7): 1279–1314, 1991.

100. Budde, W. *Analytical Mass Spectrometry.* Washington, D.C.: American Chemical Society, Oxford University Press, 2001, p. 251.

101. Loconto P.R., A. Gaind. Isolation and recovery of organophosphorous pesticides from water by solid-phase extraction with dual wide-bore capillary gas chromatography, *J Chromatogr Sci* 27: 569–573, 1989.

102. Loconto, P.R. *Interactions on hydrophobic surfaces as a means of isolating environmentally significant analytes at trace concentration levels in water,* PhD Thesis, University of Massachusetts at Lowell (formerly the University of Lowell), 1985.

103. Loconto P. *LC-GC* 20(11): 1062–1068, 2002.

104. Janák K., E. Jensen, G. Becher. *J Chromatogr B* 734: 219–227, 1999.

105. Schecter, A. et al. *J Occup. Environ Med* 47(3): 199–211, 2005

106. Loconto P.R., D. Isenga, M. O'Keefe, M. Knottnerus, Isolation and recovery of selected polybrominated diphenyl ethers from human serum and sheep serum: coupling reversed-phase solid-phase disk extraction and liquid-liquid extraction techniques with a capillary gas chromatographic electron capture negative ion mass spectrometric determinative technique, *Journal of Chromatographic Science,* 46(1): 53–60, 2008.

107. Font G., et al. *J Chromatogr A* 733: 449–471, 1996.
108. Font G., et al. *J Chromatogr* 642: 135–161, 1993.
109. Brock J., V. Burse, D. Ashley, A. Najam, V. Green, M. Korver M. Powell C. Hodge and L. Needham. An improved analysis for chlorinated pesticides and polychlorinated biphenyls (PCBs) in human and bovine sera using solid-phase extraction. *J Anal Toxicol* 20: 528–536, 1996.
110. Barker S., A. Long, C. Short. *J Chromatogr* 475: 353–361, 1989.
111. Barker S. *J Chromatogr A* 885: 115–127, 2000.
112. Barker S. In N. Simpson, Ed. *Solid-Phase Extraction: Principles, Techniques, and Applications.* New York: Marcel Dekker, 2000, chapter 13, p. 365.
113. Barker S. *LC-GC* 16: S37–S40, 1998.
114. Ling Y.-C., M.-Y. Chang, I.-P. Huang. *J Chromatogr* A 669: 119–124, 1994.
115. Ling Y.-C., I.-P. Huang. *Chromatographia* 40: 259–266, 1995.
116. Rodriguez, I. et al. *J Chromatogr A* 1072: 83–91, 2005.
117. Rodriguez, I. et al. *AnalChem* 79(4): 1675–1681, 2007.
118. Abdel-Rehim, M. *Anal Chim Acta* 701:119–128, 2011.
119. Zhang Z., M. Yang, J. Pawliszyn. *Anal Chem* 66: 844A–853A, 1994.
120. Arthur C., J. Pawliszyn. *Anal Chem* 62: 2145, 1990.
121. Ai J. *Anal Chem* 69: 1230–1236, 1997.
122. Ai J. *Anal Chem* 69: 3260–3266, 1997.
123. Lakso H., W. Ng. *Anal Chem* 69: 1866–1872, 1997.
124. Loconto P.R., Y. Pan. *1998 Pittsburgh Conference on Analytical Chemistry and Applied Spectroscopy.* New Orleans, 1998, abstract 1250.
125. David F., B. Tienport, P. Sandra. *LC-GC* 21(2): 108–118, 2003.
126. Loconto, P.R. Evaluation of automated stir bar sorptive extraction-thermal desorption-gas chromatography electron-capture negative ion mass spectrometry for the analysis of PBDEs and PBBs in sheep and human serum. *J Chromato Sci* 47: 656–669, 2009.
127. Tienport B., et al. *Anal Bioanal Chem* 373: 46–55, 2002.
128. Sandau C., et al. *Anal Chem* 75: 71–77, 2003.
129. Sjödin A., et al. *Anal Chem* 76: 1921–1927, 2004.
130. Anastassiades, M., S. Lehotay, D. Stajnbaher, F. Schrenck. *J AOAC Int.* 86: 412–431. 2003.
131. Loconto, P., Y. Pan and D. Kamdem. Isolation and Recovery of 2-aminoehtanol, N-methyl-2-aminoethanol, and N, N-dimethyl-2-aminoethanol from a copper amine aqueous matrix and from an amine-treated sawdust using LLE and LSE combined with capillary gas chromatography-ion trap mass spectrometry. *J Chrom Sci* 36: 299–305, 1998.
132. Lehotay, S., K. Mastovska, S. Yun. Evaluation of two fast and easy methods for pesticide residue analysis in fatty food matrixes. *J AOAC Int.* 88(2): 630–638. 2005.
133. Wang, X. *Determination of pesticides in coffee with QUECHERS extraction and silica gel SPE cleanup. The Application Notebook, LC-GC*, United Chemical Technologies, LLC, June 2017.
134. Burse, V. et al. Preliminary investigation of the use of dried-blood spots for the assessment of in utero exposure to environmental pollutants. Biochemical and Molecular Medicine 61: 236–239, 1997.
135. Spiethoff, H. et al. Use of newborn screening program blood spots for exposure assessment: declining levels of perfluorinated compounds in New York State Infants. *Enviro Sci Technol* 42: 5361–5367, 2008.
136. Rocio-Bautista, P., V. Pino. *Analytical Separation Science: Sample Preparation Method Validation and Analytical Applications*, Chapter 11, Volume 5 Weinheim: Germany: Wiley-VCH, 2015, pp. 1681–1724.
137. Rocio-Baustista, P., et al. *J Chrom A* 1436: 42–50, 2016.
138. Rios, A., M. Zougagh. *Trends in Anal. Chem.* 84: 2016, 72–83.
139. Rocio-Baustista, P., et al. Determination of polycyclic aromatic hydrocarbons in environmental waters using heterogeneous magnetic materials based on metal-organic frameworks, *LC-GC NA* 36: 464–471, 2018.
140. Wagner R., Ed. *Guide to Environmental Analytical Methods,* 3rd ed. Schenectady, NY: Genium Publishing, 1996, pp. 57–79.
141. Jenniss S., et al. *Applications of Atomic Spectrometry to Regulatory Compliance Monitoring,* 2nd ed. New York: Wiley VCH, 1997.
142. Van Loon J. *Selected Methods of Trace Metal Analysis,* Vol. 80. New York: Wiley Interscience, 1985, pp. 100–101.

143. Butcher D., J. Sneddon. *A Practical Guide to Graphite Furnace Atomic Absorption Spectrometry.* New York: Wiley Interscience, 1998,

144. Skoog D., J. Leary. *Principles of Instrumental Analysis,* 4th ed. Philadelphia, PA: Saunders, 1992, pp. 214–215.

145. Parsons M. In G. Ewing, Ed. *Analytical Instrumentation Handbook.* New York: Marcel Dekker, 1997, pp. 321–322.

Harris D. *Quantitative Chemical Analysis.* San Francisco: Freeman, 1991, pp. 594–595.

146. Sytris D., D. Redfield. *Anal Chem* 59: 289, 1987.

Butcher D., J. Sneddon. *A Practical Guide to Graphite Furnace Atomic Absorption Spectrometry.* New York: Wiley Interscience, 1998.

147. Schlemmer G., B. Welz. *Spectrochim Acta B* 41B: 1157, 1986.

148. Method 3015A, Revision 1 2007, *Microwave Assisted Acid Digestion of Aqueous Samples and Extracts*, and Method 3051A, Revision 1, 2007, *Microwave Assisted Acid Digestion of Sediments, Sludges, Soils, and Oils. Physical/Chemical Methods,* EPA Office of Solid Waste, SW-846 Methods.

149. Butz J., M. Burritt. *Heavy Metal Screen and Individual Analysis for As, Cd, Hg, Pb, and Tl in Urine and in Blood.* Rochester, MN: Mayo Clinic, Department of Laboratory Medicine and Pathology, 2001.

150. Nixon D., T. Moyer. *Spectrochim Acta Part B* 51: 13–25, 1996.

151. Freiser H. In B. Karger, L. Snyder, C. Horvath. *An Introduction to Separation Science.* New York: Wiley Interscience, 1973, p. 259.

Day R., Jr, A. Underwood. *Quantitative Analysis,* 5th ed. Englewood Cliffs, NJ: Prentice Hall, 1986, pp. 510–517.

152. Sturgeon W., et al. *Talanta* 29: 167, 1982.

153. Martell A., L. Sillen. *Stability Constants of Metal Ion Complexes*, 2nd ed. London: The Chemical Society, 1964.

Loconto P.R. *Interactions on Hydrophobic Surfaces as a Means of Isolating Environmentally Significant Analytes at Trace Concentrations in Water.* Ph.D. Thesis, University of Massachusetts at Lowell (formerly the University of Lowell), 1985, Tables XXIII, XXVI, and XXVII.

154. Emsley J. *The Elements.* Oxford: Clarendon Press, 1989, pp. 114–115.

155. Carro A., I. Neira, R. Rodil, R. Lorenzo. *Chromatographia* 56: 733–738, 2000.

156. Pereiro I., A. Diaz. *Anal Bioanal Chem* 372: 74–90, 2002.

157. Krachler M., K. Falk, H. Emons. *American Laboratory News Edition*, March 2002, pp. 10–14.

158. Zoorob G., J. McKiernan, J. Caruso. *Mikrochim Acta* 128: 145–168, 1998.

159. Sutton K., R. Sutton, J. Caruso, *J Chromatogr A* 789: 85–126, 1997.

160. Csuros, M. *Environmental Sampling and Analysis: Lab Manual.* Boca Raton: Lewis Publishers, CRC Press LLC, 1997, pp. 251–257.

161. Bulletins on chemical derivatization from (formerly) Pierce Chemical Co. and Regis Chemical Co. were used to develop these concepts. Several books related to chemical derivatization, among others, include: Knapp D., Ed. *Handbook of Analytical Derivatization Reactions.* New York: Wiley, 1979

162. Drozd J., J. Novak. *Chemical Derivatization in Gas Chromatography.* New York: Elsevier, 1981.

163 Shriner R, et al. *The Systematic Identification of Organic Compounds,* 7th ed. New York: Wiley, 1998.

164. Gyllenhaal O., H. Brotell, P. Hartvig. *J Chromatogr* 129: 295–302, 1976.

165. Kawahara F. *Anal Chem* 40: 2073–2075, 1968.

166. Kawashima, et al. *J Chromatogr B* 720: 1–7, 1998.

167. Dünges W. *Anal Chem* 49: 442–445, 1977.

168. Takadate A., et al. *Anal Sci* 8: 663–668, 1992.

169. Pobozy, et al. *Microchim Acta* 172(3–4): 409–417, 2011.

170. Shriner, R. et al. *The Systematic Identification of Organic Compounds,* 7th ed. New York: John Wiley and Sons, 1998.

171. Emsley, J. *The Elements*, Oxford, U.K.: Oxford University Press, 1989.

4 Determinative Techniques to Measure Organics and Inorganics

They laughed when they heard Aston say, he would weigh tiny atoms one day. But he had the last laugh—with his mass spectrograph, he "weighed" them a different way.

—Anonymous

To the observer who does not have the technical background in TEQA and walks into a contemporary environmental testing laboratory, a collection of black boxes (instruments) with cables connecting the black boxes to personal computers and other high-tech devices should be what makes the first impression. This observer will see people, some of whom wear white lab coats, running around, holding various glassware, such as vials, syringes, beakers, test tubes, or whatever else it is lab people handle when at work in the busy lab. Observers will, upon being invited to tour, see different departments within the corporate structure. Some department personnel process analytical data generated by these black boxes; some personnel prepare samples for introduction into the black boxes; and other personnel enter data into a Laboratory Information Management System (LIMS) that reads a bar code label on a given sample and tracks the status of that sample as various analytical methods and instruments are used to generate the data. Some instruments are noisy, some are silent, some incorporate robot-like arms, and some incorporate samples directly, whereas others require sample preparation; all contribute to the last and no less important step in TEQA: determination. Instruments, computers, and accessories all comprise what the Environmental Protection Agency (EPA) refers to as *determinative techniques,* hence the title of this chapter. In Chapter 2, we discussed the important outcomes of using determinative techniques to perform TEQA. In Chapter 3, we discussed the means by which environmental samples and biological specimens are made suitable and appropriate for introduction to these instruments (i.e., the science of sample preparation for TEQA). This chapter on determinative techniques therefore completes the thorough discussion of TEQA.

To the sufficiently educated observer, the contemporary environmental testing laboratory is a true testimonial to man's ingenuity, a high-tech masterpiece. However, unlike a work of art, this observer is quick to discover that this artistic endeavor is a work in progress. This observer may see a robotic arm of an autosampler depositing 5 μL of sample into the graphite tube of a graphite furnace atomic absorption spectrophotometer (GFAA). He may also peer into a monitor that reveals an electron-impact mass spectrum of a priority pollutant, semivolatile organic compound. He will become aware very quickly whether or not this particular instrument is running samples or is still running calibration standards in an attempt to meet the stringent requirements of EPA methods. If this person is interested in the progress made by a particular sample as it makes its way through the maze of methods, he can find this information by peering into the sample status section of the LIMS software.

This chapter takes the reader from the uninformed observer described above to the educated observer who can envision the inner workings of a contemporary environmental testing laboratory. This is the chapter that deals with the determinative step, a term coined by the EPA. Beginning with the SW-846 series of methods, the sample prep portion was separated from the determinative portion. This separation enabled flexibility in conceptualizing the total method objectives of TEQA in the SW-846 series. This author believes that separating the sample prep from the determinative also makes sense in the organization of this book.

Of the plethora of instrumental techniques, GC, GC-MS, GC-MS-MS, HPLC, LC-MS-MS, IC, AA, ICP-AES, and ICP-MS are the *principal determinative techniques* employed to achieve the objectives of TEQA as applied to both trace organics and trace inorganics analysis. A list of some of the classical textbooks that shaped this author's thinking on this topic early on are listed at the end of the chapter.[1–14] The separation sciences have been coupled to the optical spectroscopic and mass spectrometric sciences to yield very powerful so-called *hyphenated instruments*. These *nine* determinative techniques are also sensitive enough to give analytical information to the client that is the most relevant to environmental site remediation. For example, one way to clean up a wastewater that is contaminated with polychlorinated volatile organics (ClVOCs) is to purge the wastewater to remove the contaminants; a process known as air stripping. It is important to know that the air-stripped wastewater has a concentration of ClVOCs that meets a regulatory requirement. This requirement is usually at the level of low parts per billion. A determinative technique that can only measure as low as parts per hundred has no place in the arsenal of analytical instruments pertinent to TEQA. Recall from Chapter 2 that techniques relating the acquisition of data directly from analytical instruments provide instrumental detection limits (IDLs), whereas method detection limits (MDLs) combine the sample prep step with the determinative step. This combination serves to significantly lower the overall detection limits and is one of the prime goals of TEQA.

This chapter introduces those *nine determinative techniques* referred to earlier and adds several others. We first discuss those fundamental principles, vital to the practice of both GC and HPLC, that facilitate a more meaningful understanding of column chromatographic separations that are particularly relevant to the quantitative determination of *trace organics*. We then introduce the operational aspects of these instruments largely from a user perspective. A strong emphasis is placed on GC-MS, as this has become the dominant determinative technique for organics in TEQA. Since the early 2000s LC-MS-MS based on the *electrospray interface* is one of the more significant advances in analytical instrumentation and is introduced in this chapter. Ion chromatographic techniques (IC) as applied to trace inorganics are then introduced. This topic provides an important link to the other major class of enviro-chemical/enviro-health chemical contaminants, *trace metals*, where atomic spectroscopy, as the principal determinative technique, dominates. A link between infrared absorption spectroscopy and TEQA is made through quantitative oil and grease and total organic carbon measurements. Capillary electrophoresis is then introduced and applied to the separation, detection, and quantification of trace inorganic anions in surface water via indirect photometric detection. Finally, a relatively new determinative technique called SIFT-MS is briefly introduced and selected applications to air pollution analysis are introduced.

4.1 HOW DO YOU KNOW WHICH DETERMINATIVE TECHNIQUE TO USE?

Which determinative technique to use is dictated by the physical and chemical nature of the analyte of interest? The organics protocol flowchart introduced in Chapter 1 serves as a useful guide. Let us consider how we would determine which determinative technique to use for the following example. Ethylene glycol, 1,2-ethanediol (EG), and 1,2-dichloroethane (1,2-DCA) consist of molecules that contain a two-carbon backbone with either a hydroxyl- or chlorine-terminal functional group. The molecular structures for these are as follows:

These two molecules look alike; so, could we use the same instrument and conditions to quantitate both of these organic compounds in an environmental sample? If you build a *ball & stick model* for both molecules using correct atomic sizes, using a green ball for the chemical element chlorine, red for oxygen and white for hydrogen, you would clearly see differences between these two ball & stick models! The green ball would look much larger versus the red or white balls. This difference is due to the much larger size of the chlorine atom! Table 4.1 compares important physico-chemical properties of both organic compounds. The presence of two hydroxyl groups enables ethylene glycol to extensively hydrogen bond both *intramolecularly* (i.e., to itself) and *intermolecularly* (i.e., between solute and solvent molecules). It is well known that EG dissolves readily in polar solvents such as water (H_2O) and methanol (CH_3OH). In stark contrast to EG, 1,2-DCA interacts intramolecularly through much weaker van der Waals forces and is incapable of interacting intermolecularly with polar solvents while being miscible in nonpolar solvents such as chloroform and ether. The boiling point of EG is almost twice as large as that of 1,2-DCA. These significant differences in physical properties would also be reflected in their respective octanol–water partition coefficients (K_{ow}). 1,2-DCA can be efficiently partitioned into a nonpolar solvent or into the headspace, whereas any attempt to extract EG from an aqueous solution that contains dissolved EG is useless. Because both compounds are liquids at room temperature, they do exhibit sufficient vapor pressure. Both compounds are said to be *amenable to analysis* by gas chromatography. However, it may prove difficult to chromatograph them on the same GC chromatographic column. The fundamental differences between a hydroxyl covalently bonded to carbon and a chlorine atom covalently bonded to carbon become evident when one attempts to separate the two. We will continue to use the physico-chemical differences between EG and 1,2-DCA to develop the concept of a separation between the two organic compounds by differential migration through a hypothetical column and through a series of consecutive stages known as the Craig distribution.

4.2   WHAT IS DIFFERENTIAL MIGRATION ANYWAY?

Around 100 years ago, Mikhail Tswett, a Russian botanist, demonstrated for the first time that pigments extracted from plant leaves, when introduced into a packed column, whereby a nonpolar solvent is allowed to flow through calcium carbonate, initially separated into green and yellow rings. He called this separation phenomenon *chromatography*, derived from the Greek roots *chroma* (color) and *graphein* (to write). If additional solvent is allowed to pass through, these rings widen and separate more, and further separate into additional rings. In Tswett's own words:[15]

TABLE 4.1
Physico-Chemical Properties of Ethylene Glycol and 1,2-Dichloroethane

Compound	T (mp) (°C)	T (bp) (°C)	Soluble in
$HOCH_2CH_2OH$	−12.6	197.3	Polar solvents
$ClCH_2CH_2Cl$	−35.7	83.5	Nonpolar solvents

Like light rays in the spectrum, the different components of a pigment mixture, obeying a law, are resolved on the calcium carbonate column and then can be qualitatively and quantitatively determined. I call such a preparation a chromatogram and the corresponding method the chromatographic method.

His work in establishing the technique of liquid–solid adsorption chromatography would languish for 30 years until resurrected by Edgar Lederer in Germany.

A timeline titled *Historica Chromatographica* published recently benchmarks key advances in all of chromatography and serves to recognize those that often go unnoticed; it is summarized in tabular format in Table 4.2.[16]

TABLE 4.2
Historica Chromatographica **Benchmarks through to the Year 1990**

Year	Key Advances	Pioneers
1990	Persuasive perfusion	PerSeptive Biosystems, part of PerkinElmer, introduces perfusion chromatography, in which samples move both around and through the resin beads
1985	Superior suppression	Dionex researcher Pohl introduces micromembrane suppressors for use in ion chromatography at Pittcon
1981	Microcolumn SFC	Novotny and Lee, pioneers in microcolumn liquid chromatography, introduce capillary supercritical fluid chromatography (SFC)
1975	IC advent	Small, Stevens, and Bauman develop ion chromatography combining a cation exchange column (separator) and strongly basic resin (stripper) to separate cations in dilute HCl
1974	Capillary zone electrophoresis (CZE) under glass	Virtanen introduces commercial CZE in glass tubes, based largely on pioneering work by Hjerten
1966	I see HPLC	Horvath and Lipsky develop high-pressure liquid chromatography (HPLC) at Yale University
1966	Sugar, sugar	Green automates carbohydrate analysis, improving on the earlier efforts of Cohn and Khym, who used a borate-conjugated ion exchange column to separate mono- and disaccharides
1960	GC's heart of glass	Desty introduces the glass capillary column for GC, used in his analysis of crude petroleum; the technology was later commercialized by Hupe & Busch and Shimadzu
1958	Automating AA analysis	Stein, Moore, and Spackman automate amino acid (AA) analysis using ion exchange and Edman degradation
1955	Going to market	First gas chromatographs were introduced in the U.S. by Burrell Corp., PerkinElmer, and Podbielniak
1953	Exclusive science	Wheaton and Bauman define ion exclusion chromatography, where one solution ion is excluded from entering the resin beads and passes in the void volume
1948	Reversing phases	Boldingh develops reversed-phase chromatography when separating the higher fatty acids in methanol against a solid phase of liquid benzene supported on partially vulcanized Hevea rubber
1945	One small step to GC	Prior describes gas–solid adsorption chromatography when separating O_2 and CO_2 on charcoal column

(continued)

TABLE 4.2 (Continued)
***Historica Chromatographica* Benchmarks through to the Year 1990**

Year	Key Advances	Pioneers
1941	Protein pieces	Martin and Synge develop liquid–liquid partition chromatography when separating amino acids through ground silica gel
1938	Spotting the difference	Izmailov and Shraiber develop drop chromatography on thin horizontal sheets, a precursor to thin-layer chromatography
1937	The road to white sands	Taylor and Urey use ion exchange chromatography to separate lithium isotopes, work that eventually led to the separation of fissionable uranium for the Manhattan Project
1922	Clarifying butter	Palmer, who is later recognized for popularizing chromatography's use, separates carotenoids from butter fat
1913	Water world	First U.S. use of zeolites in water softening based on earlier work in Germany by Gans
1906	Our Father ...	Tswett develops the concept of chromatography while attempting to purify chlorophylls from plant extracts; this discovery gained him the cognomen "Father of Chromatography"
1903	Food for thought	Goppelsroeder develops theory of capillary analysis when using paper strips to examine alkaloids, dyes, milk, oils, and wine, improving on the earlier work of his mentor, Schoenbein

If a mixture containing EG and 1,2-DCA is introduced into a column, in which the mixture is allowed to flow through the column packing, it is possible to conceive of the notion that the molecules that make up each compound would *migrate differentially* through the packed bed or stationary phase. Tswett used a packed tubular column held vertically while pouring solvent through the column to separate chlorophylls. This experiment led to liquid column chromatography. If filter paper is used, capillary action can initiate solvent flow through the porous cellulose. This experiment led to paper and thin-layer chromatography. Let us assume that this hypothetical column tends to retain the more polar EG longer. A two-dimensional graphical representation of the separation of EG from 1,2-DCA is shown in Figure 4.1.

We observe that the *dispersion* of the molecules as represented by σ^2 is found to be proportional to the distance migrated, z, according to

$$\sigma^2 = kz$$

where k, the constant of proportionality, depends on the system parameters and operating conditions. Because k is a ratio of the degree of spread to migration distance, k can be referred to

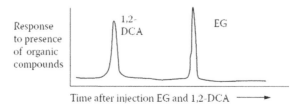

Response to presence of organic compounds

1,2-DCA   EG

Time after injection EG and 1,2-DCA ⟶

FIGURE 4.1

as a plate height. The *resolution*, R_s, between the separated peaks can be defined in terms of the distance between the apex of the peaks and the broadening of the peak according to

$$R_s = \frac{\Delta z}{\tau \sigma}$$

where τ is defined as being equal to 4. (p. 109)[17] We will have more to say on this topic later. It also becomes evident that Δz is proportional to the migration distance z and σ is proportional to the square root of the migration distance. Expressed mathematically, we have

$$\Delta z \propto z$$

and

$$\sigma \propto \sqrt{z}$$

These equations tell us that the distance between zone centers increases more rapidly than the zone widths. From the definition of R_s, this suggests that resolution improves with migration distance. We will have more to say about resolution when we take up chromatography. Differences in the rates of analyte migration, however, do not explain the fundamental basis for separating EG from 1,2-DCA. For this, we begin by discussing the principles that underline the *Craig countercurrent extraction experiment.*

4.3 WHAT CAUSES THE BANDS TO SEPARATE?

We just saw that, experimentally, EG and 1,2-DCA differentially migrate through a stationary phase when introduced into a suitable mobile phase, and that chromatography arises when this mobile phase is allowed to pass through a chemically selective stationary phase. It is not sufficient to merely state that EG is retained longer than 1,2-DCA. It is more accurate to state that EG partitions to a greater extent into the stationary phase than does 1,2-DCA, largely based on "like dissolves like." The stationary phase is more like EG than 1,2-DCA due to similar polarity. This is all well and good, yet these statements do not provide enough rationale to establish a true physical-chemical basis for separation. In Chapter 3, we introduced liquid–liquid extraction (LLE) and also considered successive or multiple LLE. What we did not discuss is what arises when we *transfer this immiscible upper or top phase or layer to a second sep funnel* (first Craig stage or $n = 1$; see below). Prior to this transfer, the second sep funnel will already contain a fresh lower phase. Equilibration is allowed to occur in the second sep funnel, while the fresh upper phase is brought in contact with the lower phase in the first funnel. What happens if we then transfer the upper phase from this second sep funnel to a third sep funnel that already contains a fresh lower phase? This transfer of the upper phase in each sep funnel to the next stage, with subsequent refill of the original sep funnel with a fresh upper phase, can be continued so that a total of n stages and $n + 1$ sep funnels are used. It becomes very tedious to use sep funnels to conduct this so-called countercurrent extraction. A special glass apparatus developed by L.C. Craig in 1949 provides a means to perform this extraction much more conveniently. Twenty or more Craig tubes are connected in series in what is called a *Craig countercurrent apparatus.* Once connected, up to 1,000 tubes previously filled with a lower phase can participate in countercurrent extraction by a mere rotation of the tubes. Figure 4.2 is a schematic diagram of a single Craig tube of 2 mL undergoing rotation.

It is this rotational motion that removes the extracted organic phase while a fresh organic phase is introduced back into the tube. The phases separate in position A, and after settling, the tube

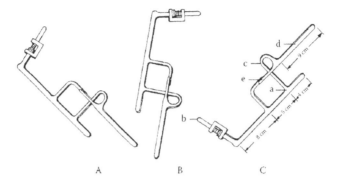

A   B   C

FIGURE 4.2

is brought to position B. Then, all of the upper phase flows into decant tube d through c, as the lower phase is at a. When the tube is brought to position C, all of the upper phases in the decant tube are transferred through e into the next tube, and rocking is repeated for equilibration. The tubes are sealed together through the transfer tube, location e in the figure, to form a unit. These units are mounted in series and form a train having the desired number of stages. (pp. 111–112)[17]

The Craig countercurrent extraction enables one to envision the concept of discrete equilibria and helps one understand how *differences in partition coefficients among solutes in a mixture can lead to separation of these solutes.* Consider a cascade of $n + 1$ stages, each stage containing the same volume of the lower phase. We seek to explain the foreboding-looking Table 4.3. Let us also assume that the total amount of solute is initially introduced into stage 0 (i.e., the first Craig tube in the cascade). The solute is partitioned between the upper and lower phases, as we saw in Equation (3.16), represented here according to

$$p = \frac{VD}{1+VD} \qquad (4.1)$$

$$q = \frac{1}{1+VD} \qquad (4.2)$$

In Table 4.3, p is the fraction of total solute that partitions into the upper phase and q is the fraction of total solute that partitions into the lower phase. Also, by definition, the following must be true:

$$p + q = 1$$

V is the ratio of the upper phase volume to the lower phase volume and is usually equal to 1 because both volumes in the Craig tubes are usually equal. D is the distribution ratio and equals K_D, the molecular partition coefficient in the absence of secondary equilibria, as discussed in Chapter 3. If we introduce a mixture of solutes to stage 0, each solute will have its own value for D, and hence a unique value for p and q. For example, for a mixture containing four solutes, we would realize a fraction p for the first solute, a fraction p' for the second solute, and so forth. In a similar manner, we would also realize a fraction q for the first solute, a fraction q' for the second solute, and so forth.

We seek now to show how the p and q values of Table 4.3 were obtained. We also wish to show how to apply the information contained in Table 4.3. We then extrapolate from the limited number of Craig tubes in Table 4.3 to a much larger number of tubes and see what effect this increase in the number of Craig tubes has on the degree of resolution, R_s.

TABLE 4.3
Countercurrent Distribution of a Given Solute in a Craig Apparatus

	Stage					Distribution
	0	**1**	**2**	**3**	**4**	
Introduce solute and equilibrate	p/q					
Total	1					1
First transfer	0/q	p/0				$(q + p)^1$
Equilibrate	p(q)/q(p)	p(p)/q(p)				
Second transfer	0/q(q)	p(q)/q(p)	p(p)/0			
Total	q^2	2pq	p^2			$(q + p)^2$
Equilibration	p(qq)/ q(qq)	p(2pq)/q(2pq)	p(pp) q(p^2)			
Third transfer	$0/q^3$	$pq^2/2\ q^2\ p$	$2p^2\ q/qp^2$	$p^3/0$		
Total	q^3	$3pq^2$	$3p^2q$	p^3		$(q + p)^3$
Equilibrate	$p(q^3)/$ $q(q^3)$	$p(3pq^2)/$ $q(3pq^2)$	$p(3p^2\ q)/$ $q(3p^2\ q)$	$p(p^3)/q(p^3)$		
Fourth transfer	$0/q^4$	$pq^3/3q^3\ p$	$3p^2\ q^2/3p^2\ q2$	$3p^3\ q/p^3\ q$	$p^4/0$	
Total	q^4	$4q^3p$	$6p^2\ q^2$	$4p^3\ q$	p^4	$(q + p)^4$

We start with a realization that once a mixture of solutes, such as our pair, EG and 1,2-DCA, is introduced into the first Craig tube, an initial equilibration occurs and this is shown at stage 0. Again, if p and q represent the fraction of EG in each phase, then p' and q' represent the fraction of 1,2-DCA in each phase. The first transfer involves moving the upper phase that contains a fraction p of the total amount of solute to stage 1. A volume of upper phase equal to that in the lower phase is now added to stage 0, and a fraction p of the total amount of solute in stage 0, p, is partitioned into the upper layer while a fraction q of the total, q, is partitioned into the lower phase. A fraction p of the total p remains in the upper layer in stage 1, and a fraction q of the total p is partitioned into the lower phase. The remainder of Table 4.3 is built by partition of a fraction p of the total after each transfer and equilibration to the upper layer and by partition of a fraction q of the total into the lower layer. The last column in Table 4.3 demonstrates that if each row labeled "total" is added, this sum is the expansion of a binomial distribution, $(q + p)^r$, where r is the number of transfer. The fraction of solute in each nth stage after the rth transfer and corresponding equilibration can then be found using

$$f = \frac{r!}{n!(r - n)!} p^n q^{r-n} \tag{4.3}$$

This fraction represents the sum of the fractions in the upper and lower phases for that stage. For example, suppose we wish to predict the fraction of EG and the fraction of 1,2-DCA in stage 3 after four transfers. Let us assume that the upper phase is the less polar phase. Let us also assume that the distribution ratio for 1,2-DCA into the less polar upper phase is favored and that $D_{1,2\text{-DCA}} = 4$. Let us also assume that EG prefers the more polar lower phase and has a distribution ratio that favors the lower phase and that $D_{EG} = 0.1$. We also assume that the volume of upper phase equals that of the lower phase (i.e., $V_{\text{upper}} = V_{\text{lower}}$) and the volume ratio is therefore 1. Substituting into Equation (4.3) without considering values for p and q yields

$$f^{3,4} = \frac{4!}{3!(4-3)!} p3q = 4q^3 q$$

A comparison of this result with that for the total fraction in the third stage after the fourth transfer (refer to Table 4.3) shows this result to be identical to that shown in the table. Next, we proceed to evaluate p and q for each of the two solutes. Using Equations (4.1) and (4.2), we find the following:

$$p(1,2-\text{DCA}) = 0.9, \quad p(\text{EG}) = 0.0909$$
$$q(1,2-\text{DCA}) \qquad q(\text{EG}) = 0.909$$

Upon substituting these values for p and q for each of the two solutes in the mixture into Equation (4.3), we obtain:

$$f^{3,4}_{1,2-\text{DCA}} = (4)(0.5)^3 (0.1) = 0.292$$

$$f^{3,4}_{\text{EG}} = (4)(0.0909)^3 (0.909) = 0.00273$$

where $f^{3,4}_{1,2-\text{DCA}}$ is the fraction of 1,2-DCA present in the third stage after four transfers and $f^{3,4}_{\text{EG}}$ is the fraction of EG present in the third stage after four transfers. Hence, after four transfers, we find the fraction of 1,2-DCA in stage 3 (i.e., upper and lower phases) to be 0.292, whereas the fraction of EG in stage 3 is only 0.00273. The fact that these fractions are so different in magnitude is the *basis for a separation of 1,2-DCA from EG!*

4.4 WHAT HAPPENS IF WE REALLY INCREASE THE NUMBER OF CRAIG TUBES?

We have just examined a relatively small number of Craig countercurrent extractions and seen that differences in D or K_D among solutes result in different distributions among the many Craig tubes. Most distributions are normally distributed. In the absence of systematic error, random error in analytical measurement is normally distributed, and this assumption formed the basis of much of the discussion in Chapter 2. A continuous random variable x has a normal distribution with certain parameters μ (mean, parameter of location) and σ^2 (variance, parameter of spread) if its density function is given by:[18]

$$f(x;\mu,\sigma) = \frac{1}{\sigma\sqrt{2\pi}} \exp\left(-\frac{1}{2}\frac{(x-\mu)^2}{\sigma^2}\right)$$

The binomial distribution in Equation (4.3) closely approximates a Gaussian distribution when r and n are large. We can write the distribution as a continuous function of the stage number as

$$f \approx \frac{1}{\sqrt{rpq}\sqrt{2\pi}} \exp\left(-\frac{1}{2}\frac{(n-rp)^2}{rpq^2}\right) \tag{4.4}$$

Equation (4.4) is a very good approximation when the total number of stages is larger than 20 or when the product rpq is greater than or equal to 3. Comparing Equation (4.4) to the above classical relationship for a Gaussian or normal distribution leads to the following observation for the standard deviation, σ, of the distribution:

$$\sigma \approx \sqrt{rpq} \qquad (4.5)$$

The mean Craig stage is also that stage with the maximum fraction. This distribution mean is given by

$$\mu = rp \qquad (4.6)$$

Equation (4.6) enables a calculation of the mean in this Craig countercurrent distribution, and Equation (4.5) yields an estimate of the standard deviation in the distribution among Craig tubes. For example, let us return to our two solutes, 1,2-DCA and EG. Earlier, we established values for p and q from a knowledge of D or, in the limit of purely molecular partitioning, K_D. Let us find μ and σ for both compounds after 100 transfers (i.e., $r = 100$) have been performed:

$$D(1,2-DCA) = 4, \qquad D(EG) = 0.1$$
$$p(1,2)-DCA = 0.8, \quad p(EG) = 0.0909$$
$$q(1,2-DCA) = 0.2, \quad q(EG = 0.909)$$

Upon substituting these values into Equation (4.5), we obtain for 1,2-DCA

$$\sigma_{1,2-DCA} = \sqrt{rpq} = \sqrt{(100)(0.8)(0.2)} = 4$$

Upon substituting these values into Equation (4.6), we obtain for 1,2-DCA

$$\mu_{1,2-DCA} = rp = (100)(0.8) = 80$$

In a similar manner, for EG we obtain

$$\sigma_{EG} = \sqrt{(100)(0.01)(0.91)} = 2.9$$

$$\mu_{EG} = (100)(0.091) = 9.1$$

It becomes obvious now that differences in D or K_D result in different stages in which the maximum fraction appears. The degree of band broadening is also larger for the solute with the larger value of μ. When the fraction of solute in a given stage is plotted against the stage number, Gaussian-like distributions are produced. A sketch of such a plot is shown in Figure 4.3.

It appears that 100 transfers using a Craig countercurrent apparatus enabled a more than adequate separation of EG and 1,2-DCA. We have therefore found a way to separate organic compounds. Before we leave the countercurrent separation concept, let us discuss the significance of Equations (4.5) and (4.6) a bit further. Equation (4.6) suggests that each solute migrates a distance equal to a constant fraction of the solvent front, and Equation (4.5) suggests that the width of the peak increases with the square root of the number of transfers. Separation is achieved as the number of transfers increase. The distance that each peak travels is proportional to r, and the width of the peak is proportional to the square root of r. It is instructive to compare these findings from countercurrent extraction to those of analyte migration discussed earlier. Differential migration and countercurrent extraction

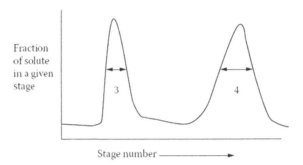

FIGURE 4.3

techniques serve to help us to begin thinking about separations. These techniques set the stage for the most powerful of separation methods, namely, chromatography.

4.5   WHAT IS CHROMATOGRAPHY?

Harris states that chromatography is a "logical extension of countercurrent distribution." (p. 627)[19] Chromatographic separation is indeed the countercurrent extraction taken to a very large number of stages across the chromatographic column. The following quotation is taken from an earlier text:[20]

> Chromatography encompasses a series of techniques having in common the separation of components in a mixture by a series of equilibrium operations that result in the entities being separated as a result of their partitioning (differential sorption) between two different phases; one stationary with a large surface and the other, a moving phase in contact with the first.

The "inventors" of partition chromatography, Martin and Synge, in 1941 first introduced gas chromatography this way:[21]

> The mobile phase need not be a liquid but may be a vapour … Very refined separations of volatile substances should therefore be possible in a column in which permanent gas is made to flow over gel impregnated with a nonvolatile solvent in which the substances to be separated approximately obey Raoult's law.

The following excerpt is titled "The King's Companions—A Chromatographical Allegory":[22]

> A great and powerful king once ruled in a distant land. One day, he decided he wanted to find the ten strongest men in his kingdom. They would be his sporting companions and would also protect him. In return, the king would give them splendid chambers in his palace and great riches.
>
> But how would these men be found? For surely thousands from his vast lands would seek this promise of wealth and power. From amongst these thousands, how would he find the ten very strongest?
>
> The king consulted his advisors. One suggested a great wrestling tournament, but that would be too time consuming and complicated. A weight-lifting contest was also rejected. Finally, an obscure advisor named *Chromos* described a plan that pleased the king.
>
> "Your majesty," said *Chromos*, "you have in your land a mighty river. Use it for a special contest. At intervals along the river, have your engineers erect poles. The ends of each pole should be anchored on opposite banks so that each pole stretches across the river. The pole must be just high enough above the surface of the river for a man being carried along by the current to reach up and grab hold of it. So strong is the current that he will not be able to pull himself out, but will just be able to hold on until, his strength sapped, the pole will be torn from his grasp. He will be carried downstream until he reaches the next pole which

he will also grasp hold of. Of course, the weakest man will be able to hold on to each pole for the shortest length of time, and will be carried downstream fastest. The strongest man will hold on the longest, and will be carried along most slowly by the river. You have only to throw the applicants into the river at one particular place and measure how long it takes each man to get to the finish line downstream (where he will be pulled out). As long as you have enough poles spaced out between the start and finish, the men will all be graded exactly according to their strength. The strongest will be those who take the longest time to reach the finish line."

So simple and elegant did this method sound, that the king decided to try it. A proclamation promising great wealth and power to the ten strongest men was spread throughout the kingdom. Men came to the river from far and wide to participate in the contest *Chromos* had devised and the contest was indeed successful. Simply and quickly, the combination of *moving* river and *stationary* poles separated all the applicants from one another according to their strength.

So the king found his ten strongest subjects, and brought them to his palace to be his companions and protectors. He rewarded them all with great wealth. But the man who received the greatest reward was his advisor, *Chromos*.

Do you see the analogy?

In Chapter 3, we have alluded to chromatographic separation as *the* most important separation technique to TEQA. Indeed, much of the innovative sample prep techniques for trace organics described in Chapter 3 are designed to enable a sample of environmental interest to be nicely introduced into a chromatograph. A chromatograph is an analytical instrument that has been designed and manufactured to perform either gas or liquid column chromatography.

Chromatography is a separation phenomenon that occurs when a sample is introduced into a system in which a mobile phase is continuously being passed through a stationary phase. Chromatography has a broad scope, in that small molecules can be separated as well as quite large ones. Of interest to TEQA are the separation, detection, and quantification of relatively small molecules. In Chapter 3, we introduced SPE as an example of frontal chromatography. In this chapter, we will discuss *elution chromatography* exclusively because this form of chromatographic separation lends itself to instrumentation. We will also limit our discussion of chromatography to column methods while being fully aware of the importance of planar chromatography, namely, paper and thin-layer chromatography, because our interest is in trace chromatographic analysis. We will further limit our discussion to the two major types of chromatography most relevant to TEQA: gas chromatography (GC) and high-performance liquid chromatography (HPLC). Figure 4.4 represents an attempt to place the major kinds of chromatographic separation science in various regions of a two-dimensional plot whereby analyte volatility increases from low to high along the ordinate, whereas analyte polarity increases from left to right along the abscissa. The reader should keep in mind that there is much overlap of these various regions and that the focus of the plot is on the use of chromatography as a separation concept without reference to the kinds of detector required. The horizontal lines denote regions where GC is not appropriate. The arrow pointing downward within the GC region serves to point out that the demarcation between volatile analytes and semivolatile ones is not clear cut. This plot reveals one of the reasons why GC has been so dominant in TEQA, while revealing just how limited GC as a determinative technique really is. This plot also reveals the *more universal nature* of HPLC in comparison to GC.

4.6 WHY IS GC SO DOMINANT IN TEQA?

There are several reasons for this, and GC is still the dominant analytical chromatographic determinative technique used in environmental testing labs today. Let us construct a list of reasons why this is so while acknowledging advances in HPLC:

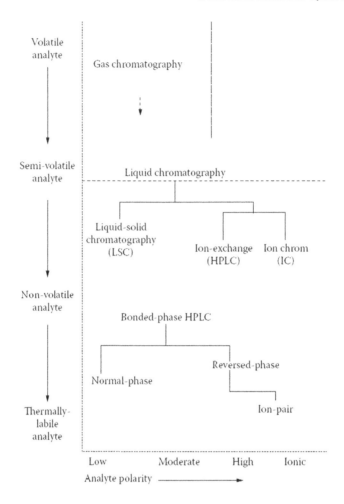

FIGURE 4.4

- Gas chromatography was historically the first instrumented means to separate organic compounds and was first applied in the petroleum industry.
- Gas chromatography continuously evolved from the packed column, where resolution was more limited to the capillary (open tubular) column. This advance occurred during the 1980s. Significant increases in chromatographic resolution R_s were realized. This achievement greatly increased the number of organic compounds that could be separated in a reasonable timeframe.
- The EPA organics protocol referred to in Figure 1.13 classifies priority pollutant organics based on the degree of volatility. Analyses that lie in the volatile and lower-molecular-weight semivolatile regions of Figure 4.4 are predominantly separated and quantitated by GC.
- Gas chromatography is simpler to comprehend in contrast to HPLC largely because the mobile phase in GC is chemically inert and contributes nothing to analyte retention and resolution.
- Detectors in GC can be of low, medium, and high sensitivity; highly sensitive GC detectors are of utmost importance to TEQA.
- The price and size of the GC-MS (transmission quadruple) instrument has declined over the past three decades. The instrument has become more sensitive and more robust. The

reduced cost and versatility of transmission quadrupole mass spectrometers in comparison to high-resolution magnetic sector mass spectrometers with respect to achieving the goals of TEQA has been realized.

- Transmission quadrupole mass spectrometers are versatile, rugged, easily accessible, and readily interfaced to both robotic autosamplers and computers for sample introduction, instrument control and for data acquisition. However, the advances made in LC-MS-MS, (commonly called "the LC triple quad") *three tandem-in-space transmission quadrupoles*, have made important in-roads into contemporary environmental laboratories. Polar analytes of environmental concern can be directly injected into such instruments without the need for chemical derivatization.

4.7 WHY IS HPLC MORE UNIVERSAL IN TEQA?

There are several reasons for this:

- High-performance liquid chromatography (HPLC) occupies a much larger region of Figure 4.4, and this fact suggests that a much larger range of organic compounds are amenable to analysis by HPLC in contrast to GC.
- HPLC can take on several different forms depending on the chemical natures of both the mobile phase and the stationary phase, respectively. This leads to a significant rise in the scope of applications.
- Chemical manipulation of the mobile phase in HPLC enables gradient elution to be conducted.
- The analyte of interest remains dissolved in the liquid phase and can be thermally labile, unlike GC, whereby an analyte must be vaporized and remain thermally stable.

4.8 CAN WE VISUALIZE A CHROMATOGRAPHIC SEPARATION?

Yes, we can. Figure 4.5 depicts a hypothetical separation of a three-component mixture. Introduction of the mixture is diagrammatically shown in snapshot 1; elution of the mixture with a mobile phase begins in snapshot 2. Snapshots 3 and 4 depict the increase in chromatographic resolution, R_s as the mixture moves through the column. Four chief parameters are

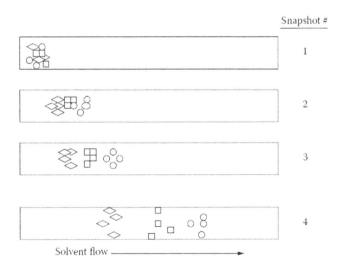

FIGURE 4.5

used to characterize a chromatographic separation: distribution coefficient, retention or capacity factor, selectivity and column efficiency, and number of theoretical plates. We will spend the next few paragraphs developing the mathematics underlying chromatographic separation.

4.9   CAN WE DEVELOP USEFUL MATHEMATICAL RELATIONSHIPS FOR CHROMATOGRAPHY?

Yes, we can. Let us begin by first recognizing that an analyte that is introduced into a chromatographic column, much like that shown in Figure 4.5, distributes itself between a mobile phase, m, and a stationary phase, s, based on the amount of analyte distributed instead of the concentration distributed. The fraction of the ith analyte in the stationary phase, ϕ_i^s, is defined as follows:

$$\phi_i^s = \frac{\text{amt}_i^s}{\text{amt}_i^{total}} = \frac{C_i^s V^s}{C_i^m V^m + C_i^s V^s} \tag{4.7}$$

where C_i^m and C_i^s are the ith analyte concentrations in both phases. V^m and V^s are the volumes of both phases. Let us define a molecular distribution constant for this ith analyte, K_i as follows:

$$K_i = \frac{C_i^s}{C_i^m} \tag{4.8}$$

Equation (4.8) can be substituted into Equation (4.7) to yield the fraction of the ith analyte in terms of the molecular distribution constant and ratio of phase volumes according to:

$$\phi_i^s = \frac{K_i V^s / V^m}{1 + K_i V^s / V^m} \tag{4.9}$$

We now define the capacity factor, k', a commonly used parameter in all of chromatography, as a ratio of the amount of analyte i in the stationary phase to the amount of analyte i in the mobile phase at any one moment:

$$k' = \frac{\text{amt}_i^s}{\text{amt}_i^m} = \frac{C_i^s V^s}{C_i^m V^m} = K_i \left(\frac{V^s}{V^m} \right) \tag{4.10}$$

As we discussed extensively in Chapter 3, when secondary equilibria are involved, K is replaced by D. Combining Equations (4.9) and (4.10) gives a relationship between the capacity factor and the fraction of analyte i distributed into the stationary phase:

$$\phi_i^s = \frac{k'}{1 + k'} \tag{4.11}$$

Also, the fraction ϕ^m of analyte i in the mobile phase is

$$\phi_i^m = \frac{1}{1 + k'} \tag{4.12}$$

Equation (4.12) gives the fraction of analyte i, once injected into the flowing mobile phase, as it moves through the chromatographic column. This analyte migrates only when in the mobile phase. The velocity of the analyte through the column, v_s, is a fraction of the mobile-phase velocity v according to

$$v_s = v\phi^m \tag{4.13}$$

Equation (4.13) suggests that the analyte does not migrate at all ($v_s = 0$) and when $\phi^m = 1$. It also suggests that the analyte moves with the same velocity as the mobile phase ($v_s = v$). The analyte velocity through the column equals the length, L, of the column divided by the analyte retention time, t_R:

$$v_s = L / t_R$$

The velocity of the mobile phase is given by the length of column divided by the retention time of an unretained component, t_0, according to

$$v = L / t_0$$

Substituting for v_s and v in Equation (4.13) yields

$$t_R = \frac{t_0}{\phi^m} \tag{4.14}$$

The mobile-phase volumetric flow rate, F, expressed in units of cubic centimeters per minute or milliliters per minute, is usually fixed and unchanging in chromatographic systems. Thus, because $t_R = V_R/F$ and $t_0 = V_0/F$, Equation (4.14) can be rewritten in terms of retention volumes:

$$V_R = \frac{V_0}{\phi^m}$$

The retention volume, V_R, for a given analyte can be seen as the product of two terms, V_0 and the reciprocal of ϕ^m:

$$= V_0 \left(\frac{1}{\phi^m} \right) \tag{4.15}$$

Substituting Equation (4.12) into the above relationship gives

$$= V_0 \left(\frac{1}{1/(1+k')} \right)$$

Upon rearranging and simplifying,

$$V_R = V_0 \left[1 + k' \right] \tag{4.16}$$

Equation (4.16) has been called *the* fundamental equation for chromatography. Each and every analyte of interest that is introduced into a chromatographic column will have its own capacity factor, k'. The column itself will have a volume V_0. The retention volume for a given analyte is then viewed in terms of the number of column volumes passed through the column before the analyte is said to elute. A chromatogram then consists of a plot of detector response

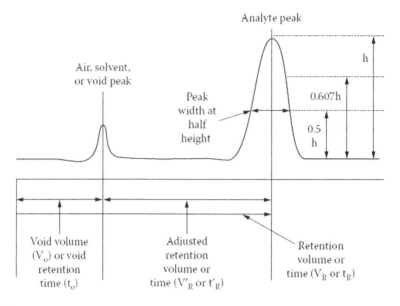

FIGURE 4.6

vs. the time elapsed after injection, where each analyte has a unique *retention time* if sufficient chromatographic resolution is provided. Hence, with reference to a chromatogram, the capacity factor becomes

$$k' = \frac{t_R - t_0}{t_0} \tag{4.17}$$

Figure 4.6 is an illustrative chromatogram that defines what one should know when examining either a GC or an HPLC chromatogram. Because the two peaks shown in Figure 4.3 have different retention times, their capacity factors differ according to Equation (4.17). The ratio of two capacity factors, α, relates to the degree that a given chromatographic column is selective:

$$\alpha = \frac{k_2'}{k_1'} \tag{4.18}$$

Let us assume that we found a column retains the more polar EG with respect to 1,2-DCA. Equation (4.13) would suggest that the $\phi_{1,2-DCA}^{m}$ is much larger than ϕ_{EG}^{m}. The unretained solute peak in the chromatogram, often called the chromatographic dead time or dead volume, might be due to the presence of air in GC or a solvent in HPLC. The adjusted retention volume, V_R', and adjusted retention time, t_R', are defined mathematically as

$$V_R' = V_R - V_0$$

$$t_R' = t_R - t_0$$

We need now to go on and address the issue of peak width. As was pointed out earlier, the longer an analyte is retained in a chromatographic column, the larger is the peak width. This is almost a universal statement with respect to chromatographic separations.

4.10 HOW DOES ONE CONTROL THE CHROMATOGRAPHIC PEAK WIDTH?

The correct answer is to minimize those contributions to peak broadening. These factors are interpreted in terms of contributions to the height equivalent to a theoretical plate (HETP). This line of reasoning leads to the need to define a HETP that in turn requires that we introduce the concept of a theoretical plate in chromatography. So let us get started.

The concept of a theoretical plate is rooted in both the theory of distillation and the Craig countercurrent extraction. A single distillation plate is a location whereby a single equilibration can occur. We already discussed the single equilibration that occurs in a single Craig stage. Imagine an infinite number of stages, and we begin to realize the immense power of chromatography as a means to separate chemical substances. An equilibration of a given analyte between the mobile phase and stationary phase requires a length of column, and this length can be defined as H. A column would then have a length L and a number of these equilibrations denoted by N, the number of theoretical plates. Hence, we define the HETP, abbreviated H for brevity here, as follows:

$$H = \frac{L}{N} \tag{4.19}$$

The number of theoretical plates in a given column, N, is mathematically defined as the ratio of the square of the retention time, t_R, or the retention volume, V_R (note that this is the apex of the Gaussian peak), of a particular analyte of interest over the variance of that Gaussian peak. Expressed mathematically,

$$N = \left(\frac{t_R}{\sigma_t}\right)^2 = \left(\frac{V_R}{\sigma_1}\right)^2 \tag{4.20}$$

The number of theoretical plates can be expressed in terms of the width of the Gaussian peak at the base. This is expressed in units of time, t_w where it is assumed that t_w approximates four standard deviations or, mathematically, $t_w = 4\sigma_t$, so that upon substituting for σ_t,

$$N = \frac{t_R^2}{\left(t_w / 4\right)^2} = 16\left(\frac{t_R}{t_w}\right)^2 \tag{4.21}$$

Columns that significantly retain an analyte of interest (i.e., have a relatively large t_R) and also have a narrow peak width at base, t_w, must have a large value for N according to Equation (4.21). Columns with large values for N, such as from 1,000 to 10, 000 theoretical plates, are therefore considered to be highly efficient. Many manufacturers prefer to cite the number of theoretical plates per meter instead of just the number of theoretical plates. The concept that the number of theoretical plates for a column (be it a GC or an HPLC column) can be calculated from the experimental GC or HPLC chromatogram is an important practical concept. In the realm of GC, when open tubular columns or capillary replaced packed columns, it was largely because of the significant difference in N offered by the former type of column. The second most useful measurement of N is to calculate N from the width of the peak at half height, $t_w^{1/2}$, using the equation

$$N = 5.55\left(\frac{t_R}{t_w^{1/2}}\right)^2 \tag{4.22}$$

It is up to the user as to whether Equation (4.21) or (4.22) is used to estimate N. Figure 4.6 defines the peak width at half height and the peak width at the base. When a GC or HPLC column is purchased from a supplier, pay close attention to how the supplier calculates N. It is also recommended that a peak be chosen in a chromatogram to calculate N whose k' is around 5. To continue our discussion of peak broadening, we need to return to the height equivalent to a theoretical plate, H.

4.11   IS THERE A MORE PRACTICAL WAY TO DEFINE H?

Yes there is. Equation (4.19) defines H in terms of column length and a dimensionless parameter N. We will now derive an expression for H in terms of the chromatogram in units of time. We start by considering a chromatographic column of length L to which a sample has been introduced. This sample will experience band broadening as it makes its way through the column. We know that the degree of band broadening denoted by σ has units of distance.

H can be defined as the ratio of the variance, in units of distance, over the column length according to

$$H = \frac{\sigma^2}{L} \tag{4.23}$$

This dispersion in distance units can be converted to time units by recognizing that

$$\sigma = \phi^m v \sigma_t$$

Upon substituting for σ and substituting into Equation (4.23), and doing some algebraic manipulation while recognizing that

$$t_R = \frac{L}{\phi^m v}$$

we now have an equation that relates H to the variance of the chromatographically resolved peak in time units according to

$$H = \frac{L \sigma_t^2}{t_R^2} \tag{4.24}$$

Equation (4.24) suggests that the height equivalent to a theoretical plate can be found from a knowledge of the length of the column, the degree of peak broadening as measured by the peak variance, in time units, and the retention time. We now discuss those factors that contribute to H because Equations (4.23) and (4.24) show that H is equal to the product of a constant and a variance. If we can identify those distinct variances, σ_i, that contribute to the overall variance, $\sigma_{overall}$, then these individual variances can merely be added. Expressed mathematically, for the ith independent contribution to chromatographic peak broadening, the statistics of propagation of error suggest that

$$\sigma_{overall}^2 = \sum_i \sigma_i^2$$

The concept that a rate theory is responsible for contributions to chromatographic peak broadening was first provided by van Deemter, Klinkenberg, and Zuiderweg. The random walk is the simplest molecular model and is due to Giddings.[23]

4.12 WHAT FACTORS CONTRIBUTE TO CHROMATOGRAPHIC PEAK BROADENING?

Contributions to the overall height equivalent to a theoretical plate abbreviated HETP or simply H! It is important for the practicing chromatographer to understand the primary reasons why the mere injection of a sample into a chromatographic column will lead to a widening of the peak width. We alluded to peak broadening earlier when we introduced band migration. We will not provide a comprehensive elaboration of peak broadening. Instead, we introduce the primary factors responsible for chromatographic peak broadening. Following this, we introduce and discuss the van Deemter equation. The concept is termed chromatographic rate theory and is adequately elaborated on in the analytical chemistry literature elsewhere. (pp. 135–155)[17,24](pp. 587–597)[25]

Equation (4.23) is the starting point for discussing those factors that broaden a chromatographic peak. By the time the solute molecules of a sample that have been injected into a column have traveled a distance L, where L is the length of the GC or HPLC column, a Gaussian profile emerges. At the end of the column where the GC or HPLC detector is located, the peak has been broadened, whereby one standard deviation has a length defined by $L - \sigma$ to the left of the peak apex at L, and $L + \sigma$ to the right of the peak apex at L. H can now be thought of as the length of column, at the end of the column, that contains a fraction of analyte that lies between $L - \sigma$ and L. (p. 586)[25] The fact that there exists a minimum H in a plot of H vs. linear flow rate, u, suggests that a complex mathematical relationship exists between H and u. The following factors have emerged:

- Multiple paths of solute molecules, the A term, are present only in packed GC and HPLC columns and absent in open tubular GC columns. This term is also called eddy diffusion.
- Longitudinal diffusion, the B term, is present in all chromatographic columns.
- Finite speed of equilibration and the inability of solute molecules to truly equilibrate in one theoretical plate, the C term, are present in all chromatographic columns. This term is also called resistance-to-mass transfer and, in more contemporary versions, consists of two mass transfer coefficients: C_S, where S refers to the stationary phase, and C_M, where M refers to the mobile phase. Equilibrium is established between M and S so slowly that a chromatographic column always operates under nonequilibrium conditions. Thus, analyte molecules at the front of a band are swept ahead before they have time to equilibrate with S and thus be retained. Similarly, equilibrium is not reached at the trailing edge of a band, and molecules are left behind in S by the fast-moving mobile phase. (p. 590)[25]

The above three factors broaden chromatographically resolved peaks by contributing a variance for each factor, starting with Equation (4.23), as follows:

$$H = \frac{1}{L}\left(\sigma^2\right) = \frac{1}{L}\left(\sum_i \sigma_i^2\right)$$

$$H = H_L + H_s + H_M$$

where H_l is the contribution to H due to *longitudinal diffusion*, H_s is the contribution to H due to *resistance-to-mass transfer to S*, and H_m is the contribution to H due to *resistance-to-mass transfer to M*. We will derive only the case for longitudinal diffusion and state the other two without derivation.

4.13   HOW DOES LONGITUDINAL DIFFUSION CONTRIBUTE TO *H*?

Molecular diffusion of an analyte of environmental interest in the direction of flow is significant only in the mobile phase. Its contribution, σ_L^2, to the total peak variance can be found by substituting the molecular diffusivity and time into the Einstein equation:

$$\sigma^2 = 2Dt$$

On average, solute molecules spend the time $t = L/u$ in the mobile phase, so that the variance in the mobile phase is given as

$$\sigma_L^2 = \frac{2D_M L}{u}$$

where D_M is the solute diffusion coefficient in the mobile phase. The plate height contribution of longitudinal diffusion, H_L, is then obtained as

$$H_L = \frac{\sigma^2}{L} = \frac{2\gamma D_M}{u}$$

$$= \frac{B}{u}$$

where γ is an obstruction factor that recognizes that longitudinal diffusion is hindered by the packing or bed structure. H_L is usually only a small contributor to H. However, when D_M is large and the mobile-phase velocity, u, is small, does H_L become significant? H_L is far more important in GC than in HPLC.

4.14   HOW DOES ALL OF THIS FIT TOGETHER?

Mass transfer into the stationary phase and the mobile-phase contribution to plate height give the terms C_S and C_M, respectively, to the total plate height in direct proportion to u. This is so because, unlike longitudinal diffusion, molecules diffuse at an angle of 90° with respect to the direction of the mobile-phase flow. The larger the mobile-phase velocity, the greater is the diffusion in this direction. Hence, we have a relationship between H and the linear mobile-phase velocity u according to

$$H = A + \frac{B}{u} + C_s u + C_M u \tag{4.25}$$

This is a more contemporary *van Deemter equation*, and this equation takes on different contributions to the terms A, B, C_S and C_M, depending on which form of chromatography is employed. We will introduce specific parameters that comprise both C terms when we discuss GC and HPLC. Equation (4.25) is plotted in Figure 4.7. This plot is obtained by measuring H as the linear velocity of the mobile phase for the solute ethyl acetate dissolved in *n*-hexane and chromatographed on a normal-phase, silica-based HPLC column. The plot includes the eddy diffusion or the A term and shows the independence of eddy diffusion with respect to the mobile-phase flow rate. The plot shows the inverse relationship between H and u with respect to longitudinal diffusion. The plot also shows the near linear relationship between H and u with respect to mass transfer in both phases. If we differentiate Equation (4.25) with respect to u and set this derivative equal to zero, upon solving for u we find

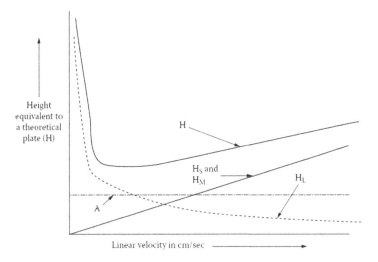

FIGURE 4.7

$$u_{optimum} = \sqrt{\frac{B}{C}}$$

In practice, the mobile-phase linear velocity is set slightly higher than $u_{optimum}$ so as to quicken the chromatographic run time (i.e., the time between injection and separation, and then detection). Figure 4.8 is a plot of H vs. u for a packed GC column and for an open tubular GC column. H is much lower for the open tubular column because multiple flow paths are eliminated. Note also that the curvature in the plot in Figure 4.8 for the open tubular column is much less than that of the packed column. This much wider range of optimal mobile-phase velocities enables the chromatographer to use a much larger range of volumetric flow rates without sacrificing H. While we are discussing flow rates, we need to point out the significant difference between linear flow and volumetric flow with respect to column chromatography.

4.15 HOW DO WE DISTINGUISH BETWEEN LINEAR AND VOLUMETRIC FLOW RATES?

Notice that Equation (4.25) examines column efficiency as a complex function of linear mobile-phase velocity. It is the linear velocity that conducts analytes of interest through a chromatographic column to the detector. Any comparison of van Deemter curves, such as that shown in Figure 4.8, must use linear velocity because the influence of the column radius is eliminated. Jennings has articulated an interesting relationship between linear and volumetric flow rates, F, incorporating the column radius, r_c, according to: (p. 18)[26]

$$u\left(cm/s\right) = \frac{1.67F\left(cm^3/min\right)}{\pi r_c\left(mm\right)^2} \tag{4.26}$$

Figure 4.8 reveals that with respect to the average linear velocity, the optimum linear velocity for a packed GC column is lower than that for an open tubular or capillary GC column. Equation (4.26) is used to calculate u for the five most commonly used GC column diameters, and these results are shown in Table 4.4. The columns are listed from the smallest commercially available

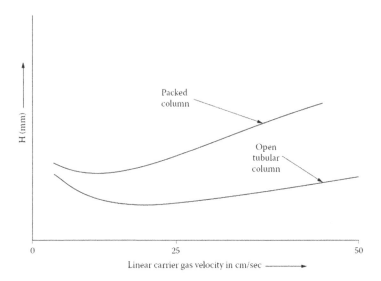

FIGURE 4.8

TABLE 4.4
Characteristics of Capillary GC Columns Having Different Internal Diameters

GC Column	r_c (mm)	F (mL/min)	u (cm/sec)
Capillary, 0.25 mm i.d.	0.125	0.75	35.6
Capillary, 0.32 mm i.d.	0.16	1.5	30.7
Capillary, 0.53 mm i.d.	0.265	5.0	38.0
Packed, 1/8 in. o.d.	3.175	25	1.3
Packed, 1/4 in. o.d.	6.35	75	1.0

---

diameter to the largest, along with a representative volumetric flow rate passing through the column. It then becomes evident that open tubular columns exhibit linear flow rates that are three to five times higher than packed columns, despite the fact that the user would have to replace compressed gas tanks less frequently. Linear flow rates can also be determined independently of Equation (4.26) by measuring the retention time of an unretained component of the injected sample. In the case of GC, injection of methane or, as this author has done, *injection of the butane vapor from a common cigarette lighter* using a 10 µL liquid-handling glass syringe gives t_0. Knowing the length of the column, L, enables the linear velocity to be calculated according to

$$u = L / t_0$$

In reversed-phase HPLC with ultraviolet absorption detection (RP-HPLC-UV), this author has used as a *source of an unretained component* the strongly absorbing and water-miscible solvent acetone. This measured retention time for acetone, commonly termed the void or dead time, can be used to calculate the linear velocity. Our understanding of what causes chromatographic peaks to broaden as they elute has led to the achievement of columns that maximize chromatographic resolution.

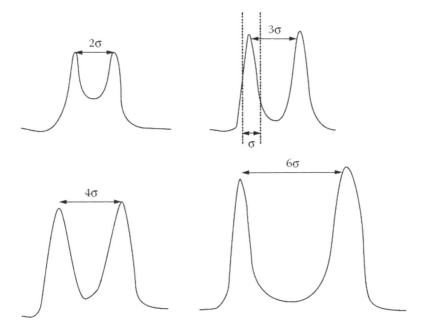

FIGURE 4.9

4.16   WHAT IS CHROMATOGRAPHIC RESOLUTION?

With reference to either a GC or HPLC chromatogram, the resolution, R_s, is defined as the ratio of the distance between retention times of two separated peaks to the average base peak width, t_w, of both peaks, 1 and 2. Expressed mathematically,

$$R_s = \frac{t_R^2 - t_R^1}{\frac{1}{2}\left(t_w^1 + t_w^2\right)}$$

$$= \frac{2\left(t_R^2 - t_R^1\right)}{t_w^1 + t_w^2} \tag{4.27}$$

Four chromatograms are presented in Figure 4.9. It becomes visually obvious that resolution increases as one goes from top to bottom in the figure. In a manner similar to that used to derive Equation (4.21), we recognize that the *peak width at the base is equal to four times the standard deviation of the Gaussian peak profile* (i.e., $t_w = 4\sigma_t$). We also assume that the variance of both peaks is equal, so that $\sigma_t^1 = \sigma_t^2 = \sigma_t$. Upon substituting this into Equation (4.27), we get

$$R_s = \frac{2\left(t_R^2 - t_R^1\right)}{8\sigma_t} = \frac{t_R^2 - t_R^1}{4\sigma_t} \tag{4.28}$$

Equation (4.28) shows that the larger the disengagement of a pair of chromatographically separated peaks, the greater the resolution. Equation (4.28) also shows that the smaller the variance of a peak, the greater the resolution. This equation can be viewed as a ratio of a change in retention time to the standard deviation, and different scenarios can be introduced. If we let Δt_R

represent a change in the retention time for a given pair of peaks that have been partly or entirely resolved chromatographically, we can rewrite Equation (4.28) to give

$$R_s = \frac{\Delta t_R}{4\sigma_t} \tag{4.29}$$

Equation (4.29) suggests that when $R_s = 0.5$, $\Delta t_R = 2\sigma_\tau$, the distance in time units between both peak apexes equals two standard deviations of the Gaussian peak profile: when $R_s = 0.75$, $\Delta t_R = 3\sigma_\tau$; when $R_s = 1.0$, $\Delta t_R = 4\sigma_\tau$; and when $R_s = 1.5$, $\Delta t_R = 6\sigma_\tau$. Again, refer to Figure 4.9, where each of these four different values for chromatographic resolution is shown. Hence, if we were to calculate a resolution whereby $R_s = 1.0$, Figure 4.9 suggests that baseline resolution has not been attained. Computerized software has enabled a partially resolved chromatographic peak to be accurately quantitated and obviates the need for baseline resolution in many cases. Equations (4.28) and (4.29) are important definitions of R_s for a given situation. These relationships do not, however, show how R_s relates to the fundamental parameters of chromatographic separation. To establish this, we need to derive *the fundamental resolution equation* in column chromatography.

4.17   HOW DO YOU DERIVE THIS FUNDAMENTAL EQUATION?

Few elementary treatments of the topic of chromatography take the reader from all that we have developed so far to a consideration of exactly how resolution depends on chromatographic efficiency, selectivity, and capacity factor. We begin to do this here by following a *derivation originally presented to the analytical literature* by Karger, Snyder, and Horvath in their classic text on separation science. (pp. 150–155)[17] Let us start with Equation (4.29), but before we do this, we need to find a way to incorporate N into this equation. Let us consider a separation of peaks 1 and 2 whose peak apexes are separated by Δt_R. Referring back to Equation (4.20), we can rewrite this equation with reference to peak 2 for a retention time for peak 2 at t_2 and standard deviation σ_2, and solve for σ_2 to give

$$\sigma_2 = \frac{t_2}{\sqrt{N}}$$

Upon substituting σ_2 back into Equation (4.29), we obtain

$$R_s = \frac{\Delta t_R}{4\left(t_2 / \sqrt{N}\right)} = \frac{\sqrt{N}}{4}\left(1 - \frac{t_1}{t_2}\right)$$

At this point, we have the resolution in terms of the number of theoretical plates and retention times. We can proceed even further. By combining Equations (4.13) and (4.15) by eliminating ϕ^m, we obtain a relationship for retention times in terms of the capacity factor, k', according to

$$\frac{t_R}{t_0} = 1 + k'$$

For peak 1, we have $t_1/t_0 = 1 + k'_1$, and likewise for peak 2, $t_2/t_0 = 1 + k'_2$, so that

$$\frac{t_1}{t_2} = \frac{1 + k'_1}{1 + k'_2}$$

We can now express resolution in terms of N and a ratio of capacity factors according to

$$R_s = \frac{\sqrt{N}}{4}\left(1 - \frac{1+k_1'}{1+k_2'}\right)$$

$$= \frac{\sqrt{N}}{4}\left(\frac{k_2' - k_1'}{1+k_2'}\right)$$

Because $\alpha = k_2'/k_1'$, we can solve for $k_1' = k_2'/\alpha$ and substitute for k_1' above to get

$$R_s = \frac{\sqrt{N}}{4}\left(\frac{k_2' - k_2'/\alpha}{1+k_2'}\right)$$

Upon simplifying, we obtain

$$R_s = \frac{\sqrt{N}}{4}\left(\frac{k_2'(1-1/\alpha)}{1+k_2'}\right)$$

Upon rearranging terms, we arrive at *the* fundamental resolution equation:

$$R_s = \frac{\sqrt{N}}{4}\left(\frac{k'}{1+k_2'}\right)\left(\frac{\alpha-1}{\alpha}\right) \tag{4.30}$$

Equation (4.30) suggests that the degree of chromatographic resolution depends chiefly on three factors: N, k', and α. N, the number of theoretical plates in a column, relates how efficient a chromatographic column is. N is independent of the chemical nature of the analyte of interest. k_2', the capacity factor of the more retained component in a given pair, depends on the chemical nature of the analyte. k', in general, is related to the product of the analyte's partition constant K and on the phase ratio β. The phase ratio in column chromatography is defined as the ratio of the stationary-phase volume, V_s, to the mobile-phase volume, V_m. The stationary-phase volume takes on different values depending on how the stationary phase is defined with respect to the column support. Jennings has stated that with respect to GC, β is between 5 and 35 for packed GC columns and between 50 and 1,000 for open tubular GC. (p. 14) [26] In any event, the capacity factor for a given analyte can be calculated by knowing the partition constant and phase ratio according to

$$k' = K\beta = K\frac{V_s}{V_m} \tag{4.31}$$

α, the chromatographic separation factor between two adjacent peaks and often called, for simplicity, the column selectivity, relates to the ratio of k' values for both peaks of interest as introduced earlier. N can be changed by increasing or decreasing the column length and adjusting the flow rate to minimize H while maximizing N for a column of fixed length. Changes in α and k' are achieved by selecting different mobile- and stationary-phase chemical compositions or by varying the column temperature or, in some cases, the column pressure. k' can also be changed for a given component by changing the relative amounts of stationary vs. mobile phase

TABLE 4.5
Chromatographic Selectivity and the Contribution of α
to the Fundamental Resolution Equation

α	$(\alpha - 1)/\alpha$
1	0
1.1	0.091
1.5	0.333
2	0.5
5	0.8
10	0.9
20	0.95
100	0.99

according to Equation (4.31). Equation (4.30) shows that these three factors enter into the resolution equation in a complex manner. We now examine the mathematical nature of each term in Equation (4.30).

4.18   WHAT IS EQUATION (4.30) REALLY TELLING US?

The first implication of Equation (4.30) is to realize that R_S approaches zero (i.e., no resolution between chromatographically resolved pairs of peaks) when N or k'_2 approaches zero or when α approaches 1. Note the nature of the third term in Equation (4.30). As the magnitude of α increases, the contribution of the third term serves to increase R_S. The effect of the α term is shown in Table 4.5. The most significant gains in chromatographic resolution occur between values of α that are greater than 1, yet diminishing returns set in if α is raised above 5. A similar argument can be made for the effect of the second term in Equation (4.30) (i.e., the k' term).

The effect of the k' term is shown in Table 4.6. It appears that very small values of k' contribute little to increased R_S, whereas, again, diminishing returns are evident as k' is increased. The greatest gains are found for a range of k' values between 2 and 5. Equation (4.30) also suggests that resolution varies with the square root of the number of theoretical plates. This is an important feature of Equation (4.30) to remember. A mere doubling of the length of the column will only increase R_S by 1.4.

To summarize the discussion that led to Equation (4.30), Karger, Snyder, and Hovath have stated it best: (pp. 150–155)[17]

Although each of these three parameters—N, α, and $k'2$—is important in controlling resolution, we usually do not attempt their simultaneous optimization in an actual separation. Experimental conditions are selected initially that favor large N values, within the practical limits of convenience plus reasonable separation times. Higher N values always provide improved resolution, other factors being equal, and this is true of analytical or preparative separations, and simple or complex mixtures. With a reasonable starting value of N, adequate resolution will be attained in most cases if we optimize k'_2 approximately. In gas chromatography an optimum value of k'_2 can be achieved by varying the temperature. In liquid chromatography it is more profitable to vary systematically the composition of the mobile phase.

TABLE 4.6
Chromatographic Capacity Factor and the Contribution
of k' to the Fundamental Resolution Equation

k'	k'/(1 + k')
0.1	0.0909
0.5	0.333
1.0	0.500
2.0	0.666
5.0	0.833
10	0.909
20	0.952
50	0.980

One other relationship is worth mentioning before we discuss the determinative techniques GC and HPLC. Equation (4.30) can be solved algebraically for the number of theoretical plates, and this gives

$$N_{\text{required}} = 16R_s^2 \left(\frac{\alpha}{\alpha-1} \right)^2 \left(\frac{k'+1}{k'} \right)^2 \qquad (4.32)$$

Equation (4.32) is of practical importance in that for a given resolution, the required number of theoretical plates can be found. To illustrate, let us calculate N_{required} for the chromatographic separation of two solutes with a given resolution $R_s = 1.5$ for two different values for the selectivity, $\alpha = 1.05$ and $\alpha = 1.10$. (p. 85)[26] The results of the application of Equation (4.32) are shown in Table 4.7. For two solutes that have a value for $k' = 0.01$, the magnitude of the k' term in Equation (4.32) is such as to require over a 100 million plates to realize a value for $\alpha = 1.05$. For two solutes that have a value for $k' = 1.0$, the magnitude of the k' term in Equation (4.32) is such as to require over 50,000 plates to realize a value for $\alpha = 1.05$. For two solutes that have a value for $k = 10$, the magnitude of the k' term in Equation (4.32) is such as to require over 10,000 plates to realize a value for $\alpha = 1.05$.

The fundamental theory of chromatographic separation has been presented in the broadest of terms. We next proceed to discuss *the* most common determinative technique for measuring trace concentration levels of organics in the environment, *gas–liquid or gas–solid chromatography*.

4.19 HOW DOES A GC WORK?

A gas chromatograph provides the proper conditions for the separation and detection of any chemical substance amenable to analysis by this determinative technique. Those organic compounds such as priority pollutants, pesticides, herbicides, rodenticides, solvents, chemical warfare agents, etc.—present and persistent chemical substances found in the environment that are amenable to analysis by GC—become the principal focus here. A GC works by properly installing the instrument in a laboratory and understanding how to optimize the separation and detection of analytes of interest. After being installed, a GC column that is appropriate to the intended application is installed into the oven of the GC. Figure 4.10 is a generalized schematic diagram of a conventional GC employing a packed column. This schematic might be one that appeared over 30 years

TABLE 4.7
Influence of k' on the Required Number of Theoretical Plates at Two Different Values for α for a Given R_s Applying Equation (4.32)

k'	$[(k'+ 1)/k']2$	$N_{required}$ ($\alpha = 1.05$)	$N_{required}$ ($\alpha = 1.10$)
0.01	10,201	162,000,000	44,000,000
0.05	441	7,000,000	1,900,000
0.10	121	1,900,000	527,000
0.15	58.8	930,000	256,000
0.5	9	143,000	39,000
1.0	4	63,500	17,500
3.0	1.8	28,600	7800
5.0	1.4	22,200	6100
10	1.1	17,500	4800
50	1.05	16,700	4600

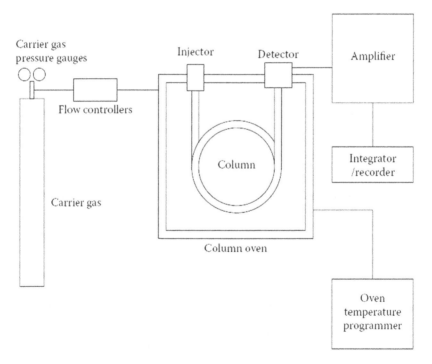

FIGURE 4.10

ago during the era of packed-column GC. Nevertheless, this schematic serves as a good starting point. We will answer the question posed above by proceeding from left to right across the diagram in Figure 4.10 and point out how each of these instrument components has evolved since the first commercial GC appeared in the year 1954. Details of how one operates a GC can be found in the appropriate experiments in Chapter 5. Before we proceed to a description of a gas chromatograph, let us view the outcome of injecting a sample into such an instrument.

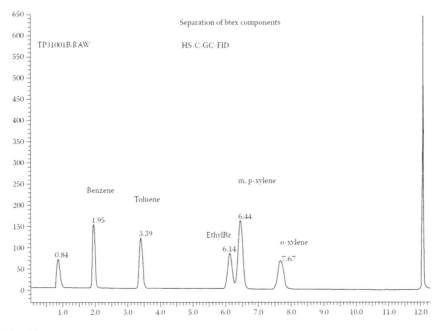

FIGURE 4.11

Figure 4.11 is an actual GC chromatogram (gas chromatogram) and shows the importance of a properly installed and optimized GC as a determinative technique. This chromatogram, if it were a real contaminated groundwater sample would serve as a good example of *enviro-chemical* TEQA. A groundwater sample that has been in contact with gasoline, perhaps from a leaking underground storage tank, was placed in a 22 mL headspace vial and sealed. The temperature of the vial was brought to 65°C and kept at that temperature for approximately ½h. The headspace was sampled using a gas-tight syringe and injected into an Autosystem® GC, manufactured by PerkinElmer Corporation. The abbreviation HS-C-GC-FID refers to *headspace capillary column-gas chromatograph-flame ionization detection*. BTEX refers to benzene, toluene, ethyl benzene, and *meta-, para-, and ortho-xylene*. Figure 4.12 shows the molecular structures for the five organic compounds that comprise the BTEX mix.

Note that the order of elution in the GC chromatogram of Figure 4.11 is from lower to higher molecular weight. In other words, the capacity factor, k', for benzene is much smaller than for *ortho*-xylene. From Equation (4.31), this suggests that *ortho*-xylene has a much larger partition constant, K, into the column stationary phase in comparison to benzene. Values of k' can also be correlated with increases in boiling point for these substituted aromatics. This particular open tubular column lacks the necessary selectivity, α, to separate the *meta*-from the *para*-xylene isomer. Also note that resolution is more than adequate for all BTEX compounds except for ethyl benzene from m, p-xylene. This chromatogram was acquired by sending the analog signal from the FID to an analog-to-digital interface. The digitized data was processed by Turbochrom® software, developed by PE-Nelson. We now return to our journey through Figure 4.10.

4.20 WHAT ARE EXTERNAL GAS PNEUMATICS AND ARE THEY IMPORTANT IN GC?

Any chromatographic separation is such that matter is forced to become more ordered. In other words, its entropy is reduced and is against the spontaneous tendency of matter to spread out. This requires an input of energy, and this input in the case of GC is provided by the potential

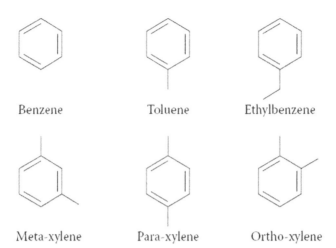

FIGURE 4.12

energy of a compressed gas. Today, the chromatographer has two options with respect to carrier gas: compressed gas cylinders or gas generators. Gas generators require a large investment;
however, the return on that investment is the elimination of gas cylinder handling. The annual
catalogs of chromatography suppliers are good sources to learn about gas generators. Carrier
gas under pressure enters the GC and provides the necessary mobile phase to enable either a
gas–liquid or gas–solid chromatographic separation to occur. Let us assume that compressed
gas cylinders are used as the source of carrier gas. A two-stage regulator is necessary to control
the gas pressure delivered to the instrument from the source. It is important to clearly distinguish between a single-stage and two-stage gas regulator. Figure 4.13 depicts schematics for
both types. Notice that in the two-stage regulator (Figure 4.13 (bottom)), the first stage is preset
and the compressed gas pressure from the tank gets reduced to 300 to 500 psi (pounds per square
inch), while the second stage reduces the pressure to the desired level, which for GC is generally
<100 psi. Bartram suggests that "if you use a single-stage regulator at the cylinder for GC, you
must constantly adjust the main line pressure as the cylinder pressure decreases." [27] In multiple-
unit GC systems, a single-stage regulator should be incorporated into the branch line to each
GC to step down the line pressure to that required by the instrument. For multiple-unit GCs, the
use of two-stage and single-stage ensures effective operation by maintaining a minimum 10 to
15 psi pressure differential across all flow- and pressure-controlling devices. This configuration
minimizes changes in the mainline pressure that might affect the operation of the individual GC.
The sketch shown in Figure 4.14 illustrates what a configuration might look like if three GCs are
plumbed into a mainline.[27]

Bartram also offers these important safety-related guidelines: [27]

Never remove a two-stage regulator from a gas line with high pressure isolated in the first
stage—the sudden release of pressure could rupture diaphragms, ruin diaphragms in downstream regulators, and create gaps in a packed column! Always depressurize a two-stage regulator through the second stage or through the GC. If your system has a single cylinder gas supply
or a gas generator, turn off the GC oven first and let the column cool. Then close the first stage
(cylinder side) valve on the regulator and leave the shutoff valve downstream from the regulator
open. Leaving this valve open will allow the gas remaining in the regulator to pass through the
regulator. Vent the pressure through the system (be sure the column is at room temperature)
through a vent installed in the gas line or a vent on the regulator itself. Finally, close the downstream pressure-control valve and remove the regulator.

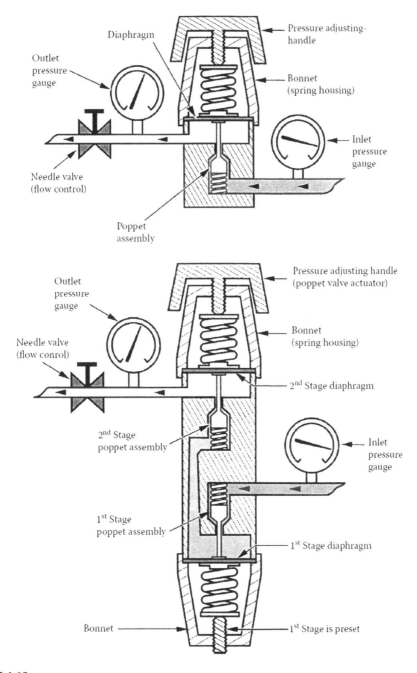

FIGURE 4.13

4.20.1 The Safe Handling of Compressed Gases is Very Important!

Figure 4.15 depicts a typical compressed gas cylinder with a two-stage gas pressure regulator. For GC, the most commonly used carrier gases are helium (He), nitrogen (N_2) and hydrogen (H_2). H_2 is much more commonly found as a GC carrier gas in Europe than in the U.S. due to the significantly higher cost of He. H_2 is also potentially flammable, unlike the other two. Today, limited supplies of He have forced chromatographers to switch to either H_2 or N_2. Most compressed gas

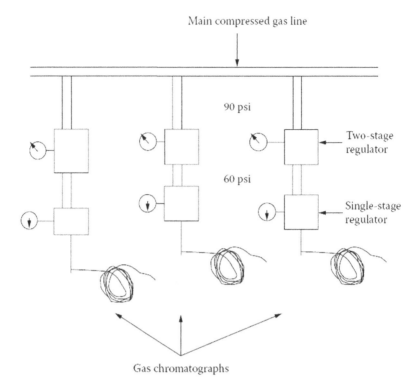

Main compressed gas line

90 psi

Two-stage
regulator

60 psi

Single-stage
regulator

Gas chromatographs

FIGURE 4.14

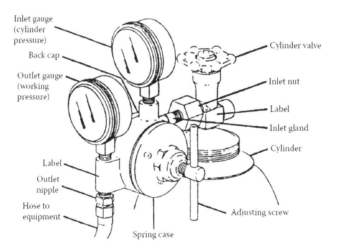

Inlet gauge
(cylinder
pressure)

Cylinder valve

Back cap

Outlet gauge
(working
pressure)

Inlet nut

Label

Inlet gland

Cylinder

Label

Outlet
nipple

Hose to
equipment

Adjusting screw

Spring case

FIGURE 4.15

cylinders are factory filled and, upon turning on the cylinder valve, should give an inlet pressure gauge reading of between 2,200 and 2,500 psi. The outlet pressure gauge is adjusted from 0 to that required, depending on whether the outlet pressure gauge reads 0 to 120 or 0 to 600 psi, by turning the adjusting screw. A constant and reproducible flow rate is essential to be able to reproduce chromatographic retention times, t_R. McNair and Miller state "a compound's retention time is a useful qualitative tool … two or more compounds may have the same t_R" (p. 17) [28] External gas pneumatics in GC consist of the following:

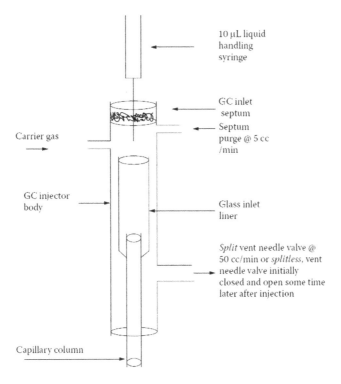

FIGURE 4.16

- Source of carrier gas and regulation of delivery pressure
- All tubing, fittings, and valves
- All gas-purifying traps.

All external gas pneumatics must be provided by the laboratory prior to installation of the gas chromatograph. Laboratories vary from simple external gas pneumatics, in which a single set of compressed gas cylinders is interfaced to a single GC, to a more complex configuration, in which a single set of cylinders is interfaced to two or more GCs. External gas pneumatics have been, in general, essentially the same over the past four decades, since compressed gas cylinders were employed. The next question to ask is how pure shall the carrier gas be that is to be delivered to the GC? The early GCs all employed a ¼-in. packed GC column and operated isothermally (i.e., the column temperature remained constant throughout the chromatographic run time). GC detectors such as the thermal conductivity detector (TCD) in those days were much less sensitive than those in use today. Gas purity was not as critical as it is today. Two options are available with respect to providing very high purity to the GC:

1. Purchase the higher-purity gas, 99.9999% research or ultrapure grade, at a significantly high cost.
2. Purchase a high-purity grade, 99.995% only, and purchase and install relatively inexpensive traps between the tank and the GC.

Most labs choose option 2 and install a series of traps that are located external to the GC, unless there is some compelling reason to go with an ultra-high- or research-grade-purity carrier gas. A hydrocarbon trap is first in line. This trap removes traces of saturated and unsaturated hydrocarbons that might be present due to residual oil from regulator diaphragms. Next should come a molecular sieve trap to remove water vapor, followed by an oxygen trap. The oxygen trap is a proprietary material designed to remove traces of oxygen from the carrier gas. The presence of oxygen in the carrier gas is detrimental to GC liquid phases. Some detectors, such as the electron-capture detector (ECD), respond to oxygen in the carrier gas and thus limit instrument sensitivity, and as we know from discussions in Chapter 2, this limits IDLs and MDLs.

4.21 WHAT ABOUT GC INLETS? WHAT DO I NEED TO KNOW?

Flow controllers are the next item in our journey through the schematic in Figure 4.10. In the early days, before GCs controlled carrier gas flow, rotameters were installed between the regulator and the inlet to the instrument. Today, GCs themselves have pressure gauges and very fine needle valves with which to control carrier, makeup, and detector gas flow rates. Contemporary GCs provide computerized control of both the column head pressure (i.e., the gas pressure at the column inlet) and column volumetric flow rate. Contemporary GCs will have internal pneumatics and additional traps. It is not readily apparent that the flow rate decreases as the column temperature is increased. The explanation for this is that the viscosity of the carrier gas increases as the column temperature is increased. A differential flow controller is used to provide a constant mass flow rate by increasing the column head pressure. Electronic pressure control is a relatively recent advance, particularly for split/splitless GC injectors that require a constant pressure. An electronic sensor is used to detect the decrease in flow rate and cause an increase in the pressure to the column, and hence to maintain a constant flow rate. The chromatographer needs to know the volumetric flow rate of an operating GC. When problems arise, the first question to ask is, "What is the GC flow rate and is it optimized?" A word about pressure settings is in order. The setting at the outlet gauge (Figure 4.15) should be at least 10 to 20 psi higher than the setting at the inlet to the GC. This is particularly important if the sequence of gas purification traps is used, as discussed earlier. The GC was once a single thermostatic unit. Contemporary GCs isolate the injector from the oven and from the detector. The temperature of each is set independent of the other. Injection ports that are usually operated isothermally can be temperature programmed in the same manner as GC ovens. A GC inlet provides the means to introduce the liquid extract (semivolatile organic solvents containing trace concentrations of semivolatile organic compounds) or a volume of gas or headspace (gases or volatile organics) to the chromatographic column. We discussed in detail the inlet interface to the GC for volatile organic compounds (VOCs) analysis in Chapter 3. Injectors have evolved from just a heated cylindrical device that was inserted into a heater block to the split/splitless injection ports used in more contemporary GCs. The objectives of injection in GC are as follows:

1. To rapidly vaporize the liquid, such as an extract, that contains the dissolved analyte of interest to TEQA.
2. To introduce all or a portion of the vapor into a GC column.
3. To split a portion of the vapor out of the injector to the atmosphere for narrow- and wide-bore capillary columns.
4. To continuously clean the septum via a purge flow with carrier gas.

Packed-column injection ports that readily accepted a ¼-in.-outer diameter (o.d.) stainless-steel or glass column during the early period are now designed to accommodate ⅛-in.-o.d. injection stainless-steel columns. This injection port can readily accommodate a 0.53 mm megabore (a term introduced by J&W Scientific) capillary provided that an adapter is used. Megabore capillary columns do not require any splitting. This enables packed-column injection ports to easily accommodate megabore capillary columns. Contemporary GCs are dual-injector and dual-detector instruments. One injector can accommodate a $\frac{1}{8}$ -in. packed column that can be adapted to a megabore capillary column, while the other injector can accommodate a split/splitless narrow- or wide-bore capillary column. The split/splitless injector requires that only a portion of the total sample volume be allowed to enter the narrow- or wide-bore capillary column. The fraction of sample that escapes to the atmosphere vs. the fraction that actually enters the capillary column is determined by the split ratio. Details are given for the Autosystem® (PerkinElmer) GC inlet in the student-tested experiments in Chapter 5.

4.22   WHAT IS A SPLIT RATIO AND WHAT DOES SPLITLESS REALLY MEAN?

The split ratio is defined as

$$\text{Split ratio} = \frac{\text{Flow rate (atmosphere)}}{\text{Flow rate (column)}}$$

A sketch of a split/splitless injector is shown in Figure 4.16. By measuring the carrier gas flow rate coming out of the vent and the carrier gas flow rate passing through the capillary column (be sure to turn off the makeup gas to the detector so that an accurate measurement of the column flow rate can be made), the split ratio can be calculated. As with optimum flow rate, the chromatographer should know at all times what the split ratio is and whether the injection was made via split or splitless. As the sketch below reveals, a splitless injection is accomplished by keeping the vent closed for t seconds after injecting an extract. McNair and Miller have recommended that splitless injection be preferred over split injection for performing TEQA (pp. 97–100).[28] These authors clearly discuss the limitations of the splitless technique. Perry states that injection without splitting produces narrower bands than injection with splitting due to the solvent effect (p. 211).[29] He explains this phenomenon by quoting the Grobs. Consider the following:

> The vapourized material is transferred on to the column essentially as a mixture. In the first stage of separation, the solvent shifts away from the sample components, leaving them on the back slope of its large peak. Thus, the moving vapour plugs of the sample components meet a liquid phase mixed with retained solvent, whereby the concentration of solvent increases rapidly in the direction of migration. Therefore, the front of every plug, in contact with stationary liquid containing more solvent, undergoes much stronger retention than the back of the plug. This effect causes the originally very broad bands of sample components to be condensed to a band width which, under properly selected conditions, may become even smaller than that which can be obtained with stream splitting.
>
> (p. 211)[29]

Jennings has argued that splitless injection is not really without split and is a term only to be used when comparing the technique to that of split injection. (p. 51[26]) More contemporary GCs enable the user to program a splitless injector (i.e., control of the time that a split vent remains closed). One other concept needs to be addressed before we move to the GC column in our journey through a gas chromatograph: sample size injected. If we are not careful, we might overload the GC column.

4.23 WHAT DOES IT MEAN TO OVERLOAD A CHROMATOGRAPHIC COLUMN?

One of the experiments in Chapter 5 asks the student to devise a way to increase the amount of analyte injected and to observe what happens to chromatographic peak shape. Increasing the amount of analyte to be injected in either GC or HPLC can be accomplished in one of two ways:

1. Inject identical volumes of a series of reference standards of increasing concentration
2. Inject increasing volumes of one reference standard.

The result of adding too great an amount of analyte is for the observed peak shape to change from a purely Gaussian peak shape to a seriously tailed peak. Figure 4.17 illustrates this loss of peak symmetry. The amount of analyte that can be injected regardless of peak shape is limited by the linear dynamic range, R_L, of a given detector. Increasing the split ratio also helps to maintain good peak shape. Peak distortion is a column problem, whereas R_L is GC detector dependent. Perry has developed a relationship between the concentration, C_{max}, corrected retention volume, V'_R, and weight of analyte injected, w_i, according to: (pp. 346–347)[29]

$$C_{max} = \frac{\sqrt{N}}{V'_R} w_i \sqrt{2\pi}$$

Using this equation, Perry has shown that if w_D is the maximum weight of a given analyte with an adjusted retention time t'_R that is injected into a GC whose column contains N theoretical plates having a detector sensitivity S and a linear dynamic range R_L, then

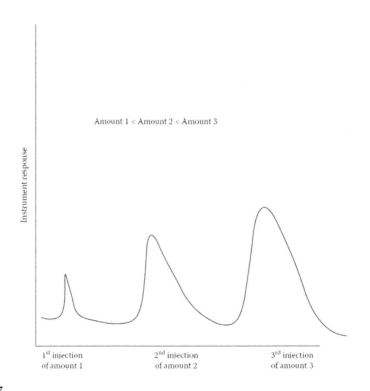

Amount 1 < Amount 2 < Amount 3

Instrument response

1st injection   2nd injection   3rd injection
of amount 1   of amount 2   of amount 3

FIGURE 4.17

$$w_D = 0.4 \frac{SR_L t_R}{\sqrt{N}}$$

For example, assume that we are using a GC with a FID. Let us assume that the FID used has a detector sensitivity $S = 1 \times 10^{-12}$ g/sec and a linear dynamic range $R_l = 1 \times 10^7$ and a column with 10,000 plates. For an analyte of interest whose adjusted retention time is 500 sec, we find

$$w_D = 0.4 \frac{\left(1 \times 10^{-12}\right)\left(10^7 \, g/s\right)\left(5 \times 10^2 \, s\right)}{\sqrt{1 \times 10^4}} = 2 \times 10^{-5} g$$

This equation applies only to isothermal operation. In TEQA, our interests tend to be directed toward the lower end of R_L, although it is good to be aware of such limitations on the high end. The fact that increasing the amount of analyte injected leads to an unsymmetrical or non-Gaussian peak shape is more formally described as moving from *linear elution to non-linear elution* column chromatography. Unsymmetrical peaks are further categorized as tailing or fronting. These correspond to convex or concave distribution isotherms whereby the distribution constant K becomes dependent upon analyte concentration. Figure 4.18 relates linear, convex and concave isotherms to chromatographic peak shapes (pp. 133–135).[17]

A convex isotherm exhibits peak tailing, while a concave isotherm exhibits peak fronting, as shown in Figure 4.18. Equation (4.16) and Equation (4.31) can be combined to yield an expression that relates a solute's (or analyte's) retention volume, V_R, to its partition constant, K, between the stationary (S) and mobile (M) phases. For the partition constant defined as

$$K \equiv C_s / C_M$$

the solute's retention volume is

$$V_R = V_0 + K V_s$$

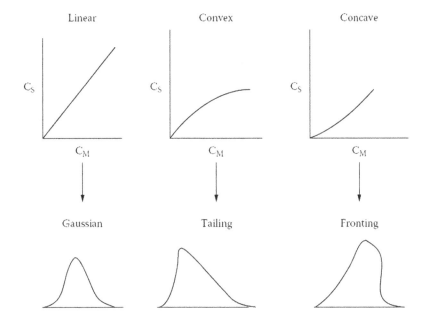

FIGURE 4.18

Since K will decrease with increasing solute concentration for a *convex isotherm*, V_R also decreases. Also, V_R increases with solute concentration for the *concave isotherm*. At the very low solute concentration, both nonlinear isotherms tend toward the linear isotherm. Consider the convex isotherm; the band center will tend to migrate more rapidly than the band extremities due to the smaller value for K at large solute concentrations. Hence, the injected plug of solute starts out at the column inlet as Gaussian and gradually becomes unsymmetrical as it migrates through the column. This results in the development of a sharp front and an extended tail. Tailed peaks become even more skewed as the amount of solute injected increases, as is shown in Figure 4.17. Unsymmetrical peaks make it nearly impossible to accurately quantitate, while some isotherm nonlinearity can be tolerated in semipreparative scale separations.

We now continue in our journey through a GC by moving on to what many chromatographers call the heart of the chromatograph: the GC column.

4.24   WHAT IS SO IMPORTANT ABOUT GC COLUMNS?

In a human, the heart is the most important organ in the body. The reasons for this are found in the field of human physiology. So too, in a GC, the chromatographic column is where separations take place, and the purpose of doing GC is to separate components in a mixture. Historically, packed GC columns dominated gas chromatographic separations up until around 1980. At that time, any work with capillary columns was accomplished using glass. This required that the ends of the coiled and brittle glass columns be heated until softened to enable an installation into the GC to occur. It took a significant level of patience and skill to install a glass capillary GC column at that time. Open tubular columns owe their origins to Marcel Golay and their present robustness from the development of fused-silica capillaries at Hewlett-Packard.[30,31]

4.25   WHAT GC COLUMNS ARE USED IN TEQA?

The significantly large values of N for *open tubular columns* when compared to packed columns have established capillary columns as essential for TEQA. For example, a 60 m capillary with 3,000 to 5,000 plates/m yields 180,000 to 300,000 plates, whereas a 2 m packed column with 2,000 plates/m yields only 4,000 plates (p. 90).[28] This difference in column efficiency was largely ignored by industry and regulatory government agencies during the 1959–1980 era, until the patent for open tubular columns expired. It was not until the introduction of the 0.53 mm *megabore column* by J&W Scientific that the EPA "allowed" the abandonment of the 6 ft × ¼-in.-o.d. packed glass column for the determination of organochlorine pesticides (OCs) via EPA Method 608. This author even heard it said in those days that "you could not connect a cap column to an ECD." The EPA was slow to abandon the packed-column mixed-liquid phase consisting of 1.5% OV-17 and 1.95% OV 210 coated on Chromosorb. This stationary phase served the regulatory agency very well when all that was needed was to measure trace residues for some dozen OCs. Three GC chromatograms from the earlier use of a packed column to separate and detect OCs are shown in Figure 4.19. During the packed-column days, where N was limited, column selectivity, α, took on much more importance and numerous liquid phases became available. The open tubular column with its very large value of N does not require as large a value for α. For this reason, the number of capillary columns with different stationary phases is much more limited. In fact, it can be argued that, today, only two different cap columns are needed to handle almost all of the volatile and semivolatile priority pollutant organic analytes. McNair and Miller present the basics of the stationary phase, packed columns, and capillary columns, and this source or an equivalent should be consulted for these topics. (chapters 4–6)[28] Narrow-, wide-, and megabore cap columns are commonplace in environmental testing labs today. A wall-coated open tubular (WCOT) column, such as a DB-624 (Agilent), or equivalent is usually a good first choice for

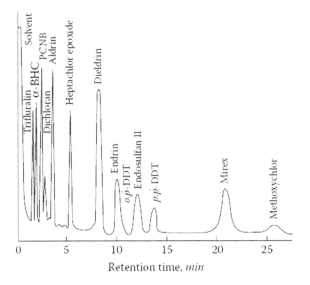

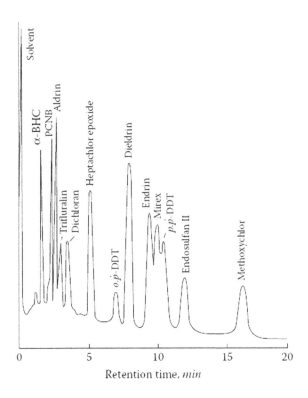

FIGURE 4.19

trace volatile organics analysis, whereas a WCOT such as DB-5 (Agilent) is a good first choice for *trace semivolatile organics* analysis. Perusal of GC consumable supply houses, such as Agilent, Millipore Sigma, Phenomenex, among others, are good starting points. Reviewing the appropriate EPA methods is also good practice in developing an understanding of applying GC to environment-related problems. Whether one is interested in *broad-spectrum* environmental

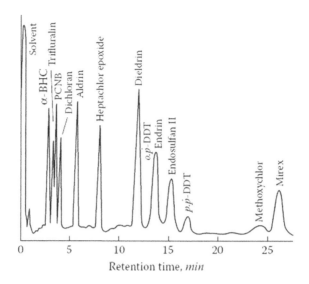

FIGURE 4.19 (Continued)

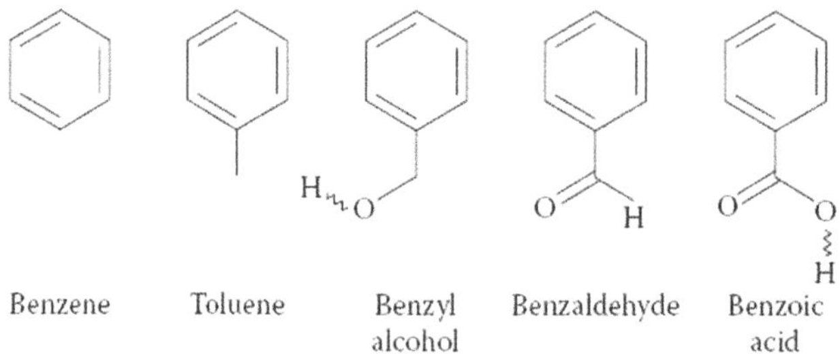

Benzene   Toluene   Benzyl alcohol   Benzaldehyde   Benzoic acid

FIGURE 4.20

monitoring or *target-specific* analysis is also an important consideration. To understand which columns to use for a given application, Farwell has considered *four basic criteria* (p. 1212):[32]

1. Polarity of analytes to be gas chromatographed
2. Polarity of the cap column stationary phase
3. Selectivity of the cap column stationary phase
4. Stationary-phase stability.

A knowledge of the basics of organic chemistry is very helpful in satisfying the *first criterion*. A scale of some kind to satisfy the *second criterion* would be most helpful. Satisfying the *third criterion* requires contributions from both the analyte and stationary phase, and *criterion 4* refers to the physical attributes of the stationary phase. Figure 4.4 serves as a reminder of the limited scope of gas chromatography. Organic compounds of environmental interest must be vaporizable and, at the same time, be thermally stable at GC inlet temperatures of 200 to 300°C. These two criteria are met only for those analytes that are neutral, relatively nonpolar, and consisting of aliphatic and aromatic carbon classes. As oxygen begins to get incorporated into the molecule, the ability of a molecule to meet these two criteria becomes more limited.

Consider the series of aromatics whose molecular structures are shown in Figure 4.20. Starting with benzene, then moving to toluene by substituting a methyl group for hydrogen. We then replace a hydrogen in the methyl group with a hydroxy group and get benzyl alcohol. We proceed to partially oxidize this benzyl group to an aldehydic group and obtain benzaldehyde. Finally, we fully oxidize the aldehydic functional group to a carboxylic group and obtain benzoic acid. The first four monoaromatics in this series are liquids at room temperature, and benzoic acid is a solid. It is possible to gas chromatograph benzoic acid; however, peak distortion and lowered sensitivity result. The polarity of these compounds takes us from the volatile, nonpolar benzene out to the semivolatile, polar benzoic acid. Benzoic acid can be moved back across the domain to the more volatile and nonpolar region (refer to Figure 4.4) by conversion of this carboxylic acid to its methyl ester by chemical derivatization techniques (refer back to Chapter 3). A more exact definition of solute polarity was developed by Kovats, who stated that "polarity cannot be expressed with one simple parameter."[33]

4.26   WHAT IS THE KOVATS RETENTION INDEX?

The Kovats retention index for a specific organic compound (i.e., the analyte of interest on a specific GC column) is nothing more than a relative retention time. The index is in relation to *aliphatic hydrocarbons*. For example, the hydrocarbon iso-octane, or 2, 2, 4-trimethyl pentane if gas chromatographed, would probably have a retention time between that of the straight-chained alkanes n-octane C_8 and n-nonane C_9. The Kovats index for n-octane is 800, and that for n-nonane is 900. Iso-octane would then have a Kovats index somewhere in between, such as 860. The Kovats index is calculated from a GC chromatogram using the retention volume for the analyte whose index is sought, $V_R^{unknown}$, and the retention volumes for the alkane with n carbon atoms per molecule, V_R^n, and for the alkane with $n + 1$ carbon atoms per molecule, V_R^{n+1} using the following relationship:

$$I = \left[\frac{\log V_R^{unknown} - \log V_R^n}{\log V_R^{n+1} - \log V_R^n} \right] + 100n \tag{4.33}$$

Note that the Kovats retention index is independent of the chemical nature of the column. If we were to chromatograph the *five monoaromatics* cited above on a purely nonpolar stationary phase such as Squalane (a C_{30} aliphatic hydrocarbon), we would generate a series of I values for each compound. If we then chromatographed these same compounds on a more polar stationary phase such as diethylene glycolsuccinate (DEGS), we would obtain an entirely different set of I values. We calculate the difference in Kovats indices as follows:

$$\Delta I^{benzene} = I_{DEGS}^{benzene} - I_{Squalane}^{benzene}$$

$$\Delta I^{benzylacohol} = I_{DEGS}^{benzylacohol} - I_{Squalane}^{benzylacohol}$$

Thus, stationary phases can have their polarities compared for a given test probe, such as benzene or benzyl alcohol. In an attempt to reduce the number of liquid phases used in packed-column GC, McReynolds published an extensive listing of Kovats retention indices using 10 probes for hundreds of liquid phases at that time.[34] Rohrschneider considered only five probes: benzene, ethanol, 2-butanone or MEK, nitromethane, and pyridine.[35] McReynolds expanded this to 10 probes and proved his assumption that liquid phases had, indeed, proliferated because many liquid phases possessed "McReynolds constants" that differed little.[36] The reasons for choosing which organic compounds are suitable probes are outlined in Table 4.8 (p. 1214).[32] Usually only the first five probes are cited. Table 4.9 lists most of the commonly used liquid phases along with recommended operating temperatures and corresponding values for McReynolds constants.

TABLE 4.8
McReynolds Probes and the Rationale for Their Selection

Symbol	Test Compound	Basic Molecular Interactions	Functional Group
X'	Benzene	Dispersion with some weak proton-acceptor properties	Aromatics, olefins
Y'	Butanol	Orientation properties with both proton-donor and proton-acceptor capabilities	Alcohols, nitriles, acids
Z'	2-Pentanone	Orientation properties with proton-acceptor capabilities	Ketone, ethers, aldehydes, esters, epoxides, dimethylamino derivatives
U'	Nitropropane	Dipole orientation properties	Nitro and nitrile derivatives
S"	Pyridine	Weak dipole orientation with strong proton-acceptor capability	Aromatic bases
H'	2-Methyl-2-pentanol		Branched alcohols
L'	1, 4-Dioxane		
M'	cis-Hydrindane		

TABLE 4.9
Most Commonly Used Liquid Phases in Capillary GC Along with Their McReynolds Constants

Liquid Phase	Commercially Available Name	Range of Operating WCOT Temperatures	Range of McReynolds Constants
Dimethyl polysiloxanes (gum, oil)	DB-1, HP-1, BP-1, OV-1, OV-101, SE-30, SP-2100	−60 to 280 (oil) −60 to 325 (gum)	220–229
5% phenyl methyl polysiloxane	DB-5, BP-5, SPB-5, SE-52, OV-73, RT_x-5, HP-5	−60 to 325	334–337
50% phenyl methyl polysiloxane	DB-17, HP-17, OV-17, GC-17, SP-2250	40 to 280	884
50% trifluoropropyl methyl polysiloxane	DB-210, OV-210, QF-1, SP-2401	40 to 240	1520–1550
25% cyanopropyl methyl	DB-225, OV-225, SP-2300		
25% phenyl methyl polysiloxane			
Poly(ethylene glycol) gum	DB-WAX, HP-20M, SP-1000, Carbowax 20M	50 to 220	2301–2309

Characterizations of GC liquid phases is an ongoing research area. Li and coworkers have developed a new set of GC-based solute parameters using solvatochromic linear solvation energy relationships. [37] McNair and Miller considered a thermodynamic view of the interaction between a solute and a stationary phase and devised a definition of the separation factor or selectivity α between solutes A and B according to (pp. 66–67):[28]

$$\alpha = \frac{K_C^A}{K_C^B} = \frac{p_A^o}{p_B^o} \frac{\overline{\gamma A}}{\overline{\overline{\gamma B}}}$$

where p^o is the vapor pressure of the solute as if it were a pure liquid and γ is the activity coefficient for a specific solute. This equation shows that the extent of separation between solutes A and B depends not only on the ratio of the vapor pressure of the pure solutes, but also on the ratio of their activity coefficients. This explains why two analytes with very similar boiling points (i.e., similar pure solute vapor pressures, p^o) have significantly different retention times.

4.27 WHY DO I NOT SEE MCREYNOLDS/ROHRSCHNEIDER CONSTANTS IN CURRENT CHROMATOGRAPHY CATALOGS?

A perusal of a current catalog of one of the largest suppliers of chromatography and related products in the U.S. makes minimal mention of McReynolds constants and no mention of Rohrschneider constants. Perusal of the current catalog of a second large supplier of chromatography and related products does not even mention these constants. Perhaps this reflects the commercial maturity of GC, in that GC has been around for some 40 years and applications have become well known. The popularity of open tubular columns has reduced the need for numerous liquid phases; hence, there is little need to distinguish liquid phases anymore. The most commonly used liquid phases for TEQA are the family of silicone polymers (also called polysiloxanes). As introduced in Chapter 1, the EPA organics protocol is divided into volatile organic compounds (VOCs) and semivolatile to nonvolatile organic compounds (SVOCs). Either a 30 or 60 m × 0.53 or 0.75 μm *open tubular or capillary column* with a stationary phase specifically designed for the separation of VOCs is commonly used. The chemical nature of these phases is considered proprietary bonded phases by most suppliers. For example, a 75 m × 0.53 mm i.d. with a 3 μm film thickness is sold by Millipore Sigma (formerly Supelco) as the SPB-624, VOCOL capillary column to separate some 93 VOCs without subambient temperatures in less than 20 min. The 624 designation refers to the EPA method number as applied to wastewater analysis. The 105 m × 0.53 mm i.d. with a 3-μm film thickness developed in the early 1980s by Restek (called the Rtx-502.2 column) is based on a diphenyl/dimethyl polysiloxane that provides low bleed and thermal stability up to 270°C. The 502.2 designation refers to the EPA method number as applied to drinking water analysis. SVOCs are commonly separated by using a WCOT column that contains a 5% phenyl methyl/dimethyl polysiloxane. The column length for adequately separating SVOCs is generally 30 m. Narrow-bore and wide-bore columns are used with a split/splitless injector, and a megabore cap column is installed on a 1/8-in. packed-column injector without split. SVOCs can be further subclassified into the 100 or so *priority pollutant base, neutral, acids* (BNAs), 30 or more organochlorine pesticides (OCs), polychlorinated biphenyls (PCB), and a miscellaneous category for many other analytes of environmental concern. VOCs can be subclassified as polychlorinated ethylenes and ethanes (ClVOCs), monoaromatics from fuel contamination of groundwater (BTEX), trihalomethanes from chlorine disinfection of drinking water (THMs), and a miscellaneous category for many other analytes of environmental concern. This classification scheme is shown in a flowchart in Figure 4.21. This author believes that a good silicone WCOT, if properly maintained and if appropriate sample prep techniques are always used, will last for years. Two other types of open tubular columns used in GC are support-coated open tubular (SCOT) and porous-layer open tubular (PLOT). SCOT columns are no longer widely used, whereas PLOT columns are essential for the *separation of permanent gases* (i.e., Ne, Ar, O_2, N_2, CO_2, Kr, Xe, and lower-molecular-weight aliphatic hydrocarbons). Let us consider these fused-silica WCOTs in more detail.

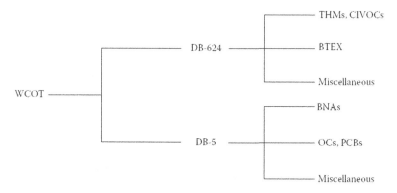

FIGURE 4.21

4.28   WHAT ARE FUSED-SILICA POLYSILOXANE WCOTs?

One of this author's earliest laboratory experiences involved placing a piece of glass tubing into a burner flame and observing the softening of the center portion of the tubing. At some point in the heating, upon pulling on both ends of the unheated portions of the tubing, fine capillary tubing can be produced. This is still the principle of *capillary column fabrication*. Figure 4.22 depicts a capillary drawing machine and some detail of the actual coating of the outside sheath.[38] These schematics are based on the original design of Desty et al.[39] A good discussion of the fabrication and technical challenges of polysiloxane WCOTs has been written by one of the founders of such cap columns. (chapters 2–4)[26] One of the more detailed reviews of liquid phases can be found in a series of three articles by Yancey.[40] Polysiloxanes are chemically bonded to the fused silica using proprietary surface treatment techniques. Jennings has summarized these approaches as follows (p. 66):[26]

> Peroxides may be added to the coating solution, and static coating procedures are generally employed to deposit a film of vinyl-containing stationary-phase oligomers on the interior wall of the tubing. Heating causes peroxide decomposition, which yields free radicals and initiates cross-linking (and, if the surface has been properly prepared, surface bonding) when the column is heated. The preferred peroxide is generally dicumyl peroxide, which decomposes to form volatile products that are dissipated during the heating step.

Let us focus on the WCOT columns used to implement EPA Method 507, "Determination of Nitrogen- and Phosphorus-Containing Pesticides in Water by GC with a Nitrogen-Phosphorous Detector." The method is summarized as follows:[41]

> A measured volume of drinking water of approximately 1 L is extracted with methylene chloride by shaking in a separatory funnel or mechanical tumbling in a bottle. The methylene chloride extract is isolated, dried and concentrated to a volume of 5 mL during a solvent exchange to methyl tert-butyl ether, MTBE. Chromatographic conditions are described that permit the separation and measurement of the analytes in the extract by capillary column GC with a nitrogen-phosphorous detector, NPD.

A primary WCOT is required to separate and detect N- and P-containing pesticides that may be present in drinking water. The separated N and P pesticides must be confirmed by injecting the MTBE extract into a confirmatory WCOT. The primary WCOT is DB-5 from J&W Scientific or an equivalent; the confirmatory WCOT is DB-1701 from J&W Scientific or an equivalent. The molecular structure for each of these polysiloxanes is shown in Figure 4.23. The liquid phases are similar in the *n* repeating monomer, whereas they differ in the substitution in the *m*

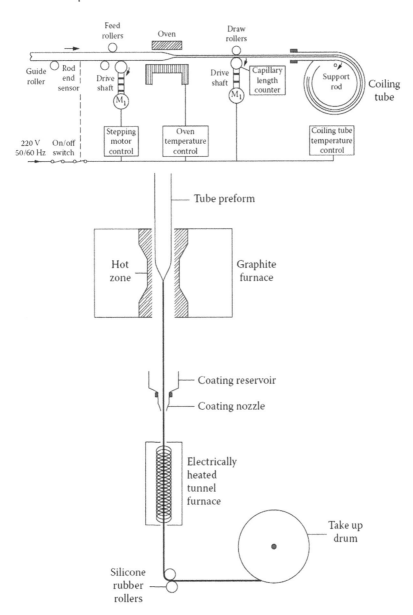

FIGURE 4.22

repeating monomer unit. The DB-5 is a 5% phenyl/dimethyl polysiloxane, and the DB-1701 is a 14% cyanopropyl/phenyl/dimethyl polysiloxane. The introduction of a cyano group increases the polarity of this WCOT, and hence shifts the retention time for the N and P pesticides that are to be monitored. An MTBE extract from an unknown drinking water sample, if injected onto both columns, will give values for t_R that differ by a fixed number of minutes. These retention times would have been previously determined by injecting standards onto both columns. The only other alternative to dual WCOT identification and confirmation is to use single WCOT and both an element-specific detector, such as a nitrogen-phosphorus detector (NPD), and a mass spectro-metric detector. Table 4.10 is a sample from the list of some 46 individual N and P pesticides and herbicides that are to be monitored in drinking water by this method. Alachlor is a substituted

Diphenyldimethylpolysiloxane (e.g. DB-5)

- non-polar
- excellent general purpose column
- wide range of applications
- low bleed
- high temperature limit
- bonded and cross-linked
- solvent rinsable
- wide range of column dimensions available
- equivalent to USP Phase G27

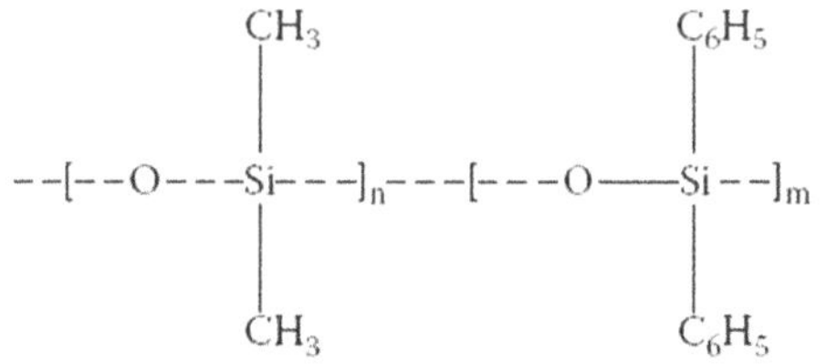

$$--[--O---Si---]_n---[---O----Si--]_m$$

with CH_3 (top) and CH_3 (bottom) on the first Si, and C_6H_5 (top) and C_6H_5 (bottom) on the second Si.

Cyanopropylphenylmethylpolysiloxane (e.g. DB 1301,1701)

- low to mid polarity
- bonded and cross-linked
- solvent rinsable
- equivalent to USP Phase G43

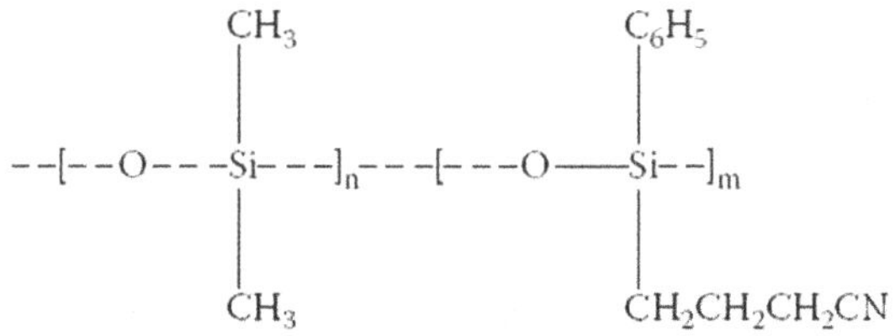

$$--[--O---Si---]_n---[---O----Si--]_m$$

with CH_3 (top) and CH_3 (bottom) on the first Si, and C_6H_5 (top) and $CH_2CH_2CH_2CN$ (bottom) on the second Si.

FIGURE 4.23

acetanilide herbicide in widespread use today, and atrazine and simazine are examples of triazine herbicides also in widespread use. Dichlorovos and hexazinone are illustrative of organophophorus pesticides. Triphenyl phosphate is used as an internal standard in the method. Note the 1 to 2 min or so shift in retention time for each N and P pesticide and herbicide listed in Table 4.10.

The differences in capacity factors among different polysiloxane WCOT columns are made quite evident in Figure 4.24. In this figure, two GC chromatograms are stacked using Turbochrom (PE-Nelson) software and obtained on gas chromatographs in the author's laboratory. A 100 ppm reference standard consisting of AR 1260 dissolved in acetone is taken and 1 μL of this solution is injected into a GC that contains DB-5 (J&W Scientific) to yield the top chromatogram, and the same solution is injected into a GC that contains DB-1 (J&W Scientific) to yield the bottom chromatogram. The liquid-phase DB-1 is a dimethyl polysiloxane.

If methyl groups are substituted for the phenyl groups in the DB-5 structure shown in Figure 4.23, a DB-1 structure is obtained. Commercially available equivalents to DB-1 include HP-1, HP-101, Ultra-1, SPB-1, Rtx-1, OV-1, SP-2100, and SE-30, among others. The influence of the phenyl groups in DB-5 creates additional aromatic group intermolecular interactions that influence the partitioning constant, and hence according to Equation (4.31), the capacity factor,

TABLE 4.10
Representative N and P Pesticides and Herbicides, Their Classification, and Retention Times on Two Chemically Different GC Columns

N/P Pesticide	Classification	$t(R)$ on DB-5 (min)	$t(R)$ DB-1701 (min)
Alachlor	Substituted acetanilide herbicide	35.96	34.10
Atrazine	Triazine herbicide	31.77	31.23
Dichlorvos	Organophosphorus pesticide	16.54	15.35
Hexazinone	Organophosphorus pesticide	46.58	47.80
Simazine	Triazine herbicide	31.49	31.32
Triphenyl phosphate	Internal standard	47.00	45.40

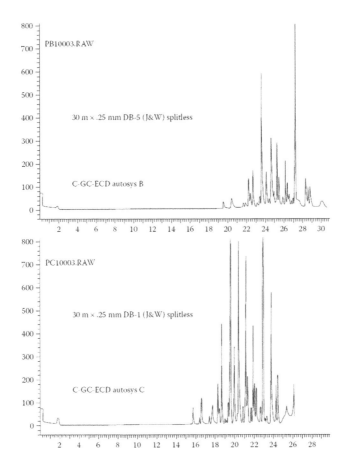

FIGURE 4.24

k'. Use of these two WCOT columns provides acceptable identification and confirmation for this complex mixture of polychlorinated biphenyl congeners. Two other significant concepts with respect to WCOTs and their use in TEQA are film thickness, df, and the influence of column temperature on k'.

4.29   WHY IS WCOT FILM THICKNESS IMPORTANT?

Because it influences the C term in the Golay equation. The Golay equation as applied to WCOT is similar to that developed earlier [e.g., Equation (4.25)]:

$$H = \frac{B}{u} + \left(C_M + C_s\right)\bar{u} \tag{4.34}$$

Note the absence of the A term and the separate resistance-to-mass transfer in the mobile phase, C_M, and in the stationary phase, C_s. This equation can be written even more specifically:

$$H = \frac{2D_M}{\bar{u}} + \left(\frac{1+6k'+11k'^2}{24\left(1+k'\right)^2}\right)\left(\frac{r_C^2}{D_M}\right)\bar{u} + \left(\frac{2k'}{3\left(1+k'\right)^2}\right)\left(\frac{d_f^2}{D_s}\right)\bar{u} \tag{4.35}$$

The Golay equation shows a dependence of H on the square of d_f in the C_s term. While we are discussing this equation, note the dependence of H on the square of the column radius, r_C, in the C_M term. The choice of GC carrier gas through the WCOT influences the diffusion of analyte in the mobile phase, D_M, and the diffusion of analyte in the stationary phase, D_s. Figure 4.25 is a Golay plot of H vs. the average linear carrier gas velocity (in cm/sec) for nitrogen, helium, and hydrogen using a WCOT with a 25 m × 0.25 mm i.d. and d_f = 0.4 μm thickness. N_2 as the carrier gas yields the lowest optimized H, whereas the lighter gases, He and H_2, do not have as severe a resistance to the mass transfer term and can be operated at a much higher linear velocity. Figure 4.25 should always be kept in mind when considering which gas to use in GC.

The B term, the contribution to H due to longitudinal diffusion, becomes evident as the two chromatograms are compared in Figure 4.26. The top GC chromatogram shows an optimized

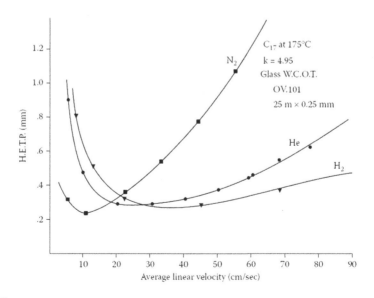

FIGURE 4.25

carrier gas flow rate and a separation of the two surrogate compounds required for the determination of various OCs and PCBs in such methods as EPA Method 8080. The molecular structures of both surrogates are as follows:

tetrachloro-m-xylene   decachlorobiphenyl

These highly chlorinated aromatics have k' values that serve to bracket the k' values for the GC elution of all OCs and PCBs. The bottom chromatogram in Figure 4.26 is at approximately half the volumetric flow rate of the top chromatogram. Both peaks are tailing, while their retention times have increased somewhat. A much stronger ECD signal is also evident. These surrogate compounds are so significantly different in their k' values that it would require a much more elevated column temperature to elute decachlorobiphenyl (DCBP). A fully chlorinated biphenyl is as nonpolar and nonvolatile as one can get. In fact, as Figure 4.26 demonstrates, it requires keeping the column temperature at 275°C. This is the recommended maximum operating temperature of the WCOT to finally elute DCBP from the column. Column temperature can significantly influence the magnitude of an analyte capacity factor, k'. All broad-spectrum monitoring EPA methods using WCOTs use GC with a column temperature that varies with time after sample injection.

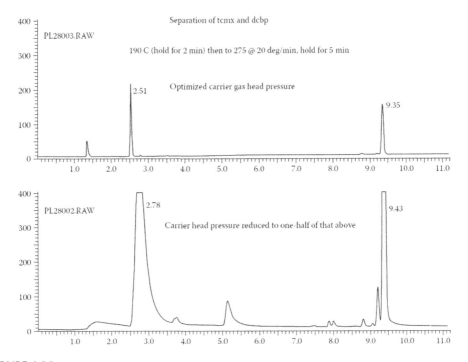

FIGURE 4.26

We know that the retention time of any organic compound is influenced by the column length, the linear velocity of the carrier gas, and the magnitude of the compound's capacity factor, k'. Revisiting Equation (4.17), rewriting it, and solving for t_R, we get

$$t_R = t_0 \left[\left(1 + k' \right) \right] = \frac{L}{u} \left(1 + k' \right)$$

In Equation (4.10), a chromatographic capacity factor, k', was defined in terms of the product of a partition constant and a phase ratio denoted by β. For a WCOT, β is related to film thickness, d_f, and column diameter, d_c, according to[42]

$$\beta = \frac{d_c}{4d_f}$$

For example, a 250 μm-i.d. WCOT column with a film thickness of 0.2 μm has a $\beta = 26.7$. Recall that $k' = K/\beta$, and if K, the partition constant for a compound between the mobile and stationary phases, is held constant, an increase in df reduces β while increasing k' and, of course, t_R. If d_c is increased, for the same film thickness, df, then a higher β results. This higher phase ratio for a fixed value of the partition constant, K, serves to decrease k' and results in a shorter retention time. In general, higher β values, such as $\beta > 100$, are more suitable for SVOCs, whereas lower β values, such as $\beta < 100$, are better for VOCs.[42]

4.30   HOW DOES COLUMN TEMPERATURE INFLUENCE K'?

Once a WCOT column is chosen, only the flow rate and column temperature can be varied so as to maximize chromatographic resolution, R_S. We have discussed the influence of flow rate on R_S earlier; what remains is to discuss the effect of column temperature, T_c, on the separation factor, α, and on R_S. Figure 4.27 shows how both k' and R_S depend on T_c. Injection of a solution that contains these four aliphatic hydrocarbons at $T_c = 45°C$ yields the top chromatogram shown in Figure 4.27. The peaks are all adequately resolved; however, the run time is excessive, over 40 min. Increasing the column temperature to 75°C reduces chromatographic run time, but also decreases chromatographic resolution. A further increase to $T_c = 95°C$ further decreases R_S to unacceptable levels. The optimum T_c that maximizes R_S for all pairs of closely spaced peaks would be somewhere between 45 and 75°C. Note the change in elution order as peak 1 is retained longer than peak 2.

If the log K is plotted against the reciprocal of absolute temperature, $1/T$, for a given organic compound, a straight line is established. The slope of the line varies somewhat with the nature of the solute. At certain temperatures, the lines can cross, and this explains the observed reversal in elution order, as shown in Figure 4.27. Such a plot is shown in Figure 4.28. At $T_c = TA$, peak 1 elutes before peak 2, whereas at $T_c = T_B$, peak 2 elutes before peak 1. Differences in the slope are caused by the difference in analyte–stationary-phase interactions. From Equation (4.30), we know that R_S depends on the separation factor, α, so that for a pair of peaks, it is of interest to know just how α varies with increasing the WCOT column temperature.

The Clausius–Clapeyron equation in physical chemistry states

$$log\, p° = -\frac{\Delta H_v}{2.3RT} + C$$

where $p°$ is the vapor pressure exerted by a substance if it were pure at an absolute temperature T. R is the ideal gas constant, C is a constant, and ΔHv is the change in the substance's enthalpy

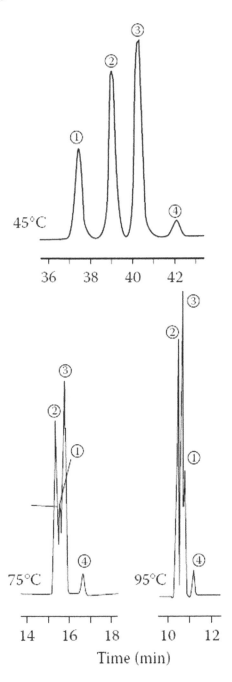

FIGURE 4.27

in moving from a liquid to a vapor and is also called the heat of vaporization. If T_c is increased, the following happens (p. 144):[28]

1. k' and t_R both decrease.
2. α can either rise; peak out, then fall; or fall, bottom out, then rise or steadily decrease.[42]
3. N slightly increases.

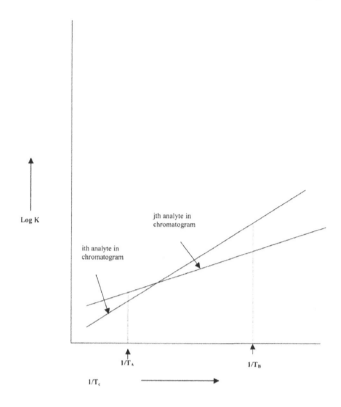

FIGURE 4.28

In the case of our two surrogates discussed earlier (TCMX and DCBP), their respective $p°$ values (alternatively, their respective boiling points) differ by such a wide range that if we set the T_c to a fixed value (e.g., 225°C), we would be able to elute TCMX but not DCBP. If the GC is capable, increasing the T_c after injection of the sample ensures that both surrogates will be eluted within a reasonable chromatographic run time, as was done in Figure 4.27. One says then that the way to elute both surrogates is to temperature program. For this reason, *programmed-temperature gas chromatography* (PTGC) is one of the most powerful features of contemporary GC instrument design. According to Perry (p. 239):[29]

> PTGC also brought about direct control of carrier gas flow rate. Before PTGC, carrier gas flow was adjusted by pressure control. However, because gas viscosity and therefore the column pressure drop, ΔP, increases with increasing T_c, the carrier gas flow tends to decrease during PTGC. Such a decrease is quickly reflected in the deviating baseline signal from a single, flow-sensitive detector. To hold the carrier gas flow constant during PTGC, and thus also the retention times (from run to run) and particularly the baseline (during each run), flow controllers were introduced. They became standard, expected components in gas chromatographs ... the prime characteristics of both GCs and stationary phases became those relevant to the demands of programmed temperature. The GC had to be able to bring about sharp but well-controlled changes in column temperature, and yet record a gas chromatogram; the stationary phases had to withstand these changes without decomposing or vaporizing to a prohibitive degree, and yet retain the selectivity relevant to each ... to permit rapid changes in column temperature, high velocity, high-power, low-mass column ovens were introduced. PTGC or no, the column oven must allow easy column installation and interchange; therefore air had to remain the heat-exchange medium. In the

newer designs, however, the air would be moved in much greater quantity and velocity, by and over much more powerful fans and heaters. Further, the column oven would become low-mass. A light inner aluminum shell of low thermal mass, insulated from the more substantial walls of the outer oven.

4.31 WHAT IS PTGC ANYWAY?

Just after a sample is introduced by either manual or autosampler syringe into the GC injection port, the oven temperature (and hence the column temperature) is increased via a previously programmed series of increases in T_c at a constant rate of change (i.e., dTc/dt is fixed), along with periods of time where T_c does not change. These periods of time are called holds or isothermal steps. GCs can be temperature programmed either in one step or in a series of steps, each step having a different value for the ramp or, as defined earlier, dTc/dt. Each step can be viewed as consisting of an initial temperature with its hold time, a ramp at a fixed value, and a final temperature with its own hold time. Contemporary GCs are temperature programmed via commercially available chromatography processing software with a precision to 0.02 min or less. This aspect of the GC oven has advanced considerably over the past 40 years to accommodate PTGC. Earlier instruments used clock motor-driven potentiometers to change the column oven set point at a constant rate. Instruments of the early 1980s vintage had electronic temperature programmers that used plugin resistors to set one or more programming rates. This author's experience with thumbwheel settings was obtained using the Model 3700 (Varian Associates). As a rule of thumb, retention times decrease by about one half for each 15 to 20°C increase in T_c.[43] McNair and Miller have pointed out that for a homologous series, analyte retention times are logarithmic, whereas under isothermal conditions, analyte t_R values are linear when temperature programmed (pp. 144–146).[28] Onuska and Karasek have pointed out in their classic text on environmental analysis that the Kovats retention index [Equation (4.33)] can be replaced with the following relationship to give a linear scale: [44]

$$I^{\mathrm{PTGC}} = 100\left(\frac{t_R^x - t_R^n}{t_R^{n+1} - t_R^n}\right) + 100n$$

4.32 CAN WE FIND HOW RETENTION TIME VARIES WITH Tc?

Yes, Perry has considered the relationship between a solute's specific retention volume, V_g (the net retention volume, V_N, per gram of liquid phase) and the column temperature, T_c^g (pp. 224–229).[29] Perry's approach is not seen in most other GC texts, so let us derive it here. Before we present his approach, we need to shore up some fundamentals first. One of the few chromatographic relationships to really commit to memory is the relationship among solute retention volume, V_R, column void volume, V_M, and the solute chromatographic capacity factor, k', according to

$$V_R = V_M \left(1 + k'\right) \tag{4.36}$$

The net retention volume, V_N, is defined as the difference between the solute retention volume and the void volume according to

$$V_N = V_R - V_M = k'V_M = \left(\frac{K}{\beta}\right)V_M$$

Upon eliminating the phase ratio, β, we get

$$V_N = K\left(\frac{V_L}{V_M}\right)V_M = KV_L$$

where V_L is the volume of the liquid phase. This is another of those important chromatographic relationships that help in understanding all this. If you double the volume of the liquid phase, you can expect a doubling of the solute net retention volume. Perry has shown earlier that V_N can be related to V_g. He defines V_g as the net retention volume per gram of stationary phase of density ρ_L and volume V_L in the column, with the temperature, T_c, of the column corrected to 0°C in the following way:

$$V_N = \left(\frac{T_c}{273}\right)V_L\rho_L V_g$$

Eliminating V_N between these two equations gives

$$K = \left(\frac{T_c}{273}\right)\rho_L V_g$$

Taking the common logarithms of both sides of this equation yields

$$\log K = \log V_g + \log \frac{\rho_L T_c}{273} \tag{4.37}$$

The Clausius–Clapeyron equation introduced earlier can be differentiated with respect to temperature, and the temperature term itself manipulated to yield

$$\frac{d}{d(1/T)}\left(\ln p^\circ\right) = -\frac{\Delta H_v}{R}$$

An analogous statement for the temperature rate of change of the Henry's law constant, K_H, discussed in Chapter 3, can be found in texts on chemical thermodynamics. The differential molar heat of vaporization of a solute from an infinitely dilute solution, ΔH_S, can be related to the Henry's law constant according to

$$\frac{d}{d(1/T)}\left(\ln K_H\right) = -\frac{\Delta H_S}{R}$$

Upon elimination of the term $d(1/T)$ from both of these differential equations, we can rewrite

$$\frac{d\left(\ln K_H\right)}{d\left(\ln p^\circ\right)} = \frac{\Delta H_S}{\Delta H_v} = a$$

so that

$$d\left(\ln K_H\right) = ad\left(\ln p^\circ\right)$$

Henry's law constant is essentially the increase of the chromatographic partition coefficient, $K = 1/KH$, and this allows us to restate this equation as

$$d\left(\ln K\right) = -ad\left(\ln p^{\circ}\right) \tag{4.38}$$

Combining Equations (4.37) and (4.38) while eliminating ln K yields a relationship between V_g and the vapor pressure of the pure solute, p°. Laub and Pecsok have also arrived at this relationship: [45]

$$\log V_g = -a \log p^{\circ} + \text{const} \tag{4.39}$$

Returning to the Clausius–Clapeyron equation and integrating, we obtain

$$\ln p^{\circ} = -\frac{\Delta H_v}{RT} + \text{const}$$

Upon multiplying through by a and converting to common logarithms, we get

$$a \log p^{\circ} = -a\left(\frac{\Delta H_v}{2.3RT}\right) + \text{const}$$

Substituting for the term "$a \log p^{\circ}$" in Equation (4.39) gives

$$\log V_g = -\left(\frac{-a\Delta H_v}{2.3RT} + \text{const}\right) + \text{const}$$

Because $\Delta H_s = a\Delta Hv$, we can now arrive at the ultimate objective of Perry's derivation:

$$\log V_g = \frac{\Delta H_s}{2.3RT_c} + \text{const}$$

The specific retention volume, a chromatographic property of a given solute of interest, is related to the column temperature such that plots of log V_g vs. $1/T_c$ should yield a straight line whose slope is related to either the heat of vaporization of the solute at infinite dilution or, if $a = 1$ (if Raoult's law holds), the heat of vaporization of the pure solute. Studies of column efficiency via a consideration of the effect of T_c on H ($N = H/L$) have shown that an optimum T_c can be found that minimizes H or maximizes N. The interested reader can refer to the excellent text by Harris and Habgood on PTGC.[46] Hinshaw has offered some additional insights into PTGC.[42]

4.33 THIS IS ALL WELL AND GOOD, BUT DO I NOT FIND TEMPERATURE PROGRAMS ALREADY PROVIDED IN EPA METHODS?

Yes, you do if you are implementing the broad-spectrum monitoring of these methods. For a shorter list of analytes of environmental interest, knowledge of the effect of T_c on various GC conditions is important. Also, be aware that there is a finite amount of time needed to cool the instrument down from the maximum T_c reached in a given temperature program. Experience in the operational aspects of PTGC using different reference standards is a great way to gain the necessary hands-on knowledge that will pay great dividends in any future laboratory work involving GCs. We would not know that we have achieved an optimized temperature program if it were not for the ability to detect the eluted analyte of interest to TEQA. The lower limit for PTGC can be extended by applying cryogenic GC techniques.

4.34 WHAT ROLE DOES CRYOGENICS PLAY IN GC?

An important one when it is desirable to either start the initial column temperature of a GC run below ambient temperature or minimize peak broadening due to the extra dead-volume effects caused by interfaces with purge-and-trap or static headspace sampling techniques as applied to VOCs. Let us digress a bit on cryogenics then and discuss the more contemporary approaches to GC today that involve *cryofocusing* techniques.

Listed in Table 4.11 are the boiling points of the most commonly used gases as cryogenic fluids.[47] Liquefied carbon dioxide and liquefied nitrogen are the two most common cryogens used with gas chromatographs. If the entire oven is to be cooled down, do not introduce the coolant until the oven has been air-cooled to ~60°C.[47] Heat pumps based on thermoelectricity (Peltier effect) and evaporation based on adiabatic expansion (Joule–Thompson effect) represent alternatives to cryogenic cooling. Cryogenic liquids are pumped into GC ovens with the use of solenoid switching valves. It is feasible to cool the GC oven to well below ambient temperature and to start a PTGC run at this subambient temperature provided that the column, usually a WCOT or PLOT, is stable at these subambient temperatures. Working with cryogenic liquids requires an increased awareness of the risks involved. It is imperative to *use liquefied CO_2 with equipment designed for liquefied CO_2, and liquefied N_2 with equipment designed for liquefied N_2.* Heavy insulation should be placed on all transfer lines and plumbing to prevent a buildup of dry ice snow around the lab. Contact between skin and a cryogenic liquid will lead to serious burns.

Cryogenic focusing, or cryofocusing, uses the cold temperatures of cryogenic coolants to focus the sample into a plug at or near the head of the GC column to improve chromatographic peak shape. Cryogenic trapping, or cryotrapping, is used to concentrate trace amounts of components rather than focusing them onto the analytical column.[48] Cryofocusing traps are commercially available units that cool the first segment of a cap column with subsequent rapid heating of the trap to 400°C at 800°/min.[49] Let us revisit Equation (4.10) or (4.31). The chromatographic capacity factor k' determines the speed of the solute through the column. Temperature influences the partition constant K, as we saw earlier. Changes in K influence k', and hence the relative speeds of the front and rear tailing with respect to peak apex. This tightening of the tailings leads to a narrowing of the solute plug at the onset of the chromatographic run. This cryofocusing might be analogous to the narrower chromatographically resolved peaks that result from manual syringe injection of a solvent that contains a dissolved solute. A firm and rapid thrust of the syringe

TABLE 4.11
Most Commonly Used Cryogens in GC

Cryogen	T_{bp} (°C)
CO_2	−78.5
CH_4	−161.4
O_2	−183.0
Ar	−185.7
F_2	−187.0
N_2	−195.8
He	−269.9

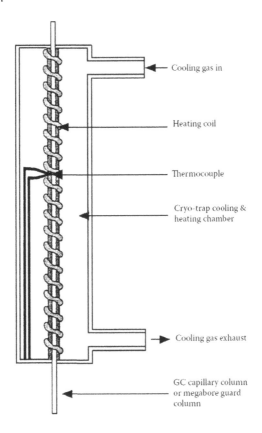

Cooling gas in

Heating coil

Thermocouple

Cryo-trap cooling &
heating chamber

Cooling gas exhaust

GC capillary column
or megabore guard
column

FIGURE 4.29

plunger usually results in narrow peaks, whereas a light and slow thrust usually yields much broader peaks.

Figure 4.29 shows a GC Cryo-Trap® (Scientific Instruments Services), which enables a GC to be converted to take advantage of cryofocusing. The Cryo-Trap consists of a small heating and cooling chamber that surrounds the first 5 inches of the GC capillary column. The unit is installed inside the GC column oven just under the injection port. A separate digital dual-temperature range controller regulates as well as measures chamber temperature. The system can be used either manually to switch between the cooling and heating cycles or automatically via an input signal from a controlling device or GC signal switch. Trace VOCs from either purge-and-trap or static headspace (Chapter 3) can be trapped at temperatures down to −70°C when liquefied CO_2 is used, or down to −180°C when liquefied N_2 is used. To release the VOCs from the Cryo-Trap, a heater coil inside the unit rapidly heats the capillary column to temperatures up to 400°C. The released VOCs are subsequently temperature programmed through the GC column.[50]

4.35 WHAT ARE THE COMMON GC DETECTORS AND HOW DO THEY WORK?

We continue now on our journey across the GC schematic drawing shown earlier (Figure 4.10) to the topic of *GC detectors*. An overview of the common detectors along with their commonly used abbreviations is found in Table 4.12. We will discuss only those detectors of relevance to TEQA and mention only some of the more recently developed GC detectors. This author was recently confronted with the notion of removing an existing detector from a GC and replacing it

TABLE 4.12

Summary of the Operating Characteristics of All of the Common GC Detectors Used in TEQA

Abbreviation	Selectivity	Flow Dependence	Destructive or Not	Detector Stability	Linear Dynamic Range	Sensitivity	Importance to TEQA	Used by Author
TCD	Universal	C	N	Good	$>10^5$	>10 ppm	Some	Yes
FID	Selective for C	M	D	Good	$>10^7$	>10 ppm	Some	Yes
PID	Selective for aromatics	C	N	Moderate	$>10^7$	>1 ppb	Much	Yes
ElCD	Selective for X, S, N	C	D	Moderate		>1 ppb	Much	Yes
ECD	Selective for X	C	N	Moderate	$>10^4$	>1 ppb	Much	Yes
TSD	Selective for N, P	M	D	Good	$>10^7$	>1 ppb	Some	Yes
FPD	Selective for N, P, S	C	D	Good	>10	>1 ppb	Some	Yes
MSD	Universal and selective	M	D	Good	$>10^6$	>10 ppb	Much	Yes

Note: C = concentration; N = nondestructive; M = mass; D = destructive.

with another type. This is not as straightforward as one might think. The one universal concept here is that all GC detectors produced an analog signal related to the analyte that has just eluted the chromatographic column. The magnitude of this signal, whether positive or negative, should always be viewed with respect to the noise that is also generated by the detector and the processing electronics. Refer to the topics involving the signal-to-noise ratio discussed in Chapter 2. One student experiment in Chapter 5 asks the student to compare organic compound specificity for a FID vs. ECD.

The GC detector is the last major instrument component to discuss. The GC detector appears in Figure 4.10 as the box to which the column outlet is connected. Evolution in GC detector technology has been as great as any other component of the gas chromatograph during the past 40 years. Among all GC detectors, the photoionization (PID), electrolytic conductivity (ElCD), electron-capture (ECD), and mass-selective detector (MSD) (or single quadrupole mass spectrometer) have been the most important to TEQA. The fact that an environmental contaminant can be measured in some cases down to concentration levels as low as *parts per trillion (ppt)* is a direct tribute to the success of these very sensitive GC detectors and to advances in electronic amplifier design! GC detectors manufactured during the packed-column era were found to be compatible with WCOTs. In some cases, makeup gas must be introduced, such as for the ECD. Before we discuss these GC detectors and their importance to TEQA, let us list the most common commercially available GC detectors and then classify these detectors from several points of view.

Most GC detectors in use today, along with the primary criteria used to compare such detectors, are listed in Table 4.12. It seems that the development of new GC detectors over the past 10 years is rare. This reflects more the maturity of gas chromatography as an analytical determinative technique rather than the lack of innovative ideas. We will discuss two of the more recent commercially developed GC detectors after we discuss the most common ones that have been around for 30 or more years. The following question is appropriate at this point: If you can detect down to ppt levels and can do this with a good signal-to-noise ratio, why change to a newly introduced detector! Each GC detector listed in the table can be further categorized as either universal or selective, concentration or mass flow dependent, and destructive or nondestructive to the analyte as it elutes from the chromatographic column. Additional questions must be asked. Does the detector have a baseline that is stable over time? Over what range of concentration or mass will the detector give a linear signal? Just how sensitive is the detector, or with respect to an entire instrument such as a GC, how low can it measure (the IDL; refer to Chapter 2)? Is the detector very important to TEQA? Has this author ever operated such a detector, or in other words, can he write from experience?

Ewing has defined a concentration-dependent GC detector as one that gives a response proportional to the concentration of sample, expressed in mole fraction. (p. 356)[51] He also defines a mass-dependent GC detector as one that depends upon the rate at which the sample is delivered to the sensing element, but the extent of dilution by carrier gas is irrelevant. The distinction between a concentration and mass-dependent GC detector has to do with understanding just what is fundamentally responsible for the signal that originates from such a detector. A Gaussian profile of a chromatographically resolved peak is shown in Figure 4.30. The nature of the signal that is responsible for any rise in the detector response when an analyte passes through is viewed in terms of an infinitesimal area, dA. For a *concentration-dependent* GC detector such as a thermal conductivity detector (TCD) or an electron-capture detector (ECD) having a mole fraction x_s that can be defined in terms of the velocity of solute, v_s, and a carrier gas velocity, v_c, we can define dA as follows:

$$dA = x_S dt = \left(\frac{v_S}{v_S + v_C} \right) dt$$

Let us assume that as the analyte of interest elutes from the column into the detector, the total gas velocity can be composed of contributions from the solute as well as from the carrier

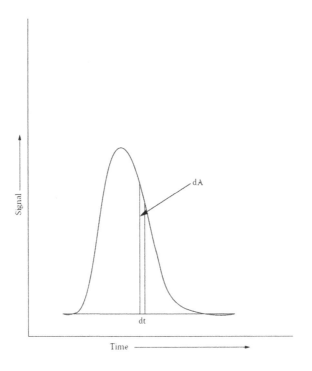

FIGURE 4.30

gas. Mathematically stated, $v = v_S + v_C$, so that upon integration over the entire peak, we can show that

$$A = \int x_S dt = \int \frac{v_S}{v} dt$$

The solute velocity is the time rate of change of mass, $v_s = dm/dt$, and we can express the integral as

$$A = \frac{1}{v} \int \frac{dm}{dt} dt = \frac{m_S}{v}$$

The above equation states that the peak area is proportional to the solute mass, m_s, only if v is held constant. We have thus shown that the peak area is indeed related to a ratio of solute mass to total flow rate.

Again, consider Figure 4.30, but in this case, we have a *mass-dependent* GC detector, and we now define the infinitesimal strip, dA, as

$$dA = v_s \, dt$$

The area under the curve is obtained as

$$A = \int v_S dt = \int \frac{dm}{dt} dt = m$$

The peak area is directly related to the mass of solute passing through it. A concentration-dependent detector does not necessarily respond to the presence of a substance in the carrier gas, but rather to a difference or change in the concentration of that substance. Hence, a TCD would have its sensitivity defined as the *smallest number of microvolt-seconds per ppm*, whereas the sensitivity of a flame ionization detector (FID) might be defined as the *smallest number of microvolt-seconds per nanogram*. Recall from the discussions in Chapter 2 that sensitivity refers to the magnitude of the slope of a plot of detector response vs. analyte concentration or analyte mass, and the detection limit of the GC detector, or as part of an instrument (IDL), refers to a signal-to-noise ratio measurement. This author believes it to be very important that practicing analysts and lab technicians have some understanding as to how GC detectors work. It helps to solve a lot of problems later. The TCD has been around for a long time. Aside from the TCD being used in environmental testing labs to measure fixed gases and any organic compounds that are present at low parts per hundred (%) levels or higher, this detector does not have a role in TEQA. The FID, however, has a somewhat limited role in TEQA, largely due to a FID detection limit in the 10 ppm realm. The FID is suggested in some EPA methods for SVOCs screening procedures. The basic design of the FID is often a starting point for explaining how other GC detectors work. For these reasons, we delve into the details of this GC detector.

4.36   HOW DOES A FID WORK?

The FID is a micro hydrogen/oxygen burner that continuously maintains a hydrogen flame. A collector electrode is set at +300 V relative to the flame tip. When an organic compound is eluted from the GC column, it is ionized and a small current is generated. It is this current that forms the basis for the analog signal from the FID whose magnitude is proportional to the mass of compound being detected. Carrier gas, being chemically inert, does not cause the formation

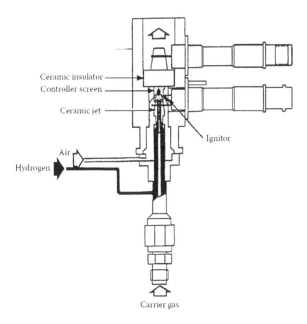

Ceramic insulator
Controller screen
Ceramic jet
Ignitor
Air
Hydrogen
Carrier gas

FIGURE 4.31

of ions. An igniter is provided to start the hydrogen flame burning. Hence, the FID looks like a chimney with two side arms. One side arm is for the electrical connections for the igniter; the other side arm provides for the polarizing voltage and analog signal output. Lighting a FID is one of the more memorable analytical lab type experiences one might confront. Recently, this author, who is accustomed to checking to see whether the FID is lit by holding the lens of his glasses above the outer barrel of the detector, changed his glasses from glass to plastic lenses. An accumulation of moisture on the lens surface is indicative of a lighted FID. A small area on the surface of the plastic lens, instead of accumulating water, melted, much to the surprise and chagrin of the author! Figure 4.31 shows a schematic of the FID. Carrier gas (either He, N_2, or H_2) enters from beneath, being joined by hydrogen and then air, and continues up through the cylinder to the jet tip. Above the jet tip is a collector cup/electrode arrangement. Coaxial cables lead out from the side of the FID and connect to instrument electronics.

The operating characteristics of the FID serve as a good frame of reference from which to view all other GC detectors. It is for this reason that we discuss these FID operating characteristics. We will then discuss the ECD, PID, ElCD, TSD, FPD, and MSD in terms of how each of these compares to the FID. This author has had numerous requests over the years to use the FID to measure down to low ppb, such as 10 to 100 ppt of an aliphatic or aromatic organic compound by direct injection of a solution that contains the dissolved compound. This it cannot do! A gas chromatograph incorporating a WCOT column and a FID, abbreviated C-GC-FID, does not have an IDL of 1 ppb or lower! Prospective clients are surprised to learn this. Let us take a look at these FID operating characteristics.

4.37 WHAT ARE THE OPTIMUM GAS FLOW RATES AND OPERATING CHARACTERISTICS OF A FID?

Compounds that do not contain organic carbon do not burn and are insensitive to the FID. These include water, inert gases, and nonmetal hydrides such as ammonia and hydrogen sulfide. The greater the number of carbon atoms per molecule in an organic compound, the greater is the FID sensitivity. For example, injection of 1 μL of a 50 ppm solution containing *n*-hexane dissolved in acetone would yield a signal that is strong, yet less than the signal from injection of 1 μL of a 50 ppm solution containing *n*-nonane dissolved in the same solvent. One says that *n*-nonane has a greater FID *relative response factor* (RRF) than *n*-hexane. Aromatic hydrocarbons have a greater RRF than aliphatic hydrocarbons having the same number of carbon atoms per molecule. As shown in one of the GC experiments in Chapter 5, a student can keep the carrier gas flow rate and the FID airflow rate fixed while varying the FID hydrogen flow rate and deduce that 30 to 50 cm^3/min for a conventional FID maximizes the GC response. Keeping the carrier and FID hydrogen flow rates fixed while varying the FID airflow rate reveals that beyond 300 cm^3/min, no further increase in GC response is observed.

McNair and Miller have characterized the FID based on the following criteria (pp. 112–116).[28] The FID is selective for all carbon-containing compounds. The FID has a minimal detectable amount of 10^{-11} g. A linear dynamic range of 10^6 enables, for a 1 μL injection volume, a detector response that is linear from 1 g of hydrocarbon all the way down to 1 μg of hydrocarbon. The next criterion is that of good baseline stability over time, and this translates into minimal baseline drift. Also, the effect of a change in flow rate or temperature on baseline stability is an important characteristic of FIDs. Finally, it is important that a GC detector be thermally stable because the analyte must remain in the gas phase. The FID can be operated up to a maximum temperature setting of 400°C. The GC chromatogram shown earlier in this chapter for the separation of BTEX components in gasoline-contaminated groundwater (Figure 4.11) was obtained using a FID. The EPA's Contract Laboratory Program utilizes a preliminary method known as "Hexadecane Screening of Water and Soil Screening for Volatiles." This method comprises a preliminary evaluation of a hazardous waste sample prior to implementing the determinative GC-MS technique in EPA Method 8270. In the author's laboratory, a C-GC-FID is a very useful instrument for conducting preliminary examinations for the presence of trace organic contaminants in drinking or wastewater samples.

As introduced in Chapter 3, for example, if gasoline-contaminated drinking water is suspected, a mini-LLE (liquid–liquid extraction) using hexadecane, with a subsequent injection of 1 μL of the hexadecane extract into the C-GC-FID, will give the analyst a quick, yet informative indication of the extent of the contamination. An example of *enviro-chemical* TEQA.

More than 30 years ago, the need to separate and detect VOCs was accomplished by placing a *PID in series with an ElCD*, in which aromatic VOCs could be detected on the PID and ClVOCs (chlorinated VOCs such as THMs) could be detected on the ElCD. This was the principal basis for EPA Methods 501 and 502 for drinking water and 601 and 602 for wastewater. The lowering of GC IDLs from the low ppm level afforded by the FID to the low ppb level was first accomplished on a routine basis with the PID, ElCD, and ECD. An instrument that incorporated a WCOT and ECD, C-GC-ECD, first detected DDT in the environment, and a cap column with a PID, C-GC-PID, first found ppb levels of BTEX in contaminated groundwater.

4.38   HOW DO A PID AND AN ElCD WORK?

Configuring both the PID and ElCD in tandem has proved to be a powerful duo for relatively clean water samples. EPA Method 8021B in the most recent SW-846 series recommends that both detectors be used in tandem or individually. The PID is an aromatics-specific detector, and the ElCD is a halogen-specific detector. When placed in tandem, both GC detectors provide for a large number of VOCs that can be separated and detected using WCOTs. The PID consists of a deuterium lamp that provides the necessary 10.2 eV of energy to ionize an aromatic molecule such as benzene, while unable to ionize many other hydrocarbons whose ionization potential is greater than 10.2 V. A neutral molecule, RH, absorbs a photon of ultraviolet (UV) energy and dissociates into a parent ion and an electron according to

$$RH + h\nu \longrightarrow RH^+ + e^-$$

The energy of the photon must be of a frequency such that this energy exceeds the ionization potential of the organic compound for photoionization to occur. The PID is equipped with a sealed UV light source that emits photons that pass through a UV transparent window made of LiF, MgF_2, NaF, or sapphire into an ionization chamber, where photons are absorbed by the eluted analyte. A positively biased high-voltage electrode accelerates the resulting ions to a collector electrode. The current produced by the ion flow is measured by the electrometer and is proportional to analyte concentration in the gas phase. The PID is a concentration-dependent, nondestructive detector. Other lamps are available that enable a wider range of hydrocarbons to be as sensitive as aromatics. A schematic of the PID is shown in Figure 4.32. The PID requires only carrier gas. Routine maintenance includes lamp window cleaning, lamp replacement, lamp window seal replacement, and positioning. With a detection limit in the picogram range coupled to a linear dynamic range of 10^7, the PID complements other detectors. The PID has also been configured in series before a FID to provide selective aliphatic hydrocarbon as well as aromatic hydrocarbon detection.

It is instructive to view the schematic of an ElCD, shown in Figure 4.33, from the perspective of the analyte of interest as it emerges from the outlet of the PID and enters the reaction tube. The analyte is pyrolyzed in the reactor. In the sulfur mode, the analyte is oxidized using oxygen reaction gas to sulfur dioxide. In the halogen and nitrogen modes, the analyte is reduced using hydrogen as a reaction gas to form HCl, HBr, HF, or NH_3. The reaction tube acts as a catalyst to hasten reaction. The chemically converted analytes of interest are swept through a transfer line into a conductivity cell. In the cell, the ionized analyte is dissolved in a flowing solvent and the change in conductivity is measured. To obtain a good response from the conductivity cell, the solvent flow and pH must be optimized. The sensitivity of the ElCD is inversely related to solvent flow rate. The pH of the solvent is controlled by passing it through an ion exchange resin located in the solvent reservoir. The proper resin mixture will provide the correct pH for the solvent. The halogen and sulfur modes are acidic and require an acidic solvent; the nitrogen mode is basic and requires a basic solvent. Let us consider what happens to ClVOCs as they enter the ElCD. The reactor is made of nickel and is maintained at a temperature of 850 to 1,000°C. The ClVOCs are reduced to haloacids by mixing them with hydrogen reaction gas. Nonhalogenated VOCs are reduced to methane, which is nonionic. The haloacids are dissolved in n-propanol and the change in solvent conductivity is measured in the cell. The ElCD is a bit more difficult to operate and

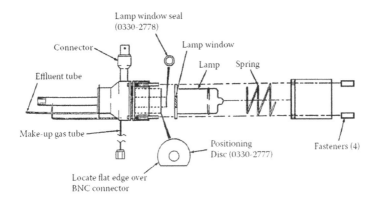

FIGURE 4.32

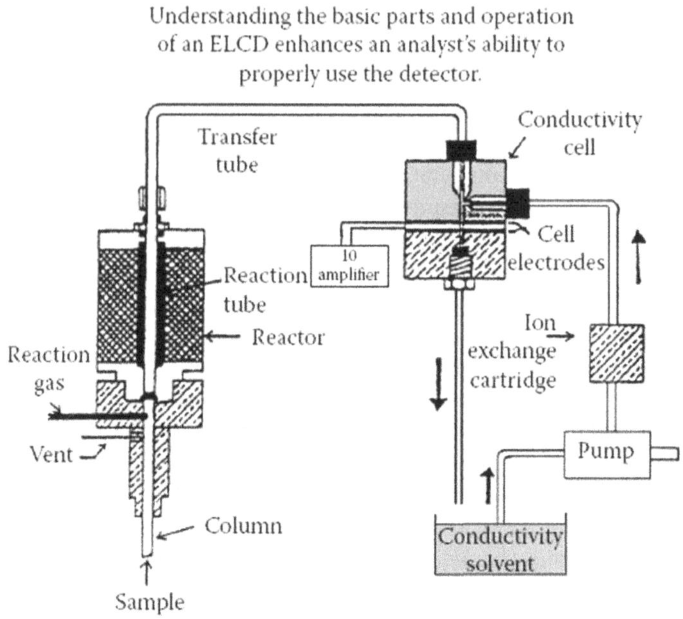

Understanding the basic parts and operation
of an ELCD enhances an analyst's ability to
properly use the detector.

FIGURE 4.33

maintain than other GC detectors. The ElCD in the halogen mode has one chief competitor, the electron-capture detector (ECD).

4.39   WHAT IS AN ECD, HOW DOES IT WORK, AND WHY IS IT SO IMPORTANT TO TEQA?

The ECD was introduced by Lovelock and Lipsky and reviewed in two parts by Lovelock and Watson.[52–54] Ewing aptly describes the ECD in the following manner (p. 357):[51]

> For many years GC made use of TCDs almost exclusively, but with the advent of capillary columns, which are limited to smaller samples, greater sensitivity was required. One method of detection that can give greater sensitivity is a modification of the ionization chamber long used for radiation detection. The effluent from the chromatographic column is allowed to flow through such a chamber, where it is subjected to a constant flux of beta ray electrons from a permanently installed radioisotope.

The inner wall of a cylindrically shaped cavity is lined with ^{63}Ni foil, and a voltage is imposed between a pair of electrodes. This generates a standing current, I_b. When N_2 or 95% Ar /5% CH_4 gas is passed through, I_b is approximately 1×10^{-8} A. The ^{63}Ni is the source of beta radiation, and thermal electrons are produced that do not recombine with either positive ions or neutral carrier gas molecules. Molecules of the analyte of interest that elute from the chromatographic column and contain electronegative heteroatoms, such as chlorine, capture electrons and become negatively charged. These negatively charged gas-phase ions quickly combine with any positive ions present. A decrease in I_b results. As molecules of the electron-capturing analyte elute and are swept through the ECD, a negative peak results. The peak is inverted by the signal-processing electronics to yield a positive peak like that obtained from all other GC detectors. The ECD requires about 30 cm^3/min of gas flow. Because a typical WCOT flow rate is more than 10 times lower, makeup gas is required. More contemporary GCs provide the necessary makeup gas. This author has used the inlet from a second GC injection port as a source of makeup gas when makeup is not "plumbed in." The advantage of ^{63}Ni over ^{3}H (tritium was an earlier choice of beta radiation) lies in the much elevated temperature enjoyed by this isotope. A ^{63}Ni ECD can be safely taken to $T_{ECD} = 400°C$. The following reactions help to explain the electron-capturing phenomena:

$$N_2 + \beta \longrightarrow N_2^+ + e^- + N_2^*$$

$$CCl_4 + e^- \longrightarrow CCl_4^- \text{?(nondissociative capture)}$$

$$CCl_4 + e^- \longrightarrow Cl + CCl_3^- \quad \text{(dissociative capture)}$$

Many of the operational problems of ECD instability that plagued these detectors for so long have been overcome with advances in design. Contemporary GCs that have ECDs enable either N_2 or Ar–CH_4 to be used as carrier or makeup gas. The low-bleed WCOTs available today are a major contributor to the good stability of ECDs. Contamination even in this era of low-bleed WCOTs and well-designed ECDs is still a problem in the day-to-day laboratory operation. It is a wrong assumption to believe that all GCs that have ECDs are ready to use all of the time to meet the objectives of TEQA. This statement is consistent with this author's experience. Onuska and Karasek have identified the sources of ECD contamination that take the form of deposition of liquid phase and of dirt on the ECD electrodes.[55] A contaminated ECD leads to loss of sensitivity, trailing peaks, and erratic baselines. The major sources of ECD contamination include the following:

1. Column bleed
2. Contaminated sample inlet, including a dirty glass insert
3. Oxygen in carrier gas or a dirty carrier gas
4. Lack of a tight WCOT connection, thus allowing oxygen to enter.

Figure 4.34 is a schematic for an ECD. EPA methods in general tend to require this detector for the determination of semivolatile to nonvolatile organochlorine analytes such as OCs and PCBs, whereby solvent extracts are introduced into the C-GC-ECD, and water has therefore been completely eliminated.

4.40   HOW RESPONSIVE ARE ECDs?

Relative response factors (RRFs) for various organic compounds of interest to TEQA reveal an answer to this question. Listed in Table 4.13 are RRFs for most of the functional groups that one might encounter.[55] RRFs differ by a factor of 100 between a trichloro-organic and a tetrachloro-organic is vividly demonstrated by referring to Figure 4.35. This is a GC chromatogram obtained on an Autosystem Gas Chromatograph (PerkinElmer) that is interfaced to an HS-40 (PerkinElmer) automated headspace analyzer. The response of carbon tetrachloride (CCl_4) is significantly higher than either chloroform ($CHCl_3$) or trichloroethylene (TCE). We close out our discussion of the ECD by showing two GC chromatograms from a C-GC-ECD in Figure 4.36. These chromatograms show a method blank extract and a sample extract from the Soxhlet extraction of a fiberglass insulation sample brought to this author's laboratory for a quantitative determination of Aroclor 1260. The envelope of peaks shows the presence of this particular commercial source of polychlorinated biphenyls (PCBs) in the insulation. This clear and almost interference-free chromatogram was obtained with no additional sample cleanup and shows the real value of an *element-specific GC detector*. Imagine what this chromatogram would look like if it were injected into a C-GC-FID!

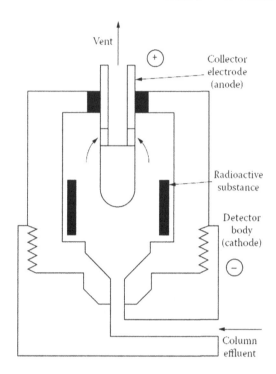

FIGURE 4.34

TABLE 4.13
Relative Response Factors and Nature of the Electron-Capturing Functional Group

RRF	Electron-Capturing Functional Group
10^0	Aliphatic hydrocarbons, aromatic hydrocarbons
10^1	Esters, ethers
10^2	Monochloro, monofluoro, aliphatic alcohols, ketones
10^3	Dichlori, difluoro, monobromo derivatives
10^4	Trichloro, anhydrides
10^5	Monoiodo, dibromo, nitro derivatives
10^6	Diiodo, tribromo, polychlorinated aromatics

4.41 ARE THERE OTHER ELEMENT-SPECIFIC DETECTORS?

Yes, indeed, and we make brief mention of these. We must ask, however, the following question: What other chemical elements are incorporated into organic molecules that are of interest to TEQA? Priority pollutant organic compounds of environmental interest that incorporate the elements *nitrogen, phosphorus*, and *sulfur* answer this question. GC detectors developed for these elements are of a more recent vintage and include: thermionic specific, flame photometric, and chemiluminescence. In addition, the atomic emission spectroscopic detector (AED) is applicable to the separation and quantitation of various *organometallics*. The AED can be tuned to a specific emission wavelength for a particular metal. Development of C-GC-AED techniques

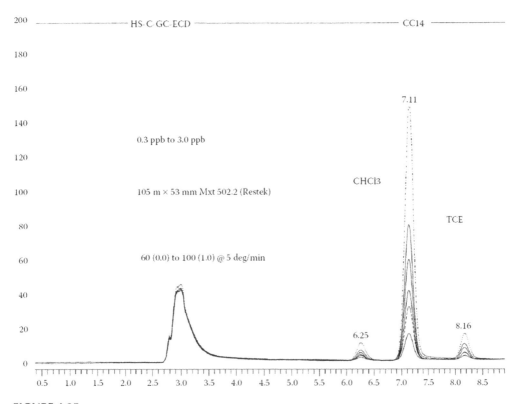

FIGURE 4.35

has advanced the analytical chemistry of *trace metal speciation*. However, the principal manufacturer of the GC-AED instrument ended production back in 2004.

The thermionic-specific detector (TSD), also called the *alkali flame ionization detector*, is really a FID with a bead of an alkali metal salt such as Rb or Cs. A schematic of the TSD is shown in Figure 4.37. The TSD shows enhanced sensitivity for organic compounds that contain the elements nitrogen and phosphorus. This author has used a GC that incorporated a TSD to measure the extent of contamination of groundwater in an aquifer somewhere in Maine with *N, N*-dimethyl formamide (DMF). A molecular structure of this groundwater contaminant is shown below:

N, N-dimethylformamide

Figure 4.37 shows that the bead can be heated to a red glow and the heating of the bead has led to increased TSD stability. A negative polarizing voltage is applied to the collector. The ion current generated by the thermionic emission can be related to the mass of, for example, DMF eluted from the WCOT.

The flame photometric detector (FPD) operates on the principle that phosphorus- and sulfur-containing organics compounds eluted from the GC column *emit characteristic green and blue*

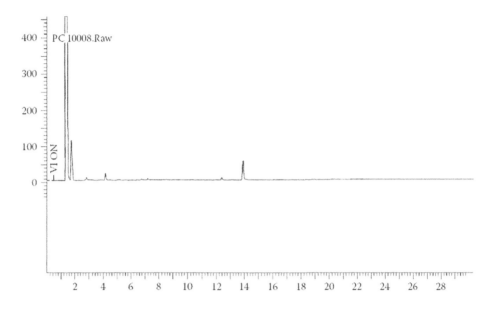

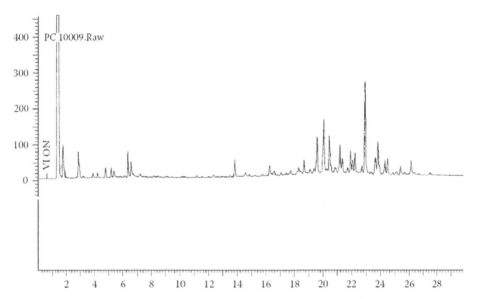

FIGURE 4.36

colorations, respectively, in hydrogen-rich H2–air rich flames. A schematic of a dual FPD is shown in Figure 4.38. This design, first developed at Varian Associates, burns the organic solute during passage through the lower flame and then excites the free radical HPO species and causes it to emit at 526 nm for phosphorus-containing organic compounds, or S_2 at 394 nm for sulfur-containing organic compounds in the higher flame.[56] These species emit characteristic wavelengths through an optical window through various cutoff filters to a photomultiplier tube, as shown in Figure 4.38. Hydrogen gas enters from the side and mixes with air and carrier gas to sustain a hydrogen flame. The design criteria in the dual-flame FPD is designed, for example, to maximize the flame photometric emission of HPO while minimizing hydrocarbon emissions and interferences due to S_2.

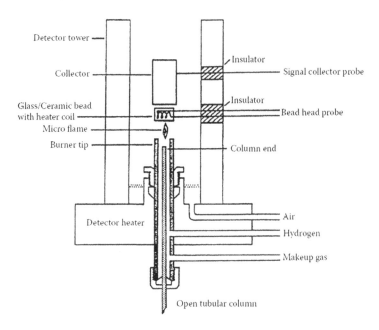

FIGURE 4.37

This author was once fortunate to have access to a gas chromatograph (Model 3400, Varian) that incorporated both a TSD and an FPD. This instrumental configuration enabled this author to conduct a comprehensive study of the isolation and recovery of representative OPs via reversed-phase solid-phase extraction (SPE).[57] Figure 4.39 is a schematic of the single-injector/dual-column/dual-detector GC configuration used to conduct the study. In order to provide the necessary makeup carrier gas to both detectors, stainless-steel tubing from the inlet of the first injector was connected to the detector inlets as shown in Figure 4.39. A dual GC chromatogram, shown in Figures 4.40A and 4.40B was obtained using the dual-column/dual-detector GC just described. A DB-5 and a DB-608 WCOT column were used to separate a mixture that contains seven OPs and the *internal standard, triphenyl phosphate*. A molecular structure for triphenyl phosphate is shown below:

Note that the elution order changes somewhat between both columns. The instrumentation used and the instrumental conditions employed in this study are given in Table 4.14. The specific operational parameters shown in Table 4.14 reveal much about how to operate both the TSD and FPD detectors. The instrumentation was of a late 1980s vintage and reveals the state of the art in GC instrumentation back then. The fact that these OPs could be isolated, recovered, and detected at trace concentration levels still is a most significant outcome toward conducting *enviro*-chemical TEQA!

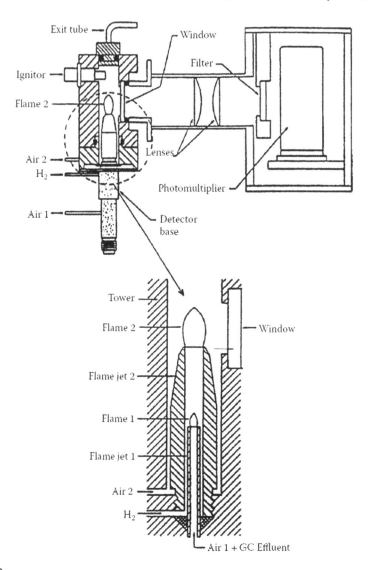

FIGURE 4.38

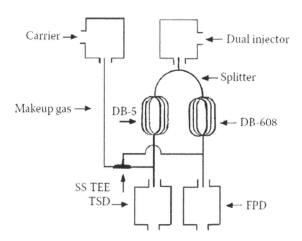

FIGURE 4.39

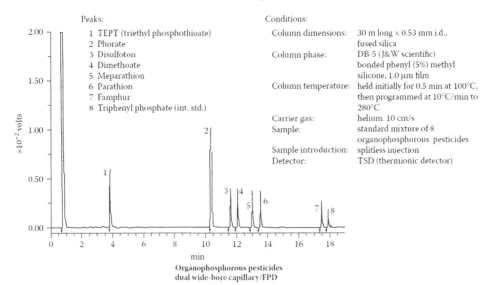

Organophosphorous pesticides
dual wide-bore capillary/TSD

Peaks:
1 TEPT (triethyl phosphothioate)
2 Phorate
3 Disulfoton
4 Dimethoate
5 Meparathion
6 Parathion
7 Famphur
8 Triphenyl phosphate (int. std.)

Conditions:
Column dimensions: 30 m long × 0.53 mm i.d., fused silica
Column phase: DB-5 (J&W scientific) bonded phenyl (5%) methyl silicone, 1.0 μm film
Column temperature: held initially for 0.5 min at 100°C, then programmed at 10°C/min to 280°C
Carrier gas: helium, 10 cm/s
Sample: standard mixture of 8 organophosphorous pesticides
Sample introduction: splitless injection
Detector: TSD (thermionic detector)

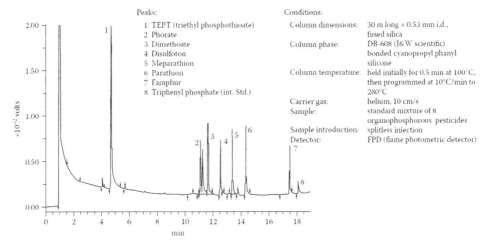

Organophosphorous pesticides
dual wide-bore capillary/FPD

Peaks:
1 TEPT (triethyl phosphothioate)
2 Phorate
3 Dimethoate
4 Disulfoton
5 Meparathion
6 Parathion
7 Famphur
8 Triphenyl phosphate (int. Std.)

Conditions:
Column dimensions: 30 m long × 0.53 mm i.d., fused silica
Column phase: DB-608 (J&W scientific) bonded cyanopropyl phanyl silicone
Column temperature: held initially for 0.5 min at 100°C, then programmed at 10°C/min to 280°C
Carrier gas: helium, 10 cm/s
Sample: standard mixture of 8 organophosphorous pesticides
Sample introduction: splitless injection
Detector: FPD (flame photometric detector)

FIGURE 4.40

TABLE 4.14
GC Operation Conditions Used to Conduct Dual Megabore Capillary GC-NP Detection

Gas chromatograph	Varian 3400 with Varian 8034 autosampler
Megabore splitter	Supelco direct injector tee, silanized with appropriate fittings
Data station	Varian 604 with Varian IIM interface using the dual-column software option
Printer	Hewlett-Packard Think Jet printer
WCOT	30 m × 0.53 mm DB-5 and 30 m × 0.53 mm DB-608 from J&W Scientific
TSD and flow characteristics	300°C, air at 175 cm³/min, H_2 at 4.5 cm³/min, carrier (N_2) at 5 cm³/min, makeup (N_2) at 25 cm³/min
FPD and flow characteristics	300°C, air 1 at 80 cm³/min, air 2 at 170 cm³/min, H_2 at 140 cm³/min, carrier (N_2) at 5 cm³/min, makeup (N_2) at 25 cm³/min

4.42   WHAT ABOUT ANY RECENTLY DEVELOPED ELEMENT-SPECIFIC GC DETECTORS?

Two such detectors will be discussed here, the pulsed-discharge detector (PDD), manufactured by VICI (Valco Instruments Company, Inc.), and the chemiluminescence detector, manufactured by Sievers Instruments. Both detectors are of a more recent vintage. The PDD has the potential of replacing either the ECD or FID, and also the PID, depending on the application to TEQA. The PDD is a nonradioactive pulsed-discharge ionization detector and was designed for ease of installation to the Model 6890 GC (Agilent Technologies, Inc.). A stable, low-power, pulsed DC discharge in helium is the source of ionization. Solutes eluting from the WCOT column flowing in the opposite direction to that of the flow of helium from the discharge zone are ionized by photons from the helium discharge above. Electrons from this discharge are focused toward the collector electrode by the two bias electrodes. A schematic of the PDD is shown in Figure 4.41. Photoionization by radiation is the principal mode of ionization and arises from the transition of diatomic helium to the dissociated monatomic He ground state. In the electron-capture mode, the PDD is a selective detector for monitoring high-electron-affinity compounds such as ClVOCs, OCs, PCBs, and so forth. GCs that incorporate the PDD would be capable of IDLs down to low ppb concentration levels. In this mode, helium and methane are introduced just upstream from the column exit. In the helium photoionization mode, the PDD is a universal, nondestructive, highly sensitive detector. In this mode, the PDD is a suitable replacement for a conventional FID. When a dopant is added to the discharge gas, the PDD also functions as a selective photoionization detector. Suitable dopants include Ar for organic compounds, Kr for unsaturated compounds, and Xe for polynuclear aromatics. A drawing of the PDD as it might appear when being installed into a 6890 GC is shown in Figure 4.42. This is a good example of how existing gas chromatographs can be retrofitted with more newly developed GC detectors as the need arises.

Because ECDs require radioactive sources, the pulsed-discharge electron-capture detector (PDECD) is a relatively recent alternative. A cross section of a PDECD that incorporates methane as a dopant is shown in Figure 4.43. Cai and coworkers provide a detailed description of the

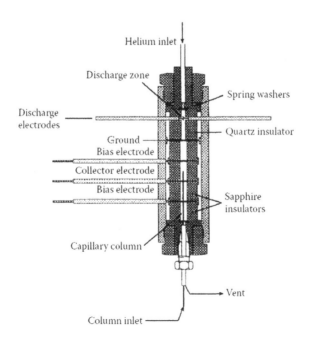

FIGURE 4.41

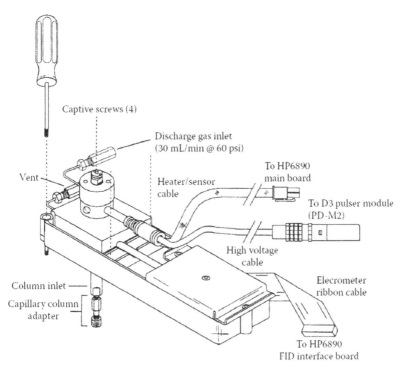

Captive screws (4)

Discharge gas inlet
(30 mL/min @ 60 psi)

To HP6890
main board

Heater/sensor
cable

Vent

To D3 pulser module
(PD-M2)

High voltage
cable

Elecrometer
ribbon cable

Column inlet

Capillary column
adapter

To HP6890
FID interface board

FIGURE 4.42

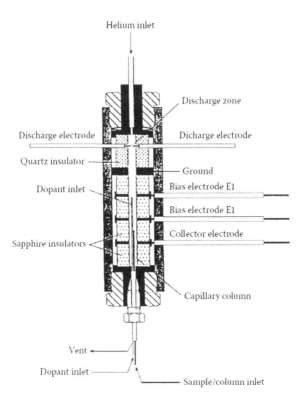

Helium inlet

Discharge zone

Discharge electrode

Discharge electrode

Quartz insulator

Ground

Bias electrode E1

Dopant inlet

Bias electrode E1

Collector electrode

Sapphire insulators

Capillary column

Vent

Dopant inlet

Sample/column inlet

FIGURE 4.43

design of this detector and described their recent developments while studying the influence of discharge current, bias and collecting voltages, and the chemical nature of the dopant gas while comparing conventional ECDs to the PDECD.[58] The pulsed discharge in pure helium produces high-energy photons according to

$$He_2 \longrightarrow 2He + h\nu$$

If methane is added downstream from the discharge, free electrons are produced. The dopant also thermalizes the electrons, making them more capturable by analytes. Analytes capture these thermalized electrons and reduce the standing current I_b according to

$$\frac{I_b - I_e}{I_e} = K_{EC}[AB]$$

where I_b is the standing current without analyte, I_e is the current with the electron-capturing analyte AB present, K_{EC} is a capture coefficient for AB, and [AB] represents the concentration of AB. This equation was first developed by Wentworth et al. for an ECD model. Thermal electrons are first formed in the discharge zone as methane undergoes photoionization according to: [59]

$$CH_4 + h\nu \longrightarrow P^+ + e^-$$

These primary electrons from photoionization become thermalized:

$$e^- + CH_4 \longrightarrow e^-_{thermal} + CH_4$$

Some electrons and positive ion species, P[+], recombine to form neutral species according to

$$e^- + P^+ \longrightarrow Neutrals$$

The dissociative and nondissociative mechanisms discussed earlier for the ECD also apply to the PDECD. The only difference in kinetic models is that of the first step of electron formation. The capture coefficient K_{EC} can then be defined in terms of the various rate constants for these various reactions. The collection of thermal electrons constitutes I_b for the PDECD. This standing current decreases during passage of an electron-capturing analyte through the detector and is responsible for the PDECD signal.[60] The use of a pulsed discharge rather than a continuous discharge serves to impart a higher energy to the cell because energy is dissipated during a very short period. The density of ions should be kept low for the charged species to be sustained in the reaction region. Pulsing also keeps the average potential to near zero, thus allowing what Wentworth et al. call a field-free condition to be realized shortly after the discharge takes place.[60] A comparison of the relative response factors (RRFs) for a selective number of ClVOCs that have been discussed earlier is given in Table 4.15.[58] The PDECD is as responsive as, if not more responsive than, the[63] Ni-ECD for aliphatic ClVOCs; however, the PDECD becomes somewhat less responsive for mono- and dichlorobenzenes. A comparison was also made between a conventional [63]NI-ECD and the PDECD with respect to satisfying the criteria posed in EPA Method 608 for OCs. GC chromatograms for the injection of a multicomponent OC standard, including the TCMX and DCBP surrogates discussed earlier, are shown for a standard sample and matrix spike for both detectors in Figure 4.44. Note that the authors plot the ratio $(I_b - I_e)/$

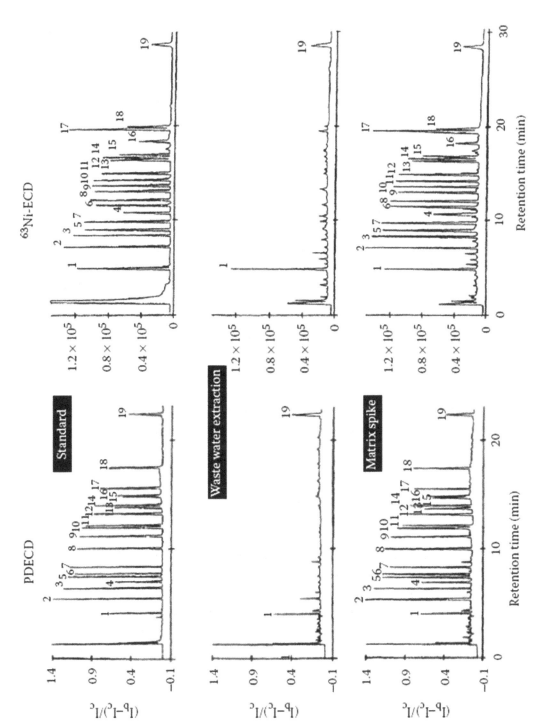

FIGURE 4.44

TABLE 4.15
Comparison of RRFs for PDECD and ^{63}Ni-ECD

CIVOC	RRF (PDECD)	RRF (^{63}Ni-ECD)
CCl$_4$	10,000	10,000
PCE	5,000	4,000
CHCl$_3$	660	470
TCE	570	380
Chlorobenzene	13	45
1,3-Dichlorobenzene	13	24
1,2-Dichloroethane	1.2	0.8
CH$_2$Cl$_2$	0.3	0.3

Note: PCE = tetrachloroethene; TCE = trichloroethene.

Ie vs. chromatographic run time in the figure.[58] The difference in WCOT column polarity is responsible for the differences in elution order. It would appear that the PDECD is a viable alternative to the ECD. One other element-specific GC detector needs to be discussed that involves chemiluminescence.

4.43   OK, BUT WHAT ABOUT CHEMILUMINESCENCE?

Chemiluminescence can be defined as the spontaneous emission of light by chemical reaction. In the sulfur chemiluminescence detector (SCD), organosulfur compounds that elute from a WCOT are combusted in a hydrogen-rich flame to produce sulfur monoxide (SO), among other products. This is the same process that occurs in a FID. These combustion products are collected and removed from the flame using a ceramic sampling tube (probe) interface and transferred under a vacuum through a flexible tube to the reaction chamber of the SCD. Sulfur monoxide is detected by an ozone / SO chemiluminescent reaction to form electronically excited sulfur dioxide (SO$_2^*$), which relaxes with emission of light in the blue and ultraviolet regions of the electromagnetic spectrum according to:[61]

$$SO + O_3 \longrightarrow SO_2^* \longrightarrow SO_2 + O_2 + h\nu$$

One model of an SCD is originally manufactured by Sievers Instruments and readily adapts to existing GCs. A more recent development from Sievers is the nitrogen chemiluminescence detector (NCD). Organonitrogen compounds that elute from a WCOT enter a ceramic combustion tube in a stainless-steel burner. The hydrogen and oxygen plasma in the combustion tubes convert all organonitrogen compounds to nitric oxide at temperatures greater than 1,800°C according to:

$$R - CH_2 - \bar{N} + H_2 + O_2 \longrightarrow NO + CO_2 + H_2O$$

Nitric oxide reacts with ozone to form electronically excited nitrogen dioxide according to

$$NO + O_3 \longrightarrow NO_2^* \longrightarrow NO_2 + h\nu$$

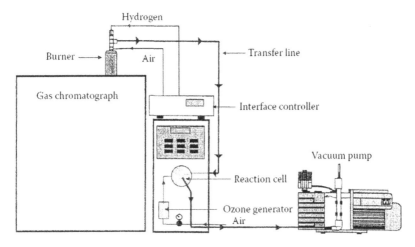

FIGURE 4.45

Excited NO_2 emits light in the red and infrared regions of the electromagnetic spectrum, from 600 to 3,200 nm, when it relaxes to its ground state. The light emitted is directly proportional to the amount of nitrogen in the sample.[62] Figure 4.45 depicts a schematic diagram for either an SCD or an NCD connected to a GC. Note that a burner head is attached to the existing detector port much like a FID. The combustion products are directed to a reaction cell where ozone can be generated and added. A photomultiplier tube, not shown in the schematic, is incorporated into the detector. Comparisons have been made between the NCD and TSD. Today, the giant instrument manufacturers such as Agilent, Shimadzu, and Thermo Fisher all offer NCDs and SCD integrated to gas chromatographs.

Element-specific GC detectors are in general very sensitive and are quite useful for target-specific determinations; however, the most widely used GC detector and the one that satisfies the most EPA methods is the mass spectrometer (MS) and the success that has been achieved in interfacing a GC with a "mass spec." This interface was *the first successful hyphenated method*, a term coined much earlier by Hirshfeld to describe how two established instruments could be married such that the whole is greater than the sum of the parts.[63] Before we delve into the mass spectrometer as a GC detector, let us digress to yet another spectroscopic GC detector, namely, the *atomic emission detector* (AED).

4.44 ATOMIC SPECTROMETRIC EMISSION AS A GC DETECTOR: WHAT WAS IT ANYWAY?

Ionization-based GC detectors have just been discussed (FIDs, ECDs, PIDs, etc.). Organic compounds that are chromatographically resolved have been introduced into a miniaturized helium *microwave-induced plasma* (MIP) discharge. The topic of atomic emission spectroscopy will be introduced as a determinative technique for trace metals analysis later in this chapter. Organic compounds entering the hot plasma are atomized and electrons are excited to higher quantized potential energy levels. Consider TCMX, a common surrogate used in the determination of various organochlorine pesticides. TCMX is retained on a WCOT column such as a DB-5. The analyte elutes from the column and enters the MIP. Look at what happens when TCMX enters the MIP:

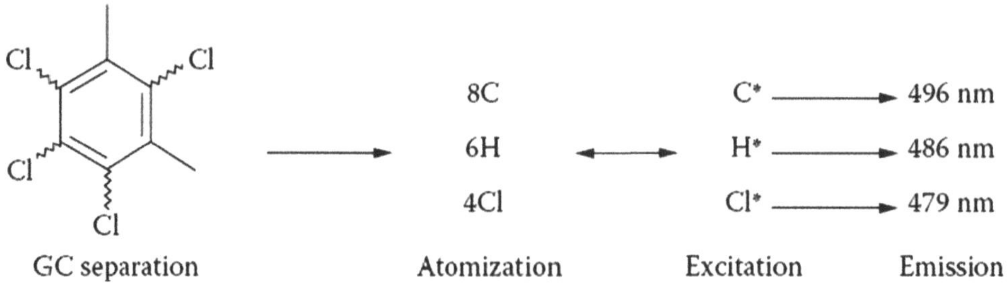

| GC separation | Atomization | Excitation | Emission |

8C	C* ⟶ 496 nm
6H	H* ⟶ 486 nm
4Cl	Cl* ⟶ 479 nm

In general, the emitted intensity is independent of the number of atoms of a given element in the molecular formula of the organic compound. Another way to express this is to say that there is an equal molar response for each element. In other words, 1 mole of TCMX with four chlorines per molecule *gives the same signal response* as would 1 mole of hexachlorobenzene with six chlorines per molecule! The physico-chemical basis of this equimolar signal response is shown by use of the *double-arrow notation*. The double arrow symbolizes a dynamic equilibrium being set up within the plasma. This equimolar response is in stark contrast to, for example, the ECD, as is demonstrated in Table 4.13, and has enabled users of C-GC-AED to accomplish *compound-independent calibration* (CIC). Software (AED ChemStation®, Agilent) has been written to enable CIC to be performed.

In addition to analytically useful atomic emission lines, there exist molecular emission bands known as broad bands and quite narrow atomic emission bands that act as interferences. A handful of elemental atomic emission lines found most useful in the practice of GC-AED are shown in Table 4.16.

One of the organophosphorous pesticides that contains several heteroatoms is chloropyrifos, whose molecular structure is shown below:

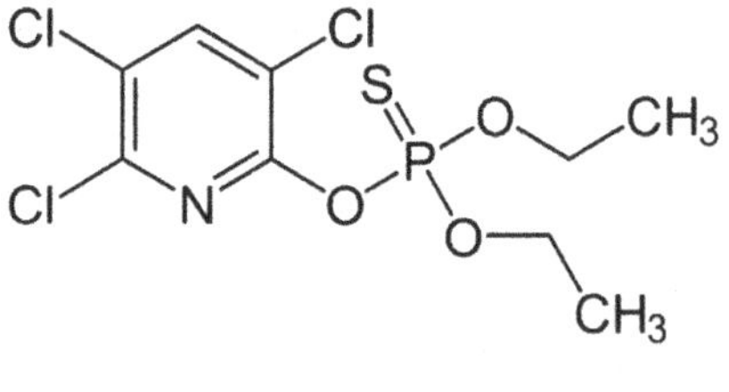

With carbon and four heteroatoms in a molecule of chloropyrifos, C, Cl, S, P, and N, it should be expected that injection of a standard that contains chloropyrifos with the appropriate software configuration should yield a peak for all five channels at the same retention time. This is indeed the case. This author, who was fortunate to have access to a GC-AED (6890GC interfaced to a G2350A® AED, Agilent Technologies) instrument, generated the series of GC chromatograms shown in Figure 4.46. One microliter of a solution containing 1.6 ppm chloropyrifos dissolved

TABLE 4.16
Most Useful Atomic Emission Wavelengths (nm)

C-193	C-496	C-179 (oils)
S-181	Cl-479	N-388 (CN-molecular band)
N-174	Br-478	C-248
P-178 requires hydrogen	O-171 (CO*)	Si-252

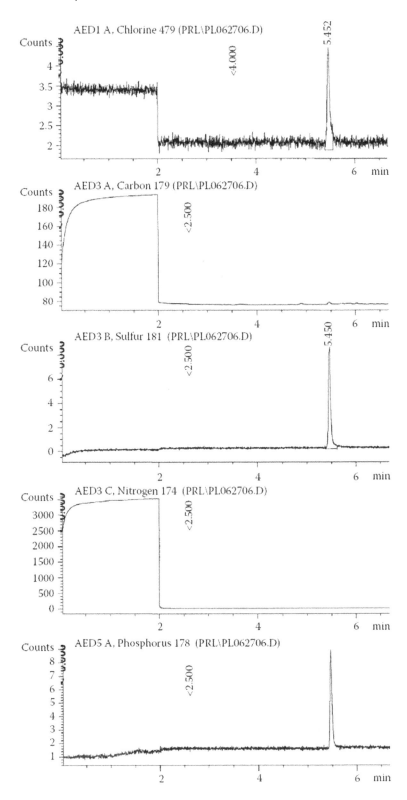

FIGURE 4.46

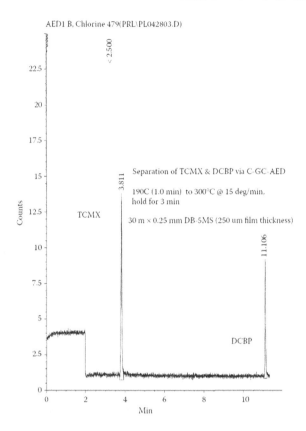

FIGURE 4.47

in iso-octane was injected into this instrument. A method within the AED ChemStation (Agilent) was developed in such a way that five different atomic emission wavelengths (called element-specific channels) are monitored simultaneously. To interpret the chromatograms in Figure 4.46, be sure to view the height of each peak with respect to the corresponding response scale.

Figure 4.47 is a GC-AED chromatogram generated in the author's laboratory. This chromatogram shows a separation of the two surrogates TCMX and DCBP monitored only at the 479 nm emission wavelength for the element chlorine. Since both compounds are polychlorinated aromatics, signals should be observed on one or more carbon channels, as well as one or more chlorine channels.

A block diagram for the GC-AED is depicted in Figure 4.48. This schematic should convey the impression that operating the instrument properly *requires a number of compressed gas sources*. Note that the 6890 GC pneumatically controls the reagent gas pressure and flow rates via Agilent Technologies' electronic pressure control (EPC). In the author's experience, helium carrier gas and nitrogen purge gas are the two most frequently replaced compressed gases in practice. A N_2 Dewar is the only practical source (short of a nitrogen generator) to use as the source for the N_2 purge. Figure 4.49 depicts the relationships between the AED cavity that incorporates a WCOT column interfaced to a capillary discharge tube, the plasma itself, and the exit through a conical aperture through an optical lens and into the monochromator, where a fixed photodiode array with the established atomic emission wavelengths for the various elements is shown that scans from 690 down to 171 nm. This fixed photodiode array for the newer model stands in contrast to the earlier model design, whereby the grating was fixed and the array moved. It is important in practice to maintain a cavity pressure of ~1.5 psi, which leads to a stable plasma. Indeed, the

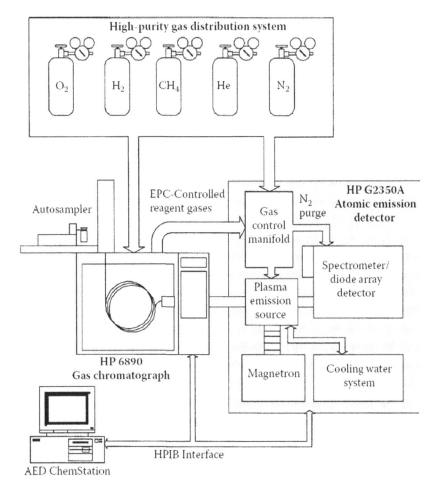

FIGURE 4.48

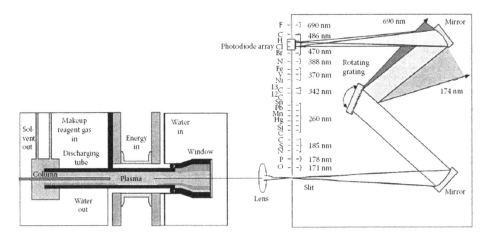

FIGURE 4.49

MIP is quite stable, and this leads to a very stable AED in practice and, subsequently, a stable baseline in the chromatogram.

The AED has witnessed a rise and fall as a popular GC detector over the past decade. The GC-AED arose from early experimental work to hyphenate GC and elemental spectroscopy. The miniaturization of the plasma was demonstrated for GC-MIP in 1965. A commercial unit was developed in Great Britain, followed by the successful development by Peter Uden at the University of Massachusetts-Amherst.[64] Agilent Technologies (formerly Hewlett-Packard) developed the first commercial instrument, HP5921A®, in 1989. The second-generation instrument, HPG2350A®, came out in 1996, and this model remains the contemporary version found most often in laboratories today.[65] Agilent decided to cease manufacture of the G2350A in 2002. The Model G2350A incorporated a number of unique features, including:

- Beeneker MIP design
- Water-cooled discharge tube
- Czerney–Turner spectrometer design
- Fixed photodiode array detector
- Utilization of EPC controlled reagent gas delivery
- Interfaces to the 5890 and 6890 GCs (Agilent)
- Computer controlled within the Windows NT® platform.

4.45   IS IT POSSIBLE TO PREPARE ENVIRONMENTAL SAMPLES AND QUANTITATE ANALYTES OF INTEREST USING GC-AED AS THE DETERMINATIVE TECHNIQUE?

Yes, indeed. The true potential for multielement quantitative analysis for TEQA may lie in the development of analytical methods that focus on the element selectivity afforded by GC-AED combined with an innovative sample prep approach. We illustrate this by describing the author's earlier attempt to couple RP-SPE with GC-AED as the determinative technique for selected *organochlorine compounds*. We then discuss some earlier literature findings on the use of SPME coupled to GC-AED to speciate the potentially *toxic chemical element Hg*. We complete this section with a description of the author's own work in adapting an EPA drinking water method and isolating, recovering, and quantitating *glyoxal* from a unique biolatex sample matrix.

This author conducted a preliminary percent recovery study for the isolation and recovery of TCMX and DCBP from spiked drinking water using RP-SPE conditions published earlier.[66] Table 4.17 compares the percent recoveries of TCMX and DCBP from two different suppliers

TABLE 4.17
RP-SPE Study Using C-GC-AED as the Determinative Technique

		% Recoveries (Mean of Triplicate SPEs)	
Sorbent	Supplier	TCMX	DCBP
C_8 (new)	A	70.9	87.1
C_2 (new)	A	51.7	90.5
C_{18} (old)	A	74.2	96.6
C_8 (new)	B	89.6	94.6
C_{18} (new)	B	81.9	80.6

(vendors A and B) of C_2, C_8 and C_{18} siloxane bonded silicas. Reversed-phase solid-phase extraction and capillary gas chromatography with atomic emission detection using a chlorine channel, $\lambda = 837.6$ nm, were employed to conduct the study.

The procedure is summarized below:

A ~70 mL tap water sample was acidified and spiked with 50 μL of 200 ppm TCMX/ DCBP in acetone. This sample was passed through a previously conditioned sorbent under reduced pressure. The cartridge was eluted with 2 × 500 μL 75:25 acetone / ethyl acetate. The eluent was transferred to a 1 mL volumetric flask and the volume adjusted to the mark with the eluent. One microliter of this eluent was injected into the C-GC-AED using a previously developed method written within the Agilent ChemStation software. A control is prepared that represents 100% recovery of the surrogates by adding 50 μL of 200 ppm TCMX/DCBP to a previously half-filled 1 mL volumetric flask, then the volume adjusted to the mark with the eluent. The ratio of the analyte peak area in the recovered sample from SPE to the corresponding peak area of the same analyte in the control ×100 gives the percent recovery for that analyte.

Percent recoveries for TCMX and DCBP varied widely among C_{18}, C_8, and C_2 chemically bonded silica sorbents obtained from the same supplier. Note that the smaller the organic moiety, the higher the recovery of the larger DCBP! A C_{18} SPE sorbent obtained over 5 years ago (compared to other sorbents used in the study) actually yielded a significantly higher percent recovery than a C_{18} SPE sorbent from a different supplier and obtained recently (new). Nevertheless, these preliminary experiments do demonstrate the importance of coupling RP-SPE with C-GC-AED toward accomplishing the objectives of TEQA.

The combination of solid-phase microextraction sampling of the headspace (SPME-HS) coupled to capillary gas chromatography and atomic emission detection (C-GC-AED) enables a speciation of Hg to be achieved. Organo-Hg compounds such as *methyl-Hg* and *dimethyl-Hg* are much more toxic than inorganic Hg. Carro and coworkers have recently demonstrated that methyl mercury, ethyl mercury, and phenyl mercury can be derivatized, volatilized, then separated and detected down to a concentration of 100 ppt levels from seawater using SPME-HS-C-GC-AED.[67] A recent review has also been published that discusses speciation of mercury, tin, and lead using C-GC-AED.[68] Methods that couple SPME techniques with determinative techniques such as C-GC-AED using a Fiber Holder for Manual Sampling (available from Milllipore Sigma) can lead to trace metal speciation of other metallic elements in addition to Hg.

The author was once approached by a company that needed to know just how much *glyoxal* was in their product, a biolatex used in paper coatings. This was a product derived from corn and is an example of the growing field of agricultural biotechnology. *Glyoxal*, a di-formaldehyde molecule is chemically derivatized based on the well-known reaction of aldehydes with hydroxylamine to form oximes. (pp. 319–325)[69] The chemical reaction is shown below:

Glyoxal   pentafluorobenzyl hydroxylamine (hydrochloride)
                                     (PFBHA)

$$F_5PhCH_2O\text{-}\bar{N}\text{=}CH\text{-}CH\text{=}O$$

PFBHA

$$F_5PhCH_2O\bar{N}\text{=}\underline{CH\text{-}CH}\text{=}\bar{N}\text{=}O\text{-}CH_2PhF_5$$

⟶ E and Z geometric isomers are possible for
the di-pentafluorobenzyl hydroxylamine oxime

Note that for a dialdehyde like glyoxal, a second molecule of PFBHA reacts with the second aldehydic functional group to yield *cis and trans geometric isomers* about the central carbon-to-carbon covalent bond underlined in the molecular structure shown above. EPA Method 556.1, a drinking water method provided guidance to the author. He adapted the method to a biolatex dispersion sample matrix while adopting the derivatization reaction, mini-LLE and extract cleanup procedures described in the method.[70] To prepare the sample, proceed as follows:

1. Weigh from ~0.25 to ~1.0 g biolatex dispersion and place in a 15 mL glass conical vial and record weight of biolatex.
2. Add ~10 mL distilled, deionized water and vortex.
3. Add ~100 mg potassium hydrogen phthalate (KHP) and vortex.
4. Add 1 mL of 15 mg/mL PFBHA, cap and vortex.
5. Place in water bath held at 35°C for 2 hours.
6. Remove, cool to ambient temperature.
7. Add 2–4 drops of concentrated sulfuric acid (H_2SO_4).
8. Add exactly 2.0 mL iso-octane (use *pesticide residue grade* iso-octane)
9. Extract using a 30 mL glass separatory funnel.
10. Let stand for 5 minutes.
11. Drain lower layer into original conical test tube.
12. Drain upper layer, still in the separatory funnel, directly into a second 30 mL separatory funnel that contains 3 mL of 0.2N H_2SO_4.
13. Shake and let stand for at least 5 minutes. (Shaking the extract with H_2SO_4 removes excess PFBHA)
14. Withdraw lower layer into a waste receptacle.
15. Carefully withdraw upper layer directly into a clean, dry 2 mL GC vial.
16. Inject 1 μL of cleaned-up extract into an optimized capillary gas chromatograph-atomic emission detector (C-GC-AED) while monitoring the 690 nm atomic emission line for fluorine.

4.45.1 CALIBRATION CONSIDERATIONS

A one-point calibration curve was used. The 40% glyoxal aqueous solution (whose concentration is known to only two significant figures) purchased from Aldrich Chemical (now part of Millipore Sigma) used as the *reference standard* was diluted as follows:

- 2.5 mL of the 40% glyoxal reference standard was added to a 10 mL volumetric flask, half-filled with distilled, deionized water (DDI). Additional DDI was added to the calibration mark. This gives a 10% glyoxal *primary reference standard*. 10% = 100,000 ppm = 100,000 ng/μL glyoxal.
- 50 μL of a 100,000 ppm glyoxal reference standard was added to a 10 mL volumetric flask half-filled with DDI. Adjust to the mark with DDI. The entire 10 mL aliquot was taken through the derivation.
- The 10 mL aliquot was extracted into 2 mL iso-octane and subsequently cleaned up using 0.2N H_2SO_4.
- The 2 mL cleaned-up extract was placed in a 2 mL GC vial.

- 1 μL of extract was injected into the C-GC-AED.
  - 50 μL × 100,000 ng/μL extracted into 2 mL iso-octane assuming a 100% recovery yields a concentration of 2,500 μg/mL = 2,500 ppm glyoxal in the extract.
  - 50 μL × 100,000ng/μL = 5×10^6 ng = 5×10^3 μg
  - 5000 μg / 2 mL extract = 2500 μg/mL glyoxal in the *single calibration standard reference*
- Since there exist two chromatographically resolved glyoxal peaks due to chemical formation of Z and E geometric isomers, the sum of both peak areas should be taken into account
- A *response factor*, RF, is established for each peak as follows:
  - For peak 1 at t_R = 10.465 min, 2,500 ppm glyoxal corresponds to an instrument response of 3,772 count-seconds to give an RF = 0.7414 ppm/count-second
  - For peak 2 at t_R = 10.588 min, 2,500 ppm corresponds to an instrument response of 4,986 count-seconds to give an RF = 0.5014 ppm/count-second

4.45.2 QUANTITATIVE ANALYSIS OF THE BIOLATEX BASED ON THE CALIBRATED RESPONSE FACTOR (RF)

- The observed instrument response for each peak is multiplied by its RF to yield the concentration in #ppm for that peak in the biolatex sample.
  - For Sample Biolatex A
    - Peak 1: 2,997 count-seconds observed × 0.7414 ppm/count-sec = 2,222.2 ppm
    - Peak 2: 4,020 count-seconds observed × 0.5014 ppm/count-sec = 2,016.0 ppm
  - For Sample Biolatex B
    - Peak 1: 2,654 count-seconds × 0.7414 ppm/count-sec = 1,967.8 ppm
    - Peak 2: 3,842 count-seconds × 0.5014 ppm/count-sec = 1,926.7 ppm
- The interpolated concentrations for both peaks are added and then multiplied by the volume of extract to give the *total amount of glyoxal* in the extract
  - For Biolatex A
    - 2,222.2 + 2,016.0 = 4,238.2 μg/mL × 2 mL = 8.476.4 μg
  - For Biolatex B
    - 1,967.8 + 1,926.7 = 3,894.5 μg/mL × 2 mL = 7,789.0 μg
- To calculate the *percent glyoxal in the biolatex sample*, we proceed as follows:
  - For Biolatex A
  - 8,476.4 μg = 8.476 mg / 0.4555 g biolatex = 18.616 mg/g biolatex = 0.0186 g / g biolatex
  - Convert to percent (%)
    - 0.0186 g / g biolatex × 100 = <u>1.86% glyoxal in Biolatex A</u>
  - For Biolatex B
    - 7,789.0 μg = 7.789 mg / 0.4200 g biolatex 18.5 mg / g biolatex = 0.0185 g / g biolatex
    - Convert to percent (%)
    - 0.0185 g / g biolatex × 100 = <u>1.85% glyoxal in Biolatex B</u>

Since the reference standard (40% glyoxal in water) was known to only *two significant figures*, both biolatex sample results were reported as having a 1.8% concentration of glyoxal in both samples of the biolatex!

Figure 4.50 shows a C-GC-AED chromatogram from injecting 1 μL of an extract containing the *derivatized glyoxal reference standard* used. Since the AED is fixed only on the 690 nm atomic emission line of fluorine, only duplicate peaks are seen due to the E and Z geometric isomers discussed earlier. Figure 4.51 shows a similar chromatogram for a starch sample which served as a sample blank and shows the absence of the dual peaks. Figure 4.52 results from injecting 1 μL of one of the unknown biolatex samples. Peak areas at t_R = 10.400 min and at t_R = 10.519 min were quantitated based on the calibrated response factor and result in the ultimate

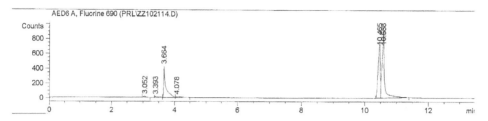

FIGURE 4.50

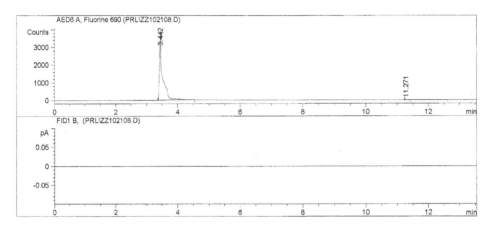

FIGURE 4.51

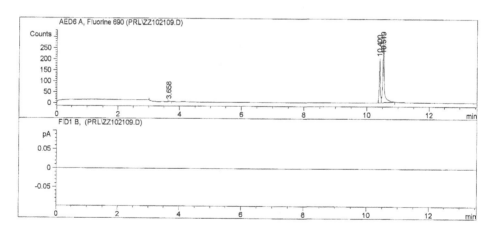

FIGURE 4.52

goal of *enviro-chemical* TEQA; the concentration of the E and Z geometric isomers of the di-pentaflurorbenzyl- oxime chemical derivative of glyoxal.

It would be utterly irresponsible for this author to leave the topic of trace organics analysis without diving into those developments that joined GC with the premier determinative technique mass spectrometry (MS). *Gas chromatography-mass spectrometry* (GC-MS) has profoundly influenced the practice of TEQA over the past 40 years! We complete our discussion of GC with a focus on MS as the *most significant and versatile* of the GC determinative techniques yet devised.

4.46   OF THE PLETHORA OF MASS SPECS, WHICH ARE MOST USEFUL TO QUANTITATE VOCs/SVOCs BY TEQA?

Because volumes have been written concerning mass spectrometry over the past 40 years, our approach here is to focus on the type of mass spectrometer instrumentation required to perform EPA methods, i.e. VOCs and SVOCs amenable to GC. Those systems can be described as being low-resolution in nature and the most affordable. The single quadrupole mass spectrometer or mass-selective detector (employees at Agilent Technologies Inc. supposedly are said to have "coined the term MSD") and the ion trap mass spectrometer (ITD) have and continue to be the "work horses" in environmental testing laboratories. This author operated a capillary column-gas chromatograph-mass selective detector (C-GC-MSD) for over 25 years! Onuska and Karasek have given a good definition and description of the importance of gas chromatography-mass spectrometry (GC-MS) to TEQA: [71]

> *In its simplest form the mass spectrometer performs three basic functions as a GC detector. These functions are to deliver a sample into the ion source at a pressure of 1×10^{-5} torr to produce ions from neutral molecules, and to separate and record a spectrum of ions according to their mass-to-charge ratios (m/z) and relative abundance. If a WCOT column represents the heart of the system, the mass spectrometric detector is the brain at the highest intelligence level. It is capable of providing both qualitative and quantitative data by means of spectral interpretation procedures developed to identify and quantify individual components in a mixture or of measuring a specific compound or group of positional isomers by means of selective ion monitoring (SIM).*

For a more complete description, as well as ease of comprehension of the broad field of mass spectrometry, including mass spectral interpretation, refer to the comprehensive treatment by Watson and Sparkman as cited earlier[13] or refer to earlier texts by Watson.[72,73] For an excellent introduction of what and how mass spectrometry has contributed to the advancements made in the broad field of environmental analysis as well as how GC, LC, and ICP determinative techniques have been interfaced to MS, refer to Budde as cited earlier.[14]

A somewhat oversimplified schematic of a single quadrupole MS is shown in Figure 4.53. This drawing depicts the trajectory of a resonant ion that makes it through the four rods of the quadrupole and a nonresonant ion that does not. The effluent from a narrow-bore WCOT can be fitted in such a way that the column outlet is positioned just in front of the electron beam. This beam is produced by "boiling off" electrons from a heated filament and accelerating the

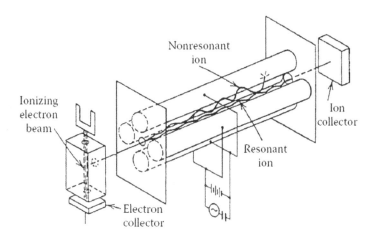

FIGURE 4.53

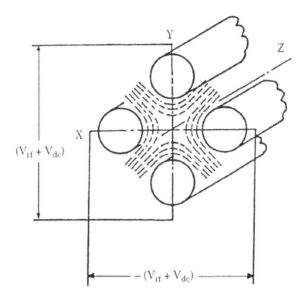

FIGURE 4.54

electrons to the collector. Molecular and fragment ions that are produced from this electron impact are accelerated through a series of electric fields and enter the center of the four rods. *Diagonally opposite rods are connected together electrically and to radio frequency (RF) and DC voltage sources,* as indicated in Figure 4.54. The applied voltages on one set of quadrupole rods are 180° out of phase with the applied voltages on the other set of rods. The quadrupole MSD is scanned by increasing the magnitude of the RF amplitude and DC voltages while maintaining a fixed ratio of the two. It is this ratio of the RF to the DC voltage that determines the resolving power of the quadrupole MSD. Sweeping the voltages from a preestablished minimum to a maximum value, while keeping the ratio of RF to DC voltage constant will provide a mass spectrum of resolution $R_S = 1$ for the molecular ion as well as any and all mass fragment ions from electron impact of the chromatographically resolved component in a mixture. This scan occurs very fast and enables a large number of scans to be recorded as the chromatographically resolved component enters the MSD. A plot of ion abundance vs. *m/z* gives a mass spectrum and serves to confirm the presence of a particular analyte. C-GC-MS therefore serves to confirm the identity of an analyte and thus complements C-GC with element-specific detection. This is one of the most powerful trace analytical instrumentation concepts devised in the 20th century.

Earlier Miller and Denton have likened the quadrupole MSD filter to a "tunable variable band-pass mass filter" and contrasted this with "true mass spectrometers" that resolve ions by dispersing them in either space, such as a magnetic sector instrument, or time, such as that of a time-of-flight mass spectrometer.[74] These authors suggest that the combination of a time-independent DC and a time-dependent RF potential applied to the four rods in essence *acts like a mass band filter*. With reference to Figure 4.55 and with the center of the quadrupoles at the origin of a set of Cartesian coordinates, the rods act as a low-pass mass filter for ions in the *x*–*z* plane, while acting as a high-pass mass filter in the *y*–*z* plane. An ion must remain stable in both the *x*–*z* and *y*–*z* planes to make it from the ion source to the detector. It must be light enough so as not to be eliminated by the low-pass filter operating in the *x*–*z* plane, yet not so light as to be eliminated by the high-pass mass filter that operates in the *y*–*z* plane. This condition of mutual stability allows a narrow mass range to make it through!

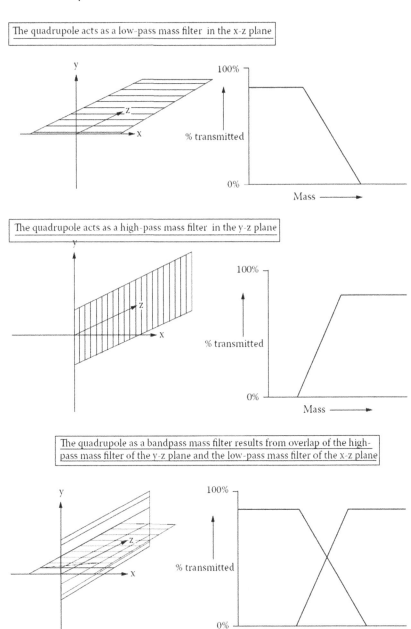

The quadrupole acts as a low-pass mass filter in the x-z plane

The quadrupole acts as a high-pass mass filter in the y-z plane

The quadrupole as a bandpass mass filter results from overlap of the high-pass mass filter of the y-z plane and the low-pass mass filter of the x-z plane

FIGURE 4.55

4.47 CAN WE PREDICT WHAT *M/Z* VALUES ARE STABLE THROUGH THE QUADRUPOLE RODS?

Yes, we can. However, we must delve into the physics of the quadrupole. For the classical hyperbolic quadrupole mass filter, the potential Φ for any time t can be described by

$$\Phi = \left[V_{DC} + V_{RF} \cos(\omega t) \right] \frac{x^2 - y^2}{2r_0^2}$$

where V_{DC} is the time-independent DC potential, V_{RF} is the magnitude of the applied RF amplitude, ω is the angular frequency (where $\omega = 2\pi f$, with f being the RF frequency), x and y are Cartesian coordinates defined with respect to Figure 4.54, and r_0 is the distance above the z-axis. The RF frequency is typically around 1.5 MHz. The partial derivative of this potential gives the magnitude of the electric field in all three directions: x, y, and z. The electric field in the x direction is obtained by taking the partial derivative with respect to x of the potential Φ, as is shown by

$$E_x = -\frac{\partial \Phi}{\partial x} = -\left[V_{DC} + V_{RF}\, \cos(\omega t)\right]\frac{x}{r_0^2}$$

The electric field in the y direction is obtained similarly as follows:

$$E_y = -\frac{\partial \Phi}{\partial y} = \left[V_{DC} + V_{RF}\, \cos(\omega t)\right]\frac{y}{r_0^2}$$

The electric field in the z direction is obtained in a similar manner as well; however, because there is no dependence of the potential in this direction, we get

$$E_z = -\frac{\partial \Phi}{\partial z} = 0$$

The force, Fx, exerted on a charged particle in the x direction by the magnitude of the electric field is equal to the product of the electric field E and the charge e such that

$$F_x = -\left[V_{DC} + V_{RF}\, \cos(\omega t)\right]\frac{ex}{r_0^2}$$

The force Fy exerted in the y direction is likewise found:

$$F_y = -\left[V_{DC} + V_{RF}\, \cos(\omega t)\right]\frac{ey}{r_0^2}$$

There is no force exerted in the z direction, so

$$F_z = 0$$

The force exerted on the ion is also equal to the product of its mass and its acceleration, expressed in terms of the x coordinate; we obtain

$$F_x = ma = m\frac{d^2 x}{dt^2}$$

Rearranging this equation, we have

$$m\frac{d^2 x}{dt^2} - F_x = 0$$

Substituting for *Fx*, we then have

$$m\frac{d^2x}{dt^2} - \left[V_{DC} + V_{RF}\cos\left(\omega t\right)\right]\frac{ex}{r_0^2} = 0$$

We can develop similar relationships for the *y* direction, and upon substituting for *Fy*, we have

$$m\frac{d^2y}{dt^2} - \left[V_{DC} + V_{RF}\cos\left(\omega t\right)\right]\frac{ey}{r_0^2} = 0$$

By letting $\xi = \omega t/2$ and designating any *x*, *y*, or *z* coordinate as *u*, the derivation as presented by March and Hughes enables us to arrive at the canonical form of the Mathieu equation: (pp. 34–35)[75]

$$\frac{d^2u}{d\xi^2} + \left[a_u + 2q_u\cos\left(2\xi\right)\right]u = 0$$

where the terms *au* and *qu* are defined as follows:

$$a_u = \frac{8V_{DC}}{\omega^2 r_0^2\left(m/e\right)}$$

$$q_u = \frac{4V_{RF}}{\omega^2 r_0^2\left(m/e\right)}$$

Solutions to the *Mathieu equation* completely describe the trajectory of an ion in terms of each ion's initial conditions. Without any force acting along the *z*-axis, the position and velocity of an ion along its *z*-axis are unaffected by any potential applied to the rods. The use of rods of a hyperbolic cross section leads to equations of motion that contain no cross-coordinate terms (pp. 34–35).[75]

Solutions to the Mathieu equation reveal regions of *a–q* space whereby the ion is bound, and therefore the ion is able to remain stable within the region between the four rods and make its way through to the conversion dynodes and electron multiplier detector. This is the so-called stability region. Note that parameters *a* and *q* are, in essence, reduced parameters and enable one to simplify the concepts. For a given mass-to-charge ratio *m/e*, a fixed distance r_0, and a fixed RF frequency ω, *a* and *q* depend only on V_{DC} and V_{RF}, respectively. In practice, quadrupoles are usually operated in a manner such that the values of parameters *a* and *q* are always related by a simple ratio. This condition is established by ensuring that the applied DC potential is always some fraction of the applied AC potential. The ratio is held constant, irrespective of the magnitude of V_{DC} or V_{RF}. Holding the ratio of V_{DC} to V_{RF} constant defines a set of points in *a–q* space called an operating line or mass scan line. If *a* is divided by *q* for a given *m/e*, a ratio equal to $2(V_{DC}/V_{RF})$ is obtained. A plot of *a* vs. *q* is shown in Figure 4.56 along with two mass scan lines that differ in their slopes. Resolution is increased by slightly changing the slope of the mass scan lines such that it is possible to achieve a desired resolution. Let us review what we mean by mass spectrometric resolution so that we do not confuse this term with chromatographic resolution.

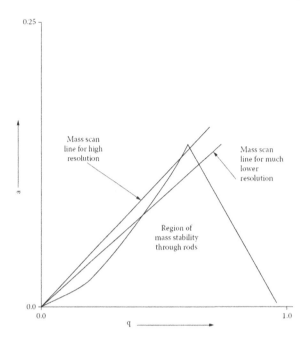

FIGURE 4.56

TABLE 4.18
Relationship between Mass Spectrometric Resolution and Mass for Low-Resolution Quadrupole Mass Spectrometers

m	R	m − Δm	m	m + Δm
10	10	9	10	11
100	10	90	100	110
10	100	9.9	10.0	10.1
100	100	99	100	101
10	1000	9.99	10.00	10.01
100	1000	99.9	100.0	100.1

4.48 WHAT IS MASS SPECTROMETRIC RESOLUTION ANYWAY?

Mass spectrometric resolution is defined as the ratio of a given mass m to the smallest discernable difference in mass, Δm, that can be measured. Mathematically, we have

$$R = \frac{m}{\Delta m}$$

A numerical expression can be obtained from this ratio where m and Δm are m/z values of two adjacent peaks in the mass spectrum. (pp. 7–8)[72] Table 4.18 suggests, for a given mass and resolution, to what extent one can resolve close m/z values.

In practice, V_{DC} can be varied from −250 to +250 V, and V_{RF} can vary from −1,500 to +1,500 V. The *magnitude of V_{RF} is approximately six times that of V_{DC}*. The mass scan line is swept across the bandpass region by changing V_{DC} and V_{RF} while keeping their ratio constant. A voltage increase is equivalent to sliding the mass scale shown in Figure 4.56 upward and to the right along the mass scan line. Sweeping the voltage applied to the rods thus provides a convenient method of scanning the bandpass region of the MSD.

4.49   THE MATH IS OK, BUT WHAT REALLY HAPPENS TO M·+ WHEN IT IS PROPELLED INTO THE CENTER OF THE QUADRUPOLE?

The mathematics just discussed provides a theoretical basis to understand how the quadrupole MSD selects specific m/z values with the desired degree of mass spectrometric resolution. For the reader who desires a more in-depth treatment on theory, consider both of March's books.[75,76] To answer the question, let us consider "riding the back" of a molecule that has lost its valence electron due to electron impact. This molecule ion can be abbreviated M·+, and taking this viewpoint will enable us to look at how this species gets knocked around by varying the applied voltages to the rods. This pedagogical approach was first used by the Finnigan MAT Institute as part of its training program.[77]

Figure 4.57 is a plot of the magnitude of the V_{DC} and V_{RF} voltages applied to the rods of an MSD. The dashed line shows how the combined DC and AC voltages vary on the first set of diagonally opposite rods, labeled 1 and 4, while the solid line represents the combined DC and AC voltages applied to the other set of rods, labeled 2 and 3. Our M·+ begins at time t_0, and we follow four subsequent time frames or "snapshots." The applied voltages for each of six snapshots are given in Table 4.19 for this exercise. The snapshots are shown in Figure 4.58. Three mass fragment ions of m/z 4, 100, and 500 are propelled into the center of the four rods.

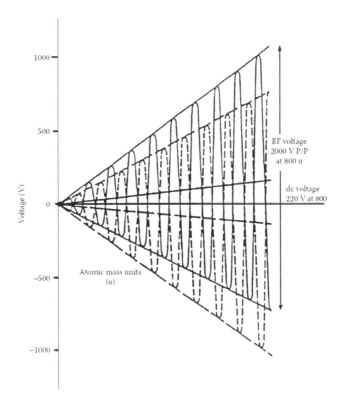

FIGURE 4.57

TABLE 4.19
Applied V_{DC} and V_{RF} to the Pair of Quadrupole Rods for Six Snapshots in Time

	$V(DC) + V(RF) \cos(\omega t)$	
Snapshot No.	Rods 1 and 4	Rods 2 and 3
t_0	+20	−20
t_1	+140	−140
t_2	+20	−20
t_3	−100	+100
t_4	+20	−20
t_5	+140	−140

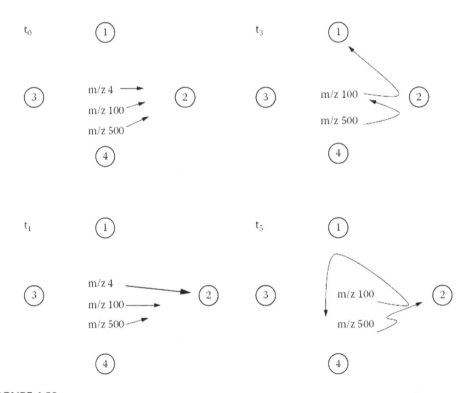

FIGURE 4.58

At time t_0 with rods 1 and 4 at 20 V and rods 2 and 3 at −20 V, all three fragment ions are attracted to rods 2 and 3, while being repelled by rods 1 and 2. At time t_1, rods 1 and 4 have become even more positive, while rods 2 and 3 have become even more negative. This change in the rod voltages accelerates all three ions to rod 2. The lightest ion, that with m/z 4, crashes into rod 2. At time t_3, rod 2 is suddenly made quite positive and ions of m/z 100 and 500 are repelled, as shown. At time t_5, rod 2 is made quite negative and m/z 500 is too heavy to change direction away from the rod and crashes into it. The ion with m/z 100 is now heading toward rod 1 and is suddenly repelled by this rod, whose voltage is +140, and is steered toward the center of the rods, as shown. Figure 4.59 shows the somewhat helical path taken by the ion

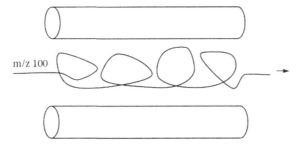

m/z 100

FIGURE 4.59

whose *m/z* is 100 through the quadrupole and to the detector. This ion would be considered bound and resides in *a–q* space. Beyond this more or less oversimplified concept, in practice, the operation can make minor changes to optimize performance. These changes with respect to operating a Finnigan Incos MSD include the following:

1. Adjusting the low resolution by keeping the V_{DC}/V_{RF} ratio, and hence the slope, constant, but changing the magnitude of the RF envelope.
2. Adjusting the high resolution by slightly changing the V_{DC}/V_{RF} ratio.
3. Adjusting the offset digital-to-analog converter by changing the zero reference of the RF start points.
4. Adjusting the offset program by changing slopes, while starting points are unaffected.

4.50 IS THE MSD THE ONLY MASS SPEC USED IN TEQA?

No, a mass spec opposite in concept to the MSD, yet utilizing similar principles, is that of the *quadrupole ion trap mass spectrometer* (ITD). The ITD is often found in environmental testing labs. Theoretically, it is the three-dimensional analog of the MSD. It is generally less expensive than the MSD and is of a more recent development. If little to no DC voltage is applied to a quadrupole, ions with any *m/z* will remain in the *a–q* stability region. The ion trap is depicted schematically in Figure 4.60. The distance from the center to the end cap, z_0, is called the axial distance. The distance from the center to the ring electrode, r_0, is called the radial distance. It helps to imagine that the plane of this paper contains the *x*, *y* coordinates and that this plane cuts through the ITD at its center. The *z*-axis is then viewed as being above the *x–y* plane, and the radial distance, *r*, rests within this plane. Ion traps are compact, with the filament positioned just above the upper end cap, in contrast to MSDs, whereby the ion source is separated. A variable RF voltage is applied to the ring electrode while the two end cap electrodes are grounded or set at either a positive or negative voltage. As the RF voltage is increased, the heavier ions remain trapped while the lighter ions become destabilized and exit the trap through the bottom end cap. This bottom cap has openings drilled into the stainless steel to allow these ions to escape. A good introduction to ion trap mass spectrometry, including some interesting historical facts, can be found in an earlier article written by Cooks and colleagues.[78]

With reference to Figure 4.60, a symmetrical electrical field allows consideration of only a radial and *z* displacement. The equations of motion of an ion in the three-dimensional quadrupole field are derived as in the case of the quadrupole mass filter:

$$\frac{d^2z}{dt^2} - \frac{4}{(m/e)r_0^2}\left[V_{DC} - V_{RF}\cos(\omega t)\right]z = 0$$

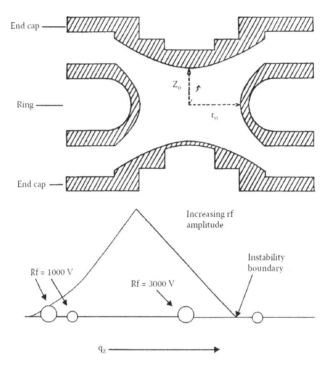

FIGURE 4.60

$$\frac{d^2r}{dt^2} - \frac{2}{(m/e)r_0^2}\left[V_{DC} - V_{RF}\cos(\omega t)\right]r = 0$$

with the following substitutions:

$$a_z = -2a_r = -\frac{16V_{DC}}{(m/e)(r_0^2)\omega_{RE}^2}$$

$$q_z = -2q_r = -\frac{8V_{RF}}{(m/e)R_0^2)\omega_{RE}^2}$$

Let u represent either r or z and let $\xi = \omega t/2$.

In a manner similar to that derived for the MSD, the canonical form of the Mathieu equation results. Note that the stability parameters for the z and r directions differ by a factor of -2. According to March and Hughes: (p. 56)[75]

The stability regions defining the a, q values corresponding to solutions of the Mathieu equation that are stable in the z direction. The regions corresponding to solutions that are stable in the r direction must be thought of as being twice the size of the z direction and then rotated about the q-axis to account for the minus sign. The region of simultaneous stability comprises the intersection of these two regions.

Radial stability expressed in terms of a_r and q_r must also be maintained with z direction stability. Ions of a given m/e are stable in the trap under operating conditions given in a–q space.

Figure 4.60 depicts a small section of the a–q stability region very near to the origin for a typical ion trap. This figure graphically demonstrates how three ions of different m/e values can be moved along the q_z-axis by simply increasing V_{RF} until the ions are sequential, according to increasing m/e ejected from the trap into the instability region of a–q space.[78] This occurs at a value of $qz = 0.908$.

4.51   AGAIN, THE MATH IS OK, BUT THIS TIME, WHAT ABOUT SOME APPLICATIONS OF C-GC-MS(ITD)?

This author had maintained a C-GC-MS(ITD) instrument for several years. One of the more interesting requests that not only utilized this instrument, but also involved some sample preparation, came from the forestry area. The *homologous series of 2-aminoethanols* (2-AEs) are organic compounds of importance to wood preservative treatments. Few methods, if any, can be found in the analytical literature that describe the qualitative and quantitative analyses of samples that contain any of the three 2-AEs. This series begins with 2-aminoethanol, followed by *N*-methyl-2-AE and, upon further substitution, *N,N*-dimethyl-2-AE. These compounds lack a UV-absorbing chromophore, and hence are not directly amenable to analysis using HPLC. The 2-AEs are quite polar, owing to both the presence of an amino functionality and being an alcohol. Molecular structures for all three are shown below:

H_2N⎯OH   ⎯$\overset{H}{N}$⎯⎯OH $\overset{CH_3}{H_3C⎯N}$⎯⎯OH

2-aminoethanol   N-methyl-2-aminoethanol   N,N-dimethyl-2-aminoethanol

The C-GC-MS(ITD) that was configured in the author's laboratory is shown in Figure 4.61. An Autosystem GC (PerkinElmer) was interfaced to a 800 Series (Finnigan, ThermoQuest) ITD. The WCOT column was passed through the transfer line into the ITD itself, as is shown. The three 2-AEs were easily separated using a DB-624 WCOT (Agilent, formerly J&W Scientific). An analytical method was developed that included novel sample preparation techniques coupled to the separation by GC, identification, and quantitative analysis by the ITD.[79]

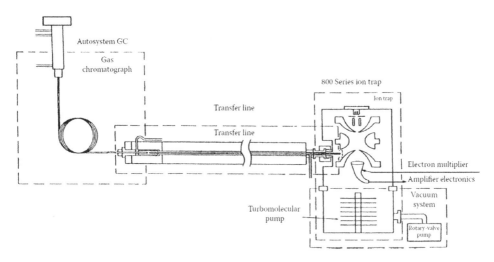

FIGURE 4.61

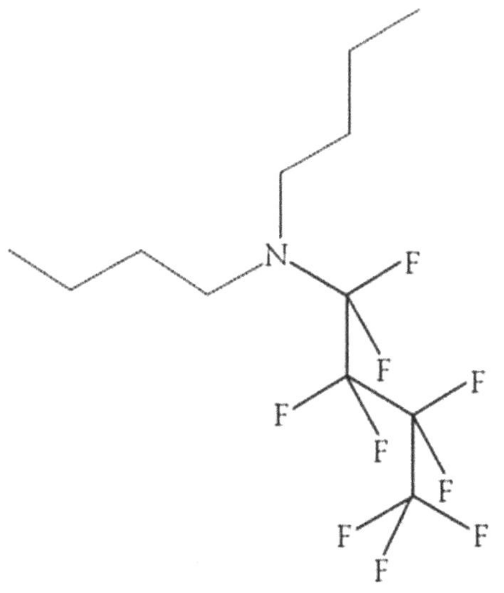

FIGURE 4.62

4.52   I HEAR THAT A MASS SPEC MUST BE TUNED: HOW IS THIS DONE?

We are most familiar with the word *tuning* as it relates to what we do each day with a radio or television. Indeed, good laboratory practice (GLP) requires that the mass spectrometer interfaced to either a GC or LC be tuned to a specific set of criteria (tuning specifications) before the instrument can be used as a determinative technique to conduct TEQA. EPA methods require that the mass spec be tuned using decafluoro-triphenyl phosphine (DFTPP) if SVOCs are to be determined and 4-bromofluorobenzene (BFB) if VOCs are to be determined. Other methods require that the mass spec be tuned to a set of criteria only for the *tuning compound intrinsic to the instrument.* In most instruments, this compound is perfluorotributyl amine (PFTBA), whose molecular structure is shown in Figure 4.62.

The three most abundant fragment ions in the EI mass spectrum for PFTBA are used for tuning. Loss of $-CF_2-CF_2-CF_3$ with the charge remaining on nitrogen, $(C_4F_9)_2N-CF_2+\cdot$, yields a fragment ion whose m/z is 502. Heteroatom cleavage of the N to C covalent bond, with the charge residing on the carbon, $C_4F_9^+\cdot$, yields a fragment ion whose m/z is 219. Cleavage of the C–C bond at the terminal carbon, with the charge residing on the carbon, $CF_3^+\cdot$, yields a fragment ion whose m/z is 69. Use of perfluoro-organic compounds as reference standards for mass spec tuning serves to eliminate multiple isotopic abundances due to hydrogen since *fluorine does not have an isotope.* The M + 1 abundance is due only to the 1.1% of all carbon that naturally exists as the ^{13}C isotope. This feature serves to simplify the EI mass spectrum of PFTBA. The m/z 69 fragment should show a ^{13}C isotopic abundance of ~1.1%, owing to a fragment containing only one carbon; the m/z 219 fragment should show a ^{13}C isotopic abundance of ~4.4%, due to this fragment having four carbons; and the m/z 502 should show a ^{13}C isotopic abundance of ~9.9% for the same reason. Because there are three N–F covalent bonds, and three C–CF_3 covalent bonds in each molecule of PFTBA, it would be expected that the m/z 69 and m/z 219 fragment ions should be the most abundant masses found, and this is indeed the case. Only one perfluoro moiety is shown in Figure 4.62 for clarity.

The Model 5973 MSD® (Agilent Technologies) at one time was considered the most widely used low-resolution mass spectrometer for single quadrupole operation in contemporary *enviro-chemical* and *enviro-health* testing laboratories. Under high vacuum, the ion source volume depicted in Figure 4.63 creates, accelerates, and focuses the intact molecular ion and fragments

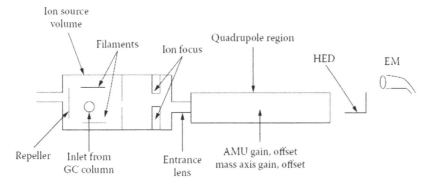

FIGURE 4.63

of the molecular ion for entrance into the quadrupole region, where mass selection occurs. Ions whose stable trajectories pass through the four rods hit the high-energy conversion dynode (HED) operating at −10,000 V, whereby electrons are created and attracted to the less negatively charged electron multipler (EM). The EM amplifies the signal output by ~100,000. Figure 4.63 is a side view of a conventional single quadrupole mass spec and shows where the various devices are located. Values assigned to these devices represent the parameters that can be adjusted in the tuning process. Referring again to the Figure 4.63, the ion source contains two filaments. This provides backup in case the filament in use burns out. The filament's emission current can be set by the user; however, it is recommended that the default setting be used. The electron energy can be set on the 5973 MSD. The optimum setting is 70 electron-volts (eV). This value yields the most vigorous ionization that results in the most abundant fragmentation patterns for organic molecules. A 70 eV setting has been used to generate most of the reference electron-impact (EI) mass spectra that comprise the large libraries available.

Referring again to Figure 4.63, a positive voltage applied to the repeller pushes these newly generated positively charged molecular ions out of the ion source. There exists an optimum repeller voltage that guarantees that sufficient ion abundance is achieved without an over-high ion velocity being created. An over-high ion velocity could cause ions to leave the ion source prematurely. The ion focus lens serves to narrow the stream of positive ions, and without this voltage, poor mass response is evident at the higher end of the mass range. The entrance lens minimizes the fringing field of the quadrupole. Increasing the entrance lens voltage increases the abundance at high mass and decreases the abundance at low mass. AMU gain and AMU offset parameters affect the V_{DC}/V_{RF} ratio of the mass filter, while the mass axis gain and mass axis offset serve to recalibrate the mass axis scale. Table 4.20 lists each major tuning parameter, possible range for programmable values, and just what influence or effect a change in a given tune setting has on PFTBA's mass spectrum for a 5973 MSD.

We now proceed to outline the *three major steps necessary to complete a manual tune* adjustment for the 5973 MSD. The manual tune should follow an autotune performed by the ChemStation® (Agilent Technologies) software. Perform a manual tune when you must meet criteria perhaps established in an analytical method being implemented.

Step 1: Minimize the peak width of PFTBA that is allowed to leak into the mass spec by adjusting the AMU gain. The *m/z* 502 is influenced more strongly than are the lower masses, 69 and 219. AMU offset moves the scan line up or down. Decreasing the offset leads to wider peaks and loss of mass spectrometric resolution, while increasing the offset leads to narrow peaks and higher resolution. The Mathieu stability diagram (Figure 4.56) is resketched in Figure 4.64 and shows all three mass-to-charge ratios for PFTBA: AMU gain affects the V_{DC}/V_{RF} ratio of the mass filter. This controls the width of the mass peaks. Increasing the gain increases the slope of

TABLE 4.20
Tuning Parameters for the 5973 MSD® [Agilent Technologies]

Tuning Parameter	5973 MSD	Effect
Filament	70 eV 300 μV emission	Energy of electron beam; number of electrons generated
Repeller	0–42.7 V	Pushes ions out of source
Draw out	Ground potential	Entrance aperture of lens stack
Ion focus	0–242.0 V	Relative abundance
Entrance lens	0–128 mV/amu	Relative abundance
Entrance lens offset	0–127.5 V	Relative abundance
AMU gain	0–4095	Peak width
AMU offset	0–255	Peak width
X-ray	HED	Optimizes sensitivity for a particular EM voltage setting
EM	0–3000 V	Sensitivity
Mass axis gain	±2047	Mass assignment
Mass axis offset	±499	Mass assignment

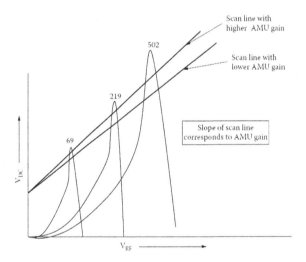

FIGURE 4.64

the scan line (refer to Figure 4.64) and more profoundly influences the 502 mass while having a much smaller effect on masses 69 and 219. AMU offset also affects the V_{DC}/V_{RF} ratio of the mass filter. This also controls the width of the mass peaks. A high offset yields narrower peaks, and this effect applies equally across the mass range.

Step 2: Perform a mass axis calibration by adjusting the mass axis gain or mass axis offset. The tuning algorithm calibrates the mass axis to within ±0.2 amu. When PFTBA is allowed to leak into the mass spec, one would expect the *m/z* for 69 to read 69.00. If 69.34 appears, the mass axis gain and offset can be adjusted to bring the mass axis back to 69.00.

Step 3: Adjust the EM voltage to match the m/z abundance criteria in terms of counts. Meeting the m/z abundance criteria, such as requiring the m/z 69 peak to exhibit an abundance of ~500,000 ± 50,000 counts, provides the requisite sensitivity that is required for a given method.

ChemStation provides a half-dozen different tune algorithms. The particular model of MSD will determine which tunes are available. Listed in Table 4.21 are the various tunes available and a brief description of each.

A properly tuned MSD yields the relative abundance ratios for one fragment ion relative to another and for carbon isotopic abundances for PFTBA as shown in Table 4.22.

4.53 WHAT'S THE DIFFERENCE BETWEEN SO-CALLED SOFT VS. HARD MASS SPEC?

GC-MS using a single quadrupole mass spectrometer can provide both "hard" electron-impact (EI) and "soft" chemical ionization (CI) mass spectra. Often CI complements EI. EI bombards neutral organic molecules with highly energetic (70 eV) electrons. These electrons are boiled off from a hot filament wire and accelerated. Energy is available from the 70 eV electrons not only to ionize the molecule, but also to cause the molecule to break apart or *fragment*. For the most part,

TABLE 4.21
Definitions for Tuning the 5973 MSD [Agilent Technologies]

Auto tune	Maximizes instrument sensitivity across the entire scan range and is a good starting point for BFB and DFTPP tuning
Std spectra auto tune	Provides for a standard response over the entire scan range; good tune to use when searching commercial libraries
Quick tune	Fast; adjusts EM voltage; resolution and mass assignments only; not recommended for beginners
User tune	Similar to auto tune except tune masses, scan settings, and repeller and entrance lens voltages are user defined
Manual tune	User controlled to a defined set of criteria
Target tune	Tunes the instrument with either BFB tune, DFTPP tune, or other targeted tune; tunes PFTBA to match specified ratios stored in *.tgt files, e.g., BFB.tgt
CI tune	Adjusts MS parameters for operation in chemical ionization (CI) mode

TABLE 4.22
Criteria for Tuning GC-MS Instruments

m/z	Relative Abundance Criteria for Proper PFTBA Tune
69	100% (base peak)
70/69	$0.5\% \leq \times \leq 1.6\%$
219/69	$70\% \leq \times \leq 250\%$
220/219	$3.2\% \leq \times \leq 5.4\%$
502/69	$3\% \leq \times$
503/502	$7.9\% \leq \times \leq 12.3\%$

the molecule fragments in well-understood ways. Ionization using CI is also achieved without the transfer of excessive energy, and this soft ionization yields a molecular adduct. Watson has articulated this difference as follows (p. 184):[73]

> *CI differs from EI in that molecules of the compound of interest are ionized by interaction or collision with ions of a reagent gas rather than with electrons. If alkyl reagent gases such as methane are employed, CI generally effects protonation of the molecule of interest. The site of protonation is most likely on a heteroatom of greatest proton affinity; some fragmentation of this protonated molecular ion may involve elimination of the heteroatom. Because fission of C–Cs bonds is rarely involved, this type of CI produces little fragmentation, which provides an insight to molecular structure.*

Consider an analyte symbolized by using the letter A being introduced into an EI ion source of a GC-MS or stand-alone MS as shown in Figure 4.65. Also consider analyte A being introduced into a second instrument, this instrument being configured with a CI ion source. Two different fragmentation pathways occur as shown in Figure 4.65. The possibility exists for fragmentation to occur as $A^{+\cdot}$ disposes of excess energy. The degree of fragmentation depends upon stabilities of various charged species. A typical EI mass spectrum for this gas-phase reaction is shown directly below the reaction for EI in Figure 4.65. Contrast this with the CI reactions whereby methane gas (CH_4) is introduced into the ion source of a mass spectrometer. When analyte A is introduced into a CI ion source, this analyte gets protonated. The predominant species is a protonated molecule with the possibility of adducts. The degree to which an analyte molecule such as A gets protonated to AH^+ is governed by the analyte's proton affinity. These values are tabulated. A typical CI mass spectrum is shown below the CI reactions in Figure 4.65.

4.54   HOW DO NEGATIVE IONS GET DETECTED IN GC-MS THAT BENEFITS TEQA?

By moving away from EI conditions, reducing energetic (70 eV) electrons to thermal (0 to 15 eV) electrons, and introducing CI conditions into the ion source! What this does is to enable the elevated gas pressure in the ion source to form secondary electrons with energies low enough (so-called thermal electrons) to yield *resonance electron capture* and *dissociative electron capture*. Reactions can be summarized in Table 4.23 where M-X represents an organic compound, with M being that moiety in the molecule that would under positive EI conditions yield a molecular ion M+· (discussed earlier) and X represents a more electronegative moiety such as Cl, F, Br, NO_2, CN, or OH.

Budde has articulated the importance of resonance electron-capture negative ion (ECNI-MS) to the broad field of GC-MS in the following manner: (p. 213)[14]

> *Electron-capture ionization is the most important negative-ion technique used in chemical analysis. High-purity methane or isobutane are the standard reagent gases used to generate the plasma of thermal electrons needed for electron capture. The ionization technique has the advantages of simplicity, high sensitivity for some analytes, high selectivity, and often little or no fragmentation of the M−· ion, which allows determination of molecular weight.*

A comprehensive review of electron-capture mass spectrometry as it relates to organic environmental contaminants up through the mid-90s is a good way to become quickly acquainted with the advantages offered by ECNI-MS.[80] After presenting the principles and practice of

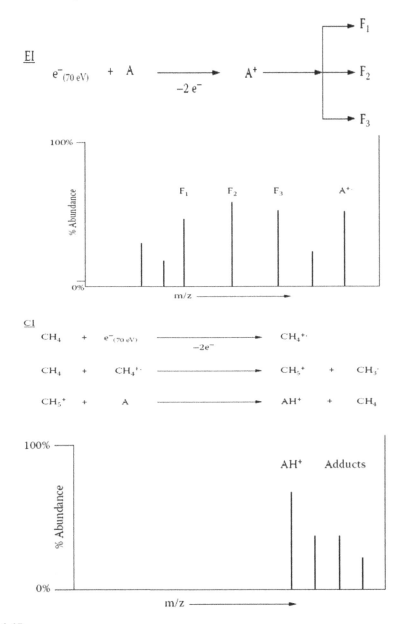

FIGURE 4.65

TABLE 4.23
Pathways to Negative Ion Formation in Mass Spectrometry

M-X + e⁻(>20 eV) M⁺ X⁻ + e⁻	Ion pair formation (observed with highly fluorinated-, poly-nitro-, and some B-, P-, S-, and metal-containing compounds)
M-X + e⁻ (~0–1 eV) MX⁻	Resonance capture or resonance electron capture or simply electron capture (most useful to TEQA)
M-X + e–(~0–15 eV) M⁻ + X⁻	Dissociative resonance capture (usually minimized or avoided in TEQA)

ECNI-MS, the authors apply this *determinative technique* to the following classes of environmental contaminants:

- polychlorinated dibenzo-p-dioxins and dibenzofurans
- polychlorinated biphenyls
- polychlorinated naphthalenes, terphenyls, and diphenyl ethers
- polychlorinated benzenes, phenols, anisoles, and styrenes
- brominated industrial compounds
- hexachlorocyclopentadiene and chlorinated diphenylethane derivatives and hexachlorocyclohexanes
- toxaphene
- nitrogen-containing herbicides
- phosphorous-containing insecticides
- polycyclic aromatic hydrocarbons and related compounds.

The author focused on polybrominated diphenyl ethers (PBDEs) and selected polybrominated biphenyls (PBBs) and converted a C-GC-MS from positive ion EI to negative ECNI. Generalized molecular structures for PBDEs and for PBBs are shown below:

Polybrominated diphenylethers
m = 1 to 5 and n = 1 to 5

polybrominated biphenyls
y = 1 - 6 and x = 1 - 6

Figure 4.66 shows two representative C-GC-ECNI chromatograms from this work focused on separating the most common PBDE congeners found in the environment.[81] Two PBB congeners were included as references. PBDEs are amenable to EI-MS fragmentation, resonance electron capture and dissociative resonance electron capture (ECNI-MS). By moving away from EI conditions, reducing energetic (70 eV) electrons to thermal (0–15 eV) electrons by introducing reagent gas into the ion source, enables the elevated gas pressure in the ion source to form secondary electrons with energies low enough (so-called thermal electrons) to yield resonance and dissociative electron capture. The decision limit x_C for these BDE/BB congeners hovered near to somewhat < 1 ppb while the detection limit x_D hovered between 0.5 to 2.5 ppb over all 12 analytes.[82] Having developed an instrumental method with specificity, the author investigated the extent to which PBDEs could be isolated and recovered from human and sheep serum comparing % recoveries using LLE, and RP-SPE sample prep techniques. He then developed a stir bar sorption extraction (SBSE) sample prep method and combined this method with automated thermal desorption C-GC-ECNI-MS to study the isolation and recovery of PBDEs from sheep and human serum.[83] These methods were extended to the following biological matrices: amniotic fluid, cord serum, placenta, and gestational membranes in support of biomonitoring efforts in Michigan. Figure 4.67 shows a C-GC-ECNI chromatogram that results from injection of 1 μL of a 100 ppb FF-1 (original Firemaster). Refer back to the discussion of the PBB crisis in Michigan in Chapter 1.

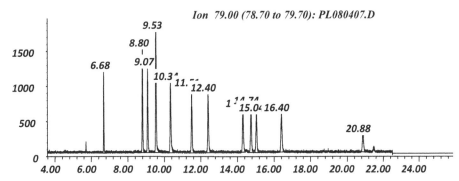

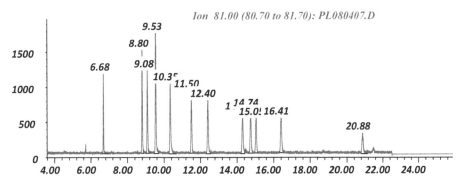

FIGURE 4.66

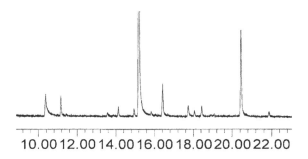

FIGURE 4.67

Having developed an instrumental method with specificity for the chemical element chlorine, the author also developed an instrumental method that isolated and recovered selected Toxaphene Parlar congeners from Great Lakes fish using C-GC-ECNI-MS.[84]

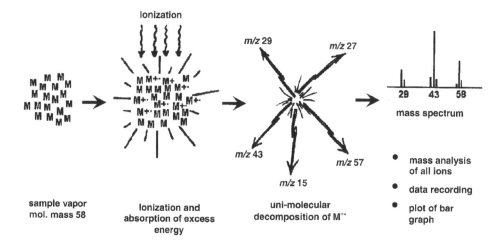

FIGURE 4.68

4.55 HOW DOES GC-MS USING A SINGLE QUADRUPOLE RELATE TO TEQA?

By separating all or most all the VOCs and/or the SVOCs present in the headspace or trapped on Tenax® or present in a liquid solvent extract from either LLE, SPE, or other trace sample prep techniques using a WCOT GC column and then letting these analytes enter the ion source of the MSD! Figure 4.68 depicts exactly what happens when a sample vapor containing a neutral molecule M subjected to a 70 eV impact. A highly energetic molecular ion whose symbol is M+· (the symbol denotes the loss of an electron while leaving an unpaired electron) is temporarily produced in the ion source. A uni-molecular decomposition occurs, often called molecular-ion fragmentation. This leads to one or more lower-molecular-weight fragments of well-defined *mass-to-charge ratios*, denoted by the symbol *m/z*.

In the words of Watson (p. 2)[73]: *The most common ionization process, electron-impact (EI), is achieved by bombardment with electrons at 70 eV of energy. The ionization process in general is nothing more than transfer of energy to the neutral molecule in the vapor state, giving the neutral molecule sufficient energy to eject one of its own electrons and thereby become charged with a residual positive charge. This process produces a molecular ion with a positive charge as represented by M+·. The molecule ion or molecular ion still has considerable excess energy, and much of that energy can be dissipated by fragmentation of its chemical bonds. The decomposition of various chemical bonds leads to the production of fragment ions whose mass is equal to the sum of atomic masses of the group of atoms retaining the positive charge during the decomposition process. It is important to realize at this stage that not all of the molecular ions decompose into fragment ions. In molecules producing a molecular ion that is stable, many of them tend to survive, or not fragment, and an intense molecular ion will be recorded. During analysis of a compound having a molecular ion that is unstable, nearly all of the molecular ions decompose into fragment ions, and in these cases, the mass spectrum contains only a small peak for the molecular ion.*

It is beyond the scope of this book to digress into the interpretation of mass spectra for qualitative analysis purposes. Several books on the interpretation of EI spectra are available.[13,85] The most abundant molecular ion or fragment ion in the mass spectrum of a given compound is usually selected for quantification. Tables at the end of each EPA method that requires GC-MS as the determinative technique will usually list these *quant* or *Q ions* for each of the analytes to be quantitated. TEQA is carried out by using an internal standard (IS) mode of instrument calibration, provided that a suitable IS can be found. If isotopically labeled organic compounds, otherwise identical to the analyte of interest, are available, the isotope dilution calibration mode is employed (Chapter 2).

We next focus on selective mass spectra themselves. Figure 4.69 shows EI mass spectra drawn from the National Institute of Standards and Technology (NIST) mass spectral library, available within the GC-MSD ChemStation® (Agilent Technologies) software, for two of the most

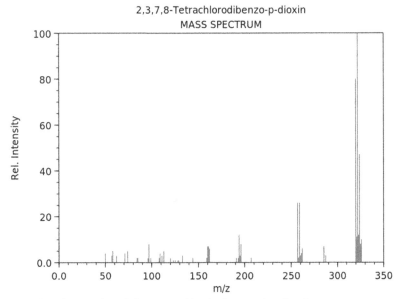

NIST Chemistry WebBook (https://webbook.nist.gov/chemistry)

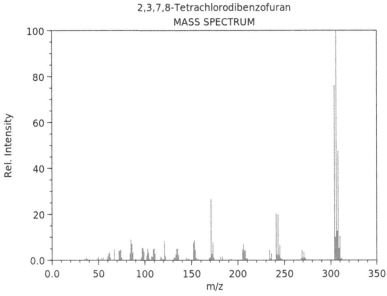

NIST Chemistry WebBook (https://webbook.nist.gov/chemistry)

FIGURE 4.69

notorious and toxic organic compounds known: 2, 3, 7, 8-tetrachlorodibenzo-*p*-dioxin (TCDD) and 2, 3, 7, 8-tetrachlorodibenzofuran (TCDF). Their molecular structures are shown below:

2, 3, 7, 8-tetrachlorodibenzo-*p*-dioxin (TCDD)   2, 3, 7, 8-tetrachlorodibenzofuran (TCDF)

n=4 & m = 4

Referring to Figure 4.69, and without any rules for EI mass spectral interpretation, we compare both mass spectra. The molecular ions (denoted by a cluster around *m/z* 322 and 306, respectively) in both spectra are the most abundant for each molecule, and the mass differs between molecules by 16 atomic mass units (amu) or Daltons (Da). This difference is explained by noting that the furan contains one less oxygen atom than the dioxin. The multiplicity of ions surrounding the molecular ion indicates polychloro substitution. The *m/z* value at 257 Da in the dioxin spectrum is attributed to loss of a COCl from the molecular ion. The ratio of the abundances of *m/z* 320 and 322, which are the molecular ions $^{35}Cl_4$-2378-TCDD and $^{35}Cl_3{}^{37}Cl$-2378-TCDD, must be within 10% of the value expected from the natural abundances of ^{35}Cl and ^{37}Cl. (p. 188)[14] This completes our discussion of the single quad mass spec, which represents a type of *scanning* mass spec. Let us digress a bit and pick up the physics that underlie one of several *batch* mass specs, namely, the time-of-flight mass spectrometer.

4.56   HOW DOES A TIME-OF-FLIGHT MASS SPECTROMETER WORK?

Watson describes a time-of-flight (TOF) mass spec this way (p. 93):[73]

> *The operating principle of the TOF mass spectrometer involves measuring the time for an ion to travel from the ion source to the detector. This process requires producing a discrete "bunch" of ions in a region near the ion source and then, through a series of synchronized events, accelerating them toward and measuring their time of arrival at a detector located 1 to 2 m from the source. All the ions receive the same kinetic energy during acceleration (e.g., 3,000 V), but because they have different masses, they separate into groups according to velocity (and hence mass) as they traverse the field-free region between the ion source and detector. The m/z value of an ion is determined by its time of arrival at the detector. Ions of low mass reach the detector before those of high mass, because these heavier ions have a lower velocity.*

A schematic for a TOF mass spec is shown in Figure 4.70 (p. 93).[73] The derivation below is adapted from a classical text on *physical chemistry* by Berry, Rice, and Ross:[86]

A molecular ion or fragment ion leaves the acceleration region with an average kinetic energy *T* of mass m_j, whose *x* component has a velocity v_{xj} according to

$$T = \frac{1}{2}m_j v_{xj}^2$$

This same molecular ion or fragment ion acquires a kinetic energy *T* whose charge is q_j under an electric field in the *x* direction E_x across a distance *d* such that

$$T = q_j E_x d$$

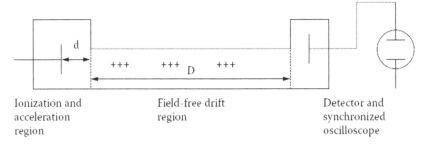

Ionization and acceleration region

Field-free drift region

Detector and synchronized oscilloscope

FIGURE 4.70

Eliminating T between both equations and solving for v_{xj} gives

$$v_{xj} = \sqrt{\frac{2q_j E_x d}{m_j}}$$

We also know that an object with a fixed velocity in the field-free drift tube of length D will arrive at the detector at a time t_j (the *time of flight*) according to

$$v_{xj} = D / t_j$$

Eliminating v_{xj} from both equations above yields an expression for the time of fight t_j according to

$$t_j = D\sqrt{\frac{m_j}{2q_j E_x d}}$$

or

$$t_j = D\sqrt{\frac{1}{2E_x d}\left(\frac{m_j}{q_j}\right)}$$

The above equation, the principal outcome of this derivation, proves that the *time of flight is directly proportional to the mass-to-charge ratio*. Two different masses, m_i and m_j, for a single charge would arrive at the detector in a ratio dictated by their ratio of masses, as follows:

$$\frac{t_i}{t_j} = \sqrt{\frac{m_i}{m_j}}$$

Interfacing a TOF to a GC gives yet another powerful hyphenated determinative technique that is currently finding its way into enviro-chemical and enviro-health laboratories. Let us digress a bit and contemplate just how GC-MS adds a third dimension to the GC chromatogram.

4.57 WHAT ARE WE REALLY SEEING WHEN WE PEER INTO A COMPUTER SCREEN WHILE ACQUIRING GC-MS DATA?

We see a three-dimensional chromatogram in two dimensions! Figure 4.71 (adapted from a training manual from Agilent Technologies) is a three-dimensional plot of ion abundance versus

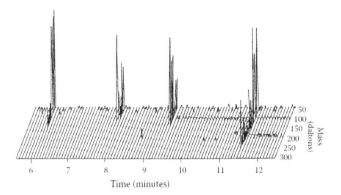

FIGURE 4.71

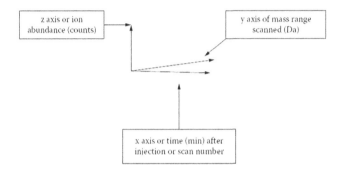

FIGURE 4.72

mass vs. chromatographic run time (i.e., time after GC injection). When the abundances at each instant of time across the mass spectrum for each chromatographically resolved analyte are added together, a *Reconstructed Ion Chromatogram* (RIC) is observed. GC-MS software will give you mass chromatograms as well as RICs. A *mass chromatogram* can be viewed in the drawing in Figure 4.71 as being represented by a plane that slices through a specific m/z value.

A plot of ion abundance for a selected m/z against chromatographic run time yields a mass chromatogram. In other words, data from a GC-MS needs to be interpreted in three dimensions even though the analyst is seeing on the PC screen only two dimensions at any moment. Figure 4.72 provides the Cartesian coordinates in an attempt to better understand the three-dimensional view shown in Figure 4.71. The RIC is viewed in the x–z plane by taking a slice at $y = 0$. A mass chromatogram is represented in the x–z plane by taking a slice at $y > 0$. Across a given chromatographically resolved peak, such as the peak that elutes near 6 min in the previous sketch, a mass spectrum at the apex of this peak is analogous in the y–z plane by taking a slice with $x > 0$.

4.58   HOW ARE THESE QUANT IONS FOUND?

TEQA using GC-MS is, in most cases, based on selecting the *base peak* from the mass spectrum of a chromatographically resolved analyte. The base peak is defined as the most abundant ion. In some organic compounds, the base peak corresponds to the molecular ion, while in other compounds, a prominent fragment ion is the base peak. To illustrate, let us focus again on the ubiquitous trihalomethanes (THMs). These volatile organics are found in municipal drinking water supplies due to disinfection processes involving chlorine and chlorine-containing chemicals. EPA Method 524.1 combines a sample prep technique (purge-and-trap, introduced in Chapter 3) with

TABLE 4.24
Molecular Weight, Primary and Secondary Quantitation Ions for the Trihalomethanes

THM	Molecular Formula	MW	Primary Q Ion	Secondary Q Ion
Bromodichloromethane	$CHCl_2Br$	162	83	85, 127
Bromoform	$CHBr_3$	250	173	175, 252
Chloroform	$CHCl_3$	118	83	85
Dibromochloromethane	$CHClBr_2$	206	129	127

a determinative technique (C-GC-MS). At the end of the method, a table of molecular weights and quantitation ions for all VOCs considered in the method is found.[87] Let us look at the four THMs tabulated from this method as shown Table 4.24.

Bromodichloromethane fragments under EI conditions as follows:

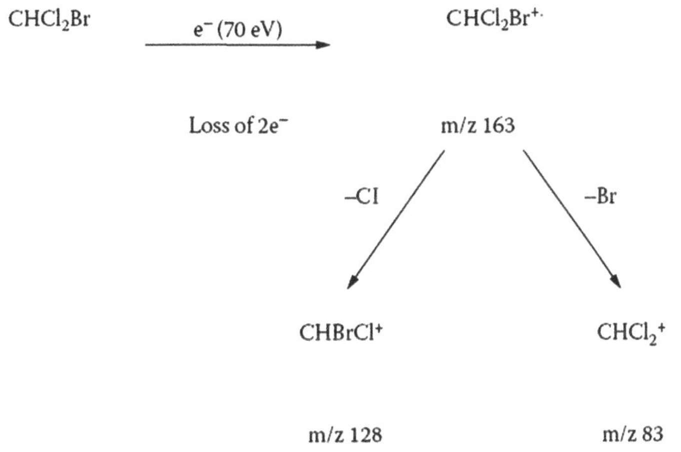

$$CHCl_2Br \xrightarrow{e^-\,(70\ eV)} CHCl_2Br^{+\cdot}$$

Loss of $2e^-$ m/z 163

$-Cl$ $-Br$

$CHBrCl^+$ $CHCl_2^+$

m/z 128   m/z 83

The other three THMs undergo heteroatom cleavage in a similar manner. The fragment ion of m/z 83 is the parent ion in the EI mass spectrum and comprises the primary quant ion. It is this *ion abundance that is proportional to the concentration* of bromodichloromethane in the original drinking water sample, and this fact leads to quantification. The secondary quant ions of m/z 85 and m/z 127 reflect the significant natural isotopic contributions of chlorine and bromine, respectively.

Tetramethylenedisulfotetramine abbreviated *tetramine* is a neurotoxin and highly poisonous semivolatile (SVOC) organic compound. Ingestion by humans of between 6 and 12 mg will lead to death. The LD_{50} is only 0.1 mg/kg. The molecular structure for tetramine is similar to the molecular structure for adamatane and is therefore considered as having "an adamantane-like structure" is shown below:

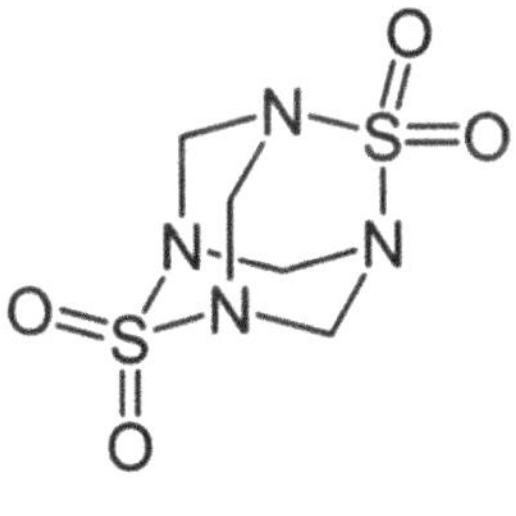

Earlier the author used C-GC-EI-MS to implement this determinative technique in order to study the isolation and recovery of tetramine using a 96-well plate RP-SPE sample prep technique (refer to Chapter 3) from human urine. This is an example of *enviro-health TEQA*. How do you start to develop such a new determinative technique? You need certified reference standards for the analyte of interest and for the isotopically labeled internal standard analyte of interest (refer to the discussion on the use of isotopically labeled internal standards found in Chapter 2). You need an optimized gas chromatographic separation and a properly operated and tuned GC-MS operated in the electron-impact (EI) mode.

First, complete the autotune algorithm. Second, with the instrument in the scan mode, record a low-resolution mass spectrum by injecting 1 μL of a higher concentration of the tetramine reference standard. In addition, record a low-resolution mass spectrum by injecting 1 μL of the internal standard. Figure 4.73 shows an EI mass spectrum for tetramine while Figure 4.74 shows an EI mass spectrum for a $^{13}C_4$ isotopically labeled tetramine. From these two mass spectra, a *qualitative m/z ion* can be set in the ChemStation® software at m/z 212 while a *quantitative ion* can be set at m/z 240 for the analyte *tetramine*. Likewise, a quantitative ion can be set at m/z 244 for $^{13}C_4$ *isotopically labeled tetramine*. Calibration standards can then be run followed by instrument calibration verification standards (ICVs) and continuing calibration standards (CCVs) as discussed in Chapter 2. Results from one representative calibration set is shown in Table 4.25. A comparison of decision (x_C) and detection limits (x_D) for tetramine was made by considering

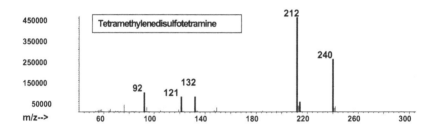

FIGURE 4.73

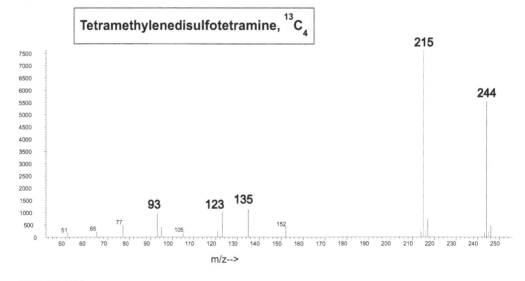

FIGURE 4.74

TABLE 4.25
Comparison of Non-Weighted vs. Weighted CBCS for Tetramine from Human Urine

Analyte	Weight	W_0	W_D	r_2	x_C (ppb)	x_D (ppb)
Tetramine	None	1	1	0.9999	2.81	5.60
Tetramine	1/x	1	0.2	0.9940	1.05	2.49

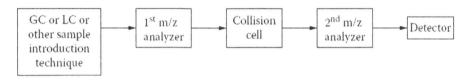

FIGURE 4.75

unweighted versus weighted least squares regressions combined with confidence band calibration statistics (to review the principles of CBCS refer to the discussion found in Chapter 2).

Interestingly, tetramine was originally thought to be amenable to liquid chromatograph via either electrospray, atmospheric pressure chemical ionization, or atmospheric pressure photoionization. Only GC-MS with EI proved quantifiable! After the GC-MS determinative technique was successfully implemented, the author transferred the GC-MS method to a newly acquired GC-MS-MS determinative technique!

4.59 WHAT IS TANDEM MASS SPECTROMETRY AND WHAT ROLE DOES IT PLAY IN TEQA?

Budde (p. 229) answers the question this way:[14]

Tandem mass spectrometry is the linking together in space or time of two independently operating m/z analyzers. Ions are formed in an ion source, separated in the first analyzer, undergo ion-molecule reactions in a collision cell, and the ionic products of these reactions are measured in the second analyzer.

Tandem MS techniques enhance analyte selectivity as well as analyte sensitivity. A fragment ion from the first *m/z* analyzer can undergo *collision-induced dissociation* (CID) with an inert gas such as He or Ar to yield product ions. The collision cell is often a quadrupole analyzer operated in the RF-only mode, thereby trapping (in a radial sense) the residual parent ions and all of the daughters in an appropriate concentration of collision gas. (p. 111)[73] The second *m/z* analyzer or third quadrupole provides a means of analyzing all of the products of CID. A block diagram of a *tandem-in-space* MS is shown in Figure 4.75.

Figure 4.76 illustrates the essence of the tandem MS technique. A molecule leaves the ion source after being ionized and fragmented. The fragment at *m/z* 129 is selected after passing through the first *m/z* analyzer and undergoes decomposition in the collision cell. The fragment at *m/z* 129 is further fragmented by CID and enters the second *m/z* analyzer to yield the product ion mass spectrum as shown.

Benchmark papers that established tandem-in-space techniques first explored double-focusing instruments consisting of magnetic and electrostatic sectors. This was followed by interfacing three quadrupole analyzers.[88–90] Benchmark papers that first explored single ion storage type analyzers, such as a Fourier-transform mass spectrometer or an ion trap by a sequence of timed ion storage, isolation, reaction, and measurement events, represent *tandem-in-time* concepts and will not be pursued further in this book.[91,92]

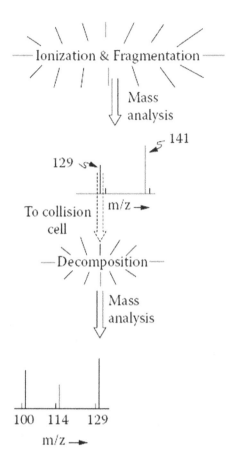

FIGURE 4.76

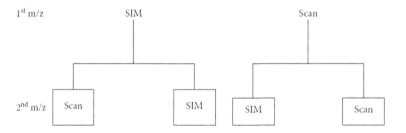

FIGURE 4.77

It is important when introducing concepts that underlie the use of tandem MS to summarize the data acquisition strategies that we have available to maximize this enhanced selectivity over that of a single quadrupole analyzer. This is best shown in Figure 4.77.

In the *product ion scan* strategy, a molecular ion or other fragment ion is selected by setting the first *m/z* analyzer to operate in the SIM mode. After collision-induced dissociation (CID), the second *m/z* analyzer is set to scan mode and a mass spectrum is generated. The chromatogram for a product ion scan would yield a series of peaks originating from a specific *m/z* selected to yield a spectrum of product ions.

In the *selected reaction monitoring* (SRM) strategy, a specific transformation such as that shown below:

$$M+ \rightarrow B+$$

provides a highly selective measure of the original molecule M. Budde cites the prime advantage of SRM as follows (p. 233):[14]

> *The advantage of SRM is tremendous selectivity for a specific substance and the exclusion of any potential interferences including chemical noise and coeluting substances with the same integer masses. Although coeluting compounds may produce multiple M+·, and these ions will be injected into the collision cell, it is unlikely that substances other than the target analyte will produce the B+ fragment ion by CID. SRM is the most widely used MS/MS technique for quantitative analysis of very complex samples.*

In the *precursor or parent ion scan* strategy, the reverse of product (daughter) ion scan occurs. In this case, the second *m/z* analyzer sits on a specific mass via SIM and the chromatogram yields a series of peaks that correspond to molecules that give precursor ions that undergo CID to produce the specific ion.

In the *neutral loss scan* strategy, both analyzers are operated in the scan mode with a constant time delay between the scans from both. This delay corresponds to a specific mass difference.

4.60   IS GC-MS-MS USED IN TEQA?

Yes, indeed. To illustrate an application, consider earlier work in *enviro-health* TEQA that utilizes a GC-MS-MS. Driskell and coworkers at the CDC's National Center for Environmental Health published a method to quantitatively determine the significant *organophosphonate* nerve agent metabolites (VX acid, GB acid, GA acid, GD acid, and GF acid) in human urine via isotope dilution (a deuterated methyl group is available) of the protonated daughter ion of each nerve agent metabolite under chemical ionization and SRM.[93] Let us focus on Sarin, the nerve agent released in an apartment located in Matsumoto, Japan, in 1994. The other nerve agents behave in a similar manner. Sarin was then released in a subway in Tokyo in 1995. Sarin is known to hydrolyze in the environment to the GB acid form. The body metabolizes Sarin to the GB acid form as well. The analytical method converts the GB acid to a methyl phosphonate ester by reacting the GB acid with diazomethane according to:

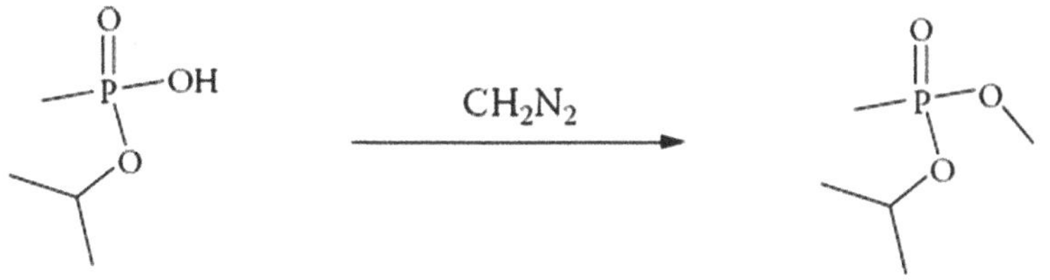

Under CI conditions using isobutane as the reagent gas and Argon as the collision gas for CID, the methyl phosphonate ester loses the isopropyl group and yields a daughter ion for both the native and deuterated isotopes as shown below:

Parent ion	Daughter ion
m/z 153	m/z 111 (native)
m/z 156	m/z 114 (deuterated)

This approach is just one of several approaches to isolating, recovering, and quantitating *organophosphonate nerve agent metabolites*. GC-MS-MS instruments have been arriving in environmental testing laboratories for close to a decade. Analytes are now quantitated under EPA Method 8270 using GC-MS-MS. The added benefit of analyte specificity to that of selectivity and sensitivity are strong driving forces.

We now move out of GC and GC-MS. Next, we introduce the other cornerstone to instrumental column chromatography deemed essential to meet the goals of TEQA: high-performance liquid chromatography (HPLC).

4.61 WHERE DOES HPLC PLAY A ROLE IN TEQA?

This author was astonished one day back in the 1980s when he obtained a sample of an aqueous phase from a Superfund waste site that had been extracted (refer to the discussions of LLE in Chapter 3) with methylene chloride. The remaining extracted aqueous phase was sent to waste. The organic extract was analyzed using EPA Method 8270. This is a C-GC-MS(MSD) determinative method that separates, identifies, and quantitates about 90 priority pollutant SVOCs. The author injected about 10 μL of this discarded aqueous phase into an HPLC in a reversed-phase mode using a UV absorbance detector. Over a half-dozen large peaks in the HPLC chromatogram were observed, and the retention times for these peaks closely matched those of phenol and some substituted phenols. This author exclaimed, "I thought all of the priority pollutant organics were extracted out!" Upon further thought, if the percent recoveries of these phenols is low, then it is reasonable to assume that these polar analytes would remain in the aqueous phase due to the appreciable hydrogen bonding between the hydroxy group on the phenol and the oxygen end of the water molecule. In Chapter 3, we discussed at length the fact that significant differences in partition coefficients between polar and nonpolar solutes lead to differences in percent recoveries. Revisiting Figure 4.4 where analyte volatility is plotted against analyte polarity, if there remained residual phenols, then we should expect the complementary determinative technique of HPLC to detect these polar analytes of environmental importance.

Current EPA methods requiring HPLC as the principal determinative technique found in SW-846 to quantitate semi- and nonvolatile analytes of *enviro-chemical TEQA* interest are listed along with their annotated titles, revision order, and date of revision below:[94]

Method 8310: Polycyclic Aromatic Hydrocarbons (PAHs) using High-Performance Liquid Chromatography (RP-HPLC-UV/FL). The current method is still Revision 0, September, 1986.

Method 8315A: Determination of Carbonyl Compounds by High-Performance Liquid Chromatography (RP-HPLC-UV). The current method is Revision 1, December, 1996.

Method 8316: Acrylamide, Acrylonitrile and Acrolein by High-Performance Liquid Chromatography (RP-HPLC-UV). The current method is Revision 0, September, 1994.

Method 8318A: N-Methylcarbamates by High-Performance Liquid Chromatography (RP-HPLC-FL with post column reaction and derivatization). The current method is Revision 1, February, 2007.

Method 8321B: Solvent Extractable Nonvolatile Compounds by High-Performance Liquid Chromatography/Thermospray/Mass Spectrometry (HPLC/ TS/MS) or Ultraviolet (UV) Detection. The current method is Revision 2, February 2007.

Method 8325: Solvent Extractable Nonvolatile Compounds by High-Performance Liquid Chromatography/Particle Beam/Mass Spectrometry (HPLC/PB/MS). Revision 0, 1996,

Method 8330B: Nitroaromatics and Nitramines by High-Performance Liquid Chromatography (RP-HPLC-UV). The current method is Revision 2, October, 2006.

Method 8331: Tetrazene by High-Performance Liquid Chromatography (RP-HPLC-UV). The current method is Revision 0, September, 1994.

Method 8332: Nitroglycerine by High-Performance Liquid Chromatography (RP-HPLC-UV). The current method is Revision 0, December, 1996.

Method 8323: Determination of Organotins by Micro-Liquid Chromatography-Electrospray Ion Trap Mass Spectrometry. The current method is Revision 0, January, 2003.

EPA Method 8323 reflects perhaps what the future will hold as LC-MS determinative techniques become increasingly accepted in the regulatory arena. Principles of LC-MS are introduced later in this chapter. Comments to this new method, found at EPA's SW-846 web site, are given below:[94]

Method 8323 is the first product of an EPA Office of Solid Waste (OSW) project to develop a series of class-specific Electrospray HPLC/MS methods to replace the obsolete Thermospray interface currently used in Method 8321. When the project is complete, OSW will issue a single integrated Electrospray HPLC/MS method. Method 8323 covers the use of solid-phase extraction (SPE) disks, solvent extractions (for biological tissues) as sample preparation methods, and micro-liquid chromatography (LC) coupled with Electrospray ion trap mass spectrometry (ES-ITMS) [this technique would also be applicable to ES-quadrupole mass spectrometry (ES-MS)] for the determination of organotins (as the cation) in waters and biological tissues.

Let us focus here on determining the various carbonyl compounds. Figure 4.78 shows an HPLC-uv (abbreviation for a high-performance liquid chromatograph with ultraviolet absorption detection) as a *determinative technique* for the separation and detection of the first six homologous series of aldehydes as their 2, 4-dinitrophenyl hydrazone derivatives, accomplished in the author's laboratory. The reaction of aldehydes and ketones with 2, 4-dinitrophenyl hydrazine under mildly acidic conditions in water to form stable hydrazones is a well-known reaction in organic chemistry (pp. 319–325).[69] The reaction of C_3 (propionaldehyde or propanal) with 2, 4-dinitrophenyl hydrazine to yield the derivative is illustrated (without showing H atoms) below:

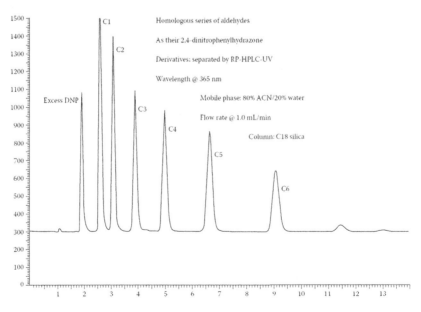

Propionaldehyde

2,4-dintrophenylhydrazine

2,4-dinitrophenylhydrazone of propionaldehyde

This method nicely illustrates the importance of HPLC to environmental testing labs and forms the basis for EPA Method 8315A, "Determination of Carbonyl Compounds by HPLC."[94] Aldehydes and ketones are, in general, too polar to be isolated from aqueous matrices and are too thermally labile or thermally reactive for the GC determinative technique. Note the wavelength setting on the UV detector to 365 nm in Figure 4.78. This wavelength is sufficiently removed from the crowded region between 200 and 300 nm, and therefore can be considered somewhat unique for 2, 4-diphenyl hydrazones. Method 8315A combines both sample prep and

FIGURE 4.78

determinative techniques and offers both LLE and SPE to isolate and recover the 2, 4-diphenyl hydrazones from an aqueous or soil matrix. The method suggests that gaseous stack samples are to be collected by Method 0011 and indoor air samples are to be collected by Method 0100. We now introduce the basics of the *HPLC determinative technique*.

4.62 WHAT IS HPLC?

Some nomenclature with respect to the broad field of liquid chromatography is in order. The determinative technique known as HPLC is an abbreviation for *high-performance liquid chromatography* because column efficiency is so greatly improved when columns contain a stationary phase whose particle size has been reduced to 10 μm, or to 5 μm, and even as low as 3 μm! Originally, HPLC was called *high-pressure liquid chromatography* to distinguish it from its low pressure or gravity fed column chromatographic precursor. The superficially porous HPLC column (currently on the market today) consists of a 1.7 μm core surrounded by a 90 Å (angstrom) porous shell. These so-called Halo® (Mac-Mod, Inc) LC columns represent state-of-the-art as of this writing. This gives a superficially porous particle with a total particle size of 2.7 μm! This reduced particle size from that used in column liquid chromatography *necessitates that the mobile phase be pumped through these columns*. The pressure drop across these low-particle-size columns required to achieve a flow rate that minimizes the height equivalent to a theoretical plate [refer to Equations (4.19) and (4.23)] exceeds 1,000 psi, and therefore necessitates pumping systems that can push liquids against this relatively high back-pressure. Instrumentation took what once was called *low-pressure* column chromatography and developed it into *high-pressure* LC. Much research and development has occurred during the last 10–20 years and has led to the development of *ultra-high-performance liquid chromatography* abbreviated UHPLC whereby pressure drops on the order of 10,000 psi are required. Chromatographic peak widths in UHPLC are comparable to capillary GC peaks widths! We will not pursue an introduction to the UHPLC determinative technique at this time.

A detailed block diagram of an HPLC is shown in Figure 4.79. This diagram is a useful way to understand how an HPLC works and shows how HPLC is clearly a modular instrument. This modular nature of HPLC enables a number of options to be considered, as introduced in Figure 4.79. The reservoir can be either left at atmospheric pressure or pressurized with inert gas. However, a mobile phase that is not pressurized must be degassed; otherwise, air bubbles are continuously squeezed out of the mobile phase and accumulate in the microflow cell of the detector. These trapped air bubbles prevent the analyst from obtaining a stable detector baseline. This is particularly true for a UV absorbance detector. Pump systems can be either those that deliver essentially a constant flow rate or those that deliver a constant pressure. Reciprocating piston pumps deliver a constant flow rate and are manufactured with single, dual, or even triple pump heads. Pumps designed for HPLC must be able to withstand fluid pressure in excess of 6,000 psi. A common safety feature built in to most HPLC pumps is a maximum pressure sensor. In practice, this value is set at approximately 4,000 psi. If the back-pressure at the column head exceeds this value, the pump will shut down automatically. The addition of a pressure transducer and a pulse dampener completes the necessary accessories for a good pump. Gradient elution HPLC is analogous to temperature programming in GC and is achieved by varying the percent organic modifier in the mobile phase after the sample has been injected. There are two different designs for performing gradient elution. One is to mix two or more different solvents at low pressure, and the other is to mix them at high pressure.

One of the great developments that led to an increase in the scope of HPLC occurred with advances in LC column stationary phase packings such as reversed-phase chemically bonded silicas. In addition, it was the realization that a reduction in particle size along with an increase in particle size homogeneity would lead to highly efficient columns. Together, these bonded-phase

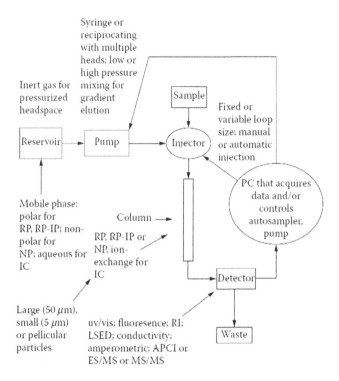

FIGURE 4.79

silicas led to the development of reversed-phase HPLC (RP-HPLC), and thus greatly expanded the scope of this instrumental technique. Referring again to a generalized plot of analyte volatility vs. analyte polarity shown in Figure 4.4, we find that for nonvolatile organics that are somewhat polar, RP-HPLC is most appropriate for their retention and separation. The fact that the analyst can make a direct aqueous injection into an HPLC that is operated under reversed-phase conditions enabled RP-HPLC to take on a major role as a determinative technique in TEQA. Snyder and Kirkland, in their classic book on HPLC, commented on how "modern LC arose" by stating the following (pp. 8–9).[95]

> *Modern LC is based on developments in several areas: equipment, special columns and column packings, and theory. High-pressure, low-dead-volume equipment with sensitive detectors plays a vital role. The new column packings that have been developed specifically for modern LC are also essential for rapid, high-performance separations. Theory has played a much more important role in the development of modern LC than for preceding chromatographic innovations. Fortunately, the theory of LC is not very different from that of GC, and a good understanding of the fundamentals of GC had evolved by the early 1960s. This GC theory was readily extended to include LC, and this in turn led directly to the development of the first high-performance column packings for LC and the design of the first model LC units … The potential advantages of modern LC first came to the attention of a wide audience in early 1969 … However, modern LC had its beginnings in the late 1950s, with the introduction of automated amino acid analysis by Spackman, Stein, and Moore and this was followed by the pioneering work of Hamilton and Giddings on the fundamental theory of high-performance LC in columns, the work of C.D. Scott at Oak Ridge on high-pressure ion exchange chromatography, and the introduction of gel permeation chromatography by J.C. Moore and Waters Associates in the mid-1960s.*

4.63 HOW DID THE NAME *REVERSED-PHASE HPLC* COME ABOUT?

A nonpolar mobile phase passing through a packed column that contains a polar stationary phase defines normal-phase HPLC (abbreviated NP-HPLC). For example, if *n*-hexane comprises the mobile phase and silica gel is used for the stationary phase, separations of non-polar organic analytes is feasible. Note the limited region in Figure 4.4 where NP-HPLC is operational. With respect to neutral organic compounds, the polar and ionic domains cannot be reached by NP-HPLC! NP-HPLC was the first high-pressure form of liquid chromatography to be developed. If the stationary phase could be made hydrophobic by chemical treatment and the mobile phase made more polar, a *reversal of mobile- / stationary-phase polarities could be achieved*. Like it or not, we are stuck with this nomenclature. RP-HPLC has certainly extended the range of analyte polarity that can be separated and detected. Reversed-phase ion pair HPLC (RP-IP-HPLC) has enabled organic cations such as quaternary ammonium salts and organic anions such as anionic surfactants to be retained, separated, and quantitated, thus extending the range of analyte polarity even further. To begin to realize the influence of mobile-phase composition on HPLC resolution, let us return to the fundamental resolution equation.

4.64 WHY DOES THE MOBILE PHASE EXERT SO MUCH INFLUENCE ON R_s?

We discussed Equation (4.30) earlier with respect to GC theory. This equation applies equally well to HPLC. Unlike GC, the HPLC mobile phase exerts considerable control over the chromatogram, and this is evident in the chromatograms shown in Figure 4.80. Consider an initial injection of two poorly resolved peaks such as in Figure 4.80A. The capacity factor can be varied as shown in Figure 4.80B, and the result either decreases R_s when k' is decreased or increases R_s when k is increased, as shown. However, sometimes an increase in k' also leads to increased peak broadening, and hence to no greater advantage with respect to R_s, as shown.

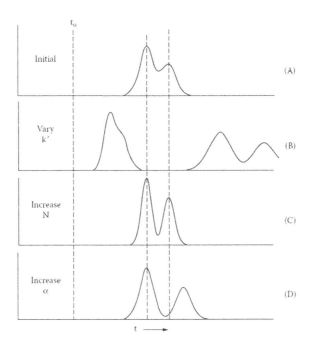

FIGURE 4.80

The column can be made more efficient by increasing the column length and thus increasing N as shown in Figure 4.80C. The chemical nature of the stationary phase can be changed to make the column more selective, and thus serve to increase α, as is shown in Figure 4.80D. To answer the question, the magnitude of the partition coefficient, K, for a neutral organic compound between a liquid mobile phase and a given stationary phase can be significantly changed by a change in the percent organic modifier. Equation (4.31) suggests that significant changes in K lead to differences in k'. As is evident in the HPLC chromatogram shown in Figure 4.81 (from the author's lab), an increase in the percent acetonitrile (solvent B by convention) from 43% in distilled deionized water (solvent A) to 50% acetonitrile in acetate buffer at pH 5.0 not only reduced k' for 3-(trifluoromethyl)-p-nitrophenol (TFM), but also improved peak shape. A well-known relationship in RP-HPLC relates the capacity factor in a given organic/aqueous mobile-phase k to the capacity factor in 100% water, kw, and the volume fraction of organics in the mobile-phase ϕ, where ϕ = %B/100 according to (pp. 235–242):[96]

$$\log k = \log k_w - S\phi$$

For low-molecular-weight compounds, $S = 4$. Thus, k increases by a factor of 2 to 3 for a decrease of 10% B. Three RP-HPLC chromatograms from the author's laboratory are shown in Figure 4.82. The software enables all three chromatograms to be stacked, as shown. The top chromatogram shows a column that has deteriorated beyond its useful life. The middle chromatogram illustrates how a change in the column can radically improve HPLC column efficiency. The isocratic test mix available from (Millipore Sigma, formerly Supelco) consists of a homologous series of substituted parabens (esters of p-hydroxybenzoic acid). The bottom chromatogram

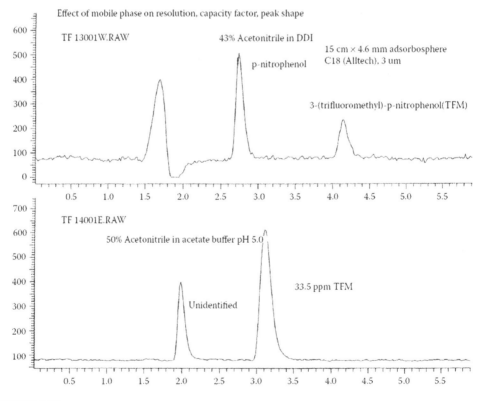

FIGURE 4.81

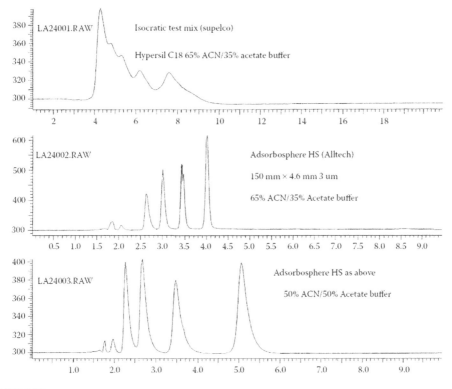

FIGURE 4.82

shows how the injection of the same mixture using the same column looks when the %B is lowered. When viewing Figure 4.82, note that the timescales have not been aligned.

4.65   ARE ALL OCTADECYL-BONDED SILICA HPLC COLUMNS THE SAME?

The stacked HPLC chromatograms shown in Figure 4.83 should provide the answer to this question. Both chromatograms were obtained on the same instrument, using the same mobile phase; however, the octadecyl-bonded silica (ODS) columns used were obtained from two different suppliers, as indicated. Note that the column dimensions differ slightly between both columns. Interpretation of the top chromatogram indicates that this column exhibits a much higher k' than the column used to generate the lower chromatogram. A narrower peak width is, however, evident in the bottom chromatogram. The number of theoretical plates for an HPLC column can be calculated using either Equation (4.21) or (4.22), depending on how the peak width is measured. When stating how good a column is, Snyder, Kirkland and Glajch have listed the following requirements (pp. 205–208):[96]

- Plate number N for a given value of k'
- Peak asymmetry
- Selectivity, α, for two different solutes
- Column back-pressure
- Retention, k', reproducibility
- Bonded-phase concentration
- Column stability.

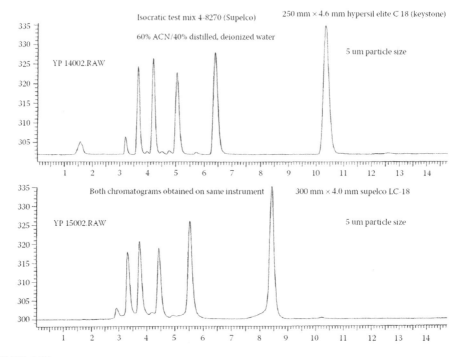

FIGURE 4.83

They also suggest that the following equation be used to estimate the column plate number for small molecules under optimum conditions for a column length L and a particle diameter d_p according to (pp. 205–208):[96]

$$N \approx \frac{3500 L\,(\text{cm})}{d_p\,(\mu\text{m})}$$

As is quite evident from interpreting the chromatograms in Figure 4.82, RP-HPLC columns have a finite lifetime. It is good practice for the analyst to keep a record of N (as calculated using the above equation) vs. either time or number of injected samples in an attempt to continuously monitor column performance. Because HPLC reciprocating pumps maintain constant flow rate, a continuous observation of the back-pressure or pressure buildup at the front on the HPLC column is an important parameter to monitor. Making sure that there are no leaks in an operating HPLC is also very important.

Another useful equation that is used to predict the back-pressure or pressure drop, ΔP, for well-packed columns having similar operating conditions, with the mobile-phase viscosity η in centipoises for a dead time t_0 for columns packed with spherical particles, can be found from (pp. 205–208):[96]

$$\Delta P\,(\text{psi}) = \frac{3000 L\,(\text{cm})\,\eta\,(\text{cP})}{t_0 d_p^2\,(\mu\text{m})}$$

Columns packed with irregular particles might yield back-pressures that are higher. A new spherical particle column should have a ΔP no greater than about 30% in excess of that predicted by the above equation. Let us return to Figure 4.79 and take a brief look at some HPLC detectors.

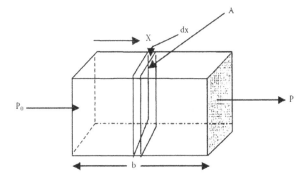

FIGURE 4.84

4.66 HOW DO HPLC DETECTORS WORK?

Depicted in Figure 4.79 are the names of the major types of HPLC detectors. It has been said that the FID is to the GC what the UV-vis-absorbance detector is to the HPLC. However, unlike GC, there is no universal HPLC non-mass spectrometric detector! The earlier advances in the interface that connects the LC eluent with a mass spectrometer and subsequent advances in LC-MS-MS have made this hyphenated technique a universal one. We will address this topic shortly. We will focus next on two detectors designed for HPLC that were in current use at one time in the author's laboratory.

The first is a UV-vis-absorption or absorbance spectrophotometric detector. (The phenomenon responsible for the signal is the *absorption of ultraviolet radiation*; however, what is actually measured is ultraviolet absorbance, hence the abbreviation HPLC-UV.) The second is a molecular fluorescence luminescence spectrophotometric detector (commonly called fluorescence, FL). Other non-mass spectrometric detectors designed for HPLC include refractive index (RI), electrochemical (EC), and, of a more recent vintage, the light-scattering evaporative detector (LSED). RI and LSED HPLC detectors are not sensitive enough to meet the needs of TEQA. Electrochemical HPLC detectors have the required sensitivity, but due to frequent fouling of electrode surfaces, they have not really found a place in TEQA. This author knows of no EPA methods as of yet that incorporate EC HPLC detectors. For this reason, EC HPLC detectors will be not considered.

Organic compounds that possess an *ultraviolet- or visible-absorbing chromophore* obey the Beer–Lambert or Beer–Bouguer law of spectrophotometry. In what is generally termed molecular absorption spectrophotometry, a cuvette (in the case of stand-alone UV-vis spectrophotometers) or a microflow cell (in the case of flow-through HPLC-UV-vis detectors) is used. We now proceed to derive the fundamental equation that relates absorbance as measured on a UV-vis HPLC detector to concentration. This relationship is important to the practice of TEQA. Similar approaches have been shown in two of the most popular textbooks in analytical chemistry. (5th ed. pp. 33–37)[1](pp. 502–506)[19] Let us start by considering what happens when UV radiation impinges onto a cuvette of path length b as shown in Figure 4.84:

Ultraviolet or visible radiation incident on the infinitesimal volume of area A and thickness dx experiences a decrease in transmitted intensity or power dP that is proportional to the incident power P, to the number of absorbing molecules Nc, where N is Avogadro's number and c is the concentration of absorbing species in moles per liter, and to the thickness dx according to

$$dP = -\beta PNcAdx$$

where β is a proportionality constant. This equation can be rearranged and integrated as follows:

$$-\frac{dP}{P} = \beta c \, dx$$

$$-\int_{P_0}^{P} \frac{dP}{P} = \beta c \int_{0}^{b} dx$$

At $x = 0$, $P = P_0$, and at $x = b$, $P = P$, and we have the limits of integration that are necessary to evaluate these integrals. Rearranging and removing the negative sign yields the desired outcome:

$$\ln\left(\frac{P_0}{P}\right) = \beta cb$$

Converting from the natural to the common logarithm while substituting $\ln x = (\ln 10)(\log x)$ yields

$$\log\left(\frac{P_0}{P}\right) = \left(\frac{\beta}{\ln 10}\right) cb$$

The absorbance, A, is defined as the logarithm of the ratio of the incident intensity, P_0, to the transmitted intensity, P, so that

$$A = \varepsilon bc$$

This equation states that the *absorbance is directly proportional to solute concentration* for a given solute/solvent, (i.e., ε, the molar absorptivity) and for a fixed length b. It must be remembered that the absorbance and the molar absorptivity are dependent on the wavelength. It becomes important in practice for an analyst to know how the molar absorptivity varies with wavelength, λ. The percent transmittance, $\%T = 100T$, a common term used with stand-alone spectrophotometers, can be related to the absorbance by manipulating the above definition for A according to

$$A = \log\left(\frac{P_0}{P}\right) = -\log\left(\frac{P}{P_0}\right)$$

$$= -\log\left(\frac{\%T}{100}\right)$$

$$= 2 - \log(\%T)$$

It is seen from this relationship that when no radiation is transmitted, $\%T = 0$ and $A = 2$, whereas if all incident radiation is transmitted through the cuvette or microflow cell, $\%T = 100$, all radiation is transmitted, and $A = 0$. The simple proportion between A and c forms the basis for TEQA using UV-vis-absorption HPLC detectors.

4.67 HOW DO YOU GO ABOUT DOING QUANTITATIVE ANALYSIS USING HPLC WITH UV DETECTION?

Because a solute's absorbance, A, and its molar absorptivity, ε, depend on the wavelength of the incident UV radiation, it is necessary that a wavelength be found that maximizes ε. It becomes important, then, to either know the UV absorption spectrum for the analyte of interest or have some means to record the UV absorption spectrum. This can be accomplished using a stand-alone scanning UV-vis spectrophotometer. An absorption spectrum is a plot of absorbance against wavelength across either the UV alone (200 to 400 nm) or the UV and visible (200 to 700 nm). Ultraviolet absorption of UV photons is enough not only to excite electrons from the ground electronic state to the first excited state, but also to excite rotational and vibrational quantized levels within each electronic state. This yields UV absorption that covers a wide range of wavelengths, and hence leads to large absorption bands. An experiment in Chapter 5 provides a number of practical details related to molecular absorption spectra. The *chlorophenoxy acids*, first introduced in Chapter 3, provide a good illustration. These organic acids are used as herbicides and their residues are likely found in the environment. Dicamba, 3, 6-dichloro-2-methoxybenzoic acid (and much in the news), although not a phenoxy acid, is included along with 2, 4-dichlorophenoxyacetic acid (2,4-D) and two structurally very similar chlorophenoxy acids, 2, 4, 5-trichlorophenoxyacetic acid (2, 4, 5-T) and 2-(2, 4, 5-trichloro)-propionic acid, commonly called Silvex. Molecular structures for all four compounds are as follows:

Dicamba   2,4,5-T   Silvex   2,4-D
(2,4,5-TP)

The UV absorption spectra were recorded at the apex of the chromatographically resolved peak using RP-HPLC with a photodiode array UV detector. This detector, unlike a fixed-wavelength detector, can record all absorbances throughout the UV region of interest simultaneously. All four compounds yield UV absorption spectra that show intense absorbance below 250 nm, while having an absorption peak maximum between 270 and 295 nm. The other factor to consider in choosing the most sensitive wavelength for trace analysis is the UV cutoff of the organic modifiers used in the mobile phase. A selected list of solvents is given in Table 4.26 (pp. 722–723).[96] The UV cutoff is the wavelength above which useful quantitative analysis can be obtained.

The solvent viscosity, boiling point, and solvent polarity are all important factors in arriving at the optimum organic solvent to be used in the mobile phase. For example, acetone is a low-viscosity liquid with a boiling point that is appropriate for RP-HPLC and a polarity that, when mixed with water, yields a mobile phase with sufficient solvent strength that would enable moderately polar solutes to have k' values between 2 and 10. However, its UV cutoff is so high as to render the solvent useless for TEQA based on molecular absorption in the

TABLE 4.26
Physical-Chemical Properties of Common Organic Solvents, Including Water

Solvent	UV Cutoff (nm)	Viscosity (cP)	Boiling Point (°C)	Polarity P'
Acetone	330	0.36	56.3	5.1
Acetonitrile	190	0.38	81.6	5.8
Cyclohexane	200	1.0	80.7	0.2
Ethyl acetate	256	0.45	77.1	4.4
Hexane	195	0.31	68.7	0.1
Methanol	205	0.55	64.7	5.1
Methylene chloride	233	0.44	39.8	3.1
Tetrahydrofuran	212	0.55	66.0	4.0
Water	190	1.00	100.0	10.2

low-UV, region where most organic compounds absorb. Acetonitrile, on the other hand, has similar physical properties combined with a low-UV cutoff. Cyclohexane has a favorable UV cutoff, viscosity, and boiling point. However, its polarity is so low that it is not even miscible in water. Cyclohexane is useless as an organic modifier for RP-HPLC, yet is appropriate as a mobile-phase additive in the normal phase (NP-HPLC) due to the extremely low solvent polarity. Acetonitrile, methanol, and tetrahydrofuran are the most commonly used organic modifiers for the laboratory practice of RP-HPLC, whereas cyclohexane and *n*-hexane are the most common solvents for NP-HPLC. Let us consider the inner workings of a typical UV absorbance HPLC detector. NP-HPLC is of limited value to TEQA and will not be considered any further. Because this book emphasizes quantitative measurement, we next focus on the HPLC-UV detector.

4.68 HOW DOES THE ABSORBANCE OF A SOLUTE GET MEASURED?

The optics for the Model 2487® Dual λ Absorbance Detector (Waters) are shown in Figure 4.85. A deuterium lamp is the source of UV radiation, and the ellipsoidal mirror collects light from the lamp and focuses it through the filter wheel onto the entrance slit. The spherical mirror directs light toward the grating. Another part of the spherical mirror focuses dispersed light of a particular wavelength, determined by the grating angle, onto the entrance of the flow cell. Prior to the flow cell, a portion of the band is split to a reference photodiode while the remaining incident intensity is transmitted through the flow cell to the sample photodiode. The preamplifier board integrates and digitizes the currents from the photodiodes for processing by the signal processing electronics and output to a computer. When a new wavelength is entered through the front panel, the detector rotates the grating to the appropriate position. The difference between a conventional flow cell and the TaperSlit® is shown in Figure 4.86. According to Waters, the TaperSlit flow cell renders the detector insensitive to changes in mobile-phase refractive index. This is significant in gradient elution HPLC where the mobile-phase composition changes with time. As shown in Figure 4.86, this tapered design minimizes the amount of radiation that is reflected off of the inner wall, thus allowing a greater UV light throughput. This is not the only non-mass spectrometric HPLC detector useful to TEQA. Let us consider measuring the UV-induced fluorescence, instead of measuring UV absorption.

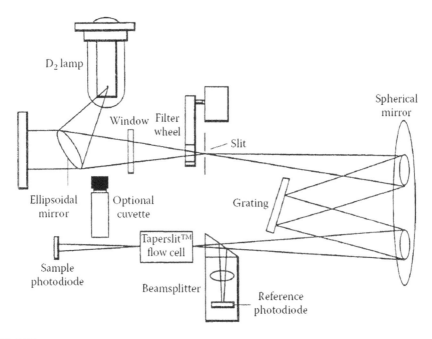

FIGURE 4.85

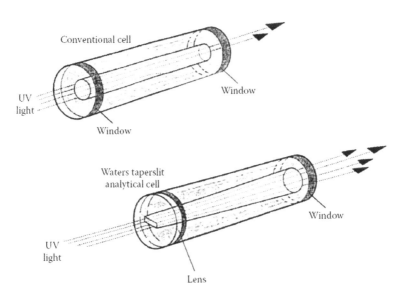

FIGURE 4.86

4.69 WHAT IS AN HPLC FLUORESCENCE DETECTOR?

The fluorescence HPLC detector (HPLC-FL) is similar to the UV absorption HPLC detector (HPLC-UV) in that a source of UV radiation is made incident to a microflow cell in which the chromatographically resolved analyte passes through. However, a photomultiplier tube (PMT) and associated optics are positioned at right angles relative to the incident UV radiation. Lakowicz has articulated just what molecular luminescence is:[97]

Luminescence is the emission of photons from electronically excited states. Luminescence is divided into two types, depending upon the nature of the ground and the excited states. In a singlet excited state, the electron in the higher-energy orbital has the opposite spin orientation as the second electron in the lower orbital. These two electrons are said to be paired. In a triplet state, these electrons are unpaired—that is, their spins have the same orientation. Return to the ground state from an excited singlet state does not require an electron to change its spin orientation. A change in spin orientation is needed for a triplet state to return to the singlet ground state. Fluorescence is the emission that results from the return to the lower orbital of the paired electron. Such transitions are quantum mechanically "allowed." ... Phosphorescence is the emission that results from transition between states of different multiplicity, generally a triplet state returns to a singlet ground state.

Ewing defines molecular luminescence that includes fluorescence, phosphorescence, and Raman spectroscopies as methods whereby radiation absorbed by molecular species is reemitted with a change in wavelength (5th ed. p. 124).[1]

Both organic and inorganic compounds exhibit fluorescence. Molecules that display significant fluorescence possess delocalized electrons due to conjugated double bonds. A fluorescence spectrum is a plot of FL emission intensity vs. wavelength of the emitted radiation. A simplified *Joblonski diagram* shown in Figure 4.87 provides a qualitative explanation of how fluorescence (FL) and phosphorescence (PHOS) spectra originate.

There exists within molecules a myriad of quantized vibrational levels in both the ground and first excited electronic states. Not shown in the diagram are the myriad of rotational quantized levels within each vibrational level. Transitions, as shown in the diagram, yield absorption and fluorescence spectra that are quite broad. The fact that these spectra are obtained for solutes that are dissolved in solvents is also a contributor to broad bands. Photons emitted due to hv(FL) are on the order of nanoseconds, whereas photons emitted due to hv(PHOS) range from milliseconds to seconds.

Fluorescence quenching is the term used to explain any decrease in the analyte-emitted fluorescence intensity due to the influence of the sample matrix. Quenching can also be attributed to the analyte itself and is called self-quenching. Ewing discusses the effect of quenching when the concentration of phenol dissolved in water is increased beyond 10 ppm, with a maximum in the emitted intensity at around 75 ppm, followed by a gradual loss in intensity due to self-quenching. (5th ed.127–129)[1] A quenching agent, if present, contributes to a loss in the emitted intensity. A good example is the quenching of PAHs by dissolved oxygen.

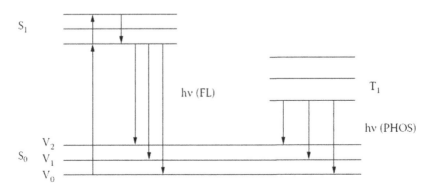

FIGURE 4.87

4.70 HOW IS IT THAT FLUORESCENCE CAN BE QUANTITATED?

For a single fluorophore, as would be the case for a chromatographically resolved compound in the HPLC chromatogram, the fluorescence is related to the incident UV radiation of power P_0 according to

$$F = P_0 K \left(1 - 10^A \right)$$

where A is the UV absorbance and K is a constant for a given system and instrument and is related to the quantum yield. This expression can be made more useful by considering a Taylor's series expansion for the exponential term:

$$F = P_0 K \left(2.30A - \frac{(2.30A)^2}{2!} + \frac{(2.30A)^3}{3!} + \cdots \right)$$

To a first approximation, for solutions of low absorbance that are most likely the case for TEQA, the higher-order terms can be neglected so that

$$F \approx P_0 K (2.30) A$$

We see that a direct proportionality exists between the emitted fluorescence intensity denoted by F and the sample absorbance denoted by A. If a stand-alone spectrofluorometer is used to conduct a quantitative analysis, the error introduced by neglecting the A^2 term should be considered. However, for a fluorescence HPLC detector where a set of working calibration standards are used (refer to Chapter 2) and where the concentration of fluorophore is relatively low, it is sufficient to ignore the higher-order terms. Upon substituting for A, we obtain

$$F \cong P_0 K (2.30)(\varepsilon b c)$$

$$\approx K' \varepsilon b c \approx K'' c$$

The linear dynamic range of this detector is limited at the high end by fluorescence quenching due to self-absorption. These equations demonstrate that a fluorescence HPLC detector is intrinsically more sensitive than a UV absorption HPLC detector because for dilute solution, the PMT senses a faint light against a dark background. For a UV absorption detector, the PMT or photodiode compares a slightly lower transmitted intensity P against a higher-incident intensity P_0, and this small difference is related to analyte concentration.

Instrumentation to measure the fluorescence of a substance can be as simple as a filter fluorimeter or as complex as a dual-excitation/emission spectrofluorometer. A schematic of the Model 474® (Waters) dual monochromatic fluorescence HPLC detector is shown in Figure 4.88. A xenon lamp is the source of UV radiation, and this light reflects off of a mirror to an excitation grating through a narrow entrance slit. This monochromator enables a narrow band of excitation wavelengths to be selected. The selected UV photons are passed through a beam splitter, whereby a portion of this excitation radiation is detected by a photodiode. This photodiode enables corrections to the lamp output to be made. The collection mirror is positioned at a right angle with respect to the direction of the excitation light, as shown in Figure 4.88 (top schematic). The fluorescence radiation is collimated and reflected through a narrow emission slit where it is diffracted off of the grating. This emission monochromator selects a narrow band of fluorescence

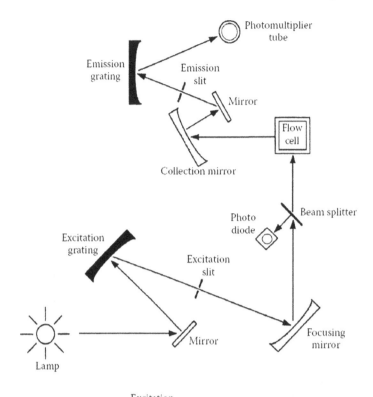

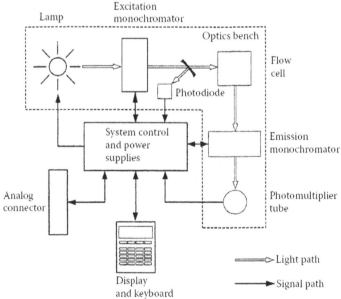

FIGURE 4.88

wavelengths whose intensity reaches a PMT. The output of the PMT is amplified and fed to an analog-to-digital (A/D) converter, and then fed into a PC for processing and quantitative analysis. Contemporary fluorescence HPLC detectors like the Model 474 are completely microprocessor controlled; this is best shown in Figure 4.88 (bottom schematic).

4.71   HOW DO UV ABSORPTION AND FLUORESCENCE EMISSION HPLC CHROMATOGRAMS COMPARE?

The answer lies in the nature of the chemical compounds that are of interest. For example, there is little difference in HPLC detector sensitivity between UV absorption and fluorescence emission for phenol, but a great difference exists for naphthalene. Naphthalene is a fused aromatic and allows for extensive delocalization of electrons and is quite sensitive when measured by RP-HPLC-FL. For both analytes of environmental concern being dissolved in water or dissolved in the RP-HPLC mobile phase, a vivid illustration of this difference in HPLC detector sensitivity is shown in Figure 4.89. An RP-HPLC-UV chromatogram is shown at the top and an RP-HPLC-FL chromatogram is shown at the bottom. The chromatograms are stacked, and both the ordinate and abscissa axes are aligned. Effluent from a reversed-phase column was fed to the Model 2478 UV absorption detector, and the effluent from this detector was made the influent to the Model 474 fluorescence detector, as shown in Figure 4.90. Analog signals from both detectors are sent through an A/D converter to a central PC workstation that uses Turbochrom (PE-Nelson) to acquire the data and to generate the chromatograms. The peaks in both chromatograms of Figure 4.89 are due to the injection of a multicomponent standard of the 16 priority pollutant PAHs. They elute in the order shown in Table 4.27.

Having completed our discussion of UV and FL detectors for HPLC, we can now introduce the *mass spec* by asking the following question.

4.72   CAN A MASS SPEC BE USED AS A DETECTOR FOR HPLC?

Yes, but look what it took! GC-MS had been developed much before LC-MS. Changing chromatographic mobile phases from gas to liquid conceptually required some type of *interface* between the exit of the LC column and the entrance to the high-vacuum region of the mass spectrometer.

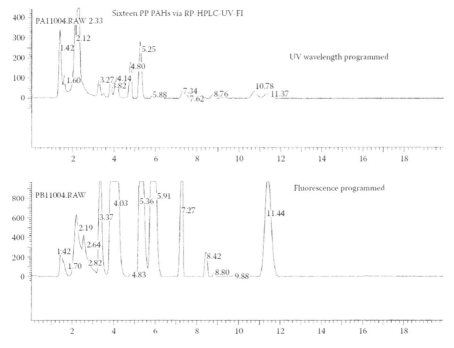

FIGURE 4.89

TABLE 4.27
Summary of the Number of Fused Rings Associated with the Priority Pollutant PAHs

No.	No. of Rings	Name
1	2	Naphthalene
2	2	Acenaphthylene
3	2	Fluorene
4	2	Acenaphthene
5	3	Phenanthracene
6	3	Anthracene
7	3	Fluoranthene
8	4	Pyrene
9	4	Benzo(*a*)anthracene
10	4	Chrysene
11	5	Benzo(*b*)fluoranthene
12	5	Benzo(*k*)fluoranthene
13	5	Benzo(*a*)pyrene
14	5	Dibenzo(*a*, *h*)anthracene
15	6	Indeno(1, 2, 3-c,*d*)pyrene
16		Benzo(g, *h*, *i*)perylene

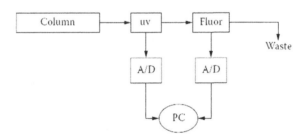

FIGURE 4.90

This was in stark contrast to GC-MS! It was also easier to work with a nonpolar mobile phase since the solvent could be more easily evaporated. However, the dilemma here as suggested again by the plot of analyte volatility vs. analyte polarity as shown in Figure 4.4, was the fact that RP- HPLC had emerged as *the* more versatile mode of column liquid chromatography when compared to NP-HPLC. In his chapter titled *Compounds Not Amenable to GC*, Budde provides an excellent historical development of the major interfaces that made LC-MS or LC-MS-MS possible.[14] Starting sometime between the years 1973 and 1984, major advances in hyphenating both single and tandem quadrupole mass spectrometers to liquid chromatographs began to occur. Of particular interest, and of a more difficult undertaking, has been to interface RP-HPLC with mass spectrometers. RP-HPLC uses aqueous and water-miscible organic solvents, while NP-HPLC uses water-immiscible organic solvents. It is much easier to deal with nonpolar organic solvents within an interface between the LC eluent and the highly evacuated mass spectrometer! Nevertheless, the successful interface technology for RP-HPLC and quadrupole-based mass spectrometry since the 1990s has established LC-MS for single quadrupole operation and

LC-MS-MS for tandem operation. In addition, the type of LC-MS interface also appears in the acronym, e.g. LC-APCI-MS, where APCI refers to *atmospheric pressure chemical ionization*, or LC-ES-MS, where ES refers to *electrospray*. Connection of a liquid–solid column chromatographic separation technique to a mass spectrometer that operates in a high vacuum requires an interface that is capable of meeting the goals of TEQA. Budde defines interfaces such as LC-MS this way (p. 246):[14]

> ... as the device that receives the condensed-phase flow from the separation system and converts this flow into gas-phase analyte molecules or ions suitable for injection into the high vacuum of the MS.

Henion and coworkers have articulated the interface in another manner:[98]

> We suggest that atmospheric pressure ionization mass spectrometry is a preferred way to simplify the coupling of liquid inlet systems such as HPLC, CE and ion pair chromatography to mass spectrometry. A key feature of this approach is that liquid effluent from a particular separation science is not directly introduced into the mass spectrometer vacuum system ... the effluent is "sprayed" in the vicinity of an in-sampling orifice ... Gas-phase ions are formed in this region by either electrospray or atmospheric pressure chemical ionization ... These ions are sampled through the ion sampling orifice into the vacuum system for mass analysis. Large excesses of solvent from the effluent do not enter the vacuum system.

A truly universal MS detector for RP-HPLC would significantly enlarge the range of analyte polarities amenable to TEQA (refer again to Figure 4.4). Commercially developed LC-MS interfaces over the past 50 years include (Chapter 6):[14]

- Direct liquid and fluid introduction
- Moving belt interface developed by McFadden at Finnigan Corp. in the mid-1970s
- Particle beam interface developed by Willoughby and Browner in the mid-1980s
- Thermospray interface developed by Vestal in the early 1980s
- The atmospheric pressure chemical ionization (APCI) interface developed by
  - Horning and associates at the Baylor College of Medicine (1970s),
  - Thomson at SCIEX (1980s)
- The electrospray (ES) interface developed by:
  - Thomson at SCIEX (1990s)
  - Fenn and colleagues at Yale University[99]
  - pneumatically assisted ES LC-MS interface by Henion at Cornell University.[100]

A cartoon-like schematic drawing of a generalized *atmospheric pressure interface* between an HPLC and a quadrupole MS is nicely shown in Figure 4.91. It is instructive to compare

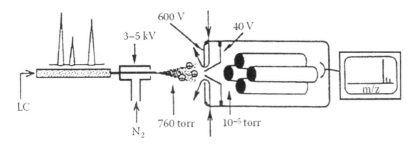

FIGURE 4.91

and contrast the two major LC-MS interfaces that are currently used in environmental laboratories: the Thomson designed APCI and the Fenn designed ES.

The Thomson APCI interface is shown in Figure 4.92 while the Fenn electrospray interface is shown in Figure 4.93 (pp. 303–317).[14] Budde compares and contrasts the advantages and limitations of both interfaces and then compares these two with a pneumatically assisted ES LC-MS interface developed by Henion (Chapter 6).[14] Combining either the APCI or ES interface with tandem mass spectrometry has proved to be a very powerful contributor to the arsenal of TEQA determinative techniques!

4.73   IS LC-MS-MS USED IN TEQA?

Yes, indeed. We discuss one application related to trace *enviro-health* quantitative analysis. Blount and coworkers at CDC earlier developed an analytical method that quantitatively

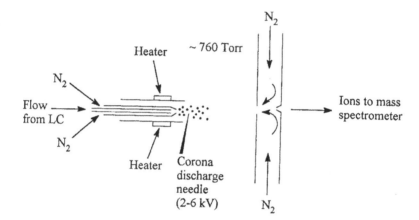

FIGURE 4.92

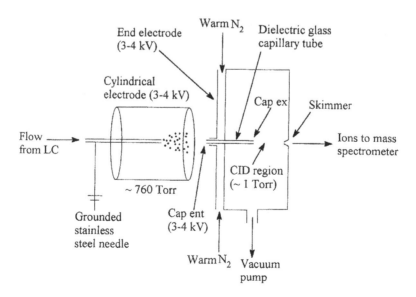

FIGURE 4.93

determines eight phthalate metabolites in human urine.[101] Enzymatic deconjugation is used to release the monoalkyl phthalate esters. Molecular structures for the environmentally ubiquitous glucuronated di(2-ethylhexyl) phthalate and its conversion to mono(2-ethylhexyl) phthalate metabolite are shown below:

Glucuronated Di (2-ethylhexyl) phthalate   Mono (2-ethylhexyl) phthalate

A glucuronide moiety is then covalently bonded to the hydroxy oxygen. Selective use of β-glucuronidase enzyme to hydrolyze glucuronide metabolites allows for quantification for both free and glucuronidated forms of each phthalate metabolite. The esters are isolated and recovered using RP-SPE techniques. $^{13}C_4$-labeled standards such as monoethyl– $^{13}C_4$ and others are commercially available and enable quantification via isotope dilution (Chapter 2). Negative-ion APCI was used to form negatively charged analyte ions. Daughter ions were formed in the collision cell using argon as the collision gas at 2.0 millitorr. The method was applied to 289 human urine samples to establish baseline levels for phthalate exposure in the American population. Detectable levels of some phthalate monoesters were found in all urine samples tested.

Today, perfluoro alkyl substances (PFAS) have now joined the organochlorine pesticides, e.g. DDT, the PBBs, the PCBs, and the PBDEs as persistent SVOC contaminants in the environment. Analytical chemists know that PFAS are not amenable to analysis by GC or GC-MS without chemical derivatization. PFAS consists of either perfluoroalkanoic acids or perfluoroalkanesulfonic acids. These unique organic compounds can repel hydrophobic *and* hydrophilic compounds. This unique property makes PFAS common ingredients in both industrial and consumer products. These products include cleaning agents, paper, paint, cosmetics, fire-retardants, and anti-wetting agents used to treat carpet, upholstery, and clothing. PFAS compounds bioaccumulate and resist environmental degradation. PFAS persist in the environment! Society wants to know what concentration of PFAS exist in drinking water and what concentration comprises the maximum contaminant level (MCL) as discussed in Chapter 1. EPA Method 537 was established in 2009 and modified in November, 2018 (EPA Method 537.1) to address this question. RP-SPE (Chapter 3) is the acceptable sample prep technique. LC-ES-MS-MS is the recommended determinative technique. Restek Corporation recommends their Raptor® C18 LC with dimensions: 100 mm x 2.1 mm ID Column with a 5 μm particle size.[102] Table 4.28 lists 20 PFAS with corresponding retention times, concentration, precursor ion, and product ion using HPLC with an ES interfaced to a triple quadrupole mass spectrometer using the MRM protocol. The reader should come away from this foreboding table of 20 PFAS compounds with a deeper appreciation of the significant advancements made in contemporary LC-MS-MS.

TABLE 4.28

Peak Identification for the Separation and Detection of 20 PFAS Compounds Using LC-MS-MS

Peak Order #	PFAS	tr_(min)	Conc (#ng/mL)	Precursor Ion	Product Ion
1	Perfluorobutanesulfonic acid (PFBS)	4.17	5	298.9	79.9
2	Perfluoro-n-[1,2^{13}C$_2$] hexanoic acid (^{13}CPFHxA)	4.90	5	312.8	269.8
3	Perfluorohexanoic acid (PFHxA)	4.91	5	362.8	318.8
4	Perfluoroheptanoic acid (PFHpA)	5.59	5	362.8	318.8
5	Perfluorohexanesulfonic acid(PFHxS)	5.65	5	398,8	79.8
6	Perfluoro-[1,2-^{13}C]octanoic acid (^{13}C-PFOA)	6.09	5	414.8	368.8
7	Perfluorootanesulfonic acid (PFOS)	6.10	5	412.7	368.8
8	Perfluoro-1-[1,2,3,4-^{13}C$_4$] octanesulfonic acid (^{13}C-PFOS)	6.51	10	502.7	79.9
9	Perfluorooctanesulfonic acid (PFOS)	6.52	5	498.7	79.9
10	Perfluorononanoic acid (PFNA)	6.52	5	462.6	418.9
11	Perfluoro-n-[1,2-^{13}C$_2$] decanoic acid (^{13}C-PFDA)	6.87	5	514.8	469.9
12	Perfluorodecanoic acid (PFDA)	6.88	5	512.7	468.8
13	N-deuteromethylperfluoro-1-octanesulfonamidoacetic acid (d3-NMeFOSAA)	7.01	10	572.7	418.9
14	N-methyl perfluorooctanesulfonami doacetic acid (NMeFOSAA)	7.02	10	569.8	418.9
15	N-deuterioethylperfluoro-1-octanesulfonamidoacetic (d5-NEtFOSAA)	7.16	10	588.8	418.9
16	N-Ethyl perfluorooctanesulfonamidoacetic acid (NEtFOSAA)	7.17	10	583.8	418.9
17	Perfluoroundecanoic acid (PFUnA)	7.17	10	562.8	518.8
18	Perfluorododecanoic acid (PFDoA)	7.44	10	612.7	568.8
19	Perfluorotridecanoic acid (PFTrDA)	7.67	10	662.8	618.9
20	Perfluorotetradecanoic acid (PFTA)	7.87	10	712.8	668.9

Column: Raptor® C18 100 mm × 2.1 mm ID, 5 μm particle size, 90 Å pore size (Restek Corporation). Column temperature: 40°C; Sample: Diluent: methanol-water (96:4); 5 μL injection volume. Mobile phase: A: 5 mM ammonium acetate in water; B: Methanol Gradient (%B):0.00 min (10%), 8.00 (95%), 8.01 min (10%), 10.0 (10%); Flow rate: 0.4 mL/min; Detector: MS/MS; Ion source: Electrospray; ion mode: ESI-; Mode: MRM.

To conclude, Watson and Sparkman have noted the following:

The modern suite of techniques available through atmospheric pressure ionization (API), which includes electrospray ionization (ESI), atmospheric pressure chemical ionization (APCI) and atmospheric pressure photoionization (APPI), has allowed LC-MS to become an effective methodology in the analysis of organic compounds that are important in environmental, pharmaceutical, forensic chemistry, and many other application areas. Of the three techniques, ESI is especially important for samples that cannot be analyzed by GC or GC-MS. APCI and APPI are limited to the analysis of volatile compounds ... many analyses using these liquid-sample interfaces are carried out without the use of an LC column. The modern LC-MS instrument is often supplied without a liquid chromatograph.

(p. 641)[13]

We have completed our discussion of the determinative techniques for the quantitative determination of organic compounds that are of concern to TEQA. What is left? Well, let us return to Figure 4.4 and consider that chromatographic region where we are interested in semi- to non-volatile compounds that in solution exist as ions. In other words, for the first time in this book we consider the analysis of *inorganic chemical compounds that ionize* when dissolved in water. We want to be able to identify ionically bonded chemical substances that are of concern to TEQA and to consider what determinative technique one would find in trace environmental testing labs today. After introducing the principle of ion exchange, we then discuss how ion chromatography (IC) became the dominant determinative technique to separate, identify, and detect trace concentrations of inorganic ions in aqueous samples of environmental interest.

4.74 HOW DID IC MAKE IT TO THE FOREFRONT IN TRACE INORGANICS ANALYSIS?

By evolving from a low-pressure (gravity fed) large-column-diameter (>1 cm) ion exchange chromatography (IEC) technique to a moderately high-pressure, narrow-column-diameter (~5 mm) high-performance IEC technique! The use of glass columns packed with either a strong or weak anion or cation exchange resin is routine in many chemical laboratories. However, the development of IC as an additional determinative technique is very useful to TEQA and requires that we discuss these principles. But first, as a way to review some principles of conventional IEC, let us discuss an ingenious use of *two chemically different ion exchange resins* and the simple experiment that can illustrate very nicely the principle of ion exchange (IE). One of the most useful IE resins is Chelex-100 whose molecular structure is shown below; it is a 1% cross-linked, 50 to 100 mesh, styrene–divinylbenzene resin containing iminodiacetate.

Conventional cation exchange resin such as AG 50W-X8, 8% cross-linked, 20 to 50 mesh, H$^+$ form (Bio-Rad), whose IE capacity is 5.1 mEq/g, is a sulfonic acid type. The structure of this resin is shown below:

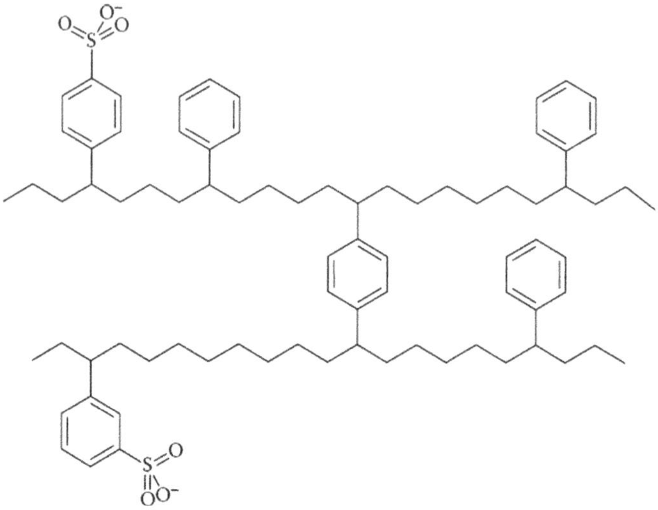

Chelex-100 in its Na$^+$ form can exchange other metal ions such as Fe^{2+}, whereas AG 50W-X8 in the H$^+$ form exchanges other metal ions such as Na$^+$ for H$^+$. The author recently packed two conventional glass columns, one containing Chelex-100 and the other containing AG 50W-X8, as shown in Figure 4.94. Suppose that we pass an aqueous solution containing Fe^{2+} through the first column in Figure 4.94. If we then take the effluent from the first column and place it on top of the second column, we discover, after enough aqueous solution has been passed through this column, that the effluent is quite acidic (i.e., its pH is <4). This solution can be titrated against a standardized base such as 0.02 M NaOH to a Bromocresol endpoint, and from the number of milliliters of titrant required, the number of millimoles of H$^+$ can be obtained. This many millimoles is also the number of millimoles of Na$^+$ that was exchanged for H$^+$. For every millimole of Fe^{2+} exchanged on the first column, 2 mmol of Na$^+$ ions must have been released. Hence, the number of milliliters of 0.02 M NaOH can be indirectly related to the number of millimoles of iron(II) present in the original sample. The *stoichiometry* for this series of chemical reactions is below:

4.74.1   CHEMICAL REACTIONS USED IN THIS DEMONSTRATION

Column to the left:

$$Fe^{2+} + 2CheR^- Na^+ \longrightarrow (CheR)_2 Fe + 2Na^+$$

Column to the right:

$$Na^+ + RSO_3^- H^+ \longrightarrow RSO_3^- Na^+ + H^+$$

At the buret:

$$H^+ + OH^- \longrightarrow H_2O$$

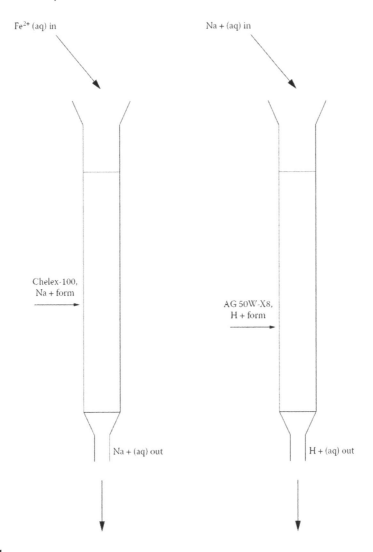

FIGURE 4.94

Bromocresol Green turns *yellow* when the pH is below 4.0 and *blue* when the pH is above 5.6.

The appearance of a gradual darkening of the head of the packed column that contains Chelex-100 is observed at first. This amount of darkening of the white resin gets larger and larger as successive aliquots of the iron (II) solution are added on top of the column. The brownish AG 50W-X8 does not change color during these repetitive additions of sodium ion to the top of this column. Let us return to our development of the principles of IC.

Two achievements that led to the acceptance of IC as an important determinative technique in TEQA are now discussed. The first was the development of surface-agglomerated low-capacity anion and cation exchange resins with particle sizes appropriate for preparing high-performance IC columns. The second was the realization that detecting trace concentration levels in a background of highly conducting IC eluent required a chemical suppression of the eluent conductivity. The research group led by Small at Dow Chemical in the early 1970s is credited with the pioneer development of suppressed IC, and a portion of their abstract to this benchmark paper follows:[103]

Ion exchange resins have a well-known ability to provide excellent separation of ions, but the automated analysis of the eluted species is often frustrated by the presence of the background electrolyte used for elution. By using a novel combination of resins, we have succeeded in neutralizing or suppressing this background without significantly affecting the species being analyzed which in turn permits the use of a conductivity cell as a universal and very sensitive monitor of all ionic species either cationic or anionic.

This benchmark publication was followed by a series of developments that continues to this day in the *evolution of IC as a trace determinative technique.* The Dionex Corporation was formed out of these developments from Dow Chemical in the mid-1970s, and this company has been quite an innovator in the manufacture of instrumentation for suppressed IC. From the earlier model, such as the 10, to the 2000I to the DX 500, and from the cation exchange column to the hollow-fiber cation suppressor to the micromembrane cation suppressor to the self-regenerate cation suppression for anion analysis and now to its "just add water" slogan, Dionex has made considerable advances in IC technology. Today, Dionex is a part of Thermo Fisher Scientific.

A nonsuppressed form of IC also developed during the 1970s. To quote from the author's abstract of this benchmark paper:[104]

The anions are separated on a column containing a macroporous anion exchange resin which has a very low exchange capacity of 0.007–0.07 milli-equivalents per gram. Because of the low resin capacity, only a very dilute solution, 1×10^{-4} M of an aromatic organic acid salt, is needed as the eluent. The eluent conductance is sufficiently low that a suppressor column is not needed, and the separated anions can be detected with a simple conductance detector.

Unfortunately, nonsuppressed IC has not been developed commercially and remains an interesting determinative technique for academic labs. It took, however, over 10 years for the EPA to approve of suppressed IC to quantitatively determine the common inorganic anions derived from moderate to strong acids such as fluoride, chloride, nitrite, phosphate, and sulfate in drinking water. Years of effort have resulted in EPA Method 300.1, "The Determination of Inorganic Anions in Drinking Water by Ion Chromatography." This method considers two different sets of inorganic anions in drinking water, reagent, surface water, and groundwater. The first set is the classical one listed earlier, and the second set consists of analytes derived from the so-called disinfection by-products (DBPs). This second set includes bromate, bromide, chlorite, and chlorate (oxyhalides). The method is replete with sufficient quality control, something that was seriously lacking in the earlier versions of Method 300.

4.75   HOW DOES AN IC WORK?

An ion chromatograph using a suppressed IC mode of operation can be viewed as a moderate-pressure-performance liquid chromatograph. A schematic of the essential components of an IC is shown in Figure 4.95. Because the trace concentration levels of various inorganic anions are of most interest to TEQA, we will focus our discussions on how anions are measured. Alkali and alkaline–earth metal ions and the ammonium ion are common applications of IC. In fact, NH_4^+ can only be measured chromatographically by cation IC. The transition metal ions can also be separated and detected; however, atomic spectroscopy, to be discussed after we complete IC, is the predominant TEQA determinative technique. We will not discuss the separation and detection of cations here.

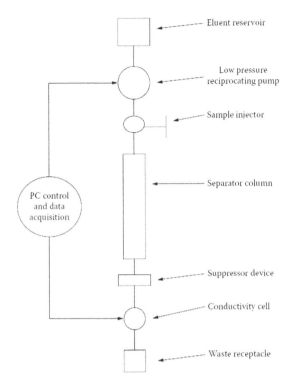

FIGURE 4.95

Referring to Figure 4.95 and focusing on trace anion analysis, the simplest ion chromatograph can be viewed as consisting of four modes:

1. A delivery mode that uses a bicarbonate/carbonate buffer as the mobile-phase eluent and a single-piston reciprocating pump with check valves and pressure gauge/dampener accessory; a means to introduce a fixed volume of sample.
2. A separation mode that consists of a column packed with a low-capacity anion exchange resin.
3. A detection mode that includes a suppressor device and a microflow conductivity cell.
4. A data mode that could be either a strip-chart recorder, an electronic integrator (ones made by either Spectra-Physics or Hewlett-Packard were popular at one time), or a PC via an A/D converter.

In the author's laboratory during the late 1980s, the analog signal from the conductivity cell is connected to a 900 interface box (PE-Nelson). The interface is connected to the input of a PC using an IEEE cable. The PC utilizes Turbochrom (PE-Nelson) *chromatography data acquisition software* in a Windows disk operating environment.

It was critical to the development of IC to achieve the IE separation of the common inorganic anions in a reasonable length of time (e.g., within 20 min of the SO_4^{2-} eluting) and with sufficient chromatographic resolution. It appeared inevitable that advances in HPLC made in the early 1970s would simultaneously advance the chromatographic separation of ions or ionizable compounds. The drastic lowering of particle size, predicted by the contributions to plate height and the realization that pellicular materials minimize the stationary-phase thickness, leads to a more chromatographically efficient IC column. Ways to lower the IE capacity were pursued, and these efforts resulted in more efficient columns and significantly lower values for k'.

TABLE 4.29
Comparison of Conventional to Surface-Agglomerated Ion Exchange Resin

Conventional Resin	Surface-Agglomerated Resin
Capacity: 3–5 mEq/g	Capacity: 0.02 mEq/g
Large degree of swelling	Very little swelling
Many diffusion paths	Limited diffusion paths
Separation by ion size	Little separation by ion size
Not easily poisoned	Easily poisoned but can be regenerated
High sample concentrations	Low sample concentrations
High eluent strength	Low eluent strength

Source: MacDonald J., *Am Lab*, 11(1): 45–55, 1979.

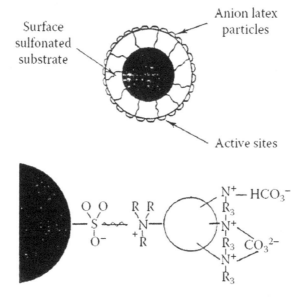

FIGURE 4.96

To achieve this increase in IC column efficiency, IE resins had to be developed that differed significantly from those of conventional resins. Table 4.29, taken from an earlier article, compares the properties of conventional IE resins to those required for IC.[105] A decrease in IE capacity by about two orders of magnitude permitted the use of eluents for IC whose concentrations are three orders of magnitude lower than those used in conventional IE analysis. Eluent concentrations in IC are routinely prepared at the millimolar level. The favorable IE properties were realized with the development of surface-agglomerated pellicular resins. Such a pellicular particle is shown in Figure 4.96. A surface-sulfonated particle is shown agglomerated with an anion latex particle. The illustration shows the bicarbonate and carbonate ions, the most common eluent ions used in anion IC, as attached to the latex particle. The chemistry involved in separating chloride from sulfate and being able to quantitate these common ions found in groundwaters is shown schematically in Figure 4.97. The eluent used in this schematic is NaOH, so that the counterion is hydroxide ion, as shown as the analyte ion symbolized as X^- is chromatographed through

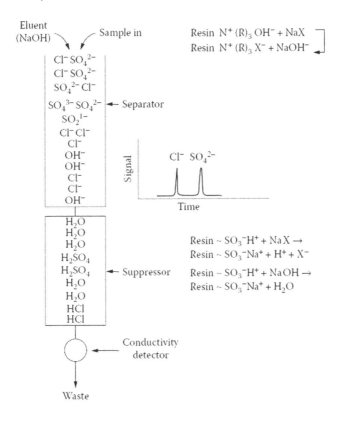

FIGURE 4.97

the separator column. As the eluent and analyte ions reach the suppressor, prior to the conductivity detector, as shown in Figure 4.97, the hydroxide ion eluent is converted to the minimally conducting water and the analyte ion is converted to the highly conducting strong acids HCl and H_2SO_4. It is well known that H^+ and OH^- are significantly more conductive than are any of the common anions.

4.76   IS THERE A THEORETICAL BASIS FOR CONDUCTIVITY MEASUREMENTS?

Let us digress a moment and discuss the relationships involving conductivity.[106] An aqueous solution containing dissolved inorganic salts or electrolytes conducts electric current. The extent of this conductance, denoted by G, is defined in terms of the electrical resistance of the electrolytic solution. The reciprocal of this resistance is the conductance such that $G = 1/R$, and the conductance is expressed in units of Ω^{-1} or siemen, denoted by S. An older unit is the mho, which is ohm spelled backward. Conductance is a bulk property and encompasses both the solute and solvent that pass through a conductivity detector. A solution has a specific conductivity κ in units of siemens per centimeter for a typical conductivity cell. This cell consists of a pair of identical electrodes, each having a surface area A and spaced 1 cm apart. An AC voltage is applied across the electrodes and the resistance measured. A cell constant k can be defined as the ratio of the spacing distance to surface area (l/A), and all of this gives

$$\kappa = \frac{Gl}{A} = Gk$$

Conductivity as an analyte-specific property can be considered by first defining the equivalent limiting conductivity, denoted by Λ, for a typical inorganic salt, MX_n, where M^{n+} is a cation of positive charge n and X is a monatomic anion (the selection of a monatomic anion serves to simplify the equations). Λ is the conductance if enough of the salt was available to equal 1 g-equivalent weight in a liter of solution. Hypothetically, this would require very large electrode surface areas while maintaining the same spacing between electrodes at 1 cm. This is summarized in the following relationship:

$$\Lambda = \frac{1000\kappa}{C}$$

For a concentration C of the salt MX_n dissolved in water (in equivalents per liter), Λ has units of siemens-square centimeters per equivalent. By substitution and rearranging, the conductance G can be related to the equivalent limiting conductivity and Λ. The use of Λ enables one to consider the separate contributions of cation and anion. The conductance can then be related to the concentration of the salt, the separate limiting conductances of the individual ions, and the cell constant k as follows:

$$G = \frac{C\Lambda}{1000k} = \frac{C_{MX}}{1000k}\left(\lambda_M + \lambda_X\right)$$

We see from the equation that each cation, M^{n+}, and each anion, X^-, can be given its own value for λ.

We now proceed to relate the bulk property G to the analyte concentration, and thus connect instrumental IC to TEQA. We focus only on anion exchange IC; however, these relationships can also be derived from the perspective of the cation using cation exchange principles. We consider any monatomic anion X, where X could be chloride, nitrate, and so forth, in a background of eluent such as E^+E^-, where E^+ might be Na^+ and E^- might be CO_3^{2-}. Let us assume that the concentration of anion is C_X and the *degree of ionization of this anion is α_X*. Because the anion displaces the anions in the eluent as it passes from the column to the detector, we can state that the concentration of eluent is

$$C_E - C_X \alpha_X$$

Let us consider the nature of the conductance G during chromatographic elution of the salt MX_n such that the following relationship holds:

$$G_{total} = G_{eluent} + G_{anion}$$

Upon substituting the equation that relates limiting equivalent conductances for separate ions to the conductance of a solution, we have

$$G_{total} = \frac{\left(\lambda_{E^+} + \lambda_{E^-}\right)\left(C_E - C_X \alpha_X\right)\alpha_E}{1000k} + \frac{\left(\lambda_{E^+} + \lambda_{X^-}\right)C_X \alpha_X}{1000k}$$

At the moment the anion is detected, there is a change in the total conductance such that

$$\Delta G = G_{total} - G_{backgd}$$

The background conductance can be expressed in terms of the eluent cation and anion, respectively, and when subtracted from the total conductance, leads to an important relationship according to

$$\Delta G = \left[\frac{\left(\lambda_{E^+} + \lambda_{X^-} \right) \alpha_X - \left(\lambda_{E^+} + \lambda_{E^-} \right) \alpha_E \alpha_X}{1000k} \right] C_X$$

This equation is the most general of all and accounts for both suppressed and nonsuppressed IC. Note that the degree of eluent ionization, α_E, has a significant effect on the magnitude of ΔG. In the case of suppressed IC, the eluent cation, E^+E^-, is converted to H^+E^-, while the analyte salt MXn is converted to highly conductive H^+X^-. The creative idea here was to recognize that if E^+E^- could be made from a moderate to strong base, such as NaOH, Na_2CO_3, or $NaHCO_3$, then H^+E^- that is formed in the suppressor is rendered minimally conductive.

If the eluent and anions are fully ionized, the above equation can be further simplified to give us a more direct relationship between the change in conductance and anion concentration:

$$\Delta G = \frac{\left(\lambda_{A^-} - \lambda_{E^-} \right) C_X}{1000k}$$

We thus see that the magnitude of the change in conductance depends not only on anion concentration, but also on the difference between the limiting equivalent conductances of the anion and the eluent ion.[106]

The idea behind eluent suppression is to convert the moderately conductive Na^+ in the eluent to the weakly conductive H_2CO_3 while converting the low-conducting analyte NaX to the high-conducting HX. Limiting equivalent conductivities, denoted by λ, of some ions in aqueous solution at 25°C are shown in Table 4.30.[107]

It is this author's opinion that the development of the micromembrane suppressor was the key advance that stabilized the background conductivity. A schematic of this suppressor is shown in Figure 4.98. The suppressor can be viewed as a "sandwich," whereby the "bread" represents eluent and regenerant flow in opposite directions and the "meat" is a semipermeable cation exchange membrane. This membrane allows H^+ to pass through and neutralize the carbonate eluent by converting it to the weakly conductive H_2CO_3. Figure 4.99 shows several ion chromatograms obtained by the author during the late 1980s. The top IC chromatogram shows

TABLE 4.30
Limiting Equivalent Conductivities for Selected Anions and Cations

Anion/Cation	λ(mho-cm²/Eq)
H^+	350
OH^-	198
SO_4^{2-}	80
Cl^-	76
K^+	74
NH_4^+	73
Pb^{2+}	71

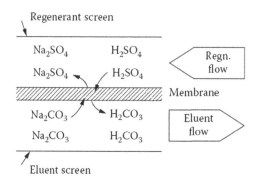

FIGURE 4.98

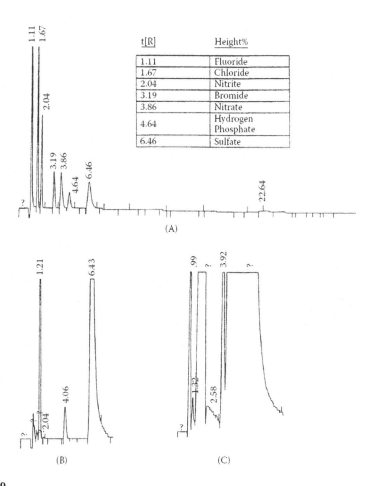

t[R]	Height%
1.11	Fluoride
1.67	Chloride
2.04	Nitrite
3.19	Bromide
3.86	Nitrate
4.64	Hydrogen Phosphate
6.46	Sulfate

FIGURE 4.99

the separation and detection of all seven of the common inorganic anions in less than 7 min using an AS-4 anion exchange column (Dionex) with an eluent consisting of 17 mM NaHCO$_3$/ 18 mM Na$_2$CO$_3$, with a flow rate of 2.0 mL/min. Referring back to Figure 4.95, this Model 2000I (Dionex) was connected to an electronic integrator at the time to generate the printout shown in Figure 4.99. The top right table lists retention times for the seven anions listed. The bottom left

IC chromatogram is that of a typical groundwater sample that identifies chloride and sulfate. These ions are very common in groundwater, revealing an unidentified peak at a retention time of 4.06 min. In the bottom right chromatogram, a second groundwater sample that happened to be preserved with sulfuric acid is shown. This IC chromatogram indicates that preserving aqueous samples from the environment for subsequent analysis by IC with H_2SO_4 is not a good idea if you want to measure the trace concentration of all the common anions found in groundwater!

Today, readers of this book will likely encounter the IC determinative technique by being asked to implement EPA Method 300.1 Revision 1 1993. The method includes two different sets of inorganic anions that are likely found in drinking water. The first set includes the common inorganic anions: bromide, chloride, fluoride, nitrite, nitrate, ortho-phosphate and sulfate. The second of analytes known as disinfection by-products includes: bromate, bromide, chlorite, and chlorate. Dichloroacetate ion serves as a surrogate for both sets of analytes. The past 30 years has seen great technological advances in the IC determinative technique. Metal ions as cations can also be separated and quantitated by IC. However, IC is not the principal determinative technique used today to quantitate metals in both enviro-chemical and in enviro-health TEQA.

4.77 IF IC IS NOT THE PRINCIPAL DETERMINATIVE TECHNIQUE TO MEASURE METALS IN THE ENVIRONMENT, WHAT IS?

Atomic emission spectroscopy (AES) and atomic absorption spectroscopy (AAS) are! In a manner similar to our discussion of molecular spectroscopy, where we compared UV absorption with UV excitation and subsequent fluorescence, these two determinative approaches are the principal ways to identify and quantitate trace concentration levels of metal contamination in the environment. As the need developed to quantitate increasing numbers of chemical elements in the periodic table, so too came advances in instrumentation that enabled this to be achieved at lower and lower IDLs. AES and AAS techniques are both complementary and competitive. Atomic fluorescence spectroscopy (AFS) is a third approach to trace metals analysis. However, instrumentation for this has not yet become widespread in environmental testing labs, and it is unlikely that one would see atomic, or what has become useful x-ray atomic, fluorescence spectroscopy. Outside of a brief mention of the configuration for AFS, we will not cover it here.

We begin this section of Chapter 4 with a discussion of *spectroscopy*, and it is useful for us to briefly survey all of molecular and atomic spectroscopy from the perspective of the electromagnetic spectrum. The entire electromagnetic spectrum is shown in Figure 4.100, along with those regions in the spectrum that are associated with different types of radiation. The various regions of the electromagnetic spectrum and their respective interactions with matter are cited below each region. The wavelength region starting from the near-UV (~200 nm) through to the near-infrared (~800 nm) is used in atomic spectroscopy, and it involves transitions of electrons between ground and excited electronic states. Spectroscopy, in general, has been defined as that branch of physical chemistry that considers the interaction of electromagnetic radiation with matter. Harris has described what it is that atomic spectroscopy does (p. 589):[19]

> *In atomic spectroscopy, samples are vaporized at very high temperatures and the concentrations of selected atoms are determined by measuring absorption and emission at their characteristic wavelengths. Because of its high sensitivity and the ease with which many samples can be examined, atomic spectroscopy has become one of the principal tools of analytical chemistry.*

The quantitative determination of trace concentration levels of the priority pollutant metals is a *major activity* in environmental testing laboratories. Instruments are designed to accept aqueous samples directly and convert these samples to aerosols or fine mists that are aspirated

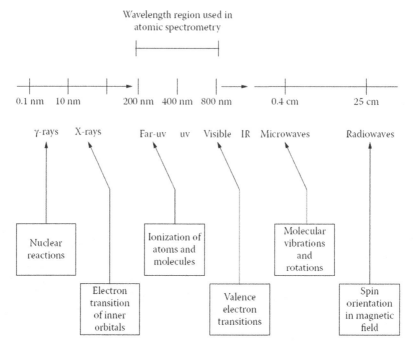

FIGURE 4.100

into flames or plasmas. Solids must be solubilized by vigorous use of strong acids and hydrogen peroxide. This oxidation results in an acidic/aqueous sample that can be subsequently aspirated into a flame or plasma or placed in a resistively heated graphite tube. A logic flowchart is shown in Figure 4.101, and the decision as to which determinative instrumental technique to use largely depends on the required IDLs, or in more familiar terms, "how low you can go."

The logic presented in Figure 4.101 should enable the reader to make the right decision given the available instrumentation. If reaching the IDLs as indicated in Figure 4.101 is not necessary, there exists a plethora of analytical methods for metal ions in aqueous solution based on metal chelation with subsequent quantitative analysis using a stand-alone UV-vis spectrophotometer. As introduced in Chapter 3, LLE or alternative column chromatographic techniques for sample prep are available to isolate or preconcentrate metal chelates from complex sample matrices. The reference work by Sandell and Onishi on the UV-vis spectrophotometry of metals is the most complete work of which this author is aware.[108] One experiment in Chapter 5 shows how one can quantitate the two oxidation states of the element iron, namely, Fe(II) and Fe(III), by forming a metal chelate and using a stand-alone UV-vis spectrophotometer.

4.78 HOW DO I CHOOSE WHICH ATOMIC SPECTRAL METHOD TO USE?

Most alkali and alkaline–earth metal ions that are found dissolved in water are readily quantitated by flame atomic emission spectroscopy (Fl-AES). This determinative technique has been in the past termed flame photometry. A simple photometer uses cutoff filters to isolate the wavelength, denoted by λ, with the strongest emission intensity that usually originates or terminates from a resonance line. Resonance transitions are those originating from or terminating to the ground electronic state. For example, the *intensity of the 589/590-nm resonance atomic emission doublet for sodium* is used to quantitate Na in aqueous samples. Contemporary atomic absorption spectrophotometers incorporate a monochromator and are used in the Fl-AES mode by merely

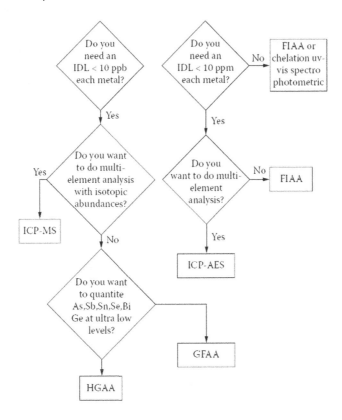

FIGURE 4.101

not using the hollow-cathode lamp. The transition or so-called heavy metal ions dissolved in water are more amenable to quantitative analysis using flame atomic absorption (FlAA), graphite furnace atomic absorption (GFAA), inductively coupled plasma-atomic emission (ICP-AES), or ICP integrated to a quadrupole mass filter. This relatively recent development is termed ICP-MS. Table 4.31 lists *instrument detection limits* (IDLs) for selective metals of interest to TEQA. The IDLs cited make it quite clear as to which atomic spectroscopic determinative technique is more appropriate.

The following example illustrates how one can decide among these several options. Consider a request from an environmental engineering firm involving the evaluation of a landfill that is suspected to contain the toxic element arsenic (As). This element is part metal and part nonmetal owing to its position in the periodic table. Atoms of the element As are usually found covalently bonded to oxygen either in a trivalent, denoted by As(III), or in a pentavalent, denoted by As(V), oxidation state. Arsenic speciation (i.e., the quantitative analysis of environmental samples) for both As(III) and As(V) is an important and current area of research interest. We will have more to say about metal speciation and its growing importance to TEQA later in this chapter. AES and AAS determinative techniques yield total As without regard to speciation. As is difficult to measure down to low IDLs by FlAA. The drinking water regulatory limit for As is set at 50 ppb. The IDLs listed in Table 4.31 for As suggest that of the several determinative techniques available, either GFAA with an IDL of 0.5 ppb, HGAA with an IDL of 0.03 ppb, or ICP-MS with an IDL of 0.006 ppb, if available, is the recommended determinative technique. Recently, considerations such as these for As have become critical in the author's laboratory. The decision as to which atomic spectroscopic instrument to choose in TEQA, as outlined in Figure 4.101, is analogous to the choice of which type of instrumental chromatographic separation determinative technique

TABLE 4.31
IDLs for the Most Common Atomic Spectrometric Determinative Techniques in ppb

Element	FlAA	GFAA	HGAA	ICP-AES	ICP-MS
Al	45	0.3		6	0.006
As	150	0.5	0.03	30	0.006
Bi	30	0.6	0.03	30	0.0005
Ca	1.5	0.03		0.15	2
Cd	0.8	0.02		1.5	0.003
Cr	3	0.08		3	0.02
Cu	1.5	0.25		1.5	0.003
Fe	5	0.3		1.5	0.04
Hg	300	15	0.009	30	0.004
K	3	0.02		75	1
La	3000			1.5	0.0005
Mg	0.15	0.01		0.15	0.007
Na	0.3	0.05		6	0.05
Pb	15	0.15		30	0.001
Sb	45	0.4	0.15	90	0.001
Se	100	0.7	0.03	90	0.06
Sn	150	0.5		60	0.002
Tl	15	0.4		60	0.0005
Zn	1.5	0.3		1.5	0.003

Note: Instruments used to generate these IDLs
• Model 5100 AA (PerkinElmer, PE)
• FIAS-200 flow injection amalgamation (PE)
• MHS-10 mercury/hydride (PE)
• Model 5100 PC AA with 5100 ZL Zeeman furnace model (PE)
• Model 4100 ZL AA
• Plasma 2000 ICP-AES (PE)
• ELAN 5000 ICP-MS (PE)
All IDLs are based on a 98% (3 σ) confidence level.

Source: Jenniss S., S. Katz, R. Lynch. *Applications of Atomic Spectrometry to Regulatory Compliance Monitoring,* 2nd ed. New York: Wiley-VCH, 1997, pp. 7–8.

to use, as outlined in Figure 4.4. In contrast to the way this author introduced column chromatographic principles and techniques earlier in this chapter, the early flame AES observations that led to principles are discussed first. Next, the ICP as the principal replacement for the flame is introduced. Following this, the principles that underlie the more recently developed ICP-MS determinative technique are discussed. Finally, we wrap up the topic of how to conduct trace metals QA with an excursion into the principles and practice of AAS.

4.79 WHAT HAPPENS WHEN VARIOUS INORGANIC SALTS ARE THRUST INTO FLAMES?

We start with the observations of Kirchoff and Bunsen in Germany in 1859. They observed the bright-line spectra for many alkali and alkaline–earth metal-based salts and are credited with the discovery of spectrochemical analysis. The so-called principal atomic emission series for the common alkali metals is shown in Figure 4.102. Note that these older atomic spectra are calibrated in terms of *wave number*, denoted by $\bar{v}$, of the emitted radiation, whose units are in reciprocal centimeters (denoted by cm⁻¹).

Wavelength is expressed in units of nanometers (denoted by nm). Frequency is expressed in units of cycles per second or Hertz (denoted by Hz). Frequency can also be viewed in terms of the number of wavelengths per second as $v = c/\lambda$. Can we convert between the two? From Planck's law, we know that the energy carried by these emitted photons is inversely related to wavelength λ, directly related to v, and directly related to $\bar{v}$; this is summarized as

$$E = hv = hc\bar{v}$$

A rule of thumb between *wavelength in nm and wave number in cm⁻¹* is given by

$$\lambda(\text{nm}) = \frac{1 \times 10^7}{v\left(\text{cm}^{-1}\right)}$$

We introduce the term *wave number* at this point because it relates directly to energy, and wave number is the principal unit used for the mid-infrared region of the electromagnetic spectrum. In Figure 4.102, the Na doublet resonance emission line occurs at $\bar{v} \sim 17{,}000$ cm⁻¹. Applying the above equation gives $v = 588$ nm, and this value is quite close to the 589/590 nm doublet.

The alkali metals all have a lone electron in an outermost s atomic orbital. The transition from the higher p atomic orbital in the excited state to this s orbital corresponds to an energy difference ΔE that can be attributed to release of a photon whose energy is hv. Quantum mechanical selection rules allow a change in the azimuthal quantum number $\Delta 1 = +1$ or $\Delta 1 = -1$. Each transition gives rise to a doublet because the spin $s = 1/2$ may couple either as $1 + s = 3/2$ or $1 - s = 1/2$ in the p state. The separation of the doublet of Na and K has been exaggerated in Figure 4.102. One of the more fascinating observations in the *introductory general chemistry laboratory* is to observe the various colors emitted when the salts NaCl, LiCl, KCl, $CaCl_2$, and $SrCl_2$ are thrust into a lab burner flame. Bright-line spectra result when this luminescence is directed through a prism or transmission grating such that the bright light is dispersed into its component λ values.

A partial energy-level diagram is shown in Figure 4.103 for the element Na. Note the two closely spaced levels for $3p$ electrons that account for the most intense 589/590 nm doublet. AE results from a relaxation of the electron from the excited $3p$ atomic orbital to the ground-state s atomic orbital, and this process is depicted as

$$3p \xrightarrow{\quad AE \quad} 3s$$

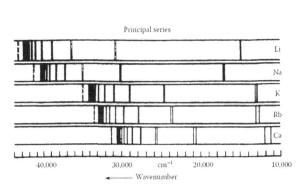

Figure 4.102

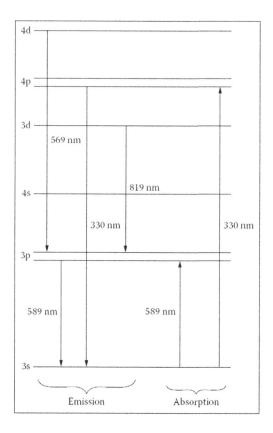

FIGURE 4.103

The energy emitted as photons according to Planck's law is depicted as

$$E_{3p} - E_{3s} = \Delta E = hv$$

Provided sodium vapor is in the path of this incident-visible radiation, these emitted photons are absorbed and a ground-state electron residing in an s atomic orbital is promoted to a $3p$ excited atomic orbital and the transition for this is depicted as

$$3s \xrightarrow{\text{AA}} 3p$$

This energy is absorbed and corresponds to

$$E_{3p} - E_{3s} = \Delta E = hv$$

This transition is also the most intense for Na among its other emission lines. In contrast to molecular spectra in solution, atomic spectra in the gas phase consist of very discreet and narrow wavelengths having very narrow bandwidths. For example, Zn has a λ at 213.9 nm and $\Delta\lambda =$ 0.002 nm.

The AES and AAS determinative techniques can be further classified according to the excitation temperatures attained. Table 4.32 provides a classification for most optical, atomic spectral methods of elemental analysis. Of the six AES and AAS determinative instrumental techniques cited, the

TABLE 4.32
Distinguishing among the Various Atomic Spectroscopic Determinative Techniques Used in Atomic Spectroscopy

Atomization	Atomization Temperature (°C)	Basis	Acronym
Flame	1700–3150	Absorption	AAS
		Emission	AES
Electrothermal	1200–3000	Absorption	GFAA
Inductively coupled argon plasma	6000–8000	Emission	ICP-AES
Direct-current argon plasma	6000–10,000	Emission	DCP
Electric arc	4000–5000	Emission	Arc-source emission
Electric spark	40,000(?)	Emission	Spark-source emission

first three are most likely to be found in environmental testing labs, and for this reason, we will consider only these. Of the three, the two most important to TEQA are ICP-AES and GFAA. FlAA is still important; however, its relatively high IDLs, particularly for the priority pollutant metals, serve to limit its usefulness. Fl-AES remains the analytical determinative technique of choice for the quantification of alkali and alkaline–earth metal ions found in aqueous samples. The degree to which a groundwater sample is categorized as *hard* is quantitatively determined by measuring the concentration of Ca and Mg via Fl-AAS as introduced in a student experiment in Chapter 5.

4.80   HOW DOES THE DESIGN OF AA, AE, AND AF INSTRUMENTS FUNDAMENTALLY DIFFER?

Figure 4.104 depicts the simplest instrument configurations for AE (top), AA (middle), and AF (bottom). Note how the positions of the flame source, optics, and detector differ among all three. The analyst should have this simple schematic in mind when approaching atomic spectroscopic instrumentation. With respect to the top schematic, instrumentation has been advanced that has sought to maximize the emitted intensity of a particular wavelength while minimizing the background emission at that wavelength. For Fl-AES or ICP-AES, it is the emitted line intensity of a particular wavelength against this background-emitted intensity that forms the basis of TEQA. This includes the analysis of environmental samples for priority pollutant metal ions that follows the process of calibration, verification of the calibration (i.e., running sufficient ICVs; see Chapter 2), interpolation of the calibration curve (quantitative analysis), and establishing the IDL for a given metal analyte. However, in today's environmental testing laboratories, one is more likely to find ICP-AES and / or ICP-MS.

4.81   WHAT IS ICP-AES?

This author has been searching for a good description of the ICP for this book. Those affiliated with the PerkinElmer Corporation, at the time, describe the ICP very well:[109]

The ICP discharge used today for optical emission spectrometry is very much the same in appearance as the one described by Velmer Fassel in the early 1970s. Argon gas is directed through a torch consisting of three concentric tubes made of quartz or some other suitable material ... A copper coil, the load coil, surrounds the top end of the torch and is connected to a radio frequency (RF) generator ... When RF power, typically 700–1500 watts, is applied

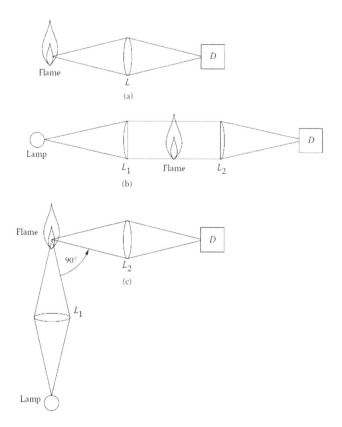

FIGURE 4.104

to the load coil, an alternating current moves back and forth within the coil, or oscillates, at a rate corresponding to the frequency of the generator. In most ICP instruments this frequency is either 27 or 40 MHz. This RF oscillation of the current in the load coil causes RF electric and magnetic fields to be set up in the area at the top of the torch. With argon gas being swirled through the torch, a spark is applied to the gas causing some electrons to be stripped from their argon atoms. These electrons are then caught up in the magnetic field and accelerated by them. Adding energy to the electrons by the use of a coil in this manner is known as inductive coupling. These high-energy electrons in turn collide with other argon atoms, stripping off still more electrons. This collisional ionization of the argon continues in a chain reaction, breaking down the gas into a plasma consisting of argon atoms, electrons, and argon ions, forming what is known as an inductively coupled plasma discharge. The ICP discharge is then sustained within the torch and load coil as RF energy is continually transferred to it through the inductive coupling process. The ICP discharge appears as a very intense, brilliant white, teardrop-shaped discharge ... At the base, the discharge is toroidal, or "doughnut shaped" because the sample-carrying nebulizer flow literally punches a hole through the center of the discharge. The body of the "doughnut" is called the induction region because this is the region in which the inductive energy transfer from the load coil to the plasma takes place ... the area from which most of the white light, the induction region, and into the center of the plasma gives the ICP many of its unique analytical capabilities.

Stanley Greenfield in England first published a report in 1964 on the use of an atmospheric pressure ICP for elemental analysis, whereas Velmer Fassel and colleagues made the earliest refinements, including nebulization of the aqueous sample.[110,111,112] The ICP-AES determinative

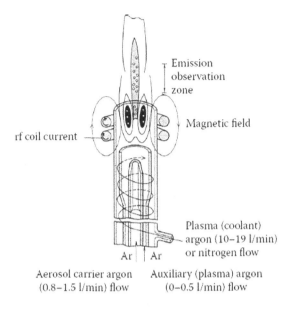

Emission observation zone

Magnetic field

rf coil current

Plasma (coolant) argon (10–19 l/min) or nitrogen flow

Ar   Ar

Aerosol carrier argon (0.8–1.5 l/min) flow

Auxiliary (plasma) argon (0–0.5 l/min) flow

FIGURE 4.105

method appeared in the early 1980s as EPA Method 200.7 for the determination of metals in drinking water. Figure 4.105 is a schematic from the original Fassel design and shows the three possible pathways up through the concentric cylindrical quartz ICP torch. An aqueous sample converted to an aerosol is taken up into the plasma via aerosol formation with Ar, and any atomic emission due to the metals in the sample is observed atop the torch, as shown. In an EPA contract lab that this author worked in during the 1980s, the ICP-AES instrument was operated continuously. A large tank containing liquefied Ar serves well when continuous operation of the ICP torch is anticipated. In the 1990s, in academia, this author used a compressed tank of Ar because of only intermittent use of the ICP-AES instrument. Figure 4.106 shows a simple schematic of a double-monochromator optical system that might be typical of mid-1980s technology for ICP-AES. The emitted radiation from the ICP is directed through a lens, L, and reflected off of a mirror, M_1. This reflected light enters slit S_1 and goes to mirror M_2, which reflects the light to a diffracting grating, G, off of M_2 a second time. This reflected light exits through slit S_2. An atomic emission spectrum for a mercury (Hg) source is also shown in Figure 4.106. A block diagram and a schematic drawing for a typical ICP-AES is shown in Figure 4.107.

4.82   HOW HAVE ICP-AES INSTRUMENTS EVOLVED?

Advances in design of ICP-AES instruments have led to systems that now incorporate an Echelle grating and a charged-couple detector as shown in Figure 4.108.[113] A comparison of the major instrumental components for an ICP-AES for instruments manufactured in the 1980s vs. those made in the 1990s, particularly during the latter part of the decade, is shown in Table 4.33.

4.83   WHAT HAPPENS TO ARGON GAS (AR) AND TO A METAL ANALYTE (M) IN THE PLASMA?

Electrons ionize the plasma gas according to

$$e^- + Ar \longrightarrow Ar^+ + 2e^-$$

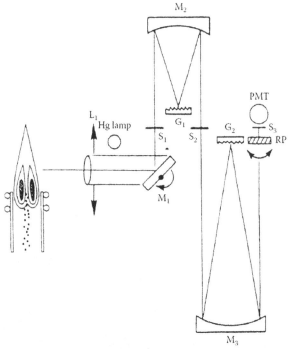

DOUBLE MONOCHROMATOR OPTICAL DIAGRAM

Emission spectrum of low-pressure mercury source
used for wavelength calibration

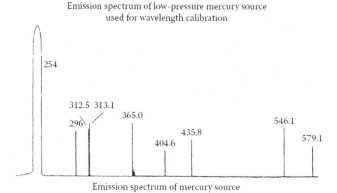

Emission spectrum of mercury source

FIGURE 4.106

In addition, recombination can occur according to

$$e^- + Ar^+ \longrightarrow Ar^* + h\nu$$

with the formation of excited Ar atoms, symbolized by Ar*, and a background emission at resonance lines of 104.8 and 106.7 nm, respectively.

The plasma source serves two roles in ICP-AES:

- To atomize the sample so as to free the metal analyte, usually in its ground state
- To partially ionize the metal analyte with excitation of both atoms and ions.

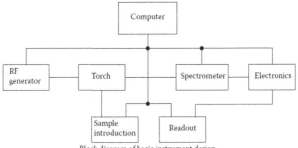

Block diagram of basic instrument design

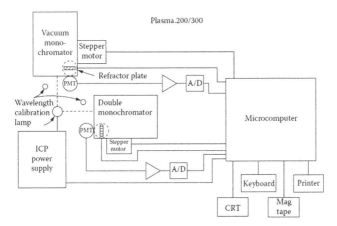

FIGURE 4.107

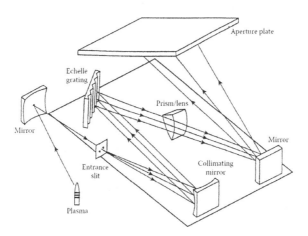

FIGURE 4.108

These processes are summarized in Table 4.34 where Ar_m represents a metastable atom and M represents the metallic analyte.[114]

When the sum of the metal analyte's ionization and excitation energies is below the ionization potential for Ar (i.e., ~16 eV), ionic lines become the most sensitive. Table 4.35 summarizes those elements that yield the most sensitive lines:[113]

TABLE 4.33
Comparison of ICP-AES Components between Decades

Component	1980s	1990s
Sample introduction	Pneumatic	Pneumatic, frit, ultrasonic, direct insert, thermospray, electrospray
RF generator	Piezoelectric	Solid state
Torch	Greenfield or Fassel type	Modified Fassel type
Plasma view	Side-on (radial)	Side-on (radial), end-on (axial), dual
Monochromator	Czerney–Turner or Ebert	Echelle, dispersion of λ values/prism, dispersion of diffraction orders
Detector	Photomultiplier	Solid state, charge transfer device
Signal processing	Analog, analog-to-digital, signal, then background	Analog-to-digital with storage of signal and background simultaneously

TABLE 4.34
Atomic Processes Occurring in the Inductively Coupled Plasma

Chief Ionization Processes	
Charge transfer	$Ar^+ + M \rightarrow M^{+^*} + Ar$
Electron-impact ionization	$e^-(fast) + M \rightarrow M^+ + 2e^-(slow)$
Penning ionization	$Ar^m + M \rightarrow M^{+^*} + Ar$
Chief Excitation Processes	
Electron-impact excitation	$e^- + M \rightarrow M^* + e^-$
Ion-electron radiative recombination	$M^+ + e^- \leftarrow M^* + hv$
Signal processing option 1	Analog, analog-to-digital, signal, then background
Signal processing option 2	Analog-to-digital with storage of signal and background simultaneously

TABLE 4.35
Sensitivity Differences between Ionic and Atomic Emission Line Intensities in ICP-AES

Elements whose ionic lines are most sensitive	Al, Ba, Be, Ca, Ce, Co, Cr, Fe, Hf, Hg, Ln, Lr, La, rare earths, Mg, Mo, Nb, Ni, Os, Pb, Sc, Sr, Ta, Th, Ti, U, V, W, Y, Zn, Zr
Elements whose atomic lines are most sensitive	Ag, As, Au, B, Bi, Ga, Ge, K, Li, Na, Rb, S, Sb, Se, Si
Elements whose atomic and ionic lines have similar sensitivities	Cu, Pd, Pt, Rh, Ni

4.84 CAN THE INTENSITY OF AN ATOMIC EMISSION LINE BE PREDICTED?

Yes, but we need to discuss relationships previously derived from statistical mechanics that relate to the population of atoms and ions among the various quantized electronic energy levels. Unless a solid sample is being directly introduced into the plasma by one of the direct insertion techniques, a metal ion dissolved in water, $M^{n+}_{(aq)}$, is most likely to be found. This hydrated metal ion becomes dehydrated in the plasma, atomized, and then undergoes excitation or ionization at the extremely high temperature of the plasma. The use of double arrows below depicts the various equilibria occurring in the plasma as shown below:

$$M^{n+}_{(aq)} \longleftrightarrow M^{n+}_{(g)} \longleftrightarrow M^{n+\ast}$$

$$M^{\ast}_{(g)} \longleftrightarrow M_{(g)}$$

The high temperature achieved in the ICP favors formation of analyte ions over formation of analyte atoms (90% conversion to ions), as indicated by the thicker arrow.

The ratio of the number of metal atoms that have their outermost electrons in the *excited quantized energy level* to the number of metal atoms that have their outermost electrons in the *ground quantized energy level* ($N_{excited}/N_{ground}$) and how this ratio is influenced by the temperature in the plasma are governed by the Boltzmann distribution. This well-known equation from statistical mechanics is

$$\frac{N_{excited}}{N_{ground}} = \left(\frac{g_{excited}}{g_{ground}}\right) e^{-\Delta E/kT}$$

where $g_{excited}$ is the degeneracy of the excited state and g_{ground} is the degeneracy of the ground state. Degeneracy is the number of states available at each quantized energy level and $g = 2J + 1$, where J is the total angular momentum quantum number. ΔE is the energy-level spacing, k is Boltzmann's constant (1.38×10^{-23} J/K) or, in terms of wave number, $k = 0.695$ cm^{-1}), and T is the absolute temperature in degrees Kelvin. The total population among all quantized atomic energy levels is given by

$$N = n_0 + n_1 + n_2 + \cdots + n_i + \cdots$$

The total number of metal analyte atoms can be considered as being distributed among or partitioned into the various quantized energy states according to a partition function denoted by Z and can be written as

$$Z = g_0 + g_1 e^{-E_1/kT} + g_2 e^{-E_1/kT} + \cdots + g_i e^{-E_1/kT} + \cdots$$

The fraction of M atoms whose electrons are in the jth excited state becomes

$$\frac{n_j}{N} = \left(\frac{g_j}{Z}\right) e^{-Ej/kT}$$

The extent that M atoms are ionized in the plasma can be predicted from applying the Saha equation, and after manipulation of the basic relationship, it can be written as follows:[113]

$$\frac{N_{ij}N_e}{N_{aj}} = \left(\frac{(2\pi m_e kT)^{2/3} 2Z_{ij}}{h^3 Z_{aj}}\right) e^{-E_i/kT}$$

TABLE 4.36
Definition of the Various Parameters When Using the Saha Equation

N_{ij} Number of ionic species j	E_i Ionization potential of atomic species j
N_{ai} Number of atomic species j	Z_{aj} Partition function of atomic species j
N_e Number of free electrons	Z_{ij} Partition function of ionic species j
M_e Mass of electron	Z_{aj} Partition function of atomic species j
h Planck's constant	T Ionization temperature
$\quad h = 6.626 \times 10^{-34}$ joule-seconds	
k Boltzmann's constant	

These parameters are defined in Table 4.36.

Consider the transition between a ground-state atom whose potential energy is E_0 and an excited state j whose potential energy is Ej. The intensity of an emitted wavelength in the plasma depends on the following factors:[113]

- The difference $E_j - E_0 = hv = \Delta E$, the number of metal atoms in the excited state, nj
- The number of possible transitions between Ej and E_0 per unit time
- The Einstein transition probability Aj such that the emitted intensity I is proportional to the product of ΔE and Aj such that

$$I \propto \Delta E A_j$$

Solving the Boltzmann distribution for the number of atoms in the excited state, N_{aj}, and substituting for N_{aj} leads to the outcome of this section and answers the question posed:

$$I = \Phi \left(\frac{hcg_j A_j N}{4\pi\lambda Z} \right) e^{-\Delta E/kT}$$

where Φ is a coefficient to account for emission being isotropic over a solid angle of 4π steradians. For the AE intensity to be directly proportional to the number of ground-state atoms, N, as shown above, all other terms in the above equation must remain constant. Therefore,

$$I \propto conc$$

and this relationship makes it possible to conduct TEQA.

We assume for purposes of TEQA that a certified reference standard such as for Cr when aspirated into the plasma exhibits nearly identical parameters as that for Cr in the environmental sample. This requires good stability of the source characteristics (i.e., forward power and gas flow rates). A calibration plot for Cr is shown in Figure 4.109; this plot was obtained using a Plasma 2000® ICP spectrometer (PerkinElmer) from the author's laboratory. The upper plot shows the raw calibration data using the *common 267 nm line for Cr*, and the lower plot shows how well these data points fit to a least squares regression line using the trend line in Excel.

4.85   WHY DO THE AE LINES BROADEN?

Two factors contribute to AE band broadening. The first is due to the *motion of M atoms* in the plasma, a so-called Doppler effect, and the second is broadening due to *collisions*. Doppler

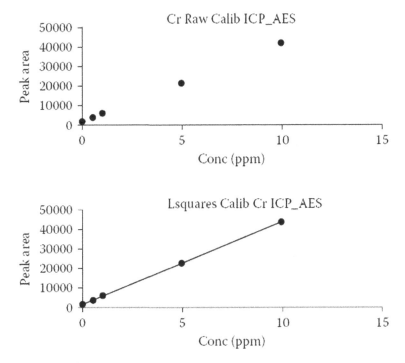

FIGURE 4.109

broadening, symbolized by $\Delta\lambda_D$, depends on the wavelength chosen, the kinetic temperature, and the molecular weight of the metal of interest and can be predicted as follows:[113]

$$\Delta\lambda_D = 7.16 \times 10^{-7} \lambda \sqrt{\frac{T_{min}}{M}}$$

Values for $\Delta\lambda D$ are low for heavy elements (e.g., Au(II) at 200.08 nm has a value of 0.8 pm), whereas $\Delta\lambda D$ values are high for light elements, such as Be(II) at 313.11 nm that has a value of 5.9 pm. Collisional broadening results from collisions among analyte ions, atoms, and neutral Ar atoms and has also been called pressure broadening. Doppler broadening is dominant near the center of the band, whereas collision broadening dominates near the tails. AE line widths dictate what resolution is needed to resolve one AE emission line from another. For each transition metal, there is a plethora of lines to consider.

4.86 WHAT IS ICP-MS AND HOW DOES THIS DETERMINATIVE TECHNIQUE COMPLEMENT AND COMPETE WITH ICP-AES?

The coupling of the ICP source with a single quadrupole mass spec via developments in the design of a suitable interface between sample introduction, the ICP torch itself, and a quadrupole mass spec "opened the flood gates" while expanding the scope of trace metals analysis. Laboratories that already have ICP-AES capability are looking to expand their analytical services to include ICP-MS. ICP-MS offers a significant *increase in sensitivity* (refer again to Table 4.31), as well as providing *isotopic abundance* data. The ICP torch is rotated to the horizontal position that enables an aqueous aerosol to be introduced into the torch in the conventional way; however, instead of energy from the plasma being used to excite ground-state electronic levels that lead to atomic emission spectra, metal ions enter a tiny orifice via a sampler/skimmer and enter the quadrupole rods, where rapid scanning enables mass-selective detection. Skoog and Leary have introduced the principle in this manner (p. 457):[25]

> *Positive metal ions, produced in a conventional ICP torch, are sampled through a differentially pumped interface linked to a quadrupole spectrometer. The spectra produced in this way, which are remarkably simple compared with conventional ICP spectra, consist of a simple series of isotope peaks. These spectra are used for quantitative measurements based upon calibration curves, often with an internal standard. Analysis can also be performed by the isotope dilution technique.*

The benchmark papers for early investigations of the ICP as an ion source for mass spec were published in the late 1970s.[114] A more detailed treatment of this topic can be found elsewhere.[115]

Figure 4.110 is a schematic diagram of a commercially available ICP-MS, with the exception of the mass spec. This drawing shows the various ways to introduce a sample into an ICP-MS instrument. Note that a GC, SFC, LC, IC, and CE can serve as important up-front separation techniques. This is in addition to the more conventional flow injection approach to sample introduction. Note also how compressed argon gas not only is used to sustain the plasma, as discussed earlier, but also is used as the source for nebulizer gas. Note also that the sample aerosol that enters the plasma gets introduced to the mass spec via a sampler under conditions of atmospheric pressure. The region between the sampler (closest to the torch) and the skimmer (closest to the mass spec) is a partially evacuated chamber in which 99.9% of unwanted species are removed to the rough pump. Figure 4.111 depicts just how an HPLC could be interfaced to an ICP-MS. This instrument was used to conduct trace arsenic speciation.[116] The drawing includes important terms used to describe the more significant features of the instrument. Budde has articulated sample introduction to the ICP-MS in the following manner (p. 331):[14]

> *Liquids are injected first into a nebulizer, which produces an aerosol spray into a spray chamber. Large droplets and condensate are removed in the spray chamber, and the fine aerosol particles are transported by the flowing Ar gas into the ICP torch. The spray chamber is similar to the desolvation chamber used with the particle beam LC interface ... Efficiencies of transport of analytes from the sample to the ICP torch are only in the 1–5% range. If a low flow rate separation column is used, for example, a microbore LC column or CE column, a supplemental flow of liquid into the nebulizer can be used or the spray chamber can be by-passed and the nebulizer spray injected directly into the ICP torch.*

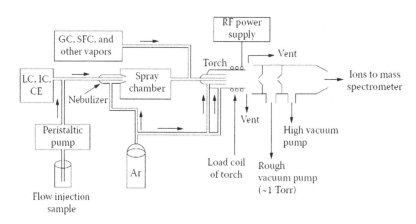

FIGURE 4.110

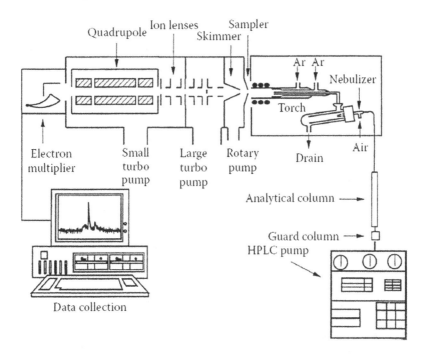

FIGURE 4.111

ICP-MS cannot (in contrast to ICP-AES) simultaneously detect and quantitate all the metal ions in a sample. However, ICP-MS provides for fast sequential acquisition using either scan or SIM modes (refer to earlier discussions). Mass spectra of metal ions are much simpler to interpret than the numerous lines in atomic emission spectra. Isotopic abundances for each element enable correct identifications to be made, and this advantage of ICP-MS has the potential to further widen the scope of trace metals analysis to, for example, the geological sciences. In addition to single quadrupole operation, both high-resolution MS (HRMS) and tandem MS are used with ICP ionization, which leads to gains in selectivity, exact *m/z* data, and structural elucidation. EPA Method 6020B Revision 2, July 2014 is the most recent and updated approach to quantitating trace metals for *enviro-chemical* TEQA. Prior to implementing EPA Method 6020B, groundwater is filtered and preserved according to methods: 3005, 3015, 3031, 3050, 3051 and 3032. For oil and grease samples a solvent dissolution approach is used as discussed in EPA Method 3040 to prepare samples prior to quantification using Method 6020B. Table 4.37 lists each chemical element with its corresponding *m/z* values for each elemental isotope. EPA Method 6800 titled: *Elemental and Molecular Speciated isotope dilution mass spectrometry* provides useful insights to the scope and application of conducting TEQA using ICP-MS. Nixon and Moyer show applicability of ICP-MS to *enviro-health* TEQA.[117]

4.87 WHAT INTERFERENCES ARE PRESENT AND HOW ARE THESE OVERCOME?

Ions from the plasma gas, matrix components, or the solvent acid used to dissolve the sample have been recognized as the major spectral interferences generated in the practice of ICP-MS. The interferences are termed *isobaric* since they have the same *m/z* as the analyte of interest. Table 4.38 lists the known isobaric interferences associated with ICP-MS.[118]

Development of collision/reactor cells placed between the interface and quadrupole rods for the most part eliminated many of these spectral interferences. Interferences can be removed in

TABLE 4.37
Recommended Elemental Isotopes for Selected Chemical Elements According to EPA Method 6020B

Aluminum	27
Antimony	121, 123
Arsenic	75
Barium	138, 137, 136, 135, 134
Beryllium	9
Bismuth (IS)	209
Cadmium	114, 112, 111, 110, 113, 116, 106
Calcium (I)	42, 43, 44, 46, 48
Chlorine (I)	35, 37, (77, 82)a
Chromium	52, 53, 50, 54
Cobalt	59
Copper	63, 65
Holmium (IS)	165
Indium (IS)	115, 113
Iron (I)	56, 54, 57, 58
Lanthanum (I)	139
Lead	208, 207, 206, 204
Lithium (IS)	6b, 7
Magnesium (I)	24, 25, 26
Manganese	55
Mercury	202, 200, 199, 201
Molybdenum (I)	98, 95, 96, 92, 97, 94, (108)a
Nickel	58, 60, 62, 61, 64
Potassium (I)	39
Rhodium (IS)	103
Scandium (IS)	45
Selenium	80, 78, 82, 76, 77, 74
Silver	107, 109
Sodium (I)	23
Terbium (IS)	159
Thallium	205, 203
Tin (I)	120, 118
Vanadium	51, 50
Yttrium (IS)	89
Zinc	64, 66, 68, 67, 70

ICP-MS by discriminating based on either kinetic energy or mass. For kinetic energy discrimination, a low reactive collision gas such as H_2 or He is bled into the cell. The cell consists of a quadrupole, hexapole, or octapole, usually operated in the RF-only mode. The *RF-only* field does not enable a mass selection to occur in contrast to a quadrupole. This difference enables a

TABLE 4.38
Isobaric Interferences in ICP-MS

Plasma Isobaric Interference	Interferes with a Determination of
$^{40}Ar^{16}O$	^{56}Fe
^{38}ArH	^{39}K
^{40}Ar	^{40}Ca
$^{40}Ar^{40}Ar$	^{80}Se
$^{40}Ar^{35}Cl$	^{75}As
$^{40}Ar^{12}C$	^{52}Cr
$^{35}Cl^{16}O$	^{51}V

focusing of ions prior to collision with reaction gas. For mass discrimination, dynamic reaction cell (DCR) technology has been developed. Thomas has described the DCR this way:[118]

In DCR technology, a quadrupole is used instead of a hexapole or octapole. A highly reactive gas, such as ammonia or methane, is bled into the cell, which is a catalyst for ion-molecule chemistry to take place. By a number of different reaction mechanisms, the gaseous molecules react with the interfering ions to convert them either into an innocuous species different from the analyte mass or a harmless neutral species. The analyte mass then emerges from the DCR free of its interference and is then steered into the analyzer quadrupole for conventional mass separation. The advantage of using a quadrupole in the reaction cell is that the stability regions are much better defined than a hexapole or octapole, so it is relatively straightforward to operate the quadrupole inside the reaction cell as a mass or bandpass filter, and not just an ion-focusing guide. Therefore, by careful optimization of the quadrupole electric fields, unwanted reactions between the gas and the sample matrix or solvent are prevented ... Every time an analyte and interfering ions enter the DCR, the bandpass of the quadrupole can be optimized for that specific problem and then changed on-the-fly for the next one.

Two DCR reactions are shown below that serve to eliminate an isobaric interference. In the first, the $^{40}Ar^+$ interference is eliminated in the determination of $^{40}Ca^+$, while in the second, the $^{56}ArO^+$ isobaric interference is eliminated in the determination of ^{56}Fe. These reactions are shown below:

$$Ar^+ + NH_3 \longrightarrow NH_3^+ + Ar$$
$$ArO^+ + NH_3 \longrightarrow NH_3^+ + O + Ar$$

We now seek to get a bit beyond the use of ICP-MS to quantitate metals in terms of a *total metal* content to the more recent developments in *trace metals speciation.*

4.88 HOW IS ICP-MS USED IN TRACE METALS SPECIATION?

The hyphenation of various HPLC techniques with ICP-MS, as shown previously in Figure 4.111, has made the largest impact. A summary of metal speciation is shown in Table 4.39 where SRM refers to the NIST's standard reference material:[116]

Figure 4.112 shows three chromatograms that illustrate the successful metal speciations obtained with HPLC-ICP-MS. In Figure 4.112A a separation of free Zn from zinc protoprophyrin

TABLE 4.39
Summary of How HPLC-ICP-MS is Used to Speciate Various Metals

Metal	Type of HPLC	Species and Sample Matrix
As	Anion exchange	As^{3+}, As^{5+}, dimethyl arsinate (DMA), monomethyl arsonate (MMA) in urine
As	Anion pairing	As^{3+}, As^{5+}, DMA, MMA, arsenobetaine in dogfish SRM
As	Micellar	As^{3+}, As^{5+}, DMA, MMA, As–betaine, As–choline in dogfish SRM
As	Anion exchange and reversed phase	As^{3+}, As^{5+}, DMA in urine
Cr	Anion exchange	Cr(III), Cr(VI)
Cr	Ion pairing	Cr(III), Cr(VI)
Sn	Reversed phase and micellar	Trimethyl tin chloride, trimethyl tin bromide, tripropyl tin chloride, monomethyl tin trichloride, dimethyl tin dichloride, trimethyl tin chloride standards
Sn	Ion pairing and ion exchange	Trimethyl tin chloride, tributyl tin chloride, triphenyl tin acetate standards
Hg	Reversed phase	Methyl mercury acetate, ethyl mercury chloride and Hg(II) chloride in tuna fish SRM, thimerosal in vaccines
Cd	Size exclusion	Cd in pig kidney
Pb	Reversed phase, cation exchange, ion pairing, size exclusion	Inorganic lead, tetraalkyl lead, tri- and dialkyl lead
P, S	Ion pairing	P and S in inorganic phosphates and sulfates, organic phosphates and amino acids
V, Ni	Anion exchange	V(III), V(V), Ni(II)
Au	Ion pairing and size exclusion	Dicyanogold (I) in blood and urine
I	Reversed phase	Iodide ion and five iodo amino acids

from the whole blood of a lead poisoned patient using RP-HPLC-ICP-MS. In Figure 4.112B, IP-RP-HPLC-ICP-MS was used to separate Hg and Pb species. A microbore PEEK® column packed with C18 chemically bonded silica was used. The mobile phase consisted of 5 mM sodium pentanesulfonate in 20:80 v/v acetonitrile-water at 1000 μL/min. Pb was detected at 0.2 pg and Hg at 7 pg. In Figure 4.112C, a separation of four arsenic standards again using IP-RP-HPLC-ICP-MS was achieved. This method has been demonstrated to speciate arsenic in urine. Emon's group in Germany report on successful speciation of As and Sb via (1) interfacing of HPLC with hydride generation and atomic absorption spectroscopy and (2) HPLC interfaced to ICP-MS.[119] Those interested in more details on speciation are referred elsewhere.[120]

"Once upon a time" in most laboratories, one or more atomic absorption spectrophotometers "dotted the landscape." Back then, EPA contract laboratories requirements were such that graphite furnace atomic absorption spectrophotometry (GFAA) was used to quantitatively determine the most toxic metals such as: As, Se, Tl, and Pb, while the emerging ICP-AES determinative technique was used to determine all the rest. Enviro-health laboratories continue to quantitate trace Pb human blood specimens using GFAA. Many state public health laboratories quantitate

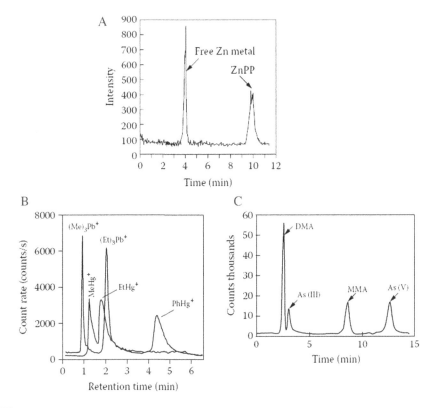

FIGURE 4.112

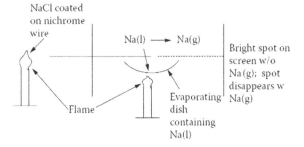

FIGURE 4.113

the presence of Pb in from 5,000 to 10,000 blood specimens annually. For this and for historical reasons, we delve into *atomic absorption spectroscopy.*

4.89   WHAT IS ATOMIC ABSORPTION?

It is the phenomenon by which resonance atomic emission lines are absorbed by atoms in the vapor phase; the phenomenon adheres to Beer's law, and hence forms the basis for TEQA involving *environmental samples that contain heavy metal residues.* Kirchoff in the mid-1800s is credited with conceiving and performing the classical experiment that first observed atomic absorption. A simple schematic of the experiment is shown in Figure 4.113. The yellow light emitted when a NaCl-coated nichrome wire is thrust into the flame was directed to a screen with a small hole. Observations of the brightness of the yellow light, reflected off of a solid back screen that disappears when elemental sodium is heated in an evaporizing dish above a second burner, must have surprised Kirchoff! The resonance emission of the characteristic 589 nm yellow light was completely absorbed by the Na vapor, explaining the absence of light at the screen and was dubbed *atomic absorption*. Contemporary sources from resonance atomic emission lines of many elements emit in the colorless UV region of the electromagnetic spectrum. The photomultiplier tube (PMT) can sense this colorless UV radiation, enabling atomic absorption to "cast a wide net" to include most *metals and metalloids*, like As and Se, in the periodic table.

Atomic absorption (AA) as an analytical technique remained dormant until Walsh in Australia and Alkemade and Milatz in the Netherlands conceived the notion and published their findings in the same year![121] Varian Associates licensed the technology from Walsh, and until the company's demise in the early 2010s, AA spectrophotometers were made in Australia. The *original experiment of Kirchoff* is repeated every time an analyst performs a quantitative analysis of a specific metal using an AA spectrophotometer. Significant modifications to the classical experiment have been made:

- The source of the resonance line has been replaced by hollow-cathode lamps (HCLs) or, in some cases, electrodeless discharge lamps (EDLs).
- The second burner has been replaced by a concentric flow nebulizer with an acetylene–nitrous oxide burner.
- The wavelength for the resonance line from the source is isolated from extraneous wavelengths due to interferences by a monochromator.
- The back screen and eyeball have been replaced by the photomultiplier tube and amplifying electronics.

This author has used AA spectrophotometers made by either PerkinElmer® or Varian® over the years. The advances in ICP-AES during the late 1970s and 1980s relegated flame atomic absorption spectrophotometers (FlAA) to the "back burner." The low IDLs enjoyed by contemporary graphite furnace atomic absorption spectrophotometers (GFAAs) have kept this technology relevant to the goals of TEQA. Before we focus our discussion on how GFAA is used to perform TEQA, let us compare AA instruments in general.

4.90   WHAT ELEMENTS OF THE AA SPECTROPHOTOMETER DO I NEED TO KNOW MORE ABOUT?

Single- vs. double-beam design and the necessity for an HCL is the answer. Figure 4.114 (top) is a schematic of a single-beam AA configuration showing the single beam of resonance line radiation emanating from an HCL, passing through a flame atomizer and into the entrance slit of a monochromator. Figure 4.114 (bottom) shows how this basic configuration is modified to accommodate a double-beam configuration. A rotating chopper is a semitransparent device that directs a portion of the HCL resonance line radiation to become diverted, pass through the atomizer and on to the entrance slit to the monochromator, and comprise the sample beam while the transparent portion bypasses the atomizer directly to the monochromator. The photo multiplier tube (PMT) in the double-beam instrument alternatively sees sample intensity/reference intensity. This comparison yields a very stable noise level over time. When an aqueous solution is aspirated into the flame, or when a microdrop is placed in the furnace, the transmitted power decreases in comparison to the reference power and a good signal-to-noise ratio can be realized. For this reason, double-beam AA instruments are preferred, but are also more expensive. Figure 4.115 shows how similar and how different the major components are for both FlAA and GFAA.

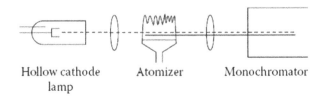

Hollow cathode lamp   Atomizer   Monochromator

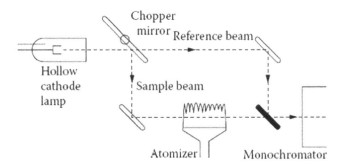

FIGURE 4.114

It would seem that an ordinary deuterium lamp that emits a continuum spectrum would make a low-cost radiation source for AA, provided the source is passed through a monochromator whereby a narrow bandpass of wavelengths can be selected. However, in order for Beer's law to be obeyed, the *bandwidth at half height, $\Delta\lambda_{1/2}$, for the monochromator should be less than $\Delta\lambda_{1/2}$ for the analyte of interest*—in this case, the atoms in the vapor phase of metallic elements. Robinson (p. 320)[122] has articulated the problem this way:

> *With the use of slits and a good monochromator, the band falling on the detector can be reduced to about 0.2 nm. If a band 0.002 nm wide were absorbed from this, the signal would be reduced 1%. Since this is about the absorption line width of atoms, even with complete absorption of radiation by atoms, the total signal would change by only 1%. This would result in an insensitive analytical procedure of little practical use. The problem of using such narrow absorption lines was solved by adopting a hollow cathode as the radiation source.*

4.91   WHY IS THE GFAA A SUITABLE TRACE DETERMINATIVE TECHNIQUE?

The relatively low IDLs listed in Table 4.31 suggest that GFAA is an ideal determinative technique for TEQA involving most metals. Contemporary GFAA instrumentation incorporates an autosampler that delivers a microdrop of sample directly into the center of a cylindrically shaped graphite tube via a predrilled hole midway up the length of the tube. Because the total amount of sample is available and when combined with the ability of the instrument to emit the resonance line associated with the HCL or EDL and then absorb this narrow band of UV radiation across a path length of a few centimeters, an appreciable absorbance is measured from a very low analyte concentration. This is in contrast to the concentric flow nebulizer used in FlAA, whereby most of the sampler volume (~90%) is drawn to waste. For the highest-quality analytical determination, it is suggested that pyrolytically coated graphite tubes be used. Prior to use, a L'vov (after the Russian scientist) platform is inserted into the graphite tube in the manner shown in Figure 4.116. The purpose of using a L'vov platform along with the additional expense is to isolate the sample

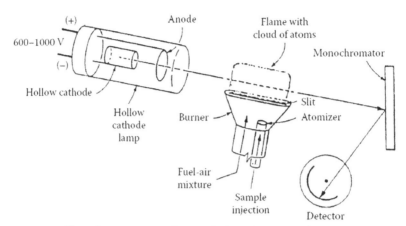

The major components of an atomic absorption spectrophotometer

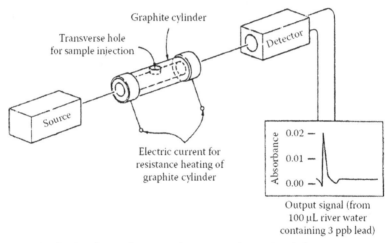

Graphite furnace for atomic absorption analysis and typical output

FIGURE 4.115

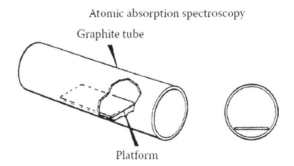

FIGURE 4.116

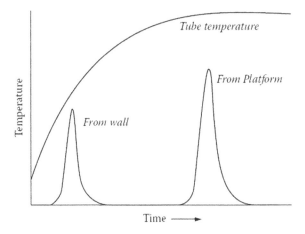

FIGURE 4.117

TABLE 4.40
Ratio of the Number of Atoms in an Excited State to the Number of Atoms in the Ground State at 3,000°K

Element	λ (nm)	N_{aj}/N_0 at 3000 K
Cs	852.1	0.007
Zn	213.8	1×10^{-10}
Na	590	4×10^{-4}

from the tube walls to allow more reproducible atomization of the sample through indirect heating. The platform enables a higher atomization temperature to be reached, as is shown in Figure 4.117, and this produces an abundance of free atoms, which reduces interferences. Pyro-coated tubes can also be purchased with the platform already built in.

4.92 WHAT DETERMINES THE STRENGTH OF ABSORPTION FOR A GIVEN METAL ANALYTE?

In a manner similar to our discussion of atomic emission, the Boltzmann distribution again needs to be considered, and the ratio of the number of atoms in an excited state to the number of atoms in the ground state is reiterated as follows:

$$\frac{N_{aj}}{N_0} = \frac{g_j}{g_0} e^{-E_{aj}/kT}$$

The maximum sensitivity obtained when 100% of all atoms are in their ground state is $N_{aj}/N_0 \sim 0$; this ratio is low when the temperature is low and E_{aj} is low. Table 4.40 examines how the Boltzmann distribution varies for three specific metals: Cs, Zn and Na.

Because most atoms are in their ground state, the atomic absorption signal depends on the number of ground-state atoms, N_0, and T has minimal influence, unlike AES, where changes in T cause significant changes in N_{aj}/N_0, as discussed earlier. Temperature-dependent chemical reactions may occur that influence the number of gaseous atoms formed, and this leads to chemical interferences.

TABLE 4.41
Oscillator Strengths for Selected Metals

Element	λ(nm)	f
Ca	422.7	1.75
Cr	357.9	0.3
Cu	324.7	0.32
Na	589	0.57
	589.6	0.655
Tl	276.8	0.3
Zn	213.9	1.5

We proceed now to take the notion of atomic absorption for GFAA a bit further. (Benchmark papers are provided in Walsh.[121]) A consideration of Einstein probability coefficients for the simple concept of a transition from the ground electronic state to the first excited state results in a relationship between the absorbance A and more fundamental parameters, assuming that $\Delta\lambda_{1/2}$ for the monochromator is less than $\Delta\lambda_{1/2}$ for the analyte of interest:

$$A = \frac{3.83 \times 10^{-13} f_{01} b N_0 (t)}{\Delta\lambda_{\text{eff}}}$$

where f_{01} is the oscillator strength, b is the path length, $N_0(t)$ is the number of ground-state atoms in the atomic vapor at time t, and $\Delta\lambda_{\text{eff}}$ is the width of a rectangular absorption profile that has the same k area and peak value as the light sources. Oscillator strengths for selected metals are given in Table 4.41.

4.93 WHY IS IT IMPORTANT TO CORRECT FOR BACKGROUND ABSORPTION?

To achieve good precision and accuracy (see Chapter 2) in quantitating trace concentration levels of various metals of interest. Robinson (pp. 338–341)[122] has suggested that unknown samples, whose residual metal content is of interest to TEQA, the calibration standards, and the initial calibration verifications (ICVs) should be as "similar as possible." To achieve this similarity, the following considerations to both FlAA and GFAA become important:

- Use identical sample matrices for samples and standards.
- Introduce a predominant anion, such as sulfate or chloride, at the same concentration in the matrix.
- For FlAA, use the same fuel–oxidant mixture, sample pressure, and flame height.
- Use the same type of background correction technique.

The metallic analyte is converted to atomic vapor in four steps:

$$M^{n+}_{(aq)}\, X^-_{(aq)} \longrightarrow MX_{n(s)} \longrightarrow M_2O_{n(s)}$$
$$M_{(g)} \longleftarrow M_{(s)} \longleftarrow$$

Butcher and Sneddon describe the conditions for GFAA in a most interesting manner (pp. 12–20):[123]

A graphite tube is typically 20 to 30 mm in length and 3 to 6 mm in diameter. The tube is surrounded by argon to prevent combustion in air at elevated temperatures. The sample is introduced into the furnace through a dosing hole (1–3 mm in diameter). In many cases, chemical compounds called chemical modifiers are also added to improve the sensitivity or accuracy for a given analyte. The temperature of the tube can be controlled from ambient up to approximately 2,700°C, with heating rates up to 1,500°C/sec. It has been shown to be beneficial for many elements to insert a platform into the tube onto which the sample is placed ... some metals may vaporize as a compound. The first step involves the relatively straightforward removal of solvent (usually water) from the sample. The remaining steps include chemical/ physical surface processes, such as homogenous or heterogeneous solid–solid interactions, solid-phase nucleation and diffusion into graphite; heterogeneous gas–solid interactions, that is, adsorption/desorption and reaction of molecules with the wall to form atoms; homogeneous gas-phase reactions; and processes by which analyte leaves the furnace.

The purpose of background correction in AAS is to accurately measure the background and subtract this absorbance value from the uncorrected signal (signal + background) to give a background-corrected signal. This is accomplished in AAS instrumentation in one of three major ways:

- By continuum source background correction, pioneered in 1965 by Koirtyohann and Pickett (see benchmark papers listed in Koirtyohann and Pickett[124].)
- By self-reversal, developed by Smith and Hieftje.[124]
- By exploiting the discovery by Zeeman that an intense magnetic field causes atomic energy levels to split, causing atomic lines to split into two or more components.[124]

Several monographs give more comprehensive discussions of these three distinct approaches to background correction.[125]

To conclude our discussion on GFAA, as we did for ICP-AES, a calibration plot for the element Cr (as total chromium) is presented in Figure 4.118, taken from the author's laboratory. We also close our atomic spectroscopy discussion by summarizing in Table 4.42 the four major techniques to measure trace levels of metals from environmental samples, and how to handle interferences. GFAA was the determinative technique in a very recent report on extracting, complexing, and quantitating the precious metal rhodium (Rh).[126] The presence of Pb in drinking water is quantitatively determined by measuring the concentration of Pb via GFAA in a student experiment shown in Chapter 5. Next, we consider two other important determinative techniques of relevance to TEQA: *infrared absorption spectroscopy* and *capillary electrophoresis*.

4.94 IN WHAT WAYS DOES IR ABSORPTION SPECTROSCOPY CONTRIBUTE TO TEQA?

Two principal applications of infrared (IR) absorption spectroscopy are useful to achieve the goals of TEQA:

- The quantitative determination of the extent to which soil, groundwater, surface water, and wastewater are contaminated with oil and grease.
- The quantitative determination of the extent to which an aqueous sample or aqueous soil leachate contains dissolved organic and inorganic forms of the element carbon.

The hyphenated instrumental technique in which a GC was interfaced to a Fourier-transform infrared (FTIR) spectrometer, via a gold-lined "light pipe," and given the acronym GC-FTIR

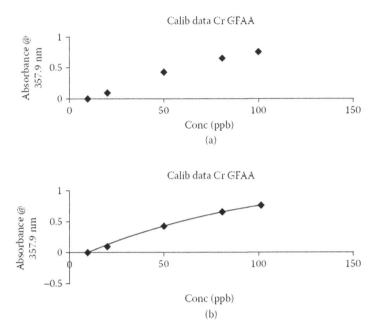

FIGURE 4.118

TABLE 4.42
Summary of Physical and Chemical Interferences in Atomic Absorption and Emission Spectroscopy

Atomic Spectral Technique	Type of Interference		Remedy
Flame AA	Ionization	⟶	Add a buffer to suppress
	Chemical	⟶	Add a releasing agent Use a C_2H_2–N_2O flame
	Physical	⟶	Dilute the sample, match the matrix, and use the standard addition mode of instrument calibration
Graphite furnace AA	Physical	⟶	Adjust furnace temperature conditions
	Molecular absorption	⟶	Zeeman or D_2 background correction
	Spectral	⟶	Zeeman background correction
ICP-AES	Spectral	⟶	Use alternative AE line or subtract out background
	Matrix	⟶	Use the internal standard mode of instrument calibration
ICP-MS	Spectral	⟶	Use a collision/reactor cell with low reactivity gases
	Spectral	⟶	Use DCR technology involving high reactivity gases

---

gained widespread acceptance in analytical chemistry during the 1980s. IDLs for the GC-FTIR determinative technique are just *not low enough to be of much use to TEQA*. We will only discuss the principles of IR absorption spectroscopy insofar as this instrumental technique pertains to the two applications cited above. We will also not address the use of IR absorption spectroscopy in air analysis. A stand-alone IR spectrometer using a conventional dispersive monochromator or an interferometer will be discussed in light of these important determinative techniques. Nondispersive IR instruments for both *oil and grease* and for *total organic carbon* are quite common and will be briefly introduced. One of the author's first laboratory tasks back in the early 1970s was to learn to operate a Total Organic Carbon (TOC) analyzer made by Beckman Instruments. During the '80s and '90s the author learned to operate an infrared absorption spectrophotometer and measured oil and grease in wastewater samples using an instrument made by the PerkinElmer Corporation.

Both determinative techniques cited above absorb radiation in the mid-IR region of the electromagnetic spectrum. The electromagnetic spectrum shown in Figure 4.100 places IR absorption with the longer wavelength, lower frequency, and lower wave number, in comparison to the visible region. IR radiation spans from just beyond the visible, known as the near-IR, beginning at $\lambda = 0.78 \, \mu m \left(\bar{\nu} = 13,000 \, cm^{-1} \right)$, through the mid-IR, 2.5 to 50 μm $\left(\bar{\nu} = 4000 - 200 \, cm^{-1} \right)$, and finally to the far-IR, beginning at $\lambda = 50 \, \mu m$ $\left(\bar{\nu} = 200 \, cm^{-1} \right)$ and ending at $\lambda = 1000 \, \mu m$ $\left(\bar{\nu} = 10 \, cm^{-1} \right)$.

Historically, mid-range IR absorption spectroscopy has been a valuable qualitative instrumental technique and has served the science of organic chemistry very well. The mid-IR spectrum is a qualitative property of an organic compound and provides important structural information. For example, if you want to distinguish between an aliphatic and an aromatic hydrocarbon, a mere record of their respective IR spectra, particularly around 3,000 cm^{-1}, reveals the difference. Those interested in a more in-depth introduction to IR absorption spectroscopy from the organic chemist's viewpoint should consult Shriner et al. (pp. 126–154)[69] or equivalent.

Photons whose λ values fall to within the mid-IR region of the electromagnetic spectrum are absorbed by organic molecules in both the solid and liquid phases and yield a characteristic absorption spectrum. If the photons of emitted IR radiation match the quantized energy-level spacing between the ground-state vibrational level denoted by ν_0 and the first excited vibrational level denoted by ν_1, a fundamental absorption is said to have occurred at that wavelength or wave number. IR absorption from the sun by glass windows in a closed automobile is a common example of how infrared radiation heats the atmosphere, and this phenomenon has helped coin the term *greenhouse effect*. Herschel performed a similar experiment in 1800 when he directed sunlight through a glass prism to a blackened thermometer as a detector, and now is credited with the discovery of IR.

4.95   HOW DO I QUANTITATE OIL AND GREASE IN WASTEWATER USING IR ABSORPTION?

The principles of sample preparation to *isolate and recover oil and grease* from a contaminated groundwater or wastewater sample involve LLE or RP-SPE sample prep techniques and were covered in Chapter 3. Both aliphatic and aromatic hydrocarbons (HCs) containing many *carbon-to-hydrogen covalent bonds* are present in oil and grease. Suppose some oil is found dispersed into surface waters due to some type of oil spill. Let us assume that a sample of this surface water has undergone either LLE or SPE and the recovered HCs are now dissolved in a solvent that lacks the C–H bond, such as 1, 1, 2-trichlorotrifluoroethane (TCTFE). The C–H bond possesses a dipole moment due to the slight difference in electronegativity between the elements carbon and hydrogen that changes upon absorbing IR radiation between 3,200 and 2,750 cm^{-1} in the mid-IR region. One such mode of vibration that absorbs IR radiation in this region is known as symmetric stretching and is depicted as follows:

$$\Delta E = hc\bar{\nu} = E_{\nu 1} - E_{\nu 0}$$

Where $\bar{\nu} = 2960 \, cm^{-1}$

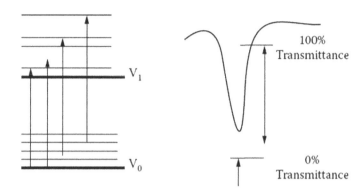

FIGURE 4.119

The energy-level spacing, ΔE, between the ground-state vibrational level and first excited-state vibrational level is depicted in Figure 4.119 along with a typical IR absorption band for the symmetric stretch just discussed. The fact that numerous quantized rotational levels exist within each vibrational level serves to help explain the broad-band nature of infrared spectra, along with the fact that samples are in condensed phases.

In contrast to UV-vis stand-alone spectrophotometers, dispersive IR spectrometers are designed so that the sample is positioned immediately after the radiation source and just before the dispersive device. This is not the case for an FTIR spectrometer. To assist in distinguishing between dispersive and FTIR instruments, a schematic of the two is presented in Figure 4.120. In a review article in 1968, Whetsel declared that "as far as instrumentation is concerned, I know of no major new developments on the horizon."[127] Then came the revolutionary development of FTIR spectroscopy based on the Michelson interferometer (he received the 1907 Nobel Prize for this invention). *FTIR designs have all but displaced dispersive designs* in infrared spectral technology today. However, the analyst is likely to find dispersive IR instruments that are dedicated to trace oil and grease determinations in environmental testing labs today. Quantitative IR analysis, as is true for UV-vis and atomic spectroscopy, is based on Beer's law. The experiment introduced in Chapter 5 provides sufficient detail to quantitatively determine the concentration of oil and grease in an environmental sample. Several texts provide good introductory discussions on IR absorption spectroscopy, including FTIR (Chapter 12),[25] (pp. 575–582),[19] (Chapter 5).[7]

This author has available to him a Model 1600® FTIR spectrometer (PerkinElmer), and this instrument utilizes software developed by PerkinElmer consistent with ASTM Method D 3921 and similar to EPA Method 431.2 ("Oil and Grease, Total Recoverable") and Method 418.1 ("Petroleum Hydrocarbons, Total Recoverable"). The idea that a nonpolar extracting solvent lacks the C–H stretch in the mid-IR originated from early work published by Gruenfeld at EPA.[128] That portion of the IR spectrum between 3,200 and 2,750 cm^{-1} of iso-octane dissolved in TCTFE is shown in Figure 4.121. This portion is sufficiently well resolved that distinct C–H stretching absorption bands can be seen for this group frequency. Almost 95% of all absorption in the region between 3,100 and 2,700 cm^{-1} is due to C–H stretching and includes the following:

- Asymmetrical stretching of a C–H bond in both a methyl and a methylene group. Both are present in iso-octane and centered at 2,962 cm^{-1}.
- Symmetrical stretching of a C–H bond in a methyl group centered at 2,872 cm^{-1}.
- Asymmetrical C–H stretching present in a methylene group at 2,926 cm^{-1}.
- Symmetrical C–H stretching present in a methylene group at 2,853 cm^{-1}.

A horizontal baseline is drawn, as shown above, to define a 100% T. The %T for an absorption as shown above is measured at the minimum of the absorption band. Beer's law is used in a

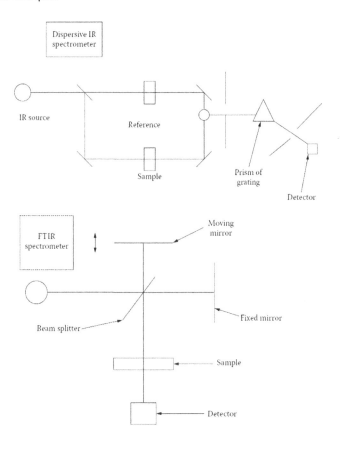

FIGURE 4.120

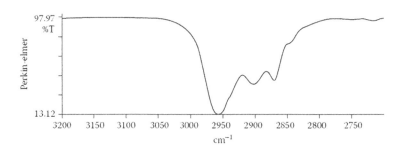

FIGURE 4.121

manner similar to that for UV-vis spectroscopy, and the measured %T is related to the absorbance according to

$$A = 2.0 - \log(\%T) = \varepsilon bc$$

Quantitative IR therefore requires a cuvette of fixed path length b. It so happens that the standard *rectangular quartz cuvette* absorbs everywhere in the mid-IR except in the C–H stretching region. This author, using a Model 567 Grating IR spectrometer (PerkinElmer) and a 10 mm rectangular quartz cuvette, developed the data and external calibration for petroleum hydrocarbons, as shown in Figure 4.122, using data shown in Table 4.43. We have used the same

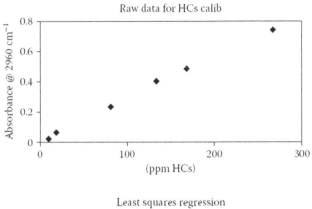

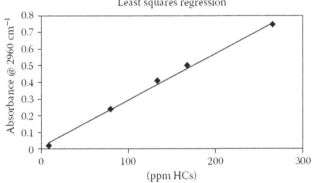

FIGURE 4.122

TABLE 4.43
Calibration Data and Quantitative Analysis for Unknown Samples for Trace Petroleum Hydrocarbons in Water

Standard No.	Concentration of HC (ppm)	Absorbance at 2960 cm⁻¹
1	7.99	0.021
2	79.9	0.235
3	133.2	0.492
4	266.4	0.735
Unknown Sample	**Found**	**Measured**
EPA reference	199.3	0.596
Blank	0.8	0.014
Groundwater	28.1	0.09
Soil A	23.0	0.076
Soil A spiked with HCs	213.7	0.607
Soil B	30.6	0.097
Soil C	21.2	0.071

Excel programming to generate the raw data calibration points in a manner similar to what we did for the GFAA and ICP-AES data discussed earlier. The analytical results for the unknown samples are real and were reported at the time. Note the influence of the spiked Soil A in comparison to the unspiked Soil A. Results for Soils B and C are given for comparison and serve to illustrate the utility of IR absorption spectroscopy toward achieving yet another goal of TEQA. One other determinative technique that utilizes IR absorption needs to be addressed: the determination of total organic carbon (TOC) in contaminated surface or ground water.

4.96   WHAT IS TOC AND HOW IS IR ABSORPTION USED?

The TOC content of a sample of water is an important and non-compound-specific parameter. It is measured by oxidizing the dissolved organic carbon (DOC) and quantitating the evolved carbon dioxide. One way to do this is to sweep the released CO_2 into a nondispersive IR (NDIR) photometer that has been tuned to measure CO_2. This is accomplished by measuring the absorbance of the characteristic C−O stretching vibration.

Two analytical approaches to the quantitative determination of TOC for aqueous samples that contain DOC have emerged over the years: *oxidation of carbon to CO2* via high-temperature catalyzed combustion and *persulfate–ultraviolet irradiation*. Teledyne-Tekmar (formerly Tekmar-Dohrmann) manufactures TOC analyzers (among others, including Beckman Coulter, OI Analytical, Shimadzu, Sievers, etc.). Teledyne-Tekmar's TOC analyzer performs *both types of oxidations*. Its most current models are the focus of this discussion. A TOC measurement involves oxidizing organic carbon in an aqueous sample, detecting and quantifying the oxidized carbon as CO_2, and presenting the results in terms of the mass of carbon per unit volume of the aqueous sample.

Let us clarify some of the terms used in this important and nonselective determinative technique:

Total carbon (TC) is the measure of all the carbon in the sample, both inorganic and organic, as a single parameter. Generally, the measurement is made by placing the sample directly into the analyzer without pretreatment.

Total organic carbon (TOC) is the sum of all the organic carbon in the sample.

TOC can be measured in one of two ways:

TOC measurement directly requires that inorganic carbon be removed by acidification and sparging. The DOC that remains is measured as TOC. Inorganic carbon and purgeable organic compounds (POCs) are lost. POCs are generally present at 1% or less of total carbon.

TOC measurement by difference requires two quantitative determinations: one to measure TC and one to measure inorganic carbon. The difference between these two measurements is rigorously TOC.

Inorganic carbon (IC) includes carbonate (CO_3^{2-}), bicarbonate (HCO_3^-), and dissolved carbon dioxide (CO_2). IC is determined in aqueous samples by acidifying with an inorganic acid to pH 3 or lower, and then sparging with a stream of inert gas. The acidification converts carbonates and bicarbonates to CO_2, which is then removed along with dissolved CO_2 by the gas stream and measured to provide an IC value.

Purgeable organic carbon (POC) is defined as the sum of volatile and semivolatile organic compounds sparged from an aqueous sample. However, these compounds are generally less than 1% of the TC in an environmental aqueous sample.

4.97   HOW DO THE TWO TYPES OF TOC ANALYZERS WORK?

UV-promoted persulfate oxidation involves exposing an aqueous sample to persulfate ions and UV radiation. This produces highly reactive sulfate and hydroxyl free radicals. The CO_2 produced from the persulfate oxidation reactions is swept by a stream of inert carrier gas, such as nitrogen, to the NDIR detector. IC and POCs are removed by acidification and sparging in the IC sparger. An aliquot of the sparged sample is then transferred to the UV reactor and persulfate reagent is added to oxidize the organic carbon based on the following chemical reactions:

$$S_2O_8^{2-} \xrightarrow{uv} 2SO_4^-$$

$$H_2O \xrightarrow{uv} H^+ + OH$$

$$SO_4^- + H_2O \xrightarrow[uv]{} SO_4^{2-} + OH + H^+$$

$$R(\text{organics}) \xrightarrow{uv} R*$$

$$R* + SO_4^- + OH \xrightarrow{uv} nCO_2 + - - -$$

These reactions show how persulfate ion and UV radiation combine to generate sulfate and hydroxyl free radicals that oxidize UV-excited organic compounds (R*) to CO_2. The UV-promoted persulfate oxidation approach to sample introduction, as developed for the Apollo 8000® TOC Analyzer (Teledyne-Tekmar), is shown in the schematic Figure 4.123. The aqueous sample is transferred to the IC sparger device, where inorganic carbon is initially removed. The sample is then moved to the UV reactor, where oxidation to CO_2 occurs and this gas product is swept into the NDIR detector.

The *high-temperature combustion* (HTC) technique uses heat (680°C or higher), in the presence of a titanium dioxide-based platinum catalyst, with a stream of hydrocarbon free compressed air or oxygen to oxidize organic carbon. DOC and particulates that contain carbon fully oxidize to CO_2 under these conditions. Following IC removal, an aliquot of the sparged aqueous sample is transferred to the combustion furnace to oxidize the organic carbon to form CO_2. The catalytic combustion oxidation products are continuously swept through the NDIR detector, which is selective to CO_2 and whose analog output signal is proportional to the concentration of CO_2 in the carrier gas, and thus in the original sample. The HTC approach, whose sketch is shown in Figure 4.124, also shows an IC sparger device and similar sample introduction technology as developed for the Apollo 9000® TOC Analyzer (Teledyne-Tekmar). The sparged sample is introduced into the injection port of a combustion furnace. An inert carrier gas continuously sweeps the CO_2 out of the furnace to the NDIR detector.

4.98   WHAT IS CAPILLARY ELECTROPHORESIS AND WHAT ROLE DOES IT HAVE IN TEQA?

We next consider a determinative technique introduced during the late 1980s and commercialized during the 1990s to the arsenal of instrumental analysis relevant to TEQA—capillary

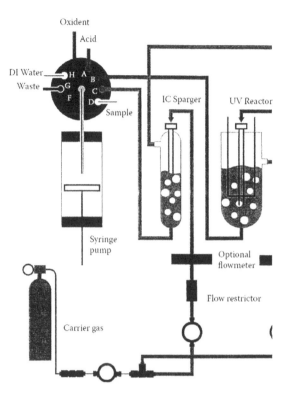

FIGURE 4.123

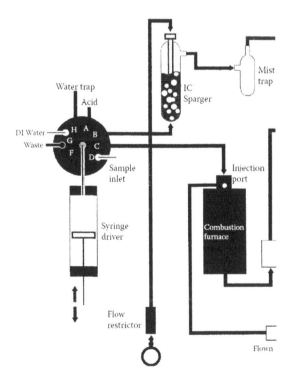

FIGURE 4.124

electrophoresis (CE). Our discussion of CE forces us to return to the separation sciences. CE encompasses a number of distinct modes:

- Capillary zone electrophoresis (CZE)
- Capillary gel electrophoresis
- Micellar electrokinetic capillary chromatography (MEKC)
- Capillary electrochromatography (CEC)
- Capillary isoelectric focusing
- Capillary isotachophoresis.

Li has provided a good definition of CE:[129]

In electrophoresis, a mixture of different substances in solution is introduced, usually as a relatively narrow zone, into the separating system, and induced to move under the influence of an applied potential. Due to differences in the effective mobilities (and hence migration velocities) of different substances under the electric field, the mixture then separates into spatially discrete zones of individual substances after a certain time.

Our focus will be on CZE in this book, although MEKC holds great potential for selective organic priority pollutant trace analysis. After we introduce the underlying principles of CZE, we will cite this author's efforts in developing a method to separate inorganic anions.

It is useful to *compare CE with the two established instrumental separation techniques* already discussed: HPLC and GC. The three schematics shown in Figure 4.125 provide a means to compare and contrast this third contributor to separation science. Moving from left to right across all three schematics reveals the following:

- All three techniques require a source of high potential energy. For CE, a high-voltage power supply is required; for HPLC, a fluid at high pressure is required; for GC, a compressed gas at high pressure is required.
- All three techniques require a column within which the separation of specific chemical compounds can be achieved. For CE, an uncoated narrow-bore cap column is used; for HPLC, a packed column; for GC, a coated-cap column or packed column.
- All three techniques require some means to detect and quantitate the separated chemical compounds. For CE, a UV absorbance detector was first used; for HPLC, several detectors, as already discussed; for GC, several detectors, as already discussed.

Electrophoresis has been known since 1886, when Lodge observed the migration of protons, H+, in a tube of phenolphthalein dissolved in a gel; Hjerten in 1967 first used a high electric field in solution electrophoresis using a 3 mm-i.d. capillary. However, Jorgensen and Lukacs are credited with writing the benchmark paper that demonstrated highly efficient separations using narrow-bore (<100 μm-i.d.) capillaries.[130] Figure 4.126 vividly depicts, in a somewhat oversimplified manner, the basic components of a CE instrument. A CE instrument can be built from simple components, as shown in the figure, and consists of two beakers joined by a filled capillary column. A high-voltage power supply that is capable of providing 30,000 V is impressed onto Pt electrodes that are placed in each beaker. Due to this extremely high-voltage requirement, an open configuration, such as is shown in Figure 4.126 is very dangerous. Safety considerations have led to the development of instruments that provide the necessary protection so that CE can be safely performed. As has been true in the historical development of GC and HPLC, the number of manufacturers of CE instruments rose and fell during the late 1980s and early 1990s, and today, only a few manufacturers remain. We digress briefly to introduce the basic theory underlying CE by asking the following question.

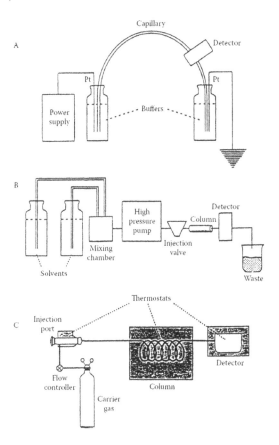

FIGURE 4.125

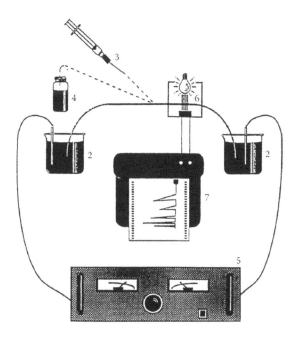

FIGURE 4.126

4.99   WHAT FACTORS INFLUENCE SEPARATIONS IN CE?

Applying such a large voltage across a capillary tube that is filled with electrolytes actually moves the fluid from anode to cathode, and this phenomenon is termed electro-osmosis. CE opens up the possibility to quantitate the presence of analytes of interest to TEQA that are either cationic, anionic, or neutral. Cations migrate to the cathode, ahead or behind the electro-osmotic velocity, depending on the polarity of the electrodes. Anions migrate to the anode, ahead or behind, and neutrals migrate with the electro-osmotic force. Electro-osmotic flow originates from the negative charges on the inner wall of the capillary tube due to deprotonizing surface silanol groups in neutral or alkaline medium. Cations from the buffer move toward this negatively charged surface. It is the water dipoles that surround the cation that get dragged when the electric field is applied.

To better explain electro-osmosis from the perspective of the analyte, consider the following analogy. What happens to the velocity (i.e., from physics, speed, *and* direction) of an individual when *he or she rides the horizontal escalator*, like those found in airport terminals? The velocity of the escalator is analogous to what is called the electro-osmotic mobility, symbolized by μ_{EO}, and the velocity of the individual riding the escalator is analogous to the electrophoretic mobility, symbolized by μ_{EP}. When you walk in a direction that is in the same direction as that of the horizontal escalator, your speed relative to a stationary observer standing on the floor next to the moving escalator is greatly increased. If you walk in a direction 180° opposite the direction of the escalator, you will still be carried in the direction of the escalator with respect to the stationary observer. You will, however, arrive at the end of the ride at a much later time. The vector notation for the mobilities of cations and anions is as follows:

$$\longrightarrow \mu_{EO}$$

$$\longrightarrow \mu +$$

$$\longrightarrow \mu -$$

The net effect to the detector (our stationary observer) is as follows:

$$+ \longrightarrow$$

$$0 \longrightarrow$$

$$- \longrightarrow$$

Ions originating from a buffer also have an electrophoretic mobility, symbolized by μ_{ion}. It should also be appreciated that the fluid flow profile through a tube is flat when the driving force is an electric field vs. the parabolic profile for flow created by a pressure difference.

The derivation shown below is adapted from earlier sources. Readers interested in a much broader introduction to CE are referred to a book by Weinberger among others.[131] The electro-osmotic velocity, v_{EO}, is defined as the product of the electro-osmotic mobility, μ_{EO}, and the applied electric field, E. For an applied voltage V and for a length of capillary L, we can write

$$v_{EO} = \mu_{EO} E = \mu_{EO} \frac{V}{L}$$

Similarly, the electrophoretic velocity, v_{EP}, is related to the electrophoretic mobility, μ_{EP}, according to

$$v_{EP} = \mu_{EP} \frac{V}{L}$$

The magnitude of μ is related to the *charge-to-size ratio of the analyte ion*. The time it takes for an analyte, once introduced into the capillary inlet, to be detected is known as the solute migration time t. The solute migration time is unique to a given analyte, and electrophoretic migration time t_{EP}, like solute retention time, t_R, in GC and HPLC, where t is defined as follows and, in turn, can be written in terms of mobilities:

$$t = \frac{L}{v_{EP} + v_{EO}} = \frac{L^2}{\left(\mu_{EP} + \mu_{EO}\right)V}$$

The above equation demonstrates that the CE solute migration time depends on the following:

- Solute charge-to-size ratio (i.e., μ_{EP})
- Applied voltage, V
- Capillary wall surface chemistry (i.e., μ_{EO})
- Buffer composition and concentration
- Length of column from inlet to detector, L.

The only factor contributing to band broadening is, unlike HPLC, longitudinal diffusion. The spread in the axial direction down the center of the capillary, for ions and molecules introduced as a tight plug at the inlet to the column, can be described by the Einstein equation:

$$\sigma^2 = 2Dt$$

where D is the molecular diffusion coefficient. The column efficiency, defined in terms of the number of theoretical plates, N, as in chromatography, is related to the capillary length, L, and the spread of molecules due to longitudinal diffusion of analyte through the capillary column, denoted by σ_L:

$$N = \frac{L^2}{\sigma_L^2} = \frac{\left(\mu_{EP} + \mu_{EO}\right)V}{2D}$$

The above equation suggests that the efficiency in CZE is related to the applied voltage and not capillary column length. Maximum efficiency and a reduction in analysis times are obtained using high voltages and short columns. This can be accomplished only by efficient heat dissipation. Control of column temperature in CE becomes very important.

To answer the question posed, as we did for GC and HPLC, we must consider those factors that influence the electrophoretic resolution, R_s. R_s can be related to the column efficiency, N, and the relative velocity difference between two zones ($\Delta v/v$), and this relationship is

$$R_s = \frac{1}{4}\sqrt{N}\left(\frac{\Delta v}{v}\right)$$

This equation can be restated in terms of electrophoretic and electro-osmotic mobilities as follows:

$$R_s = \frac{1}{4}\sqrt{N}\left(\frac{\mu_{EP2} - \mu_{EP1}}{\bar{\mu}_{EP}}\right)$$

where a mean electrophoretic mobility between separated peaks 1 and 2 is given by a simple average $\left(\bar{\mu}_{EP}\right)$ over the mobility of each analyte according to

$$\bar{\mu}_{EP} = \frac{1}{2}\left(\mu_{EP1} + \mu_{EP2}\right)$$

Eliminating N in the above equation in terms of electrophoretic ion mobilities yields

$$R_s = 0.18\left(\mu_{EP2} - \mu_{EP1}\right)\left(\frac{V}{D\left(\bar{\mu}_{EP} - \mu_{EO}\right)}\right)^{1/2}$$

This equation can be further simplified to yield

$$R_s = \frac{1}{4}\left(\frac{V}{2D}\right)^{1/2}\frac{\Delta\mu_{EP}}{\left(\bar{\mu}_{EP} - \mu_{EO}\right)^{1/2}}$$

Electrophoretic resolution, R_s, is found to depend on the following:

- $\Delta\mu_{EP}$, the difference in electrophoretic mobility between zones 1 and 2
- V, the applied voltage
- μ_{EO}, the electro-osmotic mobility
- $\bar{\mu}_{EP}$, the mean electrophoretic mobility of peaks 1 and 2.

The practice of CZE is largely influenced by the following factors:

- The chemical nature, pH, and concentration of the buffer
- Ionic strength
- Temperature, quite sensitive, needed to keep constant
- Viscosity
- Dielectric constant
- Applied voltage
- Length of capillary between the inlet and detection
- Capillary diameter, significant factor with respect to Joule heating; column inner diameters of <100 μm minimize Joule heating
- Chemical nature of buffer additives, such as the percent organic modifier, buffer ions of different μ_{EO}
- Injection time
- Internal capillary wall surface chemistry.

4.100   HOW DOES ELECTRODE POLARITY INFLUENCE CE?

Commercially available CE instruments are capable of placing either a positive or negative charge on the Pt wire at the injector, and at the same time placing a negative or positive charge on the Pt wire at the detector. A positive or negative potential can be programmed via a microprocessor. Depicted in Figure 4.127 is a schematic showing the various cation, anion, and neutral solutes introduced into a capillary, along with the anticipated electropherograms (to the right) for impressing +30 kV. In case 1, a neutral organic compound that absorbs UV radiation (this assumes that the CE instrument incorporates a UV absorption detector) is introduced into the injector end of a hypothetical column. This neutral marker will migrate with the same velocity as that of the electro-osmotic force and arrive at the detector end with a migration time that corresponds to the electro-osmotic velocity. Any anions that are present reach the detector after the neutral marker, depending on the magnitude of their anion electrophoretic mobility, as shown in cases 2 and 3. Cations that are present in the sample migrate ahead of the neutral marker, depending on their respective cation mobilities, as shown in cases 4 and 5.

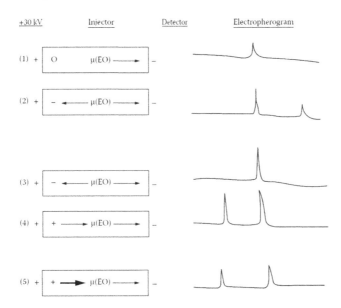

FIGURE 4.127

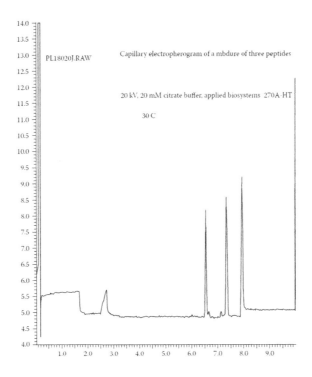

FIGURE 4.128

Cases 4 and 5 are illustrated in Figure 4.128, in which three protonated peptides are separated. This electropherogram was obtained in the author's lab using a Model 270A-HT (Applied Biosystems).

Impressing −30 kV to the injector/detector Pt wires reverses the polarity and drastically alters the order of ion migration, as shown in the second scenario depicted in Figure 4.129. The detector

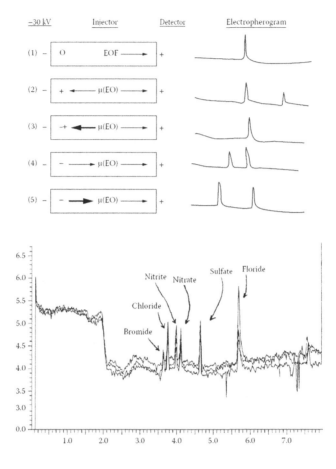

FIGURE 4.129

is now anodic, and the injector becomes cathodic. Case 1 shows the migration time for a neutral marker. Cases 2 and 3 depict two cations of different μ_{EP} values. Cases 4 and 5 show the expected migration times for two anions with different values of μ_{EP}.

To illustrate the use of a reversed polarity, consider the electropherogram, also obtained in the author's lab, and also shown in Figure 4.129. A mixture of the sodium or potassium salts of the common inorganic anions was introduced by hydrodynamic injection into the Model 270A-HT. A potential of -25 kV with a thermostatically controlled temperature of $30°C$ in a 25 mM phosphate buffer was used. The order of ion migration is shown in Figure 4.129 and differs from the ion chromatographic elution order discussed earlier. The peaks are actually negative peaks because a UV-absorbing chromophore has been added to the buffer. The lead wires from the detector were reversed and then connected to the interface.

4.101   WHAT IS VACANCY OR INDIRECT UV ABSORPTION DETECTION?

This technique has been used in HPLC to detect those analytes that either lack a UV chromophore in their molecular structure or have a weak molar absorptivity, ε. A strong UV-absorbing chromophore is added to the mobile phase in HPLC and to the buffer in CE.[132] For the separation, detection, and quantification of the common inorganic anions, vacancy or indirect UV photometric detection (IPD) is the only means that exists to quantitate all of the common anions. The detector wavelength is usually set at that λ where a maximum in the absorption spectrum for the chromophore is located. A research group at Waters Associates has made significant progress in method development for CE-IPD.[133] The methodology developed by this group has been recently applied in this author's laboratory to adapt the CE-IPD determinative technique to environmental geochemical related problems.[134] We close our discussion of CE by examining the influence of buffer pH on CE migration time and migration order.

4.102 HOW IMPORTANT IS BUFFER PH IN CE?

"Quite important" is the answer; however, it depends on the secondary equilibrium characteristics of the analyte of interest. Consider the three isomeric pyridinium carboxylates listed in Table 4.44, along with a neutral marker, mesityl oxide (MO).

The electropherograms were developed using a 25 mM phosphate buffer and an applied potential of +25 kV. A plot of the migration time vs. the buffer pH is shown in Figure 4.130. Note that the migration time for the neutral marker is not influenced by pH, in contrast to the migration times for all three isomeric pyridinium carboxylates. PA, INA, and NA all exhibit faster migrations time than the MO at pH 2. A decrease in their migration times at increasing pH becomes evident as the pH reaches the crossover pH at around 3, as indicated in the plot. The order of migration actually reverses itself after this crossover pH has been passed, as indicated in Figure 4.130.

TABLE 4.44
Name, Structure, and First and Second Acid Dissociation Constants for Compounds Discussed in Figure 4.130

Analyte	Structure	pKa1	pKa2
Mesityl oxide		–	–
PA		1.06	5.37
INA		1.70	4.89
NA		2.07	4.73

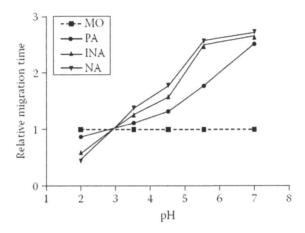

FIGURE 4.130

4.103   I UNDERSTAND THERE IS SOMETHING NEW IN ENVIRO-CHEMICAL TEQA WHEN THE SAMPLE MATRIX IS POLLUTED AIR: WHAT IS IT?

It is the application of *Selected Ion Flow Tube-Mass Spectrometry (SIFT-MS)*! Recall that near the end of Chapter 3, the topic of conventional sample preparation techniques available for sampling air pollutants was introduced. These techniques are still quite valuable and routinely used. SIFT-MS however, holds the promise of direct air sampling combined with direct quantitative analysis via quadrupole mass spectrometry, without the need for sample prep! This discussion on SIFT-MS is largely drawn from a recent book that was brought to the author's attention.[135] SIFT-MS originated in the pioneering work of Spanel and Smith in 1996 which followed from selected ion tube flow tube (SIFT) technology introduced 20 years ago by Adams and Smith. SIFT, in turn, was an extension of the early work of adapting flow tubes to investigate ion-molecule reactions begun in the late 1960s by Ferguson and collaborators. SIFT-MS is unique in that the reagent (or precursor) ions are well characterized and are mass-selected, and undergo known *ion-molecule reactions* with the analytes. In principle, an analyte such as an air pollutant can be quantitated without the need for analyte calibration!

SIFT-MS appears to complement the existing determinative techniques that measure environmental contaminants such as GC-MS and LC-MS-MS as shown in Figure 4.131 where *analyte polarity is plotted against analyte molecular weight (MW)*. SIFT-MS is seen to "fill in the gap" left by these two major determinative techniques introduced and discussed earlier. One can then envision that environmental laboratories in the future may have all three determinative techniques. Figure 4.132 shows a schematic outline of the principal operation of a SIFT-MS instrument. The ion source region is a microwave discharge of moist air. The dominant terminal ions from the discharge in air are: H_3O^+, NO^+, and O_2^+. The mixture of ions is transmitted to the lower pressure upstream quadrupole mass filter where mass selection takes place at a typical pressure in the 10^{-4} torr range (1 atmosphere of pressure = 760 torr). The selected reagent ion (for positive ion, one of the following: H_3O^+, NO^+, or O_2^+) is then transmitted into the flow tube where the reagent ion-analyte reaction occurs under controlled conditions. The flow tube pressure is 0.6 torr so that the ions from the upstream quadrupole enter the flow tube against a pressure gradient. The entry is assisted by means of a *Venturi orifice* which facilitates the transmission of ions against the pressure gradient. The reagent ions are then carried along the flow tube by the carrier gas (usually He although N_2 is being used in an increasing number of applications). It is in the flow tube where the diagnostic reagent ion-analyte reaction occurs. All ions within the flow tube are then sampled through a small orifice at the downstream end of the flow tube and are mass analyzed

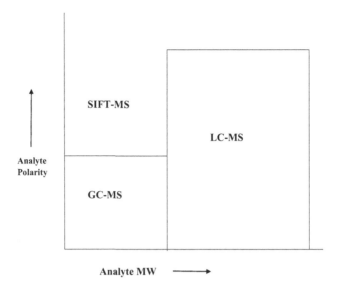

FIGURE 4.131

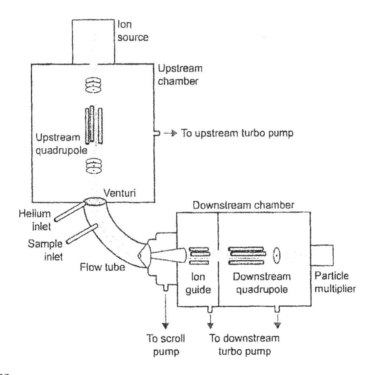

FIGURE 4.132

by the second quadrupole mass filter. The ion number densities are then counted by the pulse-counting electronics. These three reagent ions can be interchanged within a few milliseconds to obtain a complete analysis in real time of analytes (from contaminated air for example) from all three reagent ions.

Most of the commercial units made by Syft Technologies Ltd. (the Voice 200 and the Voice 200*Ultra* SIFT-MS instruments) operate at a flow tube temperature of 110°C and a carrier gas

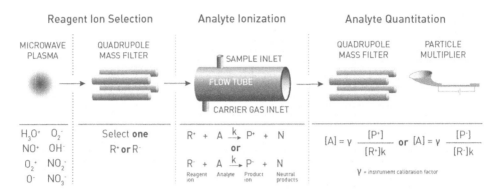

FIGURE 4.133

pressure of 0.6 torr. In some reactions where association reactions compete with electron transfer, the ratio of the product ion peaks are influenced by the conditions of temperature and pressure selected.[135]

4.104 I UNDERSTAND THERE IS SOMETHING CALLED "ION CHEMISTRY" IN SIFT-MS. WHAT IS THIS?

The term "ion chemistry" refers to the variety of reagent ions that are generated in the microwave plasma and of the chemical reactions with analytes that yield new chemical products. Figure 4.133 shows a somewhat different way to view a SIFT-MS instrument. The SIFT-MS determinative technique is seen to consist of three major parts: 1) reagent ion selection, 2) analyte ionization, and 3) analyte quantitation. Note the abundant negative ions shown in Figure 4.133 are also produced in the plasma namely O^-, O_2^-, OH^-, NO_2^-, NO_3^-. We focus here only on the three positive ions, H_3O^+, NO^+, and O_2^+ as reagent ions:

H_3O^+ Reactions

When H_3O^+ ions are injected into the flow tube, clusters of water ions are also formed. They often react in the same way as the H_3O^+ ion does with the VOC providing an exoergic pathway for proton transfer as shown below:

$$H_3O^+ + H_2O + He \rightarrow H_3O^+(H_2O) + He$$

$$H_3O^+(H_2O) + H_2O + He \rightarrow H_3O^+(H_2O)_2 + He$$

The predominant reaction of the H_3O^+ reagent ions with VOCs (e.g. from contaminant air) represented by A (see below) is *exothermic proton transfer*;

$$H_3O^+ + A \rightarrow AH^+ + H_2O$$

When proton transfer is quite exothermic (more than about 1 eV) then dissociative proton transfer may occur from AH^+ as in some reactions with organic compounds such as alcohols, aldehydes, and carboxylic acids. This leads to elimination of H_2O. The specific aldehyde shown in the example is hexanal (an aliphatic aldehyde) which serves as an example of a VOC drawn from contaminated air. The molecular structure for hexanal is shown below:

The reaction that occurs in the SIFT-MS is shown below:

$$H_3O^+ + C_5H_{11}CHO \xrightarrow[\text{hexanal}]{} C_6H_{12}O^+ + H_2O$$

$$\longrightarrow C_6H_{11}^+ + 2\,H_2O$$

Hydronium ions can also simply add to the analyte molecule either as a single product or in conjunction with other product channels:

$$H_3O^+ + A + He \longrightarrow H_3O^+(A) + He$$

In some cases the weakly bound association complexes denoted by $H_3O^+(A)$ may switch out the H_3O^+ to water and if this happens the use of H_3O^+ as a diagnostic reagent is lost.

$$H_3O^+ + A \longrightarrow H_3O^+(H_2O) + A$$

NO⁺ Chemistry

Greater variety exists in the reactions of NO^+ compared to the reactions of H_3O^+ in SIFT-MS.

There are usually one or two product ions in its reactions with a given analyte. These differences make NO^+ a very useful and important reagent ion in SIFT-MS analysis. *Electron transfer* can occur if the ionization potential of the analyte is less than the ionization potential for NO which is 9.26 eV. Numerous VOCs have lower ionization potentials when compared to NO with electron transfer as the common reaction pathway for example:

NO+ + [Phenylmethyl ether structure] $\longrightarrow$ $C_7H_8O^+$ + NO

Phenylmethyl ether

H^- (hydride ion) is also a common pathway of reactions with NO^+ with aldehydes, ethers and alcohols with the exception of tertiary alcohols when hydroxide ion transfer occurs. Consider the VOC acetaldehyde:

NO^+ + [acetaldehyde structure] $\longrightarrow$ $C_7H_8O^+$ + HNO

acetaldehyde

O₂⁺ Chemistry

O_2^+ is the most energetic of the common SIFT-MS reagent ions having an ionization potential (IP) of 12.06 eV. This value is larger than the IPs of most VOCs with the result that O_2^- generally reacts rapidly either by electron transfer (ET) or dissociative electron transfer (DET) and produces multiple product fragment in some cases. O_2^- is particularly useful in monitoring analytes that are unreactive with H_3O^+ and NO^+ reagent ions such as the lower MW hydrocarbons. The O_2^- reaction with isoprene is a typical example in that there are three product ion channels:

O_2^+ + [isoprene structure]

$\longrightarrow$ $C_5H_8^+$ + O_2 electron transfer

$\longrightarrow$ $C_5H_7^+$ + $(H + O_2)$ dissociative electron transfer

$\longrightarrow$ $C_4H_5^+$ + $(CH_3 + O_2)$ dissociative electron transfer

isoprene

Methane which is unreactive with H_3O^+ and NO^+ reagent ions, reacts with the O_2^+ reagent ion. The detection limit is not as low as for other analytes. The reaction is as follows:

$$O_2^+ + CH_4 \longrightarrow CH_3O_2^+ + H$$
$$\text{methane}$$

The small rate coefficient for methane means that the limit of detection for methane are in the mid- to high-ppb range by volume.[135]

The real-time monitoring of atmospheric VOCs has been important for years. Using the SIFT-MS determinative techniques enables Diesel engine exhaust gases such as NO, NO_2, HNO_2, aldehydes, and ketones to be quantitated. Peroxyacetyl nitrate (PAN) is known to be a precursor to photochemical smog. PAN can be measured using SIFT-MS with a limit of detection of 20 ppt by volume.[135] The molecular structure of PAN is shown below:

A bright future is ahead for SIFT-MS in direct analysis air monitoring.

4.105   WHAT DOES THE FUTURE HOLD FOR DETERMINATIVE TECHNIQUES?

That these techniques will forever be evolving, albeit ever so slightly in the future. The need for higher sensitivity and higher selectivity coupled to operational stability and ease of use serves to ensure a higher degree of precision and accuracy in trace analytical measurement. It is recognized today that attainment of high precision and accuracy is intrinsically tied to efficient sample preparation techniques as introduced in Chapter 3. The term *garbage in, garbage out* reflects this recognized synergy between sample prep and analytical instrumentation. In the future, expect to see:

- GC-MS-MS and LC-MS-MS become the dominant determinative techniques to implement both *enviro-chemical* and *enviro-health* TEQA
- More automated sample prep techniques interfaced to existing GC-MS, GC-MS-MS and LC-MS-MS instrumentation
- Elemental speciation techniques that become more routine
- Further increases in chromatographic resolution for GC and HPLC
- Autosamplers for both GC and LC become more robotic and more robust
- Greater throughput from the transducer through to a LIMS database
- SIFT-MS and other related determinative techniques get main-streamed.

This chapter has strongly emphasized GC determinative techniques and closes with the following quotation:

Gas chromatography is a volatility phenomenon in which solutes elute in an order that is mandated by the net vapor pressures of the solutes under a particular combination of operational parameters.

—Walter Jennings, Founder of J&W Scientific

REFERENCES

1. Ewing, G. *Instrumental Methods of Chemical Analysis*, 4th and 5th eds. New York: McGraw Hill, 1975, 1985.
2. Vassos, B., G. Ewing. *Analog and Digital Electronics for Scientists*, 3rd ed. New York: Wiley-Interscience, 1985.
3. Bauer H., et al. *Instrumental Analysis*. Boston: Allyn and Bacon, 1978.
4. Willard H., et al. *Instrumental Methods of Analysis*, 7th ed. San Francisco: Wadsworth, 1978.
5. Shanefield, S. *Industrial Electronics for Engineers, Chemists and Technicians*. Norwich, NY: Noyes Publications, 2001.
6. Malmstadt, H., C. Enke, S. Crouch. *Electronics and Instrumentation for Scientists*. Menlo Park, CA: Benjamin Cummings, 1981.
7. Robinson J. *Undergraduate Instrumental Analysis*, 5th ed. New York: Marcel Dekker, 1995.
8. Settle F., Ed. *Handbook of Instrumental Techniques for Analytical Chemistry*. Englewood Cliffs, NJ: Prentice Hall, 1997.
9. Strobel H., W. Heineman. *Chemical Instrumentation: A Systematic Approach*, 3rd ed. New York: Wiley, 1989.
10. Grob, R., E. Barry, Eds. *Modern Practice of Gas Chromatography*, 4th ed. Hoboken, NJ: Wiley-Interscience, 2004.
11. Snyder, L., J. Kirkland. *Introduction to Modern Liquid Chromatography*, 2nd ed. New York: Wiley-Interscience, 1979.
12. Snyder, L., J. Kirkland, J. Glajch. *Practical HPLC Method Development*, 2nd ed. New York: Wiley-Interscience, 1997.
13. Watson, J.T., O.D. Sparkman. *Introduction to Mass Spectrometry*, 4th ed. West Susssex, England: John Wiley & Sons, 2007.
14. Budde, W. *Mass Spectrometry: Strategies for Environmental and Related Applications*, Washington, D.C.: American Chemical Society, Oxford University Press, 2001.
15. Ettre, L. *LC-GC North Am* 21:458-467, 2003.
16. *Today's Chemist at Work*, February 2002.
17. Karger B., L. Snyder, C. Horvath. *An Introduction to Separation Science*. New York: Wiley, 1973.
18. Einax J., et al. *Chemometrics in Environmental Analysis*. New York: VCH, 1997, pp. 27–30.
19. Harris D. *Quantitative Chemical Analysis*, 3rd ed. San Francisco: Freeman, 1991.
20. Grob R., Ed. *Modern Practice of Gas Chromatography*. New York: Wiley, 1977, p. 5.
21. Martin A., R. Synge. *Biochem J* 35: 1358, 1941.
22. Stein A. *J Chem Educ* 53(10): 646, 1976.
23. Giddings J.C. *J Chem Educ* 35: 588–591, 1958.
24. Peters D., J. Hayes, G. Hieftje. *Chemical Separations and Measurements: Theory and Practice of Analytical Chemistry*. Philadelphia: WB Saunders, 1974, pp. 527–544.
25. Skoog D., J. Leary. *Principles of Instrumental Analysis*, 4th ed. Philadelphia: WB Saunders, 1992.
26. Jennings W. *Analytical Gas Chromatography*. San Diego: Academic Press, 1987.
27. Bartram R. *LC-GC* 15: 837–841, 1997.
28. McNair H., J. Miller. *Basic Gas Chromatography*. New York: Wiley Interscience, 1998.
29. Perry J. *Introduction to Analytical Gas Chromatography*. New York: Marcel Dekker, 1981.
30. Golay M. In D. Desty, Ed. *Gas Chromatography* (Amsterdam Symposium). London: Butterworths, 1958, pp. 36–55, 62–68.
31. Dandeneau R., E. Zerenner. *J High Resolut Chromatogr* 2: 351–356, 1979.
32. Farwell S. In G. Ewing, Ed. *Analytical Instrumental Handbook*, 2nd ed. New York: Marcel Dekker, 1997.
33. Kovats E. In *Advances in Chromatography*, Vol. 1, J. Giddings, R. Keller, Eds. New York: Marcel Dekker, 1965, pp. 229–247.
34. McReynolds W. *Gas Chromatographic Retention Data*. Preston Technical Abstracts. Niles, IL: Preston Publications, 1996.
35. Rohrschneider L. *J Chromatogr* 22: 6, 1966.
36. McReynolds W. *J Chromatogr Sci* 8: 214, 1970.
37. Li J., Y. Zhang, P. Carr. *Anal Chem* 64: 210–218, 1992.

38. Freeman R. *High Resolution Gas Chromatography*, 2nd ed. Palo Alto, CA: Hewlett-Packard, 1981, pp. 30–31.

39. Desty, D. *Anal Chem* 32: 302, 1960.

40. Yancey J. *J Chromatogr Sci* 23: 161–167, 370–377, 1985; 24: 117–124, 1986.

41. EPA Method 507, Methods for the Determination of Organic Compounds in Drinking Water, Supplement III, EPA/600/R-95/131. Office of Research and Development, Environmental Protection Agency, August 1995.

42. Hinshaw J. *LC-GC* 9: 94–98, 1996.

43. Hinshaw J. *LC-GC* 9: 470–473, 1996.

44. Onuska F., F. Karasek. *Open Tubular Column Gas Chromatography in Environmental Sciences.* New York: Plenum, 1984, p. 152.

45. Laub R., R. Pecsok. *Physicochemical Applications of Gas Chromatography.* New York: Wiley, 1978, p. 111.

46. Harris W., H. Habgood. *Programmed Temperature Gas Chromatography.* New York: Wiley, 1966.

47. Brettell T., R. Grob. *Am Lab* 17: 19–32, 1985.

48. Brettel T., R. Grob. *Am Lab* 18: 50–68, 1985.

49. Scientific Supplies Catalog, 2003–2004. Scientific Instrument Services, Inc., Ringoes, NJ, p. D32.

50. Short Path Thermal Desorption, Application Note 19. Scientific Supplies Catalog, January 1994. Scientific Instrument Services, Inc., Ringoes, NJ.

51. Ewing G. *Instrumental Methods of Chemical Analysis*, 5th ed. New York: McGrawHill, 1985.

52. Lovelock J., S. Lipsky. *J Am Chem Soc* 82: 431–433, 1960; see also Lovelock J. *J Chromatogr* 1: 35–46, 1958; Lovelock J. *Anal Chem* 33: 163, 1963; 35: 474–481, 1965.

53. Lovelock J. *J Chromatogr* 99: 3, 1974.

54. Lovelock J., A. Watson. *J Chromatogr Sci* 158: 123, 1978.

55. Onuska F., F. Karasek. *Open Tubular Column Gas Chromatography in the Environmental Sciences.* New York: Plenum, 1984, pp. 100–101.

56. Patterson P., et al. *Anal Chem* 50: 339–344, 1978.

57. Loconto P.R., A.K. Gaind. Isolation and Recovery of Organophosphorous Pesticides from Water by Solid-Phase Extraction with Dual Wide-Bore Capillary Gas Chromatography. *J Chromatogr Sci* 27: 569–573, 1989.

58. Cai H., et al. *Anal Chem* 68: 1233–1244, 1996.

59. Wentworth W., et al. *J Phys Chem* 70: 445–448, 1966.

60. Wentworth W., et al. *J Chromatogr Sci* 30: 478–485, 1992.

61. ASTM Method D 5504-94, Standard Test Method for Determination of Sulfur Compounds in Natural Gas and Gaseous Fuels by Gas Chromatography and Chemiluminescence. Philadelphia: American Society for Testing and Materials, 1994.

62. Application Note, on the Model 255 Nitrogen Chemiluminescence Detector. Sievers Instruments, 1977.

63. Hirshfeld, T. *Anal Chem* 52: 297A, 1980.

64. Uden P., Ed. *Element-Specific Detection by Atomic Emisson Spectroscopy,* ACS Symposium Series 479. Washington, DC: American Chemical Society, 1992.

65. *GC-AED Theory and Practice.* Concord, CA: Diablo Analytical, 2002, p. 16.

66. Raisgild M., M. Burke, K. Van Horne. *Am Environ Lab* 6: 24–26, 1994.

67. Carro A., I. Neira, R. Rodil, R. Lorenzo. *Chromatographia* 56: 733–738, 2002.

68. Pereiro I., A. Diaz. *Anal Bioanal Chem* 372: 74–90, 2002.

69. Shiner R., et al. *The Identification of Organic Compounds*, 7th ed. New York: John Wiley & Sons, 1998.

70. EPA Method 556.1. *Determination of Carbonyl Compounds in Drinking Water by Fast Chromatography*, Revision 1.0, September, 1999.

71. Onuska F., F. Karasek. *Open Tubular Column Gas Chromatography in the Environmental Sciences.* New York: Plenum, 1984, pp. 107–108.

72. Watson J.T. *Introduction to Mass Spectrometry*, 3rd ed. New York: Lippincott-Raven, 1997.

73. Watson J.T. *Introduction to Mass Spectrometry*, 2nd ed. New York: Raven Press, 1985. The author was fortunate to have had the opportunity to enroll in a graduate course on mass spectrometry taught by the late Professor J. Throck Watson using his 2nd edition book while the author was employed at Michigan State University!

74. Miller P., M. Denton. *J Chem Educ* 63: 617–622, 1986.
75. March R., R. Hughes. *Quadrupole Storage Mass Spectrometry.* New York: Wiley, 1989.
76. March R., J. Todd, Eds. *Practical Aspects of Ion Trap Mass Spectrometry,* Vols. I–III. Boca Raton, FL: CRC Press, 1995.
77. Incos X.L. Xle/Incos 50B/500 Course Manual. Finnigan MAT Institute, Cinncinatti, OH, 1989. The author was fortunate to have had the opportunity to be enrolled in a course offered on mass spectrometry during the early 1990s.
78. Cooks R., et al. *Chemical and Engineering News*, March 25, 1991, pp. 26–41.
79. Loconto P.R., Y. Pan, P. Kamdem. *J Chromatogr Sci* 36: 299–305, 1998.
80. Ong V., R. Hites. *Mass Spectrometry Reviews* 13: 259–283, 1994.
81. Loconto P.R. Selectivity and sensitivity improvements for selected polybrominated diphenyl ethers and polybrominated biphenyls using capillary gas chromatography/electron-capture negative ion mass selective detection: a cost-effective approach to biomonitoring. *LC-GC NA* 26: 1118–1130, 2008.
82. Loconto P.R, D. Isenga, M. O'Keefe, M. Knottnerus. Isolation and recovery of polybrominated diphenyl ethers from human and sheep serum: coupling reversed-phase solid-phase disk extraction and liquid-liquid extraction techniques with a capillary gas chromatographic electron-capture negative ion mass spectrometric determinative technique. *J Chromatogr Sci* 46: 53–60, 2008.
83. Loconto P.R. Evaluation of automated stir bar sorptive extraction-thermal desorption-gas chromatography electron-capture negative ion mass spectrometry for the analysis of PBDEs and PBBs in sheep and human serum. *J Chromatogr Sci* 47: 656–669, 2009.
84. Loconto, P.R. Quantitating Toxaphene Parlar congenrs in fish using large volume injection isotope dilution GC with electron-capture negative ion MS. *LC-GC NA* 36: 320–328, 2018.
85. Highly recommended books on mass spectral interpretation include: McLafferty F., F. Turecek. *Interpretation of Mass Spectra*, 4th ed. Sausalito, CA: University Science Books, 1993; Budzikiewicz H., C. Djerassi, D. Williams. *Mass Spectrometry of Organic Compounds.* San Francisco: Holden-Day, 1967.
86. Berry R., S. Rice, J. Ross. *Physical Chemistry*, 2nd ed. New York: Oxford University Press, 2000, pp. 14–15.
87. *Methods for the Determination of Organic Compounds in Drinking Water,* EPA/600 4-88/039. Environmental Protection Agency, December 1988, p. 309.
88. Yost R., C. Enke. *Anal Chem* 51: 1251A–1264A, 1979.
89. Cooks R., G. Glish. *Chemical Engineering News,* November 30, 1981, pp. 40–52.
90. McLafferty F., Ed. *Tandem Mass Spectrometry.* New York: Wiley, 1983.
91. Cody R., R. Burnier, B. Freiser. *Anal Chem* 54: 96–101, 1982.
92. Louris J., et al. *Anal Chem* 59: 1677–1685, 1987.
93. Driskell W., et al. *J Anal Toxicol* 26: 6–10, 2002.
94. *Test Methods for Evaluating Solid Waste, Physical/Chemical Methods*, SW-846, 3rd ed. Washington, DC: Office of Solid Waste and Emergency Response, Environmental Protection Agency.
95. Snyder L., J. Kirkland. *Introduction to Modern Liquid Chromatography*, 2nd ed. New York: Wiley, 1979.
96. Snyder L., J. Kirkland, J. Glajch. *Practical HPLC Method Development*, 2nd ed. New York: Wiley, 1997.
97. Lakowicz J. *Principles of Fluorescence Spectroscopy.* New York: Plenum, 1983, p. 1.
98. Wachs T., J. Conboy, F. Garcia, J. Henion. *J Chromatogr Sci* 29(8): 357–366, 1991.
99. Fenn J., M. Mann, C. Meng, C.S. Wong, C. Whitehouse. *Science* 246: 64–71, 1989.
100. Bruins A., T. Covey, J. Henion. *Anal Chem* 59: 2642–2646, 1987.
101. Blount B., E. Milgram, M. Silva, N. Malek. Quantitative detection of eight phthalate metabolites from human urine using HPLC-APCI-MS-MS. *Anal Chem* 72: 4127–4134, 2000.
102. *LC-GC North America the Application Notebook.* June, 2017, pp. 16–17.
103. Small H., T. Stevens, W. Baumann *Anal Chem* 47: 1801–1809, 1975.
104. Gjerde D., J. Fritz, G. Schmuckler. *J Chromatogr* 186: 509–519, 1979.
105. MacDonald J. *Am Lab* 11(1): 45–55, 1979.
106. Haddard P., P. Jandik. In J. Tarter, Ed. *Ion Chromatography.* New York: Marcel Dekker, 1987, pp. 88–91. (This is one of the more consistent treatments of the mathematics of conductance theory as applied to ion chromatography.)
107. Shpigun O., Y. Zolotov. *Ion Chromatography in Water Analysis.* London: Horwood, Halsted Press, 1988, p. 85.

108. Sandell E., H. Onishi. *Photometric Determination of Traces of Metals*, 4th ed., Part I. New York: Wiley Interscience, 1978.
109. Boss C., K. Fredeen. *Concepts, Instrumentation and Techniques in Inductively-Coupled Plasma Optical Emission Spectrometry*, 2nd ed. Norwalk, CT: PerkinElmer Corp., 1997, pp. 2.1–2.5.
110. Greenfield S., I. Jones, C. Berry. *Analyst* 89: 713–720, 1964.
111. Wendt R., V. Fassell. *Anal Chem* 37 920, 1965.
112. Ohis K., B. Bagdain. History of the ICP-AES spectral analysis from the beginning up to its coupling with mass spectrometry. *Journal of Analytical Atomic Spectrometry* 31:22–31, 2016.
113. Jenniss S., S. Katz, R. Lynch. *Applications of Atom Spectrometry to Regulatory Compliance Monitoring.* 2nd ed. New York: Wiley VCH, 1999, p. 24.
114. Benchmark papers include: Houk R., et al. *Anal Chem* 52: 2283–2289, 1980; Horlick G., et al. *Spectrochim Acta* 40B: 1555, 1985; Gray A. *Spectrochim Acta* 40B: 1525, 1986.
115. Montaser A., Ed. *Inductively Coupled Plasma Mass Spectrometry,* 3rd ed. Weinheim, Germany: Wiley VCH, 1998.
116. Zoorob G., J. McKiernan, J. Caruso. *Mikrochim Acta* 128: 145–168, 1998.
117. Nixon D., T. Moyer. *Spectrochim Acta B* 51: 13–25, 1996.
118. Thomas B. *Spectroscopy* 17: 42–48, 2002.
119. Krachler M., K. Falk, H. Emons. *Am Lab News Ed* 34: 10–14, 2002; references therein.
120. Caruso J., K. Sutton, K. Ackley, Eds. *Elemental Speciation.* Amsterdam: Elsevier, 2000.
121. Walsh A. *Spectrochim Acta* 7: 108, 1955; Alkemade C., J. Milatz. *J Opt Soc Am* 45: 583, 1955.
122. Robinson J. *Undergraduate Instrumental Analysis*, 5th ed. New York: Marcel Dekker, 1995.
123. Butcher D., J. Sneddon. *A Practical Guide to Graphite Furnace Atomic Absorption Spectrometry.* New York: Wiley Interscience, 1998.
124. Koirtyohann S., E. Pickett. *Anal Chem* 37: 601, 1965; Smith S., G. Heiftje. *Appl Spectrosc* 37: 419, 1983; deLoos-Vollegregt M., L. de Galan. *Prog Anal Atom Spectrosc* 8: 47, 1985; Yasuda K., et al. *Prog Anal Atom Spectrosc* 3: 299, 1980.
125. Butcher D., J. Sneddon. *A Practical Guide to Graphite Furnace Atomic Absorption Spectrometry.* New York: Wiley Interscience, 1998, chap. 4; Vandecasteel C., C. Block. *Modern Methods for Trace Element Determination.* New York: Wiley, 1993, chap. 5; Robinson J. *Undergraduate Instrumental Analysis*, 5th ed. New York: Marcel Dekker, 1995.
126. Yang, X. et al. Cloud-point extraction/graphite furnance atomic absorption spectromery for the determinaton of trace Rhodium in water samples. *American Laboratory* 51(3): 22–25, 2019. This issue was the last print version of this important and well-received monthly trade journal.
127. Whetsel K. Infrared spectroscopy. *Chemical Engineering News*, February 5, 1968, pp. 82–96.
128. Gruenfeld M. *Environ Sci Technol* 7: 636–639, 1973.
129. Li S. Capillary *Electrophoresis: Principles, Practice, and Applications*. Amsterdam: Elsevier, 1992, chap. 1.
130. Jorgeson J., K. Lukacs. *Anal Chem* 53: 1298, 1981.
131. Weinberger, R. *Practical Capillary Electrophoresis*, 2nd ed. San Diego: Academic Press, 2000, Chapters 1–3.
132. Foret R., et al. *J Chromatogr* 470: 299, 1989.
133. Jones W. *J Chromatogr* 640: 387–395, 1993.
134. Icopini G., P.R. Loconto, D. Long. Quantitative Trace Environmental Analysis: Inorganic Anions and Selected Organic Acid Anions via Capillary Electrophoresis with Indirect Photometric Detection. Paper presented at the *50th Pittsburgh Conference on Analytical Chemistry and Applied Spectroscopy*, Orlando, FL, 1999.
135. Murry, M. in T. Fujii, ed. *Ion/Molecule Attachment Reactions: Mass Spectrometry.* Tokyo: Springer, 2015, Chapter 8.

5 Student-Tested Laboratory Experiments

Theory guides, experiment decides.

—I.M. Kolthoff

This chapter introduces a series of laboratory experiments that attempt to show some examples of how to conduct trace environmental quantitative analysis (TEQA) in light of what has been discussed so far. These experiments address only the *enviro-chemical* aspects of TEQA. These experiments were written by the author before the first four chapters of the first edition were created! The impetus for writing these experiments was in support of a graduate-level course titled: *Environmental Analytical Chemistry Laboratory*. This course began in the mid-1990s, and the instruction followed the installation of a unique teaching laboratory coordinated by the author while employed as a laboratory manager for the *Hazardous Substance Research Center* in the Department of Civil and Environmental Engineering at Michigan State University, East Lansing, MI.

There are several options that an instructor can use to design a laboratory program that gives M.S. and Ph.D. graduate students numerous opportunities to measure environmentally significant chemical analytes. It is this author's opinion that it doesn't really matter which analytes are to be quantitated as long as an *appropriate mix of sample prep and instrumental techniques* is introduced to students who will become environmental engineers, water treatment plant operators, and environmental analytical chemists, among other occupational endeavors. A suggested laboratory schedule that was successfully implemented earlier is introduced below.

5.1   WHAT MIGHT A TYPICAL LABORATORY SCHEDULE LOOK LIKE?

Listed below is the laboratory program once implemented by the author for a graduate-level course in TEQA. These are indeed *student-tested experiments*! Beneath the title of each experiment is a statement of learning objectives and outcomes that the student should be able to achieve. The degree to which the instructor makes the course *more or less rigorous* is determined by the curriculum objectives. An experimental course in TEQA can consist of a series of experiments with everything set up beforehand for the student. Alternatively, the same experiments can be undertaken whereby the student does all the setting up. Some compromise between these two extremes might be in the best interest of the student and instructor/staff alike!

A series of actual student experiments given as *individual handouts* follows the laboratory course annotated syllabus shown below:

Experiment #	Description
1st half or full semester	Orientation to laboratory; discussion of outcomes and what is expected; definition of learning goals and student assignment to workstations; safety requirements; how to weigh properly, guidelines for safe and environmentally sound disposal of laboratory generated chemical waste
	Descriptive introductory information. Instructors should also emphasize the safety aspects of working with chemicals, laboratory equipment and sophisticated and expensive analytical instrumentation.
1	Introduction to pH measurement: estimating the degree of purity of snow; learning to measure soil pH; introduction to ion chromatography *and/or* Introduction to visible spectrophotometry *and/or* Determination of the Fe(III) / Fe(II) concentration ratio in groundwater *and/or* Determination of phosphate ion, PO_4^{3-} in eutrophicated surface water
	Quantitative analysis; emphasis on standards preparation techniques; statistical treatment of data; environmental sampling techniques; learning to operate a pH meter; learning to operate the UV-vis spectrophotometer; learning to operate a flame atomic absorption (AA) spectrophotometer; no write-up required.
2	Determination of anionic surfactants in an industrial wastewater effluent by mini-liquid–liquid (mini-LLE) extraction using ion pairing with methylene blue
	Quantitative analysis; emphasis on sample preparation, forming an ion pair, unknown sample analysis; learning to use and to calibrate a pH meter; measuring visible absorbance on a spectrophotometer; write-up required.

Experiment #	Description
3	Comparison of ultraviolet and infrared absorption spectra of chemically similar organic compounds *and/or* Determination of oil and grease and of total petroleum hydrocarbons (TPHs) via reversed-phase solid-phase extraction (RP-SPE) techniques using quantitative Fourier-Transform infrared (FTIR) spectroscopy*
	Qualitative analysis; background for understanding uv-vis and infrared absorption spectroscopy; learning to use and interpret data from molecular spectroscopic instrumentation; sampling techniques; write-up required.
4	Determination of the degree of hardness in groundwater using flame atomic absorption (FLAA) spectroscopy: measuring Ca, Mg, and Fe
	Quantitative analysis; learning to operate a FLAA spectrophotometer; calibration using external standard mode; spiked recovery; no write-up required.
5	Determination of lead (Pb) in drinking water using graphite furnace atomic absorption spectroscopy (GFAA)
	Quantitative analysis; learning to operate a GFAA spectrophotometer; learning to use the WinLab® software for GFAA spectrophotometry; calibration based on standard addition; no write-up required.
6	Comparison of soil types via a quantitative determination of the chromium content using visible spectrophotometry and FLAA spectrophotometry or inductively coupled plasma-atomic emission spectrometry
	Quantitative analysis; use of two instrumental methods to determine the Cr (III) and Cr (VI) oxidation states; digestion techniques applied to soils; write-up required.
2nd half or full semester	
7	An introduction to data acquisition and instrument control using Turbochrom®; introduction to high-performance liquid chromatography (HPLC); evaluating those experimental parameters that influence HPLC instrument performance
	Qualitative analysis; emphasis on learning to operate the HPLC instrument and the Turbochrom® computer software; no write-up required; answer questions in lab notebook.
8	Identifying the ubiquitous phthalate esters in the environment using HPLC with photodiode array detection (PDA); confirmation using gas chromatography-mass spectrometry (GC-MS)
	Qualitative analysis; interpretation of chromatograms, UV absorption spectra, mass spectra; experience with GC-MS; write-up required.
9	An introduction to gas chromatography: evaluating experimental parameters that affect gas chromatographic performance
	Qualitative analysis; emphasis on learning to operate the GC; measurement of split ratio; no write-up required; answer questions in lab notebook.
10	Screening for the presence of BTEX in gasoline-tainted water using mini-LLE and gas chromatography (GC-FID) *and/or* screening for THMs in chlorine-disinfected drinking water using static HS gas chromatography (GC-ECD)
	Semi-quantitative analysis; unknown sample analysis; statistical treatment of data; write-up required.
11	Determination of priority pollutant volatile organic compounds (VOCs) in gasoline-contaminated groundwater using static headspace (HS) and solid-phase microextraction headspace (SPME-HS) and gas chromatography
	Quantitative analysis, learning to use HS and SPME-HS sample prep techniques; learning to inject an HS syringe into a gas chromatograph; write up required.

Experiment #	Description
12	Determination of the concentration of the herbicide residue trifluralin in soil from lawn treatment by gas chromatography using reversed-phase solid-phase extraction (RP-SPE) methods
	Quantitative analysis; calibration based on internal standard mode; unknown sample analysis; statistical treatment of data; write-up required.
13	Determination of priority pollutant semivolatile (SVOC) organochlorine pesticides in contaminated groundwater: comparison of two sample preparation methods—mini-LLE vs. RP-SPE techniques*
	Quantitative analysis; emphasis on sample preparation, unknown sample analysis; calibration based on internal mode; statistical treatment of data; write-up required.
14	Determination of selected priority pollutant polycyclic aromatic hydrocarbons in oil-contaminated soil using LLE-RP-HPLC-PDA
	Quantitative analysis; sample preparation; write-up required.

* Projects are considered extra credit and thus not required. Students must make arrangements with the laboratory instructor in order to perform these experiments.

This is a very ambitious laboratory course! Instructors may wish to run the laboratory course for two semesters in one academic calendar year. It is, however, a graduate-level laboratory course of instruction. To effectively educate graduate students while delivering the course content requires a dedicated support staff, a committed faculty, sufficient laboratory glassware and accessories, and expensive analytical instrumentation, including interface of each instrument to a personal computer that operates chromatography or spectroscopy software. Each laboratory session requires a minimum of 4 hours and a maximum of 8 hours. Students must be taught not only how to *prepare environmental samples for trace analysis*, but also how to *operate and interpret the data* generated from sophisticated analytical instruments. The intensity of the lab activities starts with an initial and less rigorous laboratory session/experiment with rigor increasing as each laboratory session/experiment unfolds.

5.2   HOW IS THE INSTRUCTIONAL LABORATORY CONFIGURED?

When the laboratory experiments that follow were developed, the author had just completed coordinating the installation of an instructional laboratory that included four workstations. In addition, one ICP-MS, one FTIR spectrophotometer, one ion chromatograph and several uv-vis spectrophotometers were available for student use.

Each workstation consisted of

1. One Autosystem® (PerkinElmer Instruments) gas chromatograph incorporating dual capillary columns (one for VOCs and one for SVOCs) and dual detectors (FID and ECD).
2. One HPLC (PerkinElmer Instruments) that included a 200 Series® LC binary pump, a manual injector (Rheodyne®), a reversed-phase column and guard column, and a LC250® photodiode array (PDA) ultraviolet absorption detector.
3. One Model 3110® (PerkinElmer Instruments) atomic absorption spectrophotometer with flame and graphite furnace capability with deuterium background correction.
4. One personal computer (PC) that enabled all three instruments above to be interfaced. For GC and HPLC, *Turbochrom®* (PE Nelson) Chromatography Processing Software (now called Total Chrom; PerkinElmer Instruments) was used for the data acquisition via the 600 LINK® (PE Nelson) interface that was external to the PC. For AA, *WinLab®* (PerkinElmer Instruments) software was used via an interface board that was installed into the PC console.

5. A UV-VIS spectrophotometer Genesys 5® (Spectronic Instruments) was used. If another spectrophotometer is used, an infrared phototube is necessary to quantitate in that experiment. Multiple spectrophotometers would enhance instruction.

In addition, a Model 2000® (Dionex) ion chromatograph interfaced to the PC via a 900® interface (PE Nelson) and a Model 1600® FTIR Spectrophotometer (PerkinElmer Instruments) are available for all students to use in the instructional laboratory. Individual university and college departments will undoubtedly have their own unique laboratory configurations that include *instrumentation from other manufacturers*. In order to carry out all of the experiments introduced in this chapter, instructional laboratories must have, at a minimum, the following analytical instruments: GC-FID/ECD, HPLC-UV or HPLC-PDA, FlAA and GFAA, IC, and a UV-vis spectrophotometer (stand-alone). Access to a GC-MS is also required in one experiment. Accessories for sample preparation, as listed in each of the subsequent experiments, are also needed.

Each experiment that follows was written to be as independent of the others in the collection as possible. Beneath each experiment can be found a short list of references. Safety tips appear in each experiment as poignant reminders to students and instructors alike of the *perils associated with laboratory work*. Instructors can pick and choose to use a given experiment as written here or modify it to fit their unique laboratory situation. Students are encouraged to learn and use, in addition to mastering the software associated with each instrument, the software tool Microsoft Excel to facilitate data analysis and interpretation. Most student experiments shown below end with a list of books and/or papers cited as *suggested readings* that relate to that specific experiment while serving to broaden the scope of the given topic. As part of an introductory exercise, students should be introduced to the proper use of a contemporary analytical balance.

5.3   HOW TO WEIGH THE RIGHT WAY

The analytical balance has been and continues to be the simplest yet most profound instrument in the analytical laboratory. The precision and accuracy of the most sophisticated analytical instrument is limited by the precision and accuracy inherent in the weight of the neat form of a certified reference standard. Today's single pan electronically digitized balances are a *far cry* from the analog double pan analytical balance that required the very careful adding of various sized calibration weights placed on the right-hand pan to balance the weight of samples placed on the left-hand pan. The procedure shown below was adapted from: Ciesniewski, I. *R&D Mag* 45:30, 2003:

The contemporary electronic analytical balance must be properly maintained, cleaned frequently, calibrated with standard reference weights, and used properly. Some helpful hints on developing proper weighing skills are given below:

- Before weighing, ensure that your balance is leveled correctly.
- Periodically check, clean, and calibrate the balance.
- Zero out or tare the balance prior to weighing.
- Minimize the use of hands to place tare weights or samples in the weigh chamber. Use appropriated sized and shaped tweezers or tongs to handle weighing vessels.
- Weigh to the same side each and every time. This will maximize repeatability.
- When placing items to be weighed on to the weigh pan, open only the draft shield door on the right side on which you are weighing: e.g. if you are right-handed, open the right door.
- Understand how the balance indicates a stable weight, i.e. gives a weight that can be safely trusted. All electronic balances give a visual indication of weight stability.
- Zero out or tare weight as carefully as done for the sample. Using tweezers, place the tare weight onto the weigh pan, close all doors, press the tare button, and wait for the balance to give a stable zero.

- Introduce the sample into the weighing vessel using a long spatula, spoon, scoop, or tweezers as appropriate.
- Be aware of how the balance is affected by the working environment. Modern, busy laboratories are not ideal places for the four- or five-decimal-place balances that we need to put in them. It may be difficult to stabilize. It may take as long as 60 sec for the balance to reach stability.
- Aim for the same location on the balance weigh pan; try to aim for the center of the pan each time.
- After finishing weighing, check that the weigh chamber is clean and free of any spills. This is not just a courtesy to others, or conformance to regulations; it is helping to keep your balance working accurately by eliminating unwanted ingression that could damage the internal mechanics of the balance.

5.4 AN INTRODUCTION TO PH MEASUREMENT: ESTIMATING THE DEGREE OF PURITY OF SNOW; MEASURING SOIL PH; INTRODUCTION TO ION CHROMATOGRAPHY

5.4.1 BACKGROUND AND SUMMARY OF METHOD

This laboratory will focus on the operational aspects of pH measurement. It is appropriate that we start this course with pH because this parameter is so fundamental to the physical-chemical phenomenon that occurs in aqueous solutions. The pH of a solution that contains a weak acid determines the degree of ionization of that weak acid. Of environmental importance is an understanding of the acidic properties of carbon dioxide. The extent to which gaseous CO_2 dissolves in water and equilibrates is governed by the Henry's law constant for CO_2. We are all familiar with the carbonation of beverages. The equilibrium is

$$CO_{2(g)} \leftrightarrow CO_{2(aq)}$$

When CO_2 is dissolved in water, a small fraction of the dissolved gas exists as *carbonic acid*, an unstable substance whose theoretical chemical formula is written as H_2CO_3. The acid–base character of CO_2 is described more accurately by the following reactions and equilibrium constants:

$$CO_{2(aq)} + H_2O \leftrightarrow H^+_{(aq)} + HCO^-_{3(aq)}$$

$$K_{a1} = \frac{\left[H^+\right]\left[HCO_3^-\right]}{\left[CO_2\right]} = 4.45 \times 10^{-7}$$

$$pK_{a1} = 6.35 \tag{5.1}$$

$$HCO^-_{3(aq)} \leftrightarrow H^+_{(aq)} + CO^{2-}_{3(aq)} \tag{5.2}$$

$$pK_{a2} = 10.33$$

For example, the molar concentration of CO_2 in water saturated with this gas at a pressure of 1 atm at 25°C is 3.27×10^{-2} M. The pH, using Equation 5.1, can be calculated as 3.92. Environmental water samples generally have less than the saturated molarity value and yield a pH of approximately 5.9. Thus, the difference between *ultrapure water*, with a theoretical pH of 7, and water exposed to the atmosphere, with a measurable pH of 5.9, is attributed to dissolved CO_2 and its formation of *carbonic acid*, which in turn dissociates to *hydronium, bicarbonate*, and *carbonate* ions. The degree to which carbon dioxide dissolved water remains in its molecular form or exists as either *bicarbonate* or *carbonate* depends upon the pH of water. The fraction, α, of each of these species of the total is mathematically defined as

$$\alpha_{CO2} = \frac{[CO_2]}{[CO_2]+[HCO_3^-]+[CO_3^{2-}]} \tag{5.3}$$

$$\alpha_{HCO_3^-} = \frac{[HCO_3^-]}{[CO_2]+[HCO_3^-]+[CO_3^{2-}]} \tag{5.4}$$

$$\alpha_{CO_3^{2-}} = \frac{[CO_3^{2-}]}{[CO_2]+[HCO_3^-]+[CO_3^{2-}]} \tag{5.5}$$

Equations (5.3) to (5.5) can be reworked in terms of acid dissociation constants and hydronium ion concentrations to yield

$$\alpha_{CO2} = \frac{[H^+]^2}{[H^+]^2 + K_{a1}[H+]+ K_{a1}K_{a2}} \tag{5.6}$$

$$\alpha_{HCO_3^{2-}} = \frac{K_{a1}[H^+]}{[H^+]^2 + K_{a1}[H^+]+ K_{a1}K_{a2}} \tag{5.7}$$

$$\alpha_{CO_3^{2-}} = \frac{K_{a1}K_{a2}}{[H^+]^2 + K_{a1}[H^+]+ K_{a1}K_{a2}} \tag{5.8}$$

Equations (5.6) to (5.8) can be used to construct what are called *distribution of species diagrams*, which are plots of the fraction of each species as a function of solution pH. For the $CO_2/HCO_3^-/CO_3^{2-}$ aqueous solution, the distribution of species diagram is shown below:

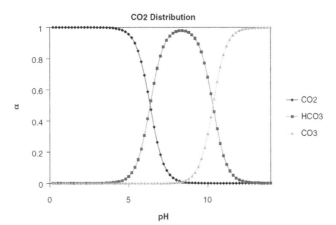

The *amount of acidic or alkaline solutes* present in a given volume of a groundwater, surface water, or wastewater is determined by conventional acid–base titration. Refer to *Standard Methods for the Examination of Water and Wastewater* for specific procedures. Titrations are performed in these methods to a specified pH. Be sure to distinguish among the concepts of acidity, alkalinity, and pH when considering the nature of environmental samples. The pH will need to be measured and adjusted prior to conducting the ion pair liquid–liquid extraction for determining methylene blue active surfactants.

The pH was originally measured by judicious choice of acid–base indicator dyes and eventually over the years gave way to potentiometric methods due to the *inherent limitations of color*. A dye serves no useful purpose when the wastewater sample is brownish in color. Advances in both the instrument and the glass electrode have taken the measurement of pH very far since the early days, when in 1935, Arnold Beckman was first asked to measure the pH of a lemon!

5.4.2   EXPERIMENTAL

5.4.2.1   Glassware Needed per Student

1 250 or 500 mL beaker to sample snow
1 50 mL beaker
1 Stirring rod.

5.4.2.2   Chemical Reagents/pH Meter Needed per Student Workstation

Each pH measurement station should consist of the following:

1. Three buffer solutions of pH 4, 7, and 10, respectively
2. One squeeze bottle containing distilled deionized water (DDI)
3. One waste beaker
4. Training guide for operating the Orion SA 720 pH/ISE® Direct Readout Meter.

5.4.2.3   Ion Chromatograph

A 2000® Dionex Ion Chromatograph will be set up by the staff for use in this experiment. As of this writing, what was the Dionex Corporation is now part of Thermo Fisher Scientific. Students are referred to the supplement titled: How to set up and operate an Ion Chromatograph.

5.4.2.4   Procedure

Use the guide located at each of four pH measurement stations and familiarize yourself with the operation of the pH meter. Each student should calibrate the meter using the pH 4 and 7 buffer solutions that are available by implementing the autocalibration. An ATC probe will not be available; therefore, samples and standards should be at the same temperature. Use a stirring rod to continuously stir the solution in the 50 mL beaker while a pH measurement is being taken.

Obtain at least three samples of snow. One sample should be obtained close to a walkway or roadway and appear visibly dirty. One sample should be obtained far away from pollutant sources and appear visibly clean. Use a relatively large beaker to collect snow and allow the snow to melt. *Record pH values in your notebook.* Draw conclusions about your observations and write these in your laboratory notebook.

Obtain one or more soil samples from a hazardous waste site. Alternatively, your instructor may have a series of fortified soils available in the laboratory. An illustration of how a series of laboratory-fortified acidic, neutral, and alkaline soils can be prepared and given to students as sample unknowns is shown in table below. Students should review how to implement EPA

Method 9045D from the EPA's SW-846 solid waste analysis methods. Today, students can quickly access most EPA Methods via a "google name search." Assume that the unknown soil is *noncalcareous*. Notice the use of flowcharts in helping to understand the procedural aspects of the method. Repeat the pH measurement for the four other samples and *record your results* in your laboratory notebook. The chemical nature of the contaminated soil samples will be revealed to you after you have completed your pH measurements. Rationalize the observed pH value for each soil based on knowledge of the chemical used to contaminate the soil. Be sure to jot down your comments in your laboratory notebook.

Sample No.	Soil Type	Chemical Added
1	Tappan B	Salicylic acid
2	Tappan B	Sodium carbonate
3	Tappan B	None
4	Hudson River sediment, possibly contaminated with PCBs	None
5	Unknown sandy soil	Citric acid
6	Unknown sandy soil	Potassium hydrogen phthalate
7	Unknown sandy soil	None
8	Unknown sandy soil	p-Nitrophenol
9	Unknown sandy soil	Tetrabutyl ammonium hydroxide

Molecular structures for the organic compounds used to adulterate the soil with the purpose of simulating a chemically polluted soil sample are given below:

Citric acid

Salicyclic acid/citric acid/potassium hydrogen phthalate/p-nitrophenol/tetrabutyl ammonium hydroxide

As an additional feature to this experiment, the environmental engineering staff will have a Dionex 2000® Ion Chromatograph in operation. The staff will engage students in the operation of the ion chromatograph. This instrument will separate and detect inorganic anions such as chloride, nitrate, and sulfate. Are these chemical species present in the various snow samples? In what manner do inorganic anions contribute to snow acidity?

You have generated hazardous waste from your soil pH measurements. Dispose of properly in accordance with the Office of Radiation, Chemical and Biological Safety (ORCBS) guidelines.

5.4.3 Suggested Readings

To develop this experiment, the author consulted the following resources:

- Many of the more lucid presentations on pH are found in *general chemistry texts*, whereas instrumental concepts of pH measurement are found in most texts on *analytical chemistry*.

- The most definitive text at an advanced level is R. Bates's *Determination of pH.* New York: John Wiley & Sons, 1964. This book is found in most university chemistry libraries.
- A useful guide to the measurement of pH in soil is found in EPA Method 9045D *Test Methods for Evaluating Solid Waste: Physical/Chemical Methods,* SW-846, Revision 4 November, 2004.
- Weiss J. *Handbook of Ion Chromatography*, E.L. Johnson ed. Sunnyvale, CA: Dionex Corporation, 1986. No better treatment of the subject at the time.

5.5 INTRODUCTION TO THE VISIBLE SPECTROPHOTOMETER

5.5.1 BACKGROUND AND SUMMARY OF METHOD

The simple visible spectrophotometer has been an important instrument in trace environmental quantitative analysis for many years. Colorless environmental contaminants must be chemically converted to a species that appears colored to the eye. The colored species will then exhibit regions of the visible electromagnetic spectrum where it will absorb photons. This absorption can be related to the Beer–Lambert or Beer–Bouguer law of spectrophotometry according to:

$$A = \varepsilon bc$$

where A is the absorbance measured in absorbance units (AUs), and in turn is related to the logarithm of the ratio of incident to transmitted intensity; ε is the molar absorptivity, a unique property of a chemical substance, viewed as a constant in the equation; b is the path length in either millimeters or centimeters, assumed to be a constant (see cuvette matching below); and c is the concentration in either molarity or weight/unit volume, if known. This forms the basis for calibration of *uv-visible and atomic absorption spectrophotometers*.

Recall that the color of a solution is the complement of the color of light that it absorbs. For example, the red cobalt chloride solution you will be using to conduct the cuvette matching (see below) actually absorbs the complement to red light, which is green light at a wavelength of 510 nm. The term *spectrophotometer* succinctly and completely describes the instrument. *Spectro* refers to the visible spectrum and the ability of the instrument to select a wavelength or, more accurately, a range of wavelengths. *Photo* refers to light, and *meter* implies a measurement process. Some instruments lack the capability of selecting a wavelength and should be called photometers. A good question to ask when using a spectrophotometer is: What is the precision in nanometers when I set the wavelength? The answer should be as follows: It depends on the *effective bandpass of the monochromator.* The effective bandpass is a measure of the band of wavelengths allowed to pass through the spectrophotometer for a given width of the exit slit of the monochromator. The bandpass is given by the product of the slit width and the reciprocal linear dispersion. The theory of light dispersion on gratings and prisms and subsequent effects on monochromator resolution has been well developed.

The major limitation on the use of visible spectrophotometric or colorimetric methods is the fact that the analyte must absorb within the visible domain of the electromagnetic spectrum. *Many analytes of environmental interest are colorless!* Some analytes can be chemically converted to a colored product by direct chemical change, formation of a metal chelate, or formation of an ion pair. An example of a chemical conversion of the analyte of interest is the colorimetric determination of nitrite–nitrogen via reaction of the nitrite ion with sulfanilic acid in acidic media to form the diazonium salt, with subsequent reaction with chromotropic acid to form a highly colored azo dye as shown:

$$NO_2^- + HO_3S - \langle\ \rangle - NH_2 + 2H^+ \longrightarrow HO_3S - \langle\ \rangle - N = N^+ + 2H_2O$$

Nitrite   Sulfanilic acid   Diazonium salt

$$HO_3S - \langle\ \rangle - N \equiv N^+ + \overset{OH\ OH}{\underset{HO_3S \qquad SO_3H}{\bigcirc\bigcirc}} \longrightarrow \overset{OH\ OH}{\underset{HO_3S \qquad SO_3H}{\bigcirc\bigcirc}} - N = N - \langle\ \rangle - SO_3H + H^-$$

Diazonium salt   Chromotropic acid   Red-orange-colored complex

An example of metal chelation can be found if one refers to *Standard Methods for the Examination of Water and Wastewater* for the recommended methods for determining the toxic metal cadmium. An atomic absorption method is listed along with a dithizone method. Cd^{2+} combines with dithizone (HDz) in aqueous media to form the neutral colored molecule $Cd(Dz)_2$. This neutral molecule can be subsequently extracted into a nonpolar solvent, thus providing an increase in the method detection limit (MDL). The extent to which a given metal ion can be chelated with HDz and subsequently extracted is defined as the distribution ratio, D_M, for a given metal ion and depends on the formation constant of the complex, the overall extraction constant, the initial concentration of HDz in the organic phase, the fraction of the total metal concentration present as the divalent cation, and, most importantly, the pH of the aqueous solution. For the equilibrium

$$M_{(aq)}^{n+} + nHDz_o \leftrightarrow MDz_{n(o)} + nH_{(aq)}^+$$

the distribution ratio is expressed mathematically as

$$D_M = \frac{\beta_n K_{ex} [HDz]_o^n}{[H^+]^n} \alpha_M$$

An example of a chelated ion association pair that when formed is subsequently extracted into a nonpolar solvent is that of the iron(II)-*o*-phenanthroline cation, which efficiently partitions into chloroform, provided a large counterion such as perchlorate is present in the chloroform.

5.5.2 Experimental

A possible major source of error in spectrophotometric measurement occurs when the path length differs from one sample to the next (i.e., the value for b (see above) is different). Hence, two solutions having identical concentrations of a colored analyte (i.e., both a and c are equal) could exhibit different absorbances due to differences in path length b. To minimize this source of systematic error, *a cuvette matching exercise is introduced*. Cuvette tubes are matched by placing a solution of intermediate absorbance in each and comparing absorbance readings. One tube is chosen arbitrarily as a reference, and others are selected that have the same reading to within 1%. Tubes should always be matched in a separate operation before any spectrophotometric experiments are begun.

5.5.2.1 Glassware Needed per Student or Group

Seven to ten 13 × 100 nm glass test tubes.

5.5.2.2   Chemical Reagents Needed per Student or Group

Ten milliliters of a 2% cobalt chloride solution; dissolve 2 g of $CoCl_2 \cdot 6H_2O$ in approximately 100 mL of 0.3 M HCl. This solution is recommended because it is stable, has a broad absorption band at about the center of the visible region, and transmits about 50% in a 1 cm cell. Each group should prepare this solution and share among members.

5.5.2.3   Miscellaneous Item Needed per Student or Group

One piece of chalk that nicely fits into a 13 × 100 mm test tube.

5.5.2.4   Spectrophotometer

Any stand-alone uv-vis absorption spectrophotometer can be used to perform this experiment. Students will be introduced to the *Spectronic 20*, commonly called a *Spec 20*. This instrument is of historical as well as educational value since it is the *most common spectrophotometer ever to be found in undergraduate college chemistry labs.* Shown below are schematic drawings of the original analog electronic instrument usually placed right on a lab benchtop followed by an optical schematic of the light path inside this "black box."

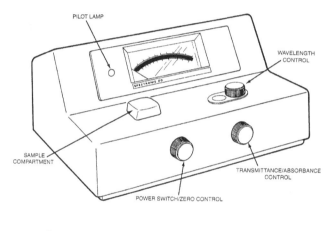

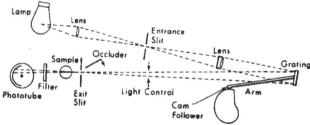

The light from a lamp is focused through a lens to an entrance slit where the light is dispersed by a rotating cam toward the exit slit whereby a narrow band of wavelengths can be absorbed or transmitted by a sample that contains a specific chemical substance. Note the location of an Occluder located just before the exit slit. Also note the role of this device in the troubleshooting section below.

5.5.2.5   Product Line History of the *Spec 20*

The original *Spec 20* was developed by Bausch & Lomb in 1953. The product line was sold to Milton Roy in 1985. Milton Roy sold its instrument group to Life Sciences International, renamed Spectronic Instruments Inc. in 1995. Spectronic Instruments was purchased by Thermo

Optek in 1997, renamed Spectronic-Unicam in 2001 and renamed Thermo-Spectronic in 2002. In 2003 the product was moved to Madison, WI and the brand renamed to Thermo Electron. In 2006 Thermo Electron merged with Fisher Scientific to become Thermo Scientific. *Spec 20* instruments found in laboratories today may bear any of these brand names.

5.5.2.6 Troubleshooting the *Spec 20*

The troubleshooting guide shown below is included in this experiment to give the student unique insights into what can go wrong when attempting to operate this or any other analytical instrument. Analytical chemists and chemical technicians working in analytical laboratories are continuously confronted with the day-to-day reality that analytical instruments will fail, will not work properly, cannot meet QA/QC minimum criteria, etc. The table below assists the student in identifying problems with the Spec 20 and how to solve these operational problems. This aspect of learning to work with analytical instruments is called troubleshooting.

	Operator's troubleshooting guide for the Spec 20 spectrophotometer	
Problem	**Possible Cause**	**Remedy**
1. Instrument does not work	a. Power line cord not connected to outlet. b. Dead power outlet. c. Source lamp burned out. d. Phototube burned out. e. Defective electronic component.	Plug in power line cord. Try a different power outlet. Replace with new lamp. Replace as required. Refer to service manual or service center.
2. Meter does not zero	a. Sample compartment cover not closed. b. Occluder binding. c. Lamp access door not tightly closed. d. Phototube defective. e. Defective electronic component.	Close cover. Check occlude action. Close door and retighten thumbscrew. Replace as required. Refer to service manual or service center.
3. Readings drift	a. Poor sampling technique. b. Fumes from sample. c. Excessive line voltage variation. d. Wrong line voltage setting. e. Source lamp defective. f. Phototube defective. g. Meter defective. (Spectronic 20 only). h. Defective electronic component.	Eliminate bubbles or particles in solution. Check voltage and grounding. Reset Line Voltage. Selection Switch. Replace with new lamp. Replace as required. Refer to service manual or service center. Refer to service manual or service center.
4. Cannot set 100%T (0.0A)	a. Occluder closed. b. Sample holder not fully inserted into adapter. c. Improper filter installed. d. Source Lamp weak. e. Wrong line voltage setting. f. Phototube weak. g. Error in wavelength calibration. h. Defective electronic component.	Install test tube in sample compartment. Insert fully. Remove of change filter. Replace with new lamp. Reset Line Voltage Selection switch. Replace as required. Check calibration. Refer to service manual or service center.

	Operator's troubleshooting guide for the Spec 20 spectrophotometer	
Problem	Possible Cause	Remedy
5. Readings are not repeatable even though meter reading is zero and 100%T control is set correctly	a. Loose lamp. b. Loose sample holder adapter. c. Poor analytical technique. d. Test tube position not repeating. e. Meter sticking (Spectronic 20 only).	Tighten thumbscrew. Tighten set screw. Clean or replace dirty test tubes; remove bubbles, etc. Always position fiducial line in exactly the sample place when test tube is inserted into adapter. Tap lightly for possible correction.

5.5.2.7 Procedure

Obtain a supply of 13 × 100 mm test tubes that are clean, dry, and free of scratches. Half-fill each tube with the 2% $CoCl_2$ solution. Set the wavelength to 510 nm on the spectrophotometer, then zero the readout. Choose one tube as a reference, place a vertical index mark near the top of the tube, and insert the tube into the sample compartment. Adjust the light control so that the meter reads 90% transmittance (T). Insert each of the other tubes and record their transmittances. If the %T is within 1% of the reference tube, place an index mark so that the tube can be inserted in the same position every time. If it is not within 1%, rotate the tube to see whether it can be brought into range. In future measurements, insert each tube in the same position relative to its index mark. Choose a set of seven tubes that have less than 1% variation in reading. Retain these tubes for subsequent photometric work and return the remainder. To compensate for variations between instruments, use the same instrument for both tube matching and experimental work.

To show that as the wavelength cam is manually rotated, different wavelengths are passed across the exit slit by the monochromator, place a piece of chalk into a test tube and insert into the sample compartment while leaving the cover open. Starting at 400 nm, scan the visible range up to 700 nm and observe the exit-slit image by looking straight down into the cell compartment as you rotate the cam.

5.5.3 For the Report

Because this is an introductory experiment and learning exercise, there is no report.

Typical calibration data obtained from a visible spectrophotometer for aqueous solutions that contain dissolved cobalt chloride ($CoCl_2$) is shown in the table below. $CoCl_2$ is a crystalline solid that readily dissolves in water to yield an aqueous solution that contains Co^{2+} and Cl^- ions. Only Co^{2+} contributes to the color of the aqueous solution.

Concentration (mg/L or ppm) $CoCl_2$ (aq)	Absorbance (absorbance units, AU)
500	0.100
600	0.235
700	0.415
800	0.610
900	0.795
1000	1.010

Use an Excel spreadsheet or equivalent to find the best straight line fit through the experimental x, y data points.

At the wavelength used in the measurement, a portion of a solution whose concentration is unknown is added to a cuvette, placed in the spectrophotometer, and the absorbance measured. Assuming that the unknown solute responsible for the color (in this case the aqueous cobalt(II) ion) is the same chemical species in which the previously prepared calibration plot was made, *find the concentration in #mg/L or ppm* from the calibration plot constructed, if the absorbance measured is 0.755 AU.

The graph shown below is a plot of the % absorption (not absorbance, be careful) versus wavelength. This is the absorption spectrum for Co^{2+} in the visible region of the electromagnetic spectrum. What wavelength should be used to conduct this quantitative analysis so as to obtain maximum sensitivity in measurement?

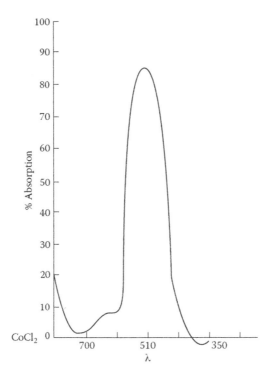

5.5.4 Suggested Readings

To develop this experiment, the author consulted the following resources:

- Harris W., B. Kratochvil. *An Introduction to Chemical Analysis*. 5th ed. New York: McGraw-Hill, 1985, pp. 23–31. The test matching test tube exercise was adapted from this excellent resource.
- Sawyer D. W. Heinemann, J. Beebe. *Chemistry Experiments for Instrumental Methods*. New York: John Wiley & Sons, 1984, pp. 163–214.

A number of good presentations can be found on the basic principles of spectrophotometry; a particularly good presentation is found in Chapter 19 of Harris D. *Quantitative Chemical Analysis*, 3rd ed. San Francisco: Freeman, 1991. Also consider Skoog D. J Leary. *Principles of Instrumental Analysis*. *4th ed*. Philadelphia: Saunders, 1992. Robinson, J. *Undergraduate Instrumental Analysis*. *5th ed*. New York: Marcel Dekker, 1995. A more comprehension treatment is found in: Ingle, J. S. Crouch. *Spectrochemical Analysis*. Englewood Cliffs, NJ: Prentice Hall,

1988. Authoritative writers have joined forces and published a recent textbook that covers the broad field of chemical analysis, consider: Skoog, D. J. Holler, S. Crouch. *Principles of Analytical Chemistry*, 7th ed. Boston: Cengage Learning. 2017.

5.6 VISIBLE SPECTROPHOTOMETRIC DETERMINATION OF TRACE LEVELS OF IRON IN GROUNDWATER

5.6.1 BACKGROUND AND SUMMARY OF METHOD

Iron (Fe), a metal in great abundance in the Earth, is a common contaminant in groundwaters in its oxidized forms, ferric ion, Fe^{3+} or iron (III) and ferrous ion, Fe^{2+} or iron (II). Common rust consists of ferric oxides and persists in groundwaters as either solubilized or particulate matter. For environmental analytical purposes, one must distinguish between total Fe and dissolved Fe that could be present in groundwaters. It is also of interest to determine the degree of metal speciation (i.e., the concentration of Fe (III) to that of Fe (II)). The hexa-aquo Fe (III) itself behaves as a weak acid and ionizes in water according to

$$Fe\left(H_2O\right)_6^{3+} \leftrightarrow Fe\left(H_2O\right)_5 OH^{2+} + H^+$$

and as a result contributes to *groundwater acidity*. Pure solutions of salts that contain the hexa-aquo Fe (III) are distinctly acidic, having a pH from about 2 to 4, depending on concentration. If either total dissolved Fe or Fe (II) is to be determined, the sample must be analyzed as soon as possible after collection. If only total Fe is to be determined, the sample should be immediately acidified with hydrochloric acid.

Fe (II), once formed by chemical reduction, forms an intensely colored complex ion with 1,10-phenanthroline according to the following reactions:

Only the Fe (II) oxidation state for iron forms the colored complex. Hence, this selectivity provides the basis for quantitatively determining the *ferric to ferrous ion concentration ratio* that characterizes the dissolved Fe portion of the iron analysis. Several organic compounds that are readily soluble in water and easily reduce Fe (III) to Fe (II) are available and include hydroquinone, ascorbic acid, and hydroxylamine hydrochloride, among others.

This exercise affords an opportunity to introduce good laboratory practices when conducting an analysis using a uv-vis spectrophotometer. Each student is given at least one unknown groundwater sample with which to measure the concentrations of Fe (III) and Fe (II).

You will need to consider how both oxidation states of iron can be quantitatively determined using the complexation with 1,10-phenanthroline method only. For a given volume of groundwater, the amount of Fe (III) and the amount of Fe (II) should approximate the amount of total Fe found independently by flame atomic absorption spectrophotometry (FLAA).

5.6.2 EXPERIMENTAL

5.6.2.1 Volumetric Glassware Needed per Student

1 Volumetric flask (100 mL)
1 Pipette (5 mL)
1 Pipette (10 mL)
1 Volumetric flask (500 mL)
1 10 mL pipette calibrated in 1/10 mL increments (10 mL).

5.6.2.2 Gravity Filtration Setup

1 Glass funnel and standard circular filter paper.

5.6.2.3 Chemical Reagents Needed per Student or Group

Note: All reagents used in this analytical method contain hazardous chemicals. Wear appropriate eye protection, gloves, and protective attire. Use of concentrated acids and bases should be done in the fume hood.

15 mL of 1.0% hydroxylamine hydrochloride. Dissolve 10 g hydroxylamine hydrochloride in every 100 mL of solution.
15 mL of 0.1% 1,10-phenanthroline. Dissolve 0.1 g 1,10-phenanthroline in enough acetone until completely solubilized, then add DDI for every 100 mL of solution.
5 mL of concentrated hydrochloric acid, HCl.
0.5 g of ferrous ammonium sulfate, $Fe_2(NH_4)_2(SO_4)_3$. Dissolve 0.35 g $Fe_2(NH_4)_2(SO_4)_3$ in a 500 mL volumetric flask half filled with DDI, then add 5 mL of concentrated HCl and adjust to the calibration mark with DDI. This gives a solution that is 100 ppm as Fe.
100 mL of saturated sodium acetate (NaOAc) solution. Dissolve enough NaOAc until crystal formation is observed.

5.6.2.4 Spectrophotometer

A stand-alone UV-vis spectrophotometer such as a GENESYS® (Spectronic Instruments) or equivalent is suitable.

An atomic absorption spectrophotometer (FLAA) or an inductively coupled plasma-atomic emission spectrophotometer (ICP-AES) should be available if the instructor chooses to include these instruments in this exercise. The instructional staff is responsible for operating these instruments and providing analytical results to students.

5.6.2.5 Procedure

1. Turn on the spectrophotometer and allow at least a 15 min warm-up time. Set the wavelength to 508 nm.
2. Prepare the calibration standards by pipetting 0 (this is called the reagent blank), 1, 2, 3, 4, and 5 mL aliquots (portions thereof) of the 100 ppm Fe stock solution into a 100 mL volumetric flask. Also prepare an instrument calibration verification (ICV) standard by pipetting 2.5 mL of the 100 ppm Fe stock solution into a 100 mL volumetric flask. Measure the ICV's absorbance after developing the color below in triplicate. The ICV is used to evaluate the precision and accuracy of any instrumental method via interpolation of the calibration curve and is essential to maintaining good quality control. *Hint:* To minimize contamination due to carryover when using only one volumetric, *prepare standards from low to high concentration.*

Standard No.	100 ppm Fe (mL)	V (total)	Concentration (ppm)
Blank	0	—	0
1	1.0	100	1.0
2	2.0	100	2.0
3	3.0	100	3.0
4	4.0	100	4.0
5	5.0	100	5.0
ICV	2.5	100	2.5

3. To 20 mL of each reference standard solution, add 10 mL of saturated NaOAc and 10 mL of 1% hydroxylamine hydrochloride. Wait 5 min and add 10 mL of the 0.1% 1,10-phenanthroline solution. Allow 10 min, then dilute to the calibration mark of the 100 mL volumetric beaker with DDI.
4. Carefully filter approximately 35 mL of the unknown groundwater sample if necessary. Pipette 20 mL of sample and place into a clean 100 mL volumetric beaker. Add reagents as done previously for the calibration standard preparation and adjust to the final volume with DDI.
5. Set the zero control and 100% transmittance controls according to the operating manual instructions using DDI. *Record transmittance values for the blank calibration standards, ICV measured in triplicate, and unknown sample.*

5.6.2.6 Determination of Total Fe by FLAA or ICP-AES

You may have an opportunity to use either FlAA or ICP-AES to quantitatively measure the total iron in the same groundwater samples that were used above. The same set of calibration standards that you prepared can be used and directly aspirated into either the flame or the plasma. If you choose FlAA, you may need to install an Fe hollow-cathode lamp.

5.6.3 FOR THE NOTEBOOK

Use a spreadsheet program such as Excel or its equivalent to construct a six-point calibration plot. The plot should show absorbance values on the ordinate and concentration as ppm Fe on the abscissa. Use a least squares regression to fit the experimental points and calculate the correlation coefficient. Calculate the percent error and the confidence interval at 95% probability for the ICV. Interpolate the calibration plot and obtain a concentration for Fe (III) and Fe (II) from the visible spectrophotometer. Obtain the concentration, C_{Fe}, for Fe(total) from the AA spectrophotometer. From these results, compare the calculated $C_{Fe(II)}/C_{Fe(III)}$ ratio from the colorimetric method against the AA method (i.e., conduct a mass balance).

5.6.4 SUGGESTED READINGS

To develop this experiment, the author consulted the following resources:

Pietryzk D., C. Frank. *Analytical Chemistry*, 2nd ed. New York: Academic Press, 1979, pp. 656–657.

Annual Book of ASTM Standards Part 31, water, 1980. Philadelphia: American Society of Testing Materials, 1980, pp. 438–442.

5.7 SPECTROPHOTOMETRIC DETERMINATION OF PHOSPHORUS IN EUTROPHICATED SURFACE WATER

5.7.1 BACKGROUND AND SUMMARY OF METHOD

The persistence of phosphates in lakes, rivers, and streams due to domestic and industrial pollution has led to elevated levels and is regarded as largely responsible for lake eutrophication. This was an even more serious problem up until some 20 years ago, prior to the ban on phosphate-containing detergents. In considering an environmental sample, one must distinguish between several chemical forms that contain the element phosphorus. Separation into dissolved and total recoverable phosphorus depends on filtration through a 0.45 μm membrane filter. Dissolved forms of phosphorus (P) include meta-, pyro-, and tripolyphosphates. The visible spectrophotometric procedure requires that all forms of P be chemically converted to the water-soluble orthophosphate ion, PO_4^{3-}. Hence, an acid hydrolysis step must be included in the method, which should account for all hydrolyzable P. A more rigorous conversion is necessary to include organo-phosphorous compounds and involves an acid-persulfate digestion of the sample. The following flowchart summarizes the analytical scheme to differentiate the various chemical forms of phosphorus:

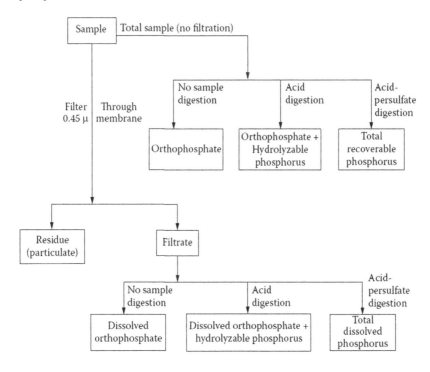

Orthophosphate forms a complex with the molybdate ion in the presence of a reducing agent such as hydrazine sulfate, amino-naphthol-sulfonic acid, tin (II) chloride, or ascorbic acid, which is commonly called *heteropoly blue* and has the molecular formula $H_3PO_4(MoO_3)_{12}$. There are two wavelengths that can be used: one between 625 and 650 nm and a more sensitive one at 830 nm. The Spectronic 21 DUV® is capable of measuring up to 1000 nm, whereas the Spectronic 20® can go only up to 600 nm with the standard phototube. An infrared phototube is necessary to widen the wavelength of this instrument.

This exercise provides an opportunity to reinforce the principles of spectrophotometry and its relationship to environmental analysis. To minimize laboratory time, the acid hydrolysis step will be eliminated, and thus only dissolved orthophosphate, PO_4^{3-}, will be measured. In addition to determining the phosphorus content of a surface water sample, each student will be given an unknown sample by the instructor whose phosphorus concentration has been previously determined by ion chromatography.

5.7.2 EXPERIMENTAL

5.7.2.1 Preparation of Chemical Reagents

Note: All reagents used in this analytical method contain hazardous chemicals. Wear appropriate eye protection, gloves, and protective attire. Use of concentrated acids and bases should be done in the fume hood.

5.7.2.1.1 5 M SULFURIC ACID

Use an appropriate size graduated cylinder and add 70.25 mL of concentrated sulfuric acid to a 250 mL volumetric flask that has been previously half filled with distilled deionized water (DDI). This solution will release heat. Allow to cool; then fill to the calibration mark with DDI. Label as appropriate. The unused 5 *M* sulfuric could be saved, diluted, and used in the exercise "Determination of Anionic Surfactants ..."

5.7.2.1.2 MOLYBDATE REAGENT

Dissolve 12.5 g of ammonium molybdate, $NH_4Mo_7O_{24}$, in approximately 100 mL of DDI in a 500 mL volumetric flask. Add 150 mL of 5 *M* sulfuric acid. Fill to the calibration mark with DDI. Label as appropriate.

5.7.2.1.3 1% ASCORBIC ACID

Dissolve 1 g of ascorbic acid into a 100 mL volumetric flask half filled with DDI. Adjust to the calibration mark with DDI. Label as appropriate.

5.7.2.1.4 PREPARATION OF STOCK PHOSPHORUS

Dissolve 0.2197 g of potassium dihydrogen phosphate, KH_2PO_4, that has been dried for 1 h at 104°C in approximately 200 mL of DDI in an appropriately sized beaker. You will need use of a correctly calibrated analytical balance. After dissolution is complete, transfer the solution to a 1 L volumetric flask and adjust to a final volume with DDI. Label as "1.0 mL = 0.05 mg P." Carefully pipette 50 mL of the stock solution into a 1 L volumetric flask half filled with DDI. Adjust to the calibration mark with DDI. Label as "1.0 mL = 0.0025 mg P."

Sample	Milliliters Std. P (1 mL = 0.0025 mg)	Milliliters Molybdate Reagent	Milliliters Ascorbic Acid Reagent	V_T (mL)	Concentration (ppm)
Blank	0	5	3	50	0
Std. 1	1	5	3	50	0.05
Std. 2	2	5	3	50	0.10
Std. 3	4	5	3	50	0.20
Std. 4	7	5	3	50	0.35
Std. 5	10	5	3	50	0.50
ICV	5	5	3	50	0.25 (expect)

5.7.2.2 Procedure

A 5 mL aliquot of sample is taken and transferred to a 50 mL volumetric flask. Add 5 mL of the molybdate reagent and 3 mL of the reducing agent. The mixture is diluted to volume, and after waiting 6 min for color development (time for development for standards and unknown should be the same), the absorbance is determined at 830 nm using the Spectronic 21 DUV or equivalent spectrophotometer. The ICV should be prepared in triplicate and its absorbance measured three times. Follow *good laboratory practices* (GLP), as discussed in Chapter 2.

5.7.3 FOR THE NOTEBOOK

Use a spreadsheet program such as Excel or its equivalent to construct a five-point calibration plot. The plot should show absorbance values on the ordinate and concentration as ppm P on the abscissa. Use a least squares regression to fit the experimental points and calculate the correlation coefficient. Calculate the percent error and the confidence interval at 95% probability for the ICV. Report on the unknown ppm P and estimate the confidence interval at 95% probability for both the surface water sample and the sample given to you by the instructor; be sure to include the code for this sample. Review this method in a resource such as *Standard Methods for the Examination of Water and Wastewater* or other sources of the colorimetric determination of phosphorus and discuss the effect of matrix interferences on the precision and accuracy for determining P in environmental samples using the visible spectrophotometric method.

5.7.4 SUGGESTED READINGS

To develop this experiment, the author consulted the following resources:

Pietryzk D., C. Frank. *Analytical Chemistry*, 2nd ed. New York: Academic Press, 1979, pp. 658. The author again drew on this excellent resource to adapt an experiment on measuring phosphorous in egg shells to measuring phosphorous in surface water based on forming the phospho-molybdate complex ($[PO_4 2MoO_3]^{3-}$).

Annual Book of ASTM Standards Part 31, Water, 1980. Philadelphia: American Society of Testing Materials, 1980, pp. 438–442. The author again drew on this well-known resource to help develop the above flowchart that outlines the numerous options available when attempting to quantitate phosphorous.

5.8 DETERMINATION OF ANIONIC SURFACTANTS BY MINI-LIQUID– LIQUID EXTRACTION (MINI-LLE) IN AN INDUSTRIAL WASTEWATER EFFLUENT USING ION PAIRING WITH METHYLENE BLUE

5.8.1 BACKGROUND AND SUMMARY OF METHOD

Synthetic detergent formulations make their way into the environment via industrial waste effluent. It is important that levels of these surfactants or surface-active substances be monitored. The classical analytical method that utilizes a visible spectrophotometer involves a consideration of the chemical nature of the surfactants. Surfactants are classified as either *anionic, nonionic,* or *cationic,* depending on the nature of the organic moiety when the substance is dissolved in water.

This mini-scale extraction method utilizes the ability of *anionic surfactants to form ion pair complexes with cationic dyes* such as methylene blue. These complexes behave as if they are neutral organic molecules. These ion pairs are easily extracted into a nonpolar solvent, thus imparting a color to the extract. The intensity of the color becomes proportional to the concentration of the surfactants in accordance with the Beer–Lambert law of spectrophotometry.

Anionic surfactant   Methylene blue

5.8.2 EXPERIMENTAL

This exercise introduces many of the techniques that are required to identify and quantitate an environmental pollutant while facilitating an understanding of the relationship between chemical principles and instrumental analysis.

5.8.2.1 Preparation of Chemical Reagents

Note: All reagents used in this analytical method contain hazardous chemicals. Wear appropriate eye protection, gloves, and protective attire. Use of concentrated acids and bases should be done in the fume hood.

5.8.2.1.1 METHYLENE BLUE (MB)
Dissolve 0.05 g of MB in 50 mL of distilled deionized water (DDI).

5.8.2.1.2 3M SULFURIC ACID
Add 16.7 mL of concentrated sulfuric acid, H_2SO_4, to a 100 mL volumetric flask half filled with DDI. Adjust to the calibration mark with DDI.

5.8.2.1.3 TO PREPARE A 0.5 M SULFURIC ACID SOLUTION
Add approximately 4 mL of the 3 M sulfuric acid to a 25 mL volumetric flask half filled with DDI, then adjust to the final volume. Transfer to a storage vial and label.

5.8.2.1.4 TO PREPARE A 0.1 M SODIUM HYDROXIDE SOLUTION
Weigh approximately 0.1 g of NaOH pellets into a 50 mL beaker that contains approximately 20 mL of DDI. Dissolve with stirring. Transfer the contents of the beaker to a 25 mL volumetric flask and adjust to the final volume. Transfer to a vial and label.

5.8.2.1.5 TO PREPARE THE WASH SOLUTION
To a 1000 mL volumetric flask half filled with DDI, add 41 mL of 3 M sulfuric acid to 5 g of $Na_2HPO_4H_2O$. Adjust to the mark with DDI.

5.8.2.1.6 TO PREPARE THE MB REAGENT
To a 500 mL volumetric flask half filled with DDI, add the following:

   15 mL of MB
   20.5 mL of 3 M sulfuric acid
   25 g of $NaH2PO_4 \cdot H_2O$.

   Shake until dissolved. Then adjust to the calibration mark with DDI. Transfer contents of volumetric flask to a glass storage bottle and label "MB Reagent."

5.8.2.2 Preparation of the 100 ppm Surfactant Stock Solution and General Comments on Standards

Dissolve approximately 0.01 g of sodium lauryl sulfate in between 5 and 10 mL of MeOH in a 50 mL beaker. Transfer the contents of the beaker to a 10 mL volumetric flask and adjust to the mark with MeOH. This yields a stock solution whose concentration is 1,000 ppm. Transfer 1 mL using a glass pipette and pipette pump to a 10 mL volumetric flask. Adjust to the mark with DDI. This yields a primary dilution reference standard whose concentration is 100 ppm. Refer to the calibration table below to prepare the blank, calibration standards and initial calibration verification (ICV) standard. A molecular structure for sodium lauryl sulfate or sodium dodecyl sulfate also known as sodium laureth sulfate is shown below:

It becomes important to *know the chemical nature of the particular surfactant* to be used to prepare the stock standard for construction of the calibration table, so that the number of parts per million (#ppm) can be related to the concentration of anionic sulfonate actually taken and extracted as an ion pair. For example, for *p*-toluene sulfonic acid, 100 mg as the *p*-toluene sulfonate ion is only 100.6 mg as the acid; however, 100 mg as the *p*-toluene sulfonate ion is 113.4 mg as its sodium salt. A stock solution is stable for no more than 1 week. Working solutions such as the 100 ppm surfactant should be prepared fresh daily. Keep in mind that approximately *0.1 g* of any pure solid dissolved in enough solvent to prepare *10 mL* of solution yields a standard whose concentration is approximately *10,000 ppm!*

Calibration Table

Sample	100 ppm Surfactant (#μL)	Surfactant Added (#μg)	MB Reagent (#μL)	DDI (#μL)	CH2Cl2 (total) (#μL)	Concentration Original Sample (#ppm)
Blank	0	0	2.5	10	10	0
Std. 1	100	10	2.5	10	10	1.0
Std. 2	250	25	2.5	10	10	2.5
Std. 3	500	50	2.5	10	10	5.0
Std. 4	1000	100	2.5	10	10	10
ICV	400	40	2.5	10	10	4.0

5.8.2.3 Operation and Calibration of the Orion SA 720A pH Meter

The pH meter must be set up and calibrated with two buffers. Use buffer solutions whose pH values are 7 and 10, as these buffers have pH values near those required in this method. Refer to the instructions for operating the specific model of the pH meter available at your workbench.

5.8.2.4 Procedure for Mini-LLE

1. Using a clean 25 mL graduated cylinder, place 10 mL of an aqueous sample whose anionic surfactant concentration is to be determined into a 50 mL beaker. The aqueous sample could be a blank (i.e., a sample with all reagents added except for the analyte of interest), calibration standard, ICV, fortified (i.e., spiked) sample, or an actual unknown wastewater effluent sample.
2. Neutralize the sample to a pH between 7 and 8 by dropwise addition of either 1 M NaOH or 0.5 M H_2SO_4.
3. Add 2.5 mL of the MB reagent and swirl; then transfer the contents of the beaker to a 30 mL glass separatory funnel. *Be sure the stopcock is closed.*
4. Add 2 mL of methylene chloride (dichloromethane) or equivalent solvent. Because methylene chloride is much denser than water, it will comprise the lower layer after the two phases separate.

5. Stopper the separatory funnel with the ground-glass top, invert the funnel, and shake. Be sure to vent the vapor. This should be done in the fume hood.

6. Withdraw the lower layer into a second clean beaker.

7. Extract the remaining aqueous phase two more times with 2 mL portions of methylene chloride. Then combine the methylene chloride extracts. Approximately 6 mL of organic solvent should be obtained at this point.

8. Wash the combined extracts with wash solution. To do this, transfer the 6 mL of extract (MeCl₂ + dissolved ion pair) to a clean 30 mL separatory funnel. Add approximately 10 mL of wash solution, shake, allow time for the two phases to separate, and then remove the lower layer directly into a clean 50 mL beaker.

9. Transfer the washed extract (approximately 6 mL) to a 10 mL volumetric flask and adjust to a final volume.

10. Transfer a portion of the 10 mL methylene chloride extract to a standard spectrophoto-metric cuvette and measure the absorbance against methylene chloride as a blank in the reference cell at 652 nm. *Record and repeat steps 1 through 10 for all standards and samples.*

11. *Discard the waste extract into the hazardous waste receptacle* located in the laboratory.

5.8.3 FOR THE REPORT (A WRITTEN LABORATORY REPORT DUE ON THIS EXPERIMENT)

Use a spreadsheet program such as Excel or equivalent to construct a four-point calibration plot. The plot should show absorbance values on the ordinate and concentration of surfactant in ppm on the abscissa. Calibration data and the least squares regression plotted in Excel for both an external standard mode and a standard addition mode for the quantitative analysis of anionic surfactants using the methylene blue colorimetric method, *are shown below.* Use a least squares regression to fit the experimental points and calculate the correlation coefficient. Calculate and report on the percent error and the confidence interval at 95% probability for the ICV. *Report on the concentrations of any unknown environmental samples to which this method was applied. Discuss* what you learned from this environmental analysis method drawing on your previous experience with spectrophotometric methods. *Comment* on the precision and accuracy afforded by the analytical method.

Two representative calibration plots comparing *the external standard* approach and *the standard addition* approach to instrument calibration for this analytical method are shown below:

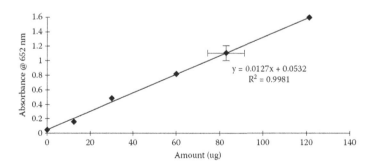

Calibration curve of anionic surfactants paired with methylene blue dye using the *external standard* mode of instrument calibration. The error bars represent the standard deviation in ICVs. ♦, standards; ■, ICV; linear regression on standards.

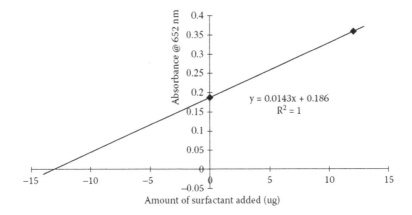

Calibration curve of anionic surfactants paired with methylene blue dye using the *standard addition* mode of instrument calibration. The graphical technique showing standard addition. The equation is for the regressed line through the points for the spiked sample and for the average of the triplicate runs of the unspiked samples.

5.8.4 SUGGESTED READINGS

To develop this experiment, the author consulted the following resources:

Sawyer D., W. Heinemann, J. Beebe. *Chemistry Experiments for Instrumental Methods.* New York: John Wiley & Sons, 1984, pp. 163–214.

Schmitt T.M. *Analysis of Surfactants, Surfactant Science Series Volume 40*, New York: Marcel Dekker, 1992, pp. 290–292.

5.9 COMPARISON OF ULTRAVIOLET AND INFRARED ABSORPTION SPECTRA OF CHEMICALLY SIMILAR ORGANIC COMPOUNDS

5.9.1 BACKGROUND AND SUMMARY OF METHOD

Because most organic compounds known to pollute the environment are colorless, it would be close to impossible to identify them and therefore to quantitate if only a visible spectrophotometer were available. Do not despair! The ultraviolet (UV) spectrophotometer enables the full UV region to be used, and these organic contaminants can be identified. The UV region is defined to be those wavelengths of electromagnetic radiation *from 190 to 400 nm*. We know that the internal energies of atoms and molecules are quantized; that is, only certain discrete energy levels are possible, and the atoms and molecules must exist at all times in one or the other of these allowed energy states. For absorption of radiation to occur, a fundamental requirement is that the energy of the photon absorbed must match exactly the energy difference between initial and final energy states within the atom or molecule. Consideration of atoms falls within the realm of atomic absorption and atomic emission energy states. In contrast, *molecular absorption* involves transitions between electronic (ultraviolet-visible absorption), vibrational (mid-infrared absorption), and rotational (microwave absorption) quantized energy states. The following diagram depicts quantized energy states for organic molecules that are dissolved in a solvent:

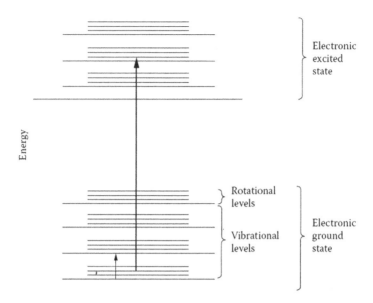

The molecular absorption phenomenon can only be accurately measured provided that the ratio of transmitted intensity, I, of the UV radiation to that of the incident intensity, I_0, is due to the presence of the dissolved solute and not to scattering of the incident beam. If the optical windows (see below) absorb UV radiation, then the absorbance that is related to the logarithm of the ratio I/I_0 would cause an increase in sample absorbance; hence, this would lead to an erroneous result. The student will encounter two types of optical window material. One consists of *glass* and is said to have a UV cutoff (UV wavelengths below cutoff value would absorb) of 300 nm (near UV), and the other consists of *quartz* with a UV cutoff of 190 nm. The rectangular cuvette is depicted as follows:

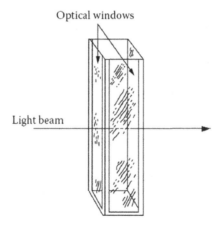

5.9.1.1   UV-Vis Absorption Spectroscopy

The light source for molecular absorption spectroscopy in the visible and UV regions (i.e., 750 nm down to 350 nm) is a tungsten filament lamp, which radiates as a black body at about 2,800° K. Below 350 nm, the hydrogen or deuterium gas discharge lamp is preferred. The *reciprocal linear dispersion* (refer to *Suggested Readings* below or other equivalent text for explanation) is around 1 nm/mm for diffraction grating spectrophotometers and between 0.5 and 5 nm/

mm for prism spectrophotometers. (*Note*: Dispersion depends on wavelength for a prism.) Air absorbs UV light of wavelengths shorter than about 180 nm, so studies at wavelengths shorter than 180 nm require the use of an evacuated spectrometer. This region is thus termed the *vacuum ultraviolet*.

The role of the solvent becomes critical in obtaining accurate UV absorption spectra. A solvent considered suitable for use in the UV-vis region must itself exhibit a low absorbance as well as dissolve the solute whose spectrum is sought experimentally. Fortunately, most common solvents are highly transparent to visible light, but all begin to absorb at some wavelength in the UV. It is not essential that the solvent have 100% transmittance, although it is desirable that as large a fraction as possible of the incident radiant energy be available for absorption by the solute. The following table lists the *approximate absorption cutoffs* for several widely used solvents. The cutoff defines a practical short-wavelength (λ) limit for the useful range of a solvent. Below this wavelength, the absorbance of the solvent, if placed in a 1 cm cell, exceeds 1.0 absorbance unit full scale (AUFS).

Solvent	Cutoff λ (nm)	Solvent	Cutoff λ (nm)
Water	200	Dichloromethane	230
Hexane	200	Chloroform	245
Heptane	200	Carbon tetrachloride	265
Cyclohexane	210	Dimethyl sulfoxide	265
Methanol	210	Dimethyl formamide	270
Ethanol	210	Benzene	275
Acetonitrile	190	Pyridine	300
Dioxane	215	Acetone	325

Different solutes exhibit different UV absorptivities. Recall that what is measured in a spectrophotometer is absorbance, and absorbance A is related to absorptivity ε via the following mathematical relationship

$$A = \varepsilon bc$$

Two chemically different solutions that contain solutes at identical concentrations in a suitable solvent using the same cuvette would not be expected to have the same absorbance due to *differences in absorptivity* (earlier terms included molar absorptivity and extinction coefficient). If logarithms are taken on both sides of the above equation, we have

$$\log A = \log \varepsilon + \log bc$$

Because the log bc term is independent of wavelength, log A will vary only as a function of absorptivity. Thus, a plot of log A vs. spectrophotometer wavelength setting will be the same even though concentrations and path lengths for individual samples may differ. In this way, a comparison of the different curves can be made.

5.9.1.2 Mid-Infrared Absorption Spectroscopy

The mid-infrared region of the electromagnetic spectrum begins on the higher energy end at 2.5 μm (4,000 cm^{-1}) and ends on the lower energy end at 6 μm (650 cm^{-1}). Dispersive type infrared absorption spectrometers have given way to *Fourier-transform infrared (FTIR) spectrometers*. Refer to *Suggested Readings* below or other equivalent texts to better understand the principles underlying infrared spectroscopy.

5.9.2 Experimental

This exercise affords students an opportunity to measure actual UV and FTIR absorption spectra of several organic solutes while comparing overall differences in UV and FTIR spectra for chemically similar and dissimilar organic solutes. The laboratory experiment focuses on the influence of delocalization of electron density and the nature of UV absorption spectra. The experiment also focuses on the relationship of organic functional group analysis and FTIR absorption spectra.

Two sets of illustrative solutions are to be studied in this exercise. The first set consists of *organo-sulfonates* and enables a comparison of UV absorption spectra of alkane vs. aromatic sulfonates that are dissolved in water. The second set consists of two representative *organic esters* that differ in their carbon backbone. The student is to engage both the UV spectrophotometer and the FTIR spectrometer in accomplishing the experimental aspects of this exercise.

5.9.2.1 Items/Accessories Needed per Student or Group

1 Pair of matching rectangular quartz cuvettes that transmit between 200 and 350 nm
1 Polyethylene (PE) disposable IR card (Type 61, 3M)
1 Polytetrafluoroethylene (PTFE) disposable IR card (Type 62, 3M)
1 Model 160-UV (Shimadzu) UV-vis scanning absorption spectrophotometer or other equivalent scanning instrument
1 Model 1600 Fourier-transform infrared spectrometer (PerkinElmer) or other equivalent instrument.

5.9.2.2 Preparation of Chemical Reagents

Note: All reagents used in this analytical method contain hazardous chemicals. Wear appropriate eye protection, gloves, and protective attire. Use of concentrated acids and bases should be done in the fume hood.

5.9.2.3 Procedure to Obtain UV Absorption Spectra for Two Sets of Chemically Similar Organic Compounds: (1) An Alkane Sulfonate vs. Alkyl Sulfate and (2) Two Esters with Different Carbon Backbones

Prepare two aqueous solutions that contain approximately 1,000 ppm each of *1-octanesulfonic acid* or *sodium dodecyl sulfate* (use whichever is available) *and* p-toluene sulfonic acid dissolved in distilled deionized water (DDI). Transfer a portion of this solution into a quartz rectangular cuvette and record the UV absorption spectrum for both organic compounds. Prepare a 100 ppm solution that contains *methyl methacrylate* and a 100 ppm solution that contains *ethyl glycolate*. You will need to use the wavelength cutoff guide to choose a suitable solvent because these organic compounds do not dissolve to any great extent in water. Record the UV absorption spectrum between 200 and 350 nm. Compare the spectra when a solvent is placed in the reference beam of the dual-beam instrument. Molecular structures for these unique organic compounds are shown below:

1-octanesulfonic acid p-toluene sulfonic acid methyl methacrylate ethyl glycolate

5.9.2.4 Procedure to Obtain FTIR Absorption (Transmission) Spectra for Various Organic Compounds

Disks that are partially to fully transparent in the infrared are available to easily prepare samples. An aqueous solution containing the anionic surfactants can be deposited directly onto either the PTFE or PE disk. Allow sufficient time for the solvent to evaporate off of the disk. Conduct a survey scan and take 16 FTIR scans. Plot the spectra. Consult the staff for assistance with the Model 1600 FTIR® (PerkinElmer). FTIR absorption (transmission) spectra for both sulfonate surfactants and for both esters can be obtained using this technique.

5.9.3 FOR THE REPORT

This exercise has introduced selected principles of molecular UV and infrared spectroscopy for qualitative chemical analysis. Interpret the significance of these spectra in terms of molecular structure. You should have obtained hard-copy printouts of UV and FTIR spectra. Please include all relevant spectra in your report.

Suppose that a client sent to you a mixture that was prepared from the ethyl benzene and styrene solutions that you used in the lab. Would benzene be a suitable solvent to use in obtaining UV spectra for ethyl benzene and styrene? Discuss how you would design a quantitative analysis to determine the concentration of ethyl benzene and of styrene in the client's sample. Assume that you have available pure ethyl benzene and pure styrene (chemists refer to these as neat forms of the liquids).

5.9.4 SUGGESTED READINGS

To develop this experiment, the author consulted the following resources:

- Thompson C. *Ultraviolet-Visible Absorption Spectroscopy*. Boston: Willard Grant Press, 1974, Chapters 2 and 3.
- Skoog D. J. Leary. *Principles of Instrumental Analysis,* 4th ed. Philadelphia: WB Saunders, 1992.
- Crooks J. *The Spectrum in Chemistry*. New York: Academic Press, 1978, pp. 100–101.
- Sawyer D., W. Heinemann, J. Beebe. *Chemistry Experiments for Instrumental Methods*. New York: John Wiley & Sons, 1984, pp. 215–221.

5.10 DETERMINATION OF OIL AND GREASE AND OF TOTAL PETROLEUM HYDROCARBONS IN WASTEWATER VIA REVERSED-PHASE SOLID-PHASE EXTRACTION TECHNIQUES (RP-SPE) AND QUANTITATIVE FOURIER-TRANSFORM INFRARED (FTIR) SPECTROSCOPY

5.10.1 BACKGROUND AND SUMMARY OF METHOD

One of the most informative and straightforward analytical methods involves the quantitative analysis of environmental samples that have been contaminated with what can be collectively termed oil and grease. The term *oil and grease* describes the extent to which an environmental sample such as wastewater or soil is contaminated. *Oil and grease* refers to any and all hydrocarbons including: lipids, high-molecular-weight fatty acids, triglycerides, higher olefinic hydrocarbons, alkanes, alkenes, monocyclic and polycyclic aromatics, and so forth. The total petroleum hydrocarbon (TPH) content of this oil and grease contamination can also be quantitatively determined by this method if the extracted *sample is placed in contact with silica gel*. The software that will be used with the FTIR spectrophotometer distinguishes between these two definitions. The absorption in the infrared region of the electromagnetic spectrum due to the presence of *aliphatic and aromatic carbon to hydrogen-stretching vibrations* occurs in the 3,100 to 2,900 cm^{-1} range. These concepts form the physicochemical basis for the instrumental measurement.

The method is adapted from Method 5520, "Oil and Grease," in *Standard Methods for the Examination of Water and Wastewater* and follows from EPA Method 413.2. EPA Method 1664 Revision B and EPA Method 413.1 replaced the extraction solvent Fluorocarbon 113 or 1,1,2-trichloro-1,2,2,-trifluoroethane with n-hexane. At present, EPA Method 1664B is called n-Hexane extractible and the sample matrix is referred to as *HEM oil and grease* while a replacement for the infrared absorption determinative technique is provided by a gravimetric determinative technique following distillation and removal of the n-hexane extraction solvent.

Primarily for educational purposes, the FTIR determinative technique will be used in this experiment because it enables students to again be introduced to the *reversed-phase solid-phase extraction sample prep technique*. We will *responsibly recover* the Fluorocarbon 113 solvent whose waste disposal is the subject of current environmental concern and debate. Software developed at PerkinElmer Corporation during the late 1980s has been downloaded to the Model 1600® FTIR spectrophotometer and will be used to provide for the quantitative analysis. The original concept was first published by and at the EPA in the early 1970s.

An aqueous sample whose oil and grease content is to be determined is first acidified, then extracted using 1, 1, 2-trichloro-1, 2, 2-trifluoroethane (TCTFE) which is a liquid at room temperature. TCTFE has a *normal boiling point* of 48°C. This relatively low boiling point classifies this solvent as a VOC. The molecular structure for this solvent is shown below:

1,2,2-trichloro-1,2,2-trifluoroethane

Molecules of this solvent lack a C–H covalent bond, and thus serve to provide an excellent background infrared spectrum because no absorption in the 3,100 to 2,900 cm^{-1} region is found. The extract is isolated from the aqueous matrix, and an aliquot (a portion thereof) is transferred to a 1 cm (path length) quartz cuvette. The FTIR absorbance vs. wavelength or wave number is graphically displayed on the screen. This absorbance, which is initially related to the concentration of oil and grease in the chemical reference standards used to calibrate the instrument yields the concentration as oil and grease in the unknown sample via interpolation of the least squares regression fit. The common concentration unit used is mg of oil and grease per 100 mL of solution (reported as a #mg oil & grease /100 mL sample) Students should be aware that the reference standards and samples after RP-SPE will be dissolved in TCTFE.

A soil/sediment or contaminated sludge with a very high solids content can be extracted via Soxhlet extraction techniques or, more conveniently, via ultrasonic probe sonication into a mixture of TCTFE and a carefully weighed amount of soil/sediment/sludge.

5.10.2 EXPERIMENTAL

5.10.2.1 Preparation of Chemical Reagents

Note: All reagents used in this analytical method contain hazardous chemicals. Wear appropriate eye protection, gloves, and protective attire. Use of concentrated acids and bases should be done in the fume hood.

5.10.2.2 Reagents Needed per Student or Group of Students

0.1 g of hexadecane, $C_{16}H_{34}$. Alternatively, dodecane, $C_{12}H_{26}$, can be substituted
0.1 g of iso-octane (2, 2, 4-trimethyl pentane), $(CH_3)_3CCH_2CH(CH_3)$.
0.1 g of benzene, C_6H_6
250 mL of TCTFE
100 mL of MeOH (methanol). Use as high a purity as is available
5 g of silica gel, needed only if TPHs are to be determined in addition to oil and grease
5 g of sodium sulfate anhydrous (Na_2SO_4)
20 mL of 1:1 hydrochloric acid (HCl)
Mix equal volumes of acid and deionized water. Remember, always add acid to water.

5.10.2.3 Apparatus Needed per Group

Vacuum manifold to conduct solid-phase extraction (SPE)
Water trap and associated vacuum tubing to be used with the Vacuum manifold
Suction vacuum pump connected to the water trap
Accessories for use of the SPE vacuum manifold, including a receiving rack
SPE cartridges packed with C_{18}-bonded silica
70 mL sample reservoirs
10 mL glass volumetric flasks.

5.10.2.4 Procedure

Refer to the oil and grease analysis method from *Standard Methods* (refer to suggested References below) or equivalent and implement the appropriate procedure. For wastewater samples, follow the procedure outlined below. If a percent recovery study is required, the procedure that immediately follows provides guidance for this study.

5.10.2.5 Percent Recovery Study

To isolate and recover TPHs from water by RP-SPE combined with quantitative FTIR:

- Spike approximately 200 mL of high-purity laboratory water that has been *acidified with 1:1 HCl* with approximately 2 mg of dodecane; note that most of the dodecane will float on top. This sample is called a *matrix spike,* using terms first developed at the EPA. Spike a second 200 mL portion of *acidified* water with approximately 2 mg of dodecane. This sample is called a *matrix spike duplicate.* Leave a third 200 mL portion of water unspiked and *acidify.* This sample is called a *method blank.*
- Set up three C_{18} SPE cartridges and condition with MeOH.
- Pass the method blank, matrix spike, and matrix spike duplicate samples through the cartridges under vacuum.
- Remove water droplets with a Kim-Wipe; apply more vacuum to remove water from the sorbent.
- Elute off the cartridge with two 500 μL portions of TCTFE into a 1 mL volumetric as receiver; use glass syringe that is available near the SPE manifold.
- Transfer contents of the 1 mL receiver to the quartz cuvette and add exactly 1.0 mL of TCTFE using a glass pipette.
- Call previous calibration standards from disk on Model 1600 FTIR.
- Run each of the eluents from SPE. Transfer each eluent to a quartz cuvette and measure the absorbance. For the sample ID, enter any number. For the initial mass of sample, enter "200," and for the volume of sample after extraction, enter "2."

For the preparation of a control (i.e., a 100% recovered sample), weigh approximately 2 mg of dodecane into a 1 mL volumetric flask half filled with TCTFE and adjust to the mark with TCTFE. Transfer to a quartz cuvette, add 1 mL of TCTFE, and measure the absorbance using the Model 1600® PerkinElmer FTIR or equivalent instrument.

5.10.2.6   Probe Sonication: Liquid–Solid Extraction

It is first necessary to estimate to what extent the solid sample is laden with oil and grease. This minimizes the necessity to dilute the extract so that the absorbance will remain on scale. This can be accomplished by taking 0.5, 5, and 15 g samples and using identical extraction volumes. Once the optimum sample weight has been estimated, proceed to the next step.

5.10.2.7   Calibration of the FTIR Spectrophotometer

Calibrate the Model 1600® FTIR (PerkinElmer) or equivalent instrument by first preparing a series of working calibration standards. The table below serves as a useful guide and yields concentrations that are compatible with the software that operates the instrument.

Prepare Blend A by obtaining a total weight of approximately 0.10 g for the pure form of the oil that is to be defined as the reference. For example, if hexadecane, iso-octane, and benzene are to be used and mixed, add approximately 0.033 g of each to obtain the desired weight. Transfer the oil to a clean, dry 10 mL volumetric flask. Add about 5 mL of TCTFE to dissolve the oil, then adjust to the calibration mark. Transfer to a clean, dry glass vial with a Teflon/silicone septum and screw cap. Label this solution "Blend A, 1,000 mg Oil and Grease/100 mL (in TCTFE)"; prepare Blends B through F according to the following table:

Blend	Blend A (mL)	Extract Volume (mL)[a]	Concentration (mg/100 mL)
F	0.01	10	1.0
E	0.1	10	10
D	0.2	10	20
C	0.3	10	30
B	0.4	10	40
ICV[b]	0.25	10	Unknown

a   Use 10 mL glass volumetric flasks and TCTFE to adjust to final volume.
b   ICV = instrument calibration verification standard; run as if it were a sample; enter a "1" for sample weight and "1" for extract volume.

When you are ready to perform the calibration, retrieve under method "og & ph" (oil, grease, and petroleum hydrocarbons) and exercise one of the six options. It is important to obtain a fresh background by placing TCTFE into a clean 1 cm quartz cuvette. If the blank reveals a large absorbance, the *quartz cuvettes must be cleaned with detergent.* Contamination of the surface of the quartz cuvettes represents a major source of error with this method.

5.10.2.8   Isolation, Recovery and Quantitation of Oil and Grease from Wastewater Samples

- Place approximately 200 mL of wastewater sample into a clean, dry 250 mL beaker using a graduated cylinder. *Record the volume of sample in your lab notebook.*
- Acidify to a pH of approximately 2 by adding sufficient 1:1 HCl.
- Set up the SPE vacuum manifold and condition the sorbent with MeOH; the sorbent surface should be wet with MeOH prior to passing the wastewater sample through the cartridge.
- Connect the 70 mL polyethylene sample reservoir to the cartridge via the adapter.

- Pass 200 mL of sample through under vacuum; observe that the sample actually flows through the SPE sorbent and watch for plugging.
- Remove the reservoir; remove the water droplets on the inner wall of the cartridge barrel.
- Elute with two 500 µL of TCTFE directly into a 1.0 mL volumetric flask.
- Add enough anhydrous sodium sulfate, if necessary, to remove residual water in the eluent and stir.
- Transfer the contents of the volumetric to the quartz cuvette, add, with a pipette, 1 mL of additional TCFFE, and measure the absorbance using the Model 1600 FTIR.
- The printer will give you a hard-copy output after a certain number of FTIR scans have been acquired. *Record all absorbance data in your laboratory notebook*. Transfer all waste TCTFE to a properly labeled waste receptacle for recovery purposes.
- If the absorbance is too large, make the appropriate dilution. Repeat the dilution step if necessary. Transfer all waste TCTFE to a properly labeled waste receptacle for recovery purposes.
- Prior to each day's FTIR measurements, the ICV standard should be measured and a log kept of its daily interpolated concentration. If the ICV value becomes a statistical outlier, rerun the calibration.

5.10.3 CALCULATIONS

The printout for a correctly measured TCTFE extract gives a direct value for the concentration of oil and grease in a solid sample, provided the weight of sample is entered in grams and the volume of extract is entered in milliliters. Because we are measuring sample volume, instead of weight, *substitute for the weight with 200 mL* and use an extract volume of *2 mL*. The calculation is based on the following:

$$\frac{(2mL)(\# \, mg \, oil \, and \, grease \, / \, 100mL)(DF)}{200mL \, sample} \, x \, \frac{1000mL}{1L}$$

$$= \# \frac{mg}{L}(oil \, and \, grease) \, in \, wastewater \, sample$$

where DF is the dilution factor. If no dilution, assume DF = 1. If a 1:25 dilution is made, DF would equal 25.

5.10.4 SUGGESTED READINGS

To develop this experiment, the author consulted the following resources:

- Eaton A., L. Clesceri, A. Greenburg, Editors. *Standard Methods for the examination of Water and Wastewater,* 19th ed. Washington, D.C., *American Public Health Association, American Water Works Association, Water Environment Federation*, 1995, pp. 5–30–5–35.
- EPA Method 413.2, Oil and Grease. Total Recoverable Hydrocarbons, Environmental Protection Agency, 1978.
- Method 1664, *N-Hexane extractable Material by Extraction and Gravimetry, EPQ-821-B-94-004.* Washington, DC: Office of Water Engineering and Analysis Division, 1995.
- McGratton, B. 1600 Method Program Automating ASTM Standard Test Method for Oil and grease and Petroleum Hydrocarbons in Water. PerkinElmer Infrared Bulletin 114.
- EPA Method 1664 Revision B February 2010. To access the method, the student should "google" the "EPA Method 1664B" and proceed to the procedure.

This experiment was adapted from: Gruenfeld, M. *Enviro Sci Technol* 7: 636–639, 1973.

5.11   DETERMINATION OF THE DEGREE OF HARDNESS IN VARIOUS SOURCES OF GROUNDWATER USING FLAME ATOMIC ABSORPTION SPECTROSCOPY

5.11.1   BACKGROUND AND SUMMARY OF METHOD

The extent to which groundwater has been rendered *hard* has been defined as the *concentration of dissolved bicarbonates containing calcium (*Ca^{2+}*) and magnesium (*Mg^{2+}*) ions present in the sam*ple. Hardness can be quantitatively measured by finding some way to measure these two alkaline–earth metal ions. The classical method that still finds widespread use is titration with EDTA. We are going to approach the problem by measuring the concentration of Ca^{2+} by *flame atomic absorption spectroscopy* (FLAA) and in addition, by a mere change of hollow-cathode lamps, by measuring the concentration of Mg^{2+} by FLAA. Several sources of groundwater will be obtained, and the concentration of the chemical elements Ca and Mg will be used to estimate the degree of water hardness.

Recall that an *analytical method's precision* is a measure of the degree to which it can be reproduced or repeated and is evaluated by calculating the standard deviation for the method's instrument calibration verification (ICV) standard following establishment of a single-point or multipoint calibration. An *analytical method's accuracy* is a measure of how close the results are to an established or authoritative value and is evaluated by calculating the percent relative error.

FLAA requires a means by which an aqueous solution containing metal ions can be aspirated into a reducing flame environment by which atomic Mg or Ca vapor is formed. Photons from the characteristic Mg emission of a hollow-cathode lamp (HCL) are absorbed by ground-state Mg atoms present in the approximately 2,300°C air–acetylene flame. The amount of radiant energy absorbed as a function of concentration of an element in the flame is the *basis of quantitative analysis using AA* and follows Beer's law. In contrast to molecular absorption in solution, atomic spectra consist of lines and originate from either atomic absorption or atomic emission processes, which are depicted schematically below:

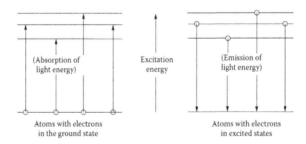

Instrument detection limits (IDLs) for most metals by FLAA are in the low-ppm realm in contrast to graphite furnace AA (GFAA) where IDLs can reach down to low-ppb concentration levels. The conventional premixed chamber type nebulizer burner is common. The sample is drawn up through the capillary by the decreased pressure created by the expanding oxidant gas at the end of the capillary, and a spray of fine droplets is formed. The droplets are turbulently mixed with additional oxidant and fuel and pass into the burner head and the flame. Large droplets deposit and pass down the drain; 85 to 90% of the sample is discarded in this way.

Flame atomic absorption spectrometry was developed in (ironically) the year 1955 independently by Walsh in Australia and by Alkemade and Milatz in the Netherlands. Because electrons in quantized energy states for atoms of alkali metal elements can easily be raised to excited states (see above), flame emission spectroscopy is a more appropriate instrumental technique, whereas plasma sources are needed for atoms of most other elements. Atomic absorption spectroscopy is unique in that it uses a flame to create the atomic vapor within which the absorption of radiation from a hollow-cathode lamp HCL can occur. This enables the quantitative determination of some 65 elements, provided a line source can be used. The source of light for AA must produce a narrow band of adequate intensity and stability for prolonged periods. An ordinary monochromator is incapable of yielding a band of radiation as narrow as the peak width of an atomic absorption line. HCLs satisfy these criteria. Refer to the following schematic:

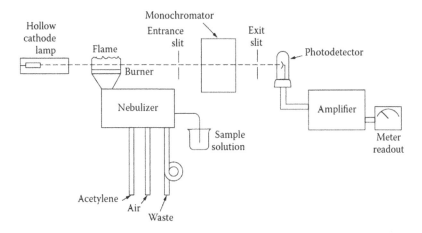

Many models of atomic absorption spectrophotometers are in use in environmental testing laboratories today. Because of this, the type of readouts that one may get might differ. Older instruments most often used *percent absorption*, whereas more contemporary instruments might read out in *absorbance or percent transmittance*. The following schematic relates all three types of AA readouts:

Absorbance	% Transmittance	% Absorption
0	100	0
0.045	90	10
0.097	80	20
0.155	70	30
0.229	60	40
0.301	50	50
0.398	40	60
0.523	30	70
0.699	20	80
1.00	10	90
∞	0	100

This experiment introduces students to quantitative trace metals analysis. The determination of the concentrations of Ca and Mg using FLAA techniques in groundwater is the objective. The PerkinElmer Model 3110® AA needs to be set up in the flame mode of operation. The FLAA will be calibrated and the calibration verified. Various sources of groundwater need to be obtained and used as unknown samples in this experiment.

5.11.2 Experimental

5.11.2.1 Preparation of Chemical Reagents

Note: All reagents used in this analytical method contain hazardous chemicals. Wear appropriate eye protection, gloves, and protective attire. Use of concentrated acids and bases should be done in the fume hood.

5.11.2.2 Chemicals/Reagents Needed per Student or Group

5 mL 1% HNO_3 (use spectroscopic-grade nitric acid only). If unavailable, prepare by placing 3.6 mL of concentrated HNO_3 into a 250 mL volumetric flask previously half filled with distilled deionized (DDI) water, adjust to the calibration mark with DDI, transfer to a plastic storage bottle, and label "1% HNO_3, spectroscopic grade."
5 mL 1,000 ppm Mg (certified stock solution).
5 mL 1,000 ppm Ca (certified stock solution).

5.11.2.3 FLAA Operating Analytical Requirement

Specification	Ca	Mg
Optimum range of concentration (minimum #ppm)	0.2	0.02
Optimum range of concentration (maximum #ppm)	7.0	0.05
Wavelength (nm)	422.7	285.2
Sensitivity (ppm)	0.08	0.007
Instrument detection limit (IDL) (ppm)	0.01	0.001

If a wavelength of 202 nm is used for Mg, a wider linear dynamic range is available (i.e., from 0 to 10 ppm). The sensitivity and IDL will be different.

5.11.2.4 Preparation of the Calibration Curve

As an illustration only, working calibration standards could be prepared as suggested below (this scheme assumes that the 1,000 ppm certified stock reference standard solution has been carefully diluted to 3 ppm).

Standard No.	#mL of 3 ppm	Final Volume (mL)[a]	Concentration of Mg (ppm)
0	0	50	0
1	4.15	50	0.25
2	12.5	50	0.75
3	25	50	1.5
ICV	10	50	0.6

a Use 1% HNO_3 for all dilutions

5.11.2.5 Procedure

Detailed instructions on how to set up and operate an FLAA using the Model 3110® AA will be made available in the laboratory. The burner–nebulizer attachment will need to be installed and the energy throughput optimized using either the Ca or Mg HCL. Proceed to aspirate the working calibration standards. An instrument calibration verification (ICV) standard should be prepared whose concentration should be approximately between the low and high standards. It should be sufficiently aspirated so that triplicate determinations of the absorbance from the same ICV solution can be made. If the precision and accuracy for the calibration and ICV are found acceptable, then any and all available unknowns can be run. A fortified sample containing both Ca and Mg should be available. This sample may have to be diluted if the absorbance is found to significantly exceed the linear dynamic range of the instrument. Be sure to *record the name or the correct code* for each unknown sample. Establish the IDL experimentally for your instrument for each of the two metals. This is accomplished by obtaining seven replicate absorbance measurements of your blank standard.

5.11.3 FOR THE LAB NOTEBOOK (NO REPORT NECESSARY)

Conduct a determination of the IDL for each metal by measuring the signal due to a blank that is repeatedly aspirated into the flame seven times. Apply the principles of blank calibrations and calculate the IDL assuming $k = 3$. Establish a least squares regression fit to the experimental calibration data points for both metals. Calculate the standard deviation in both the slope and y intercept for the calibration curve for both metals. Calculate the standard deviation in the measurement of the ICV for triplicate absorbance measurements. Calculate the confidence limits for the ICV for each metal. Calculate a percent error in the ICV for each metal. Report the concentrations for both Ca and Mg in the fortified sample and for one or more unknown groundwater samples. Comment on the differences in IDLs for Ca and Mg using FLAA.

5.11.4 SUGGESTED READINGS

To develop this experiment, the author consulted the following resources:

- Skoog D., J. Leary. *Principles of Instrumental Analysis*. 4th ed. Philadelphia: WB Saunders, 1992. Chapter 10 provides one of the better discussions of FLAA. Figure 10–15 provides an excellent drawing of the laminar flow burner for FLAA.
- Sawyer, D., W. Heinemann, J, Beebe. *Chemistry Experiments for Instrumental Methods*. New York: John Wiley & sons, 1984, pp. 242–265.
- Methods 7140 (Ca) and 7450 (Mg). In *Test Methods for Evaluating Solid Waste, SW-846, 3rd ed.* Washington, DC: EPA Office of Solid waste, 1986.
  Students who might be interested in finding the original publications that led to the development of the FLAA determinative technique will find citations to the original literature (below):
  - Walsh A. *Spectrochem Acta* 7: 108, 1955.
  - Alkemade C, J Milatz. *J Opt Soc Am* 45: 583, 1955.

5.12 DETERMINATION OF LEAD IN DRINKING WATER USING GRAPHITE FURNACE ATOMIC ABSORPTION SPECTROSCOPY (GFAA): EXTERNAL STANDARD VS. STANDARD ADDITION CALIBRATION MODE

5.12.1 BACKGROUND AND SUMMARY OF METHOD

Unfortunately, among the so-called heavy metals that have made their way into the environment, *lead (Pb) is considered extremely toxic and its presence must be identified and its concentration measured* in air, water, and soil. Two principal historical uses in which Pb was contained were in coatings e.g., house paint as $PbCO_3$ (white) and $PbCrO_4$ (yellow), in gasoline e.g., as the organolead compound tetraethyl lead. This compound was added to gasoline to boost octane ratings. Older water pipes through the U.S. were made of elemental Pb. The extreme toxicity of Pb has required that *instrumental analytical techniques that offer the lowest possible detection limits be used*. The Lead and Copper Rule established by the U.S. federal government on June 7, 1991 states the following: "To protect public health by minimizing lead (Pb) and copper (Cu) levels in drinking water, primarily by reducing water corrosivity. Pb and Cu enter drinking water mainly from corrosion of Pb and Cu containing plumbing materials." The lead and copper rule established *action levels* of 0.015 mg/L for Pb and 1.3 mg/L for Cu in drinking water. Pb joins elements such as Hg, Cd, As, and Tl as requiring determinative techniques with the lowest instrument detection limits (IDLs) possible. It is important to recognize the difference between method detection limits (MDLs) and IDLs. The MDL incorporates the IDL and is equal to the IDL if and only if there is no sample preparation involved. Sample preparation is more common in trace organics analysis in contrast to trace metals analysis. For example, the IDL for Pb using flame AA is ~0.1 ppm, whereas the IDL for Pb using graphite furnace atomic absorption spectroscopic (GFAA) methods is ~1 ppb.

When using the GFAA technique, a representative aliquot of a sample is placed in the graphite tube in the furnace, evaporated to dryness, charred, and atomized. A greater percentage of available analyte atoms is vaporized and dissociated for absorption in the tube in contrast to the flame. It becomes possible to use smaller sample volumes. Radiation from a given excited element is passed through the vapor containing ground-state atoms of that element. The intensity of the transmitted radiation decreases in proportion to the amount of the ground-state element in the vapor. The metal atoms to be measured are placed in the beam of radiation, which is nearly monochromatic (from a hollow-cathode tube), by increasing the temperature of the furnace. A monochromator isolates the wavelength of the transmitted radiation and a photosensitive device measures the attenuated intensity. Beer's Law of Spectrophotometry applies, as was the case for UV-vis absorption spectrophotometry, and the concentration of the specific element is determined by various modes of analyte calibration in a manner similar to that for UV absorption spectrophotometry. A schematic of an electrothermal atomizer follows:

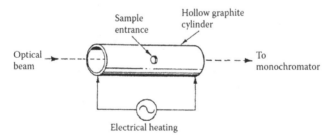

The tube is usually coated with pyrolytic graphite, which is made by heating the tube in a methane atmosphere. Pyrolytic graphite exhibits a low gas permeability and good resistance to chemical attack. This feature lengthens the lifetime (i.e., the number of successful firings) of the tube. There is, however, a finite lifetime for each tube in GFAA.

It is generally believed that the atomization mechanism for any metal M involves reduction of the solid oxide on the graphite surface according to

$$MO_{(s)} + C_{(s)} \rightarrow M_{(l)} \rightarrow M_{(g)} + 0.5M_{2(g)}$$

Most commercial electrothermal atomizers based on the L'vov furnace, as simplified by Massman, undergo vigorous changes in tube temperature. The analyte atoms volatilized from the tube wall come into a cooler gas, so that molecular species that are not detected are formed. This leads to what is termed matrix interferences. For example, equal concentrations of Pb^{2+} in a matrix of deionized water vs. one of high chloride content *would yield different absorbances.* One way to remove this chloride matrix interference is to use a matrix modifier. The addition of an ammonium salt to the chloride-containing matrix would cause volatilization of NH_4Cl with removal of chloride from the sample matrix.

Each end of the furnace tube is connected to a high-current, programmable power supply through water-cooled contacts. The power supply controlling the furnace can be programmed to dry, ash, and atomize the sample at the appropriate temperatures. The temperature and duration of each of these steps can be controlled over a wide range. The optimization of the operating conditions is very important in developing methods using GFAA. A description of the three major temperature program changes, known as steps, is important in understanding GFAA operation. In the first step (drying), the solvent is evaporated at a temperature just above its boiling point. For aqueous solutions, the temperature is held at 110°C for about 30 sec. In the second step (ashing), the temperature is raised to remove organic matter and as many volatile components from the sample matrix as possible without analyte loss. The ashing temperature varies from 350 to 1,200°C and is maintained for about 45 sec. The last step (atomization) occurs between 2,000 and 3,000°C and lasts for just a few seconds. The element of interest is atomized and the absorption of the source radiation by the atomic vapor is measured. The furnace is then cleaned by heating the atomizer to the maximum temperature for a short period. Finally, the temperature is decreased to room temperature using water coolant and inert gas flow (argon). This process is depicted graphically below. Note that the graph includes the overall transmitted intensity I_t, the furnace temperature T, and the net absorption after background correction. It becomes important to have a means to subtract out this background. The PerkinElmer Model 3110® AA uses a deuterium background correction technique.

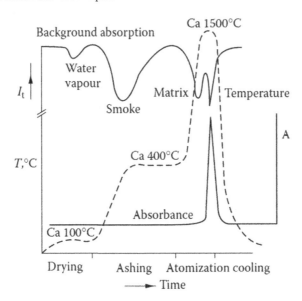

Of the three modes of calibration used in instrumental analysis, external, internal, and standard addition, the latter provides for the *most accurate analytical results for samples that exhibit a matrix interference.* The external mode of calibration is used to convert the instrumental response to concentration when matrix interferences are not considered a factor. A series of standard solutions that contain the metal of interest are prepared from careful dilution of a *certified standard* stock solution. The calibration curve is obtained and a least squares regression is performed on the *x, y* data points. The best-fit line is used to establish the calibration curve. Samples that contain the metal at an unknown concentration level in a sample matrix nearly identical to that used to prepare the serial standards can be run and the data interpolated to give the concentration. In contrast, the standard addition mode of calibration requires that calibration and analysis be performed on the sample itself. Standard addition can be used provided that (1) a linear relationship exists between the physical parameter measured and the concentration of analyte, (2) the sensitivity of the method is not changed by the additions, and (3) the method blank can be corrected for. A typical standard addition calibration follows:

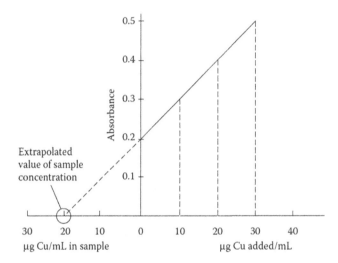

This exercise affords students an opportunity to operate a PerkinElmer Model 3110® GFAA with the objective of determining the concentration of Pb at ppb levels by calibrating the instrument *using two of the three modes: external standard and standard addition*. Following establishment of the working calibration curve, an instrument calibration verification (ICV) standard containing Pb will be prepared, run in triplicate, and the precision determined. At least one unknown sample will be provided for students to determine the concentration of Pb. *Students are asked to bring in a sample of drinking water to determine the concentration of Pb* once the GFAA has been successfully calibrated and the precision and accuracy for the ICV have been found to be acceptable.

5.12.2 EXPERIMENTAL

5.12.2.1 Preparation of Chemical Reagents

Note: All reagents used in this analytical method contain hazardous chemicals. Wear appropriate eye protection, gloves, and protective attire. Use of concentrated acids and bases should be done in the fume hood.

5.12.2.2 Reagents Needed per Student or Group

1 40 ppb reference standard containing Pb in 1% spectroscopic-grade nitric acid (HNO_3).
1 Blank reference standard in 1% spectroscopic-grade HNO_3.
1 Matrix modifier.
1 Unknown sample containing Pb. Be sure to record the code.
1 Drinking water from any source.

5.12.2.3 Procedure

The setup and operation of the GFAA involves the following sequence of activities:

1. Install and align the hollow-cathode lamp (HCL).
2. Align the furnace in the spectrophotometer.
3. Install and condition a new tube.
4. Set up an element parameter file within the PerkinElmer WinLab® software.
5. Align the autosampler.
6. Place blanks, modifiers, and the standard in appropriate locations in the autosampler carousel.
7. Run the calibration according to the programmable sequence within the WinLab® software.
8. Run any and all samples, provided that the calibration and ICV are acceptable.

5.12.2.4 Using the WinLab® Software

You should find the WinLab® (PerkinElmer) software used to control and acquire GFAA data already downloaded. Retrieve and enter the software via the keyboard. The Pb HCL should turn on as well. A stable signal is necessary in order to continue. Proceed to autozero the detector, and if not satisfied, go to the *realign lamps screen* and adjust the physical position of the Pb HCL so as to bring the energy throughput near to that for the deuterium lamp. Set up a sample sequence in the following order: run a lab blank, then a series of calibration standards for Pb, and then one or more ICVs.

Evaluate the quality of the calibration and, if satisfactory, run the ICV under the *run samples screen*. Retrieve the weight/ID screen and choose an autosampler location and run each sample in triplicate. If precision is acceptable, set up the weight/ID screen with the fortified Pb sample and any other samples of interest.

5.12.2.5 Preparation of the Stock Reference Pb Standard and Start of the Autosampler

To prepare the calibration standards for Pb, obtain the certified Pb stock standard, which should be 1,000 ppm as Pb. Using an appropriate liquid-handling syringe, take 40 μL of the 1,000 ppm Pb stock and place this aliquot into a 10 mL volumetric flask already half filled with 1% *spectroscopic-grade HNO3*. Adjust to final volume using the 1% acid. Transfer this solution to a storage container and label "4 ppm Pb." Take 100 μL of the 4 ppm Pb and place this aliquot into a clean 10 mL volumetric flask. Adjust to final volume and transfer to a clean storage container and label "40 ppb Pb." Transfer about a 1 mL aliquot of the 40 ppb Pb reference standard to a plastic GFAA autosampler vial and place in location 38. Place a clean vial filled with 1% nitric in location 36. Place a clean vial filled with matrix modifier in location 37. The programming for the autosampler is found under *element parameters \ calib*; retrieve this and write the entries into the table in your laboratory notebook.

Using the 40 ppb Pb standard and the WinLab® software, implement an *external mode* of calibration. Retrieve or create a method title for this and conduct the calibration and quantitation. Evaluate whether the calibration is free from systematic error. If so, inject the ICV in triplicate. Record the code for the unknown and inject in triplicate. Run one or more drinking water samples.

Proceed on to the implementation of the *standard addition mode* of calibration using the 40 ppb Pb standard and the WinLab® software. Retrieve the method and conduct the calibration and quantitation. Evaluate whether the calibration is free from systematic error. Inject the ICV in triplicate. Record the code for the unknown and inject in triplicate. Run one or more drinking water samples. Rerun one or more samples, provided the calibration is linear and the precision and accuracy for the ICV are acceptable.

5.12.3 For the Notebook

Include in your notebook data from both modes of calibration. Use an Excel spreadsheet or other means to calculate the confidence limits at 95% probability using the Student's t statistics for the ICV for both calibration modes. Report on the concentration of Pb in the coded unknown provided to you. Report on the Pb concentration of any unknown drinking water samples that you analyzed. Proceed to calculate the means, standard deviations, relative standard deviations, and confidence intervals at a given probability. Calculate the relative error between the mean result for the ICV and the expected result for both types of calibration modes (i.e., external standard and standard addition). Comment on the effect of the external vs. standard addition mode of GFAA calibration on the precision and accuracy in the ultra-trace determination of Pb in drinking water.

Discuss why background correction is necessary in AA. Distinguish between physicochemical interferences in AA. Explain what is meant by the term *Smith–Hieftje background correction* and discuss how it differs from the kind of correction employed in the PerkinElmer Model 3110® atomic absorption spectrophotometer.

5.12.4 Suggested Readings

To develop this experiment, the author consulted the following resources:

- Method 7000. Atomic Absorption Methods. In *Test Methods for evaluating Solid Waste. Physical/Chemical Methods.* SW-846. Revision 1. Washington, D.C. Environmental Protection Agency, 1987.
- Vandecasteele C., C. Block. *Modern Methods for trace Elemental Determination.* New York: John Wiley & Sons, 1993, p. 108.
- Massman H. *Spectrochim Acta B* 23B: 215, 1986.
- Skoog D., J. Leary. *Principles of Instrumental Analysis.* 4th ed. Philadelphia: WB Saunders, 1992, pp. 216–218.
- Sawyer, D., W. Heinemann, J. Beebe. *Chemistry Experiments for Instrumental Methods.* New York: John Wiley & Sons, 1984, pp. 242–265.
- Training Manual. Graphite Furnace Atomic Absorption with WinLab® software. PerkinElmer, 1991.
- Yang X. et al. *Am Lab* 51(3): 22–25, 2019. The authors recently reported on using a novel extraction procedure after they derivatized trace concentration levels of the rare Rh^{3+} metal ion (*rhodium*) from industrial water processing samples with 2-(5-bromo-2-pyridylazo)-5-dimethylaminoaniline. Rh^{3+} was quantitated by a GFAA determinative technique. To access this paper, visit the *American Laboratory* online website archives. Archived issues of this trade journal and also found at the *Science History Institute* located in Philadelphia, PA.

5.13 A COMPARISON OF SOIL TYPES VIA A QUANTITATIVE DETERMINATION OF THE CHROMIUM CONTENT USING VISIBLE SPECTROPHOTOMETRY AND FLAME ATOMIC ABSORPTION SPECTROSCOPY OR INDUCTIVELY COUPLED PLASMA–OPTICAL EMISSION SPECTROMETRY

5.13.1 Background and Summary of Method

Chromium exists in three oxidation states, of which Cr (III) and Cr (VI) are the most stable. Hexavalent Cr is classified as a *known human carcinogen via inhalation*, and Cr(III) is an *essential dietary element* for humans and other animals! Certain soils that exhibit a strong chemically reducing potential have been shown to convert Cr(VI) to Cr(III). It is possible for the analysis of soils from a hazardous waste site to reveal little to no Cr(VI) via the colorimetric method because this method is selective for Cr(VI) only. Because atomic absorption spectrophotometric methods yield total Cr, the *difference between analytical results* from both methods should be indicative of the Cr(III) content of a given soil type. Both methods will be implemented in this laboratory exercise and applied to one or more soil types.

This exercise affords students the opportunity to use several instrumental techniques to which they have previously been introduced in the laboratory in order to conduct a comparison of soil types with respect to *determining the ratio of Cr(III) to Cr(VI) in terms of their respective concentration* in an environmental soil matrix. The exercise includes pH measurement, calibration of a UV-vis spectrophotometer, calibration of an atomic absorption spectrophotometer in the flame mode (FLAA), and sample preparation techniques.

Cr(VI) in its dichromate form, $Cr_2O_7^{-2}$, reacts selectively with diphenyl carbazide, in acidic media to form a red–violet color of unknown composition. The molecular structure for diphenyl carbazide is shown below:

This selectivity for Cr occurs in the absence of interferences such as molybdenum, vanadium, and mercury. The colored complex has a very high molar absorptivity at 540 nm. This gives the method a very low detection limit (MDL) for Cr(VI) using a UV-vis spectrophotometer. Flame atomic absorption spectroscopy is also a very sensitive technique for determining total Cr, with instrument detection limits (IDLs) as low as 3 ppb. Inductively coupled plasma- (atomic) emission spectrometry (ICP-AES) may also be available. Your instructors may choose whether FLAA or ICP-AES is available.

A modification to EPA Method 7196 has been published and will be implemented in this lab exercise (refer to Suggested Readings below). The method uses a hot alkaline solution (pH 12) to solubilize chromates that are to be found in soils obtained from hazardous waste sites. One portion of the aqueous sample would then be aspirated into the FlAA for a determination of total Cr, whereas diphenyl carbazide dissolved in acetone will be added to another portion, and the absorbance of the red–violet complex will be measured at 540 nm using a visible spectrometer. In this manner, both total Cr and Cr(VI) can be determined on the same sample. Thus, the ratio of the concentration of Cr(III) to the concentration of Cr(VI) in a soil sample can be calculated from the data generated in this experiment.

5.13.2 Experimental

5.13.2.1 Chemical Reagents Needed per Student or Group

Note: All reagents used in this analytical method contain hazardous chemicals. Wear appropriate eye protection, gloves, and protective attire. Use of concentrated acids and bases should be done in the fume hood.

Potassium dichromate stock solution: Dissolve 141.4 mg of dried $K_2Cr_2O_7$ in distilled deionized water (DDI) and dilute to 1 L (1 mL = 50 µg of Cr).

Potassium dichromate standard solution: Dilute 10.00 mL of stock solution to 100 mL (1 mL = 5 µg of Cr).

Sulfuric acid, 10% (v/v): Dilute 10 mL of concentrated H_2SO_4 to 100 mL with DDI. Also, 1.8 M H_2SO_4 is needed. An aliquot of 10% H_2SO_4 could be used to prepare this solution.

Diphenyl carbazide (DPC) solution: Dissolve 250 mg of 1, 5-diphenyl-carbazide in 50 mL of acetone. Store in an amber bottle. Discard when the solution becomes discolored.

Acetone, CH_3COCH_3: Use the highest purity available.

Alkaline digestion reagent: 0.28 M Na_2CO_3 / 0.5 M NaOH: Use your knowledge of chemical stoichiometry to calculate the amount of each base needed to prepare a solution of the desired molarity. Recall that a 1M solution contains one mole of a pure chemical substance per liter of solution. Recall that one mole of a pure chemical substance is its formula or molecular weight in grams.

Concentrated nitric acid, HNO_3.

5.13.2.2   Procedure for Alkaline Digestion

1. Place 2.5 g of a given soil type into a 250 mL beaker. Add 50 mL of the alkaline digestion reagent. Stir at room temperature for at least 5 min and then heat on a hot plate to maintain the suspensions at 90 to 95°C, with constant stirring for about 1 h. Repeat for all other soil types to be studied whose Cr content is to be determined in this experiment. *Note:* Heating on a hot plate can cause bumping and lead to splatter, and thus loss of analyte.
2. Cool the digestates to room temperature, then filter through 0.45 µm cellulosic or polycarbonate membrane filters.
3. Adjust the pH to 7.5 using concentrated HNO_3 and dilute with DDI to a final volume of 100 mL. You now have 100 mL of digestate.

5.13.2.3   Procedure for Conducting Visible Spectrophotometric Analysis

1. Prepare six working standards from careful dilutions of the 5 µg/mL Cr standard. The range of concentrations should be from 0 to 2 µg/mL Cr. You should prepare 100 mL of each standard.
2. Prepare an initial calibration verification (ICV) standard, which should have its concentration approximately near the mid-range of the calibration. You should prepare 100 mL of the ICV.
3. Prepare a matrix spike and a matrix spike duplicate. The amount of spike should double the concentration found in the original sample. The spike recovery must be between 85 and 115% in order to verify the method.
4. To 45 mL of DDI (this is the method blank), standard, ICV, and digestate, add 1 mL of DPC solution, followed by the addition of 1.8 M H_2SO_4 until the pH reaches approximately 2. This should be done in a 125 mL beaker with stirring and immersion of the glass electrode until the desired pH is attained. After cessation of effervescence, dilute the mixture with DDI to 50 mL. Allow the solution to stand from 5 to 10 min. If the solution appears turbid after the addition of DPC, filter through a 0.45 µm membrane. Store the remaining samples and standards in properly labeled bottles. Use if it is necessary to repeat this analysis.
5. Set the spectrophotometer at 540 nm; correctly set the 0 and 100% transmittance settings. Transfer an aliquot of the 50 mL sample to a cuvette. Measure the absorbance of all blanks, standards, and samples. Construct a table in your notebook to facilitate the entry of data.

5.13.2.4 Procedure for Atomic Absorption Spectrophotometric Analysis or ICP-AES

Refer to SW-846 Methods 7000A ("Atomic Absorption Methods") and 7190 ("Chromium, Atomic Absorption, Direct Aspiration") or Method 6010D ("Inductively Coupled Plasma-Atomic Emission Spectrometry") (ICP-AES) for the quantitative determination of chromium at total Cr. In the lab, proceed to prepare calibration standards and ICVs and aspirate these into the flame using the Model 3110® (PerkinElmer) atomic absorption spectrophotometer. Your instructors may also have the ICP-AES available for you to use such as a Model 2000® (PerkinElmer) ICP-AES. Use the remaining digestates from step 3 and determine total Cr. The FlAA may need to be set up from its present configuration. Refer to previous exercises and training manuals for the necessary information.

5.13.3 For the Report

Include all calibration data, ICVs, and sample unknowns for both instrumental methods. Perform a statistical evaluation in a manner that is similar to that in previous experiments. Use Excel or an alternative to conduct a least squares regression analysis of the calibration data. Calculate the accuracy (expressed as a percent relative error for the ICV) and the precision (relative standard deviation for the ICV) from both instrumental methods. Calculate the percent recovery for the matrix spike and matrix spike duplicate. Report on the concentration of Cr in the unknown soil samples. *Be aware of all dilution factors and concentrations as you perform calculations.*

Find the ratio of the concentration of Cr (III) to that of Cr (VI) in each of the soil samples analyzed and present this value at the end of your report.

5.13.4 Suggested Readings

To develop this experiment, the author consulted the following resources:

- *Standard Methods for the Examination of Water and Wastewater.* 16th ed. Washington, DC: Association of Public Health Association. 1988, pp. 201–204.
- Method 7196A. Chromium. Hexavalent (Colorimetric). In *Test Methods for Evaluating Solid Waste Physical/Chemical Methods. SW-846.* Revision 1. Washington, DC. Environmental Protection Agency, July. 1992.
- Vitale R. et al. *Am Environ Lab* 7:1, 8–10, 1995. This paper discusses a modification to SW-846 Method 7196.
- Method 7000B. Flame Atomic Absorption Spectrophotometry. In *Test Methods for Evaluating Solid Waste, Physical/Chemical. SW-846. Revision 2.* Washington, DC: Environmental Protection Agency, 2007.
- Method 6010D Inductively Coupled Plasma Optical Emission Spectrometry. In *Test Methods for Evaluating Solid Waste, Physical/Chemical. SW-846. Revision 5.* Washington, DC: Environmental Protection Agency, July, 2018.
- Budde W. *Mass Spectrometry: Strategies for Environmental and Related Applications,* Washington, D.C.: American Chemical Society, Oxford University Press, 2001, Chapter 7 pp. 328–347 provides an excellent introduction to trace environmental elemental analysis.
- Skoog D., J. Leary. *Principles of Instrumental Analysis,* 4th ed. Philadelphia: WB Saunders, 1992, Chapter 11, pp. 233–251. This textbook and certainly subsequent editions provide the student with an excellent introduction to the principles of ICP-AES.

5.14 DATA ACQUISITION AND INSTRUMENT CONTROL USING THE TURBOCHROM CHROMATOGRAPHY SOFTWARE. AN INTRODUCTION TO HIGH-PERFORMANCE LIQUID CHROMATOGRAPHY (HPLC): EVALUATING THOSE EXPERIMENTAL PARAMETERS THAT INFLUENCE SEPARATIONS

5.14.1 BACKGROUND AND SUMMARY OF METHOD

Contemporary analytical instrumentation is said to be *interfaced to computers*. These developments commenced in the early to mid-1980s and took hold with Microsoft Windows® based software environments in the 1990s. The architecture for this technological advance can be illustrated as follows:

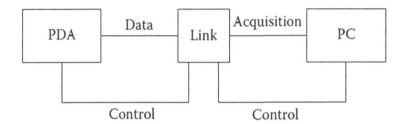

Interfaces can be either stand-alone or installed into the console of the PC. Instruments can be controlled and data acquired from a PC, or if a control is not available, only data acquisition is obtained. In our laboratory, both types of interfaces are used. With appropriate software, the control and data acquisition tasks are easily performed. If a means can be acquired to enable automatic sampling to be controlled as well, a totally automated system can be achieved! This was accomplished in our laboratory.

The HPLC within each student workstation is PC controlled, and the photodiode array detector (PDA) is interfaced to the same PC, thus enabling real-time data acquisition. Students are first asked to study the present architecture so as to gain an appreciation of contemporary HPLC-PDA-DS (data system) technology.

This experiment is designed to take students through an *initial hands-on experience* with the HPLC-PDA-DS from a first sample injection to a simple quantitative analysis. A quick method is first necessary for the software to recognize something. This is followed by optimizing the initial method, conducting a calibration, creating a customized report format, and evaluating the initial calibration verification standard (ICV).

Following completion of the initial experiment, the focus shifts to the separation of the test mixture or organic compounds using the HPLC instrument. The effect of solvent strength on k' and the effect of mobile-phase flow rate on R_S will be considered by retrieving previously developed Turbochrom® methods and making manual injections.

5.14.1.1 HPLC and TEQA

High-performance liquid chromatography (HPLC) followed GC in the early development of instrumental column chromatographic techniques that could be applied to TEQA. HPLC almost always complements and, depending on the analyte, sometimes duplicates GC. For example, *polycyclic aromatic hydrocarbons* (PAHs) can be separated and quantitated by both techniques; however, *N-methyl carbamate pesticides* can be determined only by HPLC as a result of the thermal instability in a hot GC injection port. Molecular structures for one example of each organic functional group are shown:

PAH molecule   N-methyl carbamate (pesticide Sevin®) molecule

HPLC has become the dominant determinative technique for biochemists, pharmaceutical and medicinal chemists, yet has continued to take a secondary role with environmental chemists until technological advances led to the establishment, initially of HPLC-UV and HPLC-FL, followed by HPLC-PDA during the 1970s through to the 1990s. LC-MS first appeared followed by LC-MS-MS technique in and around the early 2000s. Samples that contain the more polar and thermally labile analytes are much more amenable to analysis by HPLC rather than by GC. For example, a major contaminant in a lake in California went undetected until State Department of Health chemists identified a sulfonated anionic surfactant as the chief cause of the pollution. This pollutant was found using HPLC determinative techniques. HPLC encompasses a much broader range of applicability in terms of solute polarity and molecular weight range when compared with GC.

To illustrate how these different kinds of HPLC determinative techniques might aid the analyst in the environmental testing laboratory, consider the request from an engineering firm that wishes to evaluate the degree of *phthalate ester contamination* from leachate emanating from a hazardous waste site. Reversed-phase HPLC is an appropriate choice for the separation of lower-molecular-weight phthalate esters (e.g., dimethyl from diethyl from dibutyl). Attempts to elute higher MW and much more hydrophobic (lipophilic) phthalate esters e.g., dioctyl and bis (2-ethyl hexyl) under reversed-phase conditions were unsuccessful. The separation of these higher MW PAHs under normal-phase HPLC conditions was successful.

5.14.1.2 Flow-Through Packed Columns

High-performance liquid chromatography requires that liquid be pumped across a packed bed within a tubular configuration. Snyder and Kirkland in their classic text on HPLC have used the Hagen–Poiseuille equation for laminar flow through tubes and Darcy's law for fluid flow through packed beds and derived the following relationship:

$$t_0 = \frac{15,000L^2\eta}{\Delta P d_p^2 f}$$

where t_0 is the retention time of an unretained solute (the time it takes after injection for an unretained solute to pass through the column and reach the detector), L is the length of the column, η is the viscosity of the mobile phase, ΔP is the pressure drop across the column, d_p is the particle size of the stationary-phase packing, and f is an integer and is 1 for irregular porous, 2 for spherical porous, and 4 for pellicular packings.

The importance of stationary-phase particle size is reflected in the dependence of the void retention volume $V_0 = F \cdot t_0$ where F is the mobile-phase flow rate in #cm³/min, (recall that 1 mL = 1 cm³) on the *inverse square* of d_p. Recall that the retention volume of a retained solute whose capacity factor is given by k' is

$$V_R = V_0\left(1+k'\right)$$

Hence, the smaller the d_p, the larger is V_0 and, consequently, V_R. A smaller d_p also contributes in a significant manner to a larger N.

5.14.1.3 HPLC Also Refers to an Instrument That Is a High-Pressure Liquid Chromatograph

It is quite useful to view the instrumentation for HPLC in terms of zones according to the following schematic:

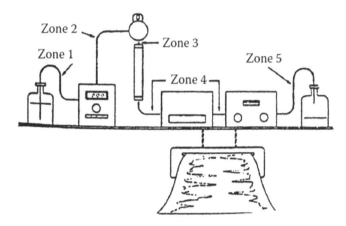

Zone 1—Low-pressure zone prior to pump. This is a *noncritical area* served by Teflon tubing. A fritted filter is placed at the inlet to prevent particulates from entering the column.

Zone 2—High-pressure zone between pump and injector. This is a *noncritical area* served by standard stainless-steel (SS) tubing usually 1/16 inch in outer diameter (o.d.). A high-surface-area 0.5 μm filter can be placed here to prevent particulates from reaching the column.

Zone 3—High-pressure area surrounding injector and column. This is a *critical area* where the sample is introduced to the separation system. The volume must be well swept and minimized. Special fittings are 0.25 mm-inner diameter (i.d.) SS tubing.

Zone 4—Low-pressure area between column and detector. In this *critical area*, separation achieved in the column can be lost prior to detection. The volume must be well swept and minimized. Special fittings and 0.25 mm SS or plastic tubing are required. The critical zone extends to all detectors or fraction collectors in series or parallel connection.

Zone 5—Low-pressure area leading to waste collector. This *noncritical area* is served by Teflon tubing. Most labs fail to fit the waste vessel with a vent line to the hood or exhaust area.

5.14.2 EXPERIMENTAL

High-performance liquid chromatograph (HPLC) instrument incorporating a UV absorption photodiode array detector (PDA) under reversed-phase conditions liquid chromatographic conditions.

5.14.2.1 Preparation of Chemical Reagents

Note: All reagents used in this analytical method contain hazardous chemicals. Wear appropriate eye protection, gloves, and protective attire. Use of concentrated acids and bases should be done in the fume hood.

5.14.2.2   Accessories to Be Used with the HPLC per Group

1 HPLC syringe. This syringe incorporates a blunt end; use of a beveled-end GC syringe would damage inner seals to the Rheodyne injector.

1 10 mL two-component mix at 1,000 ppm each. Prepare the mixture by dissolving 10 mg of ortho-phthalic acid (PhtA) and 10 mg of dimethyl ortho-phthalate (DMP) in about 5 mL of 50:50 (ACN: H_2O) in a 50 mL beaker. After dissolution, transfer to a 10 mL volumetric flask and adjust to the final mark with the 50:50 solution. Molecular structure for PhtA and DMP are shown below:

ortho-phthalic acid (PhtA)   dimethyl ortho-phthalate (DMP)

5.14.2.3   Procedure

Be sure to record your observations in your laboratory notebook.

5.14.2.3.1   INITIAL OBSERVATIONS OF A COMPUTER-CONTROLLED HIGH-PERFORMANCE LIQUID CHROMATOGRAPH

Upon approaching the HPLC-PDA-DS, conduct the following:

1. Identify each of the five zones discussed above.
2. Locate the following hardware components:
   a. The IEEE-488 cable to the LINK interface
   b. The start/stop line from the Rheodyne injector to the LINK
   c. The data acquisition line from the PDA to the LINK
   d. The keying and master key.

5.14.2.3.2   CREATING A QUICKSTART METHOD, ACQUIRING DATA, OPTIMIZING, CALIBRATING, AND CONDUCTING ANALYSIS USING THE QUICKSTART METHOD

Proceed with the Turbochrom 4 Tutorial and create a method using QuickStart. Inject an aliquot of a 100 ppm test mix reference standard. Optimize the method using the Graphic Editor. Develop the calibration and report format sections of your method. Establish a three-point calibration for DMP only between 10 and 100 ppm (always inject reference standards from low concentration to high concentration, never the reverse!) and prepare an ICV. Run the ICV in triplicate.

5.14.2.3.3   EFFECT OF SOLVENT STRENGTH ON κ′

A good practice when beginning to use a RP-HPLC instrument is to initially pass a mobile phase that contains 100% acetonitrile (ACN) so as to flush out of the reversed-phase column any nonpolar residue that might have been retained from previously running the instrument. Retrieve the Turbo method titled *100% ACN* and <u>download</u> if not already set up. *Download within "setup" using the "method" approach.*

Retrieve the method from *Turbochrom* or equivalent software titled *80% ACN* and proceed to use *Setup in the method mode* to enable you to operate the HPLC with a mobile-phase composition of 80% ACN and 20% aqueous. The use of *Setup* is called <u>downloading</u> the method and sequence file so that data acquisition can begin. The aqueous mobile phase consists of 0.05% phosphoric acid (H_3PO_4) dissolved in distilled deionized water (DDI). Carefully fill the 5 µL injection loop (the injector arm should be in the "load" position with the evaluation test mix with the HPLC syringe). Inject by moving the injector arm from the "load" position to the "inject" position. *Observe the chromatogram that results and note the retention times of the components in the mixture.* Give all members in the group the opportunity to make this initial injection.

Retrieve the method titled *60%ACN*, <u>download it</u>, and then proceed to repeat the injection procedure discussed above. *Observe the chromatogram that results and note retention times.*

Retrieve the method titled *40%ACN*, <u>download it</u>, and then proceed to repeat the injection discussed earlier. *Observe the chromatogram that results and note retention times.*

Retrieve the method titled *20%ACN*, <u>download it</u>, and then proceed to repeat the injection procedure discussed earlier. *Observe the chromatogram that results and note retention times.*

5.14.2.3.4 EFFECT OF MOBILE-PHASE FLOW RATE ON RESOLUTION
The mobile-phase flow rate will be varied and its influence on chromatographic resolution will be evaluated.

Retrieve the method titled *FlowHi* <u>download it</u>, and then proceed to use *Setup* as you did during the variation of solvent strength experiments. Allow sufficient equilibration time at this elevated mobile-phase flow rate. Notice what happens to the column back-pressure when a higher flow rate is in operation. Inject the test mix and *observe the chromatogram that results.*

Retrieve the method titled FlowLo <u>download it</u>, and then proceed to repeat the injection procedure discussed earlier. *Observe the chromatogram that results.*

5.14.3 FOR THE LAB NOTEBOOK

The following empirical relationship has been developed for RP-HPLC. Refer to a theoretical discussion on HPLC or to a more specialized monograph.

$$\log k' = \log k_W - S\Phi$$

where k' is the capacity factor for a retained peak, k_W is the capacity factor (extrapolated) k' for pure water, Φ is the volume fraction of the organic solvent in the mobile phase, and S is a constant that is approximately proportional to solute molecular size or surface area.

Choose one component in the evaluation test mix and determine whether the above equation is consistent with your observations.

Address the following:

1. Among the three major parameters upon which resolution R_s depends, which of the three is influenced by changes in mobile-phase flow rate? Explain.
2. Mr. Everett Efficient believes that he can conserve resources by operating his HPLC using a mobile phase that consists only of a 0.01 M aqueous solution containing sodium dihydrogen phosphate (NaH_2PO_4). Discuss what is seriously deficient in Mr. Efficient's fundamental assumption.
3. Assume that you could change HPLC columns in this exercise and that you installed a column that contained 3 µm particle size silica. Assume that you used the same mobile-phase composition that you used for the reversed-phase separations that you observed.

Explain what you would expect to find if the reversed-phase test mix were injected into this HPLC configuration.

4. Explain why DMP is retained longer (i.e., has the higher k') than PhtA given the same mobile-phase composition.

5.14.4 SUGGESTED READINGS

To develop this experiment, the author consulted the following resources:

- *Guide to LC.* Woburn, MA: Ranin Instruments Corporation.
- Sawyer D., W. Heineman, J. Beebe. *Chemistry Experiments for Instrumental Methods*, New York: John Wiley & Sons, 1984, pp. 344–360.
- Snyder L., J. Kirkland. *Introduction to Modern Liquid Chromatography.* 2nd ed. New York: John Wiley & Sons, 1979, pp. 36–37.
- Ahuja S. *Selectivity and Detectability in HPLC. Chemical Analysis Series of Monographs on Analytical Chemistry and Its Application*, Vol. 104, New York: Wiley Interscience. 1989. p. 28.
- Snyder L., J. Kirkland, J. Glajch. *Practical HPLC Method Development.* 2nd ed. New York: John Wiley & Sons, Inc. 1997.

5.15 IDENTIFYING THE UBIQUITOUS PHTHALATE ESTERS IN THE ENVIRONMENT USING HPLC, PHOTODIODE ARRAY DETECTION, AND CONFIRMATION BY GC-MS

5.15.1 BACKGROUND AND SUMMARY OF METHOD

The most commonly found organic contaminant in landfills and hazardous waste sites has proved to be the homologous series of aliphatic esters of phthalic acid. This author has *found phthalate esters in almost every Superfund waste site sample GC-MS report* that he reviewed during the mid-1980s while consulting for an environmental testing laboratory in New York State.

The molecular structures for two representative phthalate esters are drawn below. Dimethyl phthalate (DMP) and bis(2-ethylhexyl) phthalate (Bis) illustrate one example of a lower-molecular-weight phthalate ester versus a higher-molecular-weight ester. DMP and the higher homologs, diethyl phthalate (DEP), di-*n*-propyl (DPP), and di-*n*-butylphthalate (DBP), are the focus of this exercise.

DMP   Bis

The photodiode array UV absorption detector provides both spectral peak matching and, if desired, peak purity determinations. This is nicely illustrated below. A peak is identified in the *first drawing* and its UV absorption spectrum can be matched against a library of UV absorption spectra. Note that the UV absorption spectrum from the peak at or near a retention time $t_R = 39$ min in the HPLC chromatogram is retrieved from a stored library file. The UV spectrum for the peak and that for a reference standard are compared. The *second drawing* demonstrates how overlays of UV absorption spectra use three points across the chromatographically resolved

HPLC and are used together with an algorithm to calculate a purity match. Note the difference between the overlaid UV absorption spectra for the impure vs. the pure peak. You will not be using the peak purity algorithm in this exercise.

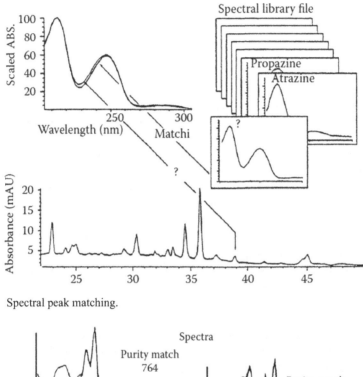

Spectral peak matching.

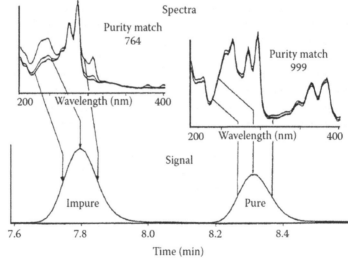

Peak purity determination by spectral overlay.

5.15.1.1 Analytical Method Development Using HPLC

Analytical method development in HPLC usually involves changing the composition of the mobile phase until the desired degree of separation of the targeted organic compounds has been achieved. One starts with a mobile phase that has a *high solvent strength* and moves downward in solvent strength to where a satisfactory resolution can be achieved. Recall the key relationship for chromatographic resolution:

$$R_s = \frac{1}{4}(\alpha - 1)(N)^{\frac{1}{2}}\left(\frac{k'}{1+k'}\right)$$

A useful illustration of the effects of selectivity, plate count, and capacity factor is shown below:

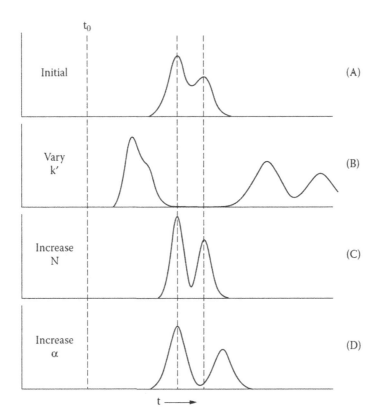

HPLC chromatogram (A) shows a partial separation of two organic compounds, e.g., DMP from DEP. This degree of resolution, R_s, could be improved by changing k', N, or α. In (B), k' is increased, which changes the retention times and shows a slight improvement in R_s. Increasing N significantly increases R_s, as shown in (C); the greatest increase in R_s is obtained by increasing α, as shown in (D). Refer to the Suggested Readings at the end of this experiment or an appropriate textbook chapter or monograph on HPLC to enlarge on these concepts.

5.15.1.2 GC-MS Using a Quadrupole Mass Spectrometer

In a manner similar to obtaining specific UV absorption spectra for chromatographically separated peaks, as in HPLC-PDA, GC-MS also provides important identification of organic compounds first separated by gas chromatography. The mass spectrometer that you will use consists of four rods arranged to form parallel sides of a rectangle, as shown below. The beam from the ion source is directed through the quadrupole section, as shown below.

The quadrupole rods are excited with a large DC voltage superimposed on a radio frequency (RF) voltage. This creates a three-dimensional, time-varying field in the quadrupole. An ion traveling through this field *follows an oscillatory path*. By controlling the ratio of RF to DC voltage, ions are selected according to their mass-to-charge ratio (*m/z*). Continuously sweeping the RF/DC ratio will bring different *m/z* ratios across the detector. An oversimplified sketch of a single quadrupole MS appears below:

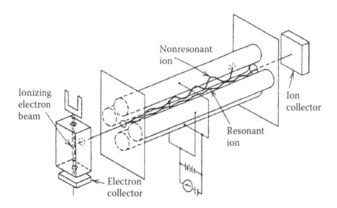

5.15.2 OF WHAT VALUE IS THIS EXPERIMENT?

The goal of this experiment is to provide an opportunity for students to engage in analytical method development by *identifying an unknown phthalate ester* provided to them. This is an example of *qualitative analysis*. The reference standard solution consists of a mixture of the four phthalate esters: dimethylphthalate (DMP), diethylphthalate (DEP), di-n-propylphthalate (DnPP) and di-n-butylphthalate (DnBP). Molecular structures for these four *phthalate esters* are shown below:

DMP   DEP   DnPP   DnBP

Each group will be given an unknown that contains one or more of these phthalate esters. A major objective would be to use available instrumentation to identify the unknown phthalate ester! Students will have available to them an HPLC in the reversed-phase mode (RP-HPLC) and also access to the department's gas chromatograph-mass spectrometer (GC-MS) system.

Students must first optimize the separation of the esters using RP-HPLC, record and store the ultraviolet absorption spectra of the separated esters, and compare the spectrum of the unknown against the stored UV spectra. In addition, staff will be available to conduct the necessary GC-MS determination of the unknown. A hard copy of the chromatogram and mass spectrum will be provided so that the student will have additional confirmatory data from which to make a successful identification of the *unknown phthalate ester*.

5.15.3 EXPERIMENTAL

High-performance liquid chromatograph with ultraviolet absorption photodiode array detection (HPLC-PDA) set up for reversed-phase liquid chromatographic separations.

Capillary gas chromatograph-mass spectrometer incorporating a quadrupole mass-selective detector (C-GC-MS). This instrument should be available for students to use or to drop off their samples at a location outside of the instrumental teaching laboratory location.

5.15.3.1 Preparation of Chemical Reagents

Note: All reagents used in this analytical method contain hazardous chemicals. Wear appropriate eye protection, gloves, and protective attire. Use of concentrated acids and bases should be done in the fume hood.

5.15.3.2 Accessories to be Used With the HPLC per Student or Group

1 HPLC syringe. This syringe incorporates a blunt end; use of a beveled-end GC syringe would damage inner seals to the Rheodyne HPLC injector.

1 Four-component phthalate ester standard. Check the label for concentration values.

1 Unknown sample that contains one or more phthalate esters. Be sure to record the code for the unknown assigned.

5.15.3.3 Procedure

Unlike other lab exercises, *no methods have been developed for this exercise.* Consult with your lab instructor regarding the details for developing a general strategy. You will be introduced to Turboscan®, software that will allow you to store and retrieve UV absorption spectra.

First, find the mobile-phase solvent strength that optimizes the separation of the four phthalate esters. *Second,* retrieve the UV absorption spectrum for each of the four and build a library. *Third,* inject the unknown sample and retrieve its UV spectrum. *Fourth,* make arrangements with the staff to get your unknown analyzed using GC-MS.

5.15.4 FOR THE REPORT

Include your unknown phthalate ester identification code along with the necessary laboratory data and interpretation of results to support your conclusions.

Please address the following in your report:

1. Compare the similarities and differences for the homologous series of phthalate esters on both UV absorption spectra and mass spectra from your data.
2. Explain what you would have to do if you achieved the optimum resolution and suddenly ran out of acetonitrile! Assume that you have only methanol available in the lab. Would you use the same mobile-phase composition in this case?
3. This exercise introduces you to the quadrupole mass filter. Briefly describe how the mass spectrum is obtained, and if you so desire, attempt to provide a brief mass spectral interpretation. You may want to review a text that discusses the principles and practice of GC-MS.

5.15.5 SUGGESTED READINGS

To develop this experiment, the author consulted the following resources:

- Snyder L., J. Kirkland. *Introduction to Modern Liquid Chromatography.* 2nd ed. New York: John Wiley & Sons, 1979.
- Snyder L., J. Glajch, J. Kirkland. *Practical HPLC Method Development.* New York: John Wiley & Sons, 1988.
- Snyder L., J. Kirkland, J. Glajch. *Practical HPLC Method Development.* 2nd ed. New York: John Wiley & Sons, Inc. 1997.

- Sawyer D., W. Heineman, J. Beebe. *Chemistry Experiments for Instrumental Methods*, New York: John Wiley & Sons, 1984, pp. 344–360.

To see how the LC-MS-MS determinative technique has significantly impacted the practice of TEQA, refer to two recent contributions:

- Shrestha P. et al. *Current Trends in Mass Spectrometry* October, 2019, pp. 7–15. The authors compared LC-MS-MS quantitation results obtained between APPI vs. ESI interface with respect to wastewater analysis for various pharmaceuticals.
- Parry E., T. Anumol. Quantitative analysis of per fluoro alkyl substances (PFAS) in drinking water using liquid chromatography tandem mass spectrometry. *Current Trends in Mass Spectrometry* October, 2019 pp. 21–24.

Current Trends in Mass Spectrometry is published (as of this writing) by MultiMedia Healthcare LLC that also publishes monthly issues of *LC-GC North America*, *LC-GC Europe* and *Spectroscopy*. Interested readers who qualify can receive the print and e-versions without cost. Issue archives are also available online.

5.16 AN INTRODUCTION TO GAS CHROMATOGRAPHY: EVALUATING EXPERIMENTAL PARAMETERS THAT INFLUENCE GAS CHROMATOGRAPHIC PERFORMANCE

5.16.1 BACKGROUND AND SUMMARY OF METHOD

Gas chromatography (GC) is *the* most widely used instrumental technique for the determination of trace concentrations of volatiles (VOCs) and semi-volatiles (SVOCs) organic pollutants found in environmental samples today! Its origins stem from the pioneering work of Martin and Synge in 1941 to the development of open tubular gas chromatographic columns advanced by Golay to the fabrication by Dandeneau and Zerenner [*J High Resol. Chromat.Chromatgr. Commun 2 351(1979)*] at Hewlett-Packard of the *fused-silica* wall-coated open tubular (WCOT) gas chromatographic column which revolutionized the practice of gas chromatography. It must be recognized, however, that approximately 20% of all of the organic compounds that exist and could possibly make their way into the environment are amenable to GC techniques without prior chemical modification! Despite this limitation, over 60 organic compounds classified as VOCs have been found in drinking water, groundwater, surface water, and wastewater and are routinely monitored. Over 100 SVOCs have also been found, which include phenols, polycyclic aromatic hydrocarbons, mono-, di-, and trichloro aromatics and aliphatics, nitro aromatics, polychlorinated biphenyls, organochlorine pesticides, organophosphorus pesticides, triazine herbicides, and phthalate esters, among others.

The theoretical principles that underlie GC are found in numerous texts and monographs. Specific methods that incorporate GC as the determinative instrumental technique are to be found in a plethora of analytical methods published by the Environmental Protection Agency (EPA), the American Public Health Association / American Water Works Association / Water Pollution Control Federation (*Standard Methods for the Examination of Water and Wastewater),* and the American Society for Testing Materials (ASTM, Part 31, *Water).*

This exercise introduces the student to those experimental GC parameters that exert a major influence on GC performance. These include (1) detector selectivity, (2) injection volume vs. chromatographic peak shape, (3) the effect of changing the carrier gas flow rate on column efficiency, and (4) the effect of column temperature on chromatographic resolution and analysis time. This exercise affords the student the opportunity to vary these parameters and assess the outcomes. *This is a qualitative analysis exercise only and involves making and recording observations and doing some calculations from information found in the chromatograms.*

5.16.2 Brief Description of Gas Chromatographs Located in the Hazardous Waste Analysis Lab at Michigan State University

The Autosystem® PerkinElmer gas chromatograph (GC) consists of a dual-injector port, dual-capillary-column configuration, and dual detectors including the flame ionization detector (FID) and the electron-capture (ECD)) connected via an analog-to-digital (A/D) interface to a personal computer (PC) workstation. The PC is driven by the Turbochrom® (PE Nelson) chromatography data processing software. You will encounter two types of A/D interfaces in the laboratory. The 600 LINK interface provides for both data acquisition and instrument control. The 900 interface provides for data acquisition only.

The *front injector* consists of a split/splitless capillary column type and is connected to a 0.25 mm (i.d.) × 30 m (length) wall-coated open tubular (WCOT) column (referred to as a narrow-bore WCOT column). The column is coated with a DB-5 liquid phase (5% phenyl dimethyl siloxane) that is chemically bonded to the inner tubing wall. This type of liquid phase is appropriate for the separation of SVOCs whose boiling points are much greater than 100°C. The optimum volumetric flow rate (i.e., the flow rate that gives a minimum in the *van Deemter* curve) is between 1 and 3 cm^3 per minute. To obtain such a low flow rate, a split vent is required to remove most of the gas. Refer to the instruction manual for *setting the split ratio*. Typical split ratios are 1:25, 1:50, or 1:100, and this ratio refers to the ratio of gas flow through the column to that through the vent. The outlet end of this column is connected to the inlet to the ECD. This detector requires an additional source of inert gas, commonly called *makeup gas*. The flow rate for the makeup should be approximately 30 cm^3/min. Using the digital flow check meter (refer to the instruction manual), measure the initial flow rate, then adjust to the optimum for operation of a narrow-bore WCOT column.

The *rear injector* consists of a *packed column adapted* for connection to a 0.53 mm (i.d.) × 30 m (length) capillary column (referred to as a megabore column). The column is coated with a cyanopropyl dimethyl polysiloxane liquid phase that is chemically bonded to the inner tubing wall. This type of liquid phase is appropriate for the separation of VOCs. The optimum volumetric flow rate is between 5 and 15 cm^3/min. The column outlet is connected to the inlet to the FID. This detector does not require makeup gas. The FID, however, requires a 10:1 ratio of airflow to hydrogen flow. Conventional flow rates are 300 cm^3/min for air and 30 cm^3/min for H$_2$. Once the air/fuel ratio has been established, the FID can be ignited. Sometimes, a slightly fuel-rich ratio is necessary to ignite the FID.

5.16.3 Principle of Separation in GC

When two compounds migrate at the same rate through a chromatographic column, no separation is possible. Two compounds that differ in retention times, t_R, or capacity factor, k', and appear to separate, do so because of differences in their equilibrium distribution constants, denoted by K_D. If K_D is independent of sample size, Gaussian elution bands (i.e., symmetrical peaks) are observed. This is the case of *linear elution chromatography*. In other words, a plot of the concentration of analyte in the stationary phase to the concentration in the mobile phase yields a straight line whose slope equals K_D. If the amount of analyte increases either by injecting equal volumes of solutions whose concentrations are increasing or by injecting increasing volumes of a solution whose concentration is fixed, nonsymmetrical chromatographic peaks result. K_D is now dependent on the amount of solute, and either peak tailing or peak fronting results. This is the case of nonlinear elution chromatography. Gaussian or symmetrical peak shape is a chief objective when GC is used to perform trace quantitative analysis. The following equations relate the parameters discussed above:

$$k' = \frac{t_R - t_0}{t_0}$$

$$K_D = \beta k'$$

where β is the volume (mobile phase)/volume (stationary phase), t_R is the retention time for a retained peak, and t_0 is the retention time for an unretained peak.

5.16.4 Experimental

Gas chromatograph interfaced to a PC that is loaded with chromatographic software. In our lab, an Autosystem® (PerkinElmer) is interfaced to a PC workstation that utilizes Turbochrom® (PE Nelson) for data acquisition, processing, and readout.

5.16.4.1 Preparation of Chemical Reagents

Note: All reagents used in this analytical method contain hazardous chemicals. Wear appropriate eye protection, gloves, and protective attire. Use of concentrated acids and bases should be done in the fume hood.

5.16.4.2 Accessories to be Used With the GC per Group

1 Digital flow check meter.
1 GC syringe with a beveled end that includes a Chaney adapter. Do not confuse with the blunt-end syringe used for HPLC.
1 GC test mix for each of the studies discussed below.

5.16.4.2.1 Summary of Turbochrom Methods to be Used in This Experiment

Order	Turbo Method	Remarks
1	FLOWRATE	Near-ambient column temperature
		Measure flow rate, split ratio
2	DETSENS	Neat acetone—FID
		Neat methylene chloride—ECD
3	INJECD	Inject increasing amounts of 10 ppm 1, 2, 4-trichlorobenzene
	INJFID	Inject increasing amounts of hexadecane at 225°C
4	PLATES	Temperature program: 265°C (0.1) to 285°C (10.0) at 6°C/min; multi-component organochlorine test mixture
5	TMAX	Isothermal at 285°C
	TMIN	Isothermal at 200°C

5.16.4.3 Procedure

Refer to *Summary of Turbochrom Methods* (above) for a definition of each of the Turbochrom methods created in support of this experiment. These previously created methods are illustrative of how chromatography-based software can be used to teach fundamental principles of GC.

5.16.4.3.1 Measurement and Adjustment of Carrier Gas Flow Rate and Split Ratio

As you approach the gas chromatograph, you will find it in an operational mode, with carrier gas flowing through both capillary columns. If not already set up, retrieve the Turbochrom file titled "FLOWRATE" and download this method. Your first task will be to measure the flow rate of the carrier gas through both capillary columns with the makeup gas off. After turning the makeup gas on, measure the split ratio through the capillary injector using the digital flow check meter. *Record flow rate data in your lab notebook.*

Once the optimum carrier flow rates have been established, the dual detector method titled "DETSENS" can be retrieved from the Turbochrom software, then transferred to the instrument via the interface (a process known as download) and the comparison of detector sensitivity can be undertaken. *Ignite the FID (refer to the operator's manual for the Autosystem GC from PerkinElmer for the specific procedure).*

5.16.4.3.2   COMPARISON OF THE FID vs. THE ECD SENSITIVITY

Allow time for the GC to equilibrate at the column temperature set in the method. Using the manual injection syringe (GC syringes are manufactured by the Hamilton Co. as well as by others), inject equal microliter (µL) aliquots of acetone into both injectors. Observe the appearance of a retained chromatographic peak found in both chromatograms. You cannot assume that t_R will be identical on both columns! Compare the peak heights from both chromatograms. Inject equal µL aliquots into both injectors, as earlier, of the specific chlorinated hydrocarbon that is available. *Record your observations* and compare the peak heights as done previously. Each member of the group should have an opportunity to make these sample injections so as to gain some experience with manual syringe injection of organic solvents.

5.16.4.3.3   INJECTION VOLUME vs. GC PEAK SHAPE

Retrieve the Turbo file titled "INJECD" and download. Inject a series of increasing µL aliquots of a reference solution labeled "10 ppm 1, 2, 4-trichlorobenzene" into the front capillary injection port. *Observe and record the changes in chromatographic peak shape as the amount of analyte is increased.* Retrieve the Turbo file titled "INJFID" and download. Repeat the series of injections as before using the reference solution labeled "hexadecane" and make these injections into the rear injector. *Observe and record the changes in chromatographic peak shape as the amount of analyte is increased.*

5.16.4.3.4   FLOW RATE vs. CAPILLARY COLUMN EFFICIENCY

A column's efficiency is determined in a quantitative manner from the chromatogram by measuring the number of theoretical plates, N. The effect of carrier flow rate on capillary column efficiency is significant in GC and will be examined under isothermal conditions (i.e., at a fixed and unchanging column temperature). Retrieve the Turbo file "PLATES" and program this method for a high flow rate by increasing the head pressure. Save this change in the method and download the method. Turn off the makeup gas and adjust the actual pressure so as to nearly match that which is set in the method and *measure* the flow rate. Turn the makeup back on. Inject 1 µL of the test mix and *observe* the chromatogram. Retrieve the method a second time and reprogram the head pressure to a much lower value. Turn off the makeup, decrease the carrier head pressure, *measure* the new flow rate, turn the makeup gas back on, and then make a second injection using the same volume.

For the carrier gas flow rate that exhibited the highest efficiency, *calculate* the number of theoretical plates using equations from your text. In addition, for the optimum carrier flow rate, choose any pair of peaks and calculate the resolution for that pair.

5.16.4.3.5   COLUMN TEMPERATURE vs. CAPACITY FACTOR

Retrieve the Turbo file titled "TMIN" and download the method. Inject approximately 1 µL of the multi- component organochlorine test mix at this column temperature of 200°C. *Observe the degree of separation among organochlorine analytes and record your qualitative comments.*

Retrieve the Turbo file titled "TMAX" and download the method. Inject the same volume of the multi-component organochlorine test mix and *observe* the chromatogram when the column temperature has been increased to 285°C.

5.16.5 FOR THE LAB NOTEBOOK

Write a brief discussion on how your experimental observations connect to the theoretical relationships for GC introduced in various textbooks and journal articles.

Address the following:

1. Explain why different GC detectors have different instrument detection limits.
2. If you operated a GC at significantly reduced carrier gas flow rates, predict what you would observe in a gas chromatogram for the injection of organic compounds. What would be the principal cause for these observations?
3. Explain why a symmetrical peak shape is important in gas chromatography.
4. What happens to K_D for a given organic compound when column temperature is varied?
5. How efficient is your GC column? That is, what is the number of theoretical plates? How many plates per meter do you have?
6. How is the phase ratio, β, determined for capillary GC columns?

5.16.6 SUGGESTED READINGS

To develop this experiment, the author consulted the following resources:

Sawyer D., W. Heineman, J. Beebe. *Chemistry Experiments for Instrumental Methods*, New York: John Wiley & Sons, 1984, pp. 321–343.

A thorough grounding in the principles and practice of the GC and GC-MS determinative techniques can be found among others in the resources shown below:

Perry J. *Introduction to Analytical Gas Chromatography.* New York: Marcel Dekker, 1981.

Jennings W. *Analytical Gas Chromatography.* San Diego: Academic Press, 1987.

McNair H., J. Miller. *Basic Gas Chromatography.* New York: Wiley Interscience, 1998.

Budde W. *Mass Spectrometry: Strategies for Environmental and Related Applications*, Washington, D.C.: American Chemical Society, Oxford University Press, 2001.

Grob R., E. Barry., Eds. *Modern Practice of Gas Chromatography*, 4th ed. Hoboken, NJ: Wiley-Interscience, 2004.

5.17 SCREENING FOR THE PRESENCE OF BTEX IN WASTEWATER USING LIQUID–LIQUID EXTRACTION (LLE) AND GAS CHROMATOGRAPHY: SCREENING FOR THMS IN CHLORINE-DISINFECTED DRINKING WATER USING STATIC HEADSPACE (HS) GAS CHROMATOGRAPHY

5.17.1 BACKGROUND AND SUMMARY OF METHOD

Two analytical *screening* approaches are introduced in this experiment. A suitable extracting solvent is experimentally selected and used to *extract suspected gasoline-tainted water samples* using a mini-LLE sample prep technique in order to detect the presence of BTEX. Alternatively, a simulated chlorinated disinfected drinking water sample is screened for *organochlorine-containing* VOCs using *static headspace gas chromatography* with electron-capture detection

(HS-GC-ECD) to detect the presence of trihalomethanes (THMs). Molecular structures together with names for the six BTEX VOCs are shown below:

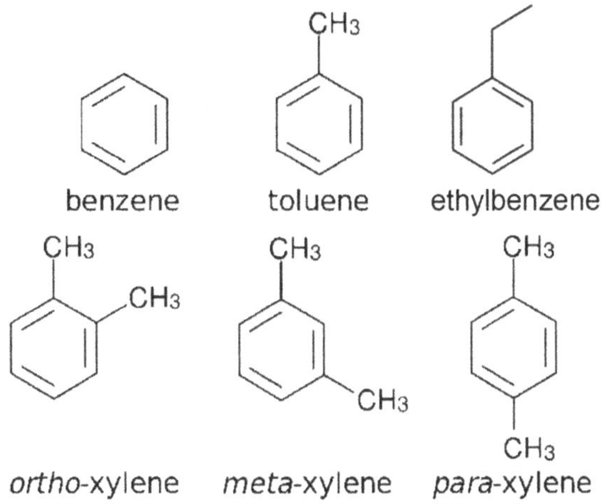

benzene
toluene
ethylbenzene
ortho-xylene
meta-xylene
para-xylene

THMs include chloroform, bromodichloromethane, dibromochloromethane, and bromoform. These toxic VOC analytes have been found in drinking water that has been disinfected using chlorine. Molecular structures for the four THMs are shown below:

Chloroform
Bromodichloromethane
Dibromochloromethane
Bromoform

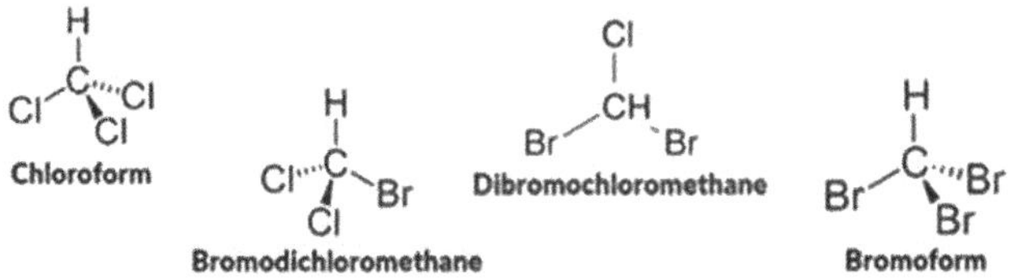

We will take a more simplified approach to trace VOCs analysis, which utilizes our limited sample preparation and instrumentation capabilities in the instructional laboratory. If a suitable extraction solvent can be found, i.e., one that does not interfere with the VOCs to be identified and quantitated by gas chromatography, then the analytes of interest can be isolated and concentrated from the environmental sample matrix via a mini-LLE technique. A 40 mL sample of wastewater that might contain BTEX is extracted with 2 mL of a suitable organic solvent. The organic solvent, being less dense than water, conveniently occupies the neck of a 40 mL vial. A 2μL aliquot of the extract is taken by a liquid-handling syringe and injected into a C-GC-FID to screen for the presence of BTEX compounds. The C-GC-FID must be previously optimized to separate most BTEX compounds. In a separate experiment, 40 mL of chlorine-disinfected drinking water is placed in a sealed HS vial, heated, and 0.5 cc of the headspace is sampled using a HS sampling syringe and injected directly into a C-GC-ECD. The C-GC-ECD must be previously optimized to separate the four THMs.

Typical levels of BTEX contamination for wastewater are in the low parts per million (ppm) concentration range. Typical levels of THM contamination for chlorine-disinfected drinking water are typically between 10 and 100 ppb for each THM. A severe limitation to LLE techniques is the possible formation of emulsions when applied to wastewaters that could have an appreciable surfactant concentration level. HS-C-GC-ECD is a very selective approach for screening chlorine-disinfected drinking water samples for THMs.

5.17.2   OF WHAT VALUE IS THIS EXPERIMENT?

This exercise affords to the student an opportunity to further utilize gas chromatography, this time as a screening tool. Two different sample preparation approaches to screening are introduced for two somewhat different sample matrices. If a method involves phase distribution equilibria either for screening or for quantification, some analyte will always be lost between phases. Volatility losses can be considerable when VOCs are dissolved in water, while these losses are not so critical for SVOCs dissolved in water.

A previously created method will be retrieved from the Turbochrom® (PE Nelson) or other chromatography processing software available in the lab. It is possible for your instructor to turn this qualitative screening experiment into a quantitative determination one. If so, external or internal standards must be prepared and run in order to create the necessary calibration plots.

5.17.3   EXPERIMENTAL

5.17.3.1   Preparation of Chemical Reagents

Note: All reagents used in this analytical method contain hazardous chemicals. Wear appropriate eye protection, gloves, and protective attire. Use of concentrated acids and bases should be done in the fume hood.

5.17.3.2   Chemicals/Reagents Needed per Group

1 Neat benzene
1 Neat toluene
1 Neat ethyl benzene
1 Neat xylene
1 Neat hexane to evaluate as a suitable screening extractant
1 Neat hexadecane to evaluate as a suitable screening extractant
1 Neat dichloromethane to evaluate as a suitable screening extractant
1 Approximately 5,000 ppm stock BTEX standard (refer to actual label for exact values)
1 40 mL of a wastewater sample for screening for BTEXs
1 40 mL of a chlorine-disinfected drinking water sample for screening for THMs
1 500 ppm reference stock standard containing THMs in MeOH.

5.17.3.3   Items/Accessories Needed per Student or per Group

1 42 mL glass vial with screw caps and PTFE/silicone septa
1 22 mL glass headspace vial with PTFE/silicone septa and crimp-top caps
1 Crimping tool for headspace vials
1 Liquid-handling syringe whose capacity is 10 μL with Chaney adapter (Hamilton or other manufacturer) for injection of liquid extracts
1 0.5 or 1.0 cc capacity gas-tight syringe for headspace sampling and direct injection (Precision Sampling, SGE, and Hamilton, among others, manufacture such syringes)
1 Heating block assembly that accepts a 22 mL HS vial and allows for measurement of the block temperature (VWR or other supply house).

5.17.3.4   Preliminary Planning

At the onset of the laboratory period, assemble as a group and decide who is going to do what. Assign specific tasks to each member of the group. Once all results are obtained, the group should reassemble and share all analytical data.

5.17.3.5   Procedure for BTEX Instrumental Analysis Using Mini-LLE Techniques

5.17.3.5.1   *Selecting the Most Suitable Extraction Solvent*
Place one small drop of each of the neat BTEX liquids into approximately 10 mL of hexane. Inject 1 μL into the GC-FID and interpret the resulting chromatogram. Repeat for dichloromethane and then for hexadecane. Methods must be previously created on Turbochrom or equivalent software. Recall, the most suitable solvent is the one that does not interfere with the GC elution of BTEXs. From these observations, select the most appropriate extraction solvent, then proceed to prepare calibration standards.

5.17.3.5.2   *Preparation of the Primary Dilution Standard and Working Calibration Standards*

1. Using a clean and dry glass pipette (volumetric), transfer 1.0 mL of the 5,000 ppm BTEX to a 10 mL volumetric flask that has been previously half filled with the most suitable solvent that you chose earlier. Adjust to the calibration mark with this solvent and label as "500 ppm BTEX," for example. This is what EPA methods call a primary dilution standard, since it is the first dilution that the analyst prepares from a given source. In this case, a 1:10 dilution has been made.
2. Prepare a series of working calibration standards according to the following table:

Standard #	500 ppm BTEX (mL)	Final Volume (mL)	Concentration of BTEX (ppm)
1	1	10	50
2	2	10	100
3	4	10	200
4	8	10	400
5	—	—	500
ICV	5	10	250

For example, to prepare standard 3, 4 mL of 500 ppm BTEX (MeOH) added to a 10 mL volumetric flask half filled with MeOH is added to bring the meniscus to the mark of the volumetric flask. This yields a calibration standard whose concentration is 200 ppm BTEX (MeOH).

3. Retrieve the method BTEX from the Turbochrom software, open a new sequence file, and name the raw data file in a manner similar to the following example: "G116" (Group 1, 16th of the month). Save the sequence file and name it in a manner similar to the following example: "G10316" (Group 1, March 16th).
4. Inject 1 μL aliquots of all calibration standards and inject the ICV in triplicate. Update the calibration method within the Turbochrom software. *Ask your lab instructor for help in updating the calibration within the method.* Observe the calibration curve and note the value of the square of the correlation coefficient. Discuss with your instructor whether this calibration is acceptable.
5. After the instrument has been properly calibrated and the ICVs quantitatively determined, proceed to inject the unknown sample extracts from the mini-LLE (refer to "Procedure to Conduct a Screen …," below). Obtain the interpolated values from the external standard mode of instrument calibration.

5.17.3.6   Procedure for THM Instrumental Analysis Using HS Techniques

Using the 500 ppm THM stock reference solution, prepare a series of calibration standards in which the THMs are present in 10 mL of DDI, which is contained in a 22 mL HS vial with PTFE/ silicone septa and aluminum crimp-top caps. *Refer to the BTEX calibration for guidance* as you prepare a series of working calibration standards for HS-GC analysis. Ask your instructor to review your calibration table for correctness. Following the development of a calibration curve, inject the ICV (only one injection per sample is acceptable in HS-GC), then inject the headspace above the aqueous samples. Following the development of a calibration curve, inject the ICV and the chlorine-disinfected drinking water samples.

5.17.3.7   Procedure to Conduct a Screen for BTEXs via Mini-LLE and Subsequent Injection into a GC-FID

Once the most appropriate extraction solvent has been selected, the wastewater sample that contains dissolved BTEX can be extracted. To a clean 42 mL glass vial with a PTFE/silicone septum and screw cap, add 40 mL of aqueous sample. Pipette 2.0 mL of extraction solvent, and place the septum and cap in place. Shake for 1 min and let stand for at least 5 min until both phases clearly separate. Using a glass transfer pipette, remove approximately 75% of the extract and place in a small test tube or vial. Inject 1 μL of extract into the GC-FID. *Discard the contents of the 42 mL glass vial into the waste receptacles that are located in the laboratory.*

5.17.3.8   Procedure to Conduct Manual Headspace Sampling and Direct Injection into a GC-ECD

If a heater block is available, place the sealed and capped 22 mL headspace vial into the block and allow time for the vial to equilibrate before sampling. A water bath, i.e., a large beaker that is half filled with water, could serve as a constant-temperature environment for headspace sampling. Insert the gas-tight syringe with the valve in the "on" position (applicable to syringes made by VICI Precision Sampling and others) by penetrating the septum seal and withdraw a 0.5 cc aliquot of headspace. Be careful not to withdraw any liquid. Immediately close the on–off valve to the syringe while positioned within the headspace. Position the syringe into the injection port, open the syringe valve, and transfer the 0.5 cc aliquot into the GC.

5.17.4   For the Report

Since this experiment involves screening only, quantification of the wastewater and chlorine-disinfected samples is unnecessary unless your instructor asks you to quantitate. If you find BTEX or THMs from the screens, discuss how you might conduct a quantitative analysis of these samples. If these samples identified additional compounds that were not BTEX or THM compounds, suggest ways that the identity of these unknown compounds could be revealed. Explain the basis on which you chose the screening extractant. Refer to literary resources that discuss the how to screen for VOCs for some help here.

5.17.5   Suggested Readings

To develop this experiment, the author consulted the following resources:

Sawyer D., W. Heineman, J. Beebe. *Chemistry Experiments for Instrumental Methods*, New York: John Wiley & Sons, 1984, pp. 321–343.

Kolb B., L. Ettre. *Static Headspace-Gas Chromatography: Theory and Practice.* New York: Wiley VCH, 1997.

Grob R., E. Barry., Eds. *Modern Practice of Gas Chromatography,* 4th ed. Hoboken, NJ: Wiley-Interscience, 2004, pp. 563–573.

5.18 DETERMINATION OF PRIORITY POLLUTANT VOLATILE ORGANIC COMPOUNDS (VOCs) IN GASOLINE-CONTAMINATED GROUNDWATER USING STATIC HEADSPACE (HS) AND SOLID-PHASE MICROEXTRACTION HEADSPACE (SPME-HS) AND GAS CHROMATOGRAPHY

5.18.1 BACKGROUND AND SUMMARY METHOD

Benzene, toluene, ethyl benzene, *para-, meta-,* and *ortho*-xylenes, collectively referred to as BTEX, constitute some of the most environmentally detrimental organic compounds that have made their way into groundwater, primarily due to gasoline spills and underground rusted storage tank leakage over time. In preparing for this experiment, this author in the 1990s (for educational purposes) contaminated (in the teaching laboratory) a small sample volume of groundwater with gasoline. The author observed not only the six BTEX compounds, however; in addition, he observed a large and early eluting peak that matches the retention time of methyl-*tert*-butyl ether (MTBE). MTBE was used as a gasoline additive designed to boost octane rating and was thought at the time to be a suitable substitute for tetraethyl lead. However, MTBE quickly polluted groundwater as well! BTEX compounds comprise about 20 to 30% of gasoline and have an appreciable solubility in water in contrast to aliphatic hydrocarbons such as n-heptane, iso-octane, n-dodecane. The *Energy Policy Act of 2005* prompted gasoline refiners to transition away from MTBE to the use of ethanol as a gasoline additive. Molecular structures for MTBE, tetraethyl lead, and ethanol are shown below:

MTBE   Tetraethyl lead   Ethanol

EPA Methods 601(purgeable halocarbons) and 602 (purgeable aromatics) comprise the real workhorse approaches to trace VOCs analyses in wastewaters. These methods use dynamic headspace sampling coupled to GC with electrolytic conductivity (EPA Method 601) and photoionization (EPA Method 602) detection. EPA Method 502.2 is a high-resolution capillary column method with both detectors cited above connected in series and provides monitoring capabilities for over 60 VOCs that could be found in municipal drinking water supplies. An alternative to dynamic headspace (commonly referred to as purge-and-trap) is static headspace capillary gas chromatographic (HS-C-GC) techniques.

Static HS techniques take advantage of the volatility exhibited by VOCs whereby the air remaining in a sealed vial above a liquid (defined as the headspace) is sampled with a gas-tight syringe and directly injected into the GC-FID for carbon-containing VOCs. This technique is called *manual HS-GC*, as distinguished from automated HS-GC techniques. A complement to static HS is *SPME-HS*. A fiber coated with a polymer such as poly-dimethyl siloxane is inserted through the septum and into the HS. VOCs partition from the HS to the polymer film. The SPME syringe-fiber assembly is removed from the vial and inserted directly into the injection port of a gas chromatograph. VOCs are thermally desorbed off of the fiber by the hot GC inlet and onto the head of a wall-coated open tubular (WCOT) column where the VOCs are gas chromatographed.

In commercial laboratories, HS-GC and SPME-HS-GC sample prep techniques have been automated! The CombiPal® (CTC Analytics) offers an HS syringe mounted on a robotic head that moves horizontally (known as a rail). This robotic autosampler can accommodate either a *gas-tight syringe* to conduct automated static HS-GC sampling or an *SPME syringe* holder to conduct automated SPME-HS-GC. The robotic autosampler called a *Multipurpose Sampler* (MPS) is complemented and controlled through hardware and software provided by Gerstel GmbH and Co. KG and others. A second rail on the MPS provides robotic automated reagent delivery to the HS vial.

5.18.2   OF WHAT VALUE IS THIS EXPERIMENT?

Students will have an opportunity in this experiment to quantitatively determine the concentration level of various BTEX compounds from gasoline-contaminated groundwater using both static HS and SPME-HS techniques. Both sampling/sample prep techniques will be performed *manually*. This experiment affords students an opportunity to operate a conventional gas chromatograph. This GC is interfaced to a personal computer that utilizes Turbochrom® (PerkinElmer) software or the equivalent for data acquisition. Hence, a student must become familiar with the sampling/sample prep technique, the GC, and the software used at the same time. This trio of techniques comprises a necessary learning experience for the student who will eventually work in fields related to trace environmental quantitative analysis.

The method titled "BTEX.mth" will be retrieved from Turbochrom or other chromatography processing software available in the lab. External standard calibration curves will be generated using both HS-C-GC-FID and SPME-HS-C-GC-FID methods. An aqueous environmental sample that has been contaminated with gasoline will be available and analyzed for traces of BTEX. Since two different analytical methods are applied to the same standards and samples, students will have the opportunity to apply *t* statistics to compare the analytical results from both methods.

5.18.3   USE OF *T* STATISTICS

Comparison of two dependent averages is a statistical procedure that helps to determine whether two different analytical methods give the same average result for a given sample. If one analyzes each of a series of samples, which could include calibration standards, ICVs, blanks, and unknowns, by the two methods, *a pair of results* for *each sample* will be obtained. The difference between these two results for each pair will reflect only the difference in the methods. The following equations are used for the *t* test on paired data:

$$t_{calc} = \frac{\overline{d}}{S_d}\sqrt{n}$$

$$S_d = \sqrt{\frac{\sum d^2 - \dfrac{\left(\sum d\right)^2}{n}}{n-1}}$$

where

d = difference in each pair of values
$\overline{d}$ = average *absolute* difference in the pairs of *values*
n = number of pairs of values
df = degrees of freedom associated with a given value for *t*
s_d = standard deviation of the differences between the pairs of observations.

A comparison of the calculated value, t_{calc}, with that from a *Table of Student's t values* is then made. If $t_{calc} > t$ (from a Student's *t* table at the desired level of significance), then both methods do not give the same result. If $t_{calc} < t$ (from a Student's *t* table at the desired level of significance), then it is statistically valid to assume that both methods are equivalent.

5.18.4 Experimental

5.18.4.1 Preparation of Chemical Reagents

Note: All reagents used in this analytical method contain hazardous chemicals. Wear appropriate eye protection, gloves, and protective attire. Use of concentrated acids and bases should be done in the fume hood.

5.18.4.2 Chemicals/Reagents Needed per Group

1 Neat benzene
1 Neat toluene
1 Neat ethyl benzene
1 Neat xylene: *ortho, meta,* and *para*
1 2,000 ppm BTEX, certified reference standard, dissolved in MeOH
1 40 mL of gasoline-contaminated groundwater for BTEX determination
1 40 mL of an unknown sample prepared by the staff for BTEX determination.

5.18.4.3 Items/Accessories Needed per Student or per Group

10 22 mL glass headspace vials with PTFE/silicone septa and aluminum crimp-top caps.
7 10 mL glass volumetric flasks to be used to prepare the calibration and ICV reference standards (refer to the calibration table below).
1 Crimping tool for headspace vials.
1 0.5 or 1.0 cc capacity gas-tight syringe for headspace sampling and direct injection (Manufacturers of syringes include: VICI Precision Sampling, SGE, Hamilton, and others)
1 Heating block assembly that accepts a 22 mL HS vial and allows for measurement of the block temperature.
1 SPME fiber holder for manual sampling (Millipore Sigma, formerly Supelco).
1 Manual SPME sampling stand (Millipore Sigma, formerly Supelco) or equivalent, including mini stir bars. This apparatus is optional; the heating block assembly can be used to conduct SPME-HS.
1 100 μm (film thickness) of poly-dimethyl siloxane (PDMS) fiber for use with the SPME holder. Instructions for installing the PDMS fiber into the SPME fiber holder are available from the manufacturer (Millipore Sigma, formerly Supelco).

5.18.4.4 Preliminary Planning

At the onset of the laboratory period, assemble as a group and decide who is going to do what. Assign specific tasks to each member of the group. Once all results are obtained, the group should reassemble and share all analytical data.

5.18.4.5 Procedure for BTEX Instrumental Analysis HS Techniques

Using the 2,000 ppm BTEX stock reference solution dissolved in MeOH, prepare a series of calibration standards in which the BTEX is present in a final volume $V_T = 10$ mL of DDI. Refer to the table below. Use of a 10 mL volumetric flask to initially prepare reference standards prior to transfer to the 22 mL headspace vial is available in this experiment but optional. Refer to the procedure stated immediately below the calibration table. Transfer each reference standard and each ICV to a 22 mL HS vial and immediately close with a PTFE/silicone septa and aluminum crimp-top cap. Refer to the calibration table below for reference as you prepare a series of working calibration standards for HS-GC analysis. Following the development of a calibration curve, inject the ICV (only one injection per vial is acceptable in HS-GC) and then one or more of the gasoline-contaminated aqueous samples.

- Prepare a series of working calibration standards and ICVs according to the following table:

Standard No.	2,000 ppm BTEX (MeOH) (µL)	V_t (mL)	Concentration of BTEX (ppm)
Blank	0	10	0
1	20	10	4
2	40	10	8
3	80	10	16
ICV 1	30	10	6
ICV 2	30	10	6
ICV 3	30	10	6

- Place the indicated aliquot of 2,000 ppm BTEX (MeOH) into a 22 mL HS vial containing 10 mL of DDI and seal promptly using the crimping tool. Place the vial into the heated block, whose temperature should be elevated above ambient. Maintain this temperature throughout the experiment.
- Retrieve the method BTEX from Turbochrom or equivalent software. Open a new sequence file and name the raw data file according to the following example: "G116" (group 1, 16th of the month). Save the sequence file and name it in a manner similar to that in the following example: "G10316" (group 1, March 16th). Download the method (BTEX.mth) and the sequence file (e.g., G10316.seq).
- Make manual injections of approximately 0.25 cc of headspace using a gas-tight syringe (refer to the technique section below). After the three calibration standards have been run, update the calibration method within Turbochrom or equivalent software. *Ask your lab instructor for help in updating the calibration within the method.* Observe the calibration curve and note the value of the square of the correlation coefficient. Discuss with your instructor whether this calibration is acceptable.
- After the instrument has been properly calibrated, proceed to inject the headspace for the three ICVs, a method blank, and unknown samples, as assigned. Your instructor may give you a sample whose concentration is unknown. Record the code on the vial label.
- Obtain the interpolated values from the least squares regression for your three ICVs, method blank, and any and all samples. *Obtain assistance from staff in getting a hard copy of your results.*

5.18.4.6   Technique to Conduct a Manual Headspace Sampling and Direct Injection Using a Gas-Tight Sampling Syringe

If a heater block is available, place the sealed and capped 22 mL headspace vial into the block and allow time for the vial to equilibrate before sampling. A water bath, i.e., a large beaker that is half filled with water, could serve as a constant-temperature environment for headspace sampling. Insert the gas-tight syringe with the valve in the "on" position (if a Precision Sampling type syringe is used) by penetrating the septum seal and withdraw a 0.25 cc aliquot of headspace. Be careful not to withdraw any liquid. Immediately close the on–off valve to the syringe while positioned within the headspace. Position the syringe into the injection port, open the syringe valve, and transfer the 0.5 cc aliquot into the GC.

5.18.4.7   Technique to Conduct an SPME Headspace Sampling and Injection/Thermal Desorption Using an SPME Syringe/Fiber Assembly

Install the 100 µm PDMS fiber into the SPME holder if this has not already been done by your instructor. Follow directions for installing the fiber. Clean the fiber by inserting the SPME holder into an unused GC injection port whose temperature is ~250°C for about ½ hour. To a sealed HS vial that contains either spiked water or an aqueous unknown sample, add a stir bar and begin magnetic stirring. Insert the retracted fiber holder through the septum. Expose the fiber by depressing the plunger and lock it in the bottom position by turning it clockwise. The PDMS fused-silica fiber that is attached to a stainless-steel rod is now exposed to the HS. The fiber should remain above the height of the liquid level. Allow the extraction to take place for ~2 min. Retrack the fiber back into the needle and pull the device out of the vial. Insert the needle of the SPME device into the injection port of the GC. This must be done carefully since SPME needles tend to be of a thinner gauge. Start the analysis by depressing the plunger and locking it in position. After 30 sec withdraw the fiber back into the needle, and pull the needle out of the injector. When the separation is completed, repeat the analysis to determine fiber carryover. Repeat this technique for every calibration standard, ICV, blank, and sample in the same general manner introduced above for the static HS technique.

5.18.5 For the Report

For each sample prep technique, include:

1. A three-point external calibration plot for each chromatographically resolved BTEX analyte with corresponding *correlation coefficient*. Note that Turbochrom finds the square of the correlation coefficient, known as a *coefficient of determination*.
2. A table that includes results for all calibration standards, ICVs, and samples from an interpolation of the regressed calibration curve. This table is to be used for the statistical comparison between both methods.
3. *The coefficient of variation for the ICVs.*
4. The *relative error* for the ICVs.
5. A representative gas chromatogram for the separation.

How do both techniques compare? Apply the *comparison of two dependent averages* to all data. Write several paragraphs using your findings to address this question. Identify those sources of error that might compromise accuracy and precision for both techniques. How might the calibration procedure be modified if an internal standard mode of instrument calibration were used? If an isotope dilution approach were used?

5.18.6 Suggested Readings

To develop this experiment, the author consulted the following resources:

- Sawyer D., W. Heineman, J. Beebe. *Chemistry Experiments for Instrumental Methods*, New York: John Wiley & Sons, 1984, pp. 321–343.
- Kolb B., L. Ettre. *Static Headspace-Gas Chromatography: Theory and Practice.* New York: Wiley VCH, 1997.
- Grob R., E. Barry., Eds. *Modern Practice of Gas Chromatography, 4th ed.* Hoboken, NJ: Wiley-Interscience, 2004, pp. 563–573.
- Pawliszyn J. *Solid Phase Microextraction: Theory and Practice.* New York: Wiley VCH, 1997, pp. 193–200. This was the first book published on SPME theory and application written by the inventor of this sample prep technique.
- Grandy J. et al. Frontiers of sampling: Design of high surface area thin-film samplers for on-site environmental analysis. *LC-GC North America* 37(9): 690–697, 2019. This paper, a recent contribution from the Pawliszyn research group at the University of Waterloo, introduces various designs developed for SPME sampling of a wide variety of environmental sample matrices. Archives of back issues of *LC-GC* are available online.

5.19 DETERMINATION OF THE HERBICIDE RESIDUE TRIFLURALIN IN CHEMICALLY TREATED LAWN SOIL BY GAS CHROMATOGRAPHY USING REVERSED-PHASE SOLID-PHASE EXTRACTION (RP-SPE) SAMPLE PREP TECHNIQUES

5.19.1 BACKGROUND AND SUMMARY OF METHOD

The persistence of trace residue levels of pesticides and herbicides in the environment has been cause for continued concern since the early 1960s, when it became apparent that these residues were detrimental to wildlife and possibly to human health. The benefits of using DDT gradually gave way to the increasing risk of continued use and eventually led to the banning of its use. Herbicides, however, do not appear to present such a high risk to the environment and continue to find widespread use. The chlorophenoxy acid herbicides are not directly amenable to GC and must first be chemically converted to their more volatile methyl esters prior to analysis using GC. Trifluralin or, according to International Union of Pure and Applied Chemistry (IUPAC) organic nomenclature, α,α,α-trifluoro-2, 6-dinitro-*N,N*-dipropyl-*p*-toluidine, is commonly one of the active pre-emergent herbicide ingredients in some lawn treatment formulations. Consider the molecular structures (shown below) of trifluralin and the *internal standard 1,2,4-trichlorobenzene*. The pre-emergent herbicide trifluralin will be extracted from herbicide treated lawn soil and quantitated against 1,2,4-trichlorobenzene in this experiment:

Trifluralin   1,2,4-trichlorobenzene

With reference to the molecular structure for trifluralin, the presence of electronegative heteroatoms, such as fluorine combined with two nitro substituents on the benzene ring, would make the organic compound highly sensitive to the electron-capture detector (ECD), provided that the substance is sufficiently vaporizable and therefore amenable to GC. With a boiling point of 139°C, trifluralin is appropriately classified as a semivolatile, neutral organic compound (SVOC) and it is thus feasible to think that trifluralin could be isolated from a soil matrix by conventional sample preparation techniques such as liquid–liquid extraction (LLE) or perhaps by the more recently developed sample prep technique, reversed-phase solid-phase extraction (RP-SPE).

The assay (on the package) for the commercially available formulation that was dispersed over lawn whose soil beneath has been sampled is given as follows:

Ingredient	%	Ingredient	%
Total nitrogen	20	Chlorine (not more than)	3
Trifluralin (N, N-dipropyl)	0.82	Ammoniacal nitrogen	1.17
Urea nitrogen	18.83	Trifluralin (N-butyl, N-ethyl)	0.43
Sulfur	1.2	Soluble potash	3
Available phosphate	3	Inert	98.5

5.19.1.1 Solid-Phase Extraction

Reversed-phase solid-phase extraction (RP-SPE) techniques provide an alternative to LLE whereby a chemically bonded silica gel is packed into 3 mL barrel-type cartridges (cylindrical) or impregnated into disks that are in turn packed in the same 3 mL barrel cartridges and used to isolate and recover SVOCs contaminants from various environmental samples. A chemically neutral (non to moderately polar) organic compound originally dissolved in water at trace concentrations is thermodynamically unstable. If an aqueous solution containing this compound is allowed to contact a hydrophobic surface, a much stronger van der Waals type of intermolecular interaction *causes the molecules of the analyte to adsorb or stick* to the surface, and thus effectively get removed from the aqueous solution. A relatively small volume of a nonpolar or even semi-polar solvent provides enough hydrophobic interaction to then remove (elute in a chromatographic sense) the analyte molecules. The following sketch is a schematic for the interaction of analyte molecules 2-naphthylamine and hexyl benzene sulfonate (isolates), with a C_8-bonded silica surface:

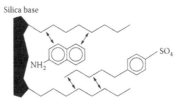

The RP-SPE technique is performed in a stepwise manner as follows:

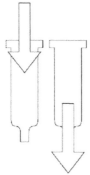

Conditioning

Conditioning the sorbent prior to sample application ensures reproducible retention of the compound of interest (the isolate).

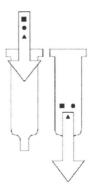

Retention

■ Adsorbed isolate

□ Undesired matrix constituents
Δ Other undesired matrix components

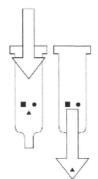

Rinse

▲ Rinse the columns to remove undesired matrix components

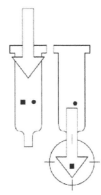

Elution

■ Purified and concentrated isolate

□ Undesired components remain

Trifluralin, which might be present in lawn-treated soil, will be initially extracted into methanol. The methanol extract will be diluted with distilled deionized water (DDI), and the aqueous solution transferred to a 70 mL SPE reservoir on top of a conditioned C_{18}-bonded silica sorbent. The sorbent cartridge will be eluted with high-purity (often called *pesticide residue grade*) iso-octane. The iso-octane eluent is dried by passing it through a second SPE cartridge that contains *anhydrous sodium sulfate* (Na_2SO_4) directly into a 1.0 mL volumetric receiver. An *internal standard* is then added and the eluent brought to a final volume of exactly 1.0 mL. A 1 to 2 µL aliquot of the eluent can then be directly injected in a C-GC-ECD (Autosystem® GC PerkinElmer). The concentration of trifluralin in the eluent can be determined after establishing and verifying the instrument calibration.

5.19.1.2 Internal Standard Mode of Calibration

In addition to external standard and standard addition, the last principal mode of calibration is the *internal standard*. This mode of calibration should be used when there exists *variability in sample injection volume* or when there is concern about the *lack of instrument stability*, or when there is *unavoidable sample loss*. Instrumental response becomes related then to the ratio of the unknown analyte X to that for the internal standard S, instead of related only to the unknown analyte. If some X is lost, one can assume that some S would be lost as well. This preserves the ratio [X]/[S]. For extraction methods, the internal standard (IS) is added to the final extract *just prior to adjusting the final volume. Selecting a suitable IS is not trivial.* The IS should possess similar physical and chemical properties to the analyte of interest and not interfere with the elution of any of the analytes that need to be identified and quantitated. The IS should be within the same concentration range as for the calibration standards and at a fixed concentration. A calibration curve for the IS mode is shown below:

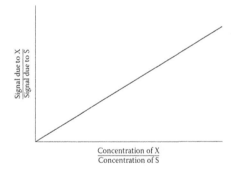

5.19.2 Experimental

5.19.2.1 Preparation of Chemical Reagents

Note: All reagents used in this analytical method contain hazardous chemicals. Wear appropriate eye protection, gloves, and protective attire. Use of concentrated acids and bases should be done in the fume hood.

5.19.2.2 Chemicals/Reagents/Accessories Needed per Group

1 10 mL of iso-octane, suitable for trace pesticide residue analysis
1 100 mL of methanol, suitable for trace pesticide residue analysis
10 SPE cartridges packed with approximately 200 mg of C_{18}-bonded silica
10 Empty SPE cartridges loosely packed with approximately 500 mg of anhydrous sodium sulfate
1 SPE vacuum manifold connected to a vacuum pump via a water trap
10 1.0 mL glass volumetric flasks with ground-glass stoppers

1 10 µL syringe (Hamilton or others) for injection into the GC

1 10 ppm trifluralin reference stock standard dissolved in *pesticide residue grade* iso-octane

1 10 ppm 1, 2, 4-trichlorobenzene (IS) dissolved in MeOH or methyl-*tert*-butyl ether (MTBE).

5.19.2.3 Preparation of the Working Calibration Standards

From the reference stock solution of trifluralin in iso-octane, prepare a series of working calibration standards that cover the range of concentration levels between 10 and 1,000 ppb of trifluralin in high-purity iso-octane. Each working standard should also contain the IS at a concentration level that should fall within the range of concentrations for the calibration standards. This level should be identical among all standards and sample extracts. Use the table below to guide you in preparing your calibration standards.

Standard No.	10 ppm Trifluralin (µL)	10 ppm IS (µL)	$V(T)$ (mL)	Concentration of Trifluralin (ppb)
0	0	50	1.0	0
1	25	50	1.0	250
2	50	50	1.0	500
3	100	50	1.0	1,000
ICV	40	50	1.0	400

5.19.2.4 Establishing the Calibration

Retrieve the method from Turbochrom® or other equivalent software titled "Triflu.mth," *create* a sequence file, and *download* the sequence. *Turn off* the nitrogen makeup gas to the ECD and *measure the split ratio. Adjust to a ratio of between 15 and 20 to 1. Turn the makeup on* after you make the split ratio measurements.

Inject 1 µL of each working calibration standard and inject the ICV in triplicate using the 10 µL liquid-handling syringe. *Update* the method titled "Triflu.mth" with the new calibration standards data using the Graphic Editor. The precision and accuracy data for the ICV can be obtained by retrieving the Graphic Editor and bringing up the raw file for each ICV. Print the tabular formatted report for that particular sample.

5.19.2.5 Isolating Trifluralin from Lawn-Treated Soil Using RP-SPE Techniques

An SPE vacuum manifold, which should be connected to a vacuum pump via a water trap, should be available at the workbench for each of the four workstations. *Condition the C_{18} sorbent by passing 2 mL of MeOH through it.* Attach a 70 mL polypropylene reservoir to the top of the SPE cartridge and fill with DDI to approximately two thirds full.

Place 0.25 g of lawn-treated soil into each of three 50 mL beakers, add 10 mL of methanol (pesticide residue grade) to each beaker, and use a glass stirring rod or magnetic stirring bar to stir this mixture for 5 min. Let stand for another 5 min; then *decant the supernatant liquid* through a Pasteur pipette, which contains non-silanized glass wool to remove large particulates, into a clean beaker. Transfer the liquid to the 70 mL reservoir. Turn on the vacuum pump and pass the contents of the reservoir through the C_{18} sorbent cartridge. After the contents of the reservoir have passed through the sorbent, rinse the reservoir and cartridge with DDI.

Remove the surface moisture with a tissue or equivalent and attach a second SPE cartridge that contains *anhydrous sodium sulfate* beneath the C_{18} sorbent cartridge that contains the retained trifluralin. Elute the sorbent with two 500 µL aliquots of high-purity iso-octane into a 1.0 mL volumetric flask. Remove the receiving volumetric flask from the apparatus and adjust to the calibration mark on the flask with iso-octane.

Inject 1 µL of the dried iso-octane eluent into the C-GC-ECD. Repeat for the other two samples.

5.19.3 FOR THE REPORT

Include all calibration plots, correlation coefficients, and precision and accuracy results of the ICV, and report on the concentration of trifluralin in the soil in #mg/kg (ppm).

5.19.4 SUGGESTED READINGS

To develop this experiment, the author consulted the following resources:

- Sawyer D., W. Heineman, J. Beebe. *Chemistry Experiments for Instrumental Methods*, New York: John Wiley & Sons, 1984, pp. 321–343.
- Perry J. *Introduction to Analytical Gas Chromatography*. New York; Marcel Dekker. 1981. This is one of the better discussions of the principles behind the operation of the ECD up through technological development at that time.
- Lee H., A. Chau, F. Kawahara. Organochlorine pesticides. In Chau A. B. Afghan Eds. *Analysis of Pesticides in Water*. Vol II. Boca Raton, FL. CRC Press. 1982. Chapter 1. This work is part of a three-volume series and is a good introduction to pesticide residue analysis even though it is outdated in the use of packed, instead of capillary GC columns.
- *Methods for the Determination of Organic Compounds in Drinking Water*. EPA-600/4–88/039. December 1988. Cincinnati: Environmental Monitoring Systems and Support Laboratory.
- Hagen E. et al *Anal Chim Acta* 236: 157–164, 1990. This paper was the first to report on a new RP-SPE sorbent, the Empore® Disk (3M Corp).
- Loconto P.R. *LC-GC* 9: 460–465, 1991; Loconto P.R. *LC-GC* 9: 752–760, 1991. The author's pioneering work that demonstrated that the multi-component EPA environmental methods for SVOCs were adaptable to a multi-modal RP-SPE approach. Archives of back issues of *LC-GC* are available online.

5.20 DETERMINATION OF PRIORITY POLLUTANT SEMIVOLATILE ORGANOCHLORINE PESTICIDES: A COMPARISON OF MINI-LIQUID–LIQUID AND REVERSED-PHASE SOLID-PHASE EXTRACTION TECHNIQUES

5.20.1 BACKGROUND AND SUMMARY OF METHOD

Organochlorine pesticides (OCs) were used widely in agriculture during the first half of the 20th century in the U.S. and were subsequently banned from use during the 1970s. Unfortunately, some of the OCs like DDT are still in widespread use around the world. Their persistence in the environment was not apparent until Lovelock introduced the electron-capture detector (ECD) in 1960. When combined with high-resolution capillary gas chromatography and appropriate sample preparation methods, the ECD provides the analytical chemist with the most sensitive means by which to identify and quantitate OCs in environmental aqueous and soil/sediment samples. As analytical chemists were seeking to identify and quantitate OCs during the early 1970s, it became apparent that many additional chromatographically resolved peaks were appearing. What were considered as unknown interfering peaks in the chromatogram were then subsequently found to be polychlorinated biphenyls (PCBs)!

The OCs and PCBs were first determined in wastewaters using EPA Method 608. This method originally required packed columns, and because of this, it necessitated extensive sample preparation and cleanup techniques, which included liquid–liquid extraction and low-pressure column liquid chromatography. Capillary GC-ECD, when combined with more contemporary methods of sample preparation, provides for rapid and cost-effective trace environmental analysis. Over the past 30 years, there have been dramatic improvements in sample preparation techniques as they relate to semivolatile and nonvolatile trace organics quantitative analyses.

In addition to external standard and standard addition, the last principal mode of calibration is called *internal standard*. This mode of calibration should be used when there exists variability in sample injection volume, when there is concern about the lack of instrument stability, and when there is unavoidable sample loss. Instrumental response then becomes related to the ratio of the unknown analyte X to that for the internal standard S, instead of related only to the unknown analyte. If some X is lost, one can assume that some S would be lost as well. This preserves the ratio [X]/[S]. For extraction methods, the internal standard (IS) is added to the final extract just prior to adjusting the final volume. *Selecting a suitable IS is not trivial.* It should possess similar physical and chemical properties to the analyte of interest and not interfere with the elution of any of the analytes that need to be identified and quantitated. The IS should be within the same concentration range as for the calibration standards and at a fixed concentration. The following is a calibration curve for the IS mode:

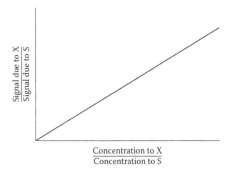

This exercise introduces the student to *reversed-phase solid-phase extraction* (RP-SPE) techniques. The *mini-LLE* method will also be implemented. The two methods will be compared. SPE in the reversed-phase (RP) mode of operation involves passing an aqueous sample over a previously conditioned sorbent that contains a chemically bonded silica gel held in place with polyethylene frits within a column configuration. A typical RP-SPE sequence follows:

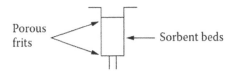

1. Activation of sorbent
2. Removal of activation solvent
3. Application of sample
4. Removal of interferences (I)
5. Elution of concentrated, purified analytes (A)

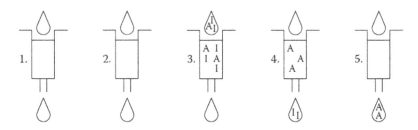

For RP-SPE, methanol is used to condition or wet the sorbent surface, thereby activating the octadecyl moiety and hence forcing it to be receptive to a van der Waals type of intermolecular interaction between the analyte and the C_{18} moiety. This phenomenon is shown below for the isolation of *n*-butyl phthalate on a C_{18} chemically bonded sorbent.

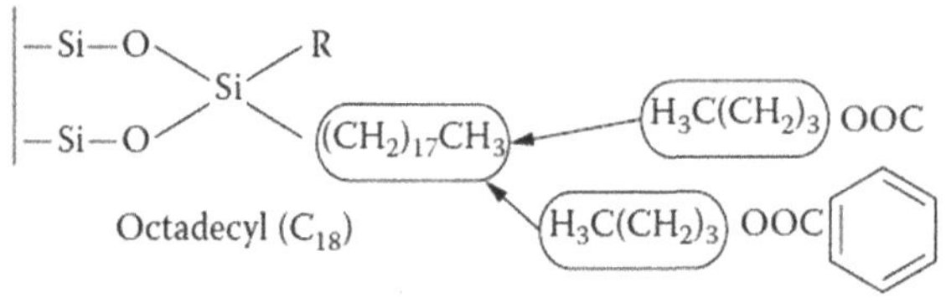

Octadecyl (C_{18})

n-Butyl phthalate

The OCs studied in this exercise are lindane, endrin, and methoxychlor. Lindane (γ-BHC) is synthesized via chlorination of benzene in the presence of ultraviolet light. This forms a mixture of BHC isomers that are identified as α, β, γ, δ, and ε. Selective crystallization isolates the γ isomer, whose aqueous solubility is 7.3 to 10.0 ppm and is the most soluble of the BHC isomers. Endrin is a member of the cyclodiene insecticides and is synthesized using Diels–Alder chemistry. Methoxychlor belongs to the *p, p'*-DDT category and structurally differs from DDT in substitution of a methoxy group in place of a chloro group *para* to the central carbon. Methoxychlor's aqueous solubility is 0.1 to 0.25 ppm and exceeds that of *p, p'*-DDT by a factor of 100. Molecular structures and correct organic nomenclature of these three representative OCs are shown below:

Gamma-BHC   Endrin

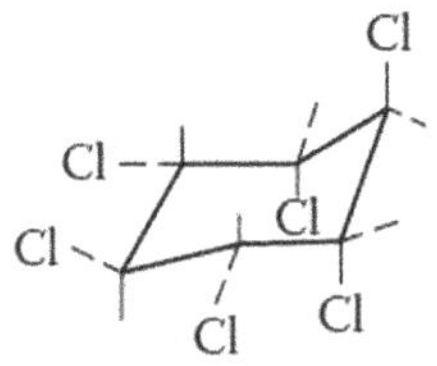

γ-1,2,3,4,5,6-Hexachlorocyclohexane

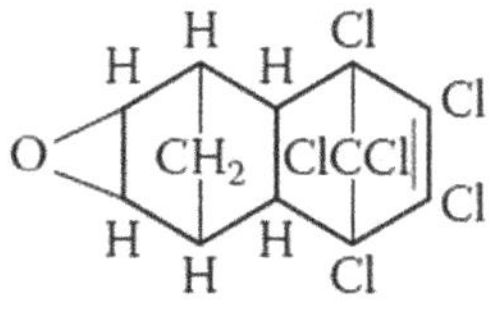

1,2,3,4,10,10-Hexachloro-exo-6,7-
epoxy-1,4,4a,5,-6,7,8,8a-octahydro-1,4
-endo, endo-5,8-dimethanonaphthalene

P,P'-Methoxychlor

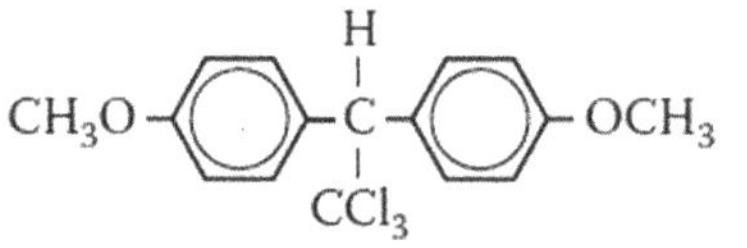

1,1,1-Trichloro-2, bis (p-methoxyphenyl) ethane

5.20.2 EXPERIMENTAL

Gas chromatograph that incorporates an electron-capture detector.

5.20.2.1 Preparation of Chemical Reagents

Note: All reagents used in this analytical method contain hazardous chemicals. Wear appropriate eye protection, gloves, and protective attire. Use of concentrated acids and bases should be done in the fume hood.

5.20.2.2 Chemicals/Reagents Needed per Group

1 1,000 ppm each of lindane, endrin, and methoxychlor stock standard solution dissolved in high-purity iso-octane. This is a solvent available in ultrahigh purity, which is an important requirement in trace environmental analysis. Look for the *pesticide-grade label* on the bottle of iso-octane to verify its purity.

1 20 ppm of an internal standard. Available candidates include 4-hydroxy-2′, 4′, 6′-trichlorobiphenyl, 3, 4, 3′, 4′-tetrachlorobiphenyl, 1, 2-dibromo-3-chloropropane, and β-BHC.

1 Vial containing approximately 30 mL of methanol for conditioning the RP-SPE sorbent.

1 Vial containing approximately 30 mL of high-purity iso-octane.

Molecular structures for the IS candidates are shown below:

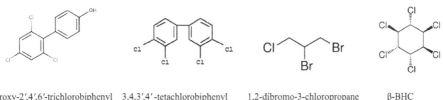

4-Hydroxy-2′,4′,6′-trichlorobiphenyl   3,4,3′,4′-tetachlorobiphenyl   1,2-dibromo-3-chloropropane   β-BHC

5.20.2.3 Preliminary Planning

Because there are two sample preparation methods to be implemented, assemble as a group at the beginning of the laboratory session and decide who does what. Once all results are obtained, the group should reassemble and share all analytical data.

5.20.2.4 Selection of a Suitable Internal Standard

The most appropriate IS needs to be selected from the above list of candidates. Consult with your laboratory instructor and proceed to inject one or more ISs and base your decision on an interpretation of the chromatogram.

5.20.2.5 Procedure for Calibration and Quantitation of the GC-ECD

1. Prepare the necessary primary and secondary dilution standards. The range of concentration levels for the unknowns is between 100 and 1,000 pg/µL (ppb). For example, take a 100 µL aliquot of the 1,000 ppm stock and add to a 10 mL volumetric flask previously half filled with iso-octane. Adjust to the calibration mark and label "10 ppm L, E, M (iso-octane), primary dilution." A 1:10 dilution of this primary dilution standard gives a *secondary dilution*, which should be labeled "1 ppm L, E, M (iso-octane), secondary dilution." Use the secondary dilution to prepare a *series of working calibrations standards* that cover the range of concentrations in the ppb domain, as discussed above.

2. Prepare a set of working calibration standards to include an ICV that brackets the anticipated range for the unknowns. To each calibration standard, add 50 μL of 20 ppm IS so that the *concentration of IS in each calibration standard is identical and at 1.0 ppm.*

3. Retrieve the method from Turbochrom or other equivalent software titled "LEMIS," which stands for *lindane, endrin, methoxychlor, internal standard* mode of calibration; allow sufficient instrument equilibration time. Write a sequence encompassing the calibration standards, ICV, and unknowns. Save the sequence as a file with e.g. the name "G#0317" (Group #, March 17th). Begin to inject a 1 μL aliquot of each working standard. Initially inject iso-octane, then inject in the order of lowest to highest concentration level. This order is important because it prevents carryover from one standard to the next.

4. Update the calibration for the LEMIS method and check with your instructor as to the acceptability of the calibration. If found acceptable, proceed to the analysis once samples have been prepared using both extraction methods. *Be sure to add the same amount of IS to each unknown extract, as was done for the calibration standards.* Because the instrument has been calibrated and updated, the report will include an accurate readout of concentration in a tabular format.

5.20.2.6   Procedure for Performing Mini-LLE and RP-SPE

5. Place exactly 40 mL of unknown sample into a 42 mL vial and extract using 2 mL of iso-octane in a manner similar to that for the BTEX/THMs exercise. This time, however, add twice the amount of IS that you added for the preparation of the calibration standards so that the concentration of IS remains identical to that for all other standards and samples.

6. Place exactly 40 mL of unknown sample into the 70 mL SPE reservoir, which sits atop a previously conditioned C_{18} sorbent, according to specific instructions given to you by your laboratory instructor. Add distilled deionized water (DDI) to the reservoir so as to fill to near capacity. Pass the aqueous sample through the cartridge, which contains approximately 200 mg of C_{18} chemically bonded silica gel. Use a wash bottle that contains DDI to rinse the residual sample from both the reservoir and the cartridge. *Place a second SPE cartridge that is filled with anhydrous sodium sulfate beneath the sorbent cartridge.* The second SPE cartridge containing anhydrous Na_2SO_4 is used to remove residual water from the eluent. Into the manifold place a 1.0 mL volumetric flask as an eluent receiver and elute with two successive 500 μL aliquots of iso-octane. Add the same amount of IS as used for the calibration standards, then adjust to a final volume of 1.0 mL. Transfer to a separate container if necessary.

7. Inject a 1 μL aliquot of the sample extract that also contains the IS into the GC-ECD. At this point, the LEMIS method should have had its calibration updated.

8. Continue to make injections into the calibrated GC-ECD until all samples have been completed. You may want to make replicate injections of a given sample extract.

5.20.3   FOR THE REPORT

Include all calibration plots and calculate the correlation coefficient for the calibration plot. Calculate the precision and accuracy for the ICV, which should have been injected in triplicate. Report on the concentration of each unknown sample. Construct a table that shows the respective concentrations for the unknowns for each of the two methods. Recall that the final extract volume from **mini**-LLE was 2 mL, and that from RP-SPE was 1 mL. Take this into account when comparing the two methods. Which sample preparation method is preferable? Give reasons for your preference and support this with analytical data.

5.20.4 SUGGESTED READINGS

To develop this experiment, the author consulted the following resources:

- Sawyer D., W. Heineman, J. Beebe. *Chemistry Experiments for Instrumental Methods*, New York: John Wiley & Sons, 1984, pp. 321–343.
- Perry J. *Introduction to Analytical Gas Chromatography*. New York; Marcel Dekker. 1981, pp. 164–175. One of the better discussions of Lovelock's development of the electron-capture detector.
- Lee H., A. Chau, F. Kawahara. Organochlorine pesticides. In Chau A., B. Afghan Eds. *Analysis of Pesticides in Water*. Vol II. Boca Raton, FL. CRC Press. 1982. Chapter 1. This work is part of a three-volume series and is a good introduction to pesticide residue analysis even though it is outdated in the use of packed, instead of capillary GC columns.
- *Methods for Organic Chemical Analysis of Municipal and Industrial Wastewaters.* EPA 600/4-82-057. July 1982. Cincinnati, OH Environmental Monitoring and Support Laboratory 1982.
- *SPE Sample Preparation.* Phillipsburg, NJ: J.T. Baker Chemical Co. 1984. p. 8.
- Thurman E., M. Mills. *Solid-Phase Extraction: Principles and Practice*. New York: Wiley-Interscience, 1998. Following an overview and theoretical background for SPE sorption and isolation, the authors discuss method development practices using SPE. This topic is followed by introducing the principles of reversed-phase, normal-phase, and ion-exchange SPE. These topics are followed by applications of SPE to environmental, pharmaceutical, food and natural product samples.
- Loconto P.R. Quantitating Toxaphene Parlar congeners in fish using large volume injection isotope dilution GC with electron-capture negative ion MS. *LC-GC North America,* 36 (5): 320–328, 2018. The author's most recent contribution discusses how to develop and validate a new analytical method to quantitate this ubiquitous organochlorine legacy pesticide down to low ppt concentration levels. Archives of back issues *LC-GC* are available online.

5.21 DETERMINATION OF PRIORITY POLLUTANT POLYCYCLIC AROMATIC HYDROCARBONS (PAHs) IN CONTAMINATED SOIL USING RP-HPLC-PDA WITH WAVELENGTH PROGRAMMING

5.21.1 BACKGROUND AND SUMMARY OF METHOD

In 1979, the EPA proposed Method 610, which, if properly implemented, would determine the 16 priority pollutant PAHs in municipal and industrial discharges. The method was designed to be used to meet the monitoring requirements of the *National Pollutant Discharge Elimination System* (NPDES). The assumption used was that a high expectation of finding some, if not all, of the PAHs was likely. The method incorporated packed-column GC in addition to HPLC, and because of the inherent limitation of packed columns, they were unable to resolve four pairs of compounds (e.g., anthracene from phenanthrene). Because RP-HPLC could separate all 16 PAHs, it become the method of choice. The method involved extracting a 1 L sample of wastewater using methylene chloride, use of *Kuderna–Danish evaporative concentrators* to reduce the volume of solvent, cleanup using a silica gel microcolumn, and a solvent exchange to acetonitrile prior to an injection into an HPLC system. The method requires that a UV absorption detector and a fluorescence emission detector be connected in series to the column outlet. This affords maximum detection sensitivity because some PAHs (e.g., naphthalene, phenanthrene, fluoranthene, among others) are much more sensitive when detected by fluorescence emission when compared to by UV absorption.

In most laboratories today, PAHs are routinely monitored under EPA Method 8270 which incorporates a WCOT GC column and includes the majority of neutrals under the base, neutral, acid (BNAs) designation of the method. This is a liquid–liquid extraction method with determination by gas chromatography-mass spectrometry (GC-MS). Careful changes in pH of the aqueous phase enable a selective extraction of bases and neutrals from acidic compounds. Examples of priority pollutant organic bases include *aniline and substituted anilines.* Examples of priority pollutant organic acids include *phenol and substituted phenols.* The most popular method of recent years has been EPA Method 525, which incorporates SPE techniques and is applicable to PAHs in drinking water.

The most common wavelength, λ, for use with aromatic organic compounds is generally 254 nm because almost all molecules that incorporate the benzene ring in their structure absorb at this wavelength. This wavelength may or may not be the most sensitive wavelength for most PAHs.

RP-HPLC chromatograms for the 16 priority pollutant PAHs in a reference standard mixture (top) and from a soil extract (bottom) are shown below. In the lower chromatogram of each figure, λ was held fixed at 255 nm, whereas for the upper chromatogram of each figure, λ was changed during the run so as to demonstrate how the wavelength influences peak height. The wavelength-programmed HPLC chromatogram shows much less background absorbance and hence increased sensitivity. This information should be used in developing the wavelength-programmed HPLC method.

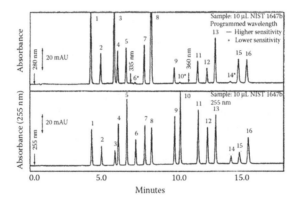

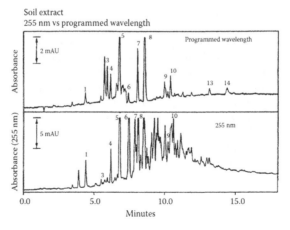

5.21.2 Of what Value is this Experiment?

The exercise affords the student an opportunity to build a new HPLC method using the chromatography data-handling software. The method will also incorporate the concept of wavelength programming, whose objective is to maximize detector sensitivity for a given analyte and which can only be performed using a photodiode array (PDA) detector and accompanying digital electronics. The following table summarizes the detection limits for each PAH in terms of nanograms (ng) injected for $\lambda = 255$ nm and for $\lambda = 280$ nm and for UV programming during the chromatographic run:

Sensitivity and Linearity Data for UV Absorption Detection

No.	PAH	$\lambda = 255$ nm (ng)	$\lambda = 280$ nm (ng)	$\lambda =$ Programmed (ng)
1	Naphthalene	0.6	0.7	0.7(280 nm)
2	Acenaphthylene	0.9	1.9	1.9
3	Acenaphthene	1.42	0.60	0.60
4	Fluorene	0.13	0.53	0.53
5	Phenanthrene	0.06	0.36	0.36
6	Anthracene	0.03	2	1.2(335 nm)
7	Fluoranthene	0.22	0.24	0.45
8	Pyrene	0.25	1.1	0.07
9	Benz(a)anthracene	0.09	0.08	0.5
10	Chrysene	0.06	0.41	6.0
11	Benzo(f)fluoranthene	0.09	0.23	0.4(360 nm)
12	Benzo(k)fluoranthene	0.14	0.31	0.6
13	Benzo(a)pyrene	0.11	0.2	0.2
14	Dibenz(a, h)anthracene	0.45	0.14	4
15	Benzo(g, h, i)perylene	0.32	0.32	0.3
16	Indeno(1, 2, 3-c, d)pyrene	0.16	0.38	0.35

5.21.3 Experimental

High-performance liquid chromatograph (HPLC) that incorporates a UV absorption detector such as a photodiode array under reversed-phase conditions.

5.21.3.1 Preparation of Chemical Reagents

Note: All reagents used in this analytical method contain hazardous chemicals. Wear appropriate eye protection, gloves, and protective attire. Use of concentrated acids and bases should be done in the fume hood.

5.21.3.2 Accessories to be Used with the HPLC per Group

1 HPLC syringe. This syringe incorporates a blunt end; use of a beveled-end GC syringe would damage inner seals to the Rheodyne injector.

1 sixteen-component PAH standard. Check the label for concentration values.

5.21.3.3 Procedure

Be sure to record your observations in your laboratory notebook.

5.21.3.3.1 CREATING THE WAVELENGTH PROGRAM METHOD

Again, you will first find the HPLC instrument in the off position; use "hands on" to activate the instrument and allow at least 15 min for the detector to warm up and stabilize. Ask your laboratory instructor for assistance if necessary. Observe the variability in baseline absorbance. Absorbance should not vary much above a $\Delta A = 0.0100$. Significant variability is most often due to trapped air bubbles because of insufficient degassing of the mobile phase. Inform your instructor if this baseline absorbance variation is significant.

Once the baseline is stable, retrieve the method titled "PAH255" and download it. This method is one previously created by the instructional staff and is a fixed wavelength (λ at 255 nm). Fill the 5 μL injection loop with the PAH standard and observe the chromatogram. The method separates the PAHs based on gradient elution. The method incorporates a one-point calibration.

Use the above tabular information and modify this method to incorporate wavelength programming as discussed earlier. Save the modified method as "PAHWP," where WP stands for "wavelength programmed." Ask your laboratory instructor for assistance in developing this software capability. Fill the 5 μL injection loop with the PAH standard. Using the "chromatograms" section in the main menu, proceed to retrieve both HPLC chromatograms that you just generated. Use the overlay capability to compare both chromatograms and print the overlay. Update the one-point calibration with this standard. You should now have a new method with an updated calibration prior to injecting the extract from the soil discussed below.

5.21.3.3.2 EXTRACTION PROCEDURE FOR SOIL

Weigh approximately 2.0 g of contaminated soil into a 50 or 125 mL glass beaker. Add 20 mL of methylene chloride and use a glass stirring rod to facilitate mixing. Let the contents of the mixture stand for at least 10 min. Decant the extract into a second beaker. It may be necessary to filter this extract if particulates become a problem. This will depend on the type of sample. Pipette 1.0 mL of the methylene chloride extract into a clean, dry 10 mL volumetric flask. Adjust to the calibration mark with *acetonitrile*. Fill the injection loop with this diluted extract. It may be necessary to use a 0.45 μm syringe filter to remove particulates from the diluted extract. Fill the HPLC syringe with about five times the loop volume to ensure a reproducible injection volume. The peak area that is found refers to the concentration of a given PAH in the diluted extract. You will be given assistance on how to allow Turbochrom® (PerkinElmer) to calculate the concentration of each PAH in the original contaminated soil. If time permits, make a second injection of the diluted extract. *Discard the excess methylene chloride extract and CH_2Cl_2/ACN diluted extract into a waste receptacle when finished.*

5.21.3.3.3 CALCULATION OF THE # PPM OF EACH PAH IN CONTAMINATED SOIL

Let us assume that upon injection of the diluted soil extract, a concentration of 225 ppm dibenzo(*a, h*)anthracene in the diluted soil extract was obtained based on a correctly calibrated instrument.

What would the original concentration of dibenzo (*a, h*) anthracene be in the contaminated soil? Please study the series of calculations shown below:

- 225 ppm means 225 μg/mL of diluted soil extract
- Thus, $225 \times 10 = 2{,}250$ μg/mL in the original 20 mL of extract before dilution
- One says that the dilution factor DF is 10
- (20 mL extract) (2250 μg/mL dibenzo (*a, h*) anthracene) = 45,000 μg total from 2 g of soil
- 45,000 μg total/2.0 g soil = 22,500 μg/g or ppm dibenzo (*a, h*) anthracene in contaminated soil

Upon properly completing the sequence file within the Turbochrom® software, the final result, 22,500 ppm, will be directly obtained in the "peak report" for that sample.

5.21.4 FOR THE REPORT

Include the overlay comparison and calibration results and list the concentration of each PAH in the contaminated soil sample. If a second sample result is available, estimate the precision of the method. One usually needs at least three replicate results to begin to use statistics to calculate an acceptable standard deviation. Comment on the advantage of using a PDA to increase sensitivity.

Address the following (refer to the table below on PAHs to assist in your answers):

1. Explain the elution order for the 16 PAHs using chemical principles.
2. The method detection limit using a UV absorption detector for some of the 16 priority pollutant PAHs could be improved if a different detector could be used. Explain.
3. Explain why this method is considered quick. Are there limitations to the use of quick methods, and if so, what are some of these?

Some representative PAHs with corresponding physicochemical properties are shown below and include: molecular weight (MW), molecular formula, molecular structure, aqueous solubility (#mg/L) and logarithm of the octanol-water partition coefficient:

Compound	Abbreviation	MW	Molecular Formula	Molecular Structure	Aqueous Solubility (#mg/L)	Log (K_{ow})
Naphthalene	NA	128	$C_{10}H_8$	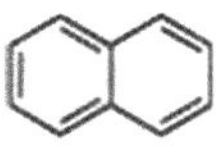	31.7	3.36
Acenaphthylene	ACY	152	$C_{12}H_8$	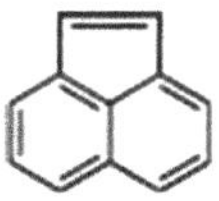	16.1	3.94
Acenaphthene	ACE	154	$C_{12}H_{10}$	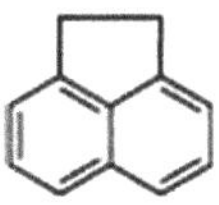	3.93	4.03
Fluorene	FLE	166	$C_{13}H_{10}$	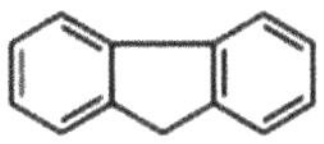	1.98	4.47
Phenanthrene	PH	178	$C_{14}C_{10}$	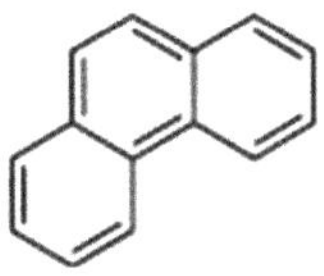	1.29	4.57
Anthracene	AN	178	$C_{14}H_{10}$	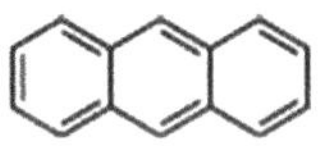	0.073	4.54
Fluoranthene	FLA	202	$C_{16}H_{10}$	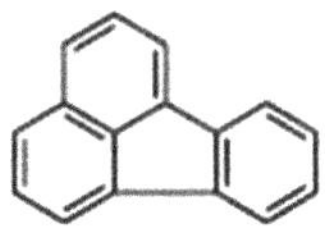	0.260	5.22

Compound	Abbreviation	MW	Molecular Formula	Molecular Structure	Aqueous Solubility (#mg/L)	Log (K_{ow})
Pyrene	PY	202	$C_{16}H_{10}$		0.135	5.18
Triphenylene	TRP	228	$C_{18}H_{12}$		0.043	5.45
Benz(a)anthracene	BaA	228	$C_{18}H_{12}$		0.014	5.91
Chrysene	CHR	228	$C_{18}H_{12}$		0.002	5.91

5.21.5 Suggested Readings

To develop this experiment, the author consulted the following resources:

- Snyder L., J. Kirkland. *Introduction to Modern Liquid Chromatography*. 2nd ed. New York: John Wiley & Sons, 1979.
- Snyder L., J. Glajch, J. Kirkland. *Practical HPLC Method Development*. New York: John Wiley & Sons, 1988.
- Snyder L., J. Kirkland. J. Glajch. *Practical HPLC Method Development*. 2nd ed. New York: John Wiley & Sons, Inc. 1997.
- Sawyer D., W. Heineman, J. Beebe. *Chemistry Experiments for Instrumental Methods*, New York: John Wiley & Sons, 1984, pp. 344–360.
- *Polynuclear aromatic hydrocarbons*. Method 610 Federal Register (233): 69514–69517, 1979.
- Method 8270E, *Gas Chromatography–Mass Spectrometry for Semi-volatile Organics: Capillary Column Technique. In Test Methods for Evaluating Solid Wastes*, Washington, DC: Office of Solid Waste, EPA, SW 846 Update VI, Revision 8, June, 2018.
- Determination of organic compounds in drinking water by liquid–solid extraction and capillary gas chromatography/mass spectrometry. In *Methods for the Determination of Organic Compounds in Drinking Water*, Method 525, EPA/600/4–88/039. Cincinnati: EMSL, 1988.
- Dong M, et al. A quick-turnaround HPLC method for PAHs in soil, water and wastewater oil. *LC-GC* 11: 802–810, 1993. Archives of back issues of *LC-GC* are available on-line.
- Stoll D., T. Lauer. Effects of buffer capacity in reversed-phase liquid chromatography, Part 1; Relationship between the sample- and mobile phase- buffers. *LC-GC North America* 38: 10–15, 2020. Exemplifies current thinking in RP-HPLC research and development. Archives of back issues of *LC-GC* are available online.

5.22   HOW TO SET UP AND OPERATE AN ION CHROMATOGRAPH

Students will notice that on the syllabus for the laboratory course (see above) for the first experiment titled: *Introduction to pH measurement: estimating the degree of purity of snow*, it was stated that an ion chromatograph was available, to be operated by the staff, for students to analyze their snow samples in an attempt to measure impurities. Below is an additional experiment/exercise that provides an opportunity for students (perhaps for extra credit) to learn to operate an additional analytical instrument known as an ion chromatograph.

5.23   DETERMINATION OF INORGANIC ANIONS USING ION CHROMATOGRAPHY (IC): ANION EXCHANGE IC WITH SUPPRESSED CONDUCTIVITY DETECTION

5.23.1   Background

An aqueous sample obtained from the environment that may be expected to contain dissolved inorganic salts. The concentration of these salts can be found by preparing and injecting aqueous samples into a properly installed and optimized ion chromatograph. The concentration of the most common inorganic anions and corresponding cations can be quantitatively determined. The instrumentation available in our laboratory is that manufactured by the Dionex Corporation in the early to mid-1980s.

Ion chromatography (IC) is a low- to moderate-pressure liquid chromatographic (LC) technique and should be clearly distinguished from that of high-pressure LC (HPLC). IC as a determinative technique has been developed to *separate and detect both cations and anions*. The instrument available to the student utilizes anion exchange IC with suppressed conductivity detection. This technique can separate the *common inorganic anions* found in aqueous environmental samples: fluoride (F⁻), chloride (Cl⁻), bromide (Br⁻), nitrite (NO_2^-), nitrate (NO_3^-), phosphate (PO_4^{3-}), and sulfate (SO_4^{2-})—under the IC conditions used here. A second set of related anions of environmental interest has emerged in recent years, and these are collectively called *inorganic disinfection by-products* and include bromate, bromide, chlorite, and chlorate. EPA Method 300.1 has been recently revised, and this method addresses both sets of analytes. Prior to the development of the micromembrane suppressor, a suppressor column was used to reduce the background conductivity of the mobile-phase eluent. The following is a schematic drawing of the classical ion chromatograph that includes the original suppressor column:

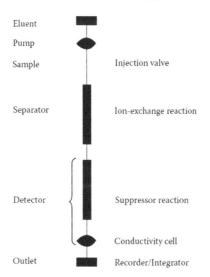

For anion exchange IC, the eluent must be a moderate to strong base (e.g., carbonate or hydroxide). Carbonate (CO_3^{2-}) or hydroxide (OH^-) ions *displace the analyte anion* through the anion exchange resin in the separator column. The separator must contain a low-capacity anion exchange resin so that the analyte ions can make it through the column in a reasonable length of time after injection. Because the conductivity detector responds to all ions, a strong signal due to the eluent would be observed, thus "swamping out" the contribution due to the analyte anions. These eluent ions can be chemically removed while the analyte ions elute from the suppressor in a low-conductivity background. This is done in the suppressor, and the conductivity of the eluent is said to be chemically suppressed. The suppressor and conductivity cell comprise the IC detector, as shown above. The suppressor must contain a cation exchange substrate whereby H^+ from the regenerant migrates across the membrane (which is itself a cation exchanger) and neutralizes the carbonate or hydroxide to form neutral carbonic acid or water. The micromembrane suppressor requires a continuous supply of regenerate solution that consists of 0.025 N sulfuric acid. The regenerant used in our laboratory utilizes mid-1980s technology and consists of a reservoir that contains the dilute H_2SO_4. This solution is made to flow into and out of the micromembrane suppressor by means of positive air pressure. More contemporary designs self-generate the H_2SO_4 electrochemically.

A typical ion chromatogram for separation and detection of a reference standard that contains all seven *common* anions follows:

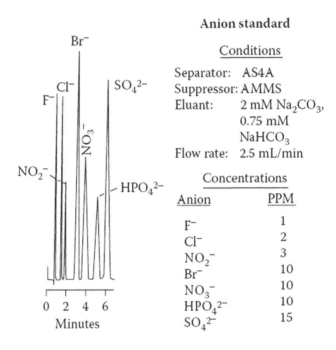

Anion standard

Conditions

Separator:	AS4A
Suppressor:	AMMS
Eluent:	2 mM Na_2CO_3, 0.75 mM $NaHCO_3$
Flow rate:	2.5 mL/min

Concentrations

Anion	PPM
F^-	1
Cl^-	2
NO_2^-	3
Br^-	10
NO_3^-	10
HPO_4^{2-}	10
SO_4^{2-}	15

5.23.2 How Do I Operate a Typical ION Chromatograph?

To operate the ion chromatograph, read the operator's manual. Alternatively, your instructor may have a written procedure. For purposes of illustration, we list below a procedure used in our lab. A Model 2000® (Dionex) instrument interfaced to a PC via a 900® (PE Nelson) interface is available:

1. Turn the compressed air valve on and adjust the main pressure gauge to 80 psi. This provides a head pressure to the eluent reservoir.
2. Adjust the small pressure gauge located adjacent to the ion chromatograph to approximately 10 psi. This provides a head pressure to the regenerant reservoir.
3. Turn on the power to the chromatography module/SP via a switch located in the rear of the module. This will start the single-piston reciprocating pump on the Model 2000® Ion Chromatograph.
4. Turn the conductivity detector cell display to "on" and monitor the output. This is located on the detector module. A reading between 10 and 20 µS with a tolerance of < 1 µS represents a stable baseline. This implies that the ion chromatograph is sufficiently equilibrated across the micromembrane suppressor. At times, the regenerate flow rate may need to be increased or decreased by adjusting the head pressure. If regenerate is not flowing through the micromembrane suppressor, the conductivity value will "skyrocket."
5. At the PC workstation, retrieve the appropriate method from the Turbochrom® (PE Nelson) software. The document "Anion.mth" is available if a more specific method has not been created.
6. Either create a sequence or use the method under "setup" and proceed to download the method. This enables the instrument and PC to communicate.
7. Once the workstation PC reaches a ready status, press inject to remove the lit LED on the Dionex Automation Interface. Fill the 50 µL injection loop, which is located at the sample port on the chromatography/SP, with a filtered, aqueous sample. Press the "inject" and immediately proceed to the 900® (PE Nelson) interface box and press "start."
8. After the chromatographic run is complete, repeat step 7 by first depressing the "inject" to remove the LED.

5.23.3 Is there a Need for Sample Prep?

Yes and no. Aqueous samples that are free of suspended matter and contain dissolved inorganic salts are the only type of sample matrix that is suitable for direct injection into the IC. Wastewater samples that contain suspended solids must be filtered prior to injection into the IC. *Wastewater samples* that have a dissolved organic content (i.e., an appreciable total organic carbon (TOC)) level should be passed through a previously conditioned reversed-phase solid-phase extraction (RP-SPE) cartridge to attempt to remove the dissolved organic matter prior to injection into the IC. Keep in mind that these RP-SPE cartridges have a finite capacity. If this capacity is exceeded, contaminants will merely pass through. If *samples come from a bioreactor*, proteins and other high-molecular-weight solutes must also be removed prior to injection into the IC.

After implementing the operations procedure, the eluent should be pumping through the separator column and the micromembrane suppressor, while the regenerant should be flowing in the opposite direction to the eluent flow, through the suppressor under building-supplied compressed air.

Also included in this experiment are specific procedures to prepare the bicarbonate/carbonate eluent, the preparation of the mixed anion reference standards, and weights of various salts to be used to prepare stock reference standard solution for all analytes of interest.

5.23.4 How Do I Prepare a Reference Stock Standard for Each Anion?

The weight of each salt that can be used to prepare a 1,000 ppm as the anion portion of the salt without any reference to a cation is listed in the following table:

Salt	Grams to dissolve to prepare 1 L of a 1,000 ppm as X
NaF	2.210
NaCl	1.646
KBr	1.488
$NaNO_2$	1.500
$NaNO_3$	1.371
K_2SO_4	1.814
$KBrO_3$	1.315
$KClO_3$	1.467
$NaClO_2$	1.341

Note: Where X = F, Cl, NO_2, etc.

5.23.5　How Do I Prepare The Bicarbonate/Carbonate Eluent from Scratch?

Dissolve 0.571 g of $NaHCO_3$ and 0.763 g of Na_2CO_3 in approximately 250 mL of distilled deionized water (DDI). Transfer this solution to a 200 mL graduated cylinder. Adjust to the mark with DDI and transfer this solution to the IC reservoir. Add 2,000 mL more of DDI to the IC reservoir for a total of 4 L. Label the IC reservoir as "1.7 mM HCO_3/1.8 mM CO_3" eluent. This eluent is used with either an AS4A® (Dionex) or equivalent anion exchange IC separator column.

5.23.6　How Do I Prepare A Mixed Anion Stock Standard for IC?

Certified reference stock standard solutions for each of the anions at 1,000 ppm each anion must be available. The following table outlines one approach to prepare a mixed-stock reference standard:

1,000 ppm Stock (mL)	Anion	Concentration (ppm)
3.0	Chloride	3.0
1.0	Nitrite	1.0
10.0	Bromide	10
2.0	Nitrate	2.0
20.0	Phosphate	20.0
10.0	Sulfate	10.0

A clean, dry 1,000 mL volumetric flask must be used. Pipette the indicated aliquot into a flask that is approximately half filled with DDI.

5.23.7　HOW DO I PREPARE A FOUR-LEVEL SET OF CALIBRATION STANDARDS FOR IC?

Using the mixed reference stock solution (above), proceed using the following table as a guide to prepare a set of working calibration standards:

Mixed-Stock Reference (mL)	V (mL)	#ppmCl	#ppmNO$_2$	#ppmBr	#ppmNO$_3$	#ppmHPO$_4$	#ppmSO$_4$
5.0 (Cal Std #1)	25	0.6	0.2	2.0	0.4	4.0	2.0
10.0 (Cal Std#2)	25	1.2	0.4	4.0	0.8	8.0	4.0
20.0 (Cal Std #3)	25	2.4	0.8	8.0	1.6	16.0	8.0
Neat(Cal Std#4)	—	3.0	1.0	10.0	2.0	20.0	10.0

To prepare a set of four calibration standards, let's consider how to prepare, for example, Calibration Standard (Cal Std #1): pipette 5.0 mL of the mixed *anion stock reference* solution and transfer this aliquot to a clean, dry 25 mL glass volumetric flask, then dilute to the calibration mark on the volumetric flask. This yields <u>Cal Std #1</u> whose concentration of *Cl- is 0.6 ppm*, whose concentration of *NO2- is 0.2 ppm*, whose concentration of *Br- is 2.0 ppm*, whose concentration of NO$_3^-$ is 0.4 ppm, whose concentration of HPO4^{2-} is 4.0 ppm and whose concentration of *SO42- is 2.0 ppm*. Repeat for <u>Cal Std #2</u> and for <u>Cal Std #3</u>. For <u>Cal Std #4</u> inject the *mixed anion stock reference* solution directly into the ion chromatograph.

5.23.8 What Does The Data Look Like?

Two ion chromatograms shown below are included in this experiment that were run on the Model 2000® Dionex instrument. A separation of six of the seven common anions (PO$_4^{3-}$ not included) only 5 min after the instrument was turned on is shown in the first in *ion chromatogram* below. Note the gradual rise in the baseline during development of the chromatogram. This same reference standard of six anions was injected long after the baseline had stabilized. The baseline stability is shown in the second ion chromatogram. A stable baseline is essential for reproducible peak areas, and hence leads to good precision and accuracy for the trace quantitative determination of trace concentration levels of the common inorganic anions.

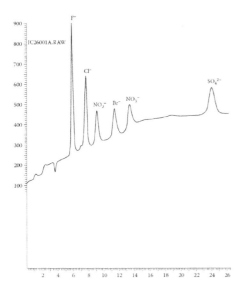
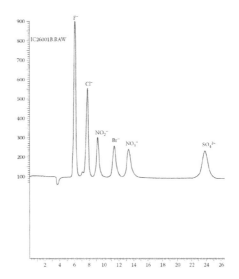

5.23.9 Suggested Readings

To develop this experiment, the author consulted the following resources:

- Weiss J. *Handbook of Ion Chromatography*, E.L. Johnson ed. Sunnyvale, CA: Dionex Corporation, 1986. No better treatment of the subject at the time.
- Shpigun O., Y. Zolotov. *Ion Chromatography in Water Analysis*. Chichester, West Sussex, England: Ellis Horwood Limited, 1988. A well-presented treatment of theory and practice of IC. Includes an excellent discussion of membrane suppression techniques.
- Small H., T. Stevens, W. Baumann *Anal Chem* 47: 1801–1809, 1975. This paper describes the principles of suppressed IC developed at the Dow Chemical Co. These concepts laid the foundation for the Dionex Corporation. Today suppressed IC technology and its numerous advances are part of Thermo Fisher Scientific Corporation.
- Loconto P.R., N. Hussain "Automated coupled ion exclusion – ion chromatography for the determination of trace anions in fermentation broth," *Journal of Chromatographic Science* 33: 75–81 1995. This paper demonstrates how IC and ion exclusion chromatography can be coupled together whereby inorganic anions such as Cl^-, NO_3^-, PO_4^{3-} and SO_4^{2-} can be detected and quantitated directly in a complex sample matrix such as fermentation broth.

Appendix A: Glossary

Books follow sciences, and not sciences, books.

—**Francis Bacon**

This appendix consists of a glossary of terms used in TEQA as well as the regulatory realm. Terms that appear in the literature of statistical evaluation of analytical data, sample preparation, and instrumental analysis are also included. The objective here is to present a wide and diverse glossary that is not unduly large. Most of the terms from the second edition were retained. In addition to a few terms added by the author, additional terms from two recently published *terminology guides* were added. The citations for these guides are shown below:

Majors, R. and J Hinshaw *Chromatography & Sample Preparation Terminology Guide. LC-GC North America* 38(s2): February 2020. (GC, HPLC, Sample Prep terms).
Workman, J. *The Molecular Spectroscopy Terminology Guide, Spectroscopy* 34(s2): February 2019 (UV, Visible and IR Spectroscopy terms)

The language for the glossary (below) was slightly modified where needed in an attempt to achieve some degree of uniformity in presentation.

96-well collection plate—A fixed-size polyethylene rectangular plate (127.8 mm × 85.5 mm) consisting of an array of 8 x 12 (96) small "test tubes" called wells; volumes of wells range from 0.5 to 2 mL.

96-well filtration plate—A fixed-size polyethylene rectangular plate (127.8 mm × 85.5 mm) consisting of an array of 8 x 12 (96) small filter tubes (volumes range from 0.5 to 2 mL): a membrane filter placed at the bottom of the well is used to filter liquid samples; sometimes a prefilter is placed above a membrane filter to prevent clogging with particulate samples.

96-well plate—A small rectangular plastic plate consisting of 96 individual wells that are basically small-volume test tubes arranged in an 8 × 12-well pattern; used for liquid handling and other such requirements.

96-well solid-phase extraction plate—A small rectangular plastic plate consisting of 96 individual flow-through SPE wells arranged in an 8 × 12 array that have top and bottom frits to contain solid particles of sorbent or resin to perform SPE on a miniaturized scale; generally 1 mg to 0.2 g of packing is placed into the well, which can have a volume of up to 2 mL; used for automated SPE with *xyz* liquid handling systems or customized workstations.

α—Separation factor of two adjacent chromatographically resolved peaks: $\alpha = k_2/k_1$.

A solvent—Usually the weaker solvent in a binary eluent or gradient elution separation. In RP-HPLC, the A solvent typically is water or a water-rich mixture.

Absorbance—Measure of the amplitude of absorbed energy in a spectrum related to the concentration of an analyte. This term is also referred to as the negative log (base 10) of transmittance $(-\log_{10}T = \log_{10}(1/T))$. This is also represented as the product of absorptivity (extinction coefficient ε), pathlength, and concentration, written as A= εbc. Absorbance = $\log_{10}(1/\text{reflectance}) = \log_{10}(1/\text{transmittance})$.

Absorptivity—Probability of energy absorbing at a particular wavelength for a specific analyte under specific conditions (i.e. a specific combination of pH, solvent and temperature). Thus a specific amount of material at specified measurement conditions will absorb a specific fraction of the energy striking it. Absorptivity is generally signified by either a lower case epsilon (ε) or *a*. It is defined as:

$$\varepsilon = \frac{A}{cl}$$

Where ε is the molar absorptivity in units of $L \times mol^{-1} \times cm^{-1}$; c is the concentration of molecules in the spectrometer beam in units of mol/L; pathlength (l) is the thickness of the sample in units of cm; A is the measured absorbance.

Absorption—The partitioning of a solute into a solvent as in LLE or the penetration of a solute into the liquid layer of a chromatographic stationary phase, to be distinguished from solute adsorption processes.

Absorption barrier—Any human exposure that may allow diffusion of an agent into a target. Examples of absorption barriers are the skin, lung tissue, and gastrointestinal tract wall. See *exposure surface*.

Accelerated solvent extraction® (ASE)—Trade name for a pressurized fluid extraction system originally introduced by the Dionex Corporation and now sold by Thermo Fisher Scientific.

Accuracy—A measure of how close a *measured* value is to a known *true* value. Accuracy is assessed by means of reference samples and percent recoveries of spiked samples.

Active site—A reactive or strongly attracting site on the surface of a chromatographic packing that may bind analytes or cause peak tailing; sometimes mobile-phase additives (such as a competing base) can negate the effects of active sites.

Activity—In adsorption chromatography, the relative strength of the surface of the packing. For silica gel, the more available the silanol groups, the more active the surface. Activity can be controlled by the addition of water or other polar modifier that hydrogen bonds to the active sites thereby reducing the surface activity; can also refer to biological activity of a biomolecule.

Acute exposure—A human contact between an agent and a target occurring over a short time, generally less than a day.

Adsorbent—Packing used in adsorption chromatography. Silica gel and alumina are the most frequently used adsorbents in chromatography and sample preparation.

Adsorption—A process of retention in which the interactions between the solute and the surface of an adsorbent dominate. The forces can be strong forces (e.g. hydrogen bonds). For silica gel, the silanol group is the driving force for adsorption and any solute functional groups which can interact with this group can be retained on silica. The term *adsorption* places emphasis on the surface versus penetration or embedding in the stationary phase coated or bonded to a surface.

Adsorption isotherm—In adsorption, a plot of the equilibrium concentration of sample in the mobile phase per unit volume versus the concentration in the stationary phase per unit weight. The shape of the adsorption isotherm can determine the chromatographic behavior of the solute such as peak tailing, peak fronting or column overload.

Aerogel—A packing prepared when the dispersing agent is removed from a gel system without collapsing the gel structure. Silica gels and glass beads used for SEC are examples of aerogels that can retain their structures even at the high pressure used in HPLC.

Affinity chromatography—A technique in which a biospecific adsorbent is prepared by coupling a specific ligand (such as an enzyme, antigen, or hormone) for the macromolecule of interest to a solid support (or carrier). This immobilized ligand will interact only with molecules that can selectively bind to it. Molecules that will not bind are eluted unretained. The retained compound can be released in a purified state. Affinity chromatography is normally practiced as an "on-off" separation technique.

Aliquot—A measured portion of a sample taken for analysis.

Alkoxysilane—A reactant used for the preparation of chemically bonded phases. It will react with silica gel to yield alkoxy siloxanes. Refer to the general reaction below:

$$R_3SiOR \equiv SiOH \rightarrow \equiv Si\text{-}OSiR_3 + ROH \qquad \text{where R is an alkyl group.}$$

Alternative hypothesis—Usually comes about from the logic of statistical testing. One example of an alternative hypothesis (refer to the definition of a null hypothesis) is to state that the precision of population A is *not* equal to that of population B. If the two variances are not equal, the nonequality may be stated mathematically in three ways: $H_A: (1)\, \sigma_A^2 \neq \sigma_B^2, (2)\, \sigma_A^2 < \sigma_B^2$, and $\sigma_A^2 > \sigma_B^2$. The first inequality is two-sided, meaning that the inequality can be approached in either direction, whereas the second and third inequalities are one-sided. If the first inequality is of interest, a two-tailed F test (for variances) or a two-tailed t test (for means) is most appropriate. For the second or third inequalities, a one-tailed F or t test is most appropriate.

Alumina—A normal-phase adsorbent used in adsorption chromatography. Aluminum oxide (Al_2O_3) is a porous adsorbent which is available with a slightly basic surface; neutral and acidic modifications can also be made. Basic alumina can have advantages over silica which is considered to have an acidic surface: alumina is seldom used as an HPLC column packing in practice; a highly purified aluminum oxide used chiefly in sample prep cleanup steps; in GC as a gas-solid chromatographic stationary phase (GSC), alumina will separate low molecular weight gases.

Amino phase—a propyl-amino phase used in normal-phase chromatography. It is a somewhat reactive phase for any solute molecule (e.g. aldehydes) or mobile-phase additive that can react with amines. The amino phase has found some applications as a weak anion exchanger and for the separation of carbohydrates using a water-acetonitrile mobile phase. It is a relatively unstable phase.

Amperometric detection—Electrochemical detection applying a constant potential to the working electrode; measured current from oxidation or reduction is proportional to the sample concentration. Very selective and sensitive method. Works with electrode reactions not changing the electrode surface (e.g. cyanide, nitrite, thiosulfate, phenols). Approximately 10% of the analyte oxidized or reduced; may be used as stand-alone as well as in series or parallel to other detectors.

Analysis—The ascertainment of the identity or the concentration of the constituents or components of a sample. Analysis is often used incorrectly in place of determination. *Only samples can be analyzed; constituents or components are determined.* Examples of correct usage are: "… analysis of fish for PCBs" and "… determination of PCBs in fish."

Analyte—The chemical element or compound an analyst seeks to determine; the chemical element of interest; the compounds of interest to be isolated from the sample matrix.

Analyte protectorant—In GC, a chemical compound added to a sample before injection to cut down on interactions between analytes that are unstable or behave poorly in the GC flow path on active sites; the protectorants are chosen so that they do not interfere with the analysis of the compounds of interest yet prevent these compounds from interacting with the active sites in the flow path; these protectorants are not generally required for LC and LC-MS.

Analytical batch—The basic unit of analytical quality control, defined as samples that are analyzed together with the same method sequence and the same host of reagents, and with the manipulations common to each sample within the same time period or in continuous sequential time periods. Samples in each batch should be of similar composition, e.g., groundwater, sludge, ash.

Analytical column—A chromatographic column used for qualitative and quantitative analysis; a typical analytical column for LC will be 50–250 cm x 4.6 mm, but columns with smaller diameters (down to 0.05 mm i.d.) can also be considered as analytical columns. GC analytical columns range in length from 1 m to as much as 60 m, with inner diameters ranging from less than 100 μm up to 2 mm. Stationary phases can be coated or bonded onto the interior of the tubing; packed GC columns are generally wider and shorter and are less frequently used nowadays. Chromatography columns can be constructed of stainless steel, glass, glass-lined stainless steel, PEEK, fused silica, and other metallic and nonmetallic materials.

Analytical sample—Any solution or medium introduced into an instrument on which an analysis is performed, excluding instrument calibration, initial calibration verification, initial calibration blank, and continuing calibration blank. The following are all analytical samples: undiluted and diluted samples, pre-digestion spike samples, duplicate samples, serial dilution samples, analytical spike samples, postdigestion spike samples, interference check samples, laboratory control samples, preparation or method blank, and linear range analysis samples.

Anion-exchange chromatography—The form of ion-exchange chromatography that uses resins or packings with functional groups that can separate anions. Synthetic resins, bonded phase silicas, and other metal oxides are available for this mode. A typical anion-exchange functional group is *tetra-alkylammonium*, making a strong anion exchanger. An *amino* group on a bonded stationary phase would be an example of a weak anion exchanger.

Appendix VIII—This list in 40 CFR 261 contains 355 compounds and classes of compounds shown to be toxic, carcinogenic, mutagenic, or teratogenic in reputable scientific studies.

Appendix IX—This list in 40 CRF 264 replaced Appendix VIII with regard to the groundwater monitoring regulations found in 40 CRF 264 and 40 CFR 270.

Area units—A term used in gas chromatography that indicates the peak area of a compound exiting a chromatographic column. The size or area of the peak is proportional to the amount of analyte in the sample.

Aroclor—Trade name (Monsanto, St. Louis, MO) for a series of commercial PCB and polychlorinated terphenyl mixtures marketed in the U.S., e.g. AR 1242 refers to numerous C_{12} polychlorinated biphenyls that are 42% chlorine by weight.

Array detector—A photoelectric detector that includes multiple, evenly spaced lines or rectangles of detectors (or pixels), each of which may independently detect a broad wavelength range of energy. These detectors may be flat or curved to accommodate a multiwavelength beam of diffracted light. They may be a linear array, or a rectangular array where different pixels form a mosaic of smaller pixels capable of collecting and reporting multidimensional images.

Askarel—A general term for a group of nonflammable, synthetic, chlorinated, aromatic hydrocarbons used as electrical insulating media.

Asymmetry—Factor describing the shape of a chromatographic peak. Theory assumes a Gaussian shape and that peaks are symmetrical. A quantitative measure is the peak asymmetry factor, which is the ratio of the distance from the peak apex to the back side of the chromatography curve over the distance from the peak apex to the front side of the chromatography curve at 10% of the peak height. Various other measures of asymmetry are also in common use.

Atomic absorption spectrophotometry (AAS)—A technique for analyzing environmental samples or human specimens for trace concentrations of metal using an element-specific lamp that emits a characteristic light spectrum. A sample is heated in a flame or graphite furnace and the light beam is passed through it. When the sample absorbs light, an energy loss is detected and is translated into a concentration of metal in the sample. This technique detects one metal at a time.

Attenuated total reflectance (ATR)—A sampling device used for surface analysis with infrared spectra where physical contact of the ATR crystal occurs with the sample surface. ATR crystal materials include: diamond, silicon (Si), and zinc selenide (ZnSe). ATR measurements can yield excellent quality spectra, provided that the contact pressures of the ATR crystal and the sample are held constant. By applying an ATR correction algorithm, ATR spectra may be compared qualitatively with transmission spectra.

Audit—A systematic check to determine the quality of some function or activity. Two basic types are *performance* audits and *system* audits. Performance audits involve a quantitative comparison of the laboratory's results to those of a proficiency sample containing known concentrations of analytes. A system audit is a qualitative evaluation that normally consists of an onsite review of a lab's quality assurance system and physical facilities.

β—Phase ratio. In GC, the ratio of mobile- to stationary-phase volumes. Thicker stationary-phase films yield longer retention times and higher peak capacities. For open tubular columns of diameter d_c and having a stationary-phase film thickness d_f, the phase ratio $\beta = V_G/V_L \sim d_c/4d_f$.

B solvent—Usually the stronger solvent in a binary eluent or gradient separation. In RP-HPLC, typically the organic modifier or modifier-rich binary mixture with water.

B term—The second term of the van Deemter equation; the first term of the Golay equation.

Back extraction—Used in LLE to perform an additional extraction to further purify a sample; initially the extraction may take place with an aqueous solvent buffered at a high pH and an immiscible organic solvent; after the initial extraction takes place and interferences are removed, then by having another aqueous solution at a low pH, one can back-extract the analyte into the organic layer based on the analyte now being in a neutral form. An example would be for the cleanup of an acidic substance containing -COOH groups; at high pH the carboxyl would be ionized and prefer the aqueous layer and impurities may migrate to the organic phase and be discarded; then the pH of the aqueous layer can be adjusted to a low value. Now the carboxyl group is in an unionized form and readily extracted into the organic layer as a purified substance.

Backflushing—Useful in chromatography to remove compounds that are held strongly at the head of a column. By reversing the flow at the conclusion of a run, analytes trapped at the head of the column can be flushed from the column entrance because they have a shorter distance to travel; sometimes a strong solvent in LC or elevated temperatures in GC will be needed to move them along. A valve or fluidic device is used to effect the change of mobile-phase flow direction. Backflushing can be used for analysis of these compounds or merely to remove them from the column.

Background correction—A technique usually employed relative to metals analysis, which compensates for variable background contribution to the instrument signal in the determination of trace elements.

Background level—The amount of an agent in a medium (e.g., water, soil) that is not attributed to the source(s) under investigation in an exposure assessment. Can be naturally occurring or anthropogenic. Natural background is the concentration of an agent in a medium that occurs naturally or is not the result of human activities.

Back-pressure regulator—In HPLC, a device placed online after the detector to maintain a positive pressure on the flow cell minimizing solvent outgassing problems in the detector. In GC, the term usually refers to a carrier gas regulator in the split vent line that maintains a constant pressure at the inlet as split flows change.

Bakeout—The process of removing contaminants from a column in GC by operation at elevated temperatures, which should not exceed a column's maximum allowable operating temperature.

Ballshmitter number—Serial numbering system for PCBs from 1 to 209, starting with 2-monochlorobiphenyl and ending with decachlorobiphenyl; also termed *IUPAC number.*

Band—Refers to a chromatographic peak as it moves along and is eluted from the column.

Band broadening—Several processes that cause solute profiles to broaden as they migrate through a column in both GC and HPLC.

Baseline—The baseline is the line drawn by the recording device representing the signal from the detector when only mobile phase is passing through, in the absence of any solutes. It also represents the point from which calculations are often made on peaks to determine peak area or peak height.

Baseline drift—Term for any regular change occurring in baseline signal from an HPLC or GC detector; it may arise from changes in flow rate of the mobile phrase or from stationary-phase bleed and may trend in a positive or negative direction. Baseline drift occurs over a *longer* period of time than baseline noise.

Baseline noise—Irregular variations (short term) in the chromatographic baseline as a result of electrical noise or temperature fluctuations, outgassing to the flow cell or poorly mixed mobile-phase solvents.

Batch—See *analytical batch*.

Beam-splitter—An optical element that is coated in such a way that ~50% of the energy striking the element is reflected and 50% is transmitted; these optical elements are often referred to as half-mirrors. A *Michelson interferometer* comprises a beam-splitter, which is an active optical element, along with one fixed and one moving mirror. Dual-beam instruments exhibiting simultaneous sample and reference beams also use a beam-splitter to separate the beams used to measure ratioed spectra as the sample and reference beams.

Bed volume—For a given mass of sorbent in SPE packed in a cylindrical bed, the volume of solvent required to fill all pores of the sorbent bed and interstitial spaces between the sorbent particles. This value is approximately 1.25 mL/g of sorbent for the most common (40 μm) silica-based sorbents.

BET method—A method for measuring surface area developed by Brunner, Emmett, and Teller (BET) that uses nitrogen adsorption-condensation in pores at liquid nitrogen temperature. Pore volume and pore size distribution can also be obtained from BET calculations.

Binary mobile phase—Mobile phase consisting of two solvents or buffers (or one of each).

Bioassay—The use of living organisms to measure the effect of a substance, factor, or condition by comparing before and after data. The term is often used to mean cancer bioassays.

Bioavailability—The degree to which an agent is capable of being absorbed by an organism and available for metabolism or interaction with biologically significant receptors. Bioavailability involves both release from a medium (if present) and absorption by an organism.

Biochemical oxygen demand (BOD)—A measure of the amount of oxygen consumed in the biochemical processes that break down organic matter in water. A larger BOD value indicates a greater degree of organic pollution. A related term *BOD5* is the amount of dissolved oxygen consumed in 5 days.

Biocompatible—A term to indicate that the column or instrument component will not irreversibly or strongly adsorb or deactivate biomolecules, such as proteins. Frequently means metal-free or ceramic surfaces and components.

Biomarker/biological marker—Indicator of changes or events in biological systems. Biological markers of exposure refer to cellular, biochemical, analytical, or molecular measures that are obtained from biological media such as tissues, cells, or fluids and are indicative of exposure to an agent.

Bit—Abbreviation of *binary digit*. A bit is the value of the least significant place of a number in the binary number system. It is the smallest unit of information a traditional computer can handle. A bit corresponds to the output of a solid-state device, which is either the "0" state (usually 0 V), or the "1" state (usually +5 V). In traditional computer circuits, intermediate states are not defined.

Blank—An artificial sample *designed to monitor* the introduction of artifacts into the measurement process. For aqueous samples, reagent water is used as a blank matrix. A universal matrix does not exist for solid samples: therefore, no matrix is routinely used. There are several types of blanks, which monitor a variety of processes:

- A *laboratory blank* is taken through sample preparation and analysis only. It is a test for contamination in sample preparation and analyses.
- A *holding blank* is stored and analyzed with samples at the laboratory. It is a test for contamination in sample storage, as well as sample preparation and analysis.
- A *trip blank* is shipped to and from the field with the sample containers. It is not opened in the field, and therefore provides a test for contamination from sample preparation, site conditions, and transport, as well as sample storage, preparation, and analysis. It is most commonly used for volatile organics.
- A *field blank* is opened in the field and tests for contamination from the atmosphere as well as those activities listed under *trip blank*.

Bleed—The loss of material from a column in GC or septum caused by high- temperature operation. Bleed can result in ghost peaks and increased detector baseline offset and noise.

BNA—Base-neutral, and acid extractable compounds. The terms *base-neutral,* and *acid* refer to the pH condition of the sample undergoing LLE. Analytes that are more efficiently extracted from alkaline of neutral conditions are referred to as "base-neutral extractables." Analytes that are more efficiently extracted from water under acidic conditions are often referred to as "acid extractables."

Bonded phase—A stationary phase in GC that has been *chemically bonded to the fused-silica inner column wall* of a capillary or wall-coated open tubular (WCOT) GC column. In HPLC, the substrate is usually a silica gel particle (a 3 µm particle size is now common) that has been chemically bonded with various alkoxysilanes. This bonded phase silica has been rendered hydrophobic and used to conduct RP-HPLC. Approximately 70% of all HPLC is carried out on chemically bonded phases. A 40 µm irregular particle size silica gel that has been *chemically bonded.* e.g. octadecyl siloxane silica is said to be a bonded phase silica. The surface of most bonded phase silicas being hydrophobic are used to isolate and recover hydrophobic analytes such as priority pollutants organics from aqueous environmental sample matrices.

Breakthrough—Undesired elution of an isolate in SPE from a sorbent bed, which occurs when the isolate is too weakly retained or sorbent capacity is exceeded.

Breakthrough volume—The volume at which a particular solute pumped continuously through a column will begin to be eluted. It is related to the column volume plus the retention factor of the solute. It is useful to determine the total sample capacity of the column for a particular solute.

4-Bromofluorobenzene (BFB)—A compound utilized in EPA gas chromatography-mass spectrometry (GC-MS) volatile methods to establish mass spectral instrument performance. It is also used as a surrogate for volatile organic analysis.

BTEX—Benzene, toluene, ethyl benzene, and the three isomeric xylenes; environmental contaminants in groundwater from leaking underground gasoline storage tanks.

Buffer—An aqueous solution containing a weak acid and its conjugate base (or a weak base and its conjugate acid) that will maintain its pH upon addition of moderate amounts of either strong bases or strong acids.

C term—The interphase mass transfer term of the van Deemter and Golay equations.

Calibration—The systematic determination of the relationship of the responses of the measurement system to the concentration of the analyte of interest. Instrument calibration performed before any samples are analyzed is called the initial calibration. Subsequent checks on the instrument calibration performed throughout analysis are called *continuing calibrations.*

California list—Created by the state of California, this list is used to determine those wastes that are restricted from land disposal. The list was incorporated by RCRA. See ***RCRA***.

Canister collection—A stainless steel vessel designed to hold vacuum to less than 1.3 Pa (10 mtorr) or pressure to 275 kPa (40 psi). Canisters are available in a range of volumes: 400 mL, 1.0 L, 3.0 L, 6.0 L, and 15 L. The size of canister used usually depends on the concentration of the analytes in the sample, the sampling time, the flow rate, and the sample volume required for the sampling period. Typically, smaller canisters are used for more concentrated samples such as soil gas collection. 3 L and 6 L canisters are used to obtain integrated (TWA) ambient air samples at sampling times of up to 24 h and large 15 L canisters are used for reference standards. Sampling time will be limited by the combination of canister size and the flow rate at which the sample is to be collected.

Capacity—The total mass of isolates or interferences in SPE that a specific sorbent mass can retain in a given solvent environment.

Capacity factor—In GC and HPLC, provides a quantitative measure of the degree to which a given analyte is retained on a stationary phase; given the symbol k or k'; where $k' = (t_R - t_0)/t_0$.

Capillary column—A long, coiled optical fiber that is the principal GC separation medium. Lengths are typically between 30 and 60 m, with column diameters defined as follows:

- Ultra-narrow bore—180 μm
- Narrow bore—250 μm
- Wide bore—320 μm
- Megabore—530 μm

The column consists of (1) an outer polyimide protective coating, (2) an inner fused-silica wall, and (3) a stationary phase such as various poly-dimethyl siloxanes either coated or chemically bonded to the inner fused-silica wall. Capillary columns are used extensively in GC and CE. See ***Wall-coated open tubular column***.

Carrier gas linear velocity—In GC, the average speed at which a molecule of carrier gas passes through a column in GC, expressed in units of *centimeters per second* and usually given the symbol $\bar{u}$. For a column of length L in *meters* and a time in *seconds* for an unretained compound to pass through the column after injection, t_0, the average linear carrier gas velocity can be found according to

$$\bar{u} = \frac{100L}{t_0}$$

This author carefully uses a butane lighter to measure $\bar{u}$ as follows: A 10 μL liquid handling glass syringe is inserted into the orifice of the lighter, the valve depressed, and ~5 μL of butane gas is removed and injected directly into the injection port of a GC-MS. The MSD is set to m/z 58, and the time it takes for the ion abundance to maximize represents t_0.

CAS registry number—Unique number assigned by the Chemical Abstracts Service to each chemical compound.

Cation-exchange chromatography—The form of ion-exchange chromatography that uses resins or packings with functional groups that can separate cations. An example of a strong cation functional group would be a *sulfonic acid*; a weak cation-exchange functional group would be a *carboxylic acid*.

CERCLA—The Comprehensive Environmental Response, Compensation and Liability Act, also known as *Superfund*. Enacted December 11, 1980, and amended thereafter, CERCLA provides for identification and cleanup of hazardous materials released over the land and into the air, waterways, and groundwater. It covers areas affected by newly released materials and older leaking or abandoned dump sites. CERCLA established the Superfund, a trust fund, to help pay for cleanup of hazardous materials sites. The EPA has authority to collect cleanup costs from those who release the waste material. Cleanup funds come from fines and penalties, taxes on chemical/petrochemical feed stocks, and the U.S. Department of the Treasury. A separate fund collects taxes on active disposal sites to finance monitoring after they close.

Chain of custody—Procedures and associated documents designed to trace the custody of a sample from the point of origin to final disposition, with the intent of legally demonstrating that custody remained intact and that tampering or substitutions were precluded.

Chalcogenide—These materials contain a group XVI element from the Periodic Table. The main use in spectroscopy is for infrared-range transmitting optics such as fiber optics and optical lens elements. The most common materials in this category include: cadmium sulfide (CdS), cadmium selenide (CdSe), cadmium telluride (CdTe), zinc selenide (ZnSe), and zinc sulfide (ZnS).

Chemical ionization MS—An alternative to electron-impact MS; uses a reagent gas such as methane to promote collision of analyte molecules with ions from this reagent gas. The result is minimum molecular fragmentation, with a heteroatom of greatest proton affinity becoming protonated. Because little fragmentation occurs, this technique provides molecular structure information. Analytes can be positive ion and negative ion. See *Electron-capture negative ion mass spectrometry*.

Chemical oxygen demand (COD)—A measure of the oxygen required to oxidize all compounds in water, both organic and inorganic.

Chemometrics—A subdiscipline of analytical chemistry involving complex overlapping molecular patterns from a sample chemistry (i.e. chromatography or spectroscopy); and the interpretation of these signal patterns (e.g. chromatograms or spectra); using a series of multivariant mathematical techniques in the usable form of computer algorithms.

Chlorinated hydrocarbons—Organic compounds containing one or more chlorine atoms. These include a class of persistent, broad-spectrum insecticides that linger in the environment and accumulate in the food chain. Among them are DDT, aldrin, dieldrin, heptachlor, chlordane, lindane, endrin, mirex, hexachlorobenzene, and toxaphene. Another example is trichloroethene (TCE), an industrial solvent.

Chronic exposure—A continuous or intermittent long-term contact between an agent and a target such as a human.

Clean Water Act (CWA)—Regulates the discharge of nontoxic and toxic pollutants into surface waters. The CWA (Public Law 92–500) became effective November 18, 1972, and has been amended significantly since then. Its ultimate goal is to eliminate all discharges into surface waters usable for fishing, swimming, and other beneficial uses. EPA and the Army Corps of Engineers have jurisdiction. EPA sets guidelines, and state agencies issue permits, e.g., National Pollutant Discharge Elimination system permits, specifying the types of control equipment and allowable discharges for each facility.

Code of Federal Regulations (CFR)—A collection of the federal regulations established by law and published by the Government Printing Office.

Cold injection—An injection in GC that occurs at temperatures lower than the final oven temperature, usually at or below the solvent boiling point.

Colorimetric—Analysis based on the measurement of the color that develops during the test-specific reaction. The intensity of color development is usually measured at a specified wavelength on a spectrophotometer.

Compound-independent calibration (CIC)—An instrument calibration model where the compound used to calibrate the instrument is not necessarily the analyte of interest. For the atomic emission detector interfaced to a gas chromatograph (GC-AED), the intensity of the spectral emission line of an element is calibrated to the concentration of the element. The analyte of interest must contain the element for which the instrument is to be calibrated, but it need not be the compound with which the instrument is calibrated. The source of the element for calibration is thus independent of the analyte of interest. This definition is adapted from EPA Method 8085. CIC is unique to GC-AED. All other GC detectors are compound dependent.

Comprehensive GC (GC × GC)—Two-dimensional technique in which all compounds experience the selectivity of two columns connected in series by a retention modulation device, thereby generating much higher resolution than that attainable with any single column.

Compressibility correction factor (j)—This factor compensates for the expansion of a carrier gas in GC as it moves along the column from the entrance at the inlet pressure, P_i, to the column exit, at the outlet pressure, P_o.

Confirmation—In GC, an unknown compound in a sample is identified on the basis of its retention time on a specific chromatographic column. Because several compounds may exhibit that exact same retention time on a given column, a secondary confirmation on a different column or detector is often recommended for additional confidence in the compound identification. This additional confirmation is often referred to as dual-column or dual-detector confirmation.

Congeners—Compounds containing different numbers and positions of chlorination (or other single-atom substituents) on the same base structure. For example, dioxin congeners all contain the same dibenzo-*p*-dioxin nucleus, but are chlorinated to different levels and in different ring positions. There are 209 PCB congeners.

Continuous liquid–liquid extraction—An extraction technique that involves boiling the extraction solvent in a flask and condensing the solvent above the aqueous sample. The condensed solvent drips through the sample and extracts the sample and the compounds of interest from the aqueous phase.

Contract Laboratory Program (CLP)—A program coordinated through the EPA to provide a wide range of analytical services by commercial laboratories in support of investigation, remediation, and enforcement actions at Superfund sites. Laboratories participating in this program are under contract to the EPA and must follow very specific analytical protocols during analyses and data delivery, as specified in the statement of work associated with the contract. Also called the *Superfund Analytical Services and Contract Laboratory Program*.

Coplanar—Those PCB congeners lacking or containing only one chlorine in the ortho positions. Chlorine substituents are larger than hydrogens and thus constrain rotation about the carbon–carbon bond bridging the two rings in the biphenyl. PCBs with zero or one ortho-chlorines can form a coplanar conformation more readily than those congeners with two, three, and four chlorines in the ortho positions. The mono-ortho-chloro congeners may or may not be included in the term *coplanar*, depending on the authors of the study.

Counterion—An ionic species associated with the ionic functional group of opposite charge on the sorbent surface.

Cross-linked phase—A stationary phase in GC that includes cross-linked polymer chains. Usually, it also is bonded to the column inner wall.

Cuvette (also cuvet)—Transparent receptacle in which sample solutions are introduced into the light path of spectrometers. Usually, the two sides are equal (e.g. 1 cm square) while the third dimension (height) is elongated, possibly as long as 15 cm. For UV work, the window material is generally low-OH quartz. For visible spectroscopy, some polymers are used such as polystyrene or polymethylmethacrylate. The z-dimension is the distance from the bottom of the cuvet holder to the center of the transmission beam of the spectrophotometer.

Data quality objective (DQO)—During the planning phase of a project requiring laboratory support, the data user must establish the quality of data required from the investigation. Such statements of data quality are known as DQOs. Qualitative and quantitative statements about the data required to support specific decisions or regulatory actions, DQOs must take into account sampling considerations as well as analytical protocols.

Data validation—A systematic effort to review data for identification of errors, thereby deleting or flagging suspect values to ensure the validity of the data for the user. This process may be done by manual or computer methods.

Dead volume—Extra volume experienced by solutes as they pass through a chromatographic system. Excessive dead volume causes additional peak broadening.

Decafluorotriphenyl phosphine (DFTPP)—An organic compound utilized in several EPA GC-MS methods to establish proper mass spectral instrument performance for semi-volatile analyses.

Deflagration—An incident involving the sudden combustion of a substance. The deflagration results in a subsonic shock wave.

DEGS—Diethylene glycol succinate; used in GC as a stationary phase.

Detector—A device sensitive to electromagnetic radiation at the wavelength of interest. The output of the device is usually an electrical signal proportional to the intensity of the electromagnetic energy input. The detector generally produces an analog signal proportional to the radiation striking the detector element. This analog signal is converted into a digital signal by means of an analog-to-digital (A/D) converter and amplifier circuit. Usually a simple silicon diode, a more sensitive photomultiplier tube (PMT), or various photodetector systems are used to detect the light energy.

Determination—The ascertainment of the quantity or concentration of a specific substance in a sample.

Detonation—An incident involving a violent explosion. Detonation results in a supersonic shock wave. For example, acetylene can detonate upon exposure to pure copper, nickel, mercury, or gold.

Digestion log—An official record of the *sample preparation procedures* used in processing a sample prior to instrumental analysis. This is most often associated with digestion of samples utilizing various acids prior to *analysis for metals*.

Digital—Refers to the behavior of certain devices (such as computers) that can only be in one of a finite number of discrete states. Computers can only be in the states associated with the output of the various electronic components exhibiting voltage levels corresponding to the 0 state or the 1 state. Each such solid-state device contains the information for one binary digit. See also *bit*.

Diol sorbent—A polar sorbent, typically used for polar extractions from nonpolar solvents.

Direct aqueous injection (DAI)—Injection of between 5 and 10 µL of an aqueous solution containing the analyte of interest; this is generally done using a packed GC column.

Direct injection—Occurs in GC when sample enters an inlet and is swept into a column by carrier gas flow. No sample splitting or venting occurs during or after the injection.

Dissolved metals—Metallic elements determined on a water sample that has been passed through a 0.45 µm filter.

Dissolved oxygen (DO)—The oxygen freely available in water. DO is vital to fish and other aquatic life and for the prevention of odors. Traditionally, the level of DO has been accepted as the single most important indicator of a water body's ability to support desirable aquatic life. Secondary and advanced waste treatment are generally designed to protect DO in waste-receiving waters.

Dissolved solids—Disintegrated organic and inorganic matter contained in water. Excessive amounts of dissolved solids make water unfit to drink or use in industrial processes.

DMCS—Dimethylchlorosilane; used for silanizing various types of glass such as GC inlet liners.

Dose—The amount of agent that enters a target (such as a human) by crossing an exposure surface. If the exposure surface is an absorption barrier, the dose is an absorbed dose/uptake dose, otherwise, it is an intake dose. See *uptake*.

Dose rate—Dose per unit time.

Double (dual) beam—This term is applied to the simultaneous measurement of optical data from both a sample beam and a stable reference beam. Although the optical radiation used to obtain the measurements may share part or all of the optical path, there must be a separation either in space or in time between the two measurements; thus, two beams of optical radiation must be used. One beam is uses as a reference, the other as a sample measurement. The *ratio of the sample intensity to that of the reference beam*, respectively provides the I/I_0 spectral measurement for use in absorbance (A) as $-\log_{10}(I/I_0)$ computations.

Dry weight—The weight of a sample based on a percent solids result. Also, the weight of a sample after drying in an oven at a specified temperature.

Efficiency—The ability of a column in GC or HPLC to produce sharp, well-defined peaks. More efficient columns have more theoretical plates N and smaller theoretical plate heights H.

Effluent—Treated or untreated wastewater that flows out of a treatment plant, sewer, or industrial outfall. Generally refers to wastes that are discharged into surface waters and are regulated under the *Clean Water Act*. Effluent limitations are restrictions on quantities, rates, and concentrations of wastewater discharges that are established by a state or EPA.

Electrolytic conductivity detector (ElCD)—A GC detector that catalytically reacts halogen-containing solutes with hydrogen (reductive mode) to produce strong acid by-products that are dissolved in a working fluid. The acids dissociate, and the detector measures increased electrolytic conductivity. Other operating modes modify the chemistry for response to nitrogen- or sulfur-containing substances.

Electromagnetic spectrum—The continuum of frequencies that contains electromagnetic radiation. Instruments measure the intensity of radiation within a defined range of the spectrum and usually present the results of their measurements as a set of values of some function, or the measured intensity at (usually) evenly spaced intervals within the range. The energy throughout the electromagnetic spectrum ranges from gamma radiation (most energetic and highest frequency) to radio waves (least energetic and lowest frequency). The spectrum is expressed in terms of wavelength, wavenumber, frequency, or energy.

Electron-capture detector (ECD)—A GC detector that ionizes solutes by collision with meta-stable carrier gas molecules produced by beta emission from a radioactive source such as ^{63}Ni. The ECD is one of the most sensitive detectors, and it responds strongly to halogenated solutes and other organic compounds with high electron-capture cross sections.

Electron-capture negative ion mass spectrometry (ECNI)—if the ionizing electron energy is reduced to the range 0–15 eV, two additional negative ion producing reactions are observed when a chemical ionization ion source is used and methane gas is introduced into the ion source: resonance capture and dissociative resonance capture. Halogenated hydrocarbons can be detected down to low ppt levels utilizing dissociative ECNI-MS.

Electron-impact mass spectrometry (EIMS)—Analyte molecules collide with 70 eV electrons boiled off of a hot wire filament in an ion source are accelerated and are fragmented; this fragmentation follows well-established patterns; low-resolution mass spectrometry operated in the electron-impact ionization mode.

Elution—Removal of a chemical species from a sorbent in SPE by changing the solvent or matrix chemistry to disrupt the analyte-sorbent interaction.

Emulsion—A stable dispersion of one liquid in a second immiscible liquid. Intractable emulsions are detrimental to efficient isolation and recovery of analytes in LLE.

Exposure—Contact between an agent and a target (human). Contact takes place at an exposure point or exposure surface over an exposure interval. For inhalation and ingestion routes, exposure is expressed as a function of exposure concentration; for the dermal route, exposure is expressed as a function of exposure loading. Related terms are cited as follows:

Exposure assessment—The process of estimating or measuring the intensity, frequency, and duration of exposure to an agent. Ideally, it describes the sources, pathways, routes, magnitude, duration, and pattern of exposure; the characteristics of the population exposed; and the uncertainties in the assessment.

Exposure concentration—The amount of agent present in the contact volume divided by the contact volume. For example, the amount of agent collected in a personal air monitor divided by volume sampled.

Exposure duration—The total period over which contacts occur between an agent and a target. For example, if an individual is in contact with an agent for 10 min a day, for 300 days over a 1-year period, the exposure duration is 1 year.

Exposure frequency—The number of exposure intervals in an exposure duration.

Exposure interval—A period of continuous contact between an agent and a target.

Exposure loading—The amount of agent present in the contact volume divided by the exposure surface area. For example, a dermal exposure measurement based on a skin wipe sample, expressed as a mass of residue per skin surface area, is an exposure loading.

Exposure mass—The amount of agent present in the contact volume. For example, the total mass of residue collected with a skin wipe sample is an exposure mass.

Exposure model—A conceptual or mathematical representation of exposure.

Exposure pathway—The course an agent takes from the source to the target.

Exposure route—The way an agent enters a human or animal after contact (e.g., by ingestion, inhalation, or dermal absorption).

Exposure scenario—A set of facts, assumptions, and inferences about how exposure takes place. Scenarios are often created to aid exposure assessors in estimating exposure.

Exposure surface—A surface on a target where an agent is present. Examples of locations of exposure surfaces include the lining of the stomach wall, the lung surface, the exterior of an eyeball, the skin surface, and a conceptual surface over the open mouth. Exposure surfaces can be absorptive or nonabsorptive.

External standards—A method of quantifying chromatographic data in which standards of known concentrations are analyzed prior to unknown samples. The chromatographic peak area (or height) of a sample component is compared to a *calibration curve* of a peak area constructed from the standard data for that component. This comparison allows the concentration of the component in the sample to be determined by interpolation.

Extraction—Distribution of an analyte of interest between two immiscible liquids; transfer of a chemical species from one phase into another; LLE obeys the Nernst distribution law.

Fast GC—Gas chromatography that uses a small-diameter capillary column (<100 μm i.d.), hydrogen carrier gas, and a fast oven temperature ramp rate to dramatically decrease chromatographic run time.

Fast HPLC—High-performance liquid chromatography that uses high flow rates and increased column temperature to maximize overall chromatographic efficiency and optimize the instrument for rapid throughput.

Fecal coliform bacteria—Bacteria found in mammals' intestinal tracts. Their presence in water or sludge is an indicator of pollution and possible contamination by pathogens.

Field screening—An investigative technique utilizing analytical chemistry at or near a work site to rapidly determine the presence or absence of environmental contaminants and the approximate concentration of specific target compounds.

Flame ionization detection (FID)—In GC where column effluent gas is mixed with hydrogen and burned in air or oxygen. The ions and electrons produced in the flame generate an electric current proportional to the amount of material in the detector. The FID responds to nearly all organic compounds, but it does not respond to air and water, which makes it exceptionally suited to environmental samples.

Flame photometric detection (FPD)—In GC where column effluent burns heteroatomic solutes in a hydrogen–air flame. This visible-range atomic emission spectrum is filtered through an interference filter and detected with a photomultiplier tube. Different interference filters can be selected for sulfur, tin, or phosphorous emission lines. The FPD is sensitive and selective.

Flammable—A substance that when mixed with air, oxygen, or other oxidant, burns upon ignition. Each flammable gas has a concentration range in air within which the gas may be ignited.

Flammable limits—The concentrations of vapor in air or oxygen in which a flame propagates on contact with a source of ignition. It is usually expressed in terms of percentage by volume of gas or vapor in air. The lower explosive limit (LEL) or lower flammable limit (LFL) is the minimum concentration of vapor below which a flame does not propagate. The upper explosive limit (UEL) is the maximum concentration above that a flame does not propagate. A change in temperature or pressure will vary the flammable limits.

Flash point—The lowest temperature at which a flammable liquid gives off sufficient vapor to form an ignitable mixture with air near its surface or within a vessel. Combustion does not continue.

Filter spectrophotometer—A spectrophotometer that uses filters to isolate narrow bands of the spectrum.

Florisil—Magnesium silicate; very polar in nature and ideal for the isolation of polar compounds from nonpolar matrices; commonly used as a sample prep cleanup adsorbent.

Fluorescence—Photons with energies in the ultraviolet (i.e. 190–360 nm) to the blue-green visible (i.e. 350–500 nm) spectral region will excite an electronic transition for atoms in molecules that fluoresce (i.e. fluorophores). Fluorescence is an electronic transition from a ground state to the excited state with the emission of a photon to return to the ground state. After a molecule is excited, it relaxes (Stokes shift) to the ground state while a photon with a femtosecond (10^{-15} s) to pico-second (10^{-12} s) timeframe. The Stokes shift indicates a lower energy of the excitation photons. The fluorescence typically has a lifetime (or duration of nanoseconds (10^{-9} s) per transition. A fluores-cence spectrophotometer normally has an excitation monochromator that defines the excitation energy, and an emission monochromator that provides a full spectrum of the fluorescence emission.

Flow rate (F)—The column outlet flow rate in GC in units of *cubic centimeters per minute* and corrected to room temperature and pressure; for example, the flow rate as measured by a flow meter. An estimate for F in cm^3/min can be calculated by knowing the column diameter in mm, d_c, the column length in m, L, and the time in minutes to elute an unretained chemical com-pound to t_0 according to

$$F = \frac{0.785 d_c^2 L}{t_0}$$

Fourier transform—A mathematical operation in which a curved repetitive function is described in terms of the sum of sine and cosine waves. A conversion from a *time domain* interferogram to a frequency *spectral domain* occurs when using this transformation. In spectroscopy, an interferogram is taken from a sample measurement in the time domain and converted into a spectrum in the frequency domain. This conversion is completed for both sample and reference interferograms which are then processed and ratioed to produce the spectrum.

Fourier transform infrared (FT-IR)—A means of measuring the electromagnetic spectrum using an interferometer, the desired spectrum is then obtained by performing a Fourier trans-form on the resulting interferometric data. Other Fourier transform measurements include Fourier transform near-infrared (FT-NIR) and Fourier transform ultraviolet (FT-UV).

Functional group—The reactive portion of an organic molecule; a group of atoms on a chemical species having properties that can be exploited in SPE for retention or sorbents. For example, the functional group for an aliphatic carboxylic acid whose general molecular formula is RCOOH is –COOH.

Gas chromatograph (GC)—An analytical instrument for detecting organic compounds by using their physical and chemical properties to separate a mixture. The compounds are identi-fied and quantified with various types of detectors as they exit the chromatograph. Selection of detectors is dependent on the particular compounds of interest.

Gas chromatography-mass spectrometry (GC-MS)—A hyphenated analytical instrument in which sample organic compounds of interest are first separated by GC and then enter the mass spectrometer. Molecules are bombarded with electrons as they exit a GC column and are fragmented into characteristic ion patterns. The mass spectrometer is the detector. It can deter-mine which fragments are present and therefore the identity of the compounds.

Gas–liquid chromatography (GLC)—In this determinative technique, solutes partition between a gaseous mobile phase and a liquid stationary phase. Selective interactions between the solutes and the liquid phase cause different retention times in the column.

Gas–solid chromatography (GSC)—In this determinative technique, solutes partition between a gaseous mobile phase and a solid stationary phase. Selective interactions between the solutes and the solid phase cause different retention times in the column.

Ghost peaks—Peaks not present in the original sample. In GC ghost peaks can be caused by septum bleed, solute decomposition, or carrier gas contamination.

GLP—Good laboratory practice.

Graphical user interface—A term invented by Xerox in Palo Alto, CA in the early 1970s that refers to a software interface between the user and the operational software code. Today, it may involve touch-screen icons and graphics that are easy to activate and understand.

Graphite furnace atomic absorption spectrophotometry (GFAA)—A technique used for the analysis of samples that contain metals. An AA spectrophotometer heats the sample within a graphite tube using an electrical current. It is also commonly called a flameless furnace and generally provides greater sensitivity for certain metals than flame or inductively coupled argon plasma techniques.

Grating—A reflective surface covered with evenly spaced, microscopic grooves whose purpose is to separate individual wavelengths from broadband energy. The distance between grooves and the angle of the faces is determined by the wavelengths to be separated. The grating (except for diode arrays) is rotated at a set angle and speed, and the desired wavelength is diffracted though an exit slit onto the sample and detector (or detectors). It is used to disperse light of various wavelengths and orders from its surface. It disperses zero order as specular reflected light, and first, second and higher orders as diffracted light. When the diffracted light interacts in a phonon effect it decreases the energy dispersed from the surface, resulting in a phenomenon termed *Wood's anomalies*. See ***Wood's anomalies***.

Grating spectrometer—A spectrometer that uses a grating for the diffraction and resulting resolution of light of various wavelengths. This is often termed a monochromator when one grating system is involved.

Gravimetric—Analyses based on the direct or indirect weighing of the analyte in question. This technique usually requires the use of an analytical balance with a sensitivity of 0.1 mg or better.

Hall electrolytic conductivity detection—Accomplished using an element-selective GC detector, primarily intended for trace analysis of organic compounds containing chlorine, nitrogen, or sulfur. In operation, this detector pyrolyzes the column effluent gas into soluble electrolytes that are dissolved in a stream of deionized liquid. The observed change in electrical conductivity, proportional to the amount of material present, is measured.

Hazardous ranking system (HRS)—The principal screening tool used by EPA to evaluate risks to public health and the environment associated with abandoned or uncontrolled hazardous waste sites. The HRS calculates a score based on the potential for hazardous substances to spread from the site through the air, surface water, or groundwater and on other factors, such as nearby population. This score is the primary factor in deciding if the site should be on the National Priorities List and, if so, what ranking it should have there.

Hazardous substance—Any material that poses a threat to human health or the environment. Typical hazardous substances are toxic, corrosive, ignitable, explosive, or chemically reactive.

Hazardous waste—Waste regulated under RCRA that can pose a substantial or potential hazard to human health or the environment when improperly managed. Such wastes possess at least one of four characteristics (ignitability, corrosivity, reactivity, or toxicity) or appear on special EPA hazardous waste lists. The term is not interchangeable with *hazardous substance or material*. See ***RCRA***.

Headspace—Any area in a container not completely filled by the sample in which gases can collect.

Heart cut—In GC, a technique in which two or more partially resolved peaks that are eluted from one column are directed onto another column of different polarity or at a different temperature for improved resolution.

Heavy metals—Metallic elements that reside in the transition metal section of the periodic table. These elements exhibit relatively high atomic weights; examples include mercury, chromium, cadmium, arsenic, and lead. They can damage the health of plants and animals at low concentrations and tend to accumulate in the food chain. Most are considered to be priority pollutants and some persist in the environment and in humans.

Height equivalent to a theoretical plate (HETP or H)—The distance along the column occupied by one theoretical plate: $H = L/N$, where L is the column length and N is the number of theoretical plates.

Heteroatoms—Organic compounds that contain atoms of elements other than carbon, hydrogen, and oxygen. This includes bromine, chlorine, fluorine, iodine, nitrogen, phosphorous, and sulfur.

High-performance liquid chromatography (HPLC)—modern liquid-phase chromatography; deteminative technique using small particles and high pressures.

Holding time—The storage time allowed between sample collection and sample analysis when the designated preservation and storage technique are employed.

Homolog—One of the 10 degrees of chlorination of PCBs ($C_{12}H_9Cl$ through $C_{12}Cl_{10}$) or other group of compound varying by systematic addition of a substituent.

Hydrocarbons (HCs)—Chemical compounds that consist entirely of carbon and hydrogen.

ICP—Inductively coupled argon plasma (also referred to as ICAP). An instrument used for determining the trace concentrations of various metals in environmental and human specimens. Because the temperature of the plasma is considerably higher ($10,000°K$) than the temperature of a flame AA spectrophotometer, it is especially useful for refractory metals. Some instruments are also capable of performing simultaneous multielement analysis. Two distinct instrument configurations incorporate the ICP: ICP-AES and ICP-MS.

Ignitable—Capable of burning or causing a fire.

Inert—A substance that does not react chemically with most material. Some gases, such as argon and helium, are inert, but can displace oxygen in air to cause asphyxiation. Nitrogen is a relatively inert gas that is responsible for many deaths by asphyxiation each year.

Inorganic chemicals—Chemical substances of mineral origin, unlike organic chemicals whose structures rely on carbon atoms.

Instrument detection limit (IDL)—The lowest concentration of an analyte of interest that is detectable within the statistical parameters defined. According to the EPA, the IDL is three times the standard deviation obtained for the analysis of a standard solution (each analyte in reagent water) at a concentration of three to five times that of the IDL on three nonconsecutive days, with seven consecutive measurements per day.

Instrument tuning—A technique used in GC-MS procedures to verify that the instrument is properly calibrated to produce reliable mass spectral information.

Interaction—Attraction or repulsion between two chemical species in a specific chemical environment.

Interferences—Undesired components in the sample matrix.

Internal standards—Compounds added to every standard, blank, matrix spike, matrix spike duplicate, sample (for volatile organics), and sample extract (for semivolatile organics) at a known concentration, prior to analysis. Internal standards are used as the basis for quantification of the target compound.

Ion exchange—Involves the interaction of an ionic isolate functional group with an ionic functional group of opposite charge on the sorbent surface.

Ionization—Utilized in mass spectrometry to fragment analyte molecules into smaller segments. These smaller mass segments are then separated and plotted to form a mass spectrum, which is used to identify the parent molecule. Electron impact is one example of ionization used in mass spectrometry. In more technical terms, ionization is the process by which neutral atoms or groups of atoms become electrically charged, either positively or negatively, by the loss or gain of electrons.

Ion trap detector—A mass spectrometric detector that uses an ion trap device to generate mass spectra.

Isomer—Chemical compounds with the same molecular weight and atomic composition, but differing molecular structure, e.g., n-pentane and 2-methyl butane.

j—Carrier gas compressibility correction factor used in GC.

k or k'—Chromatographic retention factor.

K—Partition coefficient. The relative concentration of solute in the mobile and stationary phases.

Keeper—A high-boiling solvent used to keep analyte in solution during sample evaporation. In GC methods, typically tetradecane or similar higher molecular weight hydrocarbon.

Land ban—The 1984 Resource Conservation and Recovery Act amendments mandated that by May 1990, all untreated hazardous waste must be banned from land disposal. The treatment standard and concentration levels were implemented in thirds beginning in November 1986.

Leachate—A liquid that results from water collecting contaminants as it trickles through wastes, agricultural pesticides, or fertilizers. Leaching may occur in farming areas, feedlots, and landfills, and may result in hazardous substances entering surface water, groundwater, or soil.

Library search—A technique in which an unknown mass spectrum of a compound is compared to the mass spectra of compounds contained in a computer library in an effort to identify the compound. Compounds identified in this manner are referred to as tentatively identified compounds (TICs).

Limit of decision (critical limit) x_C—Lowest concentration of analyte of interest above which indicates detection. Lowest concentration that corresponds to the critical instrumental response y_C such that the probability of not committing a type I error, α, at the y intercept, y_0, of the least squares regression (i.e., a false positive) is 95%. Represents 3 × the standard deviation in the blank. Refer to Figure 2.9 for a graphical illustration.

Limit of detection (instrumental detection limit) x_D, x_{IDL}—Lowest concentration of analyte of interest above which leads to detection. Corresponds to an instrumental detection response y_D such that the probability of not committing a type II error at y_C (i.e., a false negative) is 95%. Represents 6 × the standard deviation in the blank.

Limit of quantitation x_Q, x_{LOQ}—Lowest concentration of analyte of interest above which leads to quantification. Corresponds to an instrumental response y_Q, such that the probability of not committing a type II error at y_D (i.e., a false negative) is 99%. Represents 10 × the standard deviation in the blank.

Linear dynamic range—The range of solute concentrations or amount beyond which a GC or HPLC detector is not directly proportional to solute concentration.

Linear velocity (u)—The speed at which the carrier gas moves through the column, usually expressed as the average carrier gas linear velocity ($\bar{u}$).

Liquid chromatography (LC)—A chromatographic separation technique in which the substance to be analyzed is dissolved in a solvent and, using the same or different solvent, is eluted through a solid adsorbent exhibiting differential adsorption for the components of the substance. LC as a sample prep cleanup technique is called low or gravity pressure LC to distinguish it from high performance (HPLC) and ultrahigh performance (UHPLC) which are determinative techniques. See **HPLC** and **UHPLC**.

Liquid phase—In GC, a stationary liquid layer coated on the inner column wall (WCOT) or on a support (packed SCOT column) that selectively interacts with different solutes as they are eluted from the column.

Listed waste—Any waste listed as hazardous under the Resource Conservation and Recovery Act, but which has not been subjected to the Toxic Characteristics Listing Process because the dangers it presents are considered self-evident. See *RCRA*.

Log-in—The receipt and initial management of an environmental sample. It generally includes identifying who sent the sample, maintaining chain of custody, checking report and invoice information, recording analyses requested, including methodology and special instructions, and assigning a discreet laboratory identification, usually a number or bar code.

Mass spectrometric detector (MS, MSD)—A detector that records mass spectra of solutes (ion abundance vs. m/z); if solutes are eluted from a column such as in GC-MS, retention time becomes a third dimension.

Mass spectrum—A plot of ion mass/charge ratio, m/z, vs. intensity or ion abundance; A fragmentation pattern results from the impact upon a given molecule of a beam of electrons. The impact produces a family of charged molecular species whose mass distribution is characteristic of the parent molecule. *Qualitative information* is provided by a mass spectrum while ion abundance at a specific m/z ratio under selected ion monitoring (SIM) conditions when interfaced to a chromatographic system provides *quantitative information* as well.

Material Safety Data Sheet (MSDS)—A compilation of information required under the OSHA Communication Standard on the identity of hazardous chemicals and their associated health and physical hazards, exposure limits, and precautions.

Matrix—The physical characteristics or state of a sample (e.g., water, soil, sludge). The sample environment from which the analyte is to be extracted.

Matrix interference—The influence of the sample matrix or sample components upon the ability to qualitatively identify and quantitatively measure compounds in environmental samples.

Matrix modifiers—Chemicals added to samples for metals analysis that are used to lessen the effects of chemical interferences, viscosity, and surface tension.

Matrix spike—Aliquot of a sample fortified (spiked) with known quantities of specific compounds and subjected to the entire analytical procedure in order to indicate the appropriateness of the method for the matrix by measuring recovery of the spike.

Matrix spike duplicate—A second aliquot of the same matrix as the matrix spike that is spiked to determine the precision of the method.

Maximum allowable operating temperature—In GC, the highest continuous column operating temperature that will not damage a column, if the carrier gas is free of oxygen and other contaminants. Slightly higher temperatures are permissible for short periods during column bakeouts.

Maximum contaminant level (MCL)—The maximum permissible level of a contaminant in water delivered to any user of a public water system. MCLs are enforceable standards.

Mechanism—The nature of the chemistry leading to analyte or interference retention and elution.

Method detection limit (MDL)—The minimum concentration of a compound that can be measured using a specific method and reported with 95 or 99% confidence that the value is above zero.

Minimum detectable quantity (MDQ)—The amount of solute that produces a signal twofold that of the noise level.

Mobile phase—In GC, the carrier gas; in HPLC, a pure liquid or mixture of liquids. Both permeate a stationary phase under pressure. A *mobile phase passing across a stationary-phase* forms a fundamental definition of GC and HPLC.

Monochromator—An instrument used to provide an incident light beam with a narrow wavelength range. A monochromator becomes a spectrophotometer when it is combined with a light source, slits, grating, detectors, an amplifier, and an output energy measuring device.

Monolithic HPLC columns—Consists of a single plug (instead of the traditional beads) of an organic polymer or of silica of sufficient porosity to allow liquid throughput under pressure.

MS scan parameters—The terms listed below comprise this (adapted from ChemStation® Agilent Technologies):

Mass range—Enter the low and high masses (amu) to specify the range to be scanned by the MSD. The larger the range, the lower the number of scans per second.

MS window 1 / MS window 2—Enter the ion mass range to be plotted in real time. These fields are active only when the extracted ion plot type is selected in the real-time plot parameters section of the dialog box for the corresponding window.

Sampling—The value entered here is used to calculate the number of times the abundance of each mass is recorded before going on to the next mass. A value of 2 is suitable for most analyses. The resulting number is reported in samples and calculated as 2^N. The recommended value is 2 (giving a sample value of 4). Range is 0 to 7.

Scans/sec—An approximate value calculated from the mass range and sampling values you have entered. It does not take into account the overhead time needed to process the timed events.

Start time—The time in minutes after the start of the run at which to activate the scan parameters defined by the entries in each row of the table. A maximum of three scan ranges can be active during a run.

Threshold—Only ions with an abundance equal to or greater than this value will be retained in the mass spectrum of each scan. A threshold of 500 is typical.

Multidimensional—Separations performed with two or more columns in which peaks are selectively directed onto or removed from at least one of the columns by a timed valve system.

N—The number of theoretical plates for a given chromatographic column in GC or HPLC.

Neff—The number of effective plates. This term is an alternate measurement of theoretical plate height that compensates for the non-partitioning nature of an unretained peak.

Nreq—The number of theoretical plates required to yield a particular resolution (R_S) at a specific peak separation (α) and capacity factor (k').

Nanometer—Unit of measurement for wavelength, abbreviated nm: as one-billionth of a meter (10^{-9} m).

Narrative—In an analytical report, descriptive documentation of any problems encountered in processing the samples, along with corrective action taken and problem resolution.

National Pollutant Discharge Elimination System (NPDES)—A provision of the Clean Water Act that prohibits discharge of pollutants into waters within the U.S. unless a special permit is issued by EPA, a state (where delegated), or a tribal government on an Indian reservation.

Nitrogen-phosphorous detector (NPD)—The NPD catalytically ionizes nitrogen- and phosphorous-containing solutes on a heated rubidium or cesium surface in a reductive atmosphere. The NPD is highly selective and provides sensitivity that is somewhat better than that of a FID.

Nonpolar—Inter- and intramolecular interactions that occur between nonpolar molecules based on London dispersive forces.

Normal-phase HPLC—High-performance liquid chromatography that uses a polar stationary phase and a nonpolar mobile phase.

Null hypothesis—Usually comes about from the logic of statistical testing. One example of a null hypothesis is to state that the precision of population A is equal to that of population B. Given a symbol H_0 and stated mathematically as $\sigma_A^2 = \sigma_B^2$. To answer the question "Are my two variances really different?" is rephrased statistically as "Are the two variances equally precise?"

Nutrient—Any substance assimilated by living things that promotes growth. The term is generally applied to nitrogen and phosphorus in wastewater, but is also applied to other essential and trace elements.

Ocadecylsiloxane bonded silica sorbent—A silica gel with C_{18} moieties chemically bonded via the siloxane linkage to its surface; generally consists of 40 μm particle size with irregular shapes and used predominantly to conduct reversed-phase SPE; a much more uniformly shaped 3–5 μm particle size comprises the common reversed-phase HPLC column.

On-column injection—In GC, sample enters the column directly from the syringe and does not contact other surfaces. On-column injection usually signifies cold injection for capillary columns.

Organic—Generally, any compound that contains carbon bonded to itself and to hydrogen or halogen atoms.

Oxidant—A gas that is required for combustion of flammable materials. In some cases, the oxidant may initiate combustion. Materials that burn in air burn more vigorously or explosively in the presence of oxygen and another oxidant.

Oxidation—The general process in chemistry whereby electrons are removed from atoms or molecules.

PAH—Poly (polycyclic or polynuclear) aromatic hydrocarbon.

Part per billion—One part of solute in 10^9 parts of solution. For a low concentration of analyte in an aqueous sample, a weight/volume (w/v) basis is most commonly used; 1 ppb = 1 μg/L. For nonaqueous liquids and solids, a weight/weight (w/w) basis is most commonly used; 1 ppb = 1 μg/kg.

Part per million—One part of solute in 10^6 parts of solution. For a low concentration of analyte in an aqueous sample, a weight/volume (w/v) basis is most commonly used; 1 ppm = 1 mg/L. For nonaqueous liquids and solids, a weight/weight (w/w) basis is most commonly used; 1 ppm = 1 mg/kg.

Part per quadrillion—One part of solute in 10^{15} parts of solution. For a low concentration of analyte in an aqueous sample, a weight/volume (w/v) basis is most commonly used; 1 ppq = 1 pg/L. For nonaqueous liquids and solids, a weight/weight (w/w) basis is most commonly used; 1 ppq = 1 pg/kg.

Part per trillion—One part of solute in 10^{12} parts of solution. For a low concentration of analyte in an aqueous sample, a weight/volume (w/v) basis is most commonly used; 1 ppt = 1 ng/L. For nonaqueous liquids and solids, a weight/weight (w/w) basis is most commonly used; 1 ppt = 1 ng/kg.

PCBs—Polychlorinated biphenyls, a group of toxic and persistent chemicals once used in transformers and capacitors for insulating purposes and in gas pipeline systems as a lubricant. Sale of PCBs for new uses was banned by law in 1979.

Peak capacity—In GC or HPLC, the amount of solute that can be injected without a significant loss of column efficiency.

Peak overload—In GC or HPLC, when too much of any one solute is injected, its peak can be distorted into a triangular shape.

PEG—Polyethylene glycol.

PEL—Permissible exposure limit. This is the permissible amount of human exposure for an 8-hour workday, 40 hours per week. The PEL does not address cancer, neurological, or reproductive issues. Ten to 15% of the population may suffer acute effects at the PEL.

Percent recovery—The extent to which an analyte of interest is isolated and recovered from a given sample matrix by the specific sample prep technique employed. Percent recoveries are measured in the laboratory by using reference analytical standards that are *spiked* into various *enviro-chemical* or *enviro-health* sample matrices and then calculating the % recovery of the specific analyte.

Performance audit—A quantitative evaluation of a measurement system that involves the analysis of standard reference samples or materials that are certified as to their chemical composition or physical characteristics.

Petrochemicals—Chemicals derived from the refining of hydrocarbons (oil and natural gas). Plastics are created through processing of petrochemicals, making them valuable as a fuel in waste-to-energy incineration facilities.

Petroleum hydrocarbon fingerprinting—A method that identifies sources of oil and allows spills to be traced back to their source.

pH—A numerical designation of relative acidity and alkalinity. A pH of 7.0 indicates precise neutrality. Progressively higher values indicate increasing alkalinity and lower values increasing acidity. Mathematically, a solution's pH is defined as the negative logarithm of the hydrogen ion concentration in aqueous solution.

Photoionization detector (PID)—The PID ionizes solute molecules with photons in the UV energy range. The PID is a selective detector that responds to aromatic compounds and olefins when operated in the 10.2 eV photon range, and it can respond to other compounds with a more energetic light source.

pKa—For an acidic functional group or the cation of a basic functional group, the pH at which half of the functional groups in solution are charged and half are neutral.

Plasma vs. serum—Matrices commonly encountered......consist of body fluids, such as plasma, serum, lysed blood, urine, or sputum (saliva). These fluids are often viscous and may need some prior treatment......centrifugation, applied to blood that contains anticoagulant will consist of 55% plasma and 45% red blood cells. The plasma consists of ~90% water, ~8% proteins, ~1% organic acids, and ~1% salts. If this sample of blood did not contain anticoagulant, it would have clotted. If the clotted-blood sample were again centrifuged, the two separated phases would now be red blood cells and serum. The serum is the clear liquid above the cells, similar to plasma......Because the blood sample was allowed to clot, many of the proteins present in the serum have been used in the clot formation. The end result is that serum has approximately half the protein content of plasma and is less viscous. Typically, pH 7.0 phosphate buffer (0.01M K_2HPO_4) must be added to a plasma at a 1:1 dilution.........because plasma is so viscous....it is not necessary to add buffer to serum samples.

Polar—Inter-and intramolecular interactions that occur between polar molecules based on dipole–dipole or hydrogen bonding forces.

Pollutant—Generally, any substance introduced into the environment that adversely affects the usefulness of a resource.

Porous-layer open tubular (PLOT) column—A capillary column in GC with a modified inner wall that has been etched or otherwise treated to increase the inner surface area or to provide GSC retention behavior. Stationary phases for contemporary plot columns include:

- 5 Å molecular sieve, zeolite
- aluminum oxide, deactivated
- polystyrene-divinyl benzene
- bonded silica
- bonded monolithic carbon.

Porous polymer—A stationary-phase material that retains solutes by selective adsorption or molecular size interaction.

Potentially responsible party (PRP)—Any individual or company, including owners, operators, transporters, or generators, potentially responsible for, or contributing to, the contamination problems at a Superfund site. Whenever possible, EPA requires PRPs through administrative and legal actions, to clean up sites they have contaminated.

Practical quantitation limit (PQL)—The lowest level that can be reliably achieved within specified limits of precision and accuracy during routine laboratory operating conditions.

Precision—A measure of the ability to reproduce analytical results. It is generally determined through the analysis of replicate samples. The *standard deviation* in replicate measurements is commonly used to measure the degree of precision.

Precut—In GC, peaks at the beginning of a chromatogram are removed to vent or directed onto another column of different temperature for improved resolution.

Preservative—A chemical or reagent added to a sample to prevent or slow decomposition or degradation of a target analyte or a physical process. Physical and chemical preservation may be used in tandem to prevent sample deterioration.

Primary Drinking Water Regulation—Applies to public water systems and specifies a contaminant level that, in the judgment of the EPA administrator, will have no adverse effect on human health.

Procedure—The written directions necessary to use a method or series of methods and techniques in the laboratory.

Programmed-temperature GC (PTGC)—The column temperature changes in a controlled manner as peaks are eluted.

Programmed-temperature injection—A cold injection technique where the inlet temperature is specifically programmed from the gas chromatograph.

Programmed-temperature vaporizer—In GC, an inlet system designed to perform programmed-temperature injection.

Propagation of random error—In the computation of the *percent recovery* of the *i*th analyte, %R^i, shown in the first equation shown (see below) whose standard deviation, s^i_R, is found by propagating *relative standard deviations* for the *i*th analyte and for the control reference standard as shown in the second equation (see below). A relative standard deviation, RSD, for the percent recovery can then be found using the third equation (see below).

$$\%R^i\left(s^i_R\right) = \frac{A^i\left(s^i\right)}{A^i_c\left(s^i_c\right)} \times 100$$

$$\frac{s^i_R}{R^i} = \sqrt{\left(\frac{s^i}{A^i}\right) + \left(\frac{s^i_c}{A^i_c}\right)}$$

$$RSD = \left(\frac{s^i_R}{R^i}\right) \times 100$$

Protocol—A sampling or analysis procedure that is highly specific and from which few or no deviations are allowed.

Purgeable organic—An organic compound that is generally less than 2% soluble in water and has a boiling point at or below 200°C; a volatile organic compound; an organic compound is generally considered to be purgeable if it can be removed from water using the purging process.

Purge and trap—A technique used in the analysis of volatile organics where analytes are purged from a sample by means of an inert gas and trapped on a sorbent column. The sorbent is then flash heated, and the analytes are transferred onto a GC column for separation, identification, and quantitative analysis.

Pyrolysis GC—Sample is pyrolyzed (decomposed) in the inlet before GC analysis.

Qualitative—Having to do with establishing the presence or identity of a chemical substance.

Quality assurance (QA)—All those planned and systematic actions necessary to provide adequate confidence in laboratory results.

Quality assurance program plan—A written assembly of management policies, objectives, principles, and general procedures that outlines how the laboratory intends to generate data of known and accepted quality.

Quality control (QC)—Those quality assurance actions that provide a means to control and measure the characteristics of measurement equipment and processes in order to meet established requirements.

Quantitation limit—The minimum concentration of a compound that can be reliably quantified; very dependent upon the sample matrix. See *practical quantitation limit (PQL)*.

Quantitative—Having to do with measuring the amount or concentration of a compound in a sample.

Quantitative analysis—Describes the laboratory operations designed to *quantitatively determine* the amount or concentration of a targeted chemical substance.

Reactivity—The tendency of a chemical to explode under normal management conditions, to react violently when mixed with water, or to generate toxic gases.

Reagent water—Water in which an interference is not observed at or above the minimum quantitation limit of the parameters of interest.

Reconstructed ion chromatogram (RIC)—A mass spectral graphical representation of the separation achieved by a GC-MS or LC-MS indicating total ion current vs. retention time. All the abundances in the mass spectrum are summed and then reconstructed at the start of a chromatographically resolved peak.

Reduction—The general process in chemistry whereby electrons are acquired by atoms or molecules.

Relative response factor (RRF)—A ratio of the response factor of the ith analyte to its corresponding internal standard (IS).

Relative retention time (RRT)—Retention time of a compound on a chromatographic system, relative to an internal standard; unitless number.

Resolution—The degree of separation between peaks eluting from a chromatographic column. Sufficient resolution between peaks is required for proper quantitation of unknown analytes.

Resource Conservation and Recovery Act (RCRA)—A federal law that established a regulatory system to track hazardous substances from the time of generation to disposal. The law requires safe and secure procedure to be used in treating, transporting, storing, and disposing of hazardous substances. RCRA is designed to prevent new and uncontrolled hazardous waste sites.

Retention gap—In GC, a short piece of deactivated but uncoated column placed between the inlet and the analytical column. A retention gap often helps relieve solvent flooding. It also contains nonvolatile sample contaminants from on-column injection.

Retention index—Systematic, unitless measure of a compound's chromatographic retention as compared to a homologous series of standards, usually the n-alkanes.

Retention time—A term used in GC and HPLC describing the time elapsed from sample injection until the specific compound elutes or exits the chromatographic column at the detector. Each organic compound has a characteristic retention time on a specific column; therefore, this information is used to qualitatively identify the compound in the sample.

Reversed-phase HPLC—High-performance liquid chromatography using a nonpolar stationary phase and a polar mobile phase.

RSD—Relative standard deviation; the ratio of the *standard deviation* in the mean to the *mean* value of replicate measurements.

Sandwich technique—An injection technique in GC in which a sample plug is placed between two solvent plugs in the syringe to wash the syringe needle with solvent and obtain better sample transfer into the inlet.

Secondary Drinking Water Regulations—Unenforceable regulations that apply to public water systems and specify contamination levels that, in the judgment of EPA, are required to protect the public welfare. These regulations apply to any contaminants that may adversely affect the odor or appearance of such water and consequently may cause people served by the system to discontinue its use.

Selectivity—The fundamental ability of a stationary phase to retain analytes selectively based upon their chemical characteristics, including vapor pressure and polarity.

Sensitivity—The degree of detector response to a specified solute amount per unit time or per unit volume.

Separation number of Trennzahl—A measurement of the number of peaks that could be placed with baseline resolution between two sequential peaks z and $z + 1$ in a homologous series such as two hydrocarbons.

Septum—Silicone or other elastomeric material used in GC that isolates inlet carrier flow from the atmosphere and permits syringe penetration for injection.

Septum purge—Occurs when carrier gas in a GC is swept across the septum face to a separate vent so that material emitted from the septum does not enter the column.

Signal-to-noise ratio (S/N)—The ratio of the peak height to the noise level.

Silica gel—Common adsorbent used in sample prep cleanup techniques; composed of a network of silicon–oxygen covalent bonds with surface silanol (Si-OH) groups. Used as a support to which organic moieties can be chemically bonded through these surface silanol groups. Silica is considered to be the most polar sorbent available. In SPE, base silica is 40 µm irregularly shaped silica with a mean porosity of 60 Å.

Simulated distillation (SIMDIS)—A boiling point separation technique in GC that simulates physical distillation of petroleum products.

Skinner List—Created by John Skinner of the EPA Office of Solid Waste, a list of those compounds most often found in petroleum refining wastes.

Slit—An aperture, usually rectangular in shape with a large ban length-to-width ratio, and a fixed or adjustable shape through which radiation enters or leaves a monochromator-based instrument. The split aperture is usually quite small relative to the light source. For monochromators, there is an entrance and exit slit for each grating: for interferometers there is a J-stop aperture which in effect acts as a slit adjusting the effective resolution.

Solid wastes—Non-liquid, non-soluble materials, ranging from municipal garbage to industrial wastes, that contain complex and sometimes hazardous substances. Solid wastes include sewage, sludge, agricultural refuse, demolition wastes, mining residues, and even liquids and gases in containers.

Solvation—The process that prepares a sorbent for sample application.

Solvent—A substance, usually liquid, capable of dissolving or dispersing one or more other substances. The liquid phase involved in sorbent extraction.

Solvent effect—A solute profile sharpening technique used in GC with splitless and on-column injection. Condensed solvent in the column during and shortly after injection traps volatile solutes into a narrow band.

Sorbent—The porous, chemically modified silica used for selective extraction of chemical species from liquids.

Sorbent bed—Sorbent packed into a configuration such that solvents and liquid samples can be passed through the sorbent.

Spectrophotometer—A device used to measure and record the spectrum of a material at uniform, usually closely spaced, wavelength or wavenumber intervals. This type of instrument is sometimes referred to as a spectrometer, although correct terminology is spectrophotometer. It is specifically designed to measure the ratio of sample signal (I) versus reference signal I_0 as the ratio I/I_0.

Split injection—The sample size is adjusted to suit capillary column requirements by splitting off a major fraction of sample vapors in the inlet so that as little as 0.1% enters the column. The rest is vented.

Splitless injection—A derivative of split injection. During the first 0.4 to 4 min of sampling, the sample is not split and enters only the column. Splitting is restored afterward to purge the sample remaining in the inlet. As much as 99% of the sample enters the column.

Split ratio—The ratio of the sample amount that is vented in GC to the sample amount that enters the column during split injection. Higher split ratios place less sample on the column. The split ratio is measured as the ratio of total inlet flow to column flow; e.g., a split ratio of 40:1 suggests that for every molecule that enters the capillary column, 40 molecules are split off and exit to the atmosphere via the split vent valve.

Standard curve—A curve that plots concentrations of known analyte standards vs. the instrument response to the analyte. Calibration standards are prepared by diluting the stock analyte solution in graduated amounts that cover the expected range of the samples being analyzed. The calibration standards must be prepared by using the same type of acid or solvent at the same concentration as for the samples following sample preparation. This is applicable to organic and inorganic chemical analyses.

Standard deviation—A statistical measure, usually symbolized by s, of spread among replicate measurements assuming a Gaussian or normal distribution. The three equations below all yield the same standard deviation, s for N replicate measurements, and facilitate computation of this important statistical parameter using computer programs or calculators:

$$s = \sqrt{\frac{\sum_i \left(x_i - \bar{x} \right)^2}{N-1}}$$

$$s = \sqrt{\frac{\sum_i x_i - \frac{\left(\sum_i x_i \right)^2}{N}}{N-1}}$$

$$s = \sqrt{\frac{N \sum_i x_i^2 - \left(\sum_i x_i \right)^2}{N(N-1)}}$$

Standard operating procedure (SOP)—A written quality assurance document that describes the way an organization typically conducts a routine activity. May address instrumental operation, instrument maintenance, application of a laboratory technique, data review, or management oversight. A detailed written description of how a laboratory executes a particular procedure or method, intended to standardize its performance.

Stationary phase—Liquid or solid materials coated inside a column in GC that selectively retain solutes.

Statistical testing—Proposing a *null hypothesis* and an *alternative hypothesis*. The logic of statistical testing is given (see below):

- Formulate a null hypothesis and an alternative hypothesis.
- Define the population (or universe) and state any assumptions.
- Plan the experiment.
- Define the basis for decision making.
- Run the experiment as planned.
- Evaluate the data.
- State the conclusion.

Our decision to accept or reject a null hypothesis may be a correct one (i.e., we may accept a true null hypothesis or reject a false null hypothesis) or a wrong one. Rejecting a true null hypothesis is called committing a *Type I error*. Accepting as true a false hypothesis is called committing a *Type II error*. These possibilities can be summarized in a *truth table*. See **truth table**.

Statistically significant—When the difference between a predicted and an observed value is so large that it is improbable it could be attributed to chance.

Subchronic exposure—A contact between an agent and a target of intermediate duration between acute and chronic.

Sulfur chemiluminescence detector (SCD)—An SCD responds to sulfur-containing compounds by generating and measuring the light from chemiluminescence.

Superfund—The Response Trust Fund, established by CERCLA as a mechanism for the federal government to take emergency or remedial action to clean up both abandoned and existing disposal sites when there is a release, or potential threat of a release, of a hazardous substance presenting imminent and substantial danger to public health and welfare. See **CERCLA**.

Support-coated open tubular (SCOT)—A capillary column in GC in which the stationary phase is coated onto a support material that is distributed over the column inner wall. A SCOT column generally has a higher peak capacity than a WCOT column with the same average film thickness.

Surrogate—An organic compound similar to the analyte of interest in chemical composition, extraction, and chromatography, but not normally found in environmental samples. Primarily used in chromatography techniques, the surrogate standard is spiked into quality control blanks, calibration and check standards, samples (including duplicates and QC reference samples), and spiked samples before analysis. A percent recovery is calculated and usually reported for each surrogate.

Suspended solids—Small pollutant particles that float on the surface of, or are suspended in, sewage or other liquids. They resist removal by conventional means.

t_0—Unretained peak holdup time. The time required for one column volume (V_G) of carrier gas to pass through a column.

Target—A physical, biological, or ecological object. Examples of targets are humans, human organs, and animals.

Target Compound List (TCL)—A list of organic compounds that are determined during Superfund site remediations. Created by EPA for use in the Contract Laboratory Program, this list was formerly referred to as the Hazardous Substance List (HSL). See **Contract Laboratory Program**.

Target compounds—Specific compounds that are to be quantified in a sample, based on a standard list of potential compounds.

Technique—Scientific principle or specific operation; a skill, accomplished in the laboratory.

Tentatively identified compounds (TICs)—Compounds detected in samples when conducting *broad-spectrum* monitoring using GC-MS. TICs that are not target compounds, internal standards, system monitoring compounds, or surrogates. TICs usually consist of up to 30 peaks that are greater than 10% of the peak areas, or heights, of the nearest internal standard. They are subjected to mass spectral library searches for tentative identification. A client may specify the number of unknown peaks in its samples that it wishes the laboratory to tentatively identify.

Theoretical plate—A hypothetical entity inside a GC or HPLC column that exists by analogy to a multiple-plate distillation column. As solutes migrate through a column, they partition between the stationary phase and the mobile phase. Although this process is continuous, chromatographers often visualize a stepwise model. One step corresponds roughly to a theoretical plate.

Thermal conductivity detector (TCD)—In GC, a TCD measures the differential thermal conductivity of the carrier and reference gas flows. Solutes emerging from a column change the carrier gas thermal conductivity and produce a response. TCD is a universal detection technique with moderate sensitivity.

Thermionic-specific detector (TSD)—See **nitrogen-phosphorous detector**.

Threshold limit value time-weighted average (TLV-TWA)—The time-weighted average airborne concentration of substances for a normal 8-hour workday or 40-hour workweek, to which nearly all workers may be repeatedly exposed, day after day, without adverse effect. This is being replaced by the PEL. The TLV-TWA is subject to the same limitations as the PEL. See **PEL**.

Time-averaged exposure—The time-integrated exposure divided by the exposure duration. An example is the daily average exposure of an individual to carbon monoxide.

Time-integrated exposure—The integral of instantaneous exposures over the exposure duration. An example is the area under a daily time profile of personal air monitor readings, with units of concentration multiplied by time.

TMS—Trimethyl silyl, a chemical derivative.

Total metals—Metallic elements that have been digested prior to analysis.

Toxic—A substance that may chemically produce injurious or lethal effects. The degree of toxicity and its effects vary with the compound and time of exposure.

Toxic equivalency factor (TEF)—Toxicity values given to several of the halogenated aryl hydrocarbons relative to the most toxic congener, *2, 3, 7, 8-tetrachloridibenzo*-p-*dioxin* (2, 3, 7, 8-TCDD).

Toxic equivalent concentration (TEQ)—A toxicity-weighted concentration that accounts for both the concentration of individual congeners and their different TEF values. Calculates as TEQ = Σ (congener concentration × congener TEF) for all congeners having assigned TEF values. See **TEF**.

TPH—Total petroleum hydrocarbons.

t_R—Chromatographic retention time. The time required for a peak to pass through a column.

t_R'— Adjusted retention time; $t_R' = t_R - t_0$.

Transmittance—Transmittance (T) refers to the fraction of electromagnetic energy that passes through a sample (I), relative to the total amount of quantity of energy that strikes the sample (I_0), so that it is represented by the ratio of T = I/I_0. The T scale may be represented as 0%–100% or 0–1.0. In transmittance measurements, dry air is used as the typical reference.

Trennzahl (TZ)—See *separation number of Trennzahl*.

Trihalomethane (THM)—One of a family of organic compounds that are derivatives of methane. THMs are generally by-products of the chlorination of drinking water that contained dissolved humic and fulvic acids or other types of organic substances.

Trust Fund—A fund set up under the Comprehensive Environmental Response, Compensation and Liability Act (or equivalent state Superfund law) to help pay for cleanup of hazardous waste sites and for legal action to force those responsible for them to clean them up. See **CERCLA**.

Truth table (related to the logic of statistical testing)—The probability p of making a Type I error is α, and thus the probability of making a correct decision when the null hypothesis (H_0) is true is $1 - \alpha$. Similarly, the probability of making a Type II error when the null hypothesis (H_0) is false is β, and the probability of making a correct decision is $1 - \beta$. This logic is summarized in the *truth table* shown below:

What the Analyst Thinks	What the Scientific Facts Suggest	
Decision	H_0 is true	H_0 is false
H_0 is true	Correct	Type II error
	$p = 1 - \alpha$	$p = \beta$
H_0 is false	Type I error	Correct
	$p = \alpha$	$p = 1 - \beta$

$\bar{u}$ —Average linear carrier gas velocity; in GC, $\bar{u} = L / t_0$.

u_0—Carrier gas velocity at the column outlet; $u_0 = \bar{u} / j$.

u_{opt}—Optimum linear gas velocity. The carrier gas velocity corresponding to the minimum theoretical plate height, ignoring stationary-phase contributions to band broadening.

Ultrahigh-performance liquid chromatography (UHPLC)—UHPLC is often used loosely for any separation performed at pressures greater than provided by conventional HPLC pumps (~400 bar); original meaning was for separations in the 20,000 psi range. Commercial UHPLC is a relatively recent development. Chromatographic peak widths from UHPLC columns rival those peak widths obtained from WCOT GC columns!

Ultraviolet (UV) spectroscopy—Spectroscopy using the portion of the electromagnetic spectrum generally from 190 to 400 nm, and most often from 190 to 360 nm for the majority of ultraviolet laboratory measurements. The types of electrons that can be excited by UV-Vis light are few in number: nonbonding electrons, electrons in single bonds, and electrons involved in double bonds.

Underground storage tank—A chemical tank located entirely or partially underground. Usually associated with gasoline storage.

Uptake—The process by which an agent crosses an absorption barrier. See *dose*.

V_G—In GC, the volume of carrier gas contained in a column. For open tubular columns of length L whose diameter is d_c ignoring the stationary-phase film thickness, d_f; $V_G = L\left(\pi d_c^2 / 4\right)$.

V_L—In GC, volume of stationary phase contained in a column.

Vadose zone—Unsaturated soil, i.e., above the water table.

Vapor pressure—The pressure in mmHg characteristic of a vapor in equilibrium with its liquid or solid form. The vapor pressure of a chemical relates to its speed of evaporation. For example, since *xylene* has a higher vapor pressure than *propylene glycol*, it will evaporate faster. If equal quantities of xylene and propylene glycol were spilled, the airborne concentrations of xylene would be higher than those of propylene glycol.

Visible spectroscopy—Spectroscopic measurement using that portion of the electromagnetic spectrum detectable by human eyes—i.e. the portion of the spectrum from 360 to 780 nm. This portion is sometimes referred to as the 350 to 770 nm spectral region. This region is used for color measurements. Colors are produced by electrons in a material moving from one orbital transition to another orbital around the atoms within the molecule of the colored substance.

VOA—Volatile organic analysis; also refers to volatile organic acids.

VOA bottle—A vial used to contain samples for volatile organic analysis.

Volatile compounds—Compounds amenable to analysis by purge and trap and static headspace sample prep techniques. Synonymous with *purgeable compounds*.

Volatile organic compound (VOC)—Any organic compound that participates in atmospheric photochemical reactions, except for those designated by the EPA administrator as having negligible photochemical reactivity. A subset of VOCs that include one or more chlorine atoms covalently bonded to carbon are abbreviated, at least in this book, as ClVOCs. Examples of ClVOCs are vinyl chloride, trichloroethene (TCE), and perchloroethylene or tetrachloroethylene (PCE).

w_b and w_h—In GC, the peak width at its base and the peak width at half height (both are measured in seconds).

Wall-coated open tubular (WCOT)—In GC, a capillary column in which the stationary phase is coated directly on the inner column wall. WCOT columns are the most widely used in contemporary practice of GC. See *capillary column*.

Wet chemistry—Procedures that involve distillations, colorimetric determinations, and titrimetric measurements. Examples of analytes that are routinely quantitatively determined by wet chemical methods include cyanides, methylene blue active surfactants, and biological and chemical oxygen demands (BOD and COD).

Wide-bore open tubular—Open tubular (capillary) column with a nominal inner diameter of 320 μm; was once associated with 530 μm until J&W Scientific coined the term *megabore*. For a full description, see *Capillary column*.

Wood's anomalies—A variety of surface plasmon polarization (photon or light interaction effects) where a diffraction grating will change its efficiency relative to specific angles of incidence and the frequency (wavelength) of light interacting with the grating. A diffraction grating will dramatically change (generally decrease) the amount of energy (photon flux) dispersed from its surface, based upon the interaction of p-polarized light from the spectrometer source with the physical groove period of the grating surface (the groove or grating period of the grating is equivalent to 1/groove density). The decrease in energy dispersed from the grating is present only for p-polarized light at specific angles of incidence; noting that for p-polarized light the magnetic field of the energy is parallel to the grooves on the grating surface. Efficiency curves may be plotted for any wavelength of light incident to a diffraction grating such that the angle of incidence (x-axis) is plotted against the diffracted light flux (as percent efficiency) to yield a graphic showing the Wood's anomaly region(s).

Appendix B
QA/QC Illustrated

Read not to contradict and confute, nor to believe and take for granted, nor to talk or discourse, but to weigh and consider.

—Francis Bacon

Trace environmental quantitative analysis (TEQA) requires that a specific QA/QC document be written and available. This appendix further elaborates on QA/QC as first introduced in Chapter 2 and first written by the author (as laboratory manager) in support of the *National Institute of Environmental Health Sciences (NIEHS) Basic Superfund Research Center Analytical Core.*

B.1 WHAT DOES A SPECIFIC QC PLAN LOOK LIKE?

Many of the concepts introduced in Chapter 2 will now be applied in the example that follows. The standard operating procedure (SOP) and QC document that follow provide guidelines for the quantitative determination of trace concentrations of either organochlorine or polychlorinated biphenyl targeted analytes that were routinely performed in the past in the author's laboratory and falls within the general category of *enviro-health* TEQA.

The sample matrix was composed of rat plasma, and the sample preparation method used consisted of ultrasonic probe sonification (PS) using acetonitrile, followed by solid-phase extraction in the reversed-phase mode (RP-SPE) using either an octyl (C_8) chemically bonded silica or an octadecyl (C_{18}) chemically bonded silica. The analytes were retained on the RP silica, and the RP silica was dried and then eluted using pesticide residue-grade iso-octane. After adjustment to a precise 1 mL eluent volume, 1 μL of eluent was injected into a previously calibrated gas chromatograph.

B.2 HOW IS A QA/QC PLAN ORGANIZED?

A good QA/QC document tells the client exactly how the analysis will be conducted and how the analytical data produced by execution of the procedures will be interpreted. The document illustrated here is presented in the following sequence:

1 Title
2 Summary
3 Percent recovery study
4 Calibration, verification, and quantitative analysis
5 Establishment of the IDL
6 Establishment of the MDL
7 QC definitions
8 QC specifications
9 Reporting requirements.

B.2.1 Title: Trace Organochlorine Pesticide (OCs) / Polychlorinated Biphenyls (PCBs) Analysis via PS Coupled to RP-SPE Using Capillary Gas Chromatography with Electron-Capture Detection (C-GC-ECD): SOP and QC Protocols

B.2.2 Summary of Method

Samples of biological origin, such as plasma, serum, organ parts, and so forth, that arrive in the laboratory are immediately frozen or refrigerated as recommended by the client. Prior to beginning the analysis, the samples are thawed and brought to ambient temperature. The primary technique used in our laboratory to isolate and recover the target QC or PCB analytes from the biological matrix is *ultrasonic probe sonication* combined with *reversed-phase solid-phase extraction* (PS-RP-SPE). A 10-cartridge vacuum manifold is used. This enables up to 10 SPE extractions to be performed simultaneously. In addition to the number of samples that are to be analyzed, one sample must be selected, and two identical aliquots of this sample are needed. The first additional sample is needed to spike (matrix spike), and the second additional sample, identical to the first, is needed to spike again (matrix spike duplicate). If these additional sample matrices are not available for conducting recovery studies, matrix spikes will not be included.

B.2.3 Percent Recovery Study

A percent recovery (%) study will be conducted. A surrogate reference standard consists of a methanolic 2 ppm solution that contains tetrachloro-*m*-xylene (TCMX) and decachlorobiphenyl (DCBP). Surrogates are added to all samples, whereas matrix spikes are added to only selected samples. A matrix spike reference standard consists of a methanolic 24 ppm solution that contains PCB congeners or specific Aroclors. A *control sample* is prepared by taking the same aliquot of surrogate and matrix spike that will be added to samples for the % recovery study and placing this aliquot directly into a 1 mL volumetric flask and adjusting the final volume to 1.0 mL. A series of SPE cartridges used to conduct the % recovery study are used. In conjunction with a 10-cartridge vacuum manifold such as a Vacmaster® International Sorbent Technology, Ltd., a method development study is listed according to the following table and then discussed (see below):

SPE No.	2 ppm Surrogate (μL)	24 ppm Matrix Spike (μL)	Matrix	Description
1	100	0	ACN	Method blank
2	100	100	ACN	Blank spike 1
3	100	100	ACN	Blank spike 2
4	100	100	ACN	Blank spike 3
5	100	100	Sample	Matrix spike
6	100	100	Sample	Matrix spike duplicate
7	100	0	Sample	Sample 1
8	100	0	Sample	Sample 2
9	100	0	Sample	Sample 3
10	100	0	Sample	Sample 4
—	100	100	Iso-octane	Control

A 1 mL eluent is obtained from PS-RP-SPE that is quantitatively transferred to a clean, dry 2 mL GC autosampler glass vial with a crimped-top cap. Percent recoveries are routinely found to be between 75 and 100% for most OCs and PCBs from a biological matrix used to conduct toxicological studies. The appropriate GC method is retrieved from the chromatography software that controls and acquires data from the GC (Turbochrom®, PE Nelson). The order in which the vials are injected into the G-GC-ECD is called a sequence. A sequence file is created using the Turbochrom chromatography processing software (PE Nelson). Several QC protocols are used to consider the order in which sample eluents from PS-RP-SPE are injected into a C-GC-ECD. Blanks are run before and after standards. Usually, injection of a standard that has a high concentration of analyte is followed by injection of a blank to minimize carryover. A representative sample sequence is given in the following table:

Injection Vial No.	Description of GC Vial
1	Blank (pure iso-octane)
2	Control (100% recovery)
3	Blank (pure iso-octane)
4	SPE method blank
5	SPE blank spike 1
6	SPE matrix spike
7	Blank (pure iso-octane)
8	SPE sample unspiked 1

Injection vials 1 to 8 represent the *first cycle* in a series of several cycles in which a vial containing pure iso-octane is first injected to estimate whether there is any carryover from previous injections. Then, a control vial is injected that contains the analyte of interest and represents a 100% recovery. Pure iso-octane is then injected a second time in this cycle to prevent carryover from the high concentration of analyte in the control. A method blank is then injected that demonstrates that the method itself is free from laboratory and analyst contamination. This is followed by an injection of the blank spike eluent in order to assess the percent recovery of analyte from a relatively clean matrix. This is followed by an injection of the matrix spike eluent in order to assess the percent recovery of analyte from the sample matrix itself. A matrix effect can then be evaluated from these two spiked samples. To prevent carryover again, pure iso-octane is injected. The sample itself, *unspiked,* is then injected to provide sample analysis. A second cycle is undertaken in a manner similar to that just introduced for the first cycle. If there are sufficient samples, a third, fourth, and even a fifth cycle are also introduced.

B.2.3.1   Calculation of Analyte Percent Recovery

The integrated peak areas for all targeted analytes over replicate injections of the control and spikes are then averaged and used to calculate a mean percent recovery according to

$$\% R_i = \frac{\sum_i^N A_i^s / N}{\sum_i^n A_i^c / n} \times 100$$

where A_i^s represents the integrated peak area for the ith component in the spiked sample or spiked blank, A_i^c represents the integrated peak area for the ith component in the control sample, N represents the number of replicate spiked samples or blanks, and n represents the number of replicate control injections into the GC.

B.2.4 CALIBRATION, VERIFICATION, AND QUANTITATIVE ANALYSIS

B.2.4.1 Preparation of the Stock Solution

Weigh 0.0500 g (does not have to be exact, record in notebook) of each OC, PCB congener, or Aroclor. Place the solid or liquid into a 10 mL volumetric flask half filled with methanol (MeOH, use highest purity available). Adjust to a final volume with the bottom of the meniscus horizontally aligned with the calibration mark. This yields a 5,000 ppm stock solution. Transfer to a clean, dry glass vial with a Teflon septum and screw cap. Label the vial clearly as "5,000 ppm … (MeOH), date." Store at refrigerated temperatures.

B.2.4.2 Preparation of the Primary Dilution

Place 100 μL of stock solution into a clean, dry 10 mL volumetric flask that is half filled with MeOH. Adjust to final volume with MeOH, transfer to a storage vial, and label "50 ppm …"

B.2.4.3 Preparation of the Secondary Dilution

Place 100 μL of primary dilution into a clean, dry 10 mL volumetric flask that is half filled with MeOH. Adjust to final volume with MeOH, transfer to a storage vial, and label "500 ppb …"

B.2.4.4 Preparation of the Tertiary Dilution

Place 100 μL of secondary dilution into a clean, dry 10 mL volumetric flask that is half filled with MeOH. Adjust to final volume with MeOH, transfer to a storage vial, and label "5 ppb …"

B.2.4.5 Preparation of the Working Calibration Standards

Use the following table to prepare a series of working calibration standards using MeOH as the dilution solvent. Transfer the standards to a 2 mL GC vial, cap, and either place on autosampler or store for future use.

Standard	5 ppb (μL)	V (total)	Concentration (ppb)
1	5	1.0	0.025
2	10	1.0	0.050
3	50	1.0	0.250
4	100	1.0	0.500
5	250	1.0	1.250
6	500	1.0	2.500

A multipoint calibration curve using the ES mode is established for each OC and PCB congener or total Aroclor using either the secondary or tertiary diluted reference standard (depending on the range of concentration desired). The working calibration standards are prepared originally from either a chemically pure form of the OC or PCB or a commercially available and certified solution. A *least square regression analysis* is performed using available software such as Excel (Microsoft Corporation) or via in-house computer programs. The quality of the least squares fit is evaluated in terms of its correlation coefficient and standard deviations in both the slope and intercept. An initial calibration verification (ICV) standard is prepared and injected in triplicate to evaluate the precision and accuracy of the least squares regression. The confidence interval for the ICV is then established. Peak areas for the separated and detected OCs, PCB congeners, or total Aroclors are converted to their corresponding concentration levels expressed in either ppb or ppm of each analyte or as total Aroclor. A confidence interval about the interpolated concentration levels is also obtained. For example, the 3, 4, 3′, 4′-tetrachlorobiphenyl congener present in a sample of rat plasma is reported as 76 ± 5 ppb at the 95% significance level.

B.2.5 Establishment of the Instrument Detection Limit

The instrument detection limit (IDL) is defined as the minimum concentration or weight of analyte that can be detected at a known confidence level. The IDL depends on the ratio of the magnitude of the analytical signal to the size of the statistical fluctuations in the blank signal.

The IDL is then defined as the concentration that corresponds to the ratio of the difference between the *minimal distinguishable analytical signal* and the mean signal in a blank to the slope of the least squares regression line (calibration sensitivity). Because this minimum distinguishable signal can be defined as being equal to the sum of the mean signal from a blank, S_{blank}, and 3 or 10 times the standard deviation in the blank signal, std dev$_{\text{blank}}$, two definitions must be used: the *minimum detectable limit* C_{IDL} and the *limit of quantitation* C_{LOQ}. Expressed mathematically,

$$S_{\text{IDL}} = \bar{S}_{\text{blank}} + 3S_{\text{blank}}$$

$$C_{\text{IDL}} = \frac{S_{\text{IDL}} - S_{\text{blank}}}{\text{Slope}}$$

$$S_{\text{LOQ}} = \bar{S}_{\text{blank}} + 10S_{\text{blank}}$$

$$C_{\text{LOQ}} = \frac{S_{\text{LOQ}} - S_{\text{blank}}}{\text{Slope}}$$

Hence, C_{IDL} and C_{LOQ} yield two levels that must be considered when defining detection limits. Because the sample preparation technique involves the isolation and recovery of analytes into an extract from the original sample, a concentration factor, as well as a consideration of the efficiency of the extraction process as defined by the percent recovery, is involved. Thus, the method detection limit (MDL) will be lower than the IDL.

B.2.6 Establishment of the Method Detecton Limit

The METHOD DETECTION LIMIT (MDL) can be found only after the instrument detection limit (IDL), the concentration factor (F), and the percent recovery R (expressed as its decimal equivalent) have been established for the method. F is the ratio of eluent volume, V_{eluent}, to sample volume, V_S. The volume of rat plasma, V_S, is usually 1 or 2 mL, whereas V_{eluent} is 1.0 mL. Expressed mathematically,

$$C_{\text{MDL}} = \frac{C_{\text{IDL}} F}{R}$$

where

$$F = \frac{V_{\text{eluent}}}{V_S}$$

B.2.7 Definitions of QC Reference Standards

Calibration or working standards—A series of diluted standards prepared from dilutions made from the tertiary dilution standard. These standards are injected directly into the C-GC-ECD and used to calibrate the instrument as well as evaluate the range of ECD linearity. The range of concentrations should cover the anticipated concentration of targeted analytes in the unknown samples.

> *Control standard*—Consists of the same representative set of targeted analytes used in the matrix spike, and the same precise aliquot of this solution is added to an aliquot of the elution solvent used in PS-RP-SPE and is used in the calculation of percent recoveries.
>
> *Instrument calibration verification (ICV) standard*—Prepared in a manner similar to that of the working calibration standards; however, at a different concentration from any of the working standards. Used to evaluate the precision and accuracy for an interpolation of the least squares regression of the calibration curve.
>
> *Laboratory method blank*—A sample that contains every component used to prepare samples, except for the analyte (s) of interest, via PS-RP-SPE and is taken through the sample preparation process. It is used to evaluate the extent of laboratory contamination for the targeted analytes.
>
> *Matrix spike standard*—Consists of a representative set of the targeted analytes (i.e., OCs or PCBs dissolved in MeOH). A precise aliquot of this solution is added to a sample so that the effect of the sample matrix on the percent recovery can be evaluated.
>
> *Primary dilution standard*—Results from dilution of the stock reference standard.
>
> *Secondary dilution standard*—Results from dilution of the primary dilution standard.
>
> *Stock reference standard*—The highest concentration of target analyte obtained by either dissolving a chemically pure form of the analyte in an appropriate solvent or acquiring it commercially as a certified reference solution.
>
> *Surrogate spiking standard*—Consists of tetrachloro-*m*-xylene (TCMX) and decachlorobiphenyl (DCBP) dissolved in MeOH. These two analytes are highly chlorinated and are structurally very similar to OCs and PCB targeted analytes. TCMX elutes prior to and DCBP after the OCs and PCBs, thus eliminating any coelution interferences. A fixed volume (aliquot) of this reference solution is added to every sample, matrix spike, blank, blank spike, and control in the protocol.
>
> *Tertiary dilution standard*—Results from dilution of the secondary dilution standard.

B.2.8 Quality Control Specifications

B.2.8.1 Minimum Demonstration of Capability

Before samples are prepared and extracts injected into a GC, the laboratory must demonstrate that the instrumentation is in sound working order and that targeted analytes can be separated and quantitated. GC operating conditions must be established, and reproducibility in the GC retention times, $t(R)$, for the targeted analytes must be achieved.

B.2.8.2 Laboratory Background Contamination

Before samples are prepared and extracts injected into a GC, the laboratory must demonstrate that the sample preparation bench-top area is essentially free of trace OC and PCB residues. This is accomplished by preparing method blanks using either distilled deionized water or acetonitrile as reagent blank and observing essentially peak-free GC-ECD chromatograms.

B.2.8.3 Assessing Targeted and Surrogate Analyte Recovery

Before samples are prepared and extracts injected into a GC, the laboratory must demonstrate that the surrogate and targeted analytes can be recovered by the method to a reasonable degree.

EPA Method 508 suggests that an acceptable range in percent recovery can be found to be ±30%. Typical values for R range from 65 to 115%.

B.2.8.4 Assessing Calibration Curve Linearity

Multipoint calibration covering one or more orders of magnitude above the 0.1 ppb instrument detection limit (IDL) for C-GC-ECD based on a 1 µL injection volume with a 10:1 split ratio should satisfy a minimum correlation coefficient of 0.9500. Correlation coefficients are calculated within the Turbochrom (PE Nelson) software or by using a computer program.

B.2.8.5 Assessing ICV Precision and Accuracy

Precision can be evaluated for the triplicate injections of the ICV following establishment of the multipoint calibration by calculating a confidence limit for the interpolated value for the ICV concentration expressed in ppb.

A relative standard deviation expressed as a percent (%RSD) can also be calculated based on peak areas for the ith component according to

$$\%\text{RSD}_i = \frac{s_i}{C_i} \times 100$$

where s_i is the standard deviation in the interpolated concentration for the ith targeted component based on a multipoint calibration and triplicate measurements. %RSD is also called the coefficient of variation.

Both the mean concentration and the confidence limits about the mean concentration should be established for the ICV prior to conducting the sample analysis.

Accuracy can be evaluated based on a determination of the percent relative error according to

$$\%\text{Error} = \frac{\left| \bar{C}_i - C_i^{\text{known}} \right|}{C_i^{\text{known}}} \times 100$$

Statements of precision and accuracy should be established for the ICV, and as long as subsequent measurements of the ICV remain within the established confidence limits, no new calibration curve need be regenerated.

B.2.9 Reporting Requirements

The client generally dictates what type of reporting format to use. Software for processing raw analytical data into analytical results is usually available on the PC used with the instrument. In the author's laboratory, a sample-by-sample report can be produced and printed using Turbochrom (PE Nelson). In this *sample report*, a tabular format shows the peaks found, analyte name, analyte concentration, and analyte retention time, as follows:

Peak No.	Analyte Name	Concentration (ppb)	t(R)	Peak Area
6	Lindane	50.4	1.84	21,033
19	Endrin	153.8	4.01	30,163
24	Methoxychlor	188.1	5.56	17,202

In addition, a *summary report* is compiled in which the file name for the chromatogram is placed in the first column, the name of the sample analyzed is placed in the second column, and the concentration, retention time, and peak area for the first chromatographically resolved component are placed in the third column:

File Name	Sample Name	Analyte		
		Concentration (ppb)	$t(R)$	Peak Area
YA13001	Blank	11.0	3.77	21,245
YA13003	133 ppb ICV	140.0	3.76	298,906
YA13009	Unknown Sample #2	1316.4	3.77	2,832,169

Appendix C
A Primer on the Basics of
Probability and Statistics for
Some and a Quick Review
for Others

Some of the fundamental concepts of probability and statistics that are essential for the analytical chemist or chemical technician who is engaged in various aspects of TEQA are introduced here. Concepts presented in Chapter 2 have assumed that the reader has the background in most of the topics that will be discussed here. The author has drawn from his personal resources to help develop these concepts.[1-6]

Figure C.1 describes the fundamental measurement process involving analytical instrumentation that drives contemporary TEQA using so-called domain diagrams. The top diagram identifies the various domains of measurement available today while the bottom diagram illustrates an example of how to use the diagram. We didn't always have a digital domain! A chemical constituent in a sample can be quantitated using what was once called "chemical instrumentation" to yield a current in the analog domain. This current is converted to a phase in the time domain as is shown in the lower domain diagram. Finally, the phase is converted to a meter position in the non-electrical domain. Imagine what a domain diagram for a GC-MS instrument in which the concentration of a chemical such as for the polybrominated biphenyl, PBB-153 that might have been extracted from the groundwater located at the site of the old MCC (Chapter 1) would look like. The extract is injected into the GC-MS instrument. Referring to the lower domain diagram in Figure C.1, we would draw a line from the chemical domain to the current or voltage domain and then over to a coded count in the digital domain which would result in the computerized abundance for a specific mass to charge ratio (m/z) associated with a fragment of PBB-153. It is beneficial to compare and contrast Figure C.1 in this appendix with that of Figure 2.1. One concludes from a study of both drawings exactly how important calibration is to achieve GLP, an important ingredient in achieving and maintaining good QA/QC.

Figure C.2 is a rough sketch of instrumental baseline noise. The classical mathematical definition of a chemical analyte's limit of detection (LOD) is shown as being that concentration of chemical analyte that corresponds to a signal that is three times (3X) the standard deviation in the mean baseline noise. This statement is inherently statistical and to understand it fully we need to have a background in the basic statistics of measurement! How can we better understand what this means? Table C.1 lists some baseline noise within the digital domain. The author had injected the organochlorine pesticide lindane dissolved in iso-octane. The instrument used to generate the data was an Autosystem® (Perkin-Elmer) gas chromatograph that incorporated an open tubular GC column and an electron-capture detector (ECD). From the 13 measurements of baseline noise, we begin to calculate a mean or average value using Equation (C.1) (see below) letting N =13 and x_i refers to the ith data point. A standard deviation using either form of Equation (C.2) (see below) is then calculated:

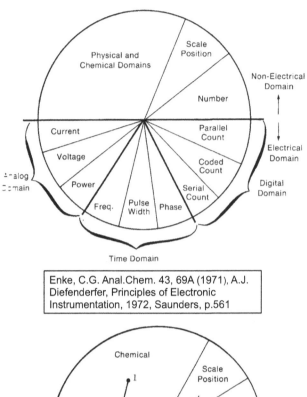

Enke, C.G. Anal.Chem. 43, 69A (1971), A.J.
Diefenderfer, Principles of Electronic
Instrumentation, 1972, Saunders, p.561

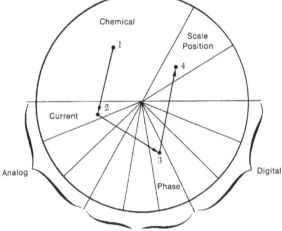

FIGURE C.1   Domain diagrams for instrumental measurement.

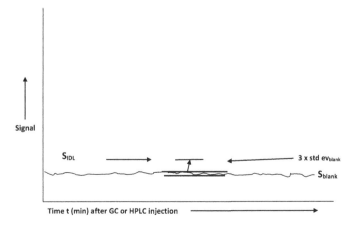

FIGURE C.2   Example chromatogram illustrating baseline noise.

TABLE C.1
Data Used to Calculate the IDL for Lindane Using C-GC-ECD

Noise (uv-sec)		
896		
897		
612		
537		
815		
285	S(bl) mean (uV-sec)	
568	984	
491		
951	std dev (bl)	
571	124.4508	
578		
856		
1072		

Instrumentation: Autosystem Gas Chromatogram incorporating a capillary GC column and an electron-capture detector (Perkin-Elmer).

$$\bar{x} = \frac{\sum_i^N x_i}{N}$$

(C.1)

$$s = \sqrt{\frac{\sum_i^N (x_i - \bar{x})^2}{N-1}}$$

(C.2)

We find that the mean signal in the baseline (bl), S(bl)is found as $\bar{x} = 984$ µV-sec with a standard deviation in the mean baseline std dev (bl) = 124.450 µV-sec. We can easily round off to 124.5 µV-sec. Refer back to Figure C.2, we can now calculate the S_{IDL} = 3 × 124.5 = 373.5 µV-sec. An amount or concentration of a chemical analyte that corresponds to 373.5 µV-sec is the value for the LOD! We have used statistics to achieve the objective! Most analytical laboratories who are not using a more advanced concept as introduced in Chapter 2 are still calculating LODs based on this blank statistical model!

It is also important to consider what is meant by a relative standard deviation (RSD) as is shown below:

$$RSD = \frac{s}{\bar{x}} \times 100$$

(C.3)

Analytical chemists often report the precision of replicate injections of a given chemical standard into a GC or an LC instrument in terms of an RSD (often referred to as a %RSD). For example, the statement "the mean of 3 replicate injections into a GC-MS of an extract obtained from one groundwater samples was reported as 41.5 ppb DDT ±0.5% RSD" is a clear statement about the precision of the method.

If we sought to calculate a population mean μ instead of a sample mean $\bar{x}$ in the baseline noise discussed above, we would need to increase N significantly. We could suppose that we can get access to many more digitized measurements of baseline noise from hundreds of GC-ECDs around the United States and beyond! In this hypothetical case, we then need to introduce a population mean μ and a population standard deviation σ as shown below:

$$\mu = \lim_{N \to \infty} \frac{\sum_{i}^{N} x_i}{N}$$

$$\sigma = \sqrt{\lim_{N \to \infty} \frac{\sum_{i}^{N} \left(x_i - \mu\right)^2}{N}}$$

It is also important to know that compounded error in measurement in general, i.e. the error inherent in the numerous steps in any analytical procedure can be found by summing the variances in each step according to:

$$\sigma_i^2 = \sigma_1^2 + \sigma_2^2 + \ldots + \sigma_n^2 \tag{C.4}$$

In the laboratory practice of TEQA, the variance due to sampling techniques (Chapter 2), sample preparation techniques (Chapter 3) and determinative techniques (Chapter 4) all contribute in an additive manner to the total variance! GLP requires that all systematic errors be eliminated. We are left with only random errors. Random errors cannot be eliminated and are interpreted by applying the mathematics of probability and statistics.

C.1 THEY SAY THAT REPLICATE ANALYTICAL DATA THAT IS FREE OF SYSTEMATIC ERROR IS NORMALLY DISTRIBUTED ABOUT THE AVERAGE VALUE: WHAT DOES THIS MEAN?

In this section we further elaborate on what Figure 2.17 is telling us. This allows us to discuss just how probable or likely a measurement is. The population average or population mean occurs at the apex of a normal (Gaussian) distribution of replicate measurements. These replicate measurements are assumed free of systematic error. Consider the author's attempt to quantitate a given concentration of nitrate ion $[(NO_3]^-$ dissolved in water. The author was equipped at the time with an ion chromatograph (ion chromatography is introduced in Chapter 4). The data is shown in Table C.2. Notice that the 50 replicate measurements of nitrate ion are organized into 10 columns of 5 replicate measurements each. Note also that an average for each of the ten columns of replicate data is also included. Next, the frequency of occurrence of identical $[(NO_3]^-$ concentrations is listed in terms of a frequency of occurrence. For example, a concentration of 0.46 ppm occurred only once in contrast to a concentration of 0.49 ppm occurring 10 times. A bar graph shows how this frequency of occurrence plotted vs. increasing nitrate concentration suggests the possibility that this data can be explained by proposing a Gaussian or normal distribution!

Consider the upper and lower graphs shown in Figure C.3. The upper graph Figure C.3(a) compares two normal distributions with different magnitudes for the standard deviation while the lower graph Figure C.3(b) clearly defines important terms to be discussed. Figure C.3(b) considers the mathematics of the normal distribution (normal error curve) in terms of the number

TABLE C.2
Results from 50 Replicate Measurements of a ~0.5 ppm Nitrate Ion in Water Using Ion Chromatography

0.51	0.51	0.51	0.5	0.51	0.49	0.52	0.53	0.5	0.47
0.51	0.52	0.53	0.48	0.49	0.5	0.52	0.49	0.49	0.5
0.49	0.48	0.46	0.49	0.49	0.48	0.49	0.49	0.51	0.47
0.51	0.51	0.51	0.48	0.5	0.47	0.5	0.51	0.49	0.48
0.51	0.5	0.5	0.53	0.52	0.52	0.5	0.5	0.51	0.51
0.51	**0.505**	**0.505**	**0.515**	**0.515**	**0.505**	**0.51**	**0.515**	**0.505**	**0.49**

Conc NO3	Frequency
0.46	1
0.47	3
0.48	5
0.49	10
0.5	10
0.51	13
0.52	5
0.53	3

Standard Deviation of the mean of 50
0.017127

Standard Deviation of the mean of 50
0.014142

of standard deviations from the mean z (refer back to Equation (2.57) and subsequent discussions). The infinitesimal ratio dL/L defines a point on the y-axis while a whole number from zero on either side of the maxima defines the x-axis. A normal or Gaussian distribution is algebraically defined as follows:

$$\frac{dL}{L} = \frac{1}{2\pi} e^{-z^2/2} dz \tag{C.5}$$

We now have the normal distribution in terms of $y = f(z)$ where $y = $ dL/L and $z = (\bar{x} - \mu)/\sigma$.

The area under the lower graph in Figure C.3 is the integral with respect to z of Equation (C.5) (p. A-8).[1] The fraction of the population with values of z between any specified limits is given by the area under the curve between these limits. For example, the area under the curve between $z = -1$ and $z = +1$ is 0.683 or 68.3% of the total area under the curve.[1] Thus, 68.3% of a population of data lies to within ± 1σ of the mean value. Furthermore, 95.5% of a population

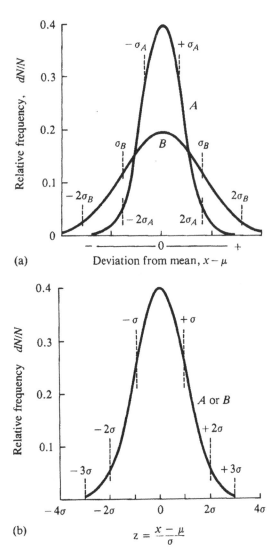

FIGURE C.3 Normal error curves. Upper curve: $\sigma_B = 2\sigma_A$ with the abscissa in units of deviations from the mean $(x-\mu)$ in the units of measurement; Lower curve: the abscissa is the deviation from the mean in the units of σ where $z = (x-\mu)/\sigma$.

lies to within $\pm 2\sigma$ of the mean value and 99.7% of a population lies to within $\pm 3\sigma$ of the mean value.[1, 4] Bear in mind we are only referring to random error in measurement. Clearly, the standard deviation is very useful in estimating and reporting the probable net random error for an analytical method.

Refer back to Table C.2 and note one other statistical concept. The standard deviation in the mean was calculated for all 50 measurement [$s = 0.0171$ (to three significant figures)]. From the ten subsets of data with each subset containing five measurements of the concentration of [$(NO_3$]$^-$ in water was quantitated using ion chromatography. The standard deviation in the mean of the 10 subsets was somewhat lower [$s = 0.0141$ (to three significant figures)]. The standard deviation in the mean of five subsets is less than the standard deviation of the population, i.e. all 50 measurements. This result is consistent with the concept of a *standard error of the mean*.[1]

C.2 WHAT IS A STANDARD ERROR OF THE MEAN?

It is the mean of a random sample of N drawn with replacement from a population with a given probability function, then (pp. 193–196)[3]:

$$\mu_{\bar{x}} = \mu_x$$

$$\sigma_{\bar{x}}^2 = \frac{\sigma_x^2}{N} \tag{C.6}$$

If the population is normally distributed, it can be proved by more advanced methods of mathematical statistics that the sample mean, $\bar{x}$ has precisely a normal distribution. The quantity obtained when we take the positive square root of Equation (C.6) is:

$$\sigma_{\bar{x}} = \frac{\sigma_x}{\sqrt{n}} \tag{C.7}$$

Equation (C.7) if often called the standard error of the mean. Normal (Gaussian) distributions of a large number N of replicate measurements denoted by x, along with the distribution of the mean for N=9 and for N=25 are shown in Figure C.4. This graph encompasses the mathematics introduced in Equation (C.6) (pp. 193–196).[3]

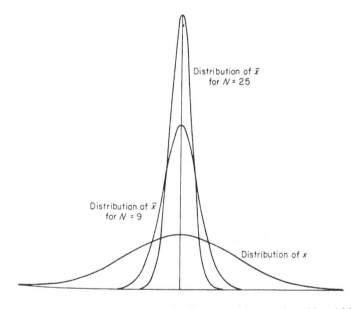

FIGURE C.4 The normal distribution of x and the distributions of for samples of 9 and 25.

C.3 LET'S RETURN TO THE NORMAL DISTRIBUTION FIRST SHOWN IN FIGURE 2–17

Equipped with some basics introduced here, let's revisit Figure 2–17 and further elaborate on confidence intervals. Figure C.5 does just that! It should become clearer now that in the absence of systematic error, replicate measurement with a mean value to within 2x the standard deviation

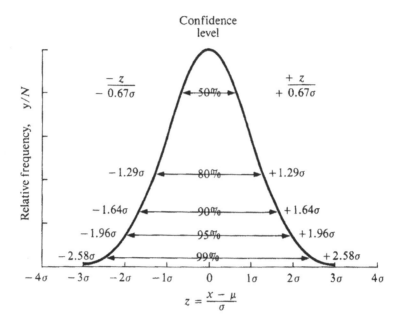

FIGURE C.5 Confidence levels for various values of z.

in the mean occurs with a probability of 95% (or is likely probable) while a mean value to within 3x the standard deviation in the mean occurs with a probability of 99% (or is likely very probable)! This should explain why ± 2s and ± 3s are so important when understanding the Westgard Rules.

C.4 WHEN DO I USE THE STUDENT–FISHER t DISTRIBUTION?

The problem of testing the significance of the deviation of a sample mean from a given population mean when N is small where N is the number of replicate samples and only the sample variance is known was first solved in 1908 by W.S. Gossett, writing under the pen name of *Student*. His method was later amended by R.A. Fisher. When the population of x is distributed normally, they proved that the variable:

$$t = \frac{\bar{x} - \mu}{s/\sqrt{N}} \quad df = N - 1 \tag{C.8}$$

has a symmetrical bell shape but non-normal distribution with $\mu_t = 0$. Its probability density function is:

$$P(t) = C\left(1 + \frac{t^2}{n}\right)^{-(n+1)/2}$$

where the parameter $n = N-1$ is the number of degrees of freedom. C is a constant depending upon n and determined so as to make the total area under the density curve equal to 1 (p. 202).[3] The curve for n = 20 together with a normal curve is shown in Figure C.6. This distribution does not fall off as rapidly in the tails as does the normal distribution (p.157).[4] For example, we find for n = 10 and P = 0.01 that t = 2.76. This is shown in Figure C.7.[3]

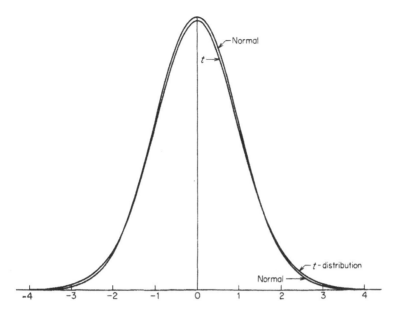

FIGURE C.6 Comparing the normal and t distributions.

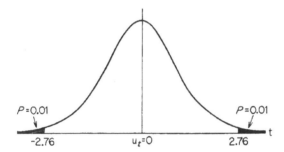

FIGURE C.7 A typical t distribution curve for $n = 10$ and a probability $P = 0.01$ such that $P(t > 2.76) = P(t < -2.76) = 0.01$.

A useful t-table is given in Appendix F. Integrals of Student's t distribution are tabulated in an inverse form, with t given as a function of $\int_{t}^{\infty} P(t)\,dt$ and n degrees of freedom [but sometimes of $2\int_{t}^{\infty} P(t)\,dt$ and n degrees of freedom or $\int_{-\infty}^{t} P(t)\,dt$ and n degrees of freedom].[4]

C.5 HOW DO I USE STUDENT'S *T* TO TEST FOR BIAS? OR DOES THE AVERAGE AGREE WITH THE THEORETICAL OR AN ACCEPTED VALUE?

Use confidence intervals (CIs) when σ (the standard deviation in the population) whose mean μ is unknown. The CI for the experimental mean $\bar{x}$ for N replicate measurements is:

$$\text{CI for } \mu = \bar{x} \pm t\, s_{\bar{x}} = \bar{x} \pm \frac{ts}{\sqrt{N}}$$

To test for bias, measure reference standards to obtain the true population μ. Then measure $\bar{x}$ over N replicate measurements. Then ask if the difference between $\bar{x}$ and μ is due to random error or due to bias?

If

$$\bar{x} - \mu > \pm \frac{ts}{\sqrt{N}}$$

then bias is likely.

If

$$\bar{x} - \mu < \pm \frac{ts}{\sqrt{N}}$$

then no bias has been demonstrated.

C.6 HOW ELSE CAN WE USE THE *T*-TABLE?

We now consider hypothesis testing and show one example that answers the question, does the average value agree with the theoretical or accepted value? A null hypothesis symbolized by H_0 suggests that there is no difference in two averages, i.e. $\mu_1 = \mu_2$. An alternative hypothesis symbolized by H_A suggests that $\mu_1 \neq \mu_2$ or H_A: $\mu_1 < \mu_2$ or H_A: $\mu_1 > \mu_2$ etc. If we are only interested in, for example, $\mu_1 < \mu_2$ or $\mu_1 > \mu_2$ we would use a one-tailed t test. If we are interested in $\mu_1 \neq \mu_2$ we would use a two-tailed t test. The null hypothesis is accepted until the test statistic exceeds the critical value. (p. 68)[6] We then apply Equation (C.8) using an approach introduced earlier (pp. 70–72).[6]

Suppose we make six replicate injections into a GC-ECD of a 50 ppb surrogate reference standard containing trichloro-m-xylene (TCMX). TCMX is often used along with deca-chlorobiphenyl (DCBP) as surrogates when implementing a method to quantitatively determine trace levels of organochlorine pesticides from an environmental matrix such as groundwater. Three scenarios are presented in Table C.3 and Equation (C.8) is applied to all three. As is evident when evaluating a two-tailed t test, only the left column results yield a $t < 2.571$ for df = 6-1= 5 degrees of freedom where 6 is the number of replicate injections. Only the first column confirms the null hypothesis that there is no difference measured mean and the accepted value, i.e. no difference between $\mu = 50$ ppb and $\bar{x} = 50.8$ since t is less than the t (critical) as shown in the table.

TABLE C.3
Hypothesis Testing Using Student's *t*

#ppb TCMX	#ppb TCMX	#ppb TCMX	
49.9	1000	100	
52.7	1500	150	
54.6	1500	150	
54.6	1400	140	
47.3	1000	50	
48.7	990	99	
51.5	880	88	
50.8	1128	1045	$\bar{x}$
50.0	50.0	50.0	μ
2.7	255	36.4	s
0.715	10.85	3.67	t

Six replicate injections of a 50 ppb TCMX surrogate reference standard. Three hypothetical scenarios are shown above.

For $df = 5$, $t(1-\alpha/2)$ where $\alpha = 0.05 = 2.571$.

REFERENCES

1. Skoog D., J. Leary. *Principles of Instrumental Analysis, 4th ed.* Philadelphia: Saunders College, 1992, pp. 5–9; Appendix 1, pp. A1–A23.
2. Harris D. *Quantitative Chemical Analysis,* 3rd ed. San Francisco: Freeman, 1991, Chapter 4 and Appendix 1.
3. Mode E. *Elements of Probability and Statistics*, Englewood Cliffs, NJ: Prentice Hall, 1966.
4. Perrin C. *Mathematics for Chemists*, New York, Wiley-Interscience, 1970.
5. Laitinen H. *Chemical Analysis*, New York, McGraw-Hill, 1960, Chapter 26.
6. Anderson R., *Practical Statistics for Analytical Chemists*, New York, Van Nostrand Reinhold, 1987.

Appendix D
Quality Control for Environmental-Health TEQA: Levey–Jennings Plots and Westgard Rules

D.1 WHAT ARE THE WESTGARD RULES AND WHEN DO YOU NEED TO APPLY THEM?

Let's assume that you have established a new quantitative analytical method on a specific instrument such as a GC-MS or LC-MS-MS for one or more chemical analytes of importance to enviro-health TEQA. Let's assume that you have completed a method validation. Let's also assume that you have completed at least 20 replicate runs. Each run consists of a set of calibration standards. Each run includes one or more quality control (QC) samples. Three QC control samples that straddle the range of analyte concentrations found in the initial calibration are common: one QC-Low (QC-L), one QC-Medium (QC-M), and one QC-High (QC-H). Multiple QCs at each of these three levels are also found, although it is less common. The analytical chemist engaged in enviro-health TEQA must then create QC charts often referred to a *Levey–Jennings plots* in clinical chemistry parlance. In 1924, Walter Shewhart at Bell Telephone Laboratories developed a statistical charting technique for monitoring a manufacturing process. In 1950, S. Levey and E.R. Jennings published a paper in the *Am J Clinical Pathology* demonstrating the use of statistical control in the clinical laboratory.

Let's first study and interpret a *Levey–Jennings Plot*. One example of a Levey–Jennings drawn from the author's work in quantitating cyanide in whole blood using Excel is shown for a QC-M in Figure D.1. Cyanide ion (CN^-) present in whole blood and after acidifying is headspace sampled as $HCN_{(g)}$. HCN is retained and eluted chromatographically using a porous layer open tubular (PLOT) column. A model 6890/5973N GC-MSD (Agilent Technologies) incorporating a split-splitless inlet with electronic pressure control and a mass-selective detector was used to generate the data. The GC system was interfaced to a workstation that used ChemStation® (Agilent) software for chromatographic control, data acquisition and storage. An MPS2® autosampler (Gerstel GmbH & Co. KG) was also used and consisted of a dual-syringe delivery system, located atop the GC-MS system. Gerstel/s Maestro® software controlled operation of the dual-rail autosampler and was integrated into the ChemStation software.

During method validation, 20 replicate QC samples (QC-M as shown in Figure D.1) were run and Levey–Jennings plots were created. The mean is then calculated and plotted as shown. Next, the [mean ±2s] is calculated and plotted as shown. In the absence of systematic error, 95% of all subsequent measurements of the CCV should fall to within ±2s *of the mean*. The [mean ±3s] *is then calculated and plotted as shown.* In the absence of systematic error, 99% of all subsequent measurements of the CCV should fall to within ±3s *of the mean*. The method validation is now complete. It is at this point in time, while reviewing the Levey–Jennings Plot, that the analyst invokes the *Westgard Rules* criteria to either accept or reject the method validation.

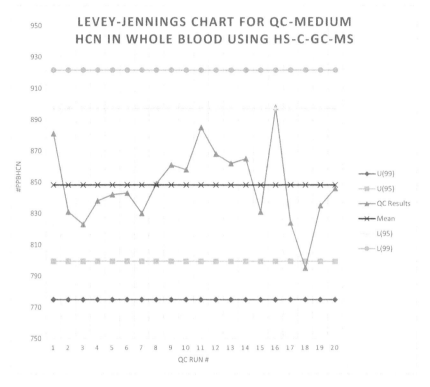

FIGURE D.1 Levey-Jennings Plot for evaluating a QC-M for the quantitative determination of trace cyanide in whole blood using HS-C-GC-MS.

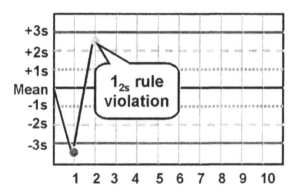

FIGURE D.2 Graphical explanation of Westgard Rule 1_{2s}.

The Westgard Rules came about from the work of Westgard and colleagues who provided a systematic interpretation of Levey–Jennings plots to easily detect systematic error in clinical chemistry.[1] Let's first define the symbolism used as well as showing a graphical interpretation of the symbolism. The symbol has the form A_L where A is the *number of control observations per run and L is the control limit*:

- 1_{2s} represents the control rule where one control observation exceeds control limits set as $\bar{u} \pm 2s$. *This is the* "warning" rule for a Shewhart chart and is interpreted in this discussion as a requirement for additional inspection of the control data, testing the data with the rules below to judge whether the analytical run should be accepted or rejected. Refer to Figure D.2 for a graphical depiction of this rule being violated.

- 1_{3s} symbolizes the control rule where a run is rejected when one control observation exceeds control limits set as $\bar{u} = \pm 3s$. Refer to Figure D.3 for a graphical depiction of this rule being violated.

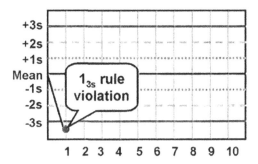

FIGURE D.3   Graphical explanation of Westgard Rule 1_{3s}.

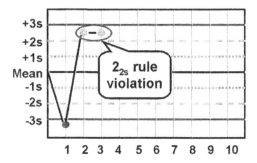

FIGURE D.4   Graphical explanation of Westgard Rule 2_{2s}.

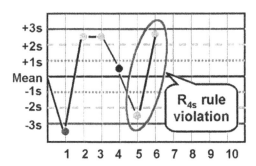

FIGURE D.5   Graphical explanation of Westgard Rule R_{4s}.

- 2_{2s} is the control rule where the run is rejected when two consecutive control observations exceed the same limit, which is either $\bar{u} + 2s$ or $\bar{u} - 2s$. This rule is initially applied to the two observations within a run, one on each of two different control materials. The run is rejected when the control observations on both materials exceed their respective +2s control limits or their respective -2s control limits. The rule can also be applied to two consecutive observations on the same control material, one for each of two consecutive runs. Refer to Figure D.4 for a graphical depiction of this rule being violated.
- R_{4s} is the control rule according to which the run is rejected when the range or difference between the two control observations within the run exceeds 4s. The rule is invoked when the observation is one control material exceeds +2s and the observation on the on the other exceeds a -2s limit, i.e. each observation is out by 2s, but in opposite directions, making a total of 4s difference between them. Refer to Figure D.5 for a graphical depiction of this rule being violated.

- 4_{1s} is the control where the run is rejected when four consecutive control observations exceed the same limit, which is either $\bar{u} + 1s$ or $\bar{u} - 1s$. These consecutive observations can occur within one control material, which would require inspecting the observations for four consecutive runs, or across control materials which would require inspecting only the present run and the one before it. Although included here, this Westgard Rule was not implemented in the author's work and therefore a graph is not shown.
- 10_{xbar} is the control rule which says the run is rejected when 10 consecutive control observations fall on the same side of the mean ($\bar{u}$). These consecutive observations can occur within one control material or across control materials. This would require inspection of 10 or five consecutive runs, respectively. Refer to Figure D.6 for a graphical depiction of this rule being violated.

Refer to the logical flow chart shown in Figure D.7 for implementing a combination of the above individual Westgard Rules. The 1_{2s} rule is used as a warning rule and prompts a more detailed inspection of the data using the other control rules. If neither control observations

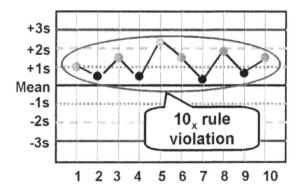

FIGURE D.6 Graphical explanation of Westgard Rule 10_{xbar}.

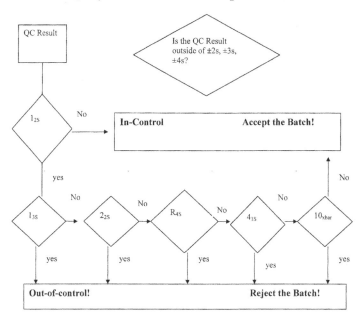

FIGURE D.7 Graphical explanation of the logic of the combined Westgard Rules.

exceeds a 2s limit, the analytical run is in-control and patients' data may be reported. It neither observation exceeds a 2s limit, the control data is tested by applying the 1_{2s}, 2_{2s}, R_{4s}, or 10_{xbar}. If none of these rules is violated, the run is *in-control*. If any one of them is violated, the run is *out-of-control*. The particular rule violated may give some indication of the type of analytical error occurring. Random error will most often be detected by the 1_{3s} and R_{4s} rules. Systematic error will usually be detected by the 2_{2s}, 4_{1a}, or 10_{xbar} and, when very large, by the 1_{3s} rule.

D.2 IS THERE AN EASIER WAY TO IMPLEMENT THE WESTGARD RULES?

The author has had to teach laboratory chemists and chemical technicians on this topic. He presents (see below) an alternate way to aid in understanding how to implement the Westgard Rules. Consider two scenarios: a) assume one QC per level and b) replicate QCs per level. Figure D.8 is this author's re-interpretation of the logic used to implement the Westgard Rules in a more easily understandable logical flow chart diagram restated as follows:

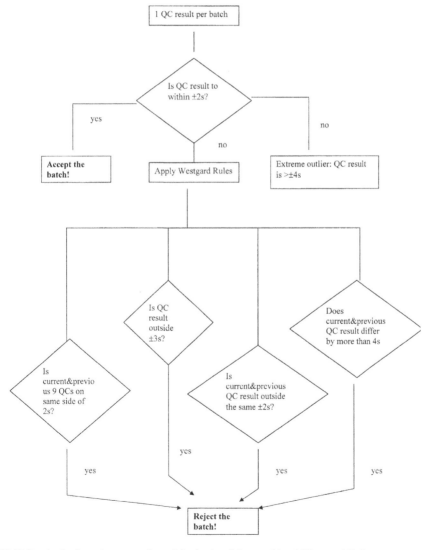

FIGURE D.8 Author's re-interpretation of the logic of the combined Westgard Rules.

For a <u>single QC per level</u>,

if the QC sample result is within ±2Si established for the analysis on the instrument used where Si is the standard deviation in the individual results, **accept the batch**

if the QC sample result is outside ±2*Si, reject the batch if*:

1 the QC run result is beyond ±4S_i of the characterization mean (extreme outlier), *if not, consider if the*
2 QC result is ±3*Si* (1_{3s}, 2_{2s}, 2_{2s}), *if not, consider if the*
3 current and previous QC run results are outside the same ±*2Si limit(22s Rule), if not, consider if the*
4 current and previous 9 results are on the same side of the mean (10xbar Rule), **if not, consider if** the
5 current and the previous result differ by more than ±4*Si(R4S), if not, accept the batch*

Note: since runs have a single QC level per batch and for this method, only one QC result per level is required, the R4s Rule is applied only across runs and not within a run.

D.3 FOR REPLICATE QCS PER LEVEL

If the QC run mean result is within ± $2S_m$ limits established or the analysis on the instrument used where S_m the standard deviation in the mean results, and individual results are within **accept the batch**

If the QC sample result is outside ± $2S_m$, *reject the batch if:*

1. the run mean is beyond ±4*Sm* of the characterization mean (extreme outlier)**, if not, consider** if the
2. the QC run mean is outside a ±3*Sm* (1_{3s} Rule), if not, consider if the
3. current and previous QC run mean are outside the same 2 S_m limit (2_{2s} Rule)**, if not, consider if** the
4. current and previous 9 results are on the same side of the mean (10X −Bar Rule)
5. if one of the two individual QC runs is outside a ±*2Si* limit, **reject the batch** if the within run range for the current and previous run exceeds 4S_w, i.e. 95% limit (R_{4S} Rule), **if not, accept the batch**

REFERENCE

1 Westgard J., P. Barry, M. Hunt. "A multi-rule shewhart chart for quality control in clinical chemistry," *ClinChem* 27(3): 493–501. 1981.

Appendix E
Innovative Sample Prep Flow Charts for TEQA

This appendix presents one well-understood and generalized *graph* that distinguishes between persistent and non-persistent pesticide residues in the environment and a number of *sample preparation flow charts* developed by the author. These logical flow chart diagrams became a useful learning tool to the author as he studied a variety of sample preparation approaches utilizing many of the LLE, LSE, and SPE approaches introduced in Chapter 3 as this relates to *enviro-health* TEQA. Flowcharts are an excellent learning tool that helps a beginning analytical chemist to quickly grasp the science underlying a given procedure. For the most part the flow charts are self-explanatory.

Figure E.1 is a generalized plot of pesticide concentration that remains in blood, urine, and meconium versus the number of days after initial exposure to the chemical. The graph clearly distinguishes between *persistent and non-persistent pesticide residue* concentration levels.

Figure E.2 shows a detailed flow chart that describes how to extract serum in order to isolate and recover persistent organic pollutants.

Figure E.3 shows a flow chart that describes a laboratory procedure to conduct a micro Florisil® cleanup of a serum extract that contains persistent organic pollutants.

Figure E.4 shows a flow chart that describes a laboratory procedure using silica gel to conduct a column fractionation of the extract from the Florisil cleanup. The hexane fraction and the benzene fraction are ready for direct injection into a C-GC-ECD or a C-GC-MS.[1]

Figures E.2, E.3, and E.4 describe laboratory procedures that are performed consecutively with the extracted and cleaned up serum which is then fractionated to yield an organic solvent extract that is subsequently injected into a C-GC-ECD or C-GC-MS.

The use of ultra-sonication combined with RP-SPE in place of conventional LLE to isolate and recover persistent organochlorine and organobromine priority pollutants is featured in Figure E.5.[2,3,4] The authors combined several sample prep approaches as shown in this this flow chart.

Coplanar PCBs can be separated from non-coplanar PCBs by implementing the sample prep procedure outlined in the flowchart shown in Figure E.6.[5] A generalized molecular structure for the biphenyl structure is shown below:

Coplanar PCBs occur when chlorine is substituted for hydrogen at the 3,4, 5 and/or the 3',4',5' positions as shown above. Non-coplanar PCBs occur when chlorine is substituted for hydrogen at the 2, 6 and/or the 2'6' as shown above. Coplanar PCBs are considered to be more highly toxic than are the non-coplanar PCBs.

An innovative use of RP-SPE using a C18 Empore® disk to isolate and recover persistent organic pollutants (POPs) from various biological fluids is outlined in a flowchart and shown in Figure E.7.[6]

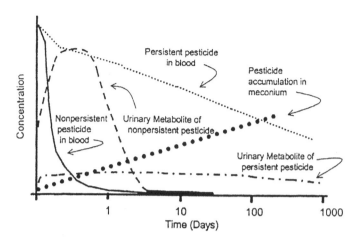

FIGURE E.1

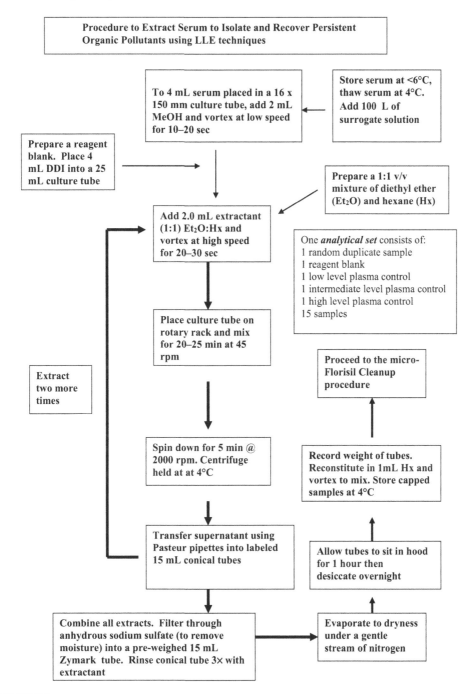

Procedure to Extract Serum to Isolate and Recover Persistent Organic Pollutants using LLE techniques

To 4 mL serum placed in a 16 x 150 mm culture tube, add 2 mL MeOH and vortex at low speed for 10–20 sec

Store serum at <6°C, thaw serum at 4°C. Add 100 L of surrogate solution

Prepare a reagent blank. Place 4 mL DDI into a 25 mL culture tube

Prepare a 1:1 v/v mixture of diethyl ether (Et₂O) and hexane (Hx)

Add 2.0 mL extractant (1:1) Et₂O:Hx and vortex at high speed for 20–30 sec

One *analytical set* consists of:
1 random duplicate sample
1 reagent blank
1 low level plasma control
1 intermediate level plasma control
1 high level plasma control
15 samples

Place culture tube on rotary rack and mix for 20–25 min at 45 rpm

Proceed to the micro-Florisil Cleanup procedure

Extract two more times

Spin down for 5 min @ 2000 rpm. Centrifuge held at at 4°C

Record weight of tubes. Reconstitute in 1mL Hx and vortex to mix. Store capped samples at 4°C

Transfer supernatant using Pasteur pipettes into labeled 15 mL conical tubes

Allow tubes to sit in hood for 1 hour then desiccate overnight

Combine all extracts. Filter through anhydrous sodium sulfate (to remove moisture) into a pre-weighed 15 mL Zymark tube. Rinse conical tube 3x with extractant

Evaporate to dryness under a gentle stream of nitrogen

FIGURE E.2

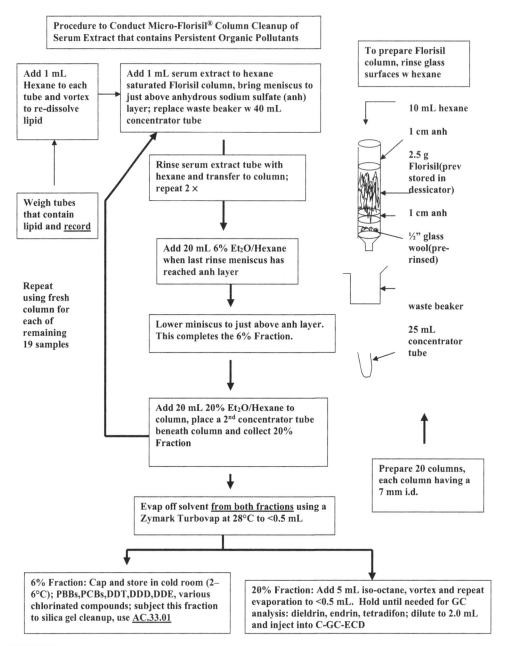

Procedure to Conduct Micro-Florisil® Column Cleanup of Serum Extract that contains Persistent Organic Pollutants

To prepare Florisil column, rinse glass surfaces w hexane

Add 1 mL Hexane to each tube and vortex to re-dissolve lipid

Add 1 mL serum extract to hexane saturated Florisil column, bring meniscus to just above anhydrous sodium sulfate (anh) layer; replace waste beaker w 40 mL concentrator tube

10 mL hexane

1 cm anh

2.5 g Florisil(prev stored in dessicator)

1 cm anh

½" glass wool(pre-rinsed)

Weigh tubes that contain lipid and record

Rinse serum extract tube with hexane and transfer to column; repeat 2 ×

Add 20 mL 6% Et₂O/Hexane when last rinse meniscus has reached anh layer

waste beaker

25 mL concentrator tube

Repeat using fresh column for each of remaining 19 samples

Lower miniscus to just above anh layer. This completes the 6% Fraction.

Add 20 mL 20% Et₂O/Hexane to column, place a 2nd concentrator tube beneath column and collect 20% Fraction

Prepare 20 columns, each column having a 7 mm i.d.

Evap off solvent from both fractions using a Zymark Turbovap at 28°C to <0.5 mL

6% Fraction: Cap and store in cold room (2–6°C); PBBs,PCBs,DDT,DDD,DDE, various chlorinated compounds; subject this fraction to silica gel cleanup, use AC.33.01

20% Fraction: Add 5 mL iso-octane, vortex and repeat evaporation to <0.5 mL. Hold until needed for GC analysis: dieldrin, endrin, tetradifon; dilute to 2.0 mL and inject into C-GC-ECD

FIGURE E.3

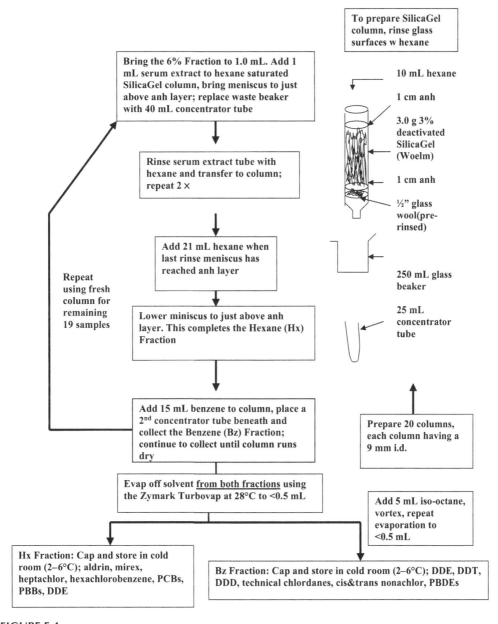

Procedure to Conduct 3% Silica Gel Column Fractionation of Extract from Florisil Cleanup that contain Persistent Organic Pollutants

To prepare SilicaGel column, rinse glass surfaces w hexane

Bring the 6% Fraction to 1.0 mL. Add 1 mL serum extract to hexane saturated SilicaGel column, bring meniscus to just above anh layer; replace waste beaker with 40 mL concentrator tube

10 mL hexane

1 cm anh

3.0 g 3% deactivated SilicaGel (Woelm)

1 cm anh

½" glass wool(pre-rinsed)

Rinse serum extract tube with hexane and transfer to column; repeat 2 ×

Add 21 mL hexane when last rinse meniscus has reached anh layer

250 mL glass beaker

Repeat using fresh column for remaining 19 samples

Lower miniscus to just above anh layer. This completes the Hexane (Hx) Fraction

25 mL concentrator tube

Add 15 mL benzene to column, place a 2ⁿᵈ concentrator tube beneath and collect the Benzene (Bz) Fraction; continue to collect until column runs dry

Prepare 20 columns, each column having a 9 mm i.d.

Evap off solvent from both fractions using the Zymark Turbovap at 28°C to <0.5 mL

Add 5 mL iso-octane, vortex, repeat evaporation to <0.5 mL

Hx Fraction: Cap and store in cold room (2–6°C); aldrin, mirex, heptachlor, hexachlorobenzene, PCBs, PBBs, DDE

Bz Fraction: Cap and store in cold room (2–6°C); DDE, DDT, DDD, technical chlordanes, cis&trans nonachlor, PBDEs

FIGURE E.4

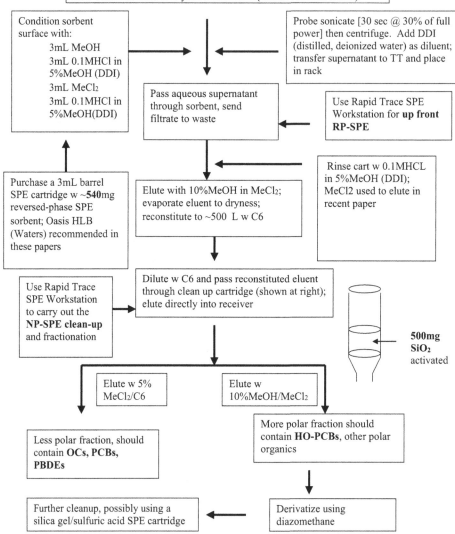

Procedure to isolate and recover PCBs, PBBs, PBDEs and HO-PCBs from human serum using probe sonication combine with RP-SPE techniques

Human Serum (1mL)
Methanolic surrogates added
Methanolic matrix spikes added, if applicable
Add formic acid: DDI to yield a **ratio1:1:1(serum : formic : DDI)**

Condition sorbent surface with:
  3mL MeOH
  3mL 0.1MHCl in 5%MeOH (DDI)
  3mL MeCl₂
  3mL 0.1MHCl in 5%MeOH(DDI)

Probe sonicate [30 sec @ 30% of full power] then centrifuge. Add DDI (distilled, deionized water) as diluent; transfer supernatant to TT and place in rack

Pass aqueous supernatant through sorbent, send filtrate to waste

Use Rapid Trace SPE Workstation for **up front RP-SPE**

Rinse cart w 0.1MHCL in 5%MeOH (DDI); MeCl2 used to elute in recent paper

Purchase a 3mL barrel SPE cartridge w ~**540**mg reversed-phase SPE sorbent; Oasis HLB (Waters) recommended in these papers

Elute with 10%MeOH in MeCl₂; evaporate eluent to dryness; reconstitute to ~500 L w C6

Use Rapid Trace SPE Workstation to carry out the **NP-SPE clean-up** and fractionation

Dilute w C6 and pass reconstituted eluent through clean up cartridge (shown at right); elute directly into receiver

500mg SiO₂ activated

Elute w 5% MeCl₂/C6

Elute w 10%MeOH/MeCl₂

Less polar fraction, should contain **OCs, PCBs, PBDEs**

More polar fraction should contain **HO-PCBs**, other polar organics

Further cleanup, possibly using a silica gel/sulfuric acid SPE cartridge

Derivatize using diazomethane

FIGURE E.5

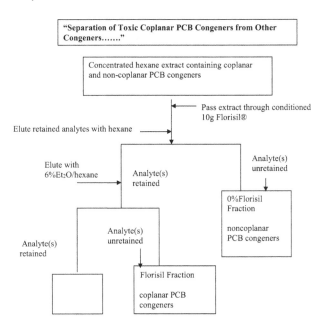

FIGURE E.6

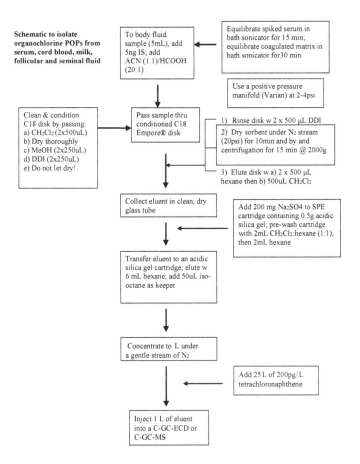

FIGURE E.7

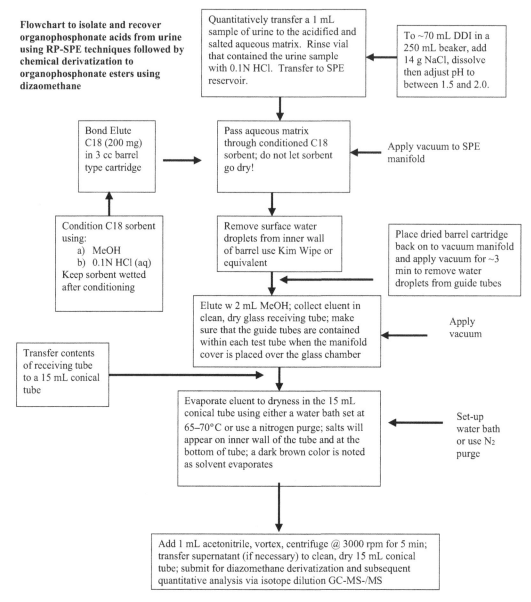

Flowchart to isolate and recover organophosphonate acids from urine using RP-SPE techniques followed by chemical derivatization to organophosphonate esters using dizaomethane

Quantitatively transfer a 1 mL sample of urine to the acidified and salted aqueous matrix. Rinse vial that contained the urine sample with 0.1N HCl. Transfer to SPE reservoir.

To ~70 mL DDI in a 250 mL beaker, add 14 g NaCl, dissolve then adjust pH to between 1.5 and 2.0.

Bond Elute C18 (200 mg) in 3 cc barrel type cartridge

Pass aqueous matrix through conditioned C18 sorbent; do not let sorbent go dry!

Apply vacuum to SPE manifold

Condition C18 sorbent using:
a) MeOH
b) 0.1N HCl (aq)
Keep sorbent wetted after conditioning

Remove surface water droplets from inner wall of barrel use Kim Wipe or equivalent

Place dried barrel cartridge back on to vacuum manifold and apply vacuum for ~3 min to remove water droplets from guide tubes

Transfer contents of receiving tube to a 15 mL conical tube

Elute w 2 mL MeOH; collect eluent in clean, dry glass receiving tube; make sure that the guide tubes are contained within each test tube when the manifold cover is placed over the glass chamber

Apply vacuum

Evaporate eluent to dryness in the 15 mL conical tube using either a water bath set at 65–70°C or use a nitrogen purge; salts will appear on inner wall of the tube and at the bottom of tube; a dark brown color is noted as solvent evaporates

Set-up water bath or use N_2 purge

Add 1 mL acetonitrile, vortex, centrifuge @ 3000 rpm for 5 min; transfer supernatant (if necessary) to clean, dry 15 mL conical tube; submit for diazomethane derivatization and subsequent quantitative analysis via isotope dilution GC-MS-/MS

FIGURE E.8

Organophosphonate nerve agent metabolites (VX acid, GB acid, GA acid, GD acid, and GF acid) can be isolated and recovered from human urine by first salting out the matrix and then eluting the sample through a conditioned C18 chemically bonded SPE sorbent. The eluent from RP-SPE as outlined in Figure E.8.[7]

A more recent and innovative use of RP-SPE to isolate and recover bisphenol and triclosan from serum and urine followed by chemical derivatization is illustrated by the flowchart shown in Figure E.9.[8]

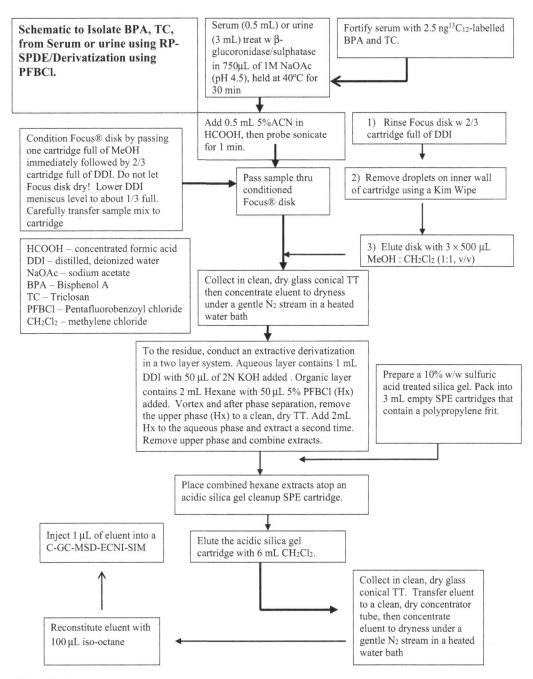

FIGURE E.9

An earlier approach developed by the author to selectively remove and quantitate the toxic weed killer herbicides (e.g. 2,4-D, 2,4,5-T, and 2,4,5-TP or Silvex) from an *enviro-chemical* aqueous sample such as aqueous agricultural runoff using RP-HPLC-PDA is shown as two flowcharts in Figures E.10.1 and E.10.2.[9] The same eluent from RP-SPE can then be solvent exchanged. The more conventional approach can then be implemented on the original sample as shown! Traditionally, chlorpheoxy acid herbicides (CPHs) were chemically derivatized and extracts were

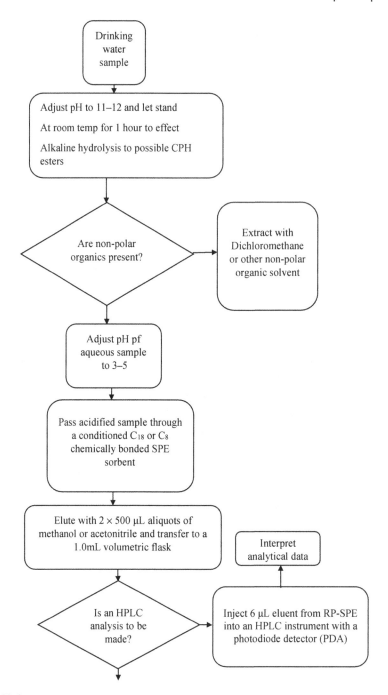

FIGURE E.10.1

quantitated as methyl esters of 2,4-D and 2,4,5-T, and 2,4,5-TP (Silvex) using either C-GC-ECD or C-GC-MS. Molecular structures for 2,4-D, 2.4.5-T and 2,4,5-TP (Silvex) are shown below:

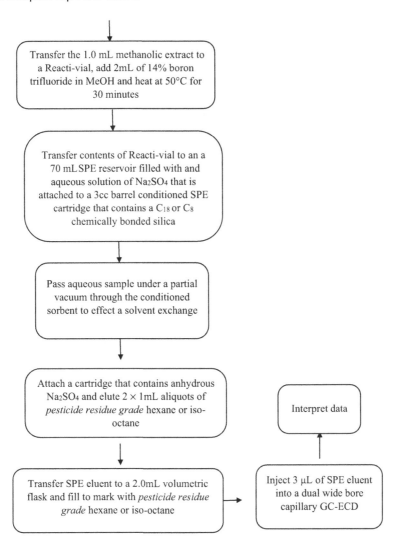

FIGURE E.10.2

A laboratory procedure that utilizes the approach shown in Figure E.10.1 flowchart is given in words as follows. Methylene chloride (dichloromethane) is added to ~100 mL of a wastewater sample whose pH has been adjusted to 11, then LLE is performed. Hydrochloric acid is used to acidify the separated aqueous phase to pH 2. The acidified aqueous sample is passed through a conditioned C_{18} or C_8 chemically bonded RP-SPE sorbent. The sorbent is washed with dilute hydrochloric acid (HCl). The analytes are eluted with two successive 1 mL aliquots of acetonitrile or methanol. The eluent is transferred to a 2.0 mL volumetric flask which is filled to the mark with elution solvent. A 6 μL HPLC injection loop is filled and the analytes are separated using isocratic reversed phase HPLC.[10]

A flowchart for a procedure developed earlier by the author that outlines an approach to isolate and recover priority pollutant base-neutral-acids (BNAs) from contaminated soil, an example of *enviro-chemical* TEQA utilizing RP-SPE techniques is shown in Figure E.11[11]. Please find examples of the plethora of BNAs whose molecular structures are shown below:

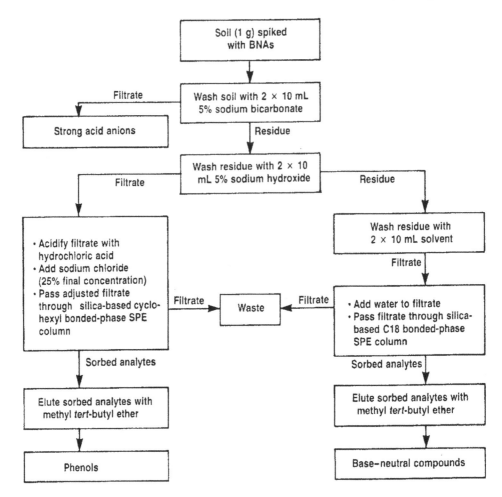

FIGURE E.11

Polycyclic aromatic hydrocarbons

Priority pollutant phenols

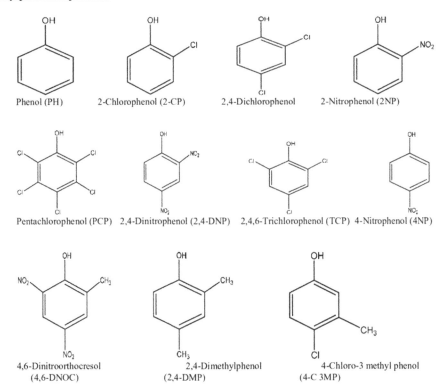

Phenol (PH)   2-Chlorophenol (2-CP)   2,4-Dichlorophenol   2-Nitrophenol (2NP)

Pentachlorophenol (PCP)   2,4-Dinitrophenol (2,4-DNP)   2,4,6-Trichlorophenol (TCP)   4-Nitrophenol (4NP)

4,6-Dinitroorthocresol (4,6-DNOC)   2,4-Dimethylphenol (2,4-DMP)   4-Chloro-3 methyl phenol (4-C 3MP)

Organo chlorine and bromine environmental contaminants

DDT   HBCD   Dioxin   PCB   BPA   PBDE

Dieldrin   Endrin

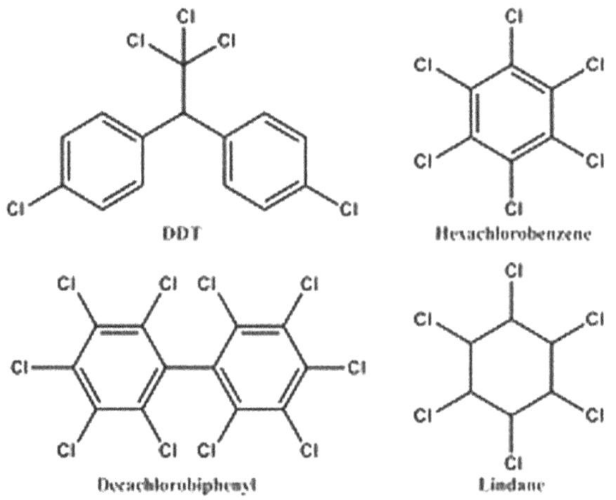

DDT

Hexachlorobenzene

Decachlorobiphenyl

Lindane

A flow chart was developed to show how RP-SPE could be incorporated into a sample preparation procedure to isolate and recover PBDEs, PCBs and other persistent organic pollutants (POPs) from *human breast milk* as shown in Figure E.12. Earlier, some excellent analytical approaches to isolating and recovering various POPs have been reported.[12,13,14]

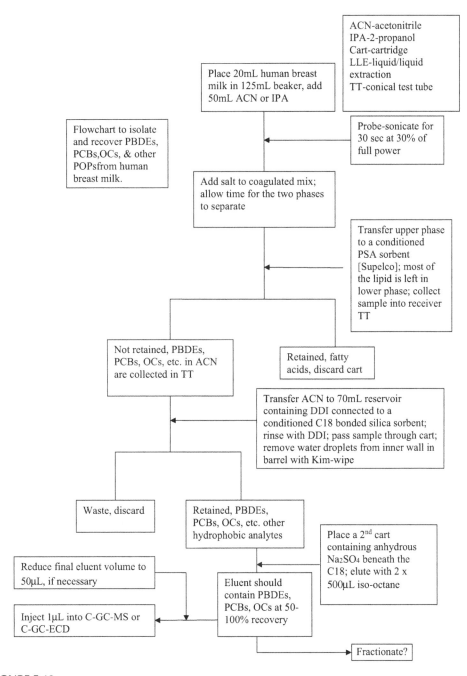

FIGURE E.12

REFERENCES

1. Figures E.2 to E.4 reflect the sample prep procedures used to isolate and recover organochlorine pesticides and PCBs and are provided courtesy of the Bureau of Laboratories, Michigan Department of Community Health, Lansing, MI.
2. Sandau, C. et al. *AnalChem* 75: 2003, 71–77.
3. Sjodin, A. et al. *AnalChem* 76: 2004, 1921–1927.
4. Sandau, C. et al. *Organohalogen Compounds 60–65, Dioxin* 2003.
5. Ricek, R. *LC-GC* 21(10): 2003, 992–1004.
6. Covaci, A. and P. Schepens, *Anal Lett* 34(9): 2001, 1449-1al 460.
7. Courtesy of George Frame, PhD, New York State Department of Health.
8. Geens, T., H. Neels, and A. Covaci, *JChromB* 877: 2009. 4042–4046.
9. Loconto, P.R. Ultra-trace determination of chlorophenoxy acid herbicides in drinking water, *American Environmental Laboratory* 4(1): 1992, 32–34.
10. Loconto, P.R. Solid-phase extraction in trace environmental analysis, current research – Part I, *LC-GC* 9(7): 1991, 460–465.
11. Loconto, P.R. Solid-phase extraction in trace environmental analysis, current research – Part II, *LC-GC* 9(11): 1991, 752–760.
12. Covaci, A., C. Hura, P. Schepens, *Chromatographia* 54: 2001, 247–252.
13. Dmitrovit, J. and S. Chan, *J Chrom B* 778: 2002, 147–155.
14. Sjodin, A. et al. *Anal Chem* 75: 2004, 4508–4514.

Appendix F
Quantitating VOCs in Serum Using Automated Headspace-SPME/Cryo-Focusing/Isotope Dilution/Capillary GC-MS

In January 2007, the author's lab received a frozen shipment of calibration standards and quality control (QC) samples to perform method validation for quantitating traces of volatile organic compounds (VOCs) in serum. Samples were supplied in 2 mL flamed-sealed glass ampoules containing 1.2 mL of calf serum previously spiked with 11 toxic VOCs (10 distinct chromatographically separated VOC peaks plus *m*- and *p*-xylene, which coelute). The validation batch consisted of 20 replicate sets of calibration standards and 20 replicate sets of QCs. Each QC set included a QC-Low (QC-L; each VOC analyte is expected to be ~250 ppb) and a QC-High (QC-H; each VOC analyte is expected to be ~8,000 ppb).

The samples arrived during a weekend when no staff was available to place them in the freezer, and thawed by the time staff returned to the office. This violated the sampling protocol of the method. After the samples were replaced, the laboratory conducted a comparative study (during the initial validation) to assess the effect of the thawed QC samples against their equivalent (the reissued batch that remained frozen throughout the shipping and receiving cycle). The quantitated results for both batches of QCs were compared by calculating a mean, standard deviation and variance over *N* replicate QC samples for each of the chromatographically separated VOCs.

This article reports the results of applying a *t* statistical comparison of two means over replicate QC samples at each of two different QC concentration levels (ppb; each VOC, except for coeluted *m*- and *p*-xylene) for each of 10 chromatographically separated VOCs.

F.1 EXPERIMENTAL

F.1.1 INSTRUMENTATION

A 6890/5973N inert GC-MSD (Agilent Technologies, Santa Clara, Calif.) incorporating a split/splitless inlet with electronic pressure control (EPC) and a single quadrupole mass selective detector were used to generate the data. The GC-MS was interfaced to a computer installed with ChemStation software (Agilent Technologies) for chromatographic control, data acquisition and storage. An MPS2 multipurpose autosampler (Gerstel GmbH & Co. KG, Mulheim an der Ruhr, Germany) was mounted above the GC-MS. Software control was via Maestro (Gerstel) integrated into ChemStation software. The robotic rail and arm perform liquid delivery, static headspace sampling and solid-phase microextraction (SPME). A CTS2 cryotrap, incubator with pulsed agitation, Peltier-cooled sample tray and 505 Controller (all from Gerstel GmbH & Co. KG) were installed as well.

This method uses only the lower rail injection unit with the Gerstel SPME Kit for the following operations: moving headspace (HS) vials from the Peltier tray to the incubator, penetrating the HS vial septum with the SPME fiber assembly during incubation, extruding and retracting the SPME fiber for analyte extraction, withdrawing the SPME assembly, inserting the SPME assembly needle into the split/splitless inlet and extruding the SPME fiber into the injector glass liner for analyte volatilization.

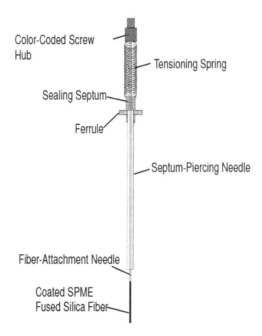

FIGURE F.1   Schematics of the SPME fiber assembly (courtesy of Millipore Sigma, formerly Supelco).

The SPME fiber remains in the inlet throughout the chromatographic run. Figure F.1 shows how the SPME fiber connects to the metal rod holder that attaches to the MPS2.

Ten of the 11 VOCs were chromatographically separated on a DB-VRX (Agilent Technologies) wall-coated open-tubular (WCOT) column. Analytes separated on the WCOT column were quantitated using the ChemStation RTE integrator. The stationary phase of the column (40 m × 0.18 mm × 1.0 μm film thickness) was designed specifically for the resolution of volatile, halogenated and aromatic hydrocarbons. A new WCOT column must be conditioned for optimal chromatographic performance. Table F.1 lists MPS2 and GC-MS operating conditions and parameters required to conduct the instrumental method: automated headspace solid-phase microextraction with cryo-focusing isotope dilution gas chromatography-mass spectrometry (HS-SPME-C-ID-GC-MS).

The VOC panel was spiked into a serum matrix at seven concentration levels for the calibrators and two QC levels (Table F.2). Since *m*- and *p*-xylene coelute, the concentration of these two VOCs is twice that of the other nine VOCs. Fixed-volume aliquots of all samples (calibrators, QCs) were added to 10 mL HS vials and were isotopically diluted with internal standards, which are analogs of the analytes labeled with ^{13}C or ^2H (Table F.3). (It is essential to use isotope dilution in GC-MS to achieve good precision and accuracy.[1]) Following isotopic dilution, the HS vials were loaded onto the Peltier-cooled sample tray, and each was sampled with the robotic sampling head using a carboxen/polydimethylsiloxane (CAR/PDMS)-coated fiber of 75 μm film thickness (SPME Fiber Assembly, MilliporeSigma).

The SPME fiber was inserted into the headspace above the serum and was exposed to each sample for 8 minutes. After the VOC analytes were extracted from the HS, thermally desorbed within the 300°C inlet, cryofocused at −110°C, and heated ballistically, they were chromatographed on the DB-VRX WCOT column and transported via carrier gas into the MSD. The MSD is a single quadrupole mass spectrometer that includes an ion source (electron impact ionization), mass analyzer (quadrupole) and detector (high-energy dynode and electron multiplier) all under high vacuum. It was operated in scan mode from a mass-to-charge ratio (*m/z*) of 35 to 250 to allow for identification of unknowns in the sample.[2]

TABLE F.1
MPS2 and GC-MS Operating Conditions Used to Calibrate and Quantitate Selected VOCs Extracted Serum Samples via HS-SPME

MPS2		GC-MS	
Runtime	14.48 min	GC inlet mode	Split
GC cool down time	2.50 min	GC inlet initial temperature	300®C
Cryo timeout	15.00 min	GC inlet split ratio	40:1
SPME injection setting	Sample preparation	Gas saver	On
Incubation temperature	35.0°C	Gas saver flow rate	20.0 mL/min
Incubation time	8.0 min	Gas saver time	14.00 min
Agitator on	5 sec	Carrier gas	Helium
Agitator off	2 sec	WCOT column	DB-VRX 40 m – 0.18 mm – 1.0 µm d_f
Agitator speed	500 rpm	GC oven	
Sample tray	VT32-10	Initial temperature	110°C
Vial penetration	22.00 mm	Initial time	1.50 min
Injection penetration	54.00 mm	Ramp 1	2.0°C/min to 130°C
Desorption time	840 sec	Ramp 2	40°C/min to 235°C
CTS2 initial temperature	–110°C	Ramp 3	0 (off)
CTS2 equilibration time	0.05 min	GC-MS transfer line	
CTS2 initial time	1.50 min	Heater	On
Cryo-cooling	On	Temperature	250°C
CTS2 inlet Ramp 1	On	MSD ionization mode	EI
CTS2 inlet ramp rate	12.0°/sec	MSD sample inlet	GC
CTS2 end temperature	235°C	EM voltage	2000
CTS2 hold time	12.5 min	Solvent delay	0.00 min
		Acquisition mode	Scan
		Fast scanning	Blank
		MS source temperature	230°C
		MS quad temperature	150°C

TABLE F.2
Calibration Levels Used to Quantitate the 14 or 15 Sets of QC Samples Studied in This Paper

Calibration Level	*m*- and *p*-xylene (ppb)	Remaining 9 VOCs (ppb)
1	200	100
2	600	300
3	1200	600
4	2000	1000
5	6000	3000
6	12000	6000
7	20000	10000

All 14 or 15 replicate QC-L and all 14 replicate QC-H samples used in this study were quantitated against a calibration that met the $r^2 = 0.9900$ criteria.

TABLE F.3
Retention Times, m/z Ratios for Quantitative and Confirmatory Ions with Corresponding Isotopically Labeled Internal Standards Needed to Quantify Ten Chromatographically Separated VOCs Including One Co-eluted Pair, m- and p-Xylene

Name	Retention Time	Target Ion	Qualifier Ion #1	Qualifier Ion #2
13C Chloroform	5.1	86	84	
Chloroform	5.1	83	85	87
D4 1,2-Dichloroethane	5.4	65	67	
1,2-Dichloroethane	5.4	62	64	49
D6 Benzene	5.6	56	54	
Benzene	5.6	78	77	51
13C Carbon Tetrachloride	5.6	117.9	119.9	
Carbon Tetrachloride	5.6	116.9	118.9	120.9
D8 Toluene	7.0	98	100	
Toluene	7.0	91	92	65
13C Tetrachloroethene	7.8	166.9	133.9	
Tetrachloroethene	7.8	164	129	94
D10 Ethylbenzene	8.8	98	116	
Ethylbenzene	9.0	91	106	77
D10 m-Xylene	9.1	98.1	116.2	
m- and *p*-Xylene	9.3	91.1	106.2	77.1
D8 Styrene	9.8	112	111	84
Styrene	9.9	104	103	77
D10 o-Xylene	9.8	98.1	116.2	
o-Xylene	10.0	91.1	106.2	77.1

F.1.2 METHODS

One ChemStation method was used to quantify the VOCs and two Maestro sample prep methods were used. One sample prep method was used to condition a new SPME fiber and the other to operate the MPS2 and run samples. The SPME fiber holder must be adjusted to accommodate the height of the 10 mL HS vial cap to prevent the bottom of the retracted coated fiber from breaking when the septum piercing needle penetrates the septum cap (Figure F.2).

Table F.3 lists each VOC, its GC retention time, its target or quantitative *m/z* ion and one or two qualifier *m/z* ions. Fourteen QC-L and 14 QC-H samples that thawed after receipt at the lab (Batch 1 [B1]) and 15 QC-L and 15 QC-H samples that did not thaw upon receipt (Batch 2 [B2]) comprise the two sets of data whose means for each of 10 VOCs will be compared statistically. Twenty sets of B1 and 20 sets of B2 were provided and analyzed by HS-SPME-C-ID-GC-MS. Each set consisted of seven calibrator vials, one QC-L vial and one QC-H vial. The sets were run according to the following sequence. First set: B1 calibrator vials, B1 QC vials, blank sample vial, B2 calibrator vials, B2 QC vials. After this run was completed, all 19 subsequent sets were prepared and run using this sequence.

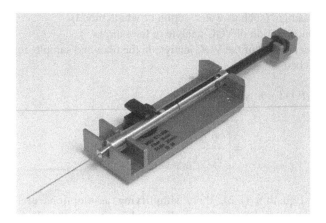

FIGURE F.2 Photo of the SPME fiber assembly attached to the MPS2 SPME holder (courtesy of Gerstel, USA).

To prepare the sample, the top of a glass ampoule was broken, and an aliquot of calf serum (previously spiked by the provider with the VOC panel) was transferred to a 10 mL headspace vial and was subsequently spiked with the VOC panel of internal standards. Of the 20 sets, 14 sets of B1 and 15 sets of B2 met the calibration curve criteria ($r^2 = 0.9900$) to calculate a mean, standard deviation and variance. These sets comprised the replicate quantitated results used in the calculations.

F.2 RESULTS

F.2.1 THEORETICAL BACKGROUND FOR HS-SPME

SPME was used to sample the headspace and preconcentrate the chemical species in the HS above a serum specimen that was continuously agitated and held at a constant temperature of 35°C. These fixed experimental conditions establish a constant partition coefficient $K^{VOC}fs(SPME)$ for a specific VOC, where f refers to the CAR/PDMS-coated SPME fiber and s is the sample. $K^{VOC}fs(SPME)$ is defined as being equal to the ratio of the concentration of a given VOC in the fiber, C_f, to that in the sample, C_s, once equilibrium is reached and assuming a negligible contribution from the HS sample interface. For HS-SPME, Pawliszyn compared extraction-time profiles for a stirred versus unstirred aqueous phase and showed that sample agitation shortens the time it takes to reach a plateau.[3] The equation that defines $K^{VOC}fs(SPME)$ can be algebraically manipulated to yield the fundamental relationship for quantitative analysis[4,5] as follows:

$$K^{VOC}_{fs(SPME)} = \frac{Cf}{Cs} = \frac{n_o / V_f}{C_o - (n_o / V_s)} \tag{F.1}$$

where:

n_o = maximum amount of VOC analyte that is partitioned into the SPME polymer coating (in moles) after sufficient incubation time and under exhaustive extraction conditions

V_f = volume of polymer coating on the SPME fiber

V_s = volume of sample (such as water, serum or whole blood)
C_o = original concentration of VOC analyte in the sample
C_f and C_s = concentration of the VOC analyte in the fiber and sample, respectively, once equilibrium is reached.

Solving Equation (F.1) for n_o yields

$$n_o = \left[\frac{K^{VOC}_{fs(SPME)} V_f V_S}{V_S + k^{VOC}_{fs(SPME)} V_F} \right] C_O \tag{F.2}$$

Upon examining Equation (F.2), three simplifying assumptions are evident: 1) If K^{VOC} $fs(SPME)$ >> V_s, as might be true for small sample volumes, then Equation (F.2) simplifies to $n_o = V_S C_o$ and demonstrates that the maximum amount of VOC analyte partitioned into the SPME fiber is directly proportional to the original VOC concentration, C_o, of that VOC in the sample. 2) If, however, a large sample volume is used, $K^{VOC} fs(SPME) V_f$ << V_s, then Equation (F.2) simplifies to $n_o = K^{VOC} fs(SPME) \cdot V_f \cdot C_o$ and demonstrates that the maximum amount of VOC analyte partitioned into the SPME fiber is also directly proportional to the original concentration, C_o, of that VOC in the sample. 3) It is not necessary to exhaustively extract VOCs from the HS to achieve quantification. It has been shown that when an aqueous solution is effectively agitated, a steady-state diffusion model can be used.[6] This model leads to a linear concentration gradient both in the bulk sample and in the PDMS-coated fiber using Fick's first law:

$$n = \left[1 - e^{-at} \right] \left[\frac{k_{fs} v_f v_s}{v_s + k_{fs} v_f} \right] C_o \tag{F.3}$$

As the sampling time t becomes very large, Equation (F.3) reduces to Equation (F.2). Hence, Equation (F.3) can be rewritten in terms of n as a function of sampling time t according to:

$$\frac{n}{n_o} = 1 - e^{-at} \tag{F.4}$$

Where a is a constant composed of various mass transfer coefficients, $K_{fs}(SPME)$, V_f and V_s. The author revised this model to include nonsteady-state considerations.[7]

All three assumptions lead to the conclusion that the amount of analyte partitioned into the SPME fiber, n, can be related to the initial concentration of analyte, C_o, in the sample. The number of moles of VOCs partitioned into the SPME fiber depends only on the original concentration of VOCs in the sample, C_o, provided that the sampling time, t, the rates of diffusion and the phase ratio all remain constant from sample to sample. The number of sorption sites on the SPME fiber is finite. As the concentration of VOCs increases, competition for sorption sites on the SPME fiber coating limits the number of sorption sites between analyte and internal standard.[2]

Figure F.3 shows a total ion chromatogram for one QC-L sample drawn from B2. This chromatogram is minimally informative since ^{13}C isotopically labeled VOCs coelute with their ^{12}C analogs, while ^{2}H isotopically labeled VOCs differ only slightly in retention time from their ^{1}H analogs.

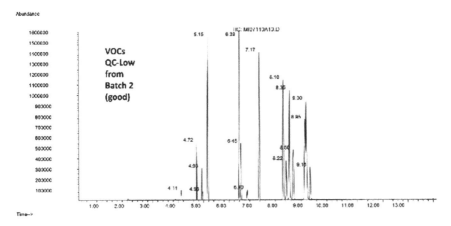

FIGURE F.3 Typical GC-MS total ion chromatogram for one QC-L sample from Batch 2. Column: DB-VRX (Agilent Technologies, Inc.), 40 m × 0.18 mm × 1.0 μm film thickness; 110°C (hold for 1.50 min) then to 130°C at 2.0°C/min; then to 235°C at 40°C/min.

F.2.2 IMPORTANCE OF THE t DISTRIBUTION

The importance of using the Student's t distribution when considering a small number of replicate measurements as is appropriate here was discussed earlier.[8] As established by the central limit theorem, for a normal distribution of x values, where N is large, 95% of the time values will be found in the confidence interval for the population mean μ_x according to

$$\bar{x} - \frac{2s_x}{\sqrt{N-1}} < \mu_x < \bar{x} + \frac{2s_x}{\sqrt{N-1}}$$ (F.5)

Where s_x is the standard deviation in the mean $\bar{x}$ of N replicate samples.

For small N (< 20), it can be shown that the variable

$$t \equiv \left(\bar{x} - \mu_x\right)/s_x/\sqrt{N-1} = \frac{\left(\bar{x} - \mu_x\right)\left(\sqrt{N-1}\right)}{s_x}$$

is not distributed normally but as a Student's t distribution:

$$P(t)dt \,\infty\, \left(1 + \frac{t^2}{v}\right)^{-(v+1)/2}$$

Thus, we find the dependence of the value of t on the number of degrees of freedom, v, where $v = N-1$. Integrals of Student's t distribution are tabulated in an inverse form with t as a function

of: $\int_t^\infty P(t)dt$ (one-sided) and v or $2\int_t^\infty P(t)dt$ (two-sided) and v. As v increases, the t distribution

begins to approximate the normal distribution.[8] The number of degrees of freedom d_f will be used instead of v in the mathematical equations shown below.

F.2.3 COMPARISON OF TWO MEANS USING *t* STATISTICS

Analytical chemists have four mathematical options to consider when *t* statistics are to be used to decide whether or not two experimental means differ: A) comparing two independent averages with known variances, B) comparing two independent averages with unknown and equal variances, C) comparing two independent averages with unknown and unequal variances and D) comparing two dependent averages (paired data). This article considers only B and C. It is assumed that both batches are independent of one another. Since the lot of VOC samples prepared and sent by the supervisory agency was the initial lot at the onset of the validation for VOCs, the variance in each mean is assumed to be unknown.

QC samples from both batches were quantitated against the B2 calibration, and then QC samples from both batches were quantitated against their respective calibrations (i.e., B1 QCs against the B1 calibration and B2 QCs against the B2 calibration) to determine if it matters whether the QCs are quantitated against their respective calibration or against only the B2 (good) calibration. Figure F.4 summarizes the tabular results below.

F.2.3.1 Applying *t* Statistics for Option B to Compare B1 and B2 Means: Both Batches Quantitated Against the B2 Calibration

When implementing option B (above), the two variances must satisfy the null hypothesis, i.e., there is no difference. An *F* test is conducted to assess whether the two variances satisfy the null

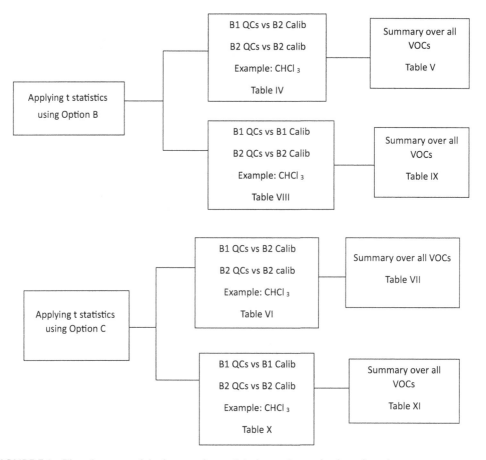

FIGURE F.4 Flowchart to explain the experimental design and organization of results.

hypothesis.[9-11] Equation (F.6) assumes that the larger variance is placed in the numerator and the smaller variance is placed as the denominator.

$$F_{calc} = \frac{s_A^2}{s_B^2}$$ (F.6)

Where s_A^2 is the larger estimated variance with $n_A - 1$ degrees of freedom and s_B^2 is the smaller estimated variance with $n_B - 1$ degrees of freedom.

If the null hypothesis is confirmed, the equations below are used to find t_{calc}, according to References 9 and 10.

$$t_{calc} = \frac{\bar{x}_1 - \bar{x}_2}{s_p} \sqrt{\frac{n_1 n_2}{n_1 + n_2}}$$ (F.7)

$$s_p = \sqrt{\frac{(n_1 = -1)s_1^2 + (n_2 - 1)s_2^2}{(n_1 + n_2 - 2)}}$$ (F.8)

$$d_f = n_1 + n_2 - 2$$ (F.9)

Where

$\bar{x}_1$ is the mean of n_1 replicate quantitated analytical results in ppb for a specific VOC such as benzene from B1

$\bar{x}_2$ is the mean of n_2 replicate quantitated analytical results in ppb for a specific VOC such as benzene from B2

s_p is the pooled standard deviation

d_f is the number of degrees of freedom.

Table F.4 shows the Excel spreadsheet created to apply the option B equations for chloroform ($CHCl_3$). Identical spreadsheets (not shown) were created for the nine other VOCs. Coeluting peaks m- and p-xylene were quantitated as one chromatographically resolved peak. B1 data (ratio of m/z 83 $CHCl_3$ abundance to m/z 86 $^{13}CHCl_3$ abundance) was quantitated against the B2 calibration for chloroform to yield the quantitated results in ppb $CHCl_3$. B2 $CHCl_3$ calibration met the $r^2 \geq 0.9900$ criterion discussed above. Each of the remaining nine ppb VOCs was calculated against its respective B2 calibration, and these VOCs also met the criterion, $r^2 \geq 0.9900$. As shown in Table F.4, the standard deviation and variance were first calculated for the N-1 degrees of freedom shown for both QC-L and QC-H. Using Equation (F.6), the F_{calc} value is shown for both QC levels and compared to the tabulated F value. Note that $F_{calc} = 2.84$ (rounded off for $CHCl_3$ from that shown in Table F.4) for the QC-L comparison of variances. This value exceeds the tabulated F value for a two-tailed chi distribution with $\alpha = 0.05$ denoted by $F_{0.05/2,13,14} = 2.51$. The null hypothesis, i.e., no difference exists when comparing the squared variances for $CHCl_3$ between B1 and B2, is refuted for the QC-Ls. The same conclusion is reached for $CHCl_3$ when comparing B1 and B2 variances for the QC-Hs, since $F_{calc} = 2.82$ vs. $F_{0.05/2,14,13} = 2.55$. Note that the value of F_{calc} for the QC-Hs is very close to that for the QC-Ls despite significant differences in concentration (~8,000 ppb $CHCl_3$ vs. ~250 ppb $CHCl_3$). Although the t test is now irrelevant, t_{calc}

TABLE F.4
Example Spreadsheet Listing Replicate Quantitated Results in ppb and Results of Calculating: x, s, s², F_{calc}, t_{calc} from Applying Option B Statistics for $CHCl_3$

Batch 1 and Batch 2 QCs are Quantitated against Batch 2 Calibration

Chloroform QC-L			Chloroform QC-H		
Batch 1	Batch 2		Batch 1	Batch 2	
267.72	259.52		8451.05	7983.64	
280.08	282.07		8145.67	7941.55	
	277.71			8146.73	
325.32	270.7		8174.05	7809.5	
289.02	280.17		8318.77	8567.87	
260.57	316.1		8276.51	6913.14	
261.27	268.94		7718.59	7532.44	
240.45	262.67		7411.71	6239.02	
262.47	299.37		8186.68	7757.24	
318.11	278.92		7975.78	6239.22	
269.2	278.73		7471.49	6732.72	
258.51	266.32		7949.65	8699.33	
226.44	291.84		6309.88	5490.35	
258.99	258.86		7615.09	7737.41	
264.22	266.31		7903.05	7109.13	
270.16929	265.97	mean	8177.05	7546.385	mean
26.490698	15.731327	std dev	546.10855	917.24219	std dev
701.75707	247.47466	[std dev]^2	298234.55	841333.24	[std dev]^2
2.8356725		F(calc)	2.8210455		F(calc)
14	15	N	14	15	N
13	14	N-1	13	14	N-1
12587.487			15655715		
27			27		
21.59174		s(p)	761.4731		s(p)
4.1992857		|x(1)-x(2)|	630.665		|x(1)-x(2)|
2.6909811		SQRT[(N(1)*N(2)/ N(1)+N(2)]	2.6909811		SQRT[(N(1)*N(2)/ N(1)+N(2)]
0.5233575		t(calc)	2.2287164		t(calc)
27		df	27		df
2.51		**F(0.05/2,13,14)***	**2.55**		**F(0.05/2,14,13)***
2.050		**t(0.05/2,27)#**	**2.050**		**t(0.05/2,27)#**

*The tabular *F* values were obtained from Anderson, R. *Practical Statistics for Analytical Chemists*, 1987, Van Nostrand Reinhold, Appendix D Table 2A.

#The tabular *t* value was obtained from Anderson, R. *Practical Statistics for Analytical Chemists*, 1987, Van Nostrand Reinhold, Appendix D Table 5 and confirmed from Dean, J. Ed. *Lange's Handbook of Chemistry*, 13th ed. New York, McGraw-Hill, 1985, pp. 1–28.

was found using Equations (F.7)–(F.9) and was compared to tabulated values for t for purposes of completeness.

Table F.5 summarizes the findings when two independent means with unknown and equal variances (option B from above) are compared based on applying t statistics for all VOCs of interest. Three significant figures are reported for the tabulated and calculated F values, and four significant figures are reported for the tabulated and calculated t values as shown in Table F.5 and all subsequent summary tables (Tables F.7, F.9 and F.11). Except for CHCl$_3$, all other VOCs for the QC-L samples confirmed the null hypothesis with respect to variances: $F_{calc} < F_{0.05/2,13,14}$. Of all the VOCs for the QC-H samples, only three VOCs confirmed the null hypothesis with respect to variances: $F_{calc} < F_{0.05/2,14,13}$. Despite some of the VOCs exhibiting unequal variances based on results from applying the F test, their respective t_{calc} values were calculated and are shown in Table F.5. The null hypothesis with respect to means was confirmed for all VOCs except for 1,2-dichloroethane in the QC-L samples. Chloroform, 1,2-dichlorethane, benzene and carbon tetrachloride failed the t test for QC-H, while the remaining VOCs passed. Failed F test values are shown in red and failed t test values in green. The assumption that the variances between the good and bad batch were unknown and unequal leads to a similar evaluation using option C.

TABLE F.5
Results of Comparing the Means (Good vs. Bad Batch) for All Ten VOCs (m & p-Xylene Co-Elute) Using Option B Statistics with Results Quantitated against B2 Calibration

VOC	F(calc) QC-L	t(calc) QC-L	F(calc) QC-H	t(calc) QC-H
Chloroform	2.84	0.523	2.82	2.229
1,2-Dichloroethane	1.71	3.500	4.86	2.460
Benzene	1.64	0.983	3.53	2.869
Carbon tetrachloride	1.64	0.935	1.83	2.375
Toluene	1.32	1.137	2.94	1.734
Tetrachloroethene	2.28	0.812	2.48	1.596
Ethylbenzene	1.33	1.69	2.71	1.149
m- & p-xylene	1.07	1.610	2.44	0.733
Styrene	2.25	1.695	2.90	0.455
o-xylene	2.51	1.520	2.66	1.119
F and *t* test criteria				
F(0.05/2,13,14) QC-L*	2.51			
t(0.05/2,27) QC-L, QC-H#		2.050		2.050
F(0.05/2, 14,13) QC-H*			2.55	

Equations taken from Anderson, R. *Practical Statistics for Analytical Chemistry*, 1987, pp. 76–78, Van Nostrand Reinhold.

*The tabular F values were obtained from Anderson, R. *Practical Statistics for Analytical Chemists*, 1987, Van Nostrand Reinhold, Appendix D Table 2A.

The tabular *t* value was obtained from Anderson, R. *Practical Statistics for Analytical Chemists*, 1987, Van Nostrand Reinhold, Appendix D Table 5 and confirmed from Dean, J. Ed. *Lange's Handbook of Chemistry*, 13th ed. New York, McGraw-Hill, 1985, pp. 1–28.

F.2.3.2 Applying *t* Statistics for Option C to Compare B1 and B2 Means: Both Batches Quantitated Against the B2 Calibration

If the *F* test finds that the variances between B1 and 2 are unknown and unequal (option C), the appropriate *t* statistics can be applied using the equation for t_{calc} and for the number of degrees of freedom d_f as shown below.[9]

$$t_{calc} = \frac{\bar{x}_1 - \bar{x}_2}{\sqrt{\dfrac{s_1^2}{n_1} + \dfrac{s_2^2}{n_2}}} \tag{F.10}$$

$$df = \frac{\left(\dfrac{s_1^2}{n_1} + \dfrac{s_1^2}{n_2}\right)^2}{\dfrac{(s_1^2)^2}{n_1 + 1} + \dfrac{(s_2^2)^2}{n_2 + 1}} - 2 \tag{F.11}$$

Where

$\bar{x}_1$ is the mean of n_1 replicate quantitated analytical results in ppb for a specific VOC such as benzene from B1

$\bar{x}_2$ is the mean of n_2 replicate quantitated analytical results in ppb for a specific VOC such as benzene from B2

s_1 is the standard deviation for the B1 QC results for a specific VOC such as benzene

s_2 is the standard deviation for the B2 QC results for a specific VOC such as benzene

d_f is the number of degrees of freedom associated with the tabulated t.

Note that d_f will always be between the smaller of $(n_1 - 1)$ and $(n_2 - 1)$ and the sum of $(n_1 + n_2 - 2)$. The absolute difference between the two means is generally used when applying Equation (F.10).[9]

Table F.6 illustrates how a spreadsheet was created to implement the *t* statistics for option C using Equations (F.10) and (F.11) for $CHCl_3$ in a manner similar to what was shown in Table F.4 for option B. Since the variances are assumed to be unequal, this spreadsheet does not consider the *F* test. For the QC-L comparison of means for $CHCl_3$, $t_{calc} < t_{(0.05/2,22)}$, confirming a null hypothesis. For the QC-H comparison of means for $CHCl_3$, $t_{calc} < t_{(0.05/2,24)}$, again confirming a null hypothesis.

Table F.7 shows a comparison of means for B1 and 2 for all VOCs using Equations (F.10) and (F.11). All VOCs at both QC levels showed that $t_{calc} < t_{(0.05/2,df)}$ with one exception. The QC results for 1,2-dichloroethane refuted the null hypothesis with respect to its means between B1 and 2 in the QC-L but not in the QC-H. Comparison of means at QC levels in B1 and 2, quantitated against the B2 calibration, proved that, for most of the VOCs studied, it is more appropriate to assume unknown and unequal variances. Consider a comparison of means using options B and C with both batches quantitated against (instead of the B2 calibration) their respective calibrations.

F.2.3.3 Applying *t* Statistics for Option B to Compare B1 and B2 Means: Both Batches Quantitated Against Their Respective Calibration

Table F.8 illustrates how a spreadsheet was created to apply option B *t* statistics when B1 QC results were quantitated against the B1 calibration for $CHCl_3$. Each B1 calibration met the $r^2 = 0.9900$ criterion as discussed above. As before, B2 QCs were quantitated against the B2

TABLE F.6
Example Spreadsheet Listing Replicate Quantitated Results in ppb and Results of Calculating: $\bar{x}$, s, s², t_{calc} from Applying Option C Statistics for CHCl$_3$

Batch 1 and Batch 2 QCs are Calculated against the Batch 2 Calibration

Chloroform QC-L			Chloroform QC-H		
Batch 1	Batch 2		Batch 1	Batch 2	
267.72	259.52		8451.05	7983.64	
280.08	282.07		8145.67	7941.55	
	277.71			8146.73	
325.32	270.7		8174.05	7809.5	
289.02	280.17		8318.77	8567.87	
260.57	316.1		8276.51	6913.14	
261.27	268.94		7718.59	7532.44	
240.45	262.67		7411.71	6239.02	
262.47	299.37		8186.68	7757.24	
318.11	278.92		7975.78	6239.22	
269.2	278.73		7471.49	6732.72	
258.51	266.32		7949.65	8699.33	
226.44	291.84		6309.88	5490.35	
258.99	258.86		7615.09	7737.41	
264.22	266.31		7903.05	7109.13	
270.16929	265.97	mean	7850.56929	7546.385	mean
26.490698	15.731327	std dev	546.108549	917.242194	std dev
701.75707	247.47466	[std dev]^2	298234.548	841333.243	[std dev]^2
14	15	N	14	15	N
50.125505	16.49831	[std dev]^2/N	21302.4677	56088.8829	[std dev]^2/N
To find Student's t			To find Student's t		
4.1992857		|x(1)-x(2)|	304.184286		|x(1)-x(2)|
8.1623413		Sqrt	278.19301		Sqrt
0.5144707		t(calc)*	1.09342893		t(calc)*
To calculate df					
4438.7328		sum (std dev)^2/N	5989421140		sum (std dev)^2/N
2512.5662	272.19424	([std dev]^2/N)^2	453795130	3145962780	([std dev]^2/N)^2
15	16	N+1	15	16	N+1
167.50442	17.01214	[(std dev)^2/N]^2/N+1	30253008.7	196622674	([std dev]^2/N]^2/N+1
184.51656			226875682		
22.056013		df	24.399573		df
Tabulated value for t			Tabulated value for t		
2.07		**t(0.05/2,22)**	**2.06**		**t(0.05/2,24)**

Refer to notes beneath Table F.4.

TABLE F.7

Results of Comparing the Means (Good vs. Bad Batch) for All Ten VOCs (m & p-Xylene Co-Elute) Using Option C Statistics with Results Quantitated against B2 Calibration

VOC	t(calc) QC-L	t(0.5/2,df) QC-L*	t(calc) QC-H	t(0.5/2,df) QC-H*
Chloroform	0.514	2.07	1.093	2.06
1,2-Dichloroethane	3.467	2.06	1.939	2.08
Benzene	0.974	2.06	1.502	2.07
Carbon tetrachloride	0.974	2.06	1.502	2.07
Toluene	1.31	2.05	0.436	2.06
Tetrachloroethene	1.746	2.05	1.571	2.06
Ethylbenzene	1.682	2.05	0.0839	2.06
m- & p-xylene	1.609	2.05	0.617	2.06
Styrene	1.671	2.06	0.413	2.06
o-xylene	1.497	2.07	0.0109	2.06

The mathematics are such that a unique df must be calculated for each analyte and for each concentration level. Refer to notes beneath Table F.5.

calibration. For QC-L $CHCl_3$, the null hypothesis was confirmed for both the F and t tests, and for QC-H $CHCl_3$, the null hypothesis was confirmed for the F test but not the t test. Table F.9 shows the results for comparing means between both batches for all VOCs assuming option B comparison of means, t statistics. For QC-L, the null hypothesis was confirmed for the F and t tests. However, for QC-H, only styrene and o-xylene (the last two eluted VOCs) have confirmation of a null hypothesis in both tests. The first four VOCs passed the F test and failed the t test, while the next four VOCs (in GC elution order) failed the F test yet passed the t test.

F.2.3.4 Applying t Statistics for Option C to Compare B1 and B 2 Means: Both Batches Quantitated Against Their Respective Calibration

Table F.10 illustrates how a spreadsheet was created to apply option C comparison of means, t statistics when B1 QC results were quantitated against the B1 calibration for $CHCl_3$. Each B1 calibration met the $r^2 = 0.9900$ criterion discussed above. As before, B2 QCs were quantitated against the B2 calibration. Again, only the t test is relevant and, in the case of $CHCl_3$, both QC levels confirmed the null hypothesis. Table F.11 shows the results for comparing means between both batches for all VOCs assuming the option C comparison of means, t statistics. For the QC-L, the first six VOCs confirmed the null hypothesis, while the last four VOCs refuted it. For the QC-H, only CCl_4 refuted the null hypothesis.

F.3 DISCUSSION

Upon comparing Tables F.5 and F.9, in which option B statistics are compared and the results quantitated against the B2 calibration versus their respective calibration, it is clear that, although the F and t tests frequently refuted the null hypotheses across the 10 VOCs, more failures are seen in the QC-H than in the QC-L. Eight of 10 VOCs (Table F.5) refuted the null hypothesis F test for QC-H when calibrated against the B2 calibration. Four of 10 VOCs (Table F.9) for the QC-H

TABLE F.8
Example Spreadsheet Listing Replicate Quantitated Results in ppb and Results of Calculating: $\bar{x}$, s, s^2, F_{calc}, t_{calc} from Applying Option B Statistics for $CHCl_3$

Batch 1 and Batch 2 are Quantitated against Their Respective Calibration

Chloroform QC-L			Chloroform QC-H		
Batch 1	Batch 2		Batch 1	Batch 2	
265.93	259.52		8321.44	7983.64	
272.66	282.07		7936.53	7941.55	
	277.71			8146.73	
275.61	270.7		6128.66	7809.5	
275.88	280.17		7844.85	8567.87	
233.29	316.1		7836.31	6913.14	
266.37	268.94		7823.48	7532.44	
254.59	262.67		8175.93	6239.02	
268.15	299.37		8207.76	7757.24	
326.91	278.92		8859.46	6239.22	
275.61	278.73		7792.28	6732.72	
263.45	266.32		7818.28	8699.33	
246.88	291.84		7582.3	5490.35	
268.67	258.86		7786.77	7737.41	
266.15	266.31		8042.04	7109.13	
268.5821	266.04	mean	8181.74	7546.385	mean
20.62016	15.73133	std dev	592.1992	917.2422	std dev
425.191	247.4747	[std dev]^2	350699.8	841333.2	[std dev]^2
1.718119		F(calc)	2.399012		F(calc)
14	15	N	14	15	N
13	14	N-1	13	14	N-1
8992.128			16337763		
27			27		
18.24943		s(p)	777.8832		s(p)
2.542143		|x(1)-x(2)|	635.355		|x(1)-x(2)|
2.690981		SQRT[(N(1)*N(2)/ N(1)+N(2)]	2.690981		SQRT[(N(1)*N(2)/ N(1)+N(2)]
0.374853		t(calc)	2.197924		t(calc)
27		df	27		df
2.51		F(0.05/2,13,14)*	**2.55**		F(0.05/2,14,13)*
2.050		t(0.05/2,27)#	**2.050**		t(0.05/2,27)#

Refer to notes beneath Table F.4.

TABLE F.9
Results of Comparing the Means (Good vs. Bad Batch) for All Ten VOCs (m & p-Xylene Co-Elute) Using Option B Statistics with Results Quantitated against Their Respective Calibration

VOC	F(calc) QC-L	t(calc) QC-L	F(calc) QC-H	t(calc) QC-H
Chloroform	1.72	0.375	2.40	2.198
1,2-Dichloroethane	0.53	1.846	1.51	2.362
Benzene	0.97	0.804	2.04	2.801
Carbon tetrachloride	1.09	0.713	2.35	2.706
Toluene	0.85	0.744	2.75	1.737
Tetrachloroethene	1.48	0.628	2.95	1.953
Ethylbenzene	0.67	0.800	2.82	1.386
m- & p-xylene	0.68	0.974	3.19	0.915
Styrene	0.60	0.629	1.87	1.485
o-xylene	0.87	0.905	2.47	1.513
F and *t* test criteria				
F(0.05/2,13,14) QC-L*	2.51			
t(0.05/2,27) QC-L, QC-H#		2.050		2.050
F(0.05/2, 14,13) QC-H*			2.55	

Refer to notes beneath Table F.5.

failed the *F* test when calibrated against their respective calibrations. However, most VOCs in the QC-L passed the *F* test irrespective of whether B1 or 2 was used to calibrate.

An overall significant decline in failures can be seen when Tables F.7 and F.11 are compared to Tables F.5 and F.9. Only 1,2-dichloroethane in the QC-L (Table F.7) failed to confirm the null hypothesis with respect to a comparison of means (failed the *t* test) when quantitated against the B2 calibration. No VOCs failed the *t* test (Table F.7) for the QC-H when quantitated against the B2 calibration. The last four eluting VOCs (Table F.11) refuted the null hypothesis for the QC-L with respect to a comparison of means (failed the *t* test) when quantitated against their respective calibration. Only carbon tetrachloride out of 10 VOCs (Table F.11) failed the *t* test for the QC-H when quantitated against their respective calibration.

F.4 CONCLUSION

Option C with results quantitated against the B2 calibration exhibited the least number of refuted null hypothesis over the 10 chromatographically separated VOCs. This conclusion was reached using the applied mathematics of *t* statistics, while illustrating significant differences in analytical chemistry outcomes.

TABLE F.10
Example Spreadsheet Listing Replicate Quantitated Results in ppb and Results of Calculating: $\bar{x}$, s, s^2, t_{calc} from Applying Option C Statistics for CHCl$_3$

Batch 1 and Batch 2 QCs (Option C) Are Calculated against Their Respective Calibration

Chloroform QC-L			Chloroform QC-H		
Batch 1	Batch 2		Batch 1	Batch 2	
265.93	259.52		8321.44	7983.64	
272.66	282.07		7936.53	7941.55	
	277.71			8146.73	
275.61	270.7		6128.66	7809.5	
275.88	280.17		7844.85	8567.87	
233.29	316.1		7836.31	6913.14	
266.37	268.94		7823.48	7532.44	
254.59	262.67		8175.93	6239.02	
268.15	299.37		8207.76	7757.24	
326.91	278.92		8859.46	6239.22	
275.61	278.73		7792.28	6732.72	
263.45	266.32		7818.28	8699.33	
246.88	291.84		7582.3	5490.35	
268.67	258.86		7786.77	7737.41	
266.15	266.31		8042.04	7109.13	
268.5821	277.2153	mean	7868.29214	7393.286	mean
20.62016	15.73133	std dev	592.19915	917.242194	std dev
425.191	247.4747	[std dev]^2	350699.834	841333.243	[std dev]^2
14	15	N	14	15	N
30.37078	16.49831	[std dev]^2/N	25049.9881	56088.8829	[std dev]^2/N
To find Student's t			**To find Student's t**		
8.63319		|x(1)-x(2)|	475.006143		|x(1)-x(2)|
6.846101		**Sqrt**	284.848856		**Sqrt**
1.261038		**t(calc)**	1.66757258		**t(calc)**
To calculate df			**To calculate df**		
2196.712		Sum (std dev)^2/N	6583516385		Sum (std dev)^2/N
922.3845	272.1942	([std dev]^2/N)^2	627501905	3145962780	([std dev]^2/N)^2
15	16	N+1	15	16	N+1
61.4923	17.01214	[(std dev)^2/N]^2/N+1	41833460.4	196622674	([std dev]^2/N]^2/N+1
78.50444			238456134		
25.98201		df	25.6089202		df
Tabulated value for t			**Tabulated value for t**		
2.06		**t(0.05/2,26)**	**2.06**		**t(0.05/2,26)**

Refer to notes beneath Table F.4.

TABLE F.11

Summary of Results: Comparing the Means (Good vs. Bad Batch) for All Ten VOCs (m & p-Xylene Co-Elute) Using Option C Statistics with Results Quantitated against Their Respective Calibration

VOC	t(calc) QC-L	t(0.5/2, df) QC-L*	t(calc) QC-H	t(0.5/2, df) QC-H*
Chloroform	1.261	2.06	1.668	2.06
1,2-Dichloroethane	0.790	2.05	1.366	2.05
Benzene	1.073	2.05	1.805	2.05
Carbon tetrachloride	1.171	2.05	2.279	2.06
Toluene	1.818	2.05	1.212	2.06
Tetrachloroethene	1.746	2.05	1.571	2.06
Ethylbenzene	2.536	2.05	0.900	2.06
m- & p-xylene	2.312	2.05	0.900	2.06
Styrene	2.883	2.05	0.659	2.05
o-xylene	2.295	2.05	0.983	2.06

The mathematics are such that a unique df must be calculated for each analyte and for each concentration level.
Refer to notes beneath Table F.5.

REFERENCES

1. Loconto, P. Use of weighted least squares and confidence band calibration statistics to find reliable instrument detection limits for trace organic chemical analysis. *American. Laboratory*. 2015, *47*(7), 34–39.
2. Chemical Terrorism Laboratory Network. VOC Analysis by SPME/GC/MS, Centers for Disease Control and Prevention, and subsequent training guides, notes and discussions.
3. Pawliszyn, J. *Solid Phase Microextraction: Theory and Practice*, Wiley-VCH: New York, N.Y., 1997, pp. 74–86.
4. Loconto, P. *Trace Environmental Quantitative Analysis. Principles, Techniques, and Applications*, 2nd ed., CRC Press: Boca Raton, Fla., 2006, pp. 255–268.
5. Snow, N. and Slack, G. In: Grob, R. and Barry, E., Eds. *Modern Practice of Gas Chromatography*, 4th ed. Wiley-Interscience: Hoboken, N.J., 2004, pp. 574–584.
6. Ai, J. Solid phase micro-extraction for quantitative analysis in non-equilibrium situations, *Anal. Chem.* 1997, 69(6), 1230–1236.
7. Ai, J., Solid-phase micro-extraction in headspace analysis. Dynamics in non-steady state mass transfer. *Anal. Chem.* 1998, *70*(22), 4822–4826.
8. Perrin, C. *Mathematics for Chemists*. Wiley-Interscience: New York, N.Y., 1970, pp. 154–157.
9. Anderson, R. *Practical Statistics for Analytical Chemists*, Van Nostrand Reinhold: New York, N.Y., 1987, pp. 72–78.
10. Mode, E. *Elements of Probability and Statistics*,. Prentice-Hall: Englewood Cliffs, N.J., 1966, p. 205.
11. Einax, J., Zwansiger, H. et al. *Chemometrics in Environmental Analysis*. VCH: Weinheim, Germany, 1997, pp. 35–40.

Appendix G
Using a Pooled Standard Deviation to Find the Uncertainty in the Percent Recovery for the Priority Pollutant Phenol Using Reversed-Phase SPE from Spiked Water by Converting Phenol to its Pyrazol-3-One Complex: Comparing *In Situ* vs. Synthesized Experimental Approaches

If you take a sample of the aqueous phase after a wastewater sample has been liquid–liquid extracted (LLE) as per EPA Method 625 and inject this sample into a high performance liquid chromatograph (HPLC) under reversed-phase conditions using a fixed wavelength ultraviolet-visible (uv-vis) absorbance detector, you might be surprised! Numerous well-separated peaks in the HPLC chromatogram become evident! If you visit the most recent version (EPA Method 625.1), the acceptable percent recovery (%R) ranges for un-, mono-, di-, and tri-substituted phenols are from a low of 17% (phenol) to a high of 167% (2-nitrophenol). [1] Phenol, mono-, di-, and even some tri-substituted phenols are sufficiently hydrophilic as to remain in the aqueous phase following both base-neutral and acid LLE. This led the author down the path of finding a different way to isolate and recover the hydrophilic phenol itself using the newly developed (at that time) reversed-phase solid phase extraction (RP-SPE) sample prep technique. This was accomplished by adapting the well-known reaction of phenol with 4-aminoantipyrene to form the 5-methyl-4-[(4-oxocyclohexa-2, 5-dien-1-ylidene) amino]-2-phenyl-1H-pyrazol-3-one (referred to here as the pyrazol-3-one complex) dye. [2,3] The fact that the pyrazol-3-one complex is hydrophobic yet water soluble made it feasible to isolate and recover this polar priority pollutant phenol using RP-SPE.

In this unpublished paper, new relative standard deviations (RSDs expressed as a percent) for each subset of experimental percent recovery (%R) results for the isolation and recovery of phenol from spiked water using RP-SPE as previously reported are re-evaluated using the concept of a pooled standard deviation first articulated by Kolthoff et al. [10]

G.1 EXPERIMENTAL

Laboratory specifications and equipment for RP-SPE that generated the data to calculate the %R of phenol as well as details on laboratory procedure were given earlier[3]. A uv-vis spectrophotometer (Hitachi) with 10 mm standard cuvettes was used to record absorbance at 460 nm. To reiterate from the earlier work, the pyrazol-3-one dye complex was developed *in-situ* in aqueous solution as well as synthesized, recrystallized, filtered, and dried. The purified solid was weighed and then dissolved in methyl t-butyl ether (MTBE) prior to conducting %R studies.

To prepare the reagents: a) NH_4^+-NH_3 buffer (dissolve 16.9 g NH_4Cl in 143 mL of concentrated NH_3 (NH_4OH) and dilute to 250 mL); b) 4-aminoantipyrene reagent (dissolve 4 g in distilled, deionized water (DDI) and dilute to 100 mL); c) potassium ferricyanide (dissolve 8 g $K_3Fe(CN)_6$ in DDI and dilute to 100 mL). To prepare the phenol reference standards used for the *two studies* whose results are shown in the accompanying tables, procedures are given as follows:

- Table G.1: The pyrazol-3-one complex was formed *in-situ* in DDI. Adding 2 mL each of buffer, 4-aminoantipyrene reagent, and K_3 $Fe(CN)_6$ for every 100 mL of sample was consistent with the standard method for total recoverable phenols at that time[4]. To prepare 500 mL of a 1 ppb phenol solution: carefully aliquot 500 µL of a 1 ppm phenol reference standard dissolved in methanol and transfer to 500 mL of sample. The sample consists of the three reagents cited above dissolved in DDI. The pH was adjusted to 10. This is in contrast to a pH of 7.9 as cited in the standard method at that time[4]. The spiked sample was passed through a RP-SPE barrel cartridge that contained a conditioned C_{18} chemically bonded silica sorbent. The cartridge was washed, dried and eluted with 2 mL MTBE. The eluent was then transferred to a 4 mL volumetric flask and adjusted to a final precise volume. Likewise, successively increasing aliquots of the 1 ppm phenol solution were taken to prepare the remaining reference standards as listed in Table G.1.
- Table G.2: The pyrazol-3-one complex was synthesized, recrystallized, filtered and dried. A 40 ppm spiking solution of the pyrazol-3-one complex in MTBE was prepared. 50 µL of the 40 ppm pyrazol-3-one complex spiking reference standard was spiked into 500 mL DDI to prepare the 1.3 ppb phenol sample. To find the concentration of phenol present in

TABLE G.1 Establishing the Statistical Variation in the Mean Absorbance for the *In-Situ* Conversion of Phenol to the Pyrazol-3-One Complex with Isolation and Recovery Using RP-SPE while Comparing Results from Applying Equation (G.2) to Results from Applying Equation (G.3) for Pooled RSDs: 500mL of Distilled, Deionized Water Spiked with Phenol

#ppb phenol	Mean absorbance[a]	RSD (Equation (G.2))[b]	RSD (Equation (G.3))[c]
1	0.086	10.8	12.5
2	0.089	17.3	20.2
4	0.130	9.2	11.0
10	0.225	6.8	7.2

a Mean uv absorbance over duplicate measurements over triplicate RP-SPEs
b Pooled relative standard deviation over triplicate RP-SPEs
c Pooled relative standard deviation over triplicate RP-SPEs

TABLE G.2 Establishing the Statistical Variation in the Percent Recovery for Isolating the Pyrazol-3-One Complex (Previously Synthesized) While Comparing Results from Applying Equation (G.2) to Results from Applying Equation (G.3) for Pooled RSDs: 500 ml of Distilled, Deionized Water Spiked with Increasing Aliquots of the 40 ppm Complex Spiking Solution

#ppb as phenol	Mean absorbance	%Recovery (Equation (G.4))[a]	RSD (Equation (G.2))[b]	RSD (Equations (G.3), (G.5), (G.6))[c]
1.3	0.037	90.2	3.8	0.21
2.6	0.075	87.6	0.97	0.16
5.2	0.142	84.0	0.50	0.11
7.8	0.205	78.5	0.27	0.37

a Percent recovery results reported are based on the ratio of the complex absorbance from RP-SPE to the complex absorbance of a control (100% recovered) that did not undergo RP-SPE

b A single RP-SPE at each phenol concentration level was conducted and compared to a single control phenol reference at the same concentration level

c Error was propagated over duplicate absorbance measurements of the control at each of the four phenol concentration levels

this sample, consider the following stoichiometric calculations (based on a 1:1 mole ratio between phenol [94 g/mole] and the pyrazol-3-one [294 g/mole] complex):

$$(50 \ \mu L) \left(40 \ ng/\mu L \ complex\right) (\frac{94 ng \ phenol}{294 ng \ complex}) = \sim 640 \ ng \ phenol$$

$$\frac{640 ng \ phenol}{500 mL \ sample} = \sim 1.3 \frac{ng}{mL} \ phenol$$

A 100 μL aliquot of 40 ppm complex spiked into 500 mL DDI gives a ~2.6 ppb phenol reference standard; a 200 μL aliquot gives a ~5.2 ppb phenol; and a 300 μL aliquot gives a ~7.8 ppb phenol. To prepare the control reference standard for the ~1.3 ppb sample, 50 μL of the 40 ppm pyrazol-3-one complex was added to a 4 mL volumetric flask, half-filled with MTBE. Enough MTBE was then added to precisely reach the calibration mark. In a similar fashion, 100, 200, and 300 μL aliquots of the pyrazol-3-one complex were taken and diluted accordingly to complete the set of four control reference standards.

G.2 RESULTS AND DISCUSSION

Following Emerson's pioneering work on a new color test for phenols using 4-aminoantipyrene[5], Martin developed a procedure for quantitating phenols in water and brine while Ettinger and colleagues first extracted the pyrazol-3-one dye from water and reported a detection limit of 1 ppb phenol[6,7]. The chemical reaction that converts the hydrophilic phenol to the hydrophobic yet water soluble pyrazol-3-one dye complex with partitioning from the aqueous phase into chloroform via LLE, forms the basis for quantifying phenols in wastewaters following simple distillation is well established.[4,8] Figure G.1 shows the chemical oxidation of phenol in the presence of 4-aminoantipyrene to form the pyrazol-3-one dye that takes place at room temperature in an alkaline pH aqueous solution containing dissolved ferricyanide ion as the oxidizing agent. The proposed molecular structure for the pyrazol-3-one dye has been confirmed by proton NMR spectroscopy[9].

RP-SPE using a barrel cartridge or 96-well plate format affords an opportunity for the analyst to easily conduct replicate RP-SPEs. For example, a given concentration with a fixed volume of a spiked aqueous sample could be prepared in five replicate samples and then passed through five identical RP-SPE devices in one vacuum manifold. This one set of quintuplet results could be compared to a control sample (either one or replicate measurements of the control) resulting in quintuplet %R values. A standard deviation over quintuplet % recovery results could be calculated. If this is repeated for higher concentrations of spiked aqueous samples, then five subsets, each subset comprising five replicates gives sufficient data to apply a more comprehensive statistical evaluation.

Earlier, Kolthoff, et.al. entertained the notion of calculating a pooled standard deviation (s_p,) from related groups of measurements. Their equation for doing this is shown as Equation (G.1) (below)[10]. In adapting Equation (G.1) to RP-SPE studies when the determinative technique is gas chromatography (GC), consider a percent recover study using RP-SPE to isolate and recover the organochlorine pesticide methoxychlor from groundwater. If j replicate GC injections of the extract are made for each of g replicate RP-SPEs conducted at a given analyte concentration in

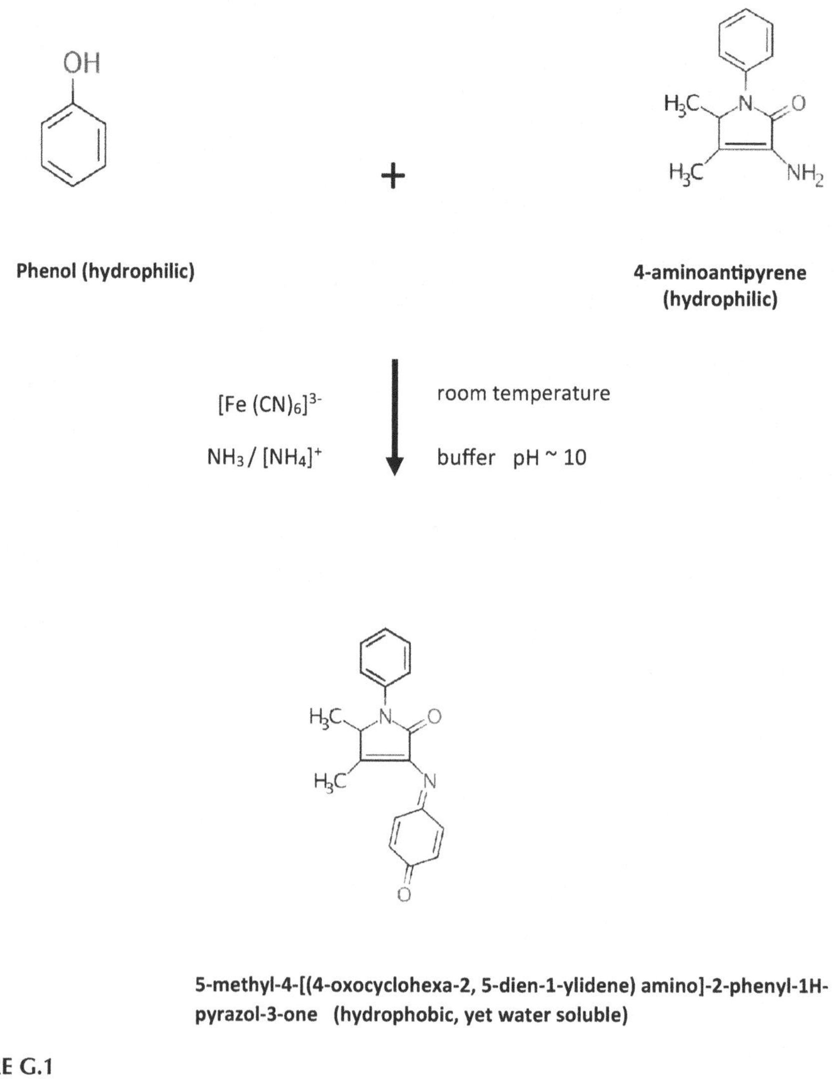

Phenol (hydrophilic)

4-aminoantipyrene
(hydrophilic)

$[Fe (CN)_6]^{3-}$

$NH_3 / [NH_4]^+$

room temperature

buffer   pH ~ 10

5-methyl-4-[(4-oxocyclohexa-2, 5-dien-1-ylidene) amino]-2-phenyl-1H-
pyrazol-3-one   (hydrophobic, yet water soluble)

FIGURE G.1

the original environmental sample, then an overall standard deviation, s_p, for this pooled data can be calculated according to:

$$s_p = \sqrt{\frac{1}{N-g}\sum_g\sum_j\left(x_j - \bar{x}\right)^2}$$ (G.1)

In the original work, x_j represents a single uv-vis absorbance measurement of the pyrazol-3-one complex that is summed over j replicate absorbance values over g replicate RP-SPEs at the same phenol concentration and thus giving a total of N measurements[3].

Revisiting Equation (G.1) with a more contemporary perspective, we find that a pooled standard deviation can be calculated from applying Equation (G.2) (below) whereby n_1 replicate RP-SPEs give a standard deviation s_1 while n_2 replicate RP-SPEs give a standard deviation s_2, all the way up through n_k replicate RP-SPEs giving a standard deviation sk:

$$s_{\text{pooled}} = \frac{\sqrt{(n_1-1)s_1^2 + (n_2-1)s_2^2 + ...(n_k-1)s_k^2}}{n_1 + n_2 + ... + n_k - k}$$ (G.2)

If $n_1 = n_2 = ... = n_k$, then Equation (G.2) simplifies to:

$$s_{\text{pooled}} = \sqrt{\frac{s_1^2 + s_2^2 + ... + s_k^2}{k}}$$ (G.3)

If unequal numbers of replicate RP-SPEs are taken, i.e. $n_1 \neq n_2 \neq ... \neq n_k$, then Equation (G.3) cannot be used and Equation (G.2) becomes equivalent to Equation (G.1). In the work reported here, since equal numbers of replicate RP-SPEs were used, Equation (G.3) was used to calculate a pooled standard deviation for a given phenol concentration level.

Of the two sets of experiments whose results are shown in the accompanying tables, only Table G.2 provides the necessary data so that a %R can be calculated. A pooled standard deviation in the pyrazol-3-one complex (phenol) absorbance denoted as $A^i(s^i)$ where superscript i refers to the ith analyte, in this case only phenol, over replicate RP-SPEs, is incorporated into the equation for the percent recovery of phenol (%R^i) along with a standard deviation (s^i_C) in the control reference standard denoted as absorbance(A^i_C). All this comes together (Equation (G.4) below) to calculate a %R with an accompanying overall precision expressed as a standard deviation in the %R denoted as s^i_R according to[11]:

$$\%R^i_{\;i}(s^i_R) = \frac{A^i\left(s^i\right)}{A^i_C\left(s^i_C\right)} \times 100$$ (G.4)

The calculation of R^i involves dividing A^i with uncertainty s^i by A^i_C with uncertainty s^i_C. To find s^i_R if given s^i and s^i_C requires a propagation of error[11,12]. The uncertainty in a %R for the ith analyte s^i_R is best stated as a relative standard deviation (RSD). It can be calculated by taking into account the propagation of error for random errors using the equation:

$$\frac{s^i_R}{R^i} = \sqrt{\left(\frac{s^i}{A^i}\right)^2 + \left(\frac{s^i_C}{A^i_C}\right)^2}$$ (G.5)

The RSD or coefficient of variation is thus calculated according to:

$$\text{RSD}\left(\text{expressed as a \%}\right) = \left(\frac{s_R^i}{R^i}\right) \times 100 \tag{G.6}$$

An Excel spreadsheet was created to incorporate Equations (G.2) through (G.6). This greatly facilitated the computational aspects whose results are discussed below.

Table G.1 summarizes the findings from a preliminary study to evaluate just how low and how reproducible phenol could be quantitated. The pyrazol-3-one complex was generated *in situ* in 500 mL of water. This scenario is closest to a *real world* application where phenol would be complexed, isolated and recovered from an aqueous matrix. The precision (RSD) from the earlier study is shown to be comparable to the precision calculated using Equations (G.2) through (G.6). Notice that the spectrophotometer reading is indistinguishable between 1 and 2 ppb phenol. This would seem to be the lowest phenol concentration that could be quantitated and represents an initial estimate of the limit of detection. In addition, this study demonstrated that phenol could be quantitated down to low ppb phenol concentration levels with adequate precision.

Converting phenol to the pyrazol-3-one complex, shown in Figure G.1, differs from most chemical derivatization reactions used in gas chromatographic analysis. *Percent recoveries depend* not only on the extent of pyrazol-3-one sorption to the chemically bonded silica surface, but on the *extent of phenol conversion to the pyrazol-3-one complex* prior to RP-SPE. To eliminate this potential second cause of low recovery, the complex was previously synthesized, recrystallized, and dried. A 40 ppm phenol spiking solution was prepared by weighing and dissolving the purified solid in MTBE and subsequently used to generate the data shown in Table G.2. 500 mL of DDI was spiked with the 40 ppm synthesized pyrazol-3-one complex in an attempt to repeat the study summarized in Table G.1 while comparing the *in situ* pyrazol-3-one complex versus the previously synthesized pyrazol-3-one complex. Mean uv-vis absorbance values along with %R and RSDs at low trace concentration levels are shown in Table G.2. A single RP-SPE at each phenol concentration was conducted with duplicate absorbance measurements taken. Replicate absorbance measurements of a control at the same phenol concentration as that for the sample were prepared. *Precision is significantly reduced* when compared to that shown in Table G.1 while *using a pooled standard deviation* and resulted in somewhat lower %RSDs using Equations (G.3), (G.5), and (G.6).

Mean absorbance values are shown in both tables. Upon comparing the mean absorbance values between the two lowest phenol standards for the *in situ* versus the previously synthesized complex, a significant gain in method sensitivity is evident. This difference might be attributable to the slow chemical kinetics for pyrazol-3-one complex formation given the extremely low reactant (phenol) concentration.

G.3 CONCLUSION

Upon applying a more contemporary equation to that shown earlier to calculate a pooled standard deviation, comparable precision was demonstrated.[13] Including a control reference and propagating error when calculating the precision in a %R yielded comparable results to that reported earlier. Gains in spectrophotometric sensitivity that could result in lower instrument detection limits for phenol can be realized if the pyrazol-3-one complex is previously synthesized and used as the reference standard versus *in situ* complex formation. Triplicate vs. duplicate absorbance measurements for each recovered pyrazol-3-one complex would have been preferable. The RP-SPE approach if combined with a simple distillation can be a suitable alternative to the conventional approach to measure total recoverable phenols in wastewater thus eliminating the need for

LLE. Further studies are needed to find the %R for mono, di, and tri-substituted phenols using the experimental approach introduced here.

G.4 ACKNOWLEDGMENT

The New York State Science and Technology Foundation supported the earlier work through its Small Business Innovation Research Matching Grants Program.

REFERENCES

1   Method 625.1 *Base Neutrals and Acids, Office of Water*, Environmental Protection Agency, 82-R-16-007, December 2016,
2   Loconto, P.R. An alternative approach to the classical chloroform extraction of total recoverable phenols at low ppb concentrations in wastewater using solid-phase extraction, Abstract #238, 1989 Pittsburgh Conference on Analytical Chemistry and Applied Spectroscopy, Atlanta, GA
3   Loconto, P.R. An alternate approach to the classical chloroform extraction of total recoverable phenols at low ppb levels in wastewaters using solid-phase extraction. *Chimica Oggi* 8(11): 21–24, 1990.
4   Total Recoverable Phenols in Wastewater by Distillation, *Standard Methods for the Examination of Water and Wastewater,* 16th ed. 1985 American Public Health Association, pp. 356–361.
5   Emerson, E. The condensation of aminoantipyrene II. A new color test for phenolic compounds. *Journal of Organic Chemistry* 8(5): 417–428, 1943.
6   Martin, R. Rapid colorimetric estimation of phenol. *Anal Chem* 21(11): 1419–1420, 1949.
7   Ettinger, M., Ruchoft, C. and Lishka, R., Sensitive 4-aminoantipyrene method for phenolic compounds. *Anal Chem* 23(12): 1783–1788, 1951.
8   SW 846 Test Method 9065 Phenolics: spectrophotometric, manual, 4-AAP with distillations, *Test Methods for Evaluating Solid Waste, Physical/Chemical Methods*, Office of Solid Waste, United States Environmental Protection Agency, September, 1986.
9   Jones, P.R. and Johnson, K.E. Estimation of phenols by the 4-Aminoantipyrine method: identification of the colored reaction products by proton magnetic resonance spectroscopy, *Canadian Journal of Chemistry* 51: 2860–2868, 1973.
10   Kolthoff, I.M., Sandell, E.B., Meehan, E.J., Bruchenstein, S. *Quantitative Chemical Analysis*, 4th ed. London: Macmillan, 1969, p. 390.
11   Loconto, P.R. *Trace Environmental Quantitative Analysis*, 2nd ed. Boca Raton, FL: CRC Press, Taylor and Francis, 2006, pp. 77–79.
12   Laitinen, H. *Chemical Analysis*, New York: McGraw-Hill, 1960, pp. 544–545.
13   Deming, S.N. Statistics in the laboratory: pooling *American Laboratory* 50(9): 2019, 16–19. This on-going series of articles on the basic principles of statistics in measurement ran from October 2015 to March 2019. Earlier and over a 10+ year contribution from January 2003 to April 2013, David Coleman and Lynn Vanatta contributed to a series titled *Statistics in Analytical Chemistry* for *American Laboratory*. Archived issues of *American Laboratory* are available on-line through April 2019. The historical archive collection can be found at the Science History Institute, Philadelphia, PA.

Appendix H
Laboratory Glass & Instrument Designs

The best education is to be found in gaining the utmost information from the simplest apparatus.
                                                                        —**A.N. Whitehead**

TEQA is constantly changing as new technologies as well as scientific advances enable the development of new analytical methods. This appendix includes photos with explanations that show innovative glass apparatus and analytical instrumentation that the author would like to share. Before we get to these, let us revisit an old, but educationally significant concept.

In Chapter 4, the *Craig countercurrent distribution* was introduced from a historical perspective as a lead into the topic of chromatography. A complete Craig countercurrent distribution apparatus is located in the lobby of the Department of Molecular and Cellular Biology at Harvard University. Figure H.1 is a photo that shows this unique laboratory glassware design. I'm indebted to my daughter Jennifer who took the photo while she was a graduate student in that department.

In Chapter 5, a student-tested experiment on the quantitative determination of Cr (VI) in a contaminated aquifer was introduced. Figure H.2 shows a photo that depicts the various instrument components used to assemble an *automated ion chromatograph* with post-column reagent delivery and UV absorbance detection in the author's laboratory at the time. Photo is courtesy of the Department of Civil and Environmental Engineering at Michigan State University.

In Chapter 3, the SPE sample prep technique was introduced and discussed. Figure H.3 is a photo that shows the author's *large-volume glass apparatus* that was designed to pass a 1 L sample of groundwater through a conditioned RP-SPE sorbent. This glass apparatus was fabricated with support from an EPA SBIR Phase I grant at the time.

In Chapter 4, the static headspace GC determinative technique was introduced and discussed. Figure H.4 shows a photo with a unique view of two HS-C-GC-MS instruments operated by the author for close to a decade in support of the CDC's Laboratory Response Network-Chemical (LRN-C). The nearer instrument shown in the photo was used to quantitate *trace cyanide in whole blood*. The farther instrument shown in the photo was used to quantitate *trace VOCs in human serum*. Human exposure to HCN and to VOCs is of great concern to public health. Photo is courtesy of the Michigan Department of Community Health, Bureau of Laboratories, Division of Chemistry and Toxicology.

In Chapter 4 the interface between an HPLC and a mass spectrometer was introduced and discussed. Figure H.5 is a photo of an LC-MS instrument in support of the CDC's Laboratory Response Network-Chemical (LRN-C). This instrument was used to *quantitate organo-phosphonates nerve agent metabolites* and the *metabolite from exposure to sulfur mustard gas in human urine*. Organic metabolites from human exposure to nerve agents and sulfur mustard gas among other important non-volatile organics is of great concern to public health. Photo is courtesy of the Michigan Department of Community Health, Bureau of Laboratories, Division of Chemistry and Toxicology.

In Chapter 4 the ICP atomic emission spectroscopic technique was introduced and discussed. Figure H.6 is a photo of an ICP-MS instrument in support of the CDC's Laboratory Response Network-Chemical (LRN-C). This instrument was used to *quantitate numerous toxic trace metals in human blood and serum*. Human exposure to toxic metals is of great concern to public health. Photo is courtesy of the Michigan Department of Community Health, Bureau of Laboratories, Division of Chemistry and Toxicology.

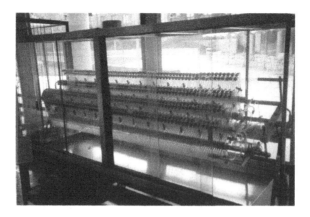

FIGURE H.1   An original Craig countercurrent distribution apparatus.

FIGURE H.2   Instrument configured from different sources to measure trace Cr(VI) via ion chromatography with post-column reagent chelation.

FIGURE H.3   Large-volume glass apparatus for solid-phase extraction (designed by author).

FIGURE H.4 A "bird's eye view" of two analytical instruments. Each instrument includes a MPS2® (Gerstel, GmbH & Co.KG) /6890 GC/5873 MSD® (Agilent Technologies).

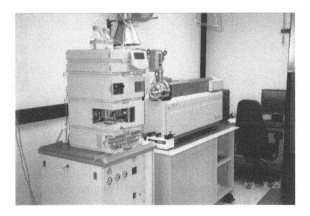

FIGURE H.5 1100® HPLC (Agilent Technologies) interfaced to an API4000® (Applied Biosystems).

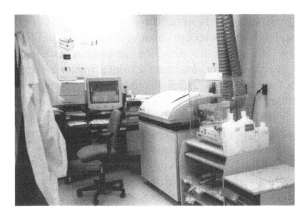

FIGURE H.6 PE SCIEX ELAN 6100 DRC Plus® ICP-MS (PerkinElmer Instruments).

FIGURE H.7   Rapid Trace® SPE Workstation. (Caliper Life Sciences, formerly Zymark Corporation.)

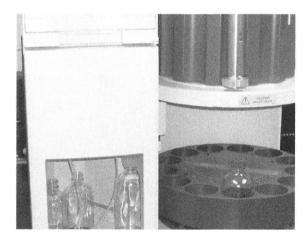

FIGURE H.8   Accelerated Solvent Extractor ASE 300® (Thermo Fisher Scientific, formerly Dionex Corporation).

In Chapter 3 the solid-phase extraction sample prep technique was introduced and discussed. Figure H.7 is a photo of an automated SPE cartridge workstation designed to isolate and recover toxic semivolatile organic compounds from human urine. Photo is courtesy of the Michigan Department of Community Health, Bureau of Laboratories, Division of Chemistry and Toxicology.

In Chapter 3 the accelerated solvent extraction sample prep technique was introduced and discussed. Figure H.8 is a photo of an automated solvent extractor used to extract priority pollutant organic pesticides from Great Lakes fish. Photo is courtesy of the Michigan Department of Community Health, Bureau of Laboratories, Division of Chemistry and Toxicology.

In Chapter 4 the Fourier Transform Infrared Spectroscopic determinative technique was introduced and discussed. Figure H.9 is a photo of a Fourier Transform Infrared Spectrometer/Microscope used to *qualitatively identify unknown organic compounds* in various solid samples by coupling magnification and infrared spectroscopy. Photo is courtesy of the Michigan Department of Community Health, Bureau of Laboratories, Division of Chemistry and Toxicology.

Figure H.10 is a photo of an X-ray Fluorescence Spectrometer designed to *quantitate trace metals* in unknown solid samples. Photo is courtesy of the Michigan Department of Community Health, Bureau of Laboratories, Division of Chemistry and Toxicology.

FIGURE H.9   Fourier Transform Infrared Spectrometer Microscope (Smith's Detection, formerly SensIR).

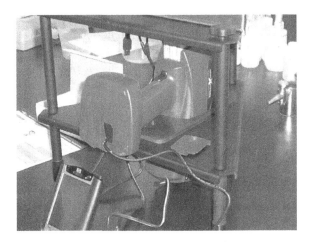

FIGURE H.10   Handheld X-Ray Fluorescence Spectrometer (Innov-X Systems).

Figure H.11 shows a photo of the original *triple quadrupole mass spectrometer*. This development was unique in its time and led to commercial advancements in tandem mass spectrometry. Photo is courtesy of Richard H. Scheel, PhD.

In Chapter 3, the solid-phase extraction sample prep technique was introduced and discussed. Figure H.12 shows a photo of an actual laboratory benchtop set up used by the author to conduct RP-SPE. A typical multi-cartridge vacuum manifold designed to conduct SPE is shown. The outlet to the manifold is connected to a small vacuum pump designed to conduct SPE. Alternatively, the vacuum manifold could be connected to a heavy walled glass side-arm flask that in turn is connected to water faucet aspirator. RP-SPE has revolutionized sample prep in the isolation and recovery of trace concentration levels of SVOCs from both *enviro*-chemical and *enviro*-health sample matrices. Early SPE manifolds evolved from the heavily walled glass development tanks used to conduct thin-layer chromatography (TLC). The TLC tanks were outfitted with multi-port covers while the bottom of the tank contained an outlet that led to a water trap connected via tubing to either a water aspirator or a vacuum pump. Photo is courtesy of the Michigan Department of Community Health, Bureau of Laboratories, Division of Chemistry and Toxicology.

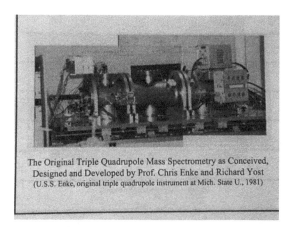

FIGURE H.11   The original triple quadrupole mass spectrometer designed and built in the Department of Chemistry at Michigan State University.

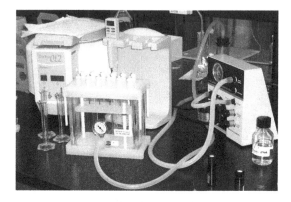

FIGURE H.12   A typical laboratory bench set up used by the author for conducting multi-sample reversed-phase solid-phase extraction (RP-SPE). Shown (left to right) are a centrifuge, an automated pipette and pipette holder stand, an SPE vacuum manifold connected to a small vacuum pump, assorted glassware and reagent bottles.

FIGURE H.13   The author is shown making a manual GC injection into a G2350A® (Agilent Technologies) Gas Chromatograph-Atomic Emission Spectrometer (GC-AED). Note the dual injection autosampler (atop the GC-AED) and the computer data system on the benchtop adjacent to the instrument (circa 2006–2008).

In Chapter 4, the atomic emission spectroscopic determinative technique was introduced and discussed. Figure H.13 shows a photo of the author making a manual injection into a GC-AED instrument sometime during the 2006–2008 time period. Photo is courtesy of the Michigan Department of Community Health, Bureau of Laboratories, Division of Chemistry and Toxicology.

A picture is worth 1,000 words.

—Anonymous

Appendix I
Useful Potpourri for
Environmental Analytical
Chemists

GRAPHICAL REPRESENTATION OF RELATIVE
ISOTOPE PEAK INTENSITIES FOR ANY GIVEN ION
CONTAINING THE INDICATED NUMBER OF
HALOGENS

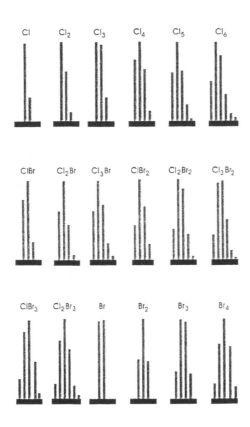

Solvent Polarity Chart

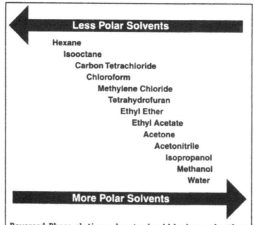

← Less Polar Solvents

Hexane
Isooctane
Carbon Tetrachloride
Chloroform
Methylene Chloride
Tetrahydrofuran
Ethyl Ether
Ethyl Acetate
Acetone
Acetonitrile
Isopropanol
Methanol
Water

More Polar Solvents →

Reversed Phase elution solvents should be less polar than wash solvents. Normal Phase elution solvents should be more polar than wash solvents.

$c = \lambda\nu$ c = velocity of light $(2.99 \times 10^{10}$ cm sec$^{-1})$
$E = h\nu$ h = Planck's Constant $(6.6 \times 1010^{-27}$ erg sec$)$
$E = mc^2$ ν = frequency E = energy m = mass λ = wavelength

The Electromagnetic Spectrum

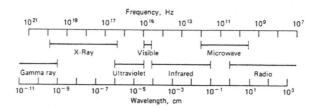

Frequency, Hz

10^{21} 10^{19} 10^{17} 10^{15} 10^{13} 10^{11} 10^{9} 10^{7}

X-Ray   Visible   Microwave

Gamma ray   Ultraviolet   Infrared   Radio

10^{-11} 10^{-9} 10^{-7} 10^{-5} 10^{-3} 10^{-1} 10^{1} 10^{3}

Wavelength, cm

PERIODIC TABLE

SOCIETY FOR APPLIED SPECTROSCOPY

GROUP IA																	VIIIA
1 H 1.008	IIA											IIIA	IVA	VA	VIA	VIIA	2 He 4.003
3 Li 6.939	4 Be 9.012											5 B 10.811	6 C 12.011	7 N 14.007	8 O 15.999	9 F 18.998	10 Ne 20.183
11 Na 22.991	12 Mg 24.312	IIIB	IVB	VB	VIB	VIIB	GROUP VIII			IB	IIB	13 Al 25.982	14 Si 28.086	15 P 30.974	16 S 32.064	17 Cl 35.453	18 Ar 39.948
19 K 39.102	20 Ca 40.08	21 Sc 44.956	22 Ti 47.90	23 V 50.942	24 Cr 51.996	25 Mn 54.938	26 Fe 55.847	27 Co 58.933	28 Ni 58.71	29 Cu 63.54	30 Zn 65.37	31 Ga 69.72	32 Ge 72.59	33 As 74.922	34 Se 78.96	35 Br 79.909	36 Kr 83.80
37 Rb 85.47	38 Sr 87.62	39 Y 88.905	40 Zr 91.22	41 Nb 92.906	42 Mo 95.94	43 Tc (98)	44 Ru 101.07	45 Rh 102.905	46 Pd 106.4	47 Ag 107.870	48 Cd 112.40	49 In 114.82	50 Sn 118.69	51 Sb 121.75	52 Te 127.60	53 I 126.904	54 Xe 131.30
55 Cs 132.905	56 Ba 137.34	57 La 138.91	72 Hf 178.49	73 Ta 180.948	74 W 183.85	75 Re 186.2	76 Os 190.2	77 Ir 192.2	78 Pt 195.09	79 Au 196.967	80 Hg 200.59	81 Tl 204.37	82 Pb 207.19	83 Bi 208.980	84 Po (210)	85 At (210)	86 Rn (222)
87 Fr (223)	88 Ra (226)	89 Ac (227)															

58 Ce 140.12	59 Pr 140.907	60 Nd 144.24	61 Pm (147)	62 Sm 150.35	63 Eu 151.96	64 Gd 157.25	65 Tb 158.924	66 Dy 162.50	67 Ho 164.930	68 Er 167.26	69 Tm 168.934	70 Yb 173.04	71 Lu 174.97
90 Th 232.038	91 Pa (231)	92 U 238.03	93 Np (237)	94 Pu (242)	95 Am (243)	96 Cm (247)	97 Bk (247)	98 Cf (249)	99 Es (254)	100 Fm (253)	101 Md (256)	102 No (254)	103 Lw (257)

I.1 STRATEGY FOR EI MASS SPECTRAL INTERPRETATION

- Attempt to recognize the molecular ion and apply the Nitrogen Rule
- Estimate the isotopic contributions particularly with respect to carbon and the halogens
- Note changes in the isotope composition between the molecular ion and higher fragment ions; this may indicate loss of certain elements, e.g., Cl vs. Br
- Look for key m/z losses from the molecular ion such as M-15 ($CH_3\cdot$), M-29 ($C_2H_5\cdot$), M-18 (H_2O)
- Check for presence of even m/z values based on McLafferty rearrangements
- Look for common m/z values at lower end of mass axis, e.g., CH_3CO^+ at m/z 43
- Be mindful of the Nitrogen Rule
  - A molecular ion will have an even m/z value for molecules that contain C, H, O, X provided there is an even (or zero) number of nitrogen atoms in the molecule.
  - A molecular ion will have an odd m/z value if an odd number of nitrogen atoms are present in the molecule.

Common Contaminants Found in GC-MS

Ions (m/z)	Compound	Possible Sources
18, 28, 32, 44 or 14, 16	H_2O, N_2, O_2 CO_2 or N, O	Residual air and water, air leaks, outgassing from Vespel® ferrules
31, 51, 69, 100, 119, 131, 169, 181, 214, 219, 264, 376, 414, 426, 464, 502, 576, 614	PFTBA, and related ions	PFTBA (tuning compound)
31	MeOH	Cleaning solvent
43, 58	Acetone	Cleaning solvent
78	Benzene	Cleaning solvent
91, 92	Toluene, xylene	Cleaning solvent
105, 106	Xylene	Cleaning solvent
151, 153	TCE	Cleaning solvent
69	Foreline pump or PFTBA	Foreline pump oil vapor or calibration valve leak
73, 147, 207, 221, 281, 295, 355, 429	Dimethylpolysiloxane	Septum bleed or methylsilicone column bleed
77, 94, 115, 141, 168, 170, 262, 354, 446	Diffusion pump fluid and related ions	Diffusion pump fluid
149	Plasticizer (phthalates)	Vacuum seals (O-rings) damaged by high temperatures, vinyl gloves
Peaks spaced 14 Da apart	Hydrocarbons	Fingerprints, foreline pump oil

I.2 LINEAR AND VOLUMETRIC FLOW RATES FOR CAPILLARY GC

Calculation of the capillary column linear carrier gas velocity:

$$\bar{u} = \frac{100L}{t_0}$$

where L = the capillary column length in *meters*

t_0 = the time it takes after injection for an unretained component to elute in *seconds*

Calculation of the volumetric capillary column flow rate:

$$F = \frac{0.785 d_c^2 L}{t_0}$$

where L = the capillary column length in *meters*
t_0 = the time it takes after injection for an unretained component to elute in *minutes*
d_c = the column diameter in *mm*

Selected Values for Student's t

df	t(1 − α) α = .05	t(1 − α/2) α = .05	t(1 − α) α = .01	t(1 − α/2) α = .01
1	6.314	12.706	31.821	63.657
2	2.920	4.303	6.965	9.925
3	2.353	3.182	4.541	5.841
4	2.132	2.776	3.747	4.604
5	2.015	2.571	3.365	4.032
6	1.943	2.447	3.143	3.707
7	1.895	2.365	2.998	3.499
8	1.860	2.306	2.896	3.355
9	1.833	2.262	2.821	3.250
10	1.812	2.228	2.764	3.169
11	1.796	2.201	2.718	3.106
12	1.782	2.179	2.681	3.055
13	1.771	2.160	2.650	3.012
14	1.761	2.145	2.624	2.977
15	1.753	2.131	2.602	2.947
16	1.746	2.120	2.583	2.921
17	1.740	2.110	2.567	2.898
18	1.734	2.101	2.552	2.878
19	1.729	2.093	2.539	2.861
20	1.725	2.086	2.528	2.845
25	1.708	2.060	2.485	2.787
30	1.697	2.042	2.457	2.750
∞	1.645	1.960	2.326	2.576

Source: Dean, J.Ed. *Lange's Handbook of Chemistry,* 13th ed. New York: McGraw-Hill, 1985, pp. 1–28.

Test for Outliers

Values of Dixon's r_{10} (Q) Parameter as Applied to a Two-Tailed Test at Various Confidence Levels

# Observations N	90% ($\alpha = 0.10$)	95% ($\alpha = 0.05$)	99% ($\alpha = 0.01$)
3	0.941	0.970	0.994
4	0.785	0.829	0.926
5	0.642	0.710	0.821
6	0.560	0.625	0.740
7	0.507	0.568	0.680
8	0.468	0.526	0.634
9	0.437	0.493	0.595
10	0.412	0.486	0.565
11	0.392	0.444	0.541
12	0.379	0.426	0.522
13	0.361	0.410	0.502

Adapted from: Rorabacher, D. *Analytical Chemistry* 63: 139–148, 1991.

Critical Values of the F Distribution, $\alpha = 0.05$

n_{lower}	n_{upper}				
	2	3	4	5	∞
2	19.00	19.16	19.25	19.30	19.50
3	9.55	9.28	9.12	9.01	8.53
4	6.94	6.59	6.39	6.26	5.63
5	5.79	5.41	5.19	5.05	4.36
∞	3.00	2.60	2.37	2.21	1.00

n_{upper} = # degrees of freedom for the upper variance.
n_{lower} = # degrees of freedom for the lower variance.
Adapted from Anderson, R. Practical Statistics for Analytical Chemists. New York: Van Nostrand Reinhold Company, 1987.

Exact Masses and Natural Isotopic Abundance Ratios

Element	Symbol	Exact Mass	Abundance	X + 1 Factor	X + 2 Factor
Hydrogen	H	1.007825	99.99		
	D or ^{2}H	2.014102	0.01		
Carbon	^{12}C	12.000000	98.9		
	^{13}C	13.003354	1.1	1.1 n_c	0.0060 n_c^2
Nitrogen	^{14}N	14.003074	99.6		
	^{15}N	15.000108	0.4	0.37 n_N	
Oxygen	^{16}O	15.994915	99.76		
	^{17}O	16.999133	0.04	0.04 n_O	
	^{18}O	17.999160	0.20		0.20 n_O
Fluorine	F	18.998405	100		
Silicon	28Si	27.976927	92.2		

Exact Masses and Natural Isotopic Abundance Ratios

Element	Symbol	Exact Mass	Abundance	X + 1 Factor	X + 2 Factor
	^{29}Si	28.976491	4.7	5.1 n_{Si}	
	^{30}Si	29.973761	3.1		3.4 n_{Si}
Phosphorous	P	30.973763	100		
Sulfur	^{32}S	31.972074	95.02		
	^{33}S	32.971461	0.76	0.8 n_S	
	^{34}S	33.967865	4.22		4.4 n_S
Chlorine	^{35}Cl	34.968855	75.77		
	^{37}Cl	36.965896	24.23		32.5 n_{Cl}
Bromine	^{79}Br	78.918348	50.5		
	^{81}Br	80.916344	49.5		98.0 n_{Br}
Iodine	I	12.904352	100		

1. Assume X = 100%; X represents the relative intensity of the first peak in a cluster of isotope peaks.

2. The factor is multiplied by the number (n) of atoms present to determine the magnitude of the abundance contribution for a given isotope. For example, the fragment ion from electron-impact ionization of the tuning compound, PFTBA at m/z 502 is due to $\left(C_4F_9\right)_2 NCF_2^{+-}$, and contains 9 carbons [$n_c = 9$] such that the contribution at X + 1 is expected to be $[1.1 \times 9] = 9.9\%$.

Adapted (with permission) from Watson J.T. *Introduction to Mass* Spectrometry, 2nd ed. New York: Raven Press, 1985.

Appendix J
Contributing Authors:
Brief Introduction

The author sought contributions from two working analytical chemists who, as of this writing, work for the Analytical Chemistry Section, Chemistry and Toxicology Division, Bureau of Laboratories, Michigan Department of Health and Human Services, Lansing, MI. The specific determinative techniques that each contributing author specializes in complements that of the author's expertise with respect to the GC-MS determinative technique. This provides the reader with a greater breadth of practical knowledge.

Both contributing authors were asked to think and articulate in writing what they *need to know* and what they *need to do* in their respective laboratories when beginning, at the bench, to conduct TEQA in each of their respective areas of expertise.

J.1 AT THE BENCH, WHAT YOU NEED TO KNOW TO SAFELY AND CORRECTLY OPERATE A CONTEMPORARY LC-MS/MS TO CONDUCT TEQA

Written by Michael Stagliano, PhD

Since the invention of electrospray ionization (ESI) in the 1980s, liquid chromatography tandem mass spectrometry (LC-MS/MS) has been an important analytical tool for the separation, detection, and quantitation of trace levels of environmental contaminants. Analyses employed are often looking for a broad range of compounds, including pesticides, plasticizers, per- and poly- fluoroalkyl substances, and small organic and inorganic molecules. Liquid chromatography offers the advantage of a wide range of mobile phases and columns (stationary phases), both of which can be tailored to specific analyses. The versatility of ESI is perfect for environmental analysis by mass spectrometry because this technique is applied directly to the analytes in solution, usually requires no derivatization, has high sensitivity, and provides reproducible results. This, along with recent advances in column chemistry, has reduced sample analysis time while improving performance, aiding in the detection of analytes from complex samples. Coupling LC to tandem mass spectrometry offers analyte specificity with sub parts-per-billion limits of detection.

When preparing to run samples on any LC-MS system, there are a series of system checks needed. The LC column(s) should be installed in the oven, and mobile phase bottles attached to the solvent lines corresponding to the appropriate pumps. A fresh preparation of aqueous mobile phases containing salts and buffers may be needed to ensure no microbial growth is introduced into the system. The solvent lines should be purged for a minimum of 2 minutes ensuring any previously used solvents are flushed from the system. After purging, the solvent lines should be inspected to ensure the lack of air bubbles, indicating an incomplete purge of solvents, or a potential pump issue. Next, the system should be equilibrated: allowing the column to reach operating temperature, and any solvents from storage or previous analyses to be flushed out. During equilibration, all solvents should be diverted from the mass spectrometer, either through the use of a divert valve or disconnection of the solvent line from the instrument. The LC system is considered ready when 10 or more column volumes of solvent have passed through the system. While this exact timeframe is dependent on the column dimensions (diameter and length) and the flow rate, 15 minutes is usually sufficient. Stabilization of the LC pump pressure (changing by less than 1%) is confirmation of system equilibration, and should be recorded to track column and system performance.

Routine maintenance and system checks are needed before the mass spectrometer is ready for use. Triple quadrupole mass spectrometers are operated under high vacuum ($\sim 10^{-6}$ torr). If the instrument is vented and not under vacuum, operating pressure can usually be reached within 1 to 2 hours. Venting, unless necessary for maintenance, is not recommended, and the system should be kept in either "operate" or "standby" mode when not in use. Most manufacturers have real-time system monitors with pressure, temperature, and voltage readbacks, and a color-coded system alerting users to the state of the system. Unlike many other analytical instruments, triple quadrupole mass spectrometers do not require daily tuning. Polypropylene glycol or reserpine are infused for tuning, which can be done yearly, as part of preventive maintenance service, or as instruments warrant.

Before performing any routine cleaning or maintenance, ensure the mass spectrometer is in "standby" mode. The ion source can then be removed for inspection of the sample orifice. This entryway for ions into the mass spectrometer should be clean of build-up from previous analyses. If needed, this area should be cleaned, per manufacturer guidelines. Inside the source, the electrospray capillary should protrude about 1–2 mm, depending on mobile phase composition and flow rate, and should appear shiny and free of debris. Once the source is reattached, the instrument can be placed in "operate" mode. The flow rate of nitrogen to the source for ionization should be checked, along with the flow rate for the collision gas (nitrogen or argon). Once all system checks are completed, and the LC system solvent flow is directed back to the mass spectrometer, sample analysis can commence.

Application: Reversed-phase HPLC Separation of Eleven PFAS compounds using LC-MS/MS according to chromatographic elution order while listing common abbreviations and MRM transitions:

A: Perfluorobutanoic acid (PFBS) m/z 299 → 80
B: Perfluorohexanoic acid (PFHxA) m/z 313 → 269
C: Perfluorohexane sulfonic acid (PFHxS) m/z 399 → 80
D: Perfluorooctanoic acid (PFOA) m/z 413 → 369
E: Perfluorooctane sulfonic acid (PFOS) m/z 499 → 80
F: Perfluorodecanoic acid (PFDA) m/z 513 → 469
G: Perfluorodecane sulfonic acid (PFDS) m/z 599 → 80
H: Perfluorododecanoic (PFDoA) m/z 613 → 569
I: Perfluorotetradecanoic acid (PFTeA) m/z 713 → 669
J: Perfluorohexadecanoic acid (PFHxDA) m/z 813 → 769
K: Perfluorooctadecanoic acid (PFODA) m/z 913 → 869

RP-HPLC Gradient Elution Profile

Time (min.)	% A	% B	Flow Rate (mL/min)
0.01	95	5	0.4
0.50	95	5	0.4
1.00	50	50	0.4
9.00	0	100	0.4
9.01	0	100	0.6
11.00	0	100	0.6
11.01	95	5	0.4
13.00	95	5	0.4

A – 2 mM ammonium acetate (NH4OAc)

B – methanol (MeOH)

A typical reversed-phase LC-MS/MS chromatogram for the separation of the eleven PFAS compounds is shown below:

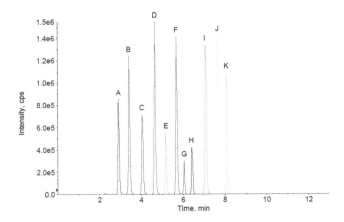

LC column: (Ascentis® Millipore Sigma, formerly Supelco, Bellefonte, PA) C8 column (2.1 x 50 mm, 3 μm particle size)

Injection: 10 μL

Temperature: 40 ºC

Mobile Phase A: 2 mM ammonium acetate

Mobile Phase B: methanol

Photo of a Triple Quad 6500® SCIEX LC-MS/MS (from left to right: mobile phase containers, auto-sampler/auto-injector/LC column heater, triple quadrupole mass spectrometer):

J.2 AT THE BENCH, WHAT YOU NEED TO KNOW TO SAFELY AND CORRECTLY OPERATE A CONTEMPORARY ICP-MS

Written by Scott Forsyth

Introduction:

It is true that inductively coupled plasma mass spectrometry is a very sensitive, and specific technique; however, it should be noted that there are significant limitations (as an example ICP-MS is not traditionally effective in quantitating C, N, O, and most of the noble gases), and problems with it (the aforementioned isobaric issues, instrument drift, detector instability, and carbon effect, to name a few). This author has worked with ICP-MS instruments for the past 9 years in a state government public health laboratory trace metals unit using either single or tandem quadrupole mass spectrometers. Our laboratory primarily determines trace quantities of various toxic metals such as Pb, As, Cd; in the following matrices: dust wipe digests, soil around residences, human whole blood, human urine, human hair, teeth/bone, and various foods. The ICP-MS analyst must consider the following factors:

- the complexity of the sample matrix since sample preparation usually involves implementing digestion/extraction procedures
- the selection of an auto-sampler
- sample introduction techniques
- instrument maintenance
- detection, calibration and quantitative analysis that meets stringent QA/QC criteria

In general, due to matrix effects, it is highly advised to matrix match calibration and QC standards to the expected matrix of the samples. In addition, if internal standards (ISTDs) are used it is advised to use ISTDs that are not found in the sample and have similar ionization energies to the target analyte(s).

This author's experience is limited to food, environmental, and clinical samples. Most of the sample preparation techniques involve acid digestion in concentrated nitric acid (HNO_3). Samples are digested using a heating block and diluted using either 1% or 5% HNO_3. Whole blood is diluted using a solution containing tetra-methyl ammonium hydroxide [$(CH_3)_4NOH$]. Other sample preparation techniques are also used. We will examine some of the issues involving sample introduction, the ICP interface, and the mass spectrometer. It should be noted here that many issues involving high background or contamination can be mitigated by using trace metals grade or electronics grade solvents and solutions.

A simplified schematic drawing of an ICP-MS (single quadrupole) is shown below. It is helpful to return to this schematic drawing as you focus on the various components that comprise a contemporary automated ICP-MS instrument:

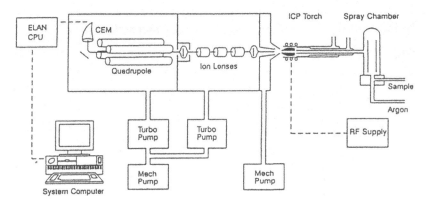

J.2.1   SAMPLE INTRODUCTION

The sample that has been digested/diluted/extracted has to be introduced to the instrument by some means. In our laboratory (and any lab concerned with high throughput) this requirement starts with an auto-sampler (for the "dilute and shoot" methods) or a separation chemistry system such as gas chromatography (GC) and/or high performance liquid chromatography (HPLC), followed by transfer of sample to a nebulizer (GC uses a gas eluate transfer line directly to the plasma torch), a spray chamber and an injector that carries the sample directly to the plasma. Each component must be optimized for the sample matrix, sample volume, acquisition time, and wash steps. Extreme patience is often required during this process!

J.2.1.1 AUTO-SAMPLER

Most vendors of ICP-MS systems contract with other manufacturers to build an auto-sampler (A/S) system. Most A/S systems that this author is familiar with consist of a robot sampling head that has a carbon fiber tube of varying internal dimensions (ID) that is connected to Teflon tubing. This tubing is then connected to a six-port or even a seven-port switching valve. To the switching valve stator several lines are connected: the A/S transfer line, a waste line, a line filled with carrier solution (fed via a peristaltic pump which will aid in pushing the sample through the loop and into the nebulizer transfer line), internal standard line (used in most applications), sample loop (taking up 2 ports), and a nebulizer transfer line. As the reader has probably surmised … the above lines can and will clog! Purchasing a large syringe with A/S fittings is recommended (vendors have kits pre-assembled for this purpose) for flushing out plugs.

The A/S system has various rack stations that can receive various size tubes, e.g. 15 mL centrifuge tubes, on its sampling deck. The analyst usually has direct control on the time in which the A/S sampling head has its probe in the sample, aspirating sample, and rate at which the sample probe is withdrawn from the sample. A visual aid such as diluted food coloring can facilitate timings (if internal standard is added to the solution, instrument times can be handled efficiently as well). The main objective is to provide enough time for the A/S to slightly overfill the sample loop connected to the six-port valve. If the sample loop is overfilled i.e. a few dozen microliters are lost to waste, not only should there be enough solution for analysis but the loop is partially flushed with each unique sample, thus minimizing carryover.

The sample probe should be of large enough diameter to quickly fill the sample loop but small enough to conserve sample and not clog. For example, for a Pb determination in whole blood, a sample probe of 500 μm ID is used with a 100 μL sample loop; this provides for at least 2 injections from a 750 μL sample.

Time is also set in the instrument's method to allow for efficient washing of the A/S and its tubing.

J.2.1.2 SAMPLE INTRODUCTION WITH HYPHENATED TECHNIQUES

It is surprisingly easy to set up a chromatographic system provided it can communicate with the ICP-MS and associated computer. This includes hyphenated instruments such as that for GC-ICP-MS and HPLC-ICP-MS. The eluate flow rate when an HPLC is interfaced to an ICP-MS is usually between 750 μL to 1,000 μL/min. At such flow rates, if the peristaltic pump is not set to increased speeds, the spray chamber could be flooded. In contrast, the GC interface directly injects the eluate into the plasma.

If the analyst is using an organic solvent to flush or to condition an HPLC column, it is advisable to not connect the eluate transfer line to the nebulizer. If concentrated organic solvent such as acetonitrile is introduced into the plasma the additional heat of combustion could melt the torch!

J.2.1.3 NEBULIZERS

There is a surprisingly large array of different nebulizer types/materials available to the analytical chemist such as a micro-flow concentric nebulizer constructed with perfluoroalkoxy (PFA) tubing. The most common nebulizer is a pneumatic type. This is a general term for a nebulizer that directs the liquid flow of the sample and disrupts it via pneumatic action into discrete liquid packets, i.e. aerosol. The most common nebulizer in our laboratory is the micro-concentric type which provides excellent aerosol formation. However, total dissolved solids greater than a few fractions of a percent of total solution will clog the nebulizer and might also clog the sample or skimmer cones downstream! Most vendors provide or purchase from a third party, a fused silica stylus of small enough diameter such that it can be inserted into the orifice of the nebulizer. It is generally not recommended to ultra-sonicate the nebulizer to clean it, especially if the nebulizer is constructed of a polymer such as PFA. The bubbles generated by the ultrasonic wave can distort the internal dimensions of the polymer constructed nebulizer. However, soaking most nebulizers in Aqua regia is usually acceptable. Always verify proper protocol for deep cleaning. Regardless of the type, the main function of a nebulizer is to generate a fine mist or aerosol from of the introduced sample. As a consequence, the main job of the analyst is not to plug the nebulizer!

J.2.1.4 Spray Chambers

The most common spray chamber types in our laboratory are either the Scott's double pass or cyclonic baffled spray chambers. The primary function of a spray chamber, regardless of type, is to select for the smallest aerosol droplets and eliminate the largest. Typically large droplets are >10 μm in diameter. The large droplets have the greatest probability of colliding with the interior wall of the spray and literally dropping out of the aerosol. The excess sample is purged via the drain tube of the spray chamber.

Many vendors recommend cooling the spray chamber to minimize solvent load onto the plasma. By minimizing solvent load the plasma can ionize sample more efficiently and by lowering the amount of solvent there is a reduced chance that oxides and hydroxides will be generated thus minimizing interferents. The final benefit of a cooled spray chamber is a stable signal.

Spray chambers are another source of contamination. Furthermore, if the interior surface of spray chamber is excessively dirty, the main function of eliminating large aerosol droplets will be compromised. To clean the spray chamber submerge it in aqua regia or if recommended by the vendor—ultrasonicate for about 15 minutes in 10% HNO_3.

J.2.1.5 Sample Injectors

The sample injectors are either built into the plasma torch or purchased separately and inserted into the torch cavity. As the name implies the sample injector introduces the small diameter aerosols directly into the plasma. If integrated into the plasma torch the sample injector will be made of quartz (often this type of injector has an extension of the same ID that transfers aerosols into the injector and then into the plasma). However, if the injector is a separate unit it can be of a different material, usually a ceramic or synthesized sapphire. Sapphire is preferred in our laboratory due to its durability and it apparently minimizes clogs.

Sample injectors can be ultra-sonicated with the tapered end pointed up in 2% HNO_3 solution. Furthermore, the injector can be cleaned with a fine tipped swab.

J.2.2 PLASMA TORCH AND LOAD COIL

The plasma torch is the ion source of the ICP-MS. The torch is most commonly made of quartz (though ceramic specialty units can be purchased) comprising two concentric tubes (three, if counting the sample injector), the outer and inner tubes. The principle of how the plasma torch functions has been explained previously. The use of liquid samples can introduce a broad range of contaminates that can disrupt the plasma flow and/or alter its ionization efficiency. It is good practice, especially if the ICP-MS has been used extensively, to check the condition of the torch prior to instrument startup. If salts or other contaminates have built up on the surface it may be a good time to clean the torch. Dismount the torch following manufacturer's instructions and either soak in aqua regia overnight or place the torch in a beaker filled with 2% HNO_3 then ultrasonicate for 15 to 30 minutes. If the ICP-MS system that is in use has the plasma torch in a cassette assembly ensure that the cassette o-rings that seal the torch are intact. It may be necessary to add a very small quantity of vacuum grade grease to the o-rings to have an adequate seal.

Conduct a torch realignment procedure via the ICP-MS software. This process may have to be conducted manually if the laboratory possesses an older instrument.

The load coil transfers the RF energy to the plasma perpetuating its formation. The coil will darken from its distinctive copper color to almost pitch black with use. It is recommended to change the load coil if it obtains a blackened, encrusted appearance. Failure to do so could lead to load coil failure which may cause extensive arcing. Electrical arcing could become so intense that the plasma torch is destroyed in the process. Obviously this is a costly and time consuming repair that might require a service engineer call! Change the load coil if the analyst has serious doubts about its integrity.

J.2.3   ICP INTERFACE

This region of the ICP-MS has its own unique set of challenges. The ion beam must successfully travel from the plasma through the sampler cone and skimmer cone prior to being focused and steered downstream for m/z determination. To reiterate, the ion beam generated by the plasma is at atmospheric pressure. As the beam travels away from the torch into the orifice of the sampler cone and skimmer cone assembly it experiences an increasingly evacuated environment. The vacuum between the cones is maintained at approximately 1–2 torr.

The orifice for the sampler cone (can be either nickel or platinum) is about 1.0 mm and that of the skimmer cone at about 0.5 mm. It should be apparent that unless the samples introduced have low percentage (about 0.2%) of total dissolved solids (TDS) the orifices of the cones could easily become clogged. Even if TDS amounts are low the orifices can become congested with salts or carbon after extended analyses. Depending on the vendor's instrument user recommendations, the analyst may have to clean the cones if they appear dirty. In practice this is not usually done unless tunes and daily performance checks fail. Sampler cone cleaning usually involves swabbing clean the cones face and orifice, front and back. A dilute nitric acid solution of 1 to 2% is commonly used, followed by an 18.2 MΩ water rinse. Depending on the instrument manufacturer the skimmer cone can receive the same treatment.

It should not be forgotten that the interface region is liquid cooled. It is highly recommended that a laboratory check chiller / heat exchanger fluid levels daily. Most current instruments will exhibit a flag if the coolant fluid is flowing too slowly or if the temperature of the interface is too high. Most instrument software control packages will shut down the plasma automatically if the interface temperature spikes or chiller solution flow rate is not detected.

J.2.4   ION OPTICS

There are several different types of ion optics and each is unique to the model and manufacturer. All ion optic systems are positioned downstream of the skimmer cone and interact with the ion beam in such a way as to direct it toward the analytical portion of the instrument. Some ion optic systems can be cleaned and replaced by the instrument analyst. Please note, always read the instrument manual prior to conducting maintenance work on the ion focusing system. The vendor may require the user to place a service call.

It is highly advised, no matter what the make and model of the ICP-MS, to conduct an ion focusing tune prior to running the daily performance check. The analyst can select various masses for this tune but the vendor has determined a set of targets (m/z) that are utilized for tuning purpose (increasing in mass), e.g. Be(9), In(115), Ce(140), Tl(205), U(238). If sensitivity benchmarks fail it could mean that the ion optics system needs to be cleaned/replaced.

J.2.5   MASS ANALYZERS

The scientific principles of how quadrupoles select m/z ratios have been discussed in length in this book so won't be reiterated here. Most ICP-MS instruments either single quadrupole or tandem have set a minimum of 0.7 amu at 10% peak height (with variable tolerances). It is recommended to conduct a mass calibration at least monthly.
Maintenance of the quadrupoles is typically done by vendor field engineers.

J.2.6   DETECTORS

This author is most familiar with the discrete dynode electron multiplier (DDEM). The DDEM is a very reliable and stable detector that usually only requires optimization once a year. However, if the ICP-MS is exhibiting signal drift for unknown reasons it may be fruitful to run a DDEM optimization.

The DDEM can be run in both pulse and analog modes. Most instrument software packages will automatically switch to analog detection if abundance of analyte exceeds a couple of million counts per second. It is advised for consistent linear calibration curves to keep your calibration standard counts below the analog detector mode.

DDEM maintenance and/or replacement is typically conducted by a vendor field engineer.

J.2.7   PASSING DAILY PERFORMANCE/INSTRUMENT TUNES

All manufacturers of ICP-MS instrumentation requires/recommends that the analyst perform a daily/startup performance verification after warm-up and/or component optimization e.g. plasma tuning. Passing criteria are vendor specific but all have sensitivity, precision, and interferent minimization parameters e.g. U must have a sensitivity of > 30,000cps, all analytes in the tune solutions must have a relative standard deviation (RSDs) <3%, and CeO/Ce and Ba++/Ba ratios below 0.02.

It is not uncommon to fail one or more verification requirements on the first attempt. A quick rerun of the performance verification may pass. However, sometimes further troubleshooting is in order. For example, if the CeO/Ce ratio is too high, this may indicate that the plasma is too cool and thus allowing the conditions for oxide formation. Alternatively, it may be the neb flow is too high. Perhaps a nebulizer flow optimization is required. Alternatively, the wattage on the load coil is too low thus cooling the plasma. Perhaps a plasma optimization is in order, etc.

The vendor's instrument user guide or hardware manual should provide some helpful tips on how to troubleshoot a failing daily performance verification.

J.2.8   PASSING THOUGHTS

If the reader is new to ICP-MS analysis please read the vendor's instrument manual! I know it is usually dry stuff but one will be surprised about how much one can learn by doing so … at the very least the new user's confidence should show significant gains. Don't forget to ask questions when the vendor's field engineer is in-house. Most field engineers are extremely knowledgeable so don't be shy and ask about what you need to know!

There is an increasing amount of information about ICP-MS principles and applications (this text book before you is an excellent example), so spend some time doing career development and avail yourself of it.

J.2.9   REFERENCES

1. Loconto P. *Trace Environmental Quantitative Analysis, Principles, Techniques and Applications,* 2nd edition. Boca Raton: CRC Press Taylor and Francis Group, 2006.
2. Thomas R. *Practical Guide to ICP-MS, A Tutorial for Beginners,* 3rd edition. Boca Raton: CRC Press Taylor and Francis Group, 2013.
3. Harris DC. *Quantitative Chemical Analysis,* 7th edition. New York: WH Freeman and Company, 2007.

A schematic drawing of an 8800–8900® (Agilent Technologies) ICP-MS/MS instrument shown below:

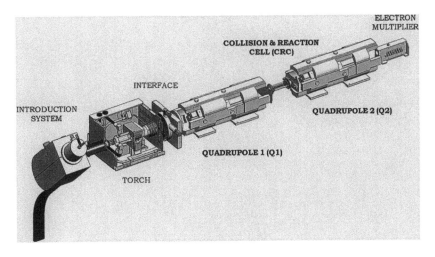

Index

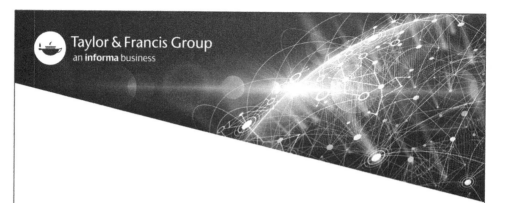

Printed in the United States
by Baker & Taylor Publisher Services